THE NEMATODE
CAENORHABDITIS
ELEGANS

COLD SPRING HARBOR MONOGRAPH SERIES

The Lactose Operon
The Bacteriophage Lambda
The Molecular Biology of Tumour Viruses
Ribosomes
RNA Phages
RNA Polymerase
The Operon
The Single-Stranded DNA Phages
Transfer RNA:
  Structure, Properties, and Recognition
  Biological Aspects
Molecular Biology of Tumor Viruses, Second Edition:
  DNA Tumor Viruses
  RNA Tumor Viruses
The Molecular Biology of the Yeast Saccharomyces:
  Life Cycle and Inheritance
  Metabolism and Gene Expression
Mitochondrial Genes
Lambda II
Nucleases
Gene Function in Prokaryotes
Microbial Development
The Nematode *Caenorhabditis elegans*

THE NEMATODE *CAENORHABDITIS ELEGANS*

Edited by

William B. Wood
and the Community of *C. elegans* Researchers

COLD SPRING HARBOR LABORATORY
1988

THE NEMATODE *CAENORHABDITIS ELEGANS*

Monograph 17

Printed in the United States of America
Book design by Emily Harste

Library of Congress Cataloging-in-Publication Data

The Nematode *Caenorhabditis elegans*

(Cold Spring Harbor monograph series; 17)
Bibliography: p.
Includes index
1. Caenorhabditis elegans. I. Wood, William Barry,
1938– II. Series [DNLM: 1. Caenorhabditis.
QX 203 N4326]
QL391.N4N375 1987 595.1′82 87-23916
ISBN 0-87969-307-X (cloth)
978-087969433-3 (pbk.)

Paperback cover: Bright-field photomicrograph of a multivulva mutant, *lin-15(n309)*. Reprinted from Ferguson and Horvitz (*Genetics*, 110: 17 [1985]).

All Cold Spring Harbor Laboratory Press publications may be ordered directly from Cold Spring Harbor Laboratory Press, 500 Sunnyside Boulevard, Woodbury, New York 11797-2924. Phone: 1-800-843-4388 in Continental U.S. and Canada. All other locations: (516) 422-4100. FAX: (516) 422-4097. E-mail: cshpress@cshl.edu. For complete catalog of Cold Spring Harbor Laboratory Press publications, visit our World Wide Web Site http://www.cshlpress.com/.

Contents

Preface

In 1965, Sydney Brenner chose the free-living nematode *Caenorhabditis elegans* as a promising model animal for a concerted genetic, ultrastructural, and behavioral investigation of development and function in a simple nervous system. Since then, with the help of a growing number of investigators, knowledge about the biology of "the worm," as it is referred to by Brenner's intellectual progeny, has accumulated at a steadily accelerating pace, to the extent that *C. elegans* is now probably the most completely understood metazoan in terms of anatomy, genetics, development, and behavior. There are presently more than 60 laboratories in several parts of the world working on the worm, and the number is growing. A *Caenorhabditis* Genetics Center has been established at the University of Missouri, Columbia, to maintain and dispense mutant strains and to collate genetic mapping data. A *C. elegans* newsletter is distributed twice yearly by the Center. The International *C. elegans* Meeting, held biennially during the past decade at Cold Spring Harbor Laboratory, now attracts over 300 participants, including an increasing number of researchers from other areas of developmental biology.

The past few years have seen the completion of two major long-term projects that provide new insights into *C. elegans* development and lay important groundwork for future investigation: completion of the cell lineages of both sexes, from zygote to adult, and description of the complete anatomy at the level of electron microscope resolution, providing a complete "wiring diagram" of cell contacts in the animal. The past 3 years have also brought the first successes in molecular cloning of developmentally interesting genes defined only by mutation, using transposon tagging as a generally applicable method for identification of desired DNA sequences. An ordered cosmid library of genomic DNA fragments, providing a physical map of the genome, as well as source of cloned sequences for further gene isolation, is well on its way to completion. The time is appropriate for a Book of the Worm that can serve as a reference source for *C. elegans* investigators as well as an introductory monograph for other biologists.

Robert Edgar, who played a seminal role in initiating the Newsletter and the series of biennial meetings, has written as follows about this enterprise: "It is somewhat unusual these days to have a book dealing exclusively with many aspects of a single multicellular organism. However, this characteristic of the book is also a distinctive feature of *C. elegans* research, which has become a field in itself. At least so far, despite the apparent diversity of studies under way ranging from molecular to behavioral, the field is united by the common goal of understanding this relatively simple organism completely and by the realization that knowledge gained in one area is likely to prove useful in another. *C. elegans* researchers are also united by a belief that understanding the organism will require understanding its development and by a common prejudice that genetics provides one of the best approaches to this understanding. As is evident from this book, the genetic control of *C. elegans* development is a focus of much current work. The ability to obtain mutant alleles of developmentally important genes and the study of their mutant phenotypes provide fruitful approaches to dealing with our current profound ignorance of the mechanisms by which genes control animal development."

Creating this book has been a group undertaking of the community of *C. elegans* researchers. Jonathan Hodgkin, John Sulston, and John White of the MRC Laboratory, Cambridge, England, and Robert Horvitz of the Department of Biology, MIT, Cambridge, Massachusetts, collaborated in the initial planning of the book and continued to provide editorial guidance during its preparation. Robert Edgar, quoted above, has had an important influence on the philosophy of the book and much of the work described in it. The authors of the 12 chapters have served throughout as an informal editorial board, helping in the early stages to formulate contents and chapter outlines, and then, toward the end of the project, critically reading all or parts of the text and offering suggestions to each other in order to make the book as complete and coherent as possible. Major contributors to the final editing process were Jonathan Hodgkin, John Sulston, John White, Donna Albertson, Martin Chalfie, Danielle Thierry-Mieg, and Robert Waterson. Numerous others, who have assisted with and commented on specific portions of the text are acknowledged at the end of each chapter. Finally, all of us are grateful to Nancy Ford, Managing Director of Publications, to production editor Dorothy Brown, and to Judy Cuddihy, Mary Cozza, and others of the staff of the Cold Spring Harbor Publications Office who helped bring the book to completion.

W.B. Wood

Foreword

This volume reviews what is known about *Caenorhabditis elegans* and will be the sourcebook on the worm for some time to come. Our field has prospered and has come of age; what was once a joke organism, often confused with the notorious flatworm of memory transfers, has now become a major experimental system for the study of development and developmental genetics. Its success is a great joy to me, and a tribute to all of those who joined a risky, unconventional research project and helped to make it what it is today. They and their students and even some of their students' students are the authors of this book.

It may be of interest to readers to learn something of the very early days of this project. In late 1962, Francis Crick and I began a long series of conversations about the next steps to be taken in our research. Both of us felt very strongly that most of the classical problems of molecular biology had been solved and that the future lay in tackling more complex biological problems. I remember that we decided against working on animal viruses, on the structure of ribosomes, on membranes, and other similar trivial problems in molecular biology. I had come to believe that most of molecular biology had become inevitable and that, as I put it in a draft paper, "we must move on to other problems of biology which are new, mysterious and exciting. Broadly speaking, the fields which we should now enter are development and the nervous system." At that time, there were extensive discussions with the Medical Research Council on building an extension to the Laboratory, and Max Perutz, the head of our laboratory, had been exploring the ground with the Council. I have recently found the correspondence on this topic, and in a letter dated 5 June, 1963 (see below), I wrote to Max and explained my views to him. Nematodes have not yet made their appearance, because I had only just started to read about them and had not yet formulated my ideas. Some people thought that our approach was too "biological" and would lead us away from molecular biology, but, in any event, we were asked to make a formal proposal, and a document was accordingly submitted to the Council in October, 1963. During the summer I had formulated my ideas, and as you will see from the excerpt from the document and the Appendix referred to, the now familiar lines of the project had emerged. Note that the paper refers to *C. briggsae*; it was some time before *C. elegans* was selected in preference.

I hope readers will enjoy the last, brief paragraph of the Appendix. They should understand that it has expanded into the contents of this book, and achieving it has taken more than 20 years and the labors of a large number of people.

Sydney Brenner
September 1987

Letter to Max Perutz

5 June, 1963

Dear Max,

These notes record and extend our discussion on the possible expansion of research activities in the Molecular Biology Laboratory.

First, some general remarks. It is now widely realized that nearly all the "classical" problems of molecular biology have either been solved or will be solved in the next decade. The entry of large numbers of American and other biochemists into the field will ensure that all the chemical details of replication and transcription will be elucidated. Because of this, I have long felt that the future of molecular biology lies in the extension of research to other fields of biology, notably development and the nervous system. This is not an original thought because, as you well know, many other molecular biologists are thinking in the same way. The great difficulty about these fields is that the nature of the problem has not been clearly defined, and hence the right experimental approach is not known. There is a lot of talk about control mechanisms and very little more than that.

It seems to me that, both in development and in the nervous system, one of the serious problems is our inability to define unitary steps of any given process. Molecular biology succeeded in its analysis of genetic mechanisms partly because geneticists had generated the idea of one gene–one enzyme, and the apparently complicated expressions of genes in terms of eye color, wing length and so on could be reduced to simple units which were capable of being analyzed. Molecular biology succeeded also because there were simple model systems such as phages which exhibited all the essential features of higher organisms so far as replication and expression of the genetic material were concerned, and which simplified the experimental work considerably. And, of course, there were the central ideas about DNA and protein structure.

In the study of development and the nervous system, there is nothing approaching these ideas at the present time. It is possible that the repressor/operator theory of Jacob and Monod will be the central clue, but there is not very much to suggest that this is so, at least in its simple form. There may well be insufficient information of the right kind to generate a central idea, and what we may require at the present is an extension of experimentation into these problems.

The experimental approach that I would like to follow is to attempt to define the unitary steps in development using the techniques of genetic analysis. At present, we are producing and analyzing conditional lethal mutants of bacteria. These are mutants which are unable to grow at 44°C but do grow normally at 37°C. The mutations affect genes controlling the more sophisticated processes of the bacterial cell, and some work which we have already done indicates that it will be possible to dissect the process of cell division into its unitary steps. We have mutants in which neither a cell membrane septum nor a cell wall is made, others in which a membrane septum is made but not a cell wall septum and so on. We have mutants in which the control of DNA replication is affected. I intend to expand this research activity in the near future.

Our success with bacteria has suggested to me that we could use the same approach to study the specification and control of more complex processes in cells of higher organisms. As a first stage, I would like to initiate studies into the control of cell division in higher cells, in particular to try to find out what determines meiosis and mitosis. In this work there is a great need to "microbiologize" the material so that one can handle the cells as one handles bacteria and viruses. Hence, like in the case of replication and transcription, one wants a model system. For cell division, in particular meiosis, the ciliates seem the likely candidates. Already, in these cells, the basic plan of meiosis is present and there is no doubt that the controlling elements must be the same in ciliates as they are in the oocytes of mammals.

Another possibility is to study the control of flagellation and ciliation. This again is a differentiation in higher cells and its control must resemble the control in amoebo-flagellates.

As a more long term possibility, I would like to tame a small metazoan organism to study development directly. My ideas on this are still fluid and I cannot specify this in greater detail at the present time.

As an even more long term project, I would like to explore the possibilities of studying the development of the nervous system using insects. . . .

Excerpts from Proposal to the Medical Research Council, October, 1963

In summary, it is probably true to say that no major discovery comparable in importance to that of, say, messenger RNA, now lies ahead in this field, but the detailed elucidiation of the mechanisms already discovered is nevertheless vital.

The *new major problem* in molecular biology is the genetics and biochemistry of control mechanisms in cellular development. We propose to start work in this field and gradually make it the Division's main research.

In the first place, control mechanisms can be studied most easily in micro-organisms, and this work has already begun. In addition we should like to start exploratory work on one or two model systems. We have in mind small metazoa, chosen because they would be suitable for rapid genetic and biochemical analysis. Proposals for such work, which we plan to begin within the next few months, are set out in Appendix I.

APPENDIX I
Differentiation in a Nematode Worm

Part of the success of molecular genetics was due to the use of extremely simple organisms which could be handled in large numbers: bacteria and bacterial viruses. The processes of genetic replication and transcription, of genetic recombination and mutagenesis, and the synthesis of enzymes could be studied there in their most elementary form, and, having once been discovered, their applicability to the higher forms of life could be tested afterwards. We should like to attack the problem of cellular development in a similar fashion, choosing the simplest possible differentiated organism and subjecting it to the analytical methods of microbial genetics.

Thus we want a multicellular organism which has a short life cycle, can be easily cultivated, and is small enough to be handled in large numbers, like a micro-organism. It should have relatively few cells, so that exhaustive studies of lineage and patterns can be made, and should be amenable to genetic analysis.

We think we have a good candidate in the form of a small nematode worm, *Caenorhabditis briggsiae,* which has the following properties. It is a self-fertilizing hermaphrodite, and sexual propagation is therefore independent of population size. Males are also found (0.1%), which can fertilize the hermaphrodites, allowing stocks to be constructed by genetic crosses. Each worm lays up to 200 eggs which hatch in buffer in twelve hours, producing larvae 80 μ in length. These larvae grow to a length of 1 mm in three and a half days, and reach sexual maturity. However, there is no increase in cell number, only in cell mass. The number of nuclei becomes constant at a late stage of development, and divisions occur only in the germ line. Although the total number of cells is only about a thousand, the organism is differentiated and has an epidermis, intestine, excretory system, nerve and muscle cells. Reports in the literature describe the approximate number of cells as follows: 200 cells in the gut, 200 epidermal cells, 60 muscle cells, 200 nerve cells. The organism normally feeds on bacteria, but can also be grown in large quantities in liver extract broth. It has not yet been grown in a defined synthetic medium.

To start with we propose to identify every cell in the worm and trace lineages. We shall also investigate the constancy of development and study its genetic control by looking for mutants.

THE NEMATODE *CAENORHABDITIS ELEGANS*

1

Introduction to *C. elegans* Biology

William B. Wood
Department of Molecular, Cellular, and Developmental Biology
University of Colorado
Boulder, Colorado 80309

I. GENERAL DESCRIPTION

Caenorhabditis elegans is a small, free-living soil nematode found commonly in many parts of the world. It feeds primarily on bacteria and reproduces with a life cycle of about 3 days under optimal conditions. The two sexes, hermaphrodites and males, are each about 1 mm in length but differ in appearance as adults, as shown in Figure 1. Hermaphrodites produce both oocytes and sperm and can reproduce by self-fertilization. Males, which arise spontaneously at low frequency, can fertilize hermaphrodites; hermaphrodites cannot fertilize each other.

A hermaphrodite that has not mated lays about 300 eggs during its reproductive life span. Juvenile worms hatch and develop through four stages (commonly referred to as larval stages, although no metamorphosis is involved), punctuated by molts. The mature adult emerging from the fourth molt is fertile for about 4 days and then lives for an additional 10–15 days.

C. elegans is a simple organism, both anatomically and genetically. The adult hermaphrodite has only 959 somatic nuclei, and the adult male has only 1031. The haploid genome size is 8×10^7 nucleotide pairs, about eight times that of the yeast *Saccharomyces* or one-half that of the fruit fly *Drosophila*.

C. elegans is easily maintained in the laboratory, where it can be grown

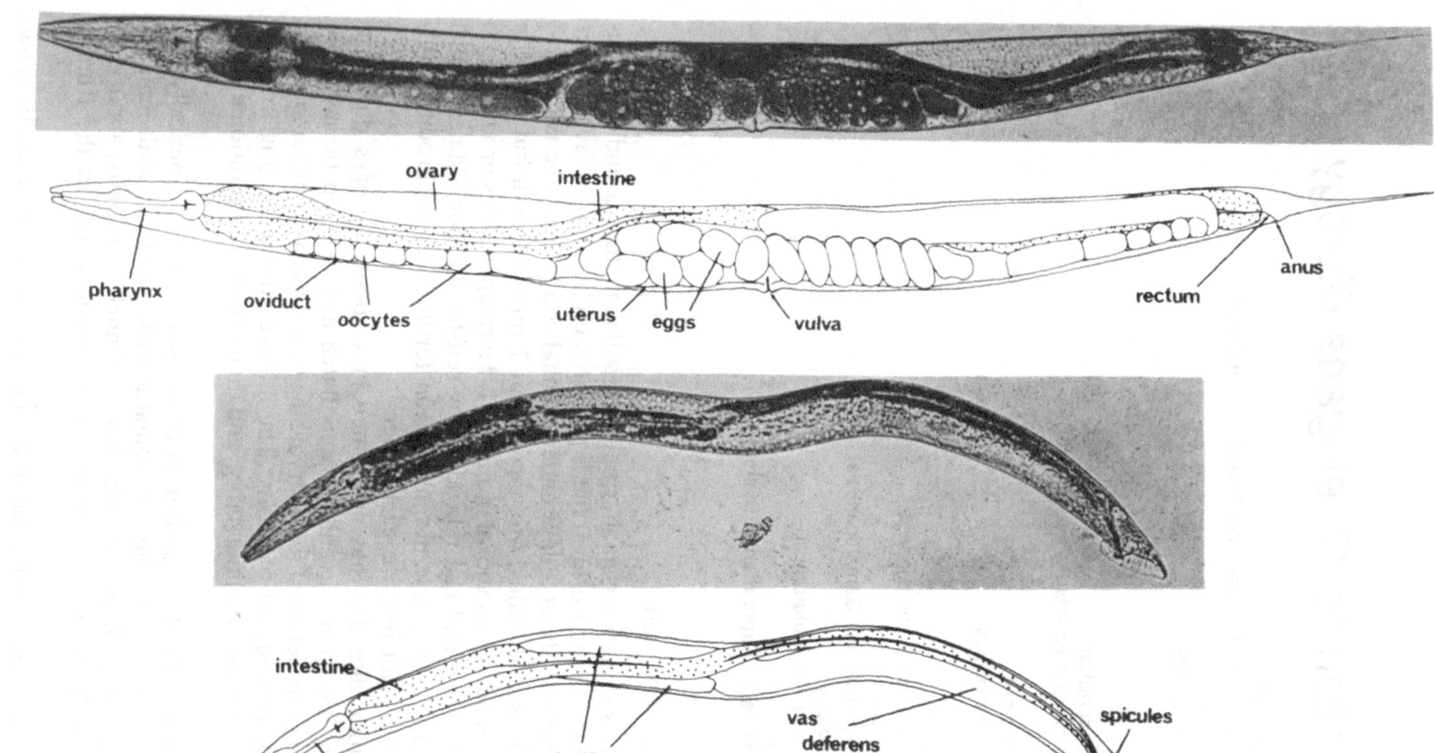

Figure 1 (See facing page for legend.)

on agar plates or in liquid culture with *Escherichia coli* as a food source (see Methods). It can also be grown axenically in liquid media. Individual animals are conveniently observed and manipulated with the aid of a dissecting microscope, and large numbers can be grown in mass culture. The animals are transparent throughout the life cycle, so that development can be followed at the cellular level in living preparations by light microscopy, preferably with differential interference contrast (Nomarski) optics. In addition, its small size allows complete anatomical description of the animal at the electron microscope level. Mutants are readily obtained following chemical mutagenesis or exposure to ionizing radiation. The simplicity, convenience of manipulation, and short life cycle of *C. elegans* make it a useful experimental organism for the study of metazoan development and behavior.

II. PHYLOGENETIC CHARACTERISTICS OF NEMATODES

Members of the phylum Nematoda[1] are numerous and diverse, having adapted to free-living existence in most terrestrial and marine environments, as well as to parasitism in a wide variety of plant and animal hosts. The parasitic forms have substantial impact on human welfare, through crop damage and diseases of both humans and domestic animals. For example, agricultural losses caused by plant parasitic nematodes in the United States alone are currently estimated at $5 billion annually, and over one quarter of the world's human population suffers from debilitating infections of nematode parasites such as hookworms and pinworms.

Evolutionarily, nematodes are thought to be of very ancient origin, arising like the other invertebrate phyla during the Precambrian era (Fig. 2). However, no early fossil record of nematodes is available, and their relationship to other invertebrates is unclear.

Taxonomically, nematodes can be divided into two classes: the Secernentea (or Phasmidia), which include most of the terrestrial free-living and parasitic species, and the Adenophorea (or Aphasmidia), which are generally marine, free-living animals. The simplified classification scheme in Figure 3 lists the orders that are recognized in each class and indicates

[1]There is still argument among taxonomists about several aspects of nematode systematics, including the question of whether the Nematoda should be regarded as a phylum or only a major class in a larger phylum (Aschelminthes or Nematelminthes), including several other morphologically related groups such as the Gastrotricha, Rotifera, and Nematomorpha. The classification used here follows that described by Nicholas (1984).

Figure 1 Photomicrographs showing major anatomical features of *C. elegans* adult hermaphrodite (*above*) and male (*below*). Lateral views; bright-field illumination. Bar represents 20 μm. (Reprinted, with permission, from Sulston and Horvitz 1977.)

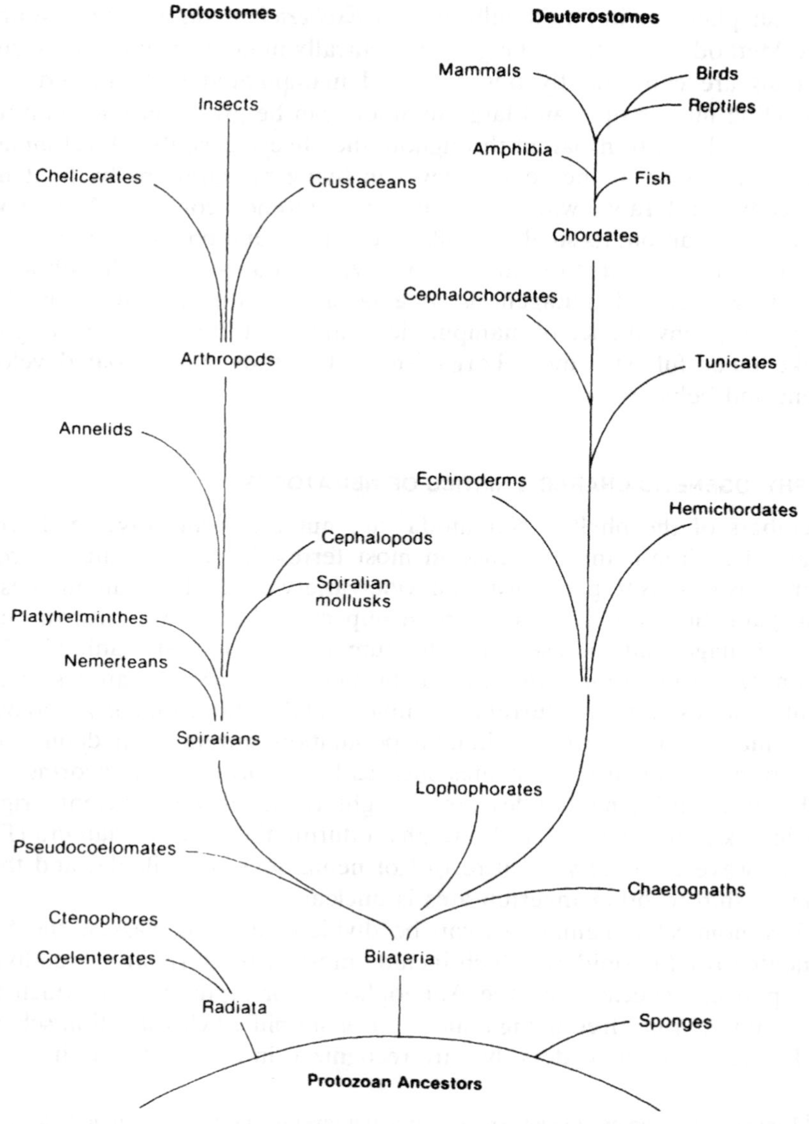

Figure 2 A metazoan phylogenetic tree. Nematodes and other Aschelminthes are included under Pseudocoelomates at lower left. (Reprinted, with permission, from Raff and Kaufman 1983.)

those genera represented in this book, as well as the major plant and animal parasites.

Despite their diverse habitats and life-styles, all nematodes are morphologically and anatomically similar. The general body plan is in the form

Phylum Nematoda

Class Adenophorea
(mostly aquatic free-living species: some plant and animal parasites)

Orders:

- **Araeolaimida**
- **Monohysterida**
- **Desmodorida**
- **Chromadorida**
- **Desmoscolecida**
- **Enoplida**
- **Dorylaimida**

Class Secernentea
(mostly terrestrial free-living species, some plant and animal parasites)

Orders:

- **Rhabditida** (mostly terrestrial microbotrophs)
  - **Family Rhabditidae**
    - **Selected genera:** *Caenorhabditis*, *Panagrellus*, *Turbatrix*
- **Tylenchida** (parasites of plants and invertebrates)
- **Aphelenchida** (parasites of plants and insects)
- **Strongylida** (parasites of vertebrates)
- **Ascaridida** (mostly parasites of vertebrates)
  - **Family Ascarididae**
    - **Selected Genera:** *Ascaris*, *Paracaris*
- **Oxyurida** (parasites of vertebrates and invertebrates)
- **Spirurida** (parasites of vertebrates and invertebrates)

Figure 3 A simplified classification of nematodes. (Adapted from Poinar 1983.)

of two concentric tubes separated by a space, the pseudocoelom. The inner tube is in the intestine; the outer tube consists of cuticle, hypodermis, musculature, and nerve cells. In the adult, the pseudocoelomic space also contains the gonad (Fig. 1).

All nematodes appear to be similar developmentally as well. The early embryonic cell divisions include a series of asymmetric asynchronous cleavages in which the germ line acts as a stem-cell lineage, giving rise sequentially to the founder cells (generally five) for the somatic tissue lineages, and finally to the germ line founder cell. As in other metazoans, the embryonic soma is clearly divided into ectoderm, mesoderm, and endoderm. Postembryonically, as typified by *C. elegans*, all nematodes develop through four larval stages characterized by different cuticle structures. Molting permits growth, but it is not so necessary as in arthropods, because nematodes increase in size between molts and after the final molt. Molting also appears to be important for adaptability. Among parasitic species, the different larval stages often have evolved to become highly specialized for survival in a particular host or host tissue, thereby making possible the complex multistage life cycles of these nematodes. Throughout development, nematodes exhibit precisely determined patterns of somatic cell division that are invariant from one individual to another. Nematodes cannot regenerate tissue if injured, and they represent one of the few phyla including organisms known to have fixed numbers of cells.

For further information on general aspects of nematology, see Nicholas (1984), Poinar (1983), Croll and Matthews (1977), and Bird (1971). An earlier classic, now reprinted, is the monograph by Chitwood and Chitwood (1974). A useful collection of review articles on the use of nematodes as biological models has been edited by Zuckerman (1980a,b).

III. GENETICS

A. The Genome

About 80% of the *C. elegans* genome is composed of single-copy sequences, and the remainder is primarily moderately repetitive sequences present in two to ten copies per genome. The repetitive sequences include a transposable element Tc1, which is present at about 30 copies per genome in some strains, including the most commonly used Bristol laboratory strain (N2), and at about 300 copies in others, including the Bergerac strain. The molecular characteristics and organization of the genome are discussed in Chapter 3.

Genes of *C. elegans* can be mapped into six linkage groups corresponding to the six haploid chromosomes. The chromosomes are holocentric, that is, kinetochores are distributed along their length rather than localized at one point. The haploid set includes five autosomes (A) and a sex chromosome (X), all roughly equal in size. Sex is determined chromosomally, depending on the X/A ratio (Chapter 9). Hermaphrodites are diploid for all six chromosomes (XX), whereas males are diploid for the autosomes but have only one X chromosome (XO). Males arise spontaneously in hermaphrodite populations by X-chromosome nondisjunction at meiosis, with a frequency of about 1 in 500 animals.

With its short life cycle and hermaphrodite mode of reproduction, *C. elegans* is well suited for genetic analysis. Chemical mutagens such as ethylmethanesulfonate induce mutations at a high frequency. The recent discovery of strains in which some transposable elements are mobile in the germ line also allows mutagenesis by transposon insertion, which is particularly useful for the molecular cloning of mutationally defined genes (Chapter 3). A mutation present in the heterozygous state in a hermaphrodite will be homozygous in one quarter of that animal's self-progeny, so that recovery of recessive alleles is convenient. Genetic crosses can be carried out using males. About 700 *C. elegans* genes, distributed among all six linkage groups, have now been identified by mutation, mapping, and complementation testing. Based on the frequency of lethal mutations and assumptions about the proportion of essential genes, the total number of genes has been estimated at less than 5000. The genetics of *C. elegans* is discussed in Chapter 2.

B. Mutant Phenotypes

An initial difficulty with *C. elegans* genetics was the limited number of fertile mutant phenotypes that could be identified under the dissecting microscope. Partially offsetting this limitation was the ability of hermaphrodites to reproduce by self-fertilization, even with severe behavioral or morphological defects that make mating impossible. A wide variety of mutant phenotypes has now been described, most of which are identifiable with a dissecting microscope, but some only with a high-power microscope equipped with Nomarski optics or polarizing optics. A list of mutationally identified genes and the corresponding mutant phenotypes is given in Appendix 4B.

Increasingly sophisticated genetic tools have become available, including balancer chromosomes for maintaining lethal mutations, unstable duplications for mosaic analysis, temperature-sensitive mutants, and nonsense suppressors. Powerful selective methods have been developed, which allow the isolation of rare mutants and revertants. A great advantage of *C. elegans* genetics is that large numbers of animals (10^6 or more) can be handled by routine laboratory methods.

Genetic analysis has been invaluable in all of the major areas of *C. elegans* research, described in Chapters 6 through 12. The genetic approach has led to explicit models of many developmental processes in the nematode, such as vulva formation (Chapter 6), sperm maturation (Chapter 7), sex determination (Chapter 9), muscle assembly (Chapter 10), neuronal differentiation (Chapter 11), and dauer larva formation (Chapter 12). Key genes identified by mutation can be cloned and subjected to molecular analysis in order to test and extend these models. The availability of mutants also allows noninvasive experiments on many aspects of nematode physiology and function, most notably in the biology of muscles and nerves.

IV. ANATOMY

C. elegans has the typical nematode body plan described in Section II above, with an outer tube that consists of cuticle, hypodermis, neurons, and muscles surrounding a pseudocoelomic space that contains the intestine and gonad. A basement membrane separates hypodermis from muscle. Figure 4 shows a schematic cross section through an adult hermaphrodite. The shape of the worm is maintained by internal hydrostatic pressure, controlled by an osmoregulatory system. The general anatomy of *C. elegans* is described in Chapter 4.

A three-layered collagenous cuticle is secreted by the underlying hypodermis. This tissue is syncytial (made up of large multinucleate cells). In adults, lateral, longitudinal cords of seam cells form treads (alae) on the cuticle surface. On solid media the worm crawls on one side, with the alae contacting the substrate.

The obliquely striated body-wall muscle cells of *C. elegans* are arranged into four strips running the length of the animal, two dorsally and two

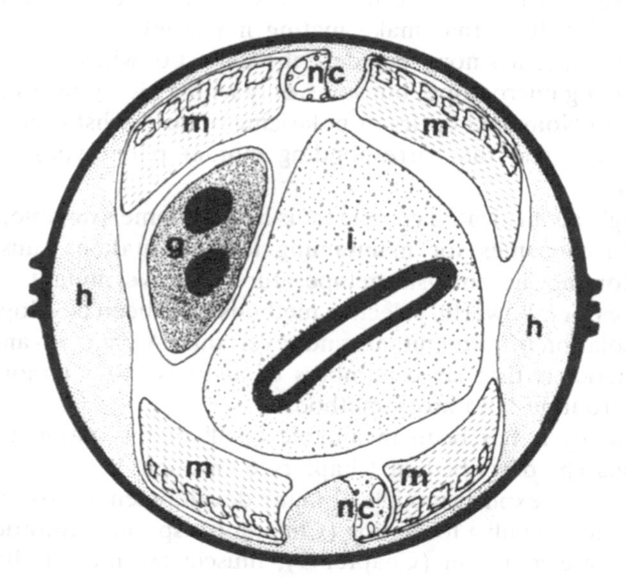

Figure 4 Diagram of a posterior cross section through the adult hermaphrodite, viewed toward the posterior. (g) Gonad; (h) hypodermal ridge; (i) intestine; (m) muscle; (nc) nerve cord. (Reprinted, with permission, from Edwards and Wood 1983.)

ventrally (Fig. 4). Most of the cells of the nervous system are found surrounding the pharynx, along the ventral midline, and in the tail. Processes from these neurons form an external ring around the pharynx (the nerve ring) or contribute to process bundles running the length of the body, the most prominent being the dorsal and ventral nerve cords. Sensory neurons run anteriorly from the nerve ring to sensory organs (sensilla) in the head.

The nerve ring receives inputs from the head region and sends its output primarily to the body-wall muscles, via motor neuron axons in the ring itself and in the dorsal and ventral cords. In nematodes, muscle cells send processes to motor neurons in the cords, rather than vice versa, as in other animals. The anatomy and function of the nervous system is described in Chapter 11 and Appendix 2.

C. elegans feeds through a bilobed pharynx, which pumps food into the intestine, crushing it as it passes through the second lobe. The intestine is formed from two rows of eight cells plus an anterior ring of four, surrounding a central lumen, which connects to the anus near the tail (Fig. 1). The simple excretory system is probably also responsible for osmoregulation. It consists of a pair of excretory canals, which are processes of a single cell that run the length of the animal, connecting to the exterior through the anteriorly located excretory pore.

The hermaphrodite reproductive system consists of a symmetrically arranged bilobed gonad, with one lobe extending anteriorly and the other posteriorly from the center of the animal. Each lobe is U-shaped, comprising a distal (to the uterus) ovary and a proximal oviduct and spermatheca (Fig. 1). The ovaries are syncytial, with germ line nuclei, partially segregated by membranes, surrounding a central cytoplasmic core. Moving proximally from the distal tip, the nuclei are first mitotic, then progress through meiotic prophase, reaching diakinesis in the oviduct prior to fertilization. At the bend in each lobe, individual nuclei become almost completely enclosed by membranes to form oocytes, which enlarge and mature as they pass down the oviduct. However, the oocytes maintain contact with the syncytium until close to the time of fertilization. The oviduct in each lobe terminates at a spermatheca carrying, in a young adult, about 150 ameboid sperm. The spermathecae connect to a common uterus, which contains fertilized eggs in early stages of embryogenesis. The uterus opens to the exterior through a vulva, which protrudes visibly from the ventral surface of the adult (Fig. 1).

The male gonad is a single-lobed, U-shaped structure, extending anteriorly from its distal end and then looping posteriorly and connecting with the cloaca near the tail (Fig. 1). At its distal end, the germ line nuclei are mitotic. Meiotic cells in progressively later stages of spermatogenesis are distributed sequentially along the gonad from the distal end to the seminal vesicle. Two meiotic divisions occur to produce the mature spermatids, which are stored in the seminal vesicle and released during copulation through a vas deferens to the cloaca.

The male tail has specialized neurons, muscles, and hypodermal structures for mating that give it quite a different appearance from that of the hermaphrodite (Figs. 1 and 5). The male tail is fan-shaped with 18 sensory rays. At the base of the tail are two spicules, which are inserted into the hermaphrodite vulva during copulation to aid in transfer of sperm. The hermaphrodite and male reproductive systems are described in more detail in Chapters 4, 7, and 9.

V. DEVELOPMENT

A. Fertilization and Embryogenesis

Mature oocytes pass through the spermatheca and become fertilized, either by the hermaphrodite's own sperm or by male sperm, which are introduced into the uterus by mating and stored in the spermatheca (Chapter 7). During the 30 minutes after fertilization, the zygote develops a tough, chitinous shell and a vitelline membrane, derived from components within the egg, which render the embryo impermeable to most solutes and able to survive outside the uterus. Normally, however, eggs are held in the uterus for the first few cleavages and are deposited through the vulva at about the time of gastrulation, approximately 3 hours after fertilization.

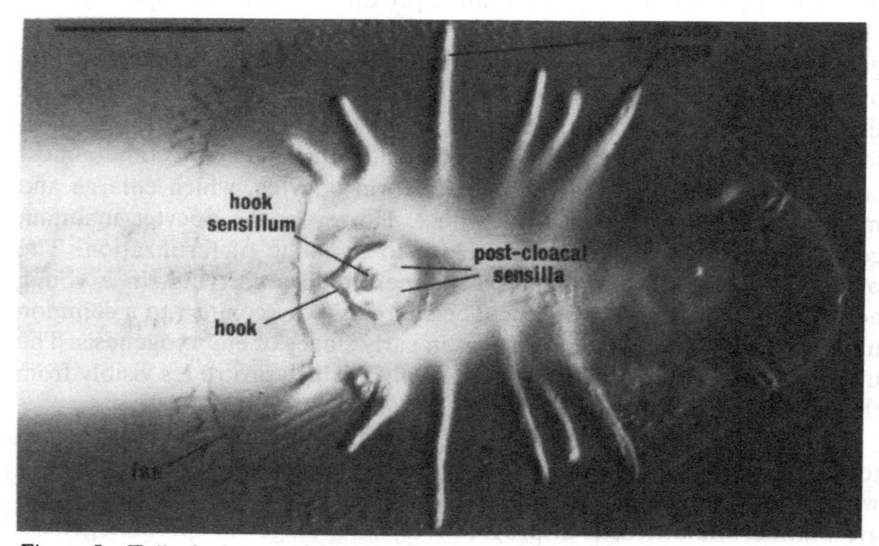

Figure 5 Tail of adult male *C. elegans*. Ventral view; Nomarski photomicrograph. Bar represents 20 μm. (Reprinted, with permission, from Sulston and White 1980.)

Embryogenesis can be divided into two phases of roughly equal duration: (1) cell proliferation and organogenesis, and (2) morphogenesis. During the proliferation phase, cell divisions, cell movements, and some cell deaths proceed according to a precise temporal and spatial pattern, invariant from one embryo to another, that gives rise to a fixed number of cells with rigidly determined fates (Chapter 5). Midway through embryogenesis, at 7 hours after fertilization, the embryo is a spheroid of approximately 550 cells, showing some differentiated characteristics. Cell proliferation then ceases almost entirely; during the next 7 hours the body elongates, neural processes grow out and interconnect and, finally, the cuticle is secreted. The first-stage larva (L1), 250 μm in length and consisting of 558 cells in the hermaphrodite and 560 in the male, hatches from the egg at about 14 hours after fertilization (Fig. 6). Embryogenesis is described further in Chapters 5 and 8.

B. Larval Development

Over the next 50 hours, larval development proceeds through three additional stages, L2, L3, and L4 (Figs. 6 and 7). About 10% of the cells in the L1 stage are somatic blast cells that undergo further cell division during larval development, contributing to the hypodermis, nervous system, musculature, and somatic gonadal structures (Chapter 5). In addition, at about 7 hours after hatching, the two germ line cells in the gonad primordium begin their proliferation, which continues through adulthood (Chapter 7). With the exception of germ line proliferation, the postembryonic cell divisions, like those in the embryo, follow precise and almost invariant temporal and spatial patterns, giving rise to fixed numbers of cells with determined fates. These patterns do differ between the sexes, however, to produce the bilobed gonad and vulva in the hermaphrodite and the single-lobed gonad and specialized tail structures for mating in the male.

The four larval stages are punctuated by molts. New cuticle is synthesized under the old, and pharyngeal pumping ceases during a brief period called lethargus, while the old cuticle is shed. The cuticles are different, both structurally and molecularly, at each of the stages, as discussed in Chapter 4. The relative sizes of larvae at each of the stages can be seen in Figure 7. Postembryonic development at the cellular level is described in Chapter 5, and its genetic control is discussed in Chapter 6.

C. Gonadogenesis and Gametogenesis

Gonadogenesis is completed during the L4 stage. In the hermaphrodite during this stage, germ cells in each arm of the gonad undergo meiosis and differentiate into about 150 mature sperm, which are stored in the spermatheca. At the L4 molt, sperm production ceases (Fig. 6), and sub-

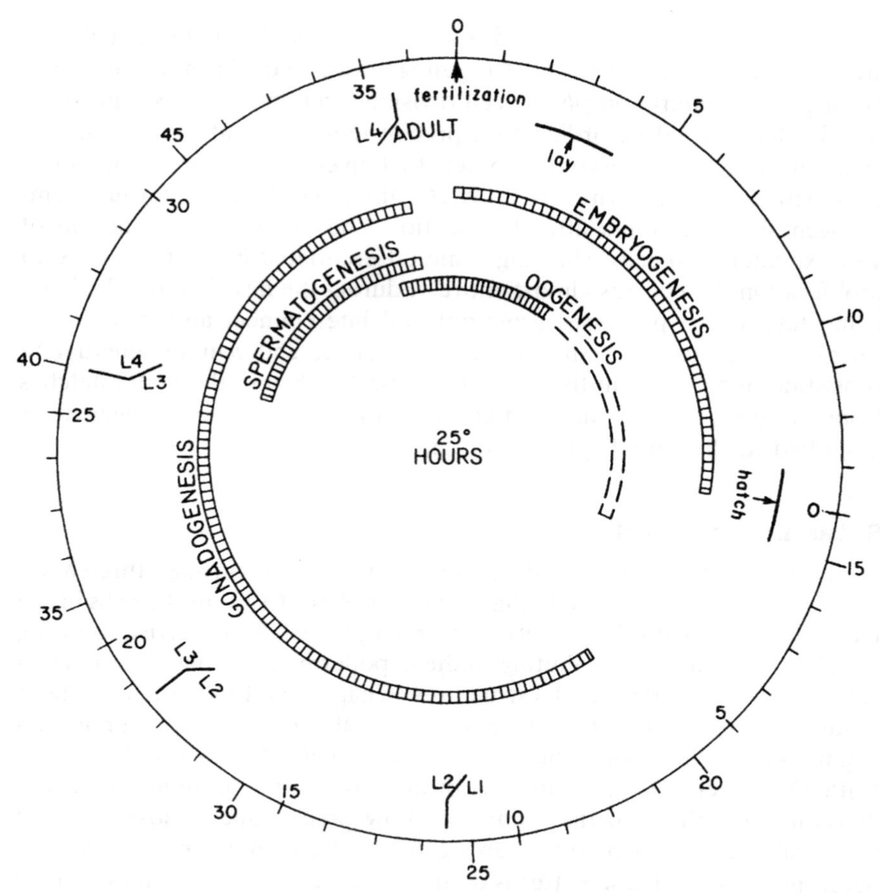

Figure 6 Diagrammatic representation of the *C. elegans* life cycle, showing durations of developmental stages. Numbers on the outside of the large circle indicate hours after fertilization at 25°C; numbers on the inside indicate approximate hours after hatching. The four larval molts are indicated by radial lines. (Reprinted, with permission, from Wood et al. 1980.)

sequent meiosis and differentiation generates only oocytes. At the same time, the vulva opens to the exterior so that mating can occur, and self-fertilization begins shortly thereafter. The number of progeny that a hermaphrodite can produce by self-fertilization is limited by the number of sperm. Oocytes are produced for about 4 days, and an unmated hermaphrodite depleted of sperm continues to produce and lay unfertilized oocytes for a day or so. A mated hermaphrodite can produce over 1000 progeny. In the male, which also becomes fertile shortly after the L4 molt, sperm production continues throughout the reproductive life span. Development of the germ line and fertilization are described in Chapter 7.

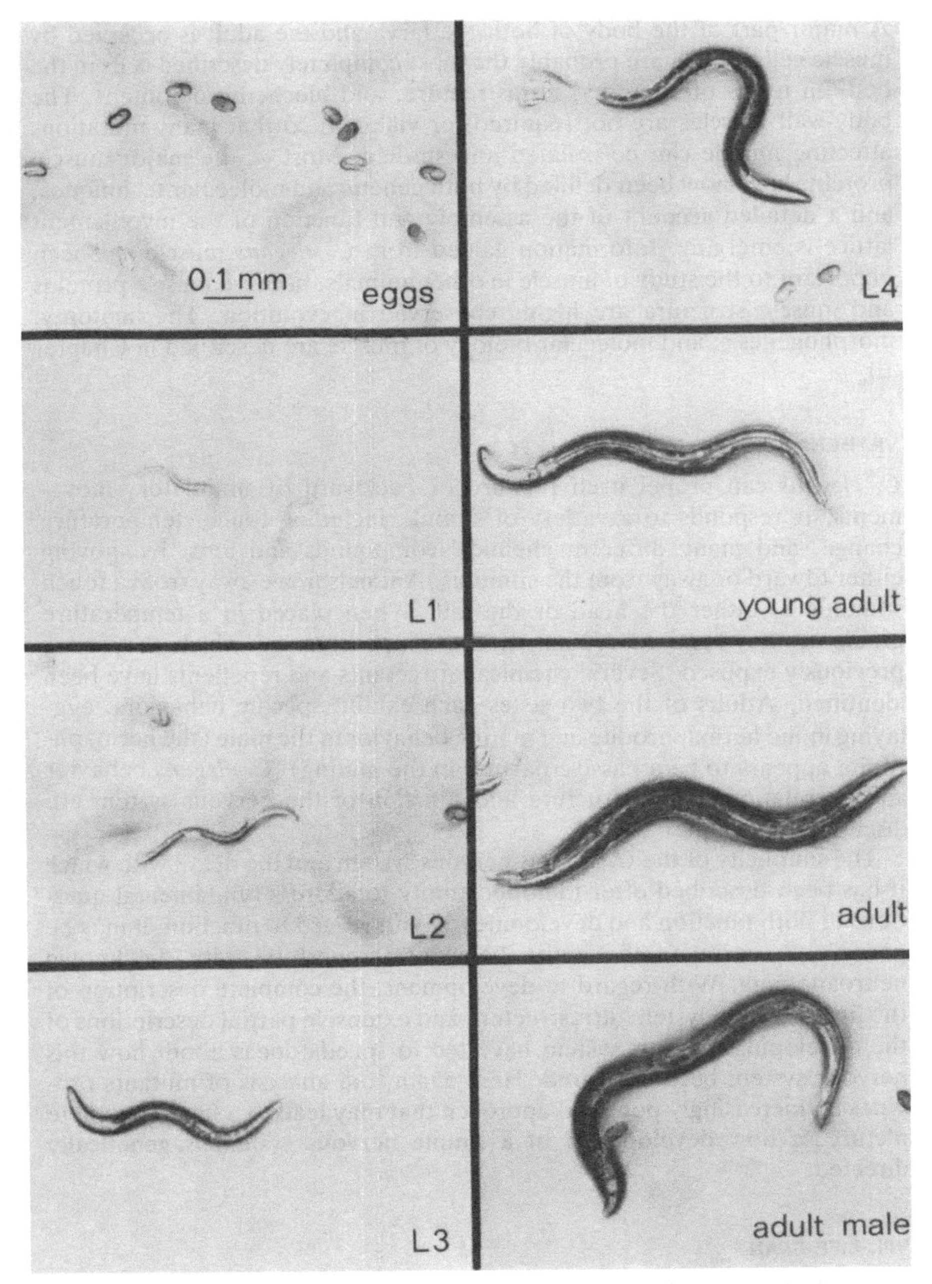

Figure 7 Eggs, larvae, and adults of *C. elegans*. All panels except lower right show hermaphrodites. Bright-field photomicrographs. (Courtesy of J. Sulston.)

VI. MUSCLE BIOLOGY

A major part of the body of both the larva and the adult is occupied by muscle cells, which are probably the most completely described cells in the body in terms of anatomy, ultrastructure, and biochemical content. The body-wall muscles are not required for viability, so that many mutations affecting muscle can be isolated and studied. Most of the major muscle proteins have now been defined by both genetic and molecular techniques, and a detailed account of the assembly and function of the myofilament lattice is emerging. Information gained from *C. elegans* muscle has been important to the study of muscle in other animals, because muscle proteins and muscle structure are highly conserved in evolution. The anatomy, morphogenesis, and molecular biology of muscle are described in Chapter 10.

VII. BEHAVIOR AND NEUROBIOLOGY

C. elegans can propel itself forward or backward by undulatory movements. It responds to a variety of stimuli, including touch, temperature change, and many different chemical compounds and ions, by moving either toward or away from the stimulus. Animals move away from a touch stimulus to either the head or the tail. When placed in a temperature gradient, they tend to remain at the temperature to which they were previously exposed. Several chemical attractants and repellents have been identified. Adults of the two sexes each exhibit specific behaviors, egg-laying in the hermaphrodite and mating behavior in the male (the hermaphrodite appears to be a passive partner in the mating). *C. elegans* behavior and its relation to the structure and function of the nervous system are discussed in Chapter 11.

The simplicity of the *C. elegans* nervous system and the detail with which it has been described offer the opportunity to address fundamental questions of both function and development. With regard to function, it may be possible to correlate the entire behavioral repertoire with the known neuroanatomy. With regard to development, the complete description of the final nervous system ultrastructure and extensive partial descriptions of the developing nervous system have led to specific ideas about how this nervous system becomes wired. Here again, the analysis of mutants provides an increasingly powerful approach that may lead to a fairly complete picture of how development of a simple nervous system is genetically directed.

VIII. LIFE SPAN

When the food supply is limited early in larval development, *C. elegans* can take an alternative development pathway at the L2/L3 molt to produce the dauer larva, a specialized L3 stage that does not feed, is resistant to desiccation, and can survive for up to 3 months without further development. If food becomes available during this period, the dauer larva molts to become an L4, which resumes normal development. The dauer larva pathway and its genetic control are described in Chapter 12.

With adequate food throughout the life cycle, under standardized laboratory conditions (liquid culture, 20°C), both hermaphrodites and males live for about 17 days after reaching adulthood, although hermaphrodites consistently show a slightly longer life span than males. Changes that occur during senescence include decreasing mobility, alterations in enzyme activities and morphology, and accumulation of the pigment lipofuscin. Quantitative genetic analysis indicates a substantial genetic component in determination of life span. Long-lived strains can be obtained by interstrain breeding or by isolation of apparent single-gene mutants; study of these strains suggests that lengths of different stages in the life cycle may be controlled separately. *C. elegans* should be a useful model for studying some aspects of aging (for review, see Zuckerman 1980b; Johnson and Wood 1982; Russell and Jacobson 1985; Johnson 1987).

IX. *C. ELEGANS* AS AN EXPERIMENTAL SYSTEM

The key attributes of *C. elegans* as an experimental system for biological studies are its simplicity, transparency, ease of cultivation in the laboratory, short life cycle, suitability for genetic analysis, and small genome size. These properties have made possible accumulation of the extensive descriptive information summarized in this volume. Now known are the complete anatomy at electron microscope resolution (Chapter 4; Appendix 2), the locations and characteristics of all somatic cells in the adult hermaphrodite and male (Appendix 1), and the complete cell lineage, that is, the timing, locations, and ancestral relationships of all cell divisions during development (Chapter 5; Appendix 3). Progressing rapidly is genomic description of the animal, including detailed genetic mapping (Chapter 2; Appendix 4), cloning and analysis of mutationally defined genes, and physical mapping of the entire genome (Chapter 3). In addition to the complete wiring diagram of the nervous system derived from the anatomical work, some knowledge of neurophysiological function has been gained from neurotransmitter analysis and electrophysiological study of the homologous nervous system in the large nematode *Ascaris suum* (Chapters 4 and 11). Because of its favorable experimental attributes and the wealth of descriptive information now available, *C. elegans* should be an increasingly useful experimental organism for investigation of a variety of problems in animal biology.

ACKNOWLEDGMENTS

I am grateful to the numerous colleagues who contributed comments, suggestions, and corrections during preparation of this chapter and, in particular, to Marty Chalfie, Jonathan Hodgkin, John Sulston, and John White for their advice and assistance.

2
Genetics

Robert K. Herman
Department of Genetics and Cell Biology
University of Minnesota
St. Paul, Minnesota 55108

I. INTRODUCTION

As discussed in Chapter 1, the philosophy behind much *Caenorhabditis elegans* research is that the analysis of mutants by a variety of methods will lead to a better understanding of animal development and behavior. The choice of *C. elegans* as an experimental organism is thus based, in part, on its tractability for genetic analysis. The purpose of this chapter is to review

The Nematode Caenorhabditis elegans. © 1988 Cold Spring Harbor Laboratory. 0-87969-307-X/88 $1 + .00.

the current status of the genetics of *C. elegans*. The discussion of identification and characterization of mutants will cite specific examples primarily for the purpose of illustrating general genetic approaches that have been used; other chapters will focus more directly on specific aspects of *C. elegans* biology. This chapter also reviews what is known about genome organization (other than at the DNA level, which will be reviewed in Chapter 3) and about chromosomal structure and stability. The system of genetic nomenclature for *C. elegans* (Horvitz et al. 1979) is discussed in Appendix 4.

II. GENOME ORGANIZATION

A. Diploid Chromosomal Complements

Using Feulgen staining and light microscopy, Nigon (1949a) showed that wild-type *C. elegans* hermaphrodites contain five pairs of autosomes and one pair of X chromosomes and that males contain five pairs of autosomes and a single X chromosome. Fluorescent dyes such as Hoechst 33258 have been used subsequently for viewing both meiotic (Herman et al. 1976) and mitotic (Albertson and Thomson 1982; Ellis and Horvitz 1986) chromosomes by fluorescence microscopy (Figs. 1 and 2). The chromosomes of oocytes at diakinesis are highly condensed, and the six bivalents present in wild-type diploid hermaphrodites are generally indistinguishable cytologically. Mitotic chromosomes are best seen in early (<50 cells) embryos (Albertson and Thomson 1982). The mitotic chromosomes appear as stiff rods, 1–2 μm in length. In early cleavage stages, one pair of chromosomes may appear longer than the others, and two pairs may appear shorter; but as development proceeds, the chromosomes all appear shorter and of uniform size at metaphase (Albertson and Thomson 1982).

B. Polyploids and Aneuploids

Nigon (1949b, 1951a,b) identified and characterized a tetraploid stock of *C. elegans* variety Bergerac following heat treatment. The hermaphrodite

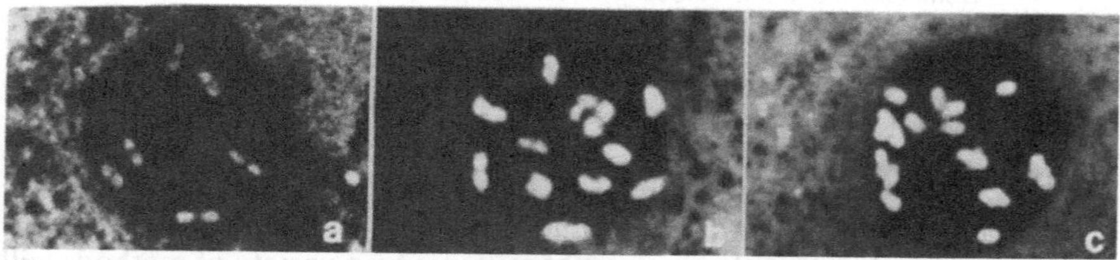

Figure 1 Fluorescence microscopy of oocytes stained with Hoechst 33258. (*a*) Diploid; (*b*) tetraploid; (*c*) triploid. Magnification, 1600 ×. (Reprinted, with permission, from Herman et al. 1979.)

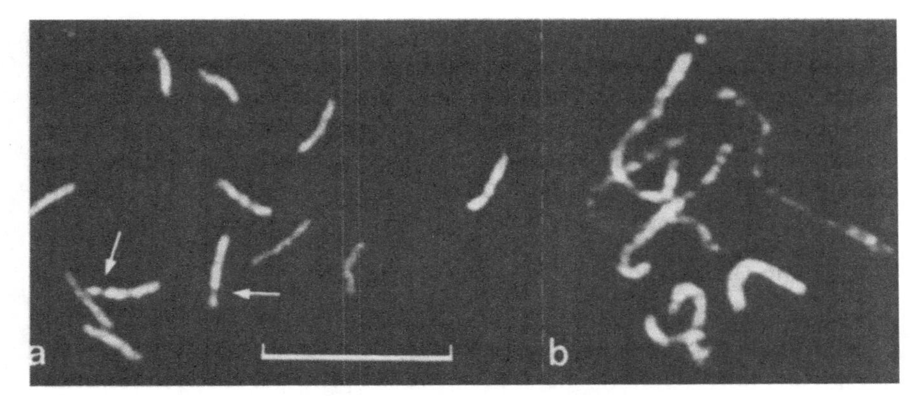

Figure 2 *C. elegans* stained with Hoechst 33258. (*a*) Mitotic chromosomes from a squashed embryo. Arrows point to a prominent dark band that is occasionally seen. (*b*) Pachytene chromosomes from a hermaphrodite. Bar represents 10 μm. (Reprinted, with permission, from Albertson and Thomson 1982.)

and male tetraploids were about 15% longer than their diploid counterparts, and the hermaphrodites fell into two classes, based on the frequencies of males among their self-progeny. We shall refer to the two classes as LFM and HFM, for low- and high-frequency male producers, respectively. Low brood sizes (averages of 17–21 animals) made some hermaphrodites unclassifiable; but omitting the unclassified animals, Nigon reported that LFM segregated about 0.6% males, 93% LFM and 7% HFM, and HFM segregated about 42% males, 41% HFM and 17% LFM. On the basis of these progeny ratios and the cytology of Feulgen-stained oocytes and spermatocytes, Nigon concluded that LFM producers were *4A;4X* (four sets of autosomes and four X chromosomes, with oocytes showing 12 bivalents), that HFM producers were *4A;3X*, and that males were *4A;2X*. Nigon also frequently observed irregular diakineses, with one or two extra or missing chromosomes; he thus suggested that the numerous sterile or weakly fertile animals he encountered were aneuploid.

Madl and Herman (1979) used heat shock to establish tetraploid stocks of the Bristol variety (Fig. 1b). The self-progeny ratios from the English tetraploid hermaphrodites were in good agreement with Nigon's results, although the broods were larger (with ~80 progeny per brood). Triploid animals were produced by mating diploids and tetraploids; oocytes of triploid hermaphrodites generally showed the 18 chromosomes as six bivalents and six univalents (Fig. 1c). Only about 15% of the eggs laid by triploid hermaphrodites hatched (compared with 87% for tetraploids), and many progeny were sterile or very weakly fertile. The genotype *3A;3X* was shown to be hermaphroditic, and *3A;2X* animals were shown to be male.

Hodgkin et al. (1979) identified *2A;3X* animals generated by mutants showing enhanced meiotic nondisjunction of the X chromosome; the *3X* animals are short, fertile hermaphrodites. Sigurdson et al. (1986) have identified triplo-IV hermaphrodites, which were fertile wild type in overall morphology. Trisomics for other chromosomes may also be viable but have not been sought. Hodgkin et al. (1979) conducted searches for monosomics for each autosome and found none, probably because they are inviable.

C. Defining Genes and Estimating Gene Number

Simple mutations (those that do not involve extensive deletions or other chromosomal aberrations) have been assigned to genes on the basis of the complementation test, which has provided the principal criterion for defining genes. In the complementation test, two recessive mutations are made *trans*-heterozygous; if the resulting phenotype is wild type, the mutations are said to complement and are generally assigned to different genes. Complexities that have sometimes been encountered in assigning mutations to genes are discussed in Section III.B. Brenner (1974) assigned about 300 mutations conferring visible phenotypes to about 100 complementation groups or genes. Because a considerable fraction of these genes was represented by more than one mutant allele, Brenner concluded that the number of genes definable by obvious visible (and fertile) phenotypes was approaching saturation; these genes do not appear to be essential for viability or fertility (Brenner 1974). During the past decade, additional apparently unessential genes have been defined, some by mutations that confer rather subtle phenotypes, detectable only by special assays (Appendix 4), but the total number so far identified is no more than about 300.

Brenner also estimated the number of essential *C. elegans* genes, defined as genes whose expression is required for viability. Based on his measured frequency of X-linked recessive lethal mutation induced by ethyl methanesulfonate (EMS) and his estimated EMS-induced mutation frequency per visible locus, Brenner calculated that the X chromosome has 300 indispensable genes. By extension to the six linkage groups of roughly equal size, a total estimate of 1800 essential genes was obtained. This estimate does not include genes that give only sterile phenotypes. Sterile mutants are generally about as frequent as lethals (Hirsh and Vanderslice 1976; Herman 1978), but a significant fraction of mutations conferring a sterile hermaphrodite phenotype appear to be weak alleles of genes essential for viability (Meneely and Herman 1981); hence, the number of genes essential for hermaphrodite viability and fertility is estimated at roughly 3000. Essential gene numbers have also been estimated from experiments in which many recessive lethal and sterile mutations were sought in a selected region of the genome and then assigned to complementation groups. Given the number of genes represented by more than one allele and certain assumptions about the distribution of mutations per locus, estimates of the total number of essential genes in the selected region were derived. This method tends to give underestimates because it overemphasizes the more mutable loci, but in general, it has agreed reasonably well with the overall genome number of roughly 3000 essential genes (Meneely and Herman 1981; Rogalski et al. 1982; Rogalski and Baillie 1985). In 235 kb of cloned DNA surrounding the five X-linked vitellogenin genes, Heine and Blumenthal (1986) identified 12 transcriptionally active genes. If this spacing of 20 kb per gene were typical of the entire genome, *C. elegans* would have about 4000 genes total (essential and unessential).

Genes classified as unessential by the criterion of not being mutable to a recessive lethal form may nonetheless contribute to essential functions. More than one gene might contribute either redundantly or additively to an essential function, for example, so that inactivation of one gene would have, at most, a slight (and nonlethal) effect. Excellent candidates for such classes of genes are members of the following gene families: vitellogenins (see Chapter 7), major sperm protein (Chapter 7), acetylcholinesterase (Chapter 11), actin (Chapter 10), collagen (Chapter 3), and tRNA (Chapter 3). Many of these genes have been defined by direct molecular analysis rather than mutation (Chapter 3). The null mutant (no gene function) phenotypes of the actin genes *act-1* and *act-3* seem to be wild type (Landel et al. 1984), but dominant mutations in these genes that affect muscle structure and animal movement, owing to altered actin or inappropriate synthesis of actin, have been identified (Waterston et al. 1984). Other examples of mutations that serve to identify genes having wild-type null phenotypes will be described in Section III.B. Such mutations seem to be rare, and it is difficult to estimate the number of genes that may be mutationally silent. Mutations in two or more different genes that are phenotypically silent by themselves may together confer an observable phenotype; examples of such synthetic mutants are discussed in Section IV.E.

D. Methods of Gene Mapping

The general genetic procedures for gene mapping were developed by Brenner (1974), who mapped about 100 genes onto six linkage groups. Methods for ascertaining linkage and measuring the map distance between two genes and for ordering genes by three-factor crosses are described in Methods.

The Bristol and Bergerac strains of *C. elegans* exhibit numerous restriction fragment length differences when analyzed by Southern blot hybridization using cloned probes (Emmons et al. 1979). Such DNA dimorphisms have been treated as standard phenotypic markers in two- and three-factor crosses (Rose et al. 1982; Files et al. 1983; Cox et al. 1985) to localize cloned genes on the genetic map (for fuller discussion, see Chapter 3).

Many recessive lethal deficiencies have been tested for complementation against recessive single-gene mutations (visibles, steriles, or lethals). Noncomplementation between a deficiency and a single-gene mutation indicates that the deficiency extends into or through the gene defined by the single-gene mutation. Such complementation tests have been used to characterize the extents of many deficiencies. Sets of overlapping deficiencies with different end points have been used to define subsegments of a region of the genetic map. New point mutations falling in the region can then be assigned to a particular segment of the genetic map by subsequent complementation tests against a small number of deficiencies; this facilitates the assignment of the point mutation to a particular gene because it needs to be complementation tested only against other point mutations occupying the same segment. This approach has been used for a portion of the X chromosome (Meneely and Herman 1979, 1981) and regions of linkage group I (LGI) (Rose and Baillie 1980), LGII (Sigurdson et al. 1984), and LGIV (Rogalski et al. 1982; Rogalski and Baillie 1985).

Albertson has developed methods for localizing genes, both repeated and single copy, by in situ hybridization of cloned probes to mitotic chromosomes (Albertson 1984b, 1985) (for a fuller discussion of the technique, see Chapter 3).

III. ANALYSIS OF SINGLE GENES

A. Isolation of Mutants

Most single-gene mutations identified in *C. elegans* have been induced with the monofunctional alkylating agent EMS. In a treatment used by Brenner (1974), the average forward mutation rate was estimated to be 5×10^{-4} per gene. EMS mutagenesis and other mutagenic treatments that have been used with *C. elegans* are described in Methods.

Many mutants have been identified by direct inspection of descendants of mutagenized hermaphrodites, but mutant selection or enrichment schemes have also been used. Various drug-resistant mutants have been isolated (e.g., Brenner 1974; Lewis et al. 1980a; Sanford et al. 1983; Rand and Russell 1984). Dusenbery (1973; Dusenbery et al. 1975) devised a countercurrent distribution technique, and Lewis and Hodgkin (1977) used an agarose plate method for enriching for mutants defective in chemotaxis. Zengel and Epstein (1980b) have described a method for selecting motility-defective mutants. The resistance of dauer larvae to sodium dodecyl sulfate (SDS) (Cassada and Russell 1975) has provided a means for enriching for mutants that form dauer larvae in the presence of food or recover slowly from the dauer state (Riddle 1977). Some useful screening methods exploit mutant interactions to make the desired mutant more easily recognizable in the background of other animals (rather than actually enriching for the mutant); these techniques, involving intragenic suppression, indirect suppression, and epistasis, are discussed in later sections. Two methods have been used to make sterile hermaphrodites identifiable by inspection. In one, animals were immobilized by exposure to levamisole; the absence of eggs near an immobilized animal was indicative of sterility (Argon and Ward 1980). In the second procedure an egg-laying-defective mutant (Trent et al. 1983), which is normally killed by internally hatching offspring, had its life prolonged by sterile (or maternal-effect embryonic lethal) mutation (J. Priess et al., pers. comm.).

B. Identifying and Analyzing Alleles

It is common to find that not all members of a set of alleles, called an allelic series, confer the same phenotype. It is often important to know the phenotype conferred by a null allele (giving no functional gene product) of a gene. The most direct criterion for identifying a null allele is to demonstrate the nature of the mutation by DNA sequencing (e.g., Dibb et al. 1985). At earlier stages in the analysis, the following criteria have been used; none of these criteria taken alone is foolproof, but cumulative evidence can make a strong case. First, a null allele is expected to give the most severe recessive phenotype in an allelic series. Second, a null allele should usually fail to complement all other alleles in the series. Third, it should give the same phenotype when homozygous as when heterozygous opposite a deficiency. Fourth, a null mutation should confer the same phenotype as a deficiency when put opposite any other allele. Fifth, null mutations are not expected to be rare, so the forward mutation frequency may be used as suggestive evidence that null mutations have been induced. Sixth, null alleles are usually not temperature sensitive. And, finally, some null alleles are suppressible by terminator codon suppressors, as discussed later in this chapter. Some unusual null phenotypes are those showing variable expressivity (Horvitz and Sulston 1980), a wild-type phenotype (Greenwald and Horvitz 1980; Park and Horvitz 1986a), and temperature sensitivity (Golden and Riddle 1984c; W. Fixsen and R. Horvitz, pers. comm.).

Mutations that confer a reduced but nonzero level of gene activity are often recognized by the following properties: They are recessive to their wild-type alleles, they do not give the strongest mutant phenotype in an allelic series, and they give a more severe phenotype when put opposite a deficiency or null mutation than when homozygous. A striking example of such a partial-loss mutation is *tra-2(e1875)*; when homozygous, this *tra* (sexual transformer) allele gives a wild-type phenotype, but when opposite a *tra-2* null mutation, leads to transformation of *XX* animals into a pseudomale phenotype (J. Hodgkin, pers. comm.). Partial-loss mutants have sometimes been referred to in the literature as hypomorphs (Muller 1932). Most mutations in *C. elegans* seem to be null or partial loss, and these two types together comprise the class 1 (loss-of-function) mutants listed in Appendix 4.

Not all loss-of-function mutations are recessive to their wild-type alleles; for example, *tra-2(null)*/+ hermaphrodites are egg-laying-defective (Trent et al. 1983), and *unc-22(null)*/+ animals display a mutant twitch in a 1% solution of nicotine (Moerman and Baillie 1979). Such loci are said to be haplo-insufficient.

The great majority of dominant or semidominant mutations appear to be gain of function, rather than loss. One class of gain-of-function mutation leads to an increased level of apparently normal gene activity (referred to as class 2 mutations in Appendix 4). The increased activity may occur in those cells that would normally contain some lower (nonzero) activity, or it may occur in cells that do not normally have any activity (inappropriate or constitutive expression). Mutations have been tentatively assigned to this class, either because their mutant phenotype is enhanced by the addition of a wild-type allele to the mutant genotype or because the dominant phenotype is the opposite of that conferred by the null allele. Examples of class 2 mutations are certain alleles of the lineage control gene *lin-12* (Greenwald et al. 1983) and *tra-1(e1575)* (Hodgkin 1983b).

Another class of dominant gain-of-function mutation leads to novel gene products (class 3 in Appendix 4). Park and Horvitz (1986a) have identified dominant mutations in nine genes with the following properties: Each mutation was non-null (the null alleles were not dominant), and each mutation opposite a null allele conferred a more severe phenotype than the mutation opposite its wild-type allele; it was concluded that each mutant produced a novel gene product and that each novel function was antagonized by the wild-type gene product. Two of the nine mutants, bearing *egl-30* (egg-laying-defective) and *unc-54* (uncoordinated, paralyzed) mutations, had the following additional property: The mutations opposite a wild-type allele resulted in a mutant phenotype similar to that of the null homozygote. Thus, the wild-type function was antagonized by the mutant gene product. Mutations that result in novel products are not invariably dominant or semidominant, for example, the recessive mutation *sup-10-(n983)* (Greenwald and Horvitz 1986).

Although a fully dominant mutation cannot be assigned to a gene on the basis of the complementation test, secondary criteria can be used. The first of these would be a related phenotype and very tight linkage. A stronger criterion is the reversion of the dominant mutation by a very closely linked *cis* mutation that confers a recessive phenotype and shows noncomplementation with another recessive allele (see Section IV.A on intragenic suppression).

It is not always possible to assign alleles to a linear series corresponding to increasing severity of their phenotypes. Hodgkin (1987b) has graded a number of *tra-1* alleles according to their effects on various aspects of sexual transformation, for example, and the allelic orders based on different aspects of the overall phenotype were different.

The property of noncomplementation can be used in mutant screens for identifying new alleles of a gene. For example, *fem-1* (*fem* = feminization) was originally defined only by a temperature-sensitive mutation, *hcl-7* (Nelson et al. 1978); Doniach and Hodgkin (1984) identified new *fem-1* alleles, which have more extreme phenotypic effects than *hcl-7*, on the basis of their showing noncomplementation with *hcl-7*.

It should be noted that, although very rare, examples of noncomplementing nonallelic mutations are known; one is described in Section IV.E.

1. Intragenic Complementation

Two mutations sometimes complement each other but are nonetheless assigned to the same gene because they both fail to complement a third allele, thought to be a single-site mutation. Such intragenic or interallelic complementation is often partial and seems to be more common for some genes than others. Three examples of intragenic complementation will be briefly noted.

Rand and Russell (1984) have described mutations in a gene called *cha-1*, which result in decreased choline acetyltransferase (ChAT) activity, uncoordinated movement, and resistance to cholinesterase inhibitors. Because a hypomorphic allele resulted in altered kinetic properties for ChAT, *cha-1* is thought to be a structural gene for the enzyme. Mutations in *unc-17*, which maps within 0.02 map unit of *cha-1*, confer both resistance to cholinesterase inhibitors and the same uncoordinated phenotype as *cha-1* mutations but do not lead to deficiency for ChAT. Complementation tests, with respect to the uncoordinated phenotype between different *cha-1* and *unc-17* mutations, gave the following results: All 4 *cha-1* alleles failed to complement each other and all 12 *unc-17* alleles failed to complement each other; 3 of the *cha-1* alleles complemented 10 *unc-17* alleles, but the fourth *cha-1* allele and the remaining 2 *unc-17* alleles failed to complement all other *cha-1* and *unc-17* mutations. Thus, it appears that *cha-1* and *unc-17* comprise a complex locus. More recently, J. Rand (pers. comm.) mapped genetically the exceptional *cha-1* mutation between the other three *cha-1* alleles and the *unc-17* mutations, and he mapped the two exceptional *unc-17* mutations to a segment of *unc-17* away from the *cha-1* side. Rand and Russell (1984) have proposed that the ChAT molecule consists of two functionally independent domains, one of which (the *cha-1* domain) contains the catalytic site and the other of which (the *unc-17* domain) is involved in a noncatalytic function, such as subcellular localization of the molecule. Intragenic complementation could be explained if the enzyme normally functions in vivo as a homodimer.

Mutations in *unc-84* disrupt two sets of nuclear migrations that occur during development. Complementation studies of 16 *unc-84* alleles have defined four classes of alleles (with 3–5 alleles in each class): Alleles in the first class fail to complement all other alleles, alleles in the second class complement alleles in the third class, and alleles in the fourth class partially complement those in the third class; thus, the *unc-84* locus may encode two independently mutable functions (W. Fixsen and R. Horvitz, pers. comm.).

Sixteen alleles of *let-2* have been identified (Meneely and Herman 1981). Fifteen are recessive embryonic lethal mutations, and 12 of these are temperature sensitive, although only 1 was isolated as such (Hirsh and Vanderslice 1976). One allele is an early larval lethal at 25°C and sterile at 20°C. The pattern of complementation between pairs of *let-2* alleles was extremely complex: The 15 alleles tested fell into 14 distinct classes on the basis of their complementation behavior. Only one allele, an amber mutation (see below), failed to complement all other alleles.

2. *Intragenic Mapping*

Mutations within *unc-22* (Moerman and Baillie 1979), *unc-13* and *unc-15* (Rose and Baillie 1980), and *unc-54* (Waterston et al. 1982a; Moerman et al. 1982) have been ordered with respect to each other by genetic mapping. The general strategy in each case was to generate hermaphrodites that were *trans*-heterozygous for the two noncomplementing alleles to be mapped and also heterozygous for closely linked markers flanking the locus. Rare wild-type recombinants were identified, and the genotypes of the recombinant chromosomes were ascertained. Most recombinant chromosomes were found to be recombinant for flanking markers (i.e., not the result of gene conversion or outside marker exchange), and the flanking marker genotypes, assuming single crossover events, were used to deduce the order of the two alleles used in the cross. In the *unc-54* work, 13 sites were ordered, and the physical positions of certain alleles were determined by protein or nucleic acid chemistry, so that parts of the map were interpretable in physical terms. The two ends of the gene, coding for the amino-terminal globular head of the myosin heavy chain and the carboxy-terminal rodlike meromyosin, were distinguished on the map, for example. The map covered more than 80% of the structural gene, and most, if not all, mapped alleles were found to lie within the structural gene.

C. Other Mutant Characteristics

Temperature-sensitive mutants can be used in temperature-shift experiments to define the temperature-sensitive period, when the gene product is either synthesized, assembled, or functional. Useful discussions of how temperature-shift experiments can be interpreted have been given by Wood et al. (1980), Miwa et al. (1980), and Swanson and Riddle (1981).

Some mutations show a maternal effect. An excellent example is mutation in *tra-3*: Homozygous mutant self-progeny of a heterozygous parent are fertile hermaphrodites, but their self-progeny are all transformed by *tra-3* to pseudomales (Hodgkin and Brenner 1977). Homozygous *tra-3* hermaphrodites, generated from *tra-3*/+ mothers, will produce hermaphrodite progeny, however, when mated with wild-type males; this indicates that zygotic expression of *tra-3(+)* is sufficient to prevent sexual transformation. Many temperature-sensitive embryonic lethal mutations and mutations affecting oogenesis and spermatogenesis have been shown to have parental effects; maternal- and paternal-effect tests for such mutations are described in Chapter 8.

IV. MUTANT INTERACTIONS

The phenotypes of double mutants have provided useful information about particular alleles or gene interactions; they have also been useful in designing screens for new mutants. Three general classes of mutant interaction will be discussed: genetic suppression, epistasis, and other interactions of nonallelic mutations. A suppressor mutation is a second mutation at a site distinct from the first mutation that reverses, at least partially, the phenotypic expression of the first mutation. Three classes of suppressor mutation will be discussed: (1) suppressors that occur in the same gene as the suppressed mutation (intragenic suppression); (2) direct or informational suppression, in which the suppressor alters the fidelity of information flow from gene to protein; and (3) indirect suppression.

The general procedure for demonstrating that a phenotypic reversion is due to an extragenic suppressor mutation is to cross the revertant to wild type and show that the original mutation can be recovered in a suppressor-free descendant; this amounts to showing that the original mutation and the suppressor mutation are separable by recombination and, therefore, must be at distinct sites. When the suppressor is intragenic, the recombination frequency is generally extremely low, so that the distinction between intragenic suppression and true reversion must be made on some other basis, such as frequency of phenotypic reversion or the types of phenotypes produced by the different mutations.

A. Intragenic Suppression

Intragenic suppressors have frequently been used as a source of new, often null, alleles. The approach usually taken makes use of mutants that have a visible phenotype by virtue of producing an altered gene product or an increased level of gene product (e.g., in an inappropriate cell type). A second mutation in the same gene that inactivates expression then suppresses the first mutation. Dominant mutations are good sources of mutations that are suppressible in this way. It is often convenient to select for phenotypic reversion of a dominant mutation when it is *trans* to a wild-type allele of the gene. The null mutation (or, more exactly, the double mutation) often turns out to have a recessive phenotype of its own, but it is only revealed following segregation of homozygotes. Examples of genes for which this approach has yielded many null alleles are *unc-54* (Anderson and Brenner 1984), the actin genes *act-1* and *act-3* (Landel et al. 1984; Waterston et al. 1984), *tra-1* (Hodgkin 1987b), *lin-12* (Greenwald et al. 1983), and *lin-14* (Ambros and Horvitz 1984, 1987). The mutation *unc-93(e1500)* is largely recessive to *unc-93(+)*, but because the homozygous *unc-93* null phenotype is wild type, null mutations are *cis*-dominant suppressors of *e1500* and are easily selected for as phenotypic revertants of *unc-93(e1500)*.

B. Informational Suppression

The *sup-5 III* and *sup-7 X* suppressors were first identified by genetic methods, which indicated they had many properties in common with tRNA nonsense suppressors of bacteria and yeast. Waterston and Brenner (1978) identified *sup-5(e1464)* as an EMS-induced suppressor of *unc-15(e1214) I*. The *unc-15* gene specifies paramyosin (Waterston et al. 1977); the *e1214* mutant contains no detectable paramyosin and is paralyzed. Rare revertants with even partially restored mobility are easily detected in a background of paralyzed animals. The *unc-15; sup-5* strain moves better than the *unc-15* strain, although not as well as wild type, and it contains some paramyosin.

Waterston and Brenner (1978) showed that the *sup-5(e1464)* mutation suppressed specific alleles of many unrelated genes; and in two genes with identified polypeptide products, suppressed alleles were invariably null, as judged by absence of polypeptide. Waterston (1981) subsequently showed that *sup-7(st5)*, isolated as a suppressor of an *unc-13* mutation, suppressed the same spectrum of mutations as did *sup-5(e1464)* but to a greater extent. Homozygous *sup-7* and *sup-5* suppressors were shown to confer as much as 39–45% and 20–25% wild-type paramyosin levels, respectively, to a null mutation in *unc-15* (Waterston 1981). Both suppressors are more efficient when homozygous, and both show more suppression at 15°C than at 25°C; associated with the increased suppression at lower temperature is slower growth and increased sterility in *sup-5* strains and inviability in *sup-7* strains. The frequency at which the *sup-7* lethality at 15°C can be reverted, apparently through *sup-7* null mutations acting as intragenic suppressors, suggests that the *sup-7* gene is dispensable (Waterston 1981).

The molecular basis of suppression by *sup-5* and *sup-7* was elucidated by two approaches (Wills et al. 1983). In one, the suppressible allele *unc-54(e1300)* was shown by DNA sequencing to have an amber codon in place of the wild-type glutamine codon at residue 1903, which, based on intragenic mapping and identification of a truncated *e1300* polypeptide, was close to the predicted nonsense site. In the second approach, the tRNA fractions from the suppressor strains, but not from wild type, were shown to promote readthrough of amber terminators of three different mRNAs in in vitro translations. More recently, DNA sequencing has shown that the wild-type *sup-7* and *sup-5* genes encode tRNAs for tryptophan and that suppressor mutations in both genes involve changes in the tRNA anticodons such that they recognize UAG (Bolten et al. 1984; K. Kondo and R. Waterston, pers. comm.).

Hodgkin (1985b) and K. Kondo et al. (pers. comm.) have identified six additional amber suppressor loci, *sup-21–sup-24* and *sup-28* and *sup-29*. The efficiencies of suppression by these suppressors were generally weaker than *sup-5* or *sup-7* suppression. Three loci, *sup-24*, *sup-28*, and *sup-29*, are $tRNA^{Trp}$ suppressors. The efficiencies of suppression of amber mutations in different genes by *sup-5* and *sup-28* were compared. The results suggested that there are tissue-specific differences in expression between these two $tRNA^{Trp}$ genes.

Showing that a particular allele of a gene is suppressed by a nonsense suppressor proves that the gene makes a polypeptide product. Thus, for example, homeotic loci such as *lin-12* (Chapter 6) and *tra-1* (Chapter 9) have been judged by this criterion to encode protein products. Most suppressible alleles can be expected to be null, although exceptions have been recognized (Ferguson and Horvitz 1985; Hodgkin 1985b, 1987b). The extent of phenotypic suppression generally is characteristically different for different genes; this may reflect the functions of the gene products, as it is known that some weakly suppressed mutants are affected in proteins that are required stoichiometrically (e.g., as part of the muscle lattice) and that a strongly suppressed mutant is affected in an enzymatic function (Waterston and Brenner 1978). An amber allele of *tra-3* was useful in developing microinjection methods; the mutant was suppressed by microinjection of suppressor tRNA (Kimble et al. 1982).

C. Indirect Suppression

An extragenic suppressor may obviate the need for the original mutant function, for example, by providing a substitute protein. Or a conformational alteration in the original mutant protein may be compensated for by the alteration of an interacting protein or by the elimination of a deleterious metabolite (for review and examples, see Hartman and Roth 1973). The search for suppressor mutations is often undertaken with the goal of identifying genes involved in just such interactions. The action of *sup-3 V* in suppressing *unc-54 I* and *unc-15 I* mutations has, for example, been taken to imply a role for the *sup-3* gene in muscle assembly (Riddle and Brenner 1978; see Chapter 10). More recently, Brown and Riddle (1985) have identified two additional mutant loci that modify a mutant *unc-15* phenotype.

Greenwald and Horvitz (1980, 1982, 1986) have identified four genes that are capable of interacting, through indirect suppression, with the mutation *unc-93(e1500) III* which, in a wild-type genetic background, appears to result in a product toxic to muscle action. Null alleles of *sup-9 II, sup-10 X*, and *sup-18 III* are recessive suppressors of *unc-93(e1500)*. A rare allele of *sup-10* by itself confers an Unc-93-like phenotype and is suppressed by null alleles of *unc-93* and apparent null alleles of *sup-9* and *sup-18*. It has been proposed that the wild-type products of the *unc-93* and *sup-10* loci may be components of a protein complex (Greenwald and Horvitz 1986). Rare dominant suppressors of *unc-93(e1500)* occur in *sup-11 I*; these confer a recessive "scrawny" phenotype and seem to be missense mutations (Greenwald and Horvitz 1982). Null mutations in *sup-11* eliminate the dominant suppressing activity and are embryonic lethal, indicating that *sup-11* is an essential gene. It has been suggested that the dominant *sup-11* suppressors may bypass the defect caused by the *unc-93(e1500)* mutation by supplying a protein alternative to the *unc-93(+)* product, the latter being able to overcome the toxic effects of *unc-93(e1500)* (Greenwald and Horvitz 1982) in *e1500/+* animals.

The actions of six mutations as dominant suppressors of the twitching phenotype of *unc-22* mutations unexpectedly turned out to define a new class of *unc-54* mutants; the phenotypes conferred by the suppressor mutations by themselves were unlike previously identified *unc-54* phenotypes (Moerman et al. 1982; see Chapter 10).

Riddle and Brenner (1978) identified a mutation that suppresses mutations in *unc-1 X* and *unc-24 IV* but not mutations in several other *unc* genes. These results led to the suggestion that the *unc-1* and *unc-24* genes, mutations that give similar uncoordinated phenotypes by themselves, may be functionally related.

Certain mutations that are thought to affect the structure of the cuticle can be suppressed indirectly. More than 30 genes have been defined by mutations that confer dumpy, squat, blister, or roller phenotypes. Blister mutants have fluid-filled blebs in their cuticles, and roller mutants rotate along their long axes as they crawl because their bodies are helically twisted. Many dumpy, blister, and roller mutants have been shown to be altered in cuticle structure, and it has been proposed that the genes defined by these mutants are involved in cuticle formation (Higgins and Hirsh 1977; Cox et al. 1980; Edgar et al. 1982). Some of the mutants show stage specificity in the expression of their mutant phenotypes; this characteristic could be related to the changes in cuticular architecture that occur during development (Cox et al. 1981b,c; Politz and Edgar 1984). A rich variety of interactions among various dumpy, blister, and roller mutations has been found (Higgins and Hirsh 1977; Cox et al. 1980; Kusch and Edgar 1986), and it has been suggested that such interactions are a consequence of interactions among different components of the cuticle, which is a very complex extracellular structure (Cox et al. 1981a). We shall cite one example of indirect suppression involving two roller mutants (Cox et al. 1980). (Other types of mutant interactions involving, dumpy, blister and roller mutants will be illustrated in the next two sections.) Homozygous *sqt-1(sc13)* animals are left-handed rollers (*sqt* = squat), and *sqt-1(sc13)/+* animals have a wild-type phenotype. On the other hand, homozygous *sqt-2(sc3)* animals do not have helically twisted bodies and do not roll, but *sqt-2(sc3)/+* animals are strong right-handed rollers. The double *trans*-heterozygote *+sqt-1(sc13)/sqt-2(sc3)+*, however, either does not roll or displays a very weak left-roller phenotype, indicating that the *sqt-1(sc13)* mutation, which by itself is a recessive roller, is acting as a dominant intergenic suppressor of *sqt-2(sc3) (sqt-1* and *sqt-2* both map on LGII, but are not closely linked). Other examples of indirect suppression involving roller genes are known (Cox et al. 1980; Kusch and Edgar 1986).

D. Epistasis

Epistasis refers to the situation in which a double mutant shows the phenotype of one of the single mutants but not the other. Epistatic interactions between pairs of mutations affecting a common developmental pathway have been used to make inferences about the order of action of the genes in the pathway. Excellent examples of this kind of analysis in *C. elegans* are the pathways for vulva development (Chapter 6), sex determination (Chapter 9), and dauer larva formation (Chapter 12).

Many double mutants involving genes implicated in cuticle development (see above) show epistasis. Mutations in certain dumpy genes are epistatic to mutations in various roller genes, for example, and in other instances, dumpy is epistatic to blister (Higgins and Hirsh 1977; Cox et al. 1980).

Epistasis has also been exploited in the development of screens for new mutants. Mutations affecting vulva development, sex determination, and dauer larva formation, for example, have been identified in screens based on the epistatic effects of the mutations on other mutations carried by the mutagenized stocks. The screen that led to the identification of *ced-3* mutations (Ellis and Horvitz 1986), which prevent programmed cell deaths, involved looking for mutations that were epistatic to a *ced-1* mutation, which prevents the engulfment of dead cells and greatly prolongs the highly refractile stage of cell death (Hedgecock et al. 1983). The epistatic effects of *unc-54* and *unc-22* mutations on *unc-105(n490)* have been taken advantage of in a very efficient screen for new *unc-54* and *unc-22* alleles (Park and Horvitz 1986b). The phenotype conferred by *unc-105* is a hypercontracted body-wall musculature and an extremely small body size. In a background of such animals, *unc-54* and *unc-22* mutations, which lead to relaxation of the hypercontraction and larger size, confer readily recognized phenotypes. Indeed, epistasis in this case can also be thought of as indirect suppression.

E. Enhancing Modifiers and Synthetic Mutants

A modifier mutation modifies the phenotype conferred by a mutation in a second gene. An intergenic suppressor is thus one type of modifier; a modifier of the opposite type enhances the mutant phenotype conferred by a second mutant gene. The mutation *sqt-1(sc13) II* confers a left roller phenotype and is normally recessive to *sqt-1(+)*; the mutation *sqt-3(e24) V* confers a dumpy nonroller phenotype and is also recessive to its wild-type allele. The genotype *sqt-1(sc13)/+; sqt-3(e24)/+*, however, confers a left roller phenotype; thus, *sqt-3(e24)* is acting as a dominant enhancing modifier of *sqt-1(sc13)*. This mutant–mutant interaction is probably the result of the interaction of gene products involved in building the cuticle (Kusch and Edgar 1986). Also note that this is an example in which the complementation test fails: Two nonallelic recessive mutations show noncomplementation.

In the following example, each of two mutations can be thought of as a recessive enhancer of the other. A temperature-sensitive allele of *fem-1 IV* (formerly called *isx-1*) causes feminization of both sexes at the restrictive temperature: Hermaphrodites (*XX*) make no sperm, and males (*XO*) produce oocytes in an intersexual gonad (Nelson et al. 1978). A temperature-sensitive allele of *fem-2 III* has a very similar phenotype (Kimble et al. 1984). The double mutant, however, displays a more extreme phenotype: *XO* animals are transformed into females at restrictive temperature (Kimble et al. 1984); thus, the two mutations together transform the sexually dimorphic nongonadal somatic structures, although neither single mutation does so. (Stronger mutant alleles of either *fem-1* or *fem-2,* when homozygous, are by themselves capable of feminizing *XO* animals; see Chapter 9.)

The extreme case of two nonallelic mutations mutually enhancing the expression of each other is the synthetic mutant, for which a mutant phenotype is observed only if both mutations are present. For example, animals homozygous for mutations in any of three acetylcholinesterase genes, *ace-1*, *ace-2*, or *ace-3*, have no apparent behavioral phenotype (Culotti et al. 1981; Johnson et al. 1981; C. Johnson, pers. comm.). However, the *ace-1 ace-2* synthetic mutant is uncoordinated (Culotti et al. 1981), and the triple mutant is arrested in development at the hatching stage (C. Johnson, pers. comm.). These synthetic mutant phenotypes reveal apparent functional overlaps of the acetylcholinesterases encoded by these genes.

A multivulva phenotype is conferred by two unlinked recessive mutations, *lin-8(n111) II* and *lin-9(n112) III*; each of the two mutations alone gives a wild-type phenotype (Horvitz and Sulston 1980: Ferguson and Horvitz 1985).

V. CHROMOSOMAL STRUCTURE AND STABILITY

A. Kinetochores

C. elegans chromosomes have diffuse kinetochores. The metaphase chromosomes lack any visible constriction, which commonly marks the position of the centromere or kinetochore of a monocentric chromosome. Albertson and Thomson (1982) have made reconstructions of *C. elegans* kinetochores, using electron micrographs of dividing nuclei in serially sectioned embryos. At prometaphase and metaphase the mitotic kinetochore is a convex plaque covering the poleward face of the chromosome and extending the length of the chromosome. In longitudinal section, the kinetochore shows a trilaminar structure. From zero to eight attached microtubules per kinetochore were counted; some may have been missed, but when several microtubules were seen attached to one kinetochore, they were distributed all along the length of the chromosome. The meiotic chromosomes, studied in the male by D.G. Albertson (pers. comm.), show spindle microtubules inserting themselves directly into the chromosomes all along their lengths in the apparent absence of a kinetochore; this feature is characteristic of certain holocentric chromosomes (for review, see Bostock and Sumner 1978).

Albertson and Thomson (1982) also saw, by electron microscopy, kinetochores as bands extending the lengths of chromosomes in whole-mount chromosomal preparations stained with ethanolic phosphotungstic acid. Finally, chromosomal fragments were observed cytologically following γ-ray treatment; it was concluded that most fragments were capable of normal segregation in embryonic mitoses, as if they carried their own functional kinetochore fragments, although some fragments displayed aberrant behavior. It was suggested that mitotic segregation of small fragments may be relatively inefficient because the probability of attachment of microtubules to the kinetochore may be proportional to kinetochore length.

The classical localized centromere serves two meiotic functions: It provides sites for attachment of spindle fibers, and it plays a role in the orderly disjunction of the meiotic chromatids of a bivalent, by keeping sister chromatids joined during meiosis I and splitting in meiosis II. If the spindle fibers can attach all along the length of a diffuse kinetochore, however, and if crossovers can also occur at variable positions along the chromosome's length, it is difficult to see how sister chromatids could remain attached throughout their lengths during meiosis I. Indeed, cytological studies of the meiotic behavior of holocentric chromosomes in other species, particularly certain insects, indicate that chiasmata are generally completely terminalized in these species by diakinesis (White 1973). Some species invariably show chiasma terminalization to one end, which results in a bivalent consisting of two pairs of chromatids connected end to end by a terminal chiasma. Furthermore, it appears that in several species, such bivalents orient equatorially with regard to the spindle at the first metaphase, so that meiosis I is an equational rather than a reductional division. The chromosomes of *C. elegans* become very small and condensed during late diakinesis, which makes interpretation of their bivalent structure very difficult, but Nigon and Brun (1955) have interpreted their Feulgen-stained preparations of oogenesis as showing end-to-end associations of homologs at diplotene–diakinesis for all six bivalents. It is not known whether the first meiotic division in *C. elegans* is equational or reductional.

A diffuse kinetochore has been reported for another nematode species (Goldstein and Triantaphyllou 1980) and may be a general property of nematode chromosomes. The early development of ascarid nematodes is characterized by elimination of chromatin; in *Parascaris equorum,* the chromosomes fragment into many small chromosomes in the early embryonic divisions of the somatic progenitor cells but remain intact in the germ line. The holocentric kinetochore may have evolved through fusion of small chromosomes, or, alternatively, the presence of a holokinetic kinetochore may have allowed nematodes to use chromatin diminution as a developmental mechanism (Albertson and Thomson 1982). Cell lineage-specific chromatin diminution has not been detected in *C. elegans*, either cytologically (D.G. Albertson, pers. comm.) or biochemically (see Chapter 3), and both somatic and germ line chromosomes in *C. elegans* have been seen to be holocentric (Albertson and Thomson 1982).

B. Recombination

Recombination occurs in males and in both the sperm and ovum lines of hermaphrodites (Brenner 1974). The recombination frequency in hermaphrodites increases with temperature and decreases with parental age (Rose and Baillie 1979b). Because spermatogenesis is completed early in development before oogenesis begins, the effect of parental age on recombination frequency is probably confined to oocytes (Rose and Baillie 1979b). Hodgkin et al. (1979) measured the frequency of double crossovers in males for a 40-map-unit interval on LGIV and found evidence for moderate interference, which is common in other organisms for map distances of this order. Hodgkin et al. (1979) also screened for double crossovers occurring in the ovum line over a 37-map-unit interval on the X chromosome and found none when the expected number was 25, suggesting very strong interference. It is not known whether high interference is limited to the X chromosome in the ovum line: Neither autosomes of the ovum line nor any linkage groups of the hermaphrodite sperm line have been checked for double crossovers. Rose and Baillie (1979a) have identified a mutation that increases the recombination frequency generally about threefold.

1. Distribution of Genes on the Genetic Map

A feature of the genetic map commented on by Brenner (1974) was the strong tendency for clustering of mutant sites on the autosomes. This feature was less apparent in the case of the X chromosome. With the addition of more genes to the map since 1974, the autosomal genes continue to show clustering, but the X-linked markers apparently do not (Appendix 4). It is known that the crossover frequency is not proportional to DNA content along the lengths of the autosomes (see Chapter 3), which may provide an explanation for the clustering.

2. Synaptonemal Complexes

Synaptonemal complexes (SCs), consisting of two lateral elements and a striated central element, have been traced in serial sections of pachytene nuclei cut for electron microscopy (Goldstein and Slaton 1982). The wild-type hermaphrodite showed six SCs, with lengths ranging from 2.5 μm to 10.4 μm. The wild-type male showed five SCs, ranging in length from 4.7 μm to 8.4 μm. This suggests that the shortest (2.5 μm) SC in the hermaphrodite, which has been reported to have more highly condensed chromatin than the other chromosomes (Goldstein and Slaton 1982), corresponds to the X chromosome; the one X in the male at pachytene appeared as a heterochromatic lump (Goldstein 1982). In *3X* hermaphrodites, generated through meiotic nondisjunction of X chromosomes in a *him-5* mutant (Hodgkin et al. 1979), six SCs with the usual range in lengths and one euchromatic chromosome, about 2.8 μm long, were found in pachytene oocytes (Goldstein 1984a); thus, the structure of the unpaired X in *XO* spermatocytes is different from the structure of the unpaired X in *3X* oocytes.

Structures composed of 6–11 granules attached to the SC have been observed and called SC knobs. Wild-type oocytes generally show six SC knobs somewhat variably distributed among the SCs, usually with zero, one, or two knobs per SC. Four wild-type male spermatocytes showed only two knobs each. Because certain *him* mutants, characterized by enhanced X-chromosome nondisjunction (Hodgkin et al. 1979), showed reduced numbers of SC knobs, Goldstein (1982, 1984a) has proposed that the knobs control X-chromosomal segregation; this interpretation makes it difficult to see why the knobs are generally found associated with autosomal SCs, however.

C. Spontaneous Mutation

Moerman and Waterston (1984) have reported that spontaneous *unc-22* germ line mutations in Bergerac BO occur at a frequency about 100-fold higher than in Bristol or in several other wild-type strains. Most of the spontaneous *unc-22* Bergerac mutants reverted to wild type at high frequency. The frequencies of both forward and reverse germ line mutations were sensitive to genetic background. The mutator activity was not localized to a discrete site in the Bergerac genome, and it did not require the Bergerac *unc-22* gene as target; a Bristol *unc-22* gene placed in the Bergerac genetic background mutated at high frequency. These workers suggested that the higher mutability of Bergerac BO might be due to transposition of the 1.6-kb genetic element Tc1. Bergerac contains over 300 dispersed copies of Tc1, and Bristol contains only about 30 copies (Tc1 is discussed more fully in Chapter 3).

Eide and Anderson (1985a,b) have characterized a large number of spontaneous *unc-54* mutants, using radiolabeled *unc-54* cloned probes to identify restriction fragment length alterations in the mutant *unc-54* genes. Among 65 independent mutants of the Bristol strain, 14 showed detectable alterations affecting 100 or more base pairs. Twelve of these mutants carried deletions; the other two were shown to contain tandem duplications. Absent from the collection were insertion mutations. In the Bergerac BO strain, however, 10 of 18 spontaneous *unc-54* mutants contained insertions of Tc1. In another high-copy-number strain, however, among 37 spontaneous *unc-54* mutations, none was a Tc1 insertion. The Tc1 insertions were shown to be genetically unstable in both the germ line and in somatic cells. Somatic reversion was about 1000-fold more frequent than germ line reversion and generated mosaic animals containing patches of wild-type muscle cells. The genes *unc-22* (Moerman et al. 1986) and *lin-12* (Greenwald 1985b) were also shown to be targets for Tc1 transposition in Bergerac BO, and the inserted Tc1 elements made possible the cloning of these genes (see Chapter 3).

It now appears that the high spontaneous mutability in Bergerac may be conferred by a very few genomic sites, and attempts are being made to identify them (I. Mori et al., pers. comm.). Collins et al. (1987) have identified an EMS-induced mutator strain showing frequencies of Tc1 excision and transposition in the germ line higher than those for Bergerac; transposable elements distinct from Tc1 are also activated in this strain (J. Collins et al., pers. comm.).

A mutant identified as hypersensitive to UV light has been reported to be a general spontaneous mutator (Hartman and Herman 1982b).

D. Chromosomal Rearrangements

1. Deficiencies

The use of overlapping deficiencies in complementation tests for rapid mapping of recessive mutations has already been described. Collections of overlapping deficiencies are also proving useful in localizing genomic clones on the genetic map; hybridization of the cloned DNA to DNA from a heterozygous deficiency strain is reduced if the cloned segment is within the deficiency (Bolten et al. 1984; E. Park and R. Horvitz, pers. comm.). A deficiency can be used as a null allele in heterozygotes; thus, for example, Hodgkin (1983b) noted that the mutation *e1575* is dominant to wild type in its effects on sex determination but that *eDf2*, which is deficient for the region in which *e1575* resides, has no such dominant effect; he therefore was able to conclude that *e1575* must not be a loss-of-function mutation. A deletion of about 300 bp within the *unc-54* gene, called *e675*, was exploited in the molecular cloning of that gene, because the mutation uniquely identified the *unc-54* mRNA, protein product, and gene as molecules of altered length in the mutant (MacLeod et al. 1981).

The most common criterion for identifying a deficiency is genetic: A mutation is generally assumed to be a deficiency if it fails to complement mutations in at least two closely linked genes. Because of the prevalance of essential genes, such deficiencies are usually recessive lethal. Homozygous fertile deletions extending into *unc-54* (Eide and Anderson 1985a), the actin gene cluster on LGV (Landel et al. 1984), and *lin-14* (G. Ruvkun et al., pers. comm.) have been identified by Southern hybridization methods using cloned probes. Albertson (1984b) has used in situ hybridization of biotin-labeled probes to chromosomes from squashed embryos to identify a partial deficiency of the ribosomal gene cluster.

The most efficient method for selecting deficiencies, which has also been employed for the selection of intragenic null alleles, has made use of mutants with a visible phenotype resulting from either an altered gene product or an increased level of gene product; inactivation of the mutation by deficiency formation can thus lead to an altered phenotype, which can be quickly identified in a background of mutant animals. A good example of this approach is provided by the work of Anderson and Brenner (1984), who screened for reversion of the dominant mutation *unc-54(e1152)* following mutagenesis. About 27% of the *unc-54* mutations induced by diepoxyoctane (DEO) were recessive lethal, and complementation tests between these mutations and EMS-induced recessive lethals confirmed that they were deficiencies extending into neighboring essential genes. Greenwald and Horvitz (1980) have made use of the fact that the null phenotype of *unc-93* is wild type to identify γ-ray-induced recessive lethal deficiencies spanning *unc-93(e1500)*. A similar strategy was used to obtain deficiencies of *sup-9* (Greenwald and Horvitz 1980).

The dominant nonsense suppressor *sup-7(st5)* confers a cold-sensitive recessive lethal phenotype. Waterston (1981) selected for reversion of this phenotype following treatment with DEO and recovered recessive lethal mutations that failed to complement a closely linked mutation in another gene.

Moerman and Baillie (1981) have taken advantage of the fact that *unc-22* heterozygotes show a characteristic twitching phenotype when placed in a 1% solution of nicotine alkaloid, whereas wild-type animals in the solution become rigid. Wild-type animals were treated with formaldehyde and their progeny were screened in 1% nicotine for *unc-22* mutations; five mutations proved to be recessive lethal and were shown to uncover certain recessive lethal mutations in the region (Moerman and Baillie 1981; Rogalski et al. 1982).

The more conventional, if less elegant, approach, to identifying deficiencies has involved treating wild-type males with mutagen, mating them with mutant hermaphrodites, and screening for mutant cross-progeny (showing noncomplementation of the putative deficiency and a visible mutation). Examples in which this general strategy has been employed include the work of Meneely and Herman (1979, 1981), Rose and Baillie (1980), and Sigurdson et al. (1984).

Finally, several recessive lethal mutations, not originally identified in direct screens for deficiencies, have subsequently been shown to be deficiencies on the basis of their showing noncomplementation with visible or lethal alleles of two or more closely linked genes (Riddle and Brenner 1978; Hodgkin 1980; Rose and Baillie 1980; Meneely and Herman 1981; Sigurdson et al. 1984; Rosenbluth et al. 1985).

2. Duplications

A chromosomal segment that is present in the genome in addition to the normal chromosomal complement is a duplication. A duplication can be tandem, transposed to another location on the same chromosome, translocated to another chromosome, or a free chromosomal fragment. Duplications in *C. elegans* have been used to vary the dosages of genes (Johnson et al. 1981; Greenwald et al. 1983), to balance recessive lethal and sterile mutations (Meneely and Herman 1979, 1981), to facilitate manipulation of X-linked markers (Herman et al. 1976), to vary the X-chromosome-to-autosome ratio (Madl and Herman 1979; Meneely and Wood 1984), to mark particular chromosomes cytologically (Albertson 1984b), and to generate genetic mosaics (Herman 1984).

The following scheme has been used to identify unlinked X duplications (Herman et al. 1976, 1979). Wild-type males are irradiated with X- or γ-rays and mated with hermaphrodites homozygous for an X-linked recessive marker. Exceptional wild-type male progeny are identified and backcrossed to mutant hermaphrodites; male progeny carrying the wild-type allele on an unlinked duplication are wild type.

Several X duplications have been shown to be translocated to autosomes. These are equivalent to half-translocations, because by the nature of the selection scheme, the chromosome that donated the duplication was not recovered. *mnDp10* and *mnDp25* map near the right end of LGI and are occasionally lost mitotically (Herman et al. 1979). Mitotic loss of translocated segments has also been reported in *Neurospora* (Newmeyer and Galeazzi 1977) and *Drosophila* (Sandler and Szauter 1978). Some, but not all, of the translocated duplications are homozygous fertile.

Many unlinked X duplications have shown no autosomal linkage; in these cases, chromosomal fragments have been observed cytologically. Several autosomal duplications have also been identified. All have been shown to be free by both genetic and cytological criteria (Herman et al. 1979; Hodgkin 1980; Anderson and Brenner 1984; Rose et al. 1984). It seems likely that the holokinetic nature of *C. elegans* chromosomes is responsible for the relatively high frequency at which free duplications have been recovered (Albertson and Thomson 1982). Generally less than half of the ova from a free duplication-bearing hermaphrodite transmit the duplication. The transmission frequencies show characteristic differences for different duplications, and it has been suggested that some of the loss is premeiotic (Herman et al. 1976). Most duplications, whether free or translocated, appear to recombine infrequently, if at all, with the homologous segments of the normal chromosomes. A notable exception is the very large duplication *sDp1(I;f)*, which shows considerable recombination with a normal LGI (Rose et al. 1984). Variants of other duplications have been recovered, but in at least some cases, the variation is due to deletion formation rather than recombination (Herman 1984).

A number of *C. elegans* duplications have been recognized by novel methods. Two were spontaneous tandem duplications within the *unc-54* gene recognized by Southern blotting and heteroduplex mapping (Eide and Anderson 1985a). A 288-bp transposed duplication in *unc-54* was studied by Southern blotting and DNA sequencing (Eide and Anderson 1985c). A duplication of the ribosomal genes translocated from LGI to LGII was identified by Albertson (1984b) by in situ hybridization of biotin-labeled probes to chromosomal squashes. Mutations in *sup-3*, which suppress certain *unc-54* and *unc-15* mutations (see Chapter 10), have been shown by DNA sequence analysis to be associated with increased copy numbers of *myo-3*, the structural gene for a second body-wall myosin heavy chain (Miller et al. 1986).

3. *Translocations and Other Rearrangements*

Translocations are rearrangements involving two (or more) nonhomologous chromosomes. Some translocations, when heterozygous, act as dominant crossover suppressors and can therefore be used to balance recessive lethal or sterile mutations in heterozygous stocks (Rosenbluth et al. 1983; Fodor and Deak 1985). When different elements of a translocation are separable, they are a source of duplications and deficiencies. Translocations have been used to mark particular linkage groups cytologically (Albertson 1984b). Like other rearrangements, translocations may prove helpful in genetically mapping genomic clones. Finally, the analysis of the meiotic behavior of translocations may help elucidate meiotic chromosome pairing and segregation in *C. elegans*.

Because of the rudimentary state of *C. elegans* cytogenetics, the primary criterion for identifying translocations has been genetic and involves showing linkage of normally unlinked markers. Consider a hermaphrodite heterozygous for a simple translocation, either reciprocal or nonreciprocal (also called insertional). If each of the two elements of the translocation tends to disjoin at meiosis from a different normal chromosome, if the two segregations occur independently, and if only balanced zygotes (diploid, with no deficiencies or duplications) are viable, then genetic markers on the two nonhomologous chromosomes will show pseudolinkage. Although pseudolinkage has been the primary criterion for identifying translocations as such, every translocation that has been studied was first identified as a possible rearrangement on the basis of another property. Attention was first focused on several translocations, for example, because they acted as dominant crossover suppressors. These include *mnT1(II;X)* (Herman 1978), *eT1(III;V)* (Rosenbluth and Baillie 1981), *szT1(I;X)* (Fodor and Deak 1985), *nT1(IV;V)* (Ferguson and Horvitz 1985), and *sT1(III;X)* and *sT2(IV;V)* (Rosenbluth et al. 1985). Several X-autosomal translocations were first selected for study because they led to X-chromosomal nondisjunction in heterozygotes (Herman et al. 1982).

Consider further the meioses occurring in a hermaphrodite heterozygous for a simple translocation. If, in addition to the assumptions made in the previous paragraph, we assume that unbalanced gametes (carrying only one of the two elements of the translocation) are functional, then 6/16 of the zygotes produced by the heterozygous hermaphrodite are expected to produce viable animals, and among the viable progeny, the expected ratios of homozygous translocation to heterozygous translocation to homozygous normal are 1:4:1. These predictions have been shown to be satisfied by at least three translocations (Herman 1978; Rosenbluth and Baillie 1981; Ferguson and Horvitz 1985).

The precise breakpoints and connections of four translocations have been proposed: *eT1(III;V)* (Rosenbluth and Baillie 1981), *mnT2*, and *mnT10* (Herman et al. 1982) are reciprocal translocations, and *mnT12* (Sigurdson et al. 1986) is a homozygous viable X-autosomal fusion chromosome.

The dominant crossover suppressor called *mnC1*, on LGII, is probably an intrachromosomal rearrangement, such as an inversion or a transposition (Herman 1978). Nearly all the eggs laid by *mnC1* heterozygous hatch to give *mnC1/mnC1*, *mnC1/+*, and *+/+* progeny in the ratios 1:2:1. *mnC1* has been used to balance recessive lethal and sterile mutations (Sigurdson et al. 1984).

E. Genetics of Meiotic Chromosomal Disjunction

1. Recessive Mutations Affecting Chromosomal Disjunction

Hodgkin et al. (1979) described 15 recessive mutations that lead to enhanced X-chromosomal nondisjunction in hermaphrodites; as a result, the mutants segregate 2–35% *XO* male self-progeny (compared with 0.2% males from wild-type hermaphrodites) and 0.8–6.7% *3X* hermaphrodites. The mutations defined ten genes, only one of which is X linked. Apart from *unc-86* (mutations of which also affect animal movement), the genes were given the general name *him*, for *h*igh *i*ncidence of *m*ales. All of eight mutants tested produced nullo-X and diplo-X ova, which were detected by crossing mutant hermaphrodites bearing an X-linked marker (and an autosomal marker to distinguish cross-progeny) and counting patroclinous males and *3X* hermaphrodites. Mutant *XO* males generally produced equal numbers of nullo-X and haplo-X sperm, as do wild-type males. On the other hand, four *him* mutations were shown to cause the production of many nullo-X and diplo-X sperm in *2X* males generated by the action of a *tra-1* mutation. Thus, it appears that these *him* mutations promote nondisjunction of X chromosomes during both oogenesis and spermatogenesis when two X chromosomes are present but generally do not affect the behavior of an unpaired X in *XO* males.

Some of the *him* mutants produced inviable zygotes at frequencies much too high to be accounted for simply by nullo-X and tetraplo-X zygotes, which are inviable. Crosses between *him* and wild-type animals indicated that dominant lethal factors were generated in the ova of four mutants and in the male sperm of three mutants. Gamete autosomal aneuploidy seemed a likely explanation for at least some of these dominant lethal effects. It was shown that a *him-6* mutation, which segregates over 75% inviable zygotes, promoted autosomal nondisjunction, which could account for the high zygote inviability. On the other hand, for other *him* mutants, particularly *him-8*, which give 35% males and less than 1% inviable zygotes, disjunction of only the X chromosomes appeared to be affected. Thus, it appears that the X chromosome is, to some degree, handled differently from the autosomes during meiosis.

In four *him* strains, including *him-8*, nondisjunction during oogenesis was shown to occur at the reductional division. Three *him* mutants showed reduced X-chromosomal recombination, with less effect on the autosomal intervals that were tested. The mutant with the greatest reduction in X-chromosomal recombination was *him-8*, which gave about one-eighth the wild-type value. Nondisjoining chromosomes showed a strong tendency to be nonrecombinant; this result suggests that exchange may be necessary for proper disjunction in *C. elegans*, as in many other organisms (Baker et al. 1976). Goldstein (1982) has shown that *him-8* hermaphrodites have six SCs, indicating that the mutant gene does not seem to affect X-chromosomal pairing. Because genetic exchange precedes disjunction, the primary defect in *him-8* mutants may be in X-chromosomal recombination which, as a consequence, leads to enhanced nondisjunction. If exchange is necessary for proper disjunction of autosomes, then mutant hermaphrodites in which recombination of all chromosomes is drastically reduced might be expected to produce almost exclusively inviable zygotes, owing to aneuploid chromosome compositions.

A mutation in the *rad-4* gene reduces meiotic X-chromosomal nondisjunction to about one-tenth the wild-type frequency (Hartman and Herman 1982b). The mutant was isolated on the basis of showing hypersensitivity to UV light and also a cold-sensitive embryogenesis defect. In addition, *rad-4* partially suppresses the effects of certain *him* mutations.

2. *Dominant Mutations Affecting Chromosomal Disjunction*

A mutation called *mn164*, recovered after X-ray treatment, was mapped near the left tip of the X genetic map and shown to promote X-chromosomal nondisjunction when either homozygous or heterozygous (Herman et al. 1982). When heterozygous, the *mn164*-bearing chromosome showed equational nondisjunction, and its homolog did not. It was proposed that *mn164* involves a structural abnormality of the X chromosome that affects chromosomal segregation.

Several dominant X-chromosomal nondisjunction mutations, generally induced by ionizing radiation, have turned out to be X-autosomal translocations. A few such translocations, such as *szT1* (Fodor and Deak 1985) and *mnT2* (Herman et al. 1982), also show very little X-chromosomal recombination in translocation heterozygotes, and it is tempting to attribute the high-frequency X nondisjunction in these cases to infrequent pairing between the X and the translocation, perhaps owing to a disruption of synapsis caused by the translocational breakpoint. Other X-autosomal translocations show nearly normal recombination, however, and still show high-frequency X nondisjunction in heterozygotes. One of these is *mnT12(IV;X)*; hermaphrodites heterozygous for *mnT12* show high-frequency nondisjunction both between *mnT12* and the X chromosome and between *mnT12* and chromosome IV. The nondisjunctions may be the result of competition between IV and X for *mnT12* as a segregating partner. Disjunction of *mnT12* in homozygotes seems to be normal (Sigurdson et al. 1986). Because the half-translocation *mnT10(X)*, which has the left end of linkage group V substituted for the left tip of the X, shows nondisjunction when homozygous or heterozygous, the nondisjunction could not be attributed to nonhomology of segregating chromosomes; it was suggested that the absence of the left tip of the X chromosome was critical (Herman et al. 1982).

The large autosomal duplication *sDp1(I;f)* promotes nondisjunction of normal LGI homologs (giving nullo-I *sDp1* gametes), which is not surprising because *sDp1* recombines with a normal I; *sDp1* also promotes the nondisjunction of X chromosomes, however, perhaps through occasional pairing with an X chromosome (Rose et al. 1984). Several free duplications, both X and autosomal, show a tendency to disjoin from the single X in male meiosis (Herman et al. 1979). This effect may be due to a process analogous to distributive pairing in female *Drosophila* (Grell 1976).

F. Genetics of Radiation and Mutagen Sensitivity

Hartman and Herman (1982b) identified and characterized nine mutants that were hypersensitive to UV light. The mutations were all recessive to their wild-type alleles and were assigned to nine separate genes, called *rad-1–rad-9*. Two of the mutants—*rad-1* and *rad-2*—were very hypersensitive to X rays, and three—*rad-2*, *rad-3*, and *rad-4*—were hypersensitive to the mutagen methylmethanesulfonate under particular conditions of exposure. It was suggested that the hypersensitivity of these mutants to more than one DNA-damaging agent may be due to defects in DNA repair. Bhat and Babu (1980) reported that *flu-2* mutants, which have reduced kynureninase activity and altered autofluorescence of the intestinal cells, have enhanced sensitivity to EMS and γ-rays but not to UV.

UV sensitivity of wild-type *C. elegans* decreases markedly during and following embryogenesis (Klass 1977; Hartman 1984b). Hartman has shown that four different *rad* mutants exhibit quite different age-dependent patterns of radiation sensitivity. The *rad-3* mutant, for example, showed *increased* radiation sensitivity during embryogenesis; it was also very hypersensitive to UV as a dauer larva (Hartman 1984a). For the four *rad* mutants analyzed, both maternal and early zygotic expression of wild-type alleles enhanced UV resistance of embryos (Hartman 1984b). Hartman (1985) has tentatively assigned the four mutants most hypersensitive to UV to two DNA repair pathways on the basis of the UV sensitivities of all combinations of double mutants: *rad-1* and *rad-2* were assigned to one epistatic group, and *rad-3* and *rad-7* were assigned to the other.

VI. GENETIC MOSAICS

A genetic mosaic is an individual having cells of two or more different genotypes. The analysis of genetic mosaics has proved to be an important tool in the developmental genetics of *Drosophila* and the mouse (for review, see Gehring 1978). For many *C. elegans* mutants affected in cell structure, cell lineage, or animal behavior, it would be useful to be able to analyze the effects of making certain particular cells mutant and the remaining cells wild type. In this way, it may be possible to ascertain the anatomical foci of mutations affecting behavior, to determine whether or not a particular mutation behaves cell autonomously, or to determine for cell nonautonomous mutations what the nature of the cell interaction might be. Genetic mosaics can also be used to set limits on the times of action of wild-type genes: If a wild-type gene is removed from a cell and, as a consequence, a descendant cell shows a recessive mutant phenotype, the implication is that the wild-type gene was needed after the time at which it was removed.

The first report of *C. elegans* mosaics was by Siddiqui and Babu (1980b), who X-irradiated (2000 rads) embryos heterozygous for a recessive *flu-3* mutation, which alters the color and intensity of the autofluorescence of intestinal cells under UV light. Some of the irradiated embryos gave rise to adults that had a patch of altered intestinal cells. The most common type of patch included either the anterior half or posterior half of the intestine. Such patches are not entirely consistent with the intestinal cell lineage (Sulston et al. 1983), and their mode of origin is unknown.

Herman (1984) has used the spontaneous loss of free chromosomal fragments or duplications to produce genetic mosaics. Duplications were used that carried one or more wild-type dominant alleles of genes that were homozygous mutant on the normal chromosomes. A duplication loss thus generated a clone of homozygous mutant cells. The analysis of mosaic animals was greatly aided by the fact that the cell lineage is completely known. Advantage was also taken of the work of Hedgecock et al. (1985), showing that a living wild-type animal when exposed to fluorescein isothiocyanate (FITC) incorporates the dye into six neurons in each of two sensilla in the head and two neurons in each of two sensilla in the tail and that mutations in a number of different genes abolish this staining. One such mutation, *osm-1* (abnormal osmotic avoidance; Culotti and Russell 1978), seemed to behave cell autonomously in mosaic animals with respect to this property. Another mutation, *daf-6* (abnormal dauer larva formation; Albert et al. 1981), was cell nonautonomous with respect to its effect on abolishing FITC staining of sensory neurons; in this case, it was concluded, on the basis of mosaic analysis and earlier electron microscopic work (Albert et al. 1981), that the genotype of a nonneuronal cell in each sensillum was probably critical. Using either *osm-1* or *daf-6* as cell markers, animals mosaic for other genes were analyzed. It was concluded that the anatomical focus for the action of the *unc-3* gene on animal movement was among descendants of cell AB.p, for example, and that expression of *sup-10* (Greenwald and Horvitz 1980) was specific to muscle cells. In more recent work, it was concluded that in animals that carry a duplication bearing *ace-1(+)* in an otherwise homozygous *ace-1 ace-2* genetic background, acetylcholinesterase synthesis directed by the *ace-1(+)* gene is required in muscle cells, but not motor neurons, for coordinated movement (Herman and Kari 1985). Several other workers have recently used the spontaneous loss of free chromosomal fragments to generate genetic mosaics. Park and Horvitz (1986b) concluded from their analysis of such mosaics that the primary effect of recessive suppressors of *unc-105(n490)* (in a gene called *sup-20 X*) is specific to muscle cells. Kenyon (1986) concluded that wild-type *unc-36* function is required by descendants of the cell AB.p (but not AB.a) and that the wild-type *mab-5* gene, the product of which affects alternative developmental decisions in many kinds of posterior cells, is cell autonomous in its action. J. Yuan and R. Horvitz (pers. comm.) deduced that *ced-3* and *ced-4*, in which mutations block programmed cell death (Ellis and Horvitz 1986), probably act cell autonomously. And E. Hedgecock (pers. comm.) concluded that *ncl-1*, in which mutations cause enlargement of the nucleoli of all cells, is also expressed autonomously; most of the *ncl-1* mosaics could be explained by duplication loss occurring in a single precursor cell, but 3 of 30 mosaics appeared to arise by loss of the duplication at two or three consecutive divisions.

Eide and Anderson (1985b) have identified genetic mosaics produced through somatic excision of Tc1 insertion mutations in *unc-54*. Inactivation of the *unc-54* myosin heavy-chain gene results in paralysis and defective egg laying. The 16 sex muscles responsible for egg laying derive from a postembryonic blast cell, M, which is also progenitor to 14 body muscles that are added postembryonically to the 81 embryonically derived body muscle cells. About 1% of the progeny of an *unc-54* mutation caused by Tc1 insertion were good egg layers, and polarized light microscopy revealed patches of revertant body-wall muscle cells in many of them.

ACKNOWLEDGMENTS

I thank Donna Albertson, David Baillie, Martin Chalfie, Jonathan Hodgkin, Robert Horvitz, and Robert Waterston for many helpful comments and suggestions.

3

The Genome

Scott W. Emmons
Department of Molecular Biology
Albert Einstein College of Medicine
Yeshiva University
Bronx, New York 10461

I. INTRODUCTION

The physical basis of the *Caenorhabditis elegans* genetic system is a genome of 8×10^7 base pairs of DNA. This DNA is organized into six nuclear chromosomes plus the mitochondrial chromosome. The small size of the genome, only 20 times that of *Escherichia coli* and half that of *Drosophila*, is consistent with the small number of genes in *C. elegans*, as measured by genetic means. The ratio of the number of base pairs of DNA to the estimated number of genetic loci, about 20,000, is similar to the value for *Drosophila* and appears to be typical for eukaryotic organisms generally. The *C. elegans* genome is typical in other respects as well. The sequences that it comprises are, in general, identically arranged in every cell. They consist of both unique sequences and repeated sequences. As in other eukaryotes, the repetitive component is a complex class made up of repeated genes, transposable elements, and diverse, short repeated sequences whose origin and function are unknown. Repeated sequences and unique sequences are intermingled throughout the DNA.

A central aim of *C. elegans* research is to understand how the information that specifies the organism is encoded in the genome and how this encoded information is read out in a temporally and spatially coordinated fashion during the life cycle. A primary aim of research on the genome, therefore, has been to provide means for isolating genes so that gene products may be identified and gene activity studied at the biochemical level. The small size of the *C. elegans* genome also provides a favorable opportunity for studying problems related to the structure of the eukaryotic genome per se. Analysis of the *C. elegans* genome should help provide insights into the reason for the large amount of apparently noncoding DNA, the relationship between the DNA sequence and the karyotype, the origin and role of short, interspersed repetitive DNA and transposable elements, and the mechanisms of genomic transmission and evolution. Molecular studies of the genome have included a general characterization of the properties of the genomic DNA, comprising the arrangement and properties of the repetitive component, and a more detailed characterization of a small number of cloned regions. To facilitate more extensive cloning studies, a coordinated community effort is under way to put together a complete set of mapped clones covering the entire genome. Methods for isolating genes, using several approaches, including transposon tagging, and for DNA-mediated transformation have also been developed and are described. Now the powerful methods of in vitro manipulation and molecular analysis of genes and gene function can be applied to *C. elegans*, and questions about the structure and transmission of the genome and the nature and expression of its code can be addressed.

II. PROPERTIES AND ORGANIZATION OF THE GENOMIC DNA

A. Physical Properties and Genome Size

The *C. elegans* genome has a nearly uniform base composition of 36% G + C and gives a single main band with one small dense satellite upon analysis by isopycnic banding in CsCl (Sulston and Brenner 1974). The satellite, which constitutes about 1% of the DNA and bands at a density corresponding to a base composition of 51% G + C, consists of the genes for the 18S and 28S rRNAs (Sulston and Brenner 1974; Files and Hirsh 1981). Density satellites consisting of simple sequence DNA have not been detected, but studies using drugs or metals have not been carried out. Methylated bases have not been detected in any of several investigations employing analysis by high-pressure liquid chromatography of nucleotides obtained by hydrolyzing DNA, and employing restriction enzymes sensitive to methylated bases (Simpson et al. 1986; J. Laufer et al., pers. comm.). In the most sensitive experiments, it would have been possible to detect one 5-methylcytosine in 10,000 cytosine residues.

Analysis by reassociation kinetics reveals two components: a single heterogeneous component consisting of repetitive sequences, and a homogeneous, slowly reannealing component, constituting 83% of the total, that reanneals at a rate corresponding to a complexity of 6.7×10^7 bp (Sulston and Brenner 1974). Assuming this slow component represents sequences present in one copy in the haploid genome (unique sequences), these values indicate a total haploid genome content of 8.0×10^7 bp, or 0.088 pg. Direct chemical analysis of the amount of DNA extracted from a known quantity of worms gives twice this value as the amount of DNA per cell, consistent with the assumption that the slow component is unique DNA and with the diploid character of most *C. elegans* cells (Sulston and Brenner 1974). The value of 0.088 pg or 8.0×10^7 bp may therefore be taken as the size of the haploid genome.

Direct analysis has confirmed that most, but not all, *C. elegans* cells are diploid. A survey of the relative amount of DNA per nucleus has been made for a number of tissues by using a microfluorometric technique employing Feulgen or Hoechst 33258 dyes (Albertson et al. 1978; Hedgecock and White 1985). Whereas most cells have twice the DNA content of sperm, as expected, nuclei of two tissues do not. The 34 nuclei of the intestinal cells of the adult have 32 times the haploid amount, and at least 98 of the 133 nuclei of the syncytium that makes up most of the hypodermis have four times the haploid amount. These polyploid nuclei arise during postembryonic development by a series of endoreduplications in which the chromosomes replicate without chromosome condensation or nuclear division. Some additional polyploid nuclei may be present in other tissues; the measurements were not exhaustive. There is no evidence for chromatin diminution in any tissue, as explained more fully below.

B. Repetitive DNA

Studies in a large number of organisms have revealed several classes of repetitive DNA in eukaryotic genomes, as well as two seemingly distinct global arrangements of the repetitive sequences among the unique sequences (for reviews, see Davidson 1976; Brutlag 1980; Long and Dawid 1980; Jelinek and Schmid 1982). The repetitive sequences that make up 17% of the *C. elegans* genome have been studied by reannealing kinetics, electron microscopy, and analysis of cloned sequences. Sequences in all previously described classes have been identified, with the exception of simple-sequence satellites, and the global arrangement of the repetitive component conforms to the so-called "short-period interspersion," or "*Xenopus*," pattern (Davidson et al. 1973; Emmons et al. 1980). This arrangement is characterized by interspersion of repetitive and unique sequences at short intervals; regions of unique sequence of 1–5 kb are separated by short regions of repetitive sequence, a few hundred nucleotides long. The genomes of most eukaryotic organisms are organized according to this pattern. The exceptions in which much longer regions of unique sequence are present uninterrupted by repetitive sequences are mostly organisms with small genomes, the most notable example being *Drosophila*, which has a genome of 1.6×10^8 bp (Spradling and Rubin 1981). This observation gave rise to the speculation that the "long-period interspersion," or "*Drosophila*," pattern was a property or consequence of small genome size. In view of the short-period arrangement found in *C. elegans*, this is not likely.

A portion of the repetitive component in *C. elegans*, as in other eukaryotes, consists of functional genes or gene-related sequences. There are approximately 70 genes for 18S and 28S rRNAs in the haploid genome and 110 genes for 5S RNA. These multiple genes are arranged in tandem at unique genomic sites (see Sections III.A.3 and III.A.4). In addition, many protein-coding genes are members of dispersed repetitive gene families. Thus, there are between 40 and 150 genes for collagen, more than 20 genes for the major sperm protein, 4 genes for myosin heavy chain, 4 genes for actin, and so forth. In addition, there are multiple copies of several transposable genetic elements.

The fastest reannealing portion of the repetitive component consists of sequences arranged as inverted repeats, which reanneal in an intramolecular "fold-back" or "snap-back" reaction (Sulston and Brenner 1974). The *C. elegans* genome contains about 2400 such regions, randomly distributed, with an average separation of 33,000 bp (Emmons et al. 1980). The lengths of several of these regions were determined from electron micrographs and found to be tightly distributed around a modal value of 300 bp. These characteristics of length and distribution are similar to those of inverted repeats studied in other eukaryotic organisms (Davidson et al. 1973; Graham et al. 1974; Wilson and Thomas 1974; Cech and Hearst 1975; Perlman et al. 1976).

A large fraction of the remaining repetitive sequences consists of isolated short repeats with a length distribution, as determined from electron micrographs, tightly distributed around a modal value of 300 bp, similar to that of the inverted repeats (Emmons et al. 1980). This length distribution is again similar to that found in studies of interspersed repetitive DNA in other organisms (see Lewin 1974; Davidson 1976). Such short, interspersed repetitive DNA makes up 10% of the *C. elegans* genome (Emmons et al. 1980). Despite this abundance and the considerable effort devoted to studying interspersed repetitive sequences in a wide variety of organisms, the origin and role of these sequences remain mysterious. A rather surprising property of this class of DNA sequences in *C. elegans* is its high complexity. The repetitive component consists of some 100–1000 probably distinct families (Emmons et al. 1980). This is not much lower than the number of families found in organisms with much larger genomes that contain a similar proportion of the genome in the short, interspersed repetitive class—in other words, with much greater total amounts of repetitive DNA. The higher total amount of repetitive DNA in organisms with larger genomes is accounted for, to some extent at least, by a higher repetition frequency of the repetitive sequences, rather than by a greater number of repetitive families. Whereas the repetition frequency of interspersed repetitive families is often above 100 and ranges up to hundreds of thousands in most organisms where they have been studied, most repetitive families in *C. elegans* contain less than 100 members, and many contain only around 10.

C. Extrachromosomal DNA

Wolstenholme et al. (1987) have determined the complete nucleotide sequence of mitochondrial DNA from *Ascaris suum*. The molecule is 14,284 nucleotide pairs in length, has a base composition of 72% A + T, and contains the genes for 12 proteins, 2 rRNAs, and 21 tRNAs. The protein-encoding genes are homologous to 12 of the 13 protein-encoding genes found in vertebrate and *Drosophila* mitochondrial DNAs. The tRNA-encoding genes, on the other hand, are of a unique structure not previously found in any other organism (see Section III.A.5). In the same study, the investigators sequenced about one third of the *C. elegans* mitochondrial DNA and showed that it was similar, containing the same genes as the corresponding segment of the *A. suum* molecule. K. McNeil and A.M. Rose (pers. comm.) have also isolated *C. elegans* mitochondrial DNA and determined a length of 15 kb.

In the Bergerac strain of *C. elegans*, extrachromosomal copies of the transposable element Tc1 have been detected in amounts of around one copy per cell (Rose and Snutch 1984; Ruan and Emmons 1984). This may be because this strain carries a large number of copies of the element in its genome, and these undergo frequent excision in somatic tissues, as described below.

Small, circular extrachromosomal DNA, which has been detected in a number of organisms, has not been studied per se in *C. elegans* (for review of circular extrachromosomal DNA in other eukaryotes, see Rush and Mishra 1985).

D. Comparison with Other Nematodes

The DNAs of seven species of nematodes have been studied (Table 1). As in other phylogenetic groups, there is a range of genome sizes. The *C. elegans* genome appears to be one of the smallest. Genome organization has been studied in *Panagrellus silusiae* (Beauchamp et al. 1979) and in *Ascaris lumbricoides* (= *A. suum*, the pig intestinal parasite) (Roth 1979; Landolt and Tobler 1980). Beauchamp et al. (1979) report a long-period interspersion pattern for the *P. silusiae* genome, as does Roth (1979) for *A. lumbricoides*, but Landolt and Tobler (1980) report a short-period interspersion pattern for *A. lumbricoides*. Similarly conflicting results have been reported for *C. elegans*: Schachat et al. (1978b) found apparent long-period interspersion, whereas Emmons et al. (1980) reported short-period interspersion. The reason for such contradictions appears to be that the method of C_0t analysis used in some of these studies is insensitive to the interspersion of repetitive DNA when the repetition frequency of the repeating elements is low. Thus, in *C. elegans,* evidence for short-period interspersion from C_0t analysis is unconvincing, but analysis of reannealed DNA by electron microscopy, as well as analysis of cloned genomic segments, shows unequivocally the presence of interspersed short repeats of low-repetition frequency (Emmons et al. 1979, 1980). In view of these results, reports of long-period interspersion based on C_0t analysis alone may not be accurate.

Interest in the genomes of the ascarid nematodes arises because the genome undergoes chromatin diminution in these nematodes, as first described 100 years ago by Boveri (1888). During the early cleavages of the embryo, the chromosomes of the presomatic blastomeres break into pieces, whereas those of blastomeres destined to contribute to the germ line remain intact. In the ensuing mitosis, some of the broken chromosomal fragments are left behind at the spindle equator, so that the daughter cells do not inherit a full genomic complement. Ultimately, the only cells in the organism having an intact and complete genome are the germ cells, the genomes of all the somatic cells being diminished. This process of chromatin diminution is found in the horse parasite, *Parascaris equorum* (known to Boveri as *Ascaris megalocephala*), in *A. lumbricoides*, and in a number of other nematode species (Walton et al. 1974; Albertson et al. 1979; Tobler 1986). The process does not occur in all nematode species, however, but it does occur in diverse plants, insects, and crustaceans (Wilson 1925; Tobler 1986). The phenomenon is therefore by no means

Table 1 DNAs of seven species of nematodes

Species	Genome size (germ line) (pg)	(kb)	Components	References
Ascaris lumbricoides (= *A. suum*)	0.32	2.9×10^5	23% satellite 77% unique plus middle repetitive	Moritz and Roth (1976)
Caenorhabditis briggsae	0.13	1.2×10^5	not determined	Searcy and MacInnis (1970)
Caenorhabditis elegans	0.088	0.80×10^5	3% fold-back 3% long repetitive 11% short repetitive 83% unique	Sulston and Brenner (1974); Emmons et al. (1980)
Panagrellus redivivus	0.076	0.70×10^5	not determined	Searcy and MacInnis (1970)
Panagrellus silusiae	0.097	0.89×10^5	9% fold-back 28% repetitive 63% unique	Beauchamp et al. (1979)
Parascaris equorum	1.2–2.1	1.1×10^6–1.9×10^6	85% satellite 15% unique plus middle repetitive	Moritz and Roth (1976)
Trichinella spiralis	0.27	2.4×10^5	not determined	Searcy and MacInnis (1970)

either peculiar to or universal among nematodes as a group. In *C. elegans*, apparently, it does not occur, as evidenced both by cytological analysis, which shows that the somatic and germinal chromosomes appear identical to each other and equal in number to that of genetic linkage groups (Nigon and Brun 1955; Brenner 1974; Albertson and Thomson 1982), and by molecular analysis of the DNA, which gives evidence for the presence of at least the majority of germ-line sequences in the DNA of somatic cells (Sulston and Brenner 1974). Analysis of cloned DNA fragments showed that developmental rearrangements, if they occur at all, can affect no more than 1% of restriction fragments or sites (Emmons et al. 1979).

Molecular analysis carried out in *P. equorum* and *A. lumbricoides* has shown that the bulk of the DNA lost from the somatic cells consists of highly reiterated satellite sequences (for review, see Tobler 1986). This is consistent with the conclusion of earlier cytologists that the lost chromatin is heterochromatin, whereas the retained chromatin is euchromatin (Walton et al. 1974). In both *P. equorum* and *A. lumbricoides,* a large fraction of the germ-line genome consists of density satellites, and this component is lacking from somatic DNA. Because *C. elegans* apparently lacks large amounts of satellite DNA, this alone could account for the absence of chromatin diminution in this species. Alternatively, the cellular processes that bring about diminution may be unique to those species in which diminution occurs.

Comparison of DNA sequences from related species can be helpful in defining coding sequences, whether genes or control regions, because coding sequences are expected to diverge more slowly than spacer DNA with no essential function. With this prospect in mind, DNA of *C. elegans* has been compared to that of the similar species *Caenorhabditis briggsae*, with the unexpected finding that the two species are highly diverged (Emmons et al. 1979). Most restriction fragments from *C. elegans* fail to hybridize to any sequence in the *C. briggsae* genome, whereas those that do hybridize are homologous to fragments of different sizes in *C. briggsae*. Butler et al. (1981) compared proteins of the two species by starch gel electrophoresis and concluded similarily that the species were highly diverged.

On morphological grounds, *C. briggsae* is the nematode most closely related to *C. elegans*. The two species are so similar that upon first isolation, *C. briggsae* was thought to be a strain of *C. elegans*. Only mating tests showed that they were distinct (Nigon and Dougherty 1949). In view of this morphological similarity, it is probable that the proteins of the two species are similar, and that those DNA sequences they do share are, in fact, genes. This supposition has been supported by results of Heine and Blumenthal (1986), who analyzed the genes in several long regions of the X chromosome containing genes for vitellogenin (see Chapter 7). They found that without exception, DNA segments for which transcripts were found in *C. elegans* were conserved in *C. briggsae*, whereas DNA segments not transcribed, with two exceptions, were diverged so completely that they failed to cross hybridize. Species comparisons of this sort can therefore provide a powerful tool for identifying the functional sequences of the DNA.

E. Overall Arrangement of Sequences

The comparison of the *C. elegans* and *C. briggsae* genomes described above suggests the general organization of sequences in the genome: Genes a few kilobases long are dispersed in the DNA at about 20-kb intervals, separated by "spacer" sequences. These figures are averages. Genes range enormously in size, the largest so far found in *C. elegans* being the *unc-22* gene, which is at least 20 kb long (Moerman et al. 1986). Similarly, the separation between genes has a wide range. Heine and Blumenthal (1986) found separations ranging from less than 2 kb to greater than 38 kb. The global average of one gene per 20 kb is calculated from the value for the total number of genes, estimated in various genetic studies as described in Chapter 2, and the total genome size. It is the same as the average value found by Heine and Blumenthal for the regions of the X chromosome they studied.

Genes of similar or related function are not necessarily found together in the DNA. Thus, multiple genes for myosin heavy chain, collagen, and vitellogenin are separated from each other, either on separate chromosomes or by many hundreds of kilobases containing unrelated genes. Coordinate regulation of these multiple genes may be brought about via sequences immediately upstream of individual genes. However, some clustering is found, for example, among the genes for actin, major sperm protein, and histones. This may reflect the existence of regulatory mechanisms affecting chromosomal domains or evolutionary events.

Genes are probably present with similar frequency and physical spacing on all the chromosomes, and it is the presence of pairing sites or other properties of the mechanism of recombination that accounts for the uneven distribution of genes on the genetic maps of the autosomes (Chapter 2). Variation in the frequency of recombination would mean that the amount of DNA per recombination map unit differs from region to region. This hypothesis is supported by recent findings. The average value is estimated to be 270 kb per map unit. This is calculated from the estimated genome size of 80 megabases and a total of 300 map units. Cloning studies and in situ hybridization give values around 1500 kb per map unit in the gene clusters at the centers of autosomes, with lower values outside the clusters. Analysis of one 600-kb region on the edge of the gene cluster of linkage group III gives a value of approximately 900 kb per map unit (Greenwald et al. 1987). Fine-structure mapping within the *unc-54* gene, which is outside a cluster near one end of linkage group I, gives a value of 500 kb per map unit (Dibb et al. 1985).

If we take 5 kb as the size of an average gene, there must be an average of 15 kb of spacer DNA between genes. Introns in *C. elegans* are small and do not account for this extra DNA (see below). Spacer DNA is A + T rich. Genes that have been sequenced so far have a G + C content of around 50%. Because the genome as a whole has a G + C content of 36% (Sulston and Brenner 1974), the difference must be due to the spacer DNA. A small amount of spacer DNA has been sequenced and is found to have a G + C content of 36%, in agreement with this prediction (E. Zucker and T. Blumenthal, pers. comm.).

The reason for the large amount of A + T-rich DNA separating genes is unknown and is one of the outstanding problems in understanding the structure of eukaryotic genomes. Not all of this DNA is expected to be functionless. Among the functions it might serve are determination of the form and stability of the karyotype; replication, pairing, recombination, and segregation of chromosomes; and determination of the X chromosome:autosome ratio for the purpose of sex determination (Chapter 8). The bulk of the DNA between genes, including the majority or all of the short, interspersed repetitive DNA, may indeed be functionless, however, at least with respect to the life of individual animals. The apparent rapid evolutionary divergence of intergenic spacer sequences and repetitive sequences supports this view. Mechanisms must exist that create and alter these sequences, and selection must act to preserve them. Selection might act either directly on the sequences themselves, as "selfish" DNA (Doolittle and Sapienza 1980; Orgel and Crick 1980), or on the animal or species that contains them.

III. PROPERTIES OF SPECIFIC GENOMIC SEQUENCES

A. Genes

The analyses of cloned genes in *C. elegans* have revealed both similarities with and differences from other eukaryotic genes. As in other eukaryotes, *C. elegans* genes are interrupted by introns, and a total of 54 different *C. elegans* introns have been sequenced as of this writing. A comparison of these sequences indicates that they share several properties atypical of most eukaryotic introns (T. Blumenthal, pers. comm.). First, many are unusually short: Three-quarters are between 45 and 59 bases long, and the majority of these short introns are either 48 or 52 bases in length. Interestingly, results of experiments with introns from higher eukaryotes have led to the suggestion that introns shorter than about 80 bases cannot be spliced (Wieringa et al. 1984). This may mean that the mechanism for splicing *C. elegans* introns is somewhat different from that used for splicing introns from higher eukaryotes. *C. elegans* introns are AT rich: They average 74% A + T, and only two of the 54 introns analyzed are under 67% A + T. Borders of *C. elegans* introns share the consensus sequences of other eukaryotic introns at most positions, but there are some significant exceptions (Fig. 1). The *C. elegans* donor consensus sequence RAG/GTAAGTT (/ represents the splice site; R is a purine) is nearly identical to the general eukaryotic consensus (C,A)AG/GTRAGT. The acceptor consensus sequence TTTCAG/R, however, differs more extensively from the general eukaryotic consensus YAG/G (Y is a pyrimidine). There is a strong preference for T at positions −4 through −6 in *C. elegans* (the last intron nucleotide before the 3′-splice junction is −1). All other positions in the *C. elegans* intron have a consensus of A/T, including the 11 positions near the 3′-splice junction, where the consensus is C/T in other eukaryotic introns.

Many *C. elegans* introns lack the internal consensus sequence CT(A/G)A(C/T), at which lariat formation occurs in the process of splicing introns in other eukaryotes. However, *C. elegans* introns have a strong preference for A at positions −15 through −17 in the intron. Because lariat formation always occurs at A residues and because the lariat consensus is only weakly adhered to in other systems, it may be that lariat formation occurs at A between −15 and −17. However, these positions are closer to the 3′ boundary than are the lariat sites of other introns (−22 through −37).

Codon usage in several *C. elegans* genes that have been analyzed is highly asymmetric, and even more so than in other metazoans. Table 2 summarizes data on 3492 codons from the genes for vitellogenin, actin, collagen, major sperm protein, myosin heavy chain, and histones. Twenty of the 61 codons are used only rarely (12 times or less), and 7 are used no more than twice. The most dramatic examples of asymmetry are proline codons, where 94% are CCA, and glycine codons, where 90% are GGA. In other coding groups, there is usually a strong selection against third-position purines. All of these genes encode abundant proteins, and genes for abundant proteins usually exhibit more biased codon usage than genes for less abundant proteins (deBoer and Kastelein 1986). Following this pattern, the *lin-12* gene, which may encode a less abundant protein, shows a similar but less extreme bias.

Although gene regulation has not yet been analyzed in detail in *C. elegans*, sequences of control regions that are available suggest that both promoters and terminators are similar to those of other eukaryotes. TATA sequences have been found about 30 bp upstream of the start sites of genes for collagen, vitellogenin, and major sperm protein. In addition, the promoter regions of each of these gene families contain short, repeated, gene-family-specific sequences that have been hypothesized to be involved in their regulation. The most thoroughly studied genes of *C. elegans* are

General

```
C                  A                  T T T T T T T T T T T      C
A  A  G | G   T    G  A  G  T.....   C C C C C C C C C C C   N  T   A    G  | G

43                 62                 51 44 50 53 60 49 49 45 45 57 58  65
40 64 73 | 100 100 29 68 84 63.....  19 25 31 21 24 30 33 28 36 36 28  31  100 100 | 52
```

C. elegans

```
A                                   T                                           G
G  A  G | G   T   A  A  G  T  T....(-)..................  T  T  T  C  A   G  |
                                    A                                           A
                                     n

47                                 44                                            45
25 63 67 | 100 100 73 77 75 62 60  33                      91 98 70 70 100 100 |  36
```

Figure 1 Comparison of general and *C. elegans* intron border consensus sequences. The general consensus is from Mount (1982); the *C. elegans* consensus includes 53 introns from the following sources: 8 from *unc-54* (Karn et al. 1983); 5 from *myo-1*, 10 from *myo-2*, and 1 from *myo-3* (Karn et al. 1985 and pers. comm.); 2 from *vit-1*, 1 from *vit-2*, 4 from *vit-5*, and 3 from *vit-6* (Spieth et al. 1985a; T. Blumenthal et al., pers. comm.); 2 from *col-1* and 1 from *col-2* (Kramer et al. 1982); 1 each from *act-1*, *act-2*, and *act-4* (Files et al. 1983); 1 each from GAPDH-1 and GAPDH-2 (P. Yarbrough and R. Hecht, pers. comm.); 1 each from *hsp-16-1* and *hsp-16-48* (Russnak and Candido 1985); 3 from myosin light-chain genes (C. Cummins and P. Anderson, pers. comm.); 2 from *lin-12* (Greenwald 1985b); and 4 from cAbl (Goddard and Capecchi 1986). The vertical lines indicate intron boundaries. The numbers under each base represent the percentage of total introns that contained the indicated bases at that position.

Table 2 Codon usage in abundantly expressed *C. elegans* genes

Codon	Amino acid	Residues	Codon	Amino acid	Residues	Codon	Amino acid	Residues	Codon	Amino acid	Residues
UUU	F	18	UCU	S	81	UAU	Y	26	UGU	C	10
UUC	F	103	UCC	S	77	UAC	Y	97	UGC	C	49
			UCA	S	27						
UUA	L	2	UCG	S	11	CAU	H	16	CGU	R	102
UUG	L	36	AGU	S	8	CAC	H	45	CGC	R	70
CUU	L	119	AGC	S	15				CGA	R	7
CUC	L	89				CAA	Q	155	CGG	R	1
CUA	L	0	CCU	P	3	CAG	Q	44	AGA	R	34
CUG	L	5	CCC	P	4				AGG	R	4
			CCA	P	238	AAU	N	38			
AUU	I	62	CCG	P	7	AAC	N	118	GGU	G	17
AUC	I	113							GGC	G	9
AUA	I	1	ACU	T	66	AAA	K	40	GGA	G	241
			ACC	T	118	AAG	K	213	GGG	G	2
GUU	V	120	ACA	T	10						
GUC	V	107	ACG	T	2	GAU	D	65			
GAU	V	5				GAC	D	82			
GUG	V	12	GCU	A	108						
			GCC	A	102	GAA	E	112			
			GCA	A	17	GAG	E	207			
			GCG	A	2						

Data are compiled from codon usage in *vit-5* (Spieth et al. 1985a); part of *vit-2* (E. Zucker and T. Blumenthal, pers. comm); an *msp* gene (Klass et al. 1984); histone genes H2A, H2B, and H4 (Roberts et al. 1987); *act-1* (Files et al. 1983); *col-1* and *col-2* (Kramer et al. 1982); and part of *unc-54* (Karn et al. 1983).

those for tRNA. These function in heterologous transcription systems, as described in more detail (Section III.A.5), where they can be demonstrated to have internal control regions, as do genes in other organisms that are transcribed by RNA polymerase III.

A remarkable recent finding is that mature mRNAs for three of the four actin genes have a *trans*-spliced leader sequence (Krause and Hirsh 1987). Twenty-two nucleotides at the 5′ end of mRNAs of *act-1*, *act-2*, and *act-3* are not found in the genomic DNA upstream of the gene (for a detailed description of these genes, see Chapter 10). The 22 nucleotides appear to come instead from the 5′ end of a 100-nucleotide RNA encoded by sequences lying in the spacer DNA between the 5S RNA genes of the 5S tandem array. The same *trans*-spliced sequence is apparently found on other *C. elegans* messages as well. Hence, the pathway for the synthesis of some *C. elegans* mRNAs is apparently similar to that for mRNA of the kinetoplastid protozoans (e.g., *Trypanosoma brucei*), in which a 35-nucleotide sequence is transferred from the 5′ end of a 140-nucleotide precursor RNA to the 5′ ends of messages (Borst 1986).

Many of the cloned genes mentioned above are discussed in other chapters in this volume. Properties of those not discussed elsewhere are summarized below.

1. Heat Shock Genes

Like other organisms, *C. elegans* responds to elevated temperature (29–35°C, normal growth temperature is 20°C), by a general reduction in overall protein synthesis and a specific induction of the transcription and translation of a small set of polypeptides (Snutch and Baillie 1983). Using the gene for a *Drosophila* 70-kD heat shock protein, Snutch and Baillie (1984) isolated three *C. elegans* genes, one of which is activated by high temperature. A curious property of the inducible gene is that it lies in a genomic segment that shows a very high degree of nucleotide sequence divergence between the Bristol and Bergerac strains. Two cDNA clones encoding 16-kD heat shock proteins have been isolated by screening a cDNA library constructed from mRNA of heat-shocked worms with labeled and size-fractionated poly(A)$^+$ heat shock RNA (Russnak et al. 1983). Using these cDNA clones as probes, Russnak and Candido (1985) and Jones et al. (1986) defined the genomic organization of the 16-kD heat shock genes. There are six genes for this class of proteins, and these define four distinct but related polypeptides. The six genes are arranged in three sets of two, the two genes of each set being divergently oriented and separated by a few hundred nucleotides. Two of the sets are closely linked in inverted orientation. They lie in a 1.9-kb genomic segment that occurs twice as a perfect inverted repeat, the two repeats being separated by 416 bp. The four proteins defined by these genes have homology at the amino acid level to *Drosophila* 16-kD heat shock proteins and to murine and bovine α-crystalline proteins. The homology is confined to a carboxy-terminal domain, which is separated from a variable amino-terminal domain by an intron located identically in the genes of both *C. elegans* and mouse. The nucleotide sequences surrounding the *C. elegans* genes contain appropriately located TATA and polyadenylation (AATAAA) sequences; in the 5′ upstream region, they also contain sequences close to the consensus sequence CT GAA TTC AG found in the 5′ region of *Drosophila* heat shock genes (Pelham 1985).

2. Histone Genes

The *C. elegans* genome contains about 10–12 copies of each of the core histone genes, H2A, H2B, H3, and H4 (Roberts et al. 1987). These genes are not present in tandem arrays, as they are in sea urchins, *Drosophila*, and some other organisms, but rather occur in separated clusters as they do in mammals. Each cluster in *C. elegans* contains one or a small number of each of the core histone genes. Several clusters have been cloned and analyzed by sequencing. The encoded histone genes have amino acid sequences nearly identical to those of the sea urchin. Genes for H1 histones have also been cloned and lie elsewhere in the genome (M. Sanicola et al., pers. comm.). Histone proteins have been analyzed by Vanfleteren and colleagues (Vanfleteren et al. 1979; Vanfleteren 1983; Vanfleteren and Van Beeumen 1983).

3. 5S RNA Genes

The *C. elegans* genome contains about 110 copies of the genes for 5S RNA organized into a single tandem array or a small number of closely linked arrays. 5S DNA in a CsCl equilibrium density gradient bands as a single homogeneous peak, at a slightly denser position (corresponding to a G + C content of 42%) than main band DNA (36% G + C) and distinct from rDNA (51% G + C) or genes for tRNA (38% G + C, with a broad range) (Sulston and Brenner 1974). The 5S genes are therefore clustered in the genome and are separated from genes for 18S and 28S RNA and for tRNA. The 5S cluster consists of a tandem array of uniform repeating units of 1.0 kb, each containing a single gene for 5S RNA surrounded by spacer DNA (Nelson and Honda 1985). The spacer contains the gene for the *trans*-spliced leader of some *C. elegans* mRNAs, as discussed above. *C. elegans* 5S RNA has been sequenced by Butler et al. (1981) and Kumazaki et al. (1982), and one 5S DNA repeat was sequenced by Nelson and Honda (1985). The sequence of the internal promoter of the *C. elegans* gene is similar to that of *Xenopus* and *Drosophila*, and the gene ends with the termination sequence TTTTT. The promoter is weakly expressed to give a 5S RNA product in heterologous pol III transcription systems in vitro. Expression in a homologous extract has also been demonstrated (Honda et al. 1986). The 5S cluster has been mapped to linkage group V by using a DNA strain–polymorphism in flanking sequences and by hybridization in situ of cloned sequences to chromosomes (Albertson 1984b; Nelson and Honda 1986a,b). Both of these mapping techniques are described more fully below (Section IV.A,D).

4. rRNA Genes

The rRNA genes of *C. elegans*, like the 5S RNA genes, are organized into a single tandem array, or a small number of closely linked tandem arrays, containing a total of about 70 repeating units, 7.0 kb in length (Sulston and Brenner 1974; Files and Hirsh 1981). They are readily isolated because they can be separated completely from the bulk of genomic DNA as a dense satellite in CsCl equilibrium density gradients. Cloned repeats have been isolated from purified material (Files and Hirsh 1981), and one repeating unit was cloned in a random selection of recombinant plasmids containing nematode DNA (pCe7; Emmons et al. 1979).

Files and Hirsh (1981) analyzed the structure of the repeating units of *C. elegans* rRNA in detail. They found that unlike rRNA repeats in most other organisms, the repeating units in *C. elegans* are almost identical. One cloned repeating unit (pCe7) has been completely sequenced, along with parts of others (Ellis et al. 1986). The repeating unit contains a single gene for each of the large rRNAs and for 5.8S rRNA, as well as a nontranscribed spacer of 1.0 kb. Overall, the repeat length of 7.0 kb is smaller than those reported for other eukaryotes. The short length is due in part to the short nontranscribed spacer and in part to the short length of the rRNAs themselves. *C. elegans* 28S rRNA is 3500 bases long, and 18S rRNA is 1750 bases long. These values are similar to those for yeast rRNAs and are several hundred bases shorter than for rRNAs in other higher eukaryotes. The *C. elegans* genes have long regions of sequence homology with other eukaryotic rRNA genes, interspersed with shorter diverged regions.

The site of the ribosomal gene cluster is the right end of linkage group I. This was determined by hybridization in situ of a cloned repeat to metaphase chromosomes (see Section IV.D) (Albertson 1984b). Analysis using this method uncovered two strains with mutations affecting the rRNA genes. One strain has a duplication of the ribosomal locus translocated to the right end of linkage group II (designated *eDp20[I;II]*), and one carries a deletion of over 90% of the ribosomal genes *(let[e2000]I)*. The latter mutation results in late embryonic lethality, possibly owing to a deficiency of rRNA.

5. tRNA Genes

There are about 300 genes for tRNA in the *C. elegans* haploid genome, as determined from saturation hybridization experiments (Sulston and Brenner 1974). Several have been cloned by Cortese et al. (1978) and used extensively to study the structure and expression of nuclear tRNA genes. The gene for a mutant $tRNA^{Trp}$ with amber suppressor activity, the *sup-7* locus, has also been cloned (Bolten et al. 1984). The 300 tRNA genes are dispersed in the genome. *C. elegans* tRNAs can be fractionated into six size classes on urea–polyacrylamide gels. When these classes are isolated, labeled, and separately hybridized to genomic *Eco*RI restriction fragments, a broad smear of fragments hybridizes (Cortese et al. 1978). Two cloned *Eco*RI fragments, 5.5 kb and 5.0 kb, identified as having homology to labeled nematode tRNA, each carry a single tRNA gene, whereas a third, 2.5 kb, carries two. A λ recombinant phage similarly isolated carries three tRNA genes. These results suggest that though they are generally dispersed in the genome, there may be some degree of clustering of the genes. Five of the cloned wild-type genes have been sequenced. They encode sequences for a $tRNA^{Pro}$, a $tRNA^{Asp}$, a $tRNA^{Leu}$, and two $tRNA^{Lys}$. All the sequences and predicted tRNA structures are similar to those that have been found in other higher eukaryotes (Tranquilla et al. 1982).

When cloned nematode tRNA genes are injected into *Xenopus* oocytes, they are correctly transcribed and processed, and the processed products appear in the cytoplasm (Cortese et al. 1978; Melton and Cortese 1979). Using this assay, Cortese and his colleagues carried out an extensive investigation of the control sequences required for this activity (Cortese et al. 1980; Ciampi et al. 1982; Ciliberto et al. 1982a,b, 1983; Traboni et al. 1984). They found that like other genes, such as 5S RNA genes, that are transcribed by RNA pol III, *C. elegans* tRNA genes have promoters that are internal to the coding sequence. In *C. elegans* tRNA genes, the promoter consists of two regions of about 10 nucleotides, separated by 30–40 nucleotides. These two regions are denoted, in 5′–3′ order, box A and box B. Initiation of transcription occurs at a fixed distance 5′ of box A, and terminates at the first run of four or more Ts, 3′ of box A (not necessarily 3′ of box B). The distance between the two box sequences is not important for correct initiation. This general organization of promoter sequences holds for all genes transcribed by RNA pol III. The box A and box B sequences occur in regions of the tRNA sequence that are highly conserved, both from one tRNA species to the next and from one organism to another. Detailed analysis of the promoter region of a *Xenopus* tRNA gene gave essentially identical results to those in *C. elegans* (Hofstetter et al. 1981). In the 5S RNA genes, the two regions, box A and box B, are adjacent in the 5S sequence. Nevertheless, Ciliberto et al. (1983) showed that the two halves of the *Xenopus* 5S RNA promoter region were functionally interchangeable with box-A and box-B regions of *C. elegans* tRNA genes in experiments in which hybrid 5S RNA–tRNA genes were injected into oocytes.

Wolstenholme et al. (1987) have reported the sequences of 21 apparent tRNA genes identified in the complete sequence of the mitochondrial DNA from *A. lumbricoides*. These sequences show most of the characteristics expected of tRNA genes, except that all are distinctive in lacking the usual TψC arm and variable loop, which are replaced with a single loop of between 4 and 12 nucleotides. These investigators also reported the sequence of a portion of the *C. elegans* mitochondrial DNA, containing six tRNA gene sequences that showed the same distinctive structure. This feature is not shared with any other tRNA structure described so far but may eventually provide information on the evolutionary origin of nematode mitochondria.

6. *Collagen Genes*

The *col-1* and *col-2* genes were first isolated from *C. elegans* by hybridization with a chicken collagen cDNA probe (Kramer et al. 1982). Using these isolates to probe genomic Southern blots and recombinant phage libraries under low stringency conditions (which allow all collagen genes to cross hybridize), Cox et al. (1984) demonstrated that there are between 40 and 150 collagen genes dispersed throughout the *C. elegans* genome, mostly singly, which have been mapped to several linkage groups. In contrast, vertebrate genomes contain approximately ten collagen genes (Bornstein and Sage 1980), and *Drosophila* has fewer than ten (Natzle et al. 1982).

Six of the *C. elegans* genes have been sequenced (*col-1*, *col-2*, *col-6*, *col-8*, *col-14*, and *col-19*), and all six have similar, but not identical, structures (Kramer et al. 1982; G. Cox et al., pers. comm.). All are less than 1.5 kb in size and contain only one or two small introns. They encode proteins of 28–32 kD, of which approximately one-half is the Gly-X-Y repeat characteristic of collagen proteins. The Gly-X-Y repeat regions contain several short interruptions (2–21 amino acids) that do not have glycine residues at every third position. The *C. elegans* collagens are quite different from the vertebrate fibrillar collagens, which are proteins of 100–200 kD in size, have uninterrupted Gly-X-Y repeat regions, and are encoded by 50-kb genes with 40–50 introns (Boedtker et al. 1983). The *C. elegans* collagens resemble the vertebrate nonfibrillar collagens, such as the chondrocyte-specific type-IX (Ninomiya and Olsen 1984; Lozano et al. 1985b) and type-X collagens (Schmid and Conrad 1982; Schmid and Linsenmayer 1985).

The six sequenced *C. elegans* genes appear typical. Restriction mapping and hybridization analysis of 24 collagen-containing phage clones showed that 20 of the phage contained a single small collagen-hybridizing region, similar in size to the hybridizing regions seen for the sequenced genes. The four other phage contained two or possibly three small collagen genes (Cox et al. 1984; J. Kramer, pers. comm.). Genomic Southern blot hybridization experiments using various stringency conditions and comparisons between the sequenced collagen genes have shown that there is a wide range in the degree of sequence homology among different members of the collagen gene family. Some genes share greater than 90% sequence homology, whereas others are less than 67% homologous. Most, and possibly all, are unique and do not cross hybridize with other collagen genes under appropriately stringent conditions (Cox et al. 1984; Cox and Hirsh 1985).

Northern blot analyses of *C. elegans* RNAs have shown that the majority (90%) of the collagen mRNAs are between 1.1 and 1.4 kb in length, the size expected for transcripts from the sequenced genes (Cox et al. 1984). Two of the sequenced genes, *col-1* and *col-2*, have been shown to produce transcripts within this size range (Kramer et al. 1985). Collagen mRNAs of approximately 4.0 and 5.0 kb have also been detected in smaller amounts and could encode collagens similar to the vertebrate fibrillar or basement membrane types.

Several lines of evidence suggest that the large family of small collagen genes encodes the cuticle collagens: (1) Transcripts from the small collagen genes are highly abundant at molts, when the cuticle is being synthesized. (2) The small collagen transcripts are the most abundant collagen transcripts in the animal, and localization by in situ hybridization has shown that collagen transcripts are most abundant in the hypodermis (Edwards and Wood 1983), the tissue that synthesizes the cuticle. (3) The collagens isolated from cuticles range from 60 kD to over 200 kD in size, much larger than the proteins encoded by the family of small collagens. However, during the L4 adult molt, antibodies raised against adult cuticle proteins detect a class of small (38- to 52-kD) collagenous proteins that may be precursors of the larger collagens found in mature cuticles (Politz et al. 1986). The large (60- to 200-kD) collagens may therefore arise by the formation of nonreducible covalent cross-links between the small collagen polypeptides, as is known to occur with other collagens. In vitro translation products from stage-specific RNAs consisted of numerous small (37- to 52-kD) collagenase-sensitive polypeptides (Politz and Edgar 1984). Over 60 distinct collagenous products could be identified, setting a lower limit to the number of different collagen genes that must be expressed.

Expression of the *C. elegans* collagen genes is regulated during development. Using small, gene-specific probes, it was demonstrated that *col-1* transcripts are highly abundant during late embryogenesis (when the L1 cuticle is formed) and the L2 dauer molt, and at low levels during the L4 dauer and L4 adult molts. *col-2* transcripts are detectable only during the L2 dauer molt (Kramer et al. 1985). The expression patterns of a large number of collagen genes were studied by hybridizing 70 collagen-containing phage clones with stage-specific labeled cDNAs under high-stringency conditions (Cox and Hirsh 1985). In general, a sequential activation of collagen genes during development was noted. Once a gene is activated it remains active during subsequent molts, although its level of expression can change.

B. Transposable Elements

The *C. elegans* genome contains several families of transposable elements, denoted Tc elements (for transposon *Caenorhabditis*). Two of these, Tc1 and Tc3, have been shown to transpose in the germ line, causing mutations by inserting into genes. Several other repetitive families are thought to consist of transposable elements because of their structure or highly polymorphic arrangement in strains. It is likely that additional families remain to be identified. The structure of Tc1 resembles that of P elements of *Drosophila* and Ac/Ds elements of maize in having short, terminal inverted repeats. Two of the putative transposon families have an inverted repeat structure, possibly resembling that of fold-back transposons in *Drosophila*, sea urchins, and *Dictyostelium*. Elements have been identified in *Panagrellus redivivus* and *A. lumbricoides* that have structures resembling retrotransposons such as *copia* of *Drosophila* and Ty of yeast. Therefore, although research on transposons of nematodes is at an early stage, it is already clear that, as in other groups, this is a complex class of sequences.

1. Tc1

Tc1 elements are 1610 bp long, are highly uniform in structure, and are located at a number of dispersed sites in the genome (Emmons et al. 1983; Liao et al. 1983). All of one and part of two other cloned elements have been sequenced (Rosenzweig et al. 1983a,b; Rose et al. 1985). They have 54-bp perfect terminal inverted repeats and an open reading frame capable of encoding a polypeptide of 273 amino acids. Most Tc1 elements have the same structure, in contrast to P transposable elements in *Drosophila* or Ac/Ds elements in maize, many of which contain deletions (Fedoroff et al. 1983; O'Hare and Rubin 1983). Tc1 elements do exhibit microheterogeneity, however. A minority of elements contains a *Hin*dIII site (Rose et al. 1985); a different subset contains an *Eco*RI site (Eide and Anderson 1985b; A. Otsuka, pers. comm.), and there is some sequence variation near the inverted repeat termini (D. Eide and P. Anderson, pers. comm.).

Tc1 elements may cause a 2-base target-site duplication upon insertion, or, alternatively, the elements themselves may be 1612 bp long, have 55-bp inverted terminal repeats, and duplicate no target-site sequences. This ambiguity arises because all Tc1 elements so far examined are inserted at a TA dinucleotide. A TA duplication that occurs at each end of the element could be a duplication of this dinucleotide, but it could also be part of the element itself. Sequences at ten Tc1 insertion sites give the following consensus target sequence (number in parentheses is number of sites with the given base at this position; insertion occurs within the underlined region): A(7)T(6)R(8)T(10)A(10)Y(8)NT(7).

Initial evidence for transposition of Tc1 came from comparisons of the arrangement of insertion sites in various strains. Active transposition of Tc1, resulting in insertional mutation at defined loci, has since been observed in several strains.

C. elegans strains fall into two classes with respect to the number of Tc1 elements in their haploid genome: low-copy-number strains and high-copy-number strains. Tc1 was discovered because low- and high-copy-number strains were included in initial surveys of cloned *C. elegans* DNA (Emmons et al. 1979; Files et al. 1983). Tc1 elements are so plentiful in high-copy-number strains that strain polymorphisms due to them are frequently observed when low- and high-copy-number strains are compared (Rose et al. 1982; Blumenthal et al. 1984). So far, no strain completely lacking Tc1 elements has been found. Low-copy-number strains contain about 30 copies of Tc1 per haploid genome. These strains include the standard laboratory strain Bristol (N2), with which most genetic and cell lineage studies have been carried out, and the majority of new wild isolates that have been obtained from various parts of the world. The arrangement of Tc1 elements in all these strains is nearly the same, suggesting that transposition is infrequent, if it occurs at all. High-copy-number strains have around 300 copies of Tc1 inserted singly at new genomic sites, strongly suggesting that a process of transposition resulting in amplification has taken place during the descent of these strains (Emmons et al. 1983; Liao et al. 1983). At present, three high-copy-number strains are known, including the laboratory strain Bergerac and two newly isolated wild strains from California (DH424; Liao et al. 1983) and Wisconsin (TR403; P. Anderson, pers. comm.).

Germ-line transposition of Tc1 has been detected by analysis of spontaneous mutations at specific genetic loci in a Bergerac genetic background. Moerman and Waterston (1984) found a high spontaneous mutation frequency at the *unc-22 IV* locus in Bergerac. The frequency, 10^{-4} (per gamete), is at least 100-fold greater than the spontaneous mutation frequency at *unc-22* found in several other strains, and, furthermore, spontaneous *unc-22* mutations isolated in the Bergerac background are unstable and revert to wild type at a frequency of 10^{-3} or greater. These unstable, spontaneous mutations are due to disruption of the *unc-22* gene by insertion of a Tc1 element, and revertants are due to excision of the element (Moerman et al. 1986). To prove this, mutant alleles isolated in the Bergerac background were outcrossed to Bristol to remove most of the 300 endogenous Bergerac elements. An element newly transposed into the *unc-22* locus was then detected in a Southern hybridization experiment by using a cloned Tc1 element as a probe. The *unc-22* gene is the first *C. elegans* gene to have been isolated by transposon tagging. It encodes a previously unknown component of muscle thick filaments, as described in Chapter 10.

Similarly, Eide and Anderson (1985b) have reported that approximately two thirds of spontaneous *unc-54 I* mutations isolated in Bergerac are due to insertion of a Tc1 element. The frequency of spontaneous *unc-54* mutants in this strain is approximately 10^{-6}. Many Tc1-induced mutations revert by excision at frequencies between 10^{-4} and 10^{-7}. Analysis of *unc-54* mutations was facilitated by the prior availability of the cloned gene (MacLeod et al. 1981). In both *unc-22* and *unc-54*, spontaneous mutations can be studied because a sensitive and specific screen is available for detecting mutant worms. In *unc-22*, heterozygous or homozygous mutant worms are detected as worms that twitch or vibrate in a solution of 1% nicotine, which causes wild-type worms to become rigidly paralyzed (Moerman and Baillie 1979). *unc-54* mutants are isolated as mutations that give rise to a thin, paralyzed animal in a genetic background containing a second mutation (*unc-105[n490]II*) that causes hypercontraction of the body-wall muscle and, hence, a Dumpy phenotype (Eide and Anderson 1985a; Park and Horvitz 1986b). Both screens make it possible to identify mutations occurring at a frequency of 10^{-6} or lower.

A striking observation made in these studies is that transposition activity of Tc1 is restricted to the high-copy-number Bergerac strain. As already mentioned, the forward mutation frequency of *unc-22* is 100-fold higher in Bergerac than in other strains. Reversion of Tc1-induced *unc-22* mutations by excision of Tc1 is also strain dependent. The reversion frequency falls from 10^{-3} to less than 10^{-6} when *unc-22*::Tc1 alleles isolated in Bergerac are crossed into a Bristol background (Moerman and Waterston 1984). The spontaneous mutation frequency of *unc-54* is threefold higher in Bergerac than in Bristol, the increase being accounted for by Tc1-induced mutations. Analysis of 65 spontaneous *unc-54* mutations isolated in the Bristol background has proved that transposition of Tc1 is very rare in this strain, if it occurs at all. None of the 65 spontaneous mutants was due to insertion of Tc1 or of any other transposable element (Eide and Anderson 1985a).

These observations provide an opportunity to examine the genetic basis for activation of transposition. Experiments carried out so far have shown that Tc1 is under the control of activators in Bergerac, rather than repressors in Bristol: In hybrid strains containing chromosomes from both Bergerac and Bristol and in heterozygous Bristol/Bergerac hybrids, Tc1 elements are active, that is, activity is dominant over inactivity. There are several or many such activators in the Bergerac background, because activators can be mapped to more than one Bergerac chromosome. On placing Bristol chromosomes into a Bergerac background, Moerman and Waterston (1984) found that no single Bergerac chromosome was responsible for the high frequency of *unc-22* mutation in this strain. They have mapped an activator to linkage group I in one strain and to linkage group IV in another (D.G. Moerman and R.H. Waterston, pers. comm.). Anderson has mapped an activator to linkage group II (D.G. Anderson, pers. comm.). Activators might be a subset of Tc1 elements, but activity is not caused simply by a high level of ordinary Tc1 elements. Both Moerman and Waterston (1984) and Eide and Anderson (1985b) found high-copy-number strains in which transposition of Tc1 could not be detected. These strains not only included the newly isolated (1970) wild-type strain DH424 (Liao et al. 1983) but also Bergerac stocks held by some laboratories. In this and in other respects, it has become evident that not all Bergerac lines are identical to each other, even though all of them originally came from the *C. elegans* laboratory at Lyon, France, and all of them contain a large number of Tc1 elements. Because of this diversity, stocks maintained by different laboratories are separately designated by a two-letter code. The strain in which Tc1 is transpositionally active is designated Bergerac (BO).

Collins et al. (1987) found that the frequency of Tc1 excision and transposition can be elevated to even higher levels by mutations occurring in the Bergerac(BO) background. Such mutants were isolated by screening ethylmethanesulfonate (EMS)-mutagenized worms that carry an *unc-54::Tc1* mutation for derivatives that exhibit elevated levels of germ-line reversion. These mutations define several genetic loci, designated *mut*. In strains carrying *mut* mutations, the frequencies of Tc1 germ-line excision and transposition are elevated 10- to 100-fold. The *unc-22* gene, for example, mutates spontaneously by insertion of a Tc1 element at a frequency of greater than 10^{-3} in mutator strains. These strains are presently being exploited by many laboratories for cloning genes by transposon tagging, and the nature of the changes occurring at the *mut* loci, as well as the nature of the endogenous Bergerac(BO) activators themselves, are under intensive investigation.

Biochemical evidence for mobility of Tc1 elements has been obtained by analysis of individual sites of Tc1 insertion in Southern hybridization experiments. Such experiments show that at Tc1 insertion sites, DNA from populations of worms is a mixture of sequences, some containing the Tc1 element and some lacking it (Emmons et al. 1983; Eide and Anderson 1985b; Moerman et al. 1986). This is because Tc1 elements undergo frequent excision in somatic tissues, leaving behind a perfectly restored or nearly perfectly restored empty insertion site (Emmons and Yesner 1984; Eide and Anderson 1985b; Ruan and Emmons 1987). Excision proceeds throughout development, and DNA sequences lacking the element accumulate as the worms mature and eventually amount to some 5% or more of the DNA at each site. Somatic excision of Tc1 elements inserted into the *unc-54* gene gives rise to mosaic animals with patches of revertant muscle tissue (Eide and Anderson 1985b).

In germinal tissue, excision of Tc1 occurs but is much less frequent than in somatic cells, as shown by the genetic experiments described above. Tc1 excision is therefore under tissue-specific regulation. Emmons et al. (1986)

have investigated the basis of this regulation. They showed that elements at five different genomic sites in Bergerac (BO) underwent excision at similar frequencies during several developmental stages. This suggests that regulation acts at the level of transposition factors that affect all genomic elements, rather than resulting from the effects of flanking sequences acting in *cis*. In the latter event, differences in the behavior of elements at different genomic sites might have been expected. Emmons et al. (1986) also found that somatic excision occurs in Bristol, albeit at a lower frequency. Tc1 elements are therefore not completely quiescent in strains where germ-line transposition has not been detected.

Somatic excision could account for the observed presence of extrachromosomal copies of Tc1 in Bergerac DNA preparations. Linear, as well as relaxed and supercoiled circular, monomers of the element can be detected at a level corresponding roughly to one copy per cell when whole DNA preparations are analyzed by using a Tc1-specific probe (Rose and Snutch 1984; Ruan and Emmons 1984). Analysis of these molecules with restriction endonucleases revealed no differences from chromosomal copies. Circular copies appear to result from circularization of the genomic sequence, including the ambiguous terminal nucleotides mentioned earlier. Linear copies have fixed ends nearly, if not exactly, equivalent to the ends of genomic elements, strongly suggesting that they are excision products.

At present, the relationship(s) that the four phenomena—germ-line transposition, germ-line excision, somatic excision, and extrachromosomal elements—bear to one another is not known. The most attractive hypothesis is that they are related: Maybe somatic excision represents the first step in a transposition pathway active in somatic tissues. Extrachromosomal elements could be transposition intermediates in this pathway. The same pathway might be operative in the germ line in Bergerac (BO) and in mutator strains. Alternatively, the four phenomena may reflect unrelated properties of the Tc1 sequence.

2. Tc3

Tc3 transposable elements were identified in three spontaneous mutant alleles of the *unc-22* gene (J. Collins et al., pers. comm.). These 2.5-kb elements are present in 10–15 copies in the genomes of most *C. elegans* strains examined and in relatively constant arrangement, indicating that transposition in these strains is rare. In mutator strains, however, additional Tc3-containing restriction fragments can be detected in genomic Southern hybridization experiments. This suggests that the same genetic change that has caused an elevation of Tc1 transposition has activated Tc3 as well. Tc3 does not hybridize to Tc1, and details of its structure and relationship to Tc1 remain to be determined.

3. *Additional Possible* C. elegans *Transposons*

Restriction fragment length polymorphisms have led to the identification of three additional repetitive families that are likely to be transposable elements. One family is present in some 30 copies in several of the strains that have large numbers of Tc1 elements in their genomes (e.g., Bergerac-[BO]) and in about 5 copies in low-copy-number Tc1 strains (Emmons et al. 1985). This striking difference is taken as evidence that the family is transposable, and the element has been termed Tc2. The correlation of numbers of Tc2 elements and numbers of Tc1 elements suggests that Tc2, like Tc3, may be in some way coregulated with Tc1.

Another Bristol–Bergerac restriction fragment length polymorphism (identified by plasmid pCe14; Emmons et al. 1979) has been shown to be due to a family of 1.6-kb elements with an inverted repeat structure (D. Dreyfus and S.W. Emmons, in prep.). There are approximately 20 such sequences in the genome, and their arrangement is not highly polymorphic. One sequenced element shares a sequence of nine consecutive nucleotides at its ends with the ends of Tc1. Another family of about 20 elements with an inverted repeat structure has been identified because two members are found in spontaneous mutant alleles of *unc-86* (M. Finney and R. Horvitz, pers. comm.).

4. *Transposons of Other Nematode Species*

Link et al. (1987) used an approach similar to that utilized in the demonstration of Tc1 transposition in *C. elegans* to identify a transposon of *P. redivivus*. They examined the molecular structure of spontaneous mutations of the *P. redivivus* homolog of the *unc-22* gene. In the strain they used (*C15*), spontaneous twitcher mutations of the type expected for alleles of *unc-22* arose at a frequency comparable to that in Bergerac(BO) and at least two orders of magnitude greater than the spontaneous frequency in Bristol. The cloned *C. elegans* gene was used to isolate the homologous gene from wild type and two spontaneous mutants. One of the mutant genes was found to contain a 4.8-kb insertion. The insertion defines a repetitive family of from 10 to 50 members in various *P. redivivus* strains and has direct repeats of 170 bp at its ends. These properties suggest that this is a transposable element of the *copia* or retrotransposon class. These elements have been designated PAT. They are not present in *C. elegans*.

A transposon of *A. lumbricoides*, also with *copia*like structure, was identified in studies of satellite DNA-containing clones (Aeby et al. 1986). This element, termed TAS, is 7.5 kb long and has 256-bp terminal direct repeats. The terminal repeats have been sequenced and have features expected for long terminal repeats of a retroviral provirus or a retrotransposon. There are about 50 TAS elements in the *A. lumbricoides* genome, and their arrangement varied somewhat among five individual animals examined.

C. Short, Interspersed Repetitive DNA and Segregator Sequences

Although a large fraction of the genome, up to 10%, consists of short, interspersed repeats, this component of the genome has been characterized minimally. Individual families that have been analyzed using cloned sequences are not conserved in either of the two closely related species, *C. briggsae* or *Caenorhabditis remanei*, suggesting that they are a nonessential component of the genome (Emmons et al. 1979; Felsenstein and Emmons 1987).

As discussed earlier in this chapter, possible roles for nongene repeats include chromosome counting for sex determination, and chromosome segregation. *C. elegans* kinetochores are diffuse (Chapter 2), which may be because sequences with centromeric properties are scattered throughout the DNA. *C. elegans* DNA sequences exist that can stabilize autonomous plasmids in yeast cells, and these might be the postulated dispersed centromeric sequences. Stinchcomb et al. (1985a) isolated such sequences active in yeast, denoted segregator (SEG) sequences, by direct selection. They estimated that about 30 SEG sequences are present in the *C. elegans* genome. Felsenstein and Emmons (1988) have described an interspersed repetitive family that has replication and segregator function in yeast. SEG sequences stabilize yeast plasmids during mitosis and meiosis and also lower their copy number. These are also characteristics of true yeast centromere sequences (Blackburn and Szostak 1984), but the *C. elegans* sequences are not as effective. When included on DNA segments microinjected into nematodes, SEG sequences do not stabilize the extrachromosomal concatemers formed (see Section IV.F). The function of SEG sequences in the *C. elegans* genome, if any, therefore remains to be established.

IV. FINDING AND ANALYZING SINGLE GENES

Several methods are available for cloning and characterizing single genes. These include the genetic mapping of restriction fragment length polymorphisms, transposon tagging, hybridization in situ to chromosomes, and transformation. To provide a comprehensive correlation between the genetic and physical maps, a cooperative project involving many laboratories is under way to create an overlapping set of clones covering the entire genome. The ordered set is positioned on the genetic map by using the existing genetically identified clones.

A. Mapping Restriction Fragment Length Polymorphisms

Comparisons of genomic restriction fragments of the Bristol and Bergerac strains in Southern hybridization experiments with cloned genomic probes showed that differences between the strains occur, affecting about one out of ten fragments (Emmons et al. 1979; Rose et al. 1982). Some of these differences are due to Tc1 elements, present in about 300 copies in Bergerac but in only 30 copies in Bristol. A polymorphism due to Tc1 would be expected to occur roughly once every 300 kb and, hence, should affect less than 1 fragment in 60. Other differences between the strains at the DNA level evidently exist, but these have not been characterized.

A DNA polymorphism may be used as a genetic marker and mapped in a manner identical to conventional mutant phenotypes. Experiments carried out to date have utilized the Bristol and Bergerac strains. Typically, linkage to a chromosome is determined by analyzing the DNA of populations of worms known to be homozygous Bristol or Bergerac for one of the linkage groups but of mixed parentage for the other linkage groups. These populations are grown from appropriate segregants of hybrid worms constructed by mating a Bristol strain carrying a morphological marker on one linkage group to an unmarked Bergerac strain.

Several genes have been assigned to linkage groups by this method, including genes for actin (Hirsh et al. 1979; Files et al. 1983), vitellogenin (Blumenthal et al. 1984; Heine and Blumenthal 1986), 5S RNA (Nelson and Honda 1986a), collagen (Cox et al. 1985), and heat shock proteins (T.P. Snutch and D.L. Baillie, pers. comm.), as well as SEG sequences (S. Carr et al., pers. comm.) and randomly cloned restriction fragments (Rose et al. 1982).

Once the linkage of a particular fragment is known, the position of the fragment on the linkage group can be determined by means of three-factor crosses in which the DNA polymorphism is introduced as the unselected marker (see Chapter 2). Using such an approach, Files et al. (1983) positioned actin genes to a 2% recombination interval on linkage group V. This interval was known to contain a gene, *unc-92*, identified by mutations that caused disruption of body-wall muscle. Subsequent analysis of DNA sequences in mutants and revertants of this locus confirmed that it is the site of three of the four actin genes of *C. elegans* (Landel et al. 1984).

B. Construction of Prelocalized Restriction Fragment Length Polymorphisms

The existence of *C. elegans* strains with high- and low-copy numbers of the transposable element Tc1 has made possible a mapping and gene isolation method involving the construction of new strains with prelocalized restriction fragment length polymorphisms, identifiable by hybridization to Tc1.

The new strains are constructed in such a way that a defined segment of the genome of a low-copy-number strain is replaced by DNA from a high-copy-number strain. Comparison of a Southern hybridization of such a congenic strain using a Tc1 probe with a similar Southern hybridization of the low-copy-number parental strain reveals the additional Tc1-containing restriction fragments due to the high-copy-number DNA. The specific segment of the low-copy-number genome to be replaced with high-copy-number DNA is predetermined by the use of genetic markers in the construction of the congenic strain. For example, the congenic strain may be constructed by a repeated series of crosses to a mutant Bristol strain, in which a wild-type Bergerac strain is used in the first cross to Bristol, and progeny carrying the wild-type Bergerac allele are selected and used in each subsequent cross. Additional markers flanking the gene of interest may be included to select recombination events that remove linked high-copy-number DNA.

A number of *C. elegans* laboratories are successfully using congenic strains of this type to clone specific genes of interest. Strains that have been constructed and examined so far have about the expected number of additional Tc1-containing restriction fragments, giving evidence for a random distribution of Tc1 elements in the Bergerac genome. Subsequent three-factor crosses have shown that the additional Tc1-containing restriction fragments are indeed linked to the marker used during the construction, implying that no extensive mobilization of the element has occurred during the strain construction and that Tc1 elements may be used as stable markers of specific DNA segments. Three-factor crosses are used to identify the Tc1 element most closely linked to the gene of interest, which must then be cloned by "walking."

C. Gene Isolation by Transposon Tagging

The most straightforward method for isolating a gene of interest is to take advantage of a transposon-induced allele. Genes can be tagged by transposable elements Tc1, Tc3, or other elements in the Bergerac(BO) strain and in mutator strains. Most spontaneous mutations in these strains are due to insertion of Tc1. A suitable selection or screen is required to isolate a spontaneous mutation in the gene of interest. Once isolated, the mutation must be outcrossed to Bristol to reduce the number of endogenous Tc1 elements, so that the newly inserted element can be visualized in a Southern hybridization.

D. Gene Mapping by In Situ Hybridization to Chromosomes

Despite the small size and uniform appearance of the six *C. elegans* chromosomes, it is possible to locate cloned DNA sequences in specific chromosomal regions by hybridization in situ (Albertson 1984b, 1985). The problem of the uniformity of the chromosomes is partially overcome by using strains carrying rearrangements that alter the karyotype, making linkage groups distinguishable.

The sequence to be mapped and a cloned ribosomal gene repeat are hybridized together to embryos of a strain carrying three mutations or rearrangements: a partial deletion of the tandemly reiterated rDNA locus on linkage group I (*let[e2000]I*), a translocation of a wild-type ribosomal gene locus to the right end of linkage group II (*eDp20[I;II]*), and a translocation chromosome consisting of linkage groups IV and X joined together (*mnT12[IV;X]*). In this hybridization, the right end of linkage group I is identified by a small hybridization signal to the ribosomal repeat, the right end of linkage group II is identified by a large hybridization signal to the ribosomal repeat, and linkage groups IV and X are conjointly identified as the double-length translocation chromosome. A second hybridization, including appropriately linked probes, is required to assign an unknown probe to linkage groups III, IV, V, or X. Mapping genes by hybridization in situ in *C. elegans* is relatively straightforward because cosmids can be used as the probe DNA. The large genomic inserts in the probes provide large hybridization signals that are easily distinguished from background, so that few chromosome spreads need to be analyzed.

The results of mapping genes by this method are given in Appendix 4D. Map locations can be determined to within 10–20% of the length of a chromosome, which corresponds to 1000–2500 kb of DNA. The physical map determined in this way is colinear with the genetic map. For genes within the genetic clusters, the cytological map reveals a greater physical distance between these genes than that implied by the recombination frequency (Appendix 4D). Location of cloned sequences by hybridization in situ has also led to the identification of associated mutations in the genes for rRNA (*let[e2000]*) and a gene for myosin heavy chain (*sup-3*). In addition, hybridization in situ has been useful in determining the structure of translocations that do not recombine and are therefore difficult to analyze genetically.

E. Cloning and Mapping the Entire Genome

As research on *C. elegans* moves more and more into the area of molecular cloning, many individual cloning projects will focus on particular loci or regions of the genome. Ultimately, given the small size of the genome and the growing interest in problems to which *C. elegans* lends itself, the entire genome may come under study. In anticipation of this likelihood, the *C. elegans* community has undertaken coordination of its diverse cloning efforts to speed the process of gene isolation and avoid duplication of effort.

The program is being coordinated at the Medical Research Council Laboratory of Molecular Biology in Cambridge, U.K., where an overlapping set of clones covering the entire genome is being assembled by a method that allows the recognition of a diagnostic set of restriction fragments from the 20- to 40-kb inserts of cosmid or λ clones. The patterns of fragments from individual inserts are compared by computer, and clones are placed into overlapping sets known as "contigs" (Coulson et al. 1986). Up to 400 clones can be analyzed per week.

The goal, expected to be achieved within a few years, is to join all the contigs together into a complete physical map. Contigs are placed on the genetic map by using one of the methods for correlating the physical and genetic maps described above (Section IV.A–D). If a genetically located clone falls within a previously defined contig, genomic sequences flanking the original fragment become immediately available. This obviates the need for genome walking and can be one immediate benefit of this communal effort. Ultimately, separate gene isolation projects will become unnecessary as the physical map nears completion. All that will then be required to isolate a genetic locus will be to first determine its genetic map position with sufficient accuracy to allow the identification of the appropriate clones covering the region and then to locate the gene itself physically within this region. For this second step, deficiencies, transposable element insertions, or other physical breakpoints mapped with respect to the gene are required. Alternatively, it may be possible to identify the gene by transformation of mutant worms with wild-type DNA.

F. Transformation

1. Microinjection of RNA

The feasibility of introducing nucleic acids into *C. elegans* by microinjection was first demonstrated in experiments in which a nonsense suppressor tRNA was injected into a strain carrying an amber mutation (Kimble et al. 1982). Total tRNA was isolated from the *sup-7* strain that suppresses amber mutations (Waterston 1981; Wills et al. 1983). From genetic experiments, *sup-7* was known to suppress the mutant phenotype of the *e1107* allele of the *tra-3* gene (Hodgkin 1985b). This transformer mutation displays a maternal-effect phenotype in which homozygous *XX* mutant progeny of a heterozygous mother develop as fertile hermaphrodites, but their homozygous mutant *XX* progeny develop as sterile pseudomales. Thus, the wild-type *tra-3* gene is expressed in the mother, and the product is probably present in oocytes and utilized during embryogenesis.

The phenotype of the homozygous *tra-3(e1107)* mutant provided a strong screen for suppression of the amber allele by the microinjected tRNA. Indeed, when *sup-7* tRNA was microinjected into the ovary of homozygous *e1107* hermaphrodites, fertile hermaphrodite progeny were produced. These hermaphrodites, in turn, produced pseudomale progeny, verifying that they were of the *tra-3*/*tra-3* genotype. Thus, injection of *sup-7* tRNA suppressed the *tra-3* amber mutation, allowing normal development.

2. *Integrative Transformation*

Cloned genes can be reintroduced into the *C. elegans* germ line by microinjecting DNA into the syncytial ovary of the hermaphrodite or into oocyte nuclei. The injected DNA can either integrate (apparently randomly) into a chromosome or form unstable, extrachromosomal tandem arrays. In either form, reintroduced genes can be expressed and can suppress the mutant phenotype that results from a mutation in the endogenous gene.

The highest frequencies of stably transformed transgenic animals are obtained when DNA is microinjected into the nuclei of oocytes. Oocytes inside the hermaphrodite gonad, between the bend in the ovary and the spermatheca (see Chapter 7), are injected using standard microinjection equipment. Multiple oocytes in each animal are injected, and a frequency of one stable transgenic offspring (an animal in which the microinjected DNA has integrated into a chromosome) per 5–20 injected animals can be obtained.

To identify transformed individuals, a selectable genetic marker is either incorporated on the injected DNA or coinjected with it (Fire 1986). In the latter case, the unselected DNA is frequently found in transformants containing the selected DNA. Cloned genes that have been used to select for transformants include the *sup-7* amber tRNA gene, mentioned above, and the *unc-54* gene for a body-wall-muscle myosin heavy chain. The *sup-7* gene is injected into a homozygous *tra-3* mutant, and stably suppressed self-fertile hermaphrodites are selected. The myosin gene is injected into animals homozygous for a null mutation at the *unc-54* locus, and progeny displaying improved movement or restored egg laying are selected. DNA carrying the *myo-3* gene, which encodes a second body-wall-muscle myosin heavy chain, has been introduced after ligation to a plasmid containing the *sup-7* gene. In this case, suppression of a *myo-3* mutation was demonstrated by crossing mutant and transgenic animals (A. Fire and R. Waterston, pers. comm.).

In the stably transformed animals that have been analyzed so far, one to ten copies of the injected DNA are found, integrated at a particular chromosomal site (Fire 1986). The multiple copies appear to have been linearized randomly and rejoined or are tandemly arrayed as if by homologous recombination or rolling-circle replication. Several integration sites have been mapped genetically. A given transforming DNA can be found on a number of chromosomes at locations unrelated to the site of the homologous locus. Nevertheless, restoration of wild-type or nearly wild-type phenotype by the transgenic DNA indicates that it is expressed. In the cases of the myosin heavy chain genes discussed above, additional evidence for correct expression has been provided by hybridization in situ to specific antibodies, which has shown that the genes are only expressed in the appropriate muscle cells.

3. Extrachromosomal Transformation

When populations derived from injected animals are screened directly for the injected DNA (by Southern hybridization), transformed animals of a different type are detected. These contain hundreds of copies of the transforming DNA in a long, extrachromosomal tandem array (Stinchcomb et al. 1985c). Transgenic animals of this type occur with high frequency: One or more progeny from one out of four injected hermaphrodites may be a transgenic animal of this type. Such transformed animals may also be detected by expression of an included gene. So far, two genes have been shown to be expressed from extrachromosomal arrays. Extrachromosomal DNA containing the wild-type mechanosensory gene *mec-3* has been shown to suppress a homozygous *mec-3* mutation (J. Way and M. Chalfie, pers. comm.), and, similarly, a wild-type copy of the *unc-22* gene can suppress the uncoordinated phenotype of a homozygous *unc-22* mutation (A. Fire, pers. comm.). The tandem arrays are large enough to be visualized cytologically as accessory DNA bodies in oocyte nuclei of transformants. The injected DNA can be joined in a head-to-head or head-to-tail manner, and the arrays may contain other rearranged structures as well.

It is not clear which experimental factors determine whether integrants or extrachromosomal tandem arrays are obtained, although selection scheme and point of injection seem to be important. DNA from many different sources has been injected, including plasmids of entirely bacterial origin, with similar results. Tandem extrachromosomal arrays can be formed, regardless of whether or not nematode sequences are included. On average, 50% of the worms at each generation lose the extrachromosomal foreign DNA, but there is a wide range of stabilities in different transformed lines. The instability observed is similar to that found for small chromosomal free duplications (Herman et al. 1979). Injection of nematode DNA containing *C. elegans* SEG sequences, which stabilize replicating plasmids containing such sequences in yeast (see Section III.C), did not increase the stability of the extrachromosomal arrays. It appears likely that the worm is simply promiscuous in replicating and segregating foreign DNA, or in ligating worm replication origins to it.

ACKNOWLEDGMENTS

Portions of the chapter were written by Thomas Blumenthal (University of Indiana, Bloomington), David Hirsh (University of Colorado, Boulder), and James Kramer (University of Illinois, Chicago). We thank the numerous colleagues who read drafts of the chapter and made helpful suggestions.

4
The Anatomy

John White
MRC Laboratory of Molecular Biology
Cambridge CB2 2QH, England

I. INTRODUCTION

The complete cellular architecture of *Caenorhabditis elegans* is now known. Much of this knowledge was obtained from reconstructions of electron micrographs of serial sections and is therefore quite detailed. The invariance of cell number and cell fate within the somatic tissues of *C. elegans* has enabled every cell in the animal to be identified and assigned a

The Nematode Caenorhabditis elegans. © 1988 Cold Spring Harbor Laboratory. 0-87969-307-X/88 $1 + .00.

unique label. This information has been related to the cell lineages (Sulston et al. 1983), thereby providing, for the first time, a comprehensive description of the ontogeny and ultimate differentiated state of all the various cells that comprise a metazoan.

One of the original motivations for undertaking ultrastructural reconstructions of *C. elegans* was to deduce the structure and connectivity of the nervous system, which is dealt with more fully in Chapter 11. This chapter presents a general anatomical description of the somatic tissues of the animal. It begins by describing some features of the somatic tissue types and then goes on to describe the major organs of the body.

II. TISSUES

A. Epithelia

Epithelial cells, together with the cuticle that they secrete, establish the basic body form of *C. elegans*. In common with all epithelial cells, those of *C. elegans* have two distinct regions in their plasma membranes: the basolateral surface and the apical surface. Cuticle is always laid down adjacent to the apical surface, and there is usually a basement membrane adjacent to the basal surface. It is likely that the apparatus for secreting the material of the cuticle is localized at the apical surface. Such a localization is obviously necessary for epithelial cells, because without it they would be incarcerated in their own secreted cuticle.

Epithelial cells in *C. elegans* are connected by belt desmosomes. These structures are seen in thin sections as regions of closely apposed membrane from two adjacent cells that are more darkly staining than surrounding regions. Sometimes, darkly staining material can be seen extending into the cytoplasm (Fig. 1). Serial section reconstruction reveals that these structures are always in the form of closed loops that wrap around cells. At the points where three cells come into contact, Y-shaped structures are seen. Belt desmosomes demarcate the apical and basolateral regions of epithelial cells; they are always situated at the boundary of the region of the cell adjacent to the cuticle. It therefore seems likely that belt desmosomes have at least three functions: (1) to mechanically attach epithelial cells, (2) to provide a seal between the two sides of a sheet of epithelial cells, and (3) to maintain the segregation of the basolateral and apical regions of the cell plasma membrane.

The apical surface of a sheet of connected epithelial cells is usually flat, perhaps suggesting that cells tend to minimize the surface area of the apical face. Thus, belt desmosomes probably play a major role in the establishment of the smooth outer surface of an epithelium such as the hypodermis of *C. elegans*. The same minimization of apical surface area seems to occur in the development of internal epithelia. The epithelial cells of the pharynx have their apical faces adjacent to the lumen. The lumen is small during

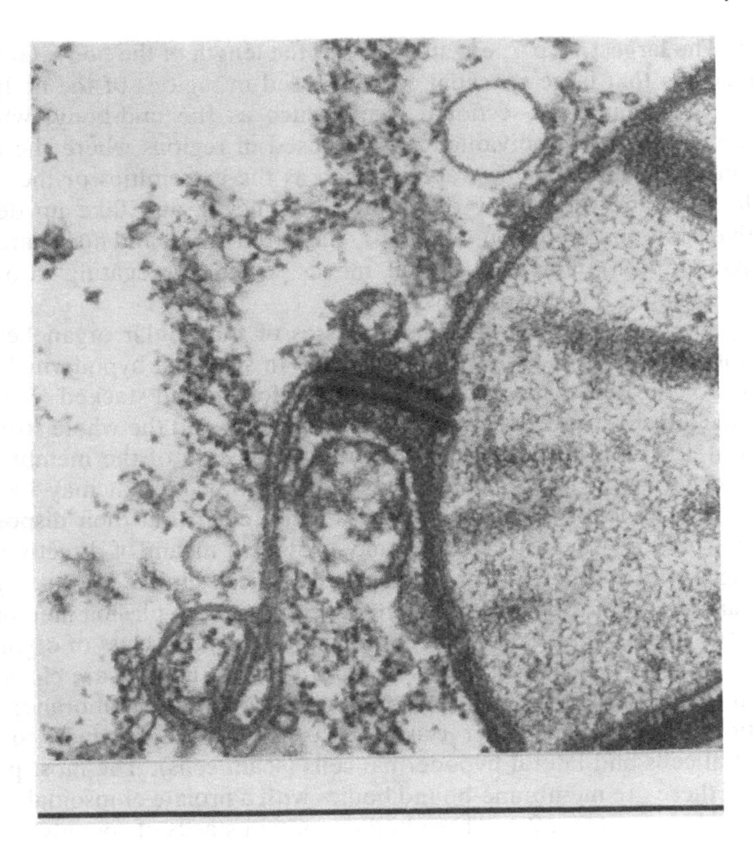

Figure 1 Desmosomal connection between intestinal cells. Epithelial and intestinal cells are connected together by belt desmosomes. These structures appear as darkly staining regions in the apposed membranes of the connected cells. Belt desmosomes form closed loops around cells, segregating their surfaces into basal and apical domains. In the case of epithelial cells, cuticle is always adjacent to the apical face. Scale bar represents 1 μm.

development, and so the epithelial cells are crowded around it and are consequentially wedge-shaped.

An interesting problem arises in the course of the postembryonic development of the gonad, where a newly developed internal epithelium (the uterus or the vas deferens) must connect to a preexisting epithelium (the hypodermis or the rectum). In both cases, a single cell makes the initial connection between the two epithelial layers. This cell is subsequently removed, opening a passageway between the two epithelia.

1. *The Hypodermis*

We refer to the external epithelium as the hypodermis. Many of the hypodermal cells are multinucleate, arising by cell fusion during development. The largest, hyp-7, extends most of the length of the body. Generally, it seems that large syncytial cells are used in regions of the body that have a constant cross-sectional shape, such as the mid-body, whereas smaller syncytial or individual cells are used in regions where the cross-sectional shape is changing rapidly, such as the extremities or the vulva. Nuclei are fairly uniformly distributed in syncytia and take up defined relative positions. In the mutant *anc-1*, this order is lost and nuclei are seen to change position as the animal moves, often aggregating into rafts (Hedgecock and Thomson 1982).

Hypodermal cells contain several classes of subcellular organelles that are not seen in other tissues of *C. elegans*. In syncytial hypodermal cells, regions of the apical plasma membrane are folded and stacked (Fig. 2a). The stacked sheets are separated by about 60 nm, and the whole structure is about 1000 nm in diameter. The cytoplasmic side of the membrane is covered with a layer of small darkly staining objects that may be ribosomes. The function of these structures is not clear, but their disposition and morphology suggest that they may provide a means of directly transporting newly synthesized products to the apical surface, bypassing the internal system of directed transport of vesicles to the plasma membrane. Syncytial hypodermal cells also contain another unique class of organelle, known as multivesicular bodies (Fig. 2b). These structures are clusters of 50-nm vesicles that are encapsulated in a bounding membrane; their function is unknown. Other classes of organelle are common to both syncytial cells and lateral hypodermal cells (seam cells). The most prominent of these are membrane-bound bodies with a prolate ellipsoidal shape, the Ward bodies (Fig. 2c). These bodies contain stacks of roughly parallel sheets of membrane. Again their function is unclear. Vesicles with dark cores are seen in both syncytial and seam cells just prior to a molt (Fig. 2b). These vesicles sometimes are seen to be fusing with the apical surface of the plasma membrane and so are likely to be involved with the secretion of cuticle precursors.

It is convenient to group the hypodermal cells of *C. elegans* into four general categories: (1) the main body syncytium (hyp-7), (2) seam cells, (3) the hypodermal cells of the head and tail, and (4) interfacial hypodermal cells.

The Main Body Syncytium. This tissue is present at hatching and contains 23 nuclei. At this stage, it makes up only the dorsal hypodermis. The ventral hypodermis is made up of a set of 12 P blast cells. In the course of development, these cells divide, and some of their daughters join the hypodermal syncytium, which eventually replaces the P cells on the ventral side. Other cells derived from divisions of the seam cells also join the body syncytium. In all, a total of 110 cells fuse and join the main hypodermal syncytium during postembryonic development (Sulston and Horvitz 1977).

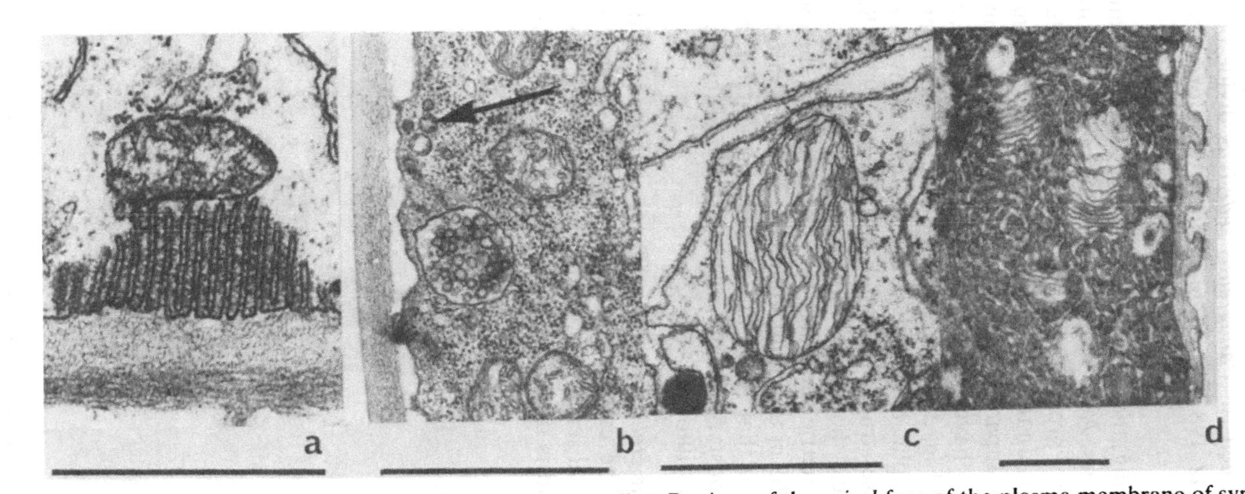

Figure 2 Epithelial cells contain a variety of characteristic organelles. Regions of the apical face of the plasma membrane of syncytial cells are folded into stacked sheets or, possibly, fingers (*a*). Similar structures are also seen on the luminal face of the excretory duct cell. Membrane-bound multivesicular bodies (*b*) are found only in syncytial hypodermal cells. Stacked sheets of membrane, also membrane-bound (*c*), are found in both syncytial and seam hypodermal cells. Seam cells contain many prominent Golgi bodies (*d*) just prior to molt. Dark-cored vesicles are often seen near the apical surface of epithelial cells (arrow in *b*), particularly at times when cuticle synthesis is taking place. Scale bars represent 1 μm.

The cells that join the syncytium become tetraploid before they fuse, whereas the nuclei present at hatching are diploid (Hedgecock and White 1985). The equivalent cells in *Panagrellus redivivus* divide before fusing (Sternberg and Horvitz 1982). In *Panagrellus*, therefore, the syncytial nuclei may be diploid.

Seam Cells. At hatching, these cells are arranged as rows of ten cells that run along each lateral line. All of them (except the most anterior) are blast cells, contributing many progeny to the set of cells added during post-embryonic development. In the adult, the seam cells fuse to form two seam syncytia, which remain separate from the main body syncytium. The L1, dauer, and adult stages have cuticle specializations, the alae, which run along the lateral lines (Fig. 3). Laser ablation studies and observations of certain mutants have implicated the seam cells as the only hypodermal cells capable of making these structures (Singh and Sulston 1978). Seam cells, or the specialized cuticle that they produce, have also been shown to be responsible for the diametric shrinkage that occurs during the dauer molt. The posterior seam cells in the adult male (the set cells) differ from their anterior counterparts in that there are no alae present on the adjacent posterior cuticle. The posterior seam cells are syncytial but separate from the anterior seam syncytium.

Hypodermal Cells of the Head and Tail. These cells are fairly diverse and almost certainly could be categorized further; however, they have been lumped together in the absence of criteria other than their positions. The hypodermis in the head is made up of a series of six concentric annular

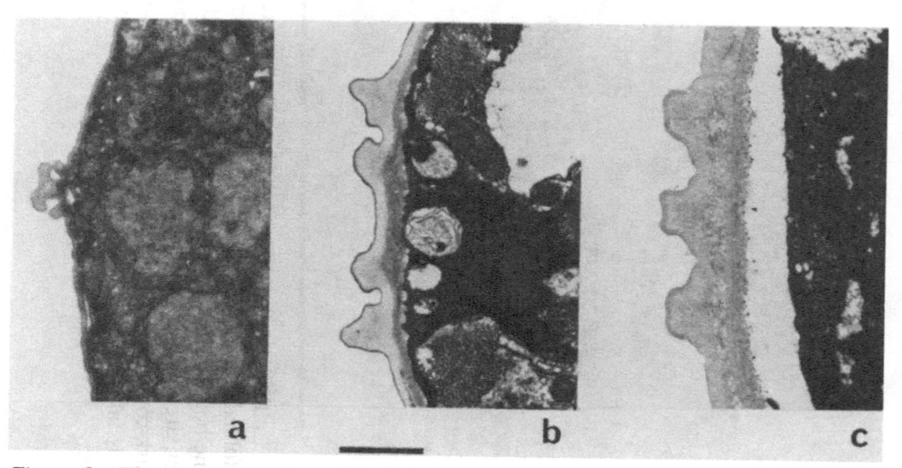

Figure 3 The seam hypodermal cells lay down longitudinal ridges of cuticle along the lateral lines in the L1 (*a*), dauer (*b*), and adult (*c*) stages. Scale bars represent 1 μm.

syncytia, hyp-1 through hyp-6 (Fig. 4), containing 3, 2, 2, 3, 2, and 6 nuclei, respectively. The tip of the head is richly endowed with sensory endings, and all the sensilla are embedded in the cuticle of hyp-2 and hyp-3. The first hypodermal cell in the sequence, hyp-1, is attached via desmosomes to the anterior cells of the buccal cavity (arcade cells). Two cells (XXX) form part of the hypodermis in the head during embryogenesis but eventually become detached from it and appear to be free-standing undifferentiated cells in the adult. Two other hypodermal cells are also transiently deployed in the construction of the tail spike, but they subsequently undergo programmed cell death (Sulston et al. 1983). Four additional cells make up the tail hypodermis (hyp-8 through hyp-11). These are mononucleate, with the exception of hyp-10, which is binucleate.

Interfacial Hypodermal Cells. The hypodermis is not a closed sheet of cells; it is pierced at the points where the pharynx, the anus, the vulva, and the excretory duct are connected to it. In addition, many sensory receptors form sensilla that are attached to the hypodermis. Several of these (most notably the amphids) also pierce the hypodermis and are open to the outside. In all of these cases, specialized cells apparently act as interfaces between the organ and the hypodermis. These interfacial cells are usually arranged in a toroidal conformation, achieved by one of several strategies. For sensilla, the interfacial cells (socket cells) are single, mononucleate cells that form a toroid by surrounding the orifice with a process that seals to itself by means of a tight junction or a desmosome (Ward et al. 1975). The pharynx is interfaced to the body hypodermis by means of two syncytial arcade cells, each of which is a complete toroid (Fig. 4). The anus is also interfaced to the hypodermis with a sequence of two toroidal structures, each of which is made up of two mononucleate cells coupled together with desmosomes (Fig. 4).

Interfacial cells are often intermediate in function and morphology between the adjacent structures to which they are connected. All have epithelial characteristics of apical regions delineated by desmosomes that are adjacent to cuticle. The arcade cells of the pharynx secrete the cuticular lining of the anterior part of the buccal cavity, whereas pharyngeal cells secrete the lining in the posterior part of the cavity (Wright and Thomson 1981). The socket cells of sensilla are generally rather small cells with long, neuronlike processes. Laser ablation studies have indicated that socket cells of particular sensilla are capable of accommodating neurons that are normally components of different sensilla (Sulston et al. 1983).

Multiple Functions of Hypodermal Cells. The hypodermal cells of *C. elegans* have functions other than secretion of cuticle. One of these functions is the elimination of the many cells that undergo programmed cell death during development. Dead cells are phagocytosed predominantly

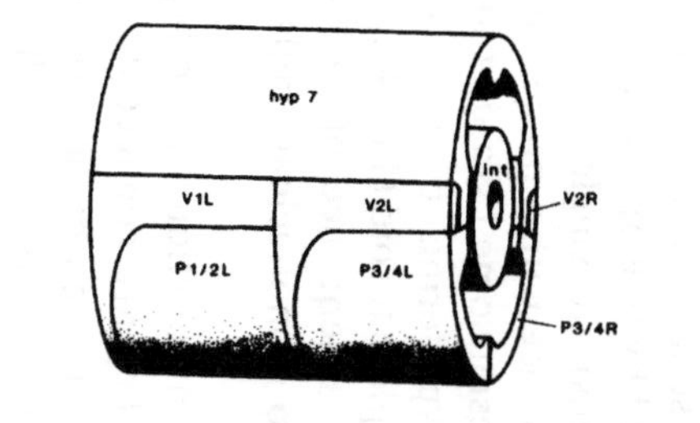

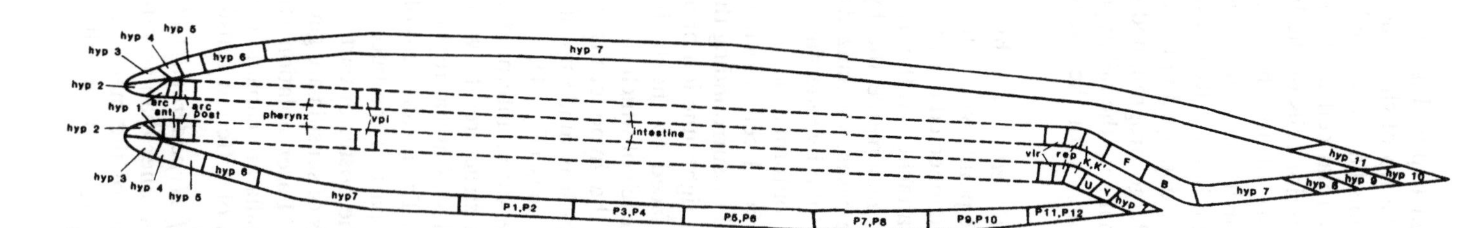

Figure 4 Schematic diagram of a longitudinal section of a newly hatched L1, showing the arrangement of hypodermal cells. (*Inset*, top) Three-dimensional organization of a region of mid-body. Commas indicate two cells meeting in the plane of the drawing. In the course of postembryonic development, hyp-7 enlarges by cell fusion and spreads over most of the ventral region. (arc) Arcade; (vpi) pharyngeal–intestinal valve; (vir) intestinal–rectal valve; (rep) rectal epithelial cell. (Reprinted, with permission, from Sulston et al. 1983.)

(but not exclusively) by hypodermal cells (Robertson and Thomson 1982; Sulston et al. 1983). Mutations in the *ced-1* and *ced-2* loci cause defects in this process, so that dead cells persist in the body for a considerable time (Hedgecock et al. 1983). Hypodermal cells also function as major storage depots and often contain many storage granules and lipid droplets.

Several classes of hypodermal cells act as blast cells, giving rise to most of the new cells that are produced during postembryonic development. These cells generally divide as stem cells, with one of the daughters taking on the attributes of the mother. The nonstem daughter may be either a neuroblast or a different type of hypodermal cell. When epithelial blast cells divide, the cytokinetic furrow runs through the belt desmosome around the cell, and both daughters inherit a belt desmosome. When one of the daughters is destined to become a neuroblast, its belt desmosome rapidly shrinks in diameter until it disappears and the cell becomes detached from the epithelium (C. Kenyon, pers. comm.).

A few interesting exceptions to the stem cell mode of division occur in certain hypodermal cells. The W blast cell has all the characteristics of a hypodermal cell on hatching, but all of its progeny are neurons. Even more striking is the Y blast cell, which again has the appearance of a hypodermal cell when the animal hatches but subsequently becomes the neuron PDA without dividing, in the hermaphrodite. In the male, the equivalent cell is a blast cell, one daughter of which differentiates into PDA (Sulston et al. 1980).

One of the most striking examples of multifunctional cells in *C. elegans* is the seam cells of the tail (T cells). There are two chemoreceptive sensilla in the tail, the phasmids. In the L1 larva, a process extends out of each of the T cells, wraps around the orifice of a phasmid sensillum, and seals to itself with a tight junction (Sulston et al. 1980). The process is coupled to the sheath cell of the sensillum and to the adjacent syncytial hypodermis with desmosomes. Part of a T cell's function in the L1 is therefore to act as a socket cell for a phasmid. In the course of postembryonic development, one of the daughters of each T cell ultimately becomes the phasmid socket. This cell has the appearance of a normal socket cell, with no seam characteristics. Thus, a function that was originally performed by one region of the T cell becomes compartmentalized into a single specialized descendant in the course of development.

B. Cuticle

The cuticle of *C. elegans* is an extracellular structure that covers the outermost surfaces of all hypodermal cells and also lines the pharynx and rectum. The surface of the external cuticle is divided into numerous regularly spaced circumferential ridges (annuli). In the L1, dauer, and adult stages, elevated longitudinal ridges (alae) mark the lateral surfaces of the cuticle and overlie the seam cells (Fig. 3). A new cuticle is synthesized prior to each molt, at which time the cuticle from the previous larval stage is shed.

Although some nematodes, such as *P. silusiae,* apparently secrete larger versions of the same cuticle at each molt, *C. elegans* appears to secrete at least four different types of cuticle. The cuticles of the L1, dauer, L4, and adult stages differ in ultrastructure and protein composition. At least four distinct layers within the cuticle can be seen in the electron microscope: (1) the outermost external cortical layer, (2) the inner cortical layer, (3) the cortical layer, and (4) the innermost basal layer (Cox et al. 1981a). The cortical and basal layers show the greatest variation in structure between stages. Both the L1 and dauer cuticles contain a highly organized "striated zone" within the basal layer. The striated zone appears to consist of intersecting sets of longitudinally and circumferentially oriented fibrils. Animals mutant at the *sqt-3* locus have marked defects in the striated zone of the L1 cuticle and hatch as short, fat larvae (Priess and Hirsh 1986).

The basal layer of the adult cuticle consists of two layers of highly organized fibers that appear to spiral in opposite direction around the body. In the adult cuticle, the basal layer is connected to the cortical layer by columns of material, termed struts. The struts bridge a space separating the two layers, which is probably filled with fluid. This space is often grossly expanded in the adult stage of "blister" mutants, which have prominent, blisterlike bubbles along their bodies.

Cuticles are separated easily from body tissues by sonication and treatment with the detergent SDS (Cox et al. 1981a). Many of the cuticle proteins can be made soluble by treating isolated cuticles with reducing agents such as β-mercaptoethanol. Most of the soluble proteins from adult cuticles are degraded by collagenase and are thus likely to be collagens. Several other soluble proteins, including most of the soluble proteins from L1 cuticles, are not affected by collagenase treatment and are thus not collagens or are modified such that they are resistant to digestion by this enzyme. The collagens and collagen genes of *C. elegans* are described in Chapter 3.

C. Musculature

The detailed structure and biochemistry of *C. elegans* muscle are discussed in Chapter 10. This section describes some of the general features of *C. elegans* musculature and its organization within the body.

Perhaps the most unusual feature of nematode muscle is the manner in which it is innervated. Motor neurons do not extend axons peripherally to innervate muscles; rather, muscles send neuronlike processes to the neuropil in which the axons of motor neurons reside (Chitwood and Chitwood 1974). Neuromuscular junctions, directed to the distal terminals of the muscle processes, are distributed along the length of motor neuron axons, which are generally located on the surfaces of process bundles. Muscle arms are often seen to converge and extensively interdigitate at the presynaptic specializations of motor neurons. As in most other organisms, there is a basement membrane separating neural tissue from muscle tissue, so that neuromuscular junctions are constrained to lie in the plane of the basement membrane.

Muscles in *C. elegans* are of two general types: single sarcomere muscles with focal attachment points at the extremities and obliquely striated muscles containing several sarcomeres that have no substantial focal attachment points but probably have distributed attachments to the hypodermis along their length. All the muscles associated with the alimentary system and the sex muscles belong to the former class, whereas all the body-wall muscles belong to the latter class. Muscle cells are mononucleate, with the exception of four cells in the pharynx.

1. Body-wall Muscles

The body musculature is arranged as longitudinal bands of muscle cells, with one band running in each quadrant of the body. Two rows of rhomboid-shaped muscle cells make up each band (Fig. 5). There are a total of 95 body muscle cells in the adult: 23 in the left ventral quadrant and 24 in each of the remaining quadrants (Sulston and Horvitz 1977). The four bands of muscle lie in grooves in the hypodermis. The hypodermal cells are remarkably thin (50 nm) in the regions where they are overlaid by muscle cells. Fibrous elements, seen extending through the cytoplasm of the hypodermal cells in these regions, presumably act to anchor the body muscles to the cuticle.

Body muscles can be classified according to their source of innervation (Fig. 5). The anterior four muscles in each quadrant are innervated by motor neurons in the nerve ring. The next four muscles are dually innervated by motor neurons of the nerve ring and the ventral cord. The remaining muscles are exclusively innervated by motor neurons of the ventral cord. Individual motor neurons of the ventral cord innervate either dorsal muscles or ventral muscles. The body is therefore limited to making flexures in the dorsoventral plane only. The head, on the other hand, often makes lateral as well as dorsoventral movements when the animals are foraging. This behavior is almost certainly a consequence of the pattern of innervation of these muscles. In the head, innervation by individual motor neurons is restricted to two adjacent rows, not necessarily in the same band, thereby allowing differential activation of muscles in adjacent bands and possibly even in adjacent rows (White et al. 1986).

2. Single Sarcomere Muscles

The pharynx has a total of 20 muscle cells, several of which are multinucleate (Fig. 9). These cells are organized into eight distinct muscle layers,

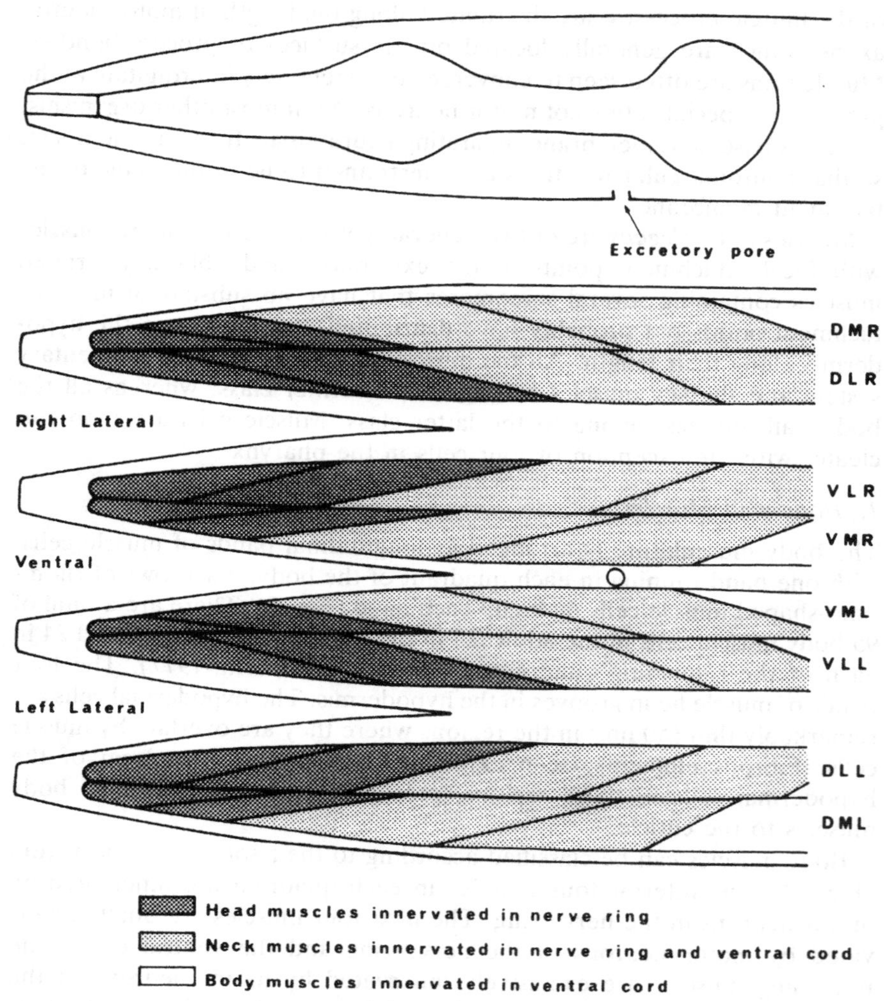

Figure 5 "Orange peel" projection of the muscles in the head. Muscles are organized as longitudinal bands in each of the four body quadrants. Each band contains two adjacent rows of muscle cells. The first (anterior) four muscles in each band are innervated by motor neurons in the nerve ring; the next four are dually innervated by motor neurons in the nerve ring and ventral nerve cord; the rest of the muscles are innervated solely by ventral cord motor neurons. (DMR) Dorsomedial right; (VLL) ventrolateral left. (Reprinted, with permission, from White et al. 1986.)

each one (except the most posterior) having threefold radial symmetry (Albertson and Thomson 1976). The pharyngeal muscles are myoepithelial cells with clearly defined apical regions bounded by belt desmosomes.

These regions are adjacent to the lumen cuticle, which they probably secrete (Albertson and Thomson 1976). Pharyngeal muscles are also unusual in that they have discrete attachments to the bounding basement membrane in the form of hemidesmosomes.

The remaining single sarcomere muscles are linear muscles that are attached at either end to epithelial cells. The rectal muscle, the intestinal muscles, and the sex-specific muscles in the male tail and hermaphrodite vulva are of this type. The anal sphincter muscle is unusual in that it is made up of a single toroidal cell containing a continuous ring of contractile filaments. There are no obvious focal attachment points for the filaments, and in this respect the sphincter muscle is analogous to vertebrate smooth muscle.

3. Head Mesodermal Cell

This single cell in the head lies in the pseudocoelom with the body muscles and makes extensive gap junctions with them, although it does not contain contractile filaments itself (Sulston and Horvitz 1977). From its cell body, situated on the dorsal midline, a short, broad process extends posteriorly and makes gap junctions with arms from dorsal muscles. A second process extends ventrally and splits at the pharynx. The two branches extend along either side of the pharynx and fuse together on the ventral midline. The resulting single process then runs posteriorly, making gap junctions with arms from ventral body muscles (E.M. Hedgecock and J.G. White, unpubl.). The function of this cell is unknown, although it seems likely that it could provide electrical coupling between the dorsal and ventral muscles in the neck region.

D. Basement Membranes

A thin (20-nm) basement membrane lines the pseudocoelomic cavity and effectively segregates the musculature from the hypodermal and nervous tissues. Striations with a spacing of 30 nm can sometimes be seen when they are oriented at a grazing angle to the plane of sectioning (White et al. 1976). The gonad and the gut are ensheathed by similar basement membranes, whereas the pharynx is ensheathed in its own, rather thicker (45-nm), basement membrane (Albertson and Thomson 1976).

The arrangement of the basement membrane lining the pseudocoelom suggests that it may be instrumental in defining some of the major tracts for nervous system processes, which are constrained to run alongside the membrane. Nerve processes that travel from the ventral side to the dorsal run around the animal, traveling under the muscle quadrants instead of taking a more direct internal route. In the main part of the body, the dorsal and ventral hypodermal ridges are quite small, but toward the head they enlarge as they become filled with the cell bodies of neurons (Fig. 6c). Eventually, the basement membrane bounding the four ridges meets and fuses (Fig. 6b), opening an internal tract that allows neuronal processes to run around inside the muscle quadrants, forming the nerve ring. This organization is maintained up to the tip of the head, with the four muscle quadrants running inside tubes of basement membrane (Fig. 6a). The central ring of membrane remaining after fusion of the ridges ends next to the pharynx in the vicinity of the nerve ring. It appears to terminate onto a cylinder made up of the six sheetlike processes of the GLR glial cells, which line the layer of muscle arms on the inside surface of the nerve ring (White et al. 1986).

E. Nervous Tissue

The nervous system is described in detail in Chapter 11; only a few of its general features will be mentioned here. The 302 neurons of the adult *C. elegans* hermaphrodite all have simple, relatively unbranched morphologies. Processes of neurons are organized into ordered bundles that tend to run either anteroposteriorly or circumferentially. Synaptic contacts are made en passant between adjacent processes within bundles. Many synaptic contacts are multiple, with two or more postsynaptic elements. Most of the sensory receptors have their sensory endings in the tip of the head. Processes from these neurons project into a large toroidal bundle (the circumpharyngeal nerve ring), which encircles the central region of the pharynx. Most of the sensory integration takes place in the nerve ring. Its inside surface is lined with the axons of motor neurons, which are arranged in a precisely ordered, circumferential pattern. Arms from muscle cells in the neck and head run past the outside of the ring and then turn and distribute themselves around the inside surface of the ring where they receive their synaptic input.

A prominent bundle of processes (the ventral cord) emanates from the nerve ring on the ventral side and extends the length of the animal, eventually terminating in the preanal ganglion where the outputs of some of the posteriorly located sensory receptors are integrated. A series of motor neurons that innervate body muscles are distributed along the length of the ventral cord. They receive their synaptic inputs either from interneurons running in the cord or from other motor neurons. Motor neurons that innervate ventral muscles have axons that run in a defined location adjacent to the basement membrane bounding the ventral cord. Motor neurons that innervate dorsal muscles send out commissures that leave the ventral cord and run circumferentially around the animal to the dorsal midline. Here they turn and run anteriorly or posteriorly, forming the dorsal cord. Dorsal body muscles send out arms that synapse with motor neuron axons in the dorsal cord.

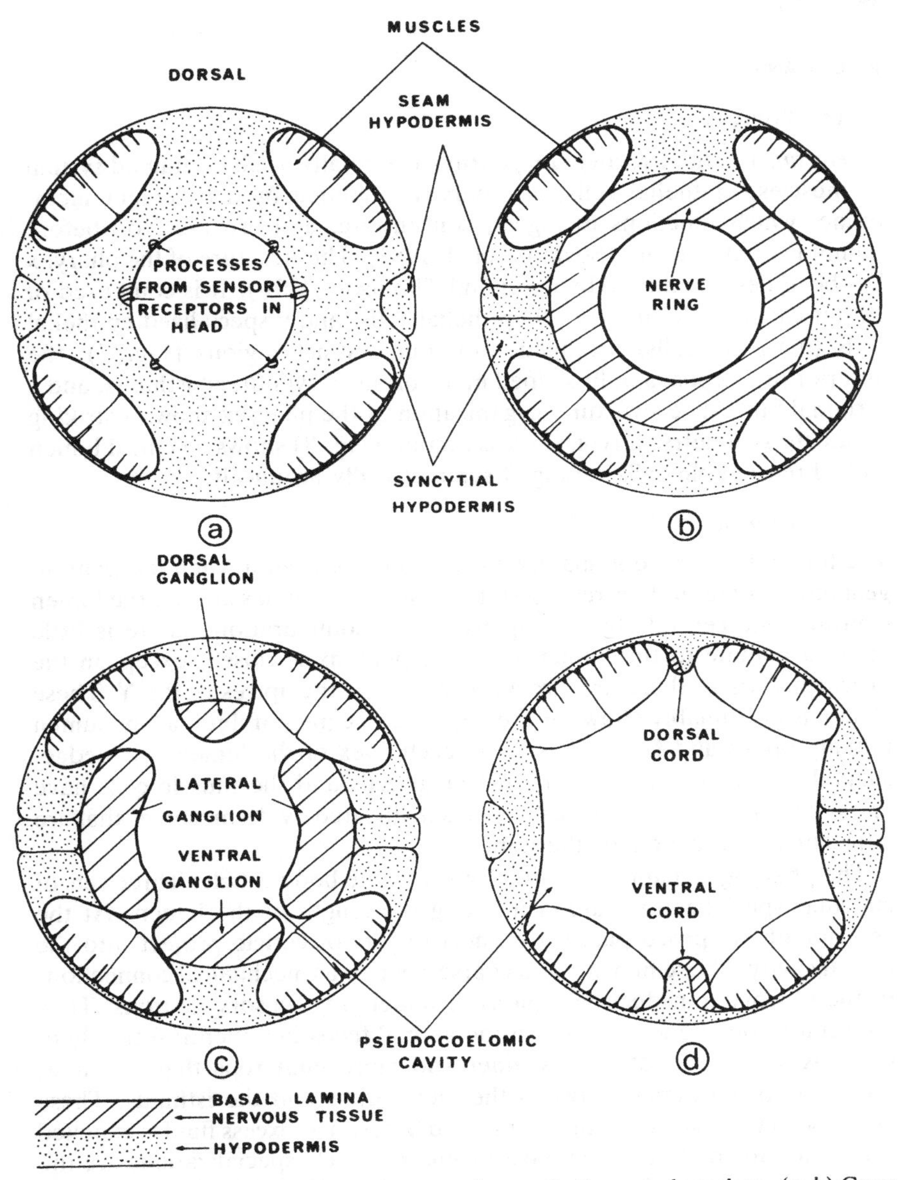

Figure 6 Arrangement of basement membranes in the pseudocoelom. (*a*,*b*) Cross sections anterior to and through the nerve ring, respectively. At the level of the nerve ring, lobes of basement membrane fuse inside the muscle bands, opening an internal route for the processes of the nerve ring. (*c*) Cross section at the anterior end of the pseudocoelom near the nerve ring; the ventral and lateral cords enlarge where they become filled with the cell bodies of neurons. (*d*) Cross section near the center of the animal. The pseudocoelom is lined with basement membrane; muscles are situated inside the coelomic cavity. (Reprinted, with permission, from White et al. 1986.)

III. ORGANS

A. The Pharynx

C. elegans is a filter feeder. Its pharnyx is virtually a self-contained system of muscles, epithelial cells, and nerves, bounded by a basement membrane, which functions to ingest, concentrate, and process food before pumping it into the gut (Seymour et al. 1983). The structure of the pharynx has been described by Albertson and Thomson (1976). It is composed of 20 muscle cells, 20 neurons, 9 epithelial cells, and 9 specialized epithelial cells (marginal cells) arranged into four distinct regions (Fig. 7): the anterior procorpus, a bulb-shaped metacorpus, a cylindrical isthmus, and a terminal bulb. The structural organization of the pharynx shows a striking triradiate symmetry; a typical cross section (Fig. 8) shows a central lumen with three muscle cells arranged symmetrically around it.

1. The Lumen

The lumen has a predominantly Y-shaped cross section when the pharyngeal muscles are in their resting state. When the muscles around the lumen contract, the central channel opens into Δ configuration. There is little internal volume to the lumen when the pharynx is closed, except in the procorpus where there are circular channels at the apices of the Y. These channels presumably allow for the ingress and egress of fluid, as the lumen initially opens and finally closes. At each apex of the lumen is a wedge-shaped marginal cell. Both the marginal cells and the muscle cells have epithelial characteristics: apical regions delineated by desmosomes that are adjacent to the cuticle of the lumen.

The pharyngeal cuticle is continuous with the body cuticle. Three sets of cuticular specializations are seen along the length of the lumen. At the anterior of the procorpus, three short fingers of cuticle project into the buccal cavity. These may act as a coarse sieve or as mechanical connections to the sensory transduction regions of underlying sensory neurons. They are seen to open and close when the animal feeds in a plentiful supply of food. A second set of fingers, finer and more numerous than the first, project into the lumen between the metacorpus and the isthmus. These may also act as a sieve, trapping ingested bacteria as excess fluid is expelled when the lumen closes. The most extensive cuticle specializations are the knobbed structures that grind up food in the terminal bulb. Contractile elements in the muscle cells adjacent to these structures are arranged both radially and longitudinally, allowing grinding movements to be generated.

2. Pharyngeal Musculature

Of the eight layers of muscle in the pharynx, three are in the procorpus, one in the metacorpus, one in the isthmus, and three in the terminal bulb (Fig. 9). The first layer consists of a single cell with six nuclei. Each of the

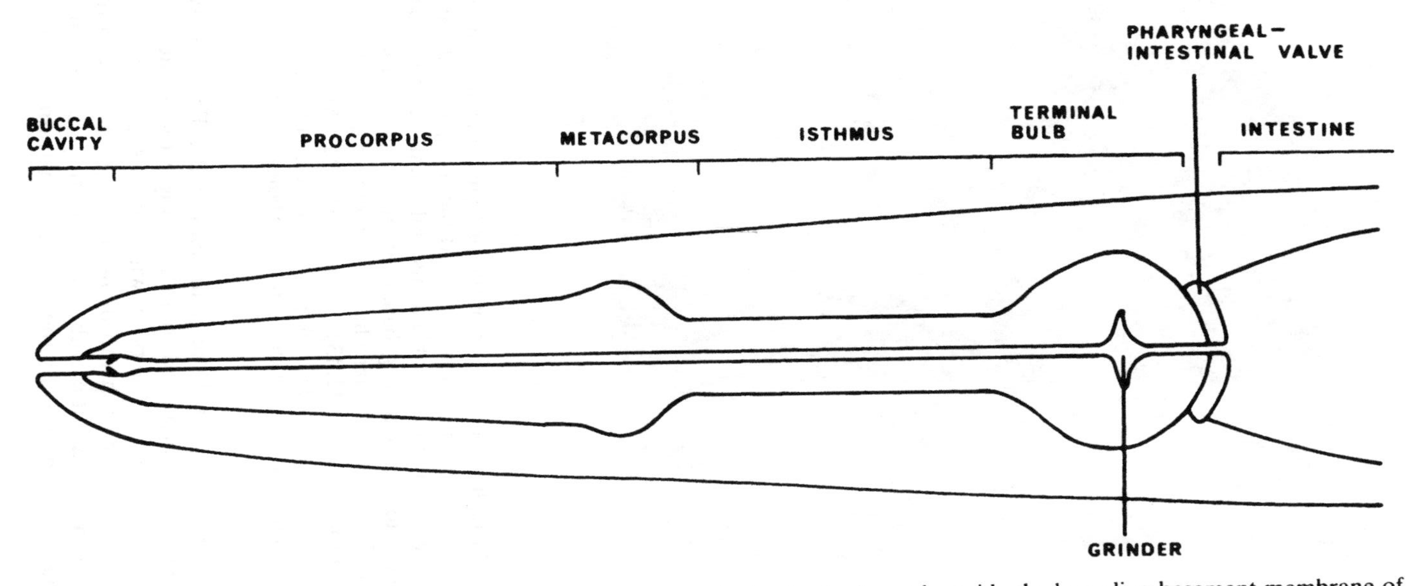

Figure 7 Principal regions of the pharynx. The pharyngeal–intestinal valve is situated outside the bounding basement membrane of the pharynx.

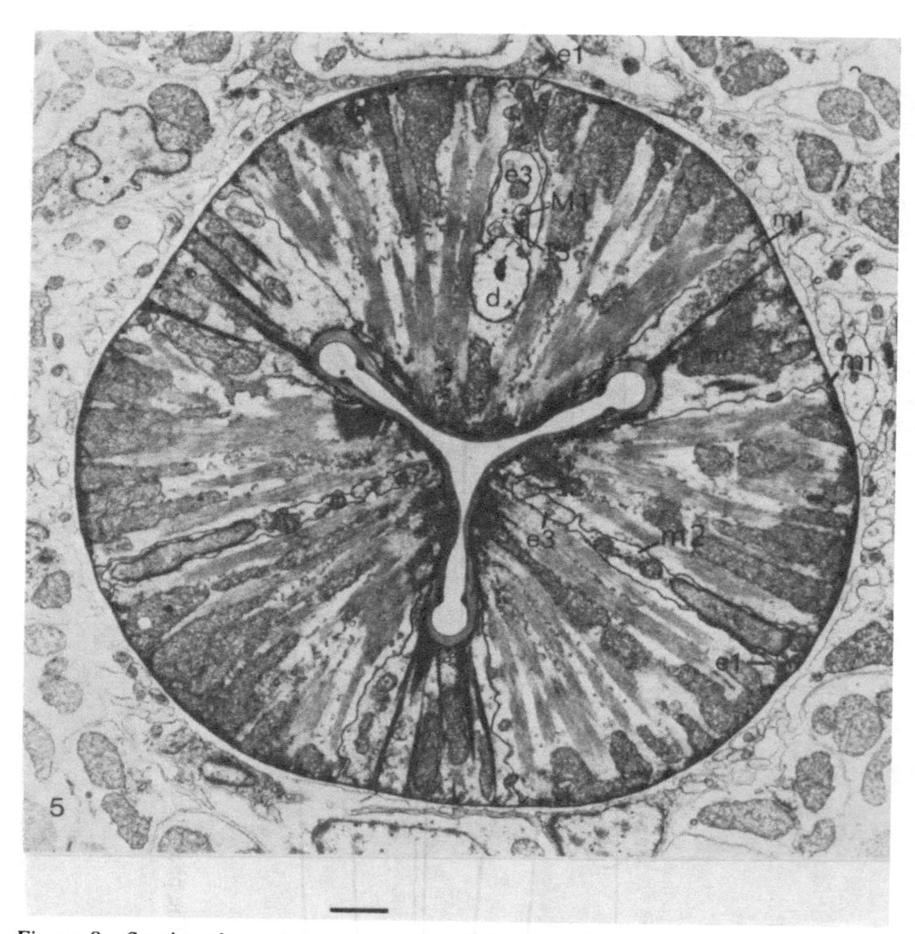

Figure 8 Section through the procorpus of the pharynx, showing the lumen, the muscle cells with radially oriented filaments, and the marginal cells (mc), which contain darkly staining intermediate filaments. Filaments in both muscle and marginal cells attach to the bounding basement membrane of the pharynx by hemi-desmosomes. The three circular channels at the apices of the lumen are present only in the procorpus. (I1, I2, I3) Interneurons; (M1, M2) motor neurons; (d) gland cell duct; (e1, e3) epithelial cell processes. Scale bar represents 1 μm. (Reprinted, with permission, from Albertson and Thomson 1976.)

next four layers is made up of three binucleate cells. The next two have three mononucleate cells each, and there is a single mononucleate cell at the posterior extremity. Unlike the body-wall muscles, pharyngeal muscles do not have arms synapsing with motor neurons, perhaps relating to the absence of an internal basement membrane separating nerves and muscles. Without this separation, there is no mechanical constraint on the arrange-

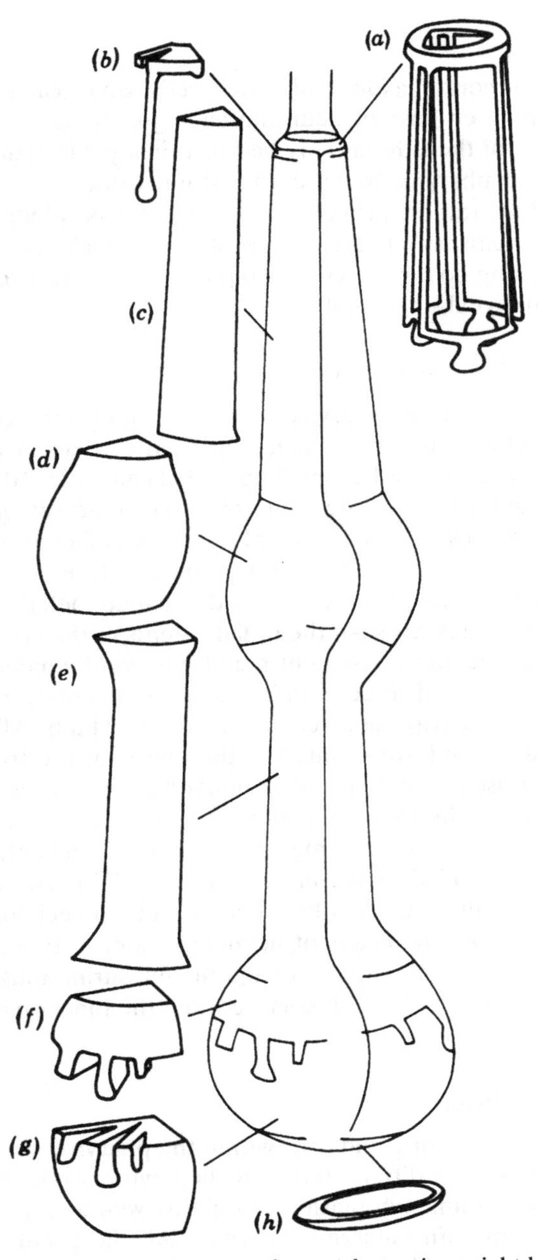

Figure 9 Organization of the pharyngeal musculature into eight layers (*a–h*). (*a*) Single cell with six nuclei; (*b–e*) each made up of three binucleate cells; (*f–g*) made up of three mononucleate cells and a saucer-shaped mononucleate cell layer; (*h*) lines the posterior wall of the pharynx. (Reprinted, with permission, from Albertson and Thomson 1976.)

ments of motor neuron axons and muscle cells, which can lie alongside one another, making en passant neuromuscular junctions.

The muscles of the isthmus may be functionally different from the other muscles of the pharynx because the isthmus does not open and close periodically but, rather, propels aggregates of food along its length with peristaltic movements. The single posterior muscle is also unusual in apparently having no direct synaptic input. It is closely associated with the adjacent pharyngeal–intestinal valve.

3. *Pharyngeal Epithelial Cells*

The anterior lumen of the pharynx has a fixed triangular cross section. This structure, together with the connections to the adjacent arcade cells, is formed by a set of nine rather small epithelial cells (Fig. 10), which can be conveniently divided into three sets of three. One set of three cells is situated at the apices of the triangular lumen and is similar in this respect to the marginal cells. The other two sets form two layers of epithelial cells occupying the space that is taken up by the myoepithelial cells in the bulk of the pharynx. Filaments span the radial extents of the epithelial cells and connect to the bounding basement membrane via hemidesmosomes.

There are two sets of three marginal cells in the anterior regions of the pharynx and a single trinucleate cell in the terminal bulb. All marginal cells are wedge-shaped and are situated at the apices of the triangular lumen (Fig. 8). The posterior cell in the terminal bulb has three wedge-shaped regions. Marginal cells are lighter staining than the anterior epithelial cells and contain bundles of dark filaments that stain with an anti-intermediate-filament antibody (Hedgecock and Thomson 1982). As in the anterior cells, these filaments are radially oriented and connect to the bounding basement membrane by means of hemidesmosomes. It seems likely that these filaments secure the apices of the lumen during muscle activation. Some of the marginal cells are innervated, but the function of this innervation is not known.

4. *Pharyngeal Glands*

There are two classes of gland cells within the pharynx, designated g1 and g2. The two classes are differentiated by their cytoplasmic structure; g1 has a more darkly staining cytoplasm than g2, as well as a prominent endoplasmic reticulum with enlarged cisternae. All the gland cell bodies are situated in the terminal bulb. At the rear of the bulb are two subventral g2 cell bodies, which have processes that connect to the lumen in the region of the grinder. There are three g1 gland cell bodies. One, situated dorsally at the rear of the terminal bulb, sends a process anteriorly, which connects to the lumen behind the buccal cavity. The other two, situated subventrally at the front of the bulb, have processes that connect to the lumen, just behind

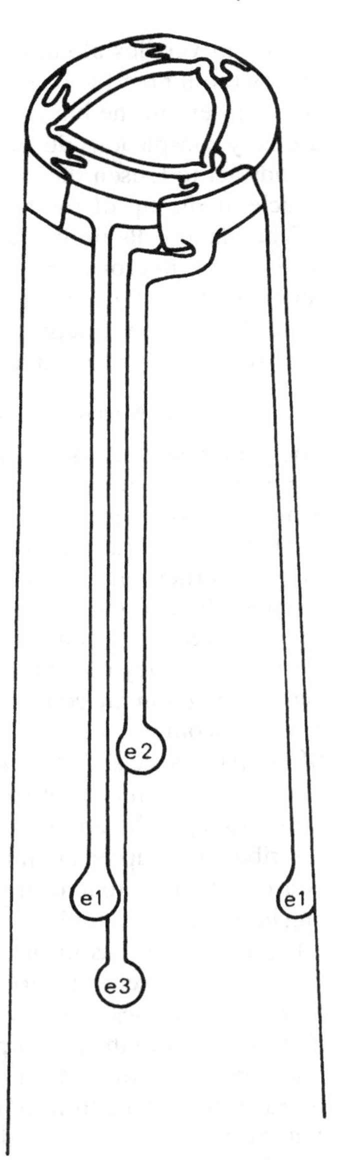

Figure 10 Nine epithelial cells, arranged as three groups (e1, e2, e3) of three cells, ring the anterior margin of the pharynx. The three e2 cells are similar to the marginal cells in that they are situated at the apices of the lumen. The other two groups, e1 and e3, make up two layers of epithelia in the regions between marginal cells. (Reprinted, with permission, from Albertson and Thomson 1976.)

the metacorpus. The right-hand subventral and the dorsal g1 cells are usually fused.

Both types of gland lack internal ducts but have long cytoplasmic processes that connect to epithelial cells via desmosomes. Unlike epithelial cells, the gland cells do not secrete cuticle on their apical surfaces, which are therefore directly accessible to the lumen. Secreted material is transported in vesicles along the processes of the gland cells and presumably enters the lumen by exocytosis of the vesicles at the apical surface of the cell adjacent to the lumen. Movements of vesicles along the g1 gland cells are very conspicuous during molting, suggesting that secretions from these glands act to loosen the old pharyngeal cuticle lining and also possibly the cuticle at the tip of the head (Singh and Sulston 1978).

The gland cells probably receive some synaptic input from the pharyngeal motor neurons M4 and M5, because some of the neuromuscular junctions from these neurons are dyadic, with the gland cells as corecipients. This arrangement may allow activation of the glands to secrete digestive enzymes coordinately with the activation of muscles.

5. Pharyngeal Nervous System

The pharyngeal nervous system has a predominantly bilateral symmetry, in contrast to the threefold symmetry of the musculature and epithelium. Motor neurons are generally present as either a single dorsal cell with symmetrically arranged processes or paired subventral cells. Several classes of interneuron (I1, I2, and I3) have specialized endings that are situated close to the lumen. These endings are attached to adjacent nonneuronal cells by means of desmosomes. The disposition of these endings suggests that they may function in sensory transduction, perhaps responding to changes in the conformation of the lumen or even to chemical stimuli from the contents of the lumen. Other classes of neuron (M3, MC, and NSM) have specialized attachments to muscle cells, suggesting that they could act as proprioceptors.

The topography and connectivity of the pharyngeal nervous system are described in Chapter 11 and Appendix 2. Most of the neuropil is organized as a nerve ring situated in the metacorpus. There is also a smaller nerve ring in the terminal bulb. The arrangement of processes within the neuropil is highly ordered, as in the rest of the nervous system. The only connections with the central nervous system are made via the two RIP interneurons. Processes from these neurons penetrate the basement membrane of the pharynx in the procorpus and make gap junctions with two classes of pharyngeal neuron (M1 and I1). RIP is exclusively postsynaptic in its interactions with neurons of the nonpharyngeal nervous system, suggesting that information flow is unidirectional, from the central nervous system to the pharynx.

Several of the pharyngeal neurons have unusual features. The interneuron I5 has two completely closed loops of process, one in the terminal bulb and the other in the main pharyngeal nerve ring. The serotonergic motor neurons, NSM, are probably neurosecretory, based on their prominent, varicosed processes that run underneath the basement membrane and contain both large and small vesicles. The MC neurons synapse exclusively onto the marginal cells, which seem to be epithelial in nature. Laser ablation studies have shown that with the exception of the M4 motor neuron, the pharyngeal nervous system is not required for pumping (L. Avery, pers. comm.). It is possibly used for the sensory-mediated modulation and inhibition of pumping.

B. The Intestine

The pharynx connects to the intestine via the pharyngeal–intestinal valve. This structure is made up of six cells, all situated outside the bounding basement membrane of the pharynx. The cells are arranged into three layers: a single toroidal cell connected to the pharynx, a pair of cells connected to the anterior cells of the intestine, and a layer of three cells sandwiched in the middle (Fig. 11). All except the pair coupled to the intestine have microvilli on their apical surfaces. Although it is closely associated with the single saucer-shaped muscle cell at the posterior extremity of the pharynx, the pharyngeal–intestinal valve is probably not a true valve, because there is no sphincter muscle to close it. The cells of the valve are coupled together and to the intestine and pharynx by a system of desmosomes.

The main body of the intestine consists of a tube of 20 cells, each bearing a dense layer of microvilli on its apical surface (Fig. 11). The intestine twists 180° along its length, with most of the twist occurring in the middle region of the body where the gonad primordium resides in the L1 larva. The lumen of the intestine is surrounded by a quartet of cells (int1) at the anterior end and thereafter by eight pairs of cells, designated int2–int9 (Fig. 11). All but the six most anterior intestinal cells of int1 and int2 undergo nuclear divisions at the beginning of L1 lethargus and become binucleate. This process is somewhat variable; individual nuclei often fail to divide (Sulston and Horvitz 1977). The intestinal cells also undergo endoreduplications of their DNA without nuclear divisions at each lethargus, resulting in C values of 32 in the adult (Hedgecock and White 1985).

Like many of the cell types in *C. elegans*, the intestinal cells are multifunctional. Their primary function is probably to secrete digestive enzymes into the lumen and to absorb the processed nutrients. One of the enzymes secreted by the gut or pharyngeal glands is an endodeoxyribonuclease. A mutant defective in the gene *nuc-1 X* lacks this activity and accumulates DNA from digested bacteria in the intestine (Sulston 1976).

The intestine also seems to be one of the main storage organs of the body, based on observations that intestinal cells contain numerous storage granules of diverse appearance (Fig. 11). Just prior to a molt, it contains dense accumulations of granules that disperse during lethargus. Some of these granules are lipid; others appear darkly stained in electron micrographs and are probably protein and/or carbohydrate. Some intestinal granules are strongly autofluorescent when irradiated with 300- to 400-nm light and are refractile when viewed with polarized light. These granules appear early in the ontogeny of the intestine and have proved to be useful

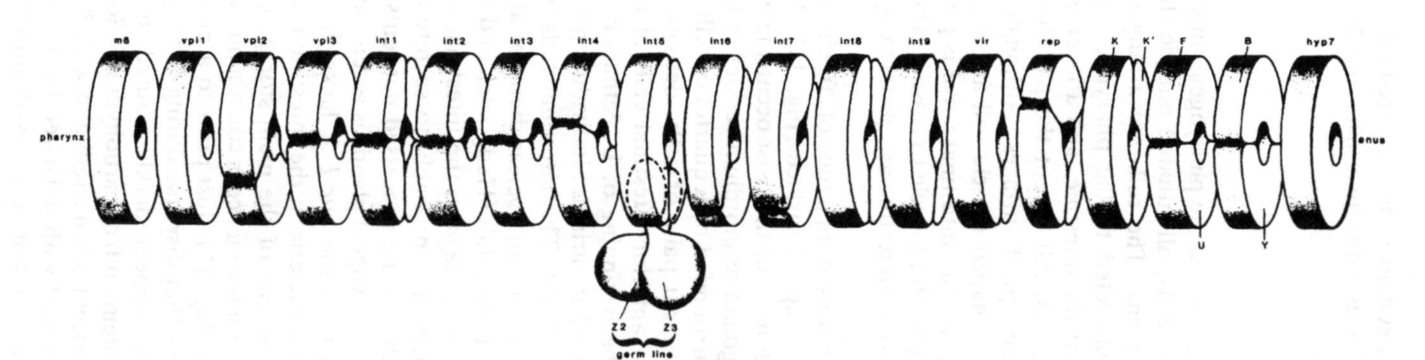

Figure 11 Schematic diagram of part of the alimentary tract of a 430-min-old embryo. The L1 has a similar arrangement, except that there are no germ cell lobes into int5. Microvilli line the apical surfaces of the intestinal cells (int), the vp1 and vp2 cells of the pharygeal–intestinal valve, and the rectal epithelial cell (rep). *(Inset)* A section of the adult intestine. (Reprinted, with permission, from Sulston et al. 1983.)

cell-type-specific markers (Laufer et al. 1980). Mutations in the *flu* genes cause alterations in the intensity or wavelength of the fluorescence, as a result, at least for some *flu* genes, of defects in tryptophan catabolism (Babu 1974).

The intestine plays a major role in the nurture of germ cells. Yolk proteins produced in intestinal cells of the adult hermaphrodite are transported to the oocytes, as described in Chapter 7 (Kimble and Sharrock 1983). If the gonad of a hermaphrodite is prevented from developing by laser ablation, yolk accumulates in the pseudocoelomic cavity. In addition, two of the embryonic intestinal cells seem to act as nurse cells for the primordial germ cells Z2 and Z3, which extend prominent pseudopodia into the two int5 cells of the intestine (Fig. 11). These pseudopodia disappear before hatching, as though this function of the intestine were no longer required or could be taken over by the somatic gonad cells during subsequent development.

C. The Rectum

The intestine is connected to the rectum by an intestinal–rectal valve, which is similar to the pharyngeal–intestinal valve in some respects. In both cases, a ring of two cells, which do not bear microvilli, is attached to the intestine. This ring is, in turn, attached to a second ring of three cells that do bear microvilli. The intestinal–rectal valve is connected to the main body syncytium (hyp-7) through three layers of two endothelial cells each, which comprise the rectum (Fig. 11). Most of these endothelial cells also act as blast cells during postembryonic development of the male (Sulston and Horvitz 1977).

Three sets of muscle are associated with the rectum: the anal depressor muscle, the sphincter muscle, and two intestinal muscles (Fig. 12). The anal depressor is a large H-shaped muscle that lifts the roof of the anus when it contracts. The sphincter muscle wraps around the intestinal–rectal valve and acts to close it. The intestinal muscles have longitudinally oriented filaments localized in the ventral regions of the cells. The dorsal regions flatten out into thin sheets, which wrap around the posterior ventral regions of the intestine and are probably attached to it. The three sets of muscles are coupled together by gap junctions and send arms to the dorsal surface of the preanal ganglion. Here they receive synaptic input from a single neuron (DVB).

D. The Excretory/Secretory System

The fine structure of the excretory/secretory system has been described by Nelson et al. (1983). The excretory system consists of three cells: a single large excretory cell, a duct cell, and a pore cell that interfaces the duct to

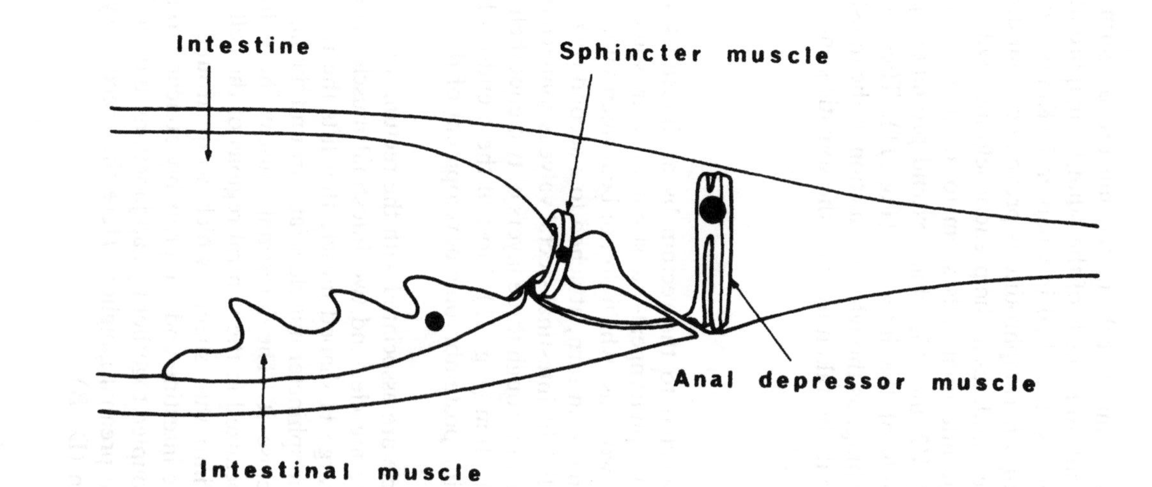

Figure 12 Organization of the defecation muscles in the hermaphrodite. The anal depressor is a large H-shaped muscle, which lifts the posterior dorsal surface of the rectum. The intestinal muscles have longitudinally oriented filaments and attach to the intestine and the body wall. The intestinal and depressor muscles send arms to the preanal ganglion, where they receive synaptic input from the neuron DVB, along with the sphincter muscle. (Reprinted, with permission, from White et al. 1986.)

the main body hypodermis (hyp-7). In addition, a binucleate gland cell is coupled to the excretory cell and the duct cell. All four cells are situated on the hypodermal side of the pseudocoelomic basement membrane (Fig. 13).

The excretory cell is the largest mononucleate cell in *C. elegans*. Its cell body is situated ventrally, at the level of the terminal bulb of the pharynx.

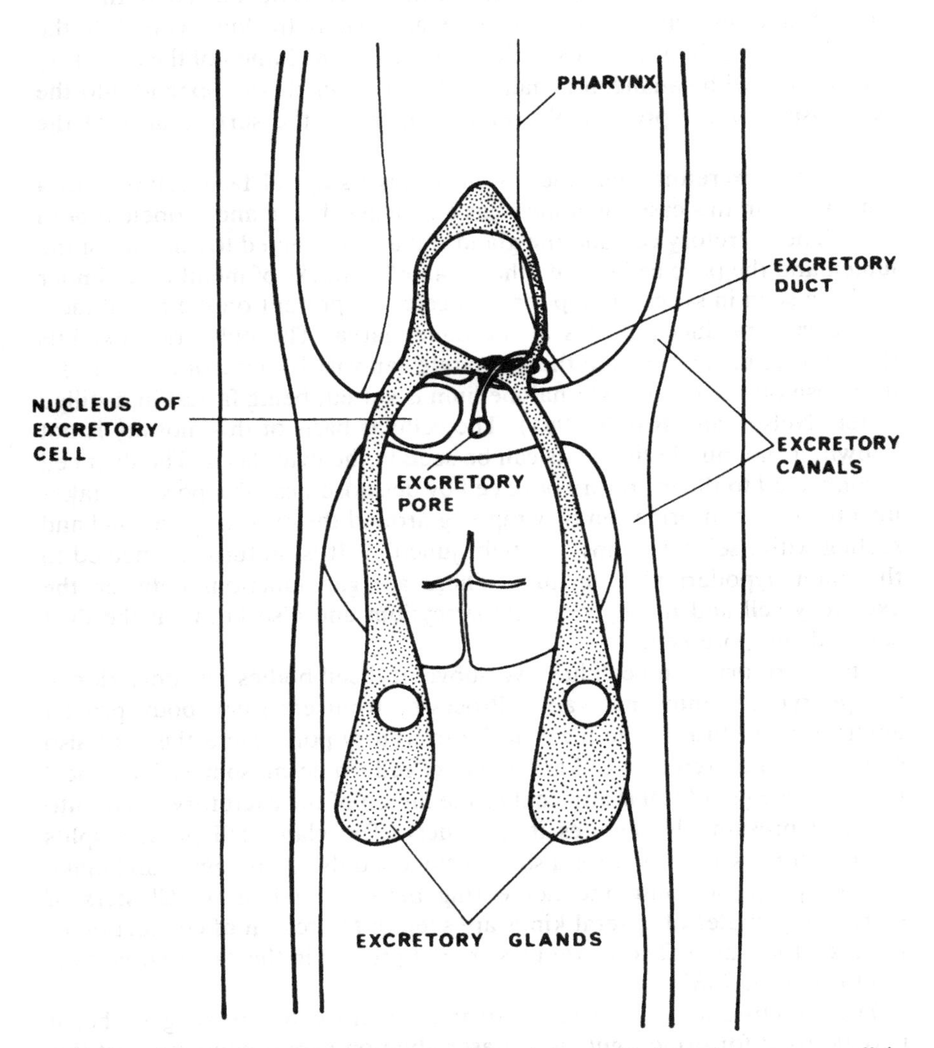

Figure 13 The excretory/secretory system. A large H-shaped excretory cell is connected to a bilobed gland cell and the excretory duct cell which, in turn, is connected to the excretory pore cell. All of these connections are made via desmosomes. The lumen of the duct cell is continuous with that of the excretory cell. The excretory gland appears to secrete into this lumen.

The cell itself is H-shaped, with the two side arms extending along the lateral lines for most of the length of the animal. These arms are slightly dorsal of lateral and run in close association with the processes of three neurons (CAN, PVD, and ALA). The excretory cell is polarized, with basal and apical faces. The apical face is adjacent to a lumen, which runs inside the arms and cross-bridge; the basal face is on the outside of the cell. There is a desmosome between the two faces where the lumen opens to the outside of the cell, at the cross-bridge of the H. The lumen of the excretory cell is lined with canaliculi (small dead-end channels that extend into the cytoplasm), which presumably serve to increase the surface area of the apical face.

Like the excretory cell, the duct cell has its apical face adjacent to a lumen, but in this case the lumen has a cuticular lining and is open at both ends. The excretory cell and the gland cell are connected to one end of the lumen and the pore cell to the other. Lamellar stacks of membrane similar to those seen in syncytial hypodermal cells are present on the apical face. The excretory duct pulsates in the dauer larva. The pulse rate exhibits short-term fluctuations, but the mean rate can vary by a factor of five or six if the osmolarity of the external medium is varied, being fastest in distilled water (Nelson and Riddle 1984). The cellular basis of this motility is not known, as no muscle filaments can be seen in the cytoplasm. The duct cell is connected to the excretory pore cell by desmosomes. The pore cell takes up a toroidal conformation by wrapping around the excretory channel and sealing with itself by means of a tight junction. It is, in turn, connected to the main hypodermal syncytium. There are gap junctions between the excretory cell and the hypodermal syncytium and also between the duct cell and the pore cell.

The excretory gland cell has two subventral cell bodies just posterior to the pharyngeal–intestinal valve. Processes from each cell body project anteriorly and fuse at the level of the excretory pore. Here they are also coupled to the excretory cell and the duct cell via desmosomes. The apical region of the gland cell is adjacent to the lumen of the excretory canal, into which it presumably secretes its products. The gland cell process splits again as it runs anteriorly for a short distance, only to fuse again and make a short projection into the nerve ring before terminating. Clusters of secretory granules of several kinds are seen in the region of connection of the excretory canal. These granules are not present in the dauer larva stage (Nelson et al. 1983).

The structure and organization of the excretory system suggest that it may be used for osmoregulation. Laser ablation experiments support this notion; animals in which the duct cell has been ablated become bloated with fluid, if in a hypotonic medium, and eventually die (Nelson and Riddle 1984). The function of the excretory glands is less clear, however. Removal of the glands by means of laser surgery has no obvious consequences (Nelson and Riddle 1984).

E. Coelomocytes

There are six coelomocytes in the adult hermaphrodite and five in the male. They are situated in the pseudocoelomic cavity and have a glandlike morphology (Chitwood and Chitwood 1974). During larval development, these cells become filled with granules and vacuoles (Sulston and Horvitz 1977). In the adult, the cytoplasm contains large vacuoles and many coated pits and vesicles (Fig. 14). Laser ablation experiments have shown that

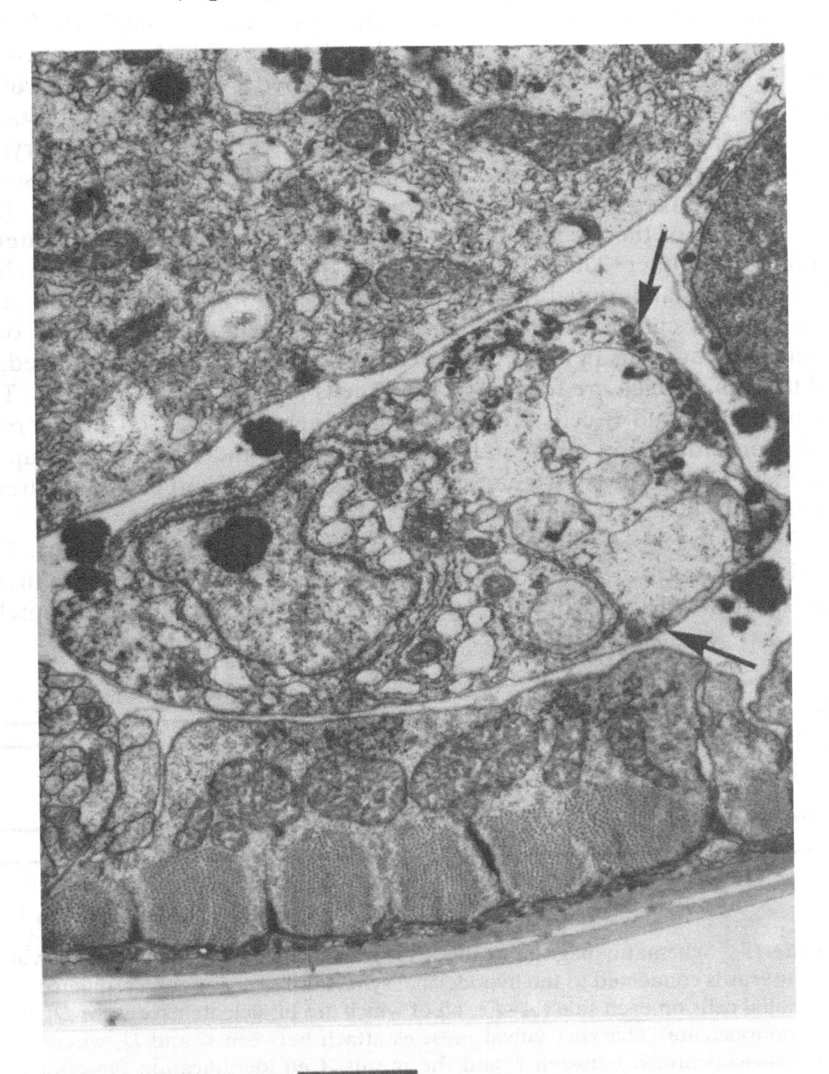

Figure 14 Coelomocytes. These cells are situated in the pseudocoelomic cavity and contain large membrane-bound vesicles. They also have many coated pits and coated vesicles (arrows). Scale bar represents 1 μm.

these cells are not essential for survival (J. Sulston, pers. comm.); however they have been seen to accumulate injected dyes, suggesting that they could provide a scavenging function (J. Kimble, pers. comm.).

F. Sex-specific Structures—Hermaphrodite

1. Vulva

The vulva consists of a transverse slit in the hypodermis through which eggs are laid. It is structured in such a way that the internal hydrostatic pressure of the animal keeps it closed until it is opened by means of the vulval muscles, so that it probably also acts as a valve. It is made up of a total of 12 epithelial cells and 8 muscles and has twofold rotational symmetry, as does the rest of the hermaphrodite gonad. In each half of the vulva, a series of six cells links the uterus to the hypodermal syncytium, hyp-7 (Fig. 15). All but one of these cells are binucleate. The F cells attach to the uterus and to each other. The E cells are next in the sequence; they are attached to the F cells and probably to each other. These two pairs of cells produce the internal cuticle of the vulva. The two D cells, which are the only mononucleate cells in the series, occupy prominent positions on the edges of the labia, which are formed by the D cells and the adjacent C cells. The remaining A and B cells produce a thickening of the ventral hypodermal ridge on either side of the vulva. The A, B, C, and D cells are coupled laterally to the body syncytium, hyp-7, whereas the E and F cells apparently do not have such connections.

There are two sets of four vulval muscles, vm1 and vm2 (Fig. 16). The vm1 muscles are attached to the body wall subventrally, insinuating themselves between the two ventral rows of body muscles. The vm1 muscles

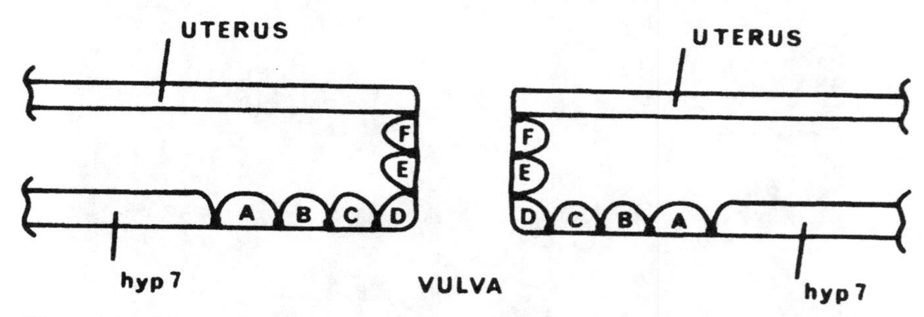

Figure 15 Schematic diagram of the arrangement of epithelial cells in the vulva. The uterus is connected to the hypodermal syncytium by means of a sequence of six epithelial cells on each side (*A–F*), all of which are binucleate except for *D*, which is mononucleate. The vm1 vulval muscles attach between *C* and *D*, whereas the vm2 muscles attach between *F* and the uterus. Cell identification (anterior half-vulva): (*A*) P5paaa, P5paap; (*B*) P5papa, P5papp; (*C*) P5ppal, P5ppar; (*D*) P5ppp; (*E*) P6paal, P6paar; (*F*) P6papl, P6papr.

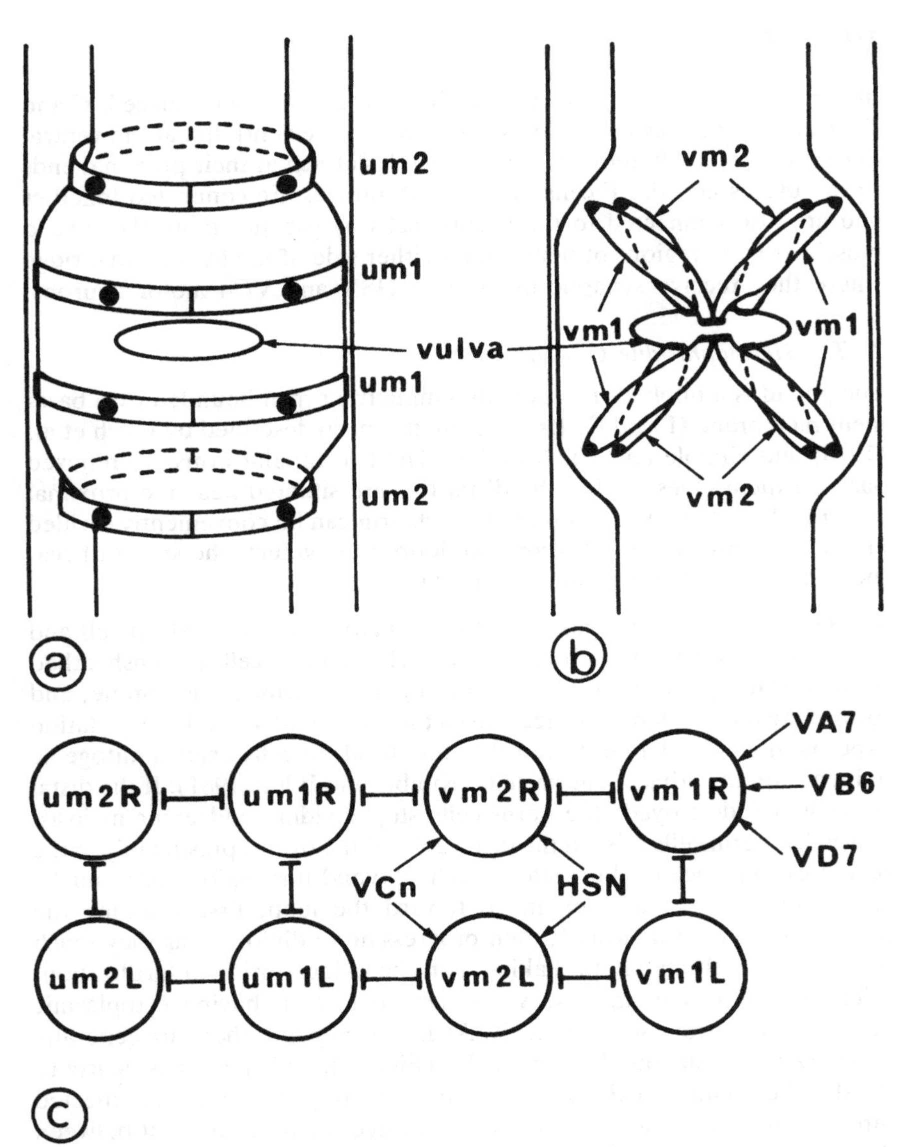

Figure 16 Arrangement of the 16 muscle cells that control egg laying. Eight are associated with the uterus (*a*) and eight with the vulva (*b*). Uterine muscles (um2) form circumferential bands of muscle around the uterus distal to the vulva. The um1 muscles attach to the body wall laterally (Fig. 20) and extend to the proximal regions of the ventral uterus. The vm1 muscles are attached to the vulval opening, and the vm2 muscles are attached to the junction between the vulva and the uterus. The vulval and uterine muscles are coupled together by gap junctions as shown in *c*. Synaptic input is only onto the vm2 muscles. (Reprinted, with permission, from White et al. 1986.)

attach at their proximal ends to the labia of the vulva, between cells C and D. The vm2 muscles attach to the body wall more ventrally, at the ventral margins of the body muscles. These muscles attach at their proximal ends to the uterus and the F cells. The vulval muscles are connected together and are also connected to the uterine cells by gap junctions. They send muscle arms to regions of neuropil on either side of the hypodermal ridge where they receive synaptic input from HSN and VCn motor neurons.

2. *The Hermaphrodite Gonad*

The gonad is a bilobed, rotationally symmetric organ bounded by a basement membrane (Fig. 17). Its structure has been described by Hirsh et al. (1976) and Kimble and Hirsh (1979). The two tubular arms are reflexed back on themselves so that the distal tips are situated near the proximal region, where the vulva is situated. Each arm can be conveniently divided into six regions: the distal ovary, the loop, the oviduct, the spermatheca, the sphermathecal valve, and the uterus.

The Ovary. The ovary has only three somatic cells, a distal tip cell and two ventrally located sheath cells. The enclosed germ cells are ensheathed in the distal region by the distal tip cell and a basement membrane, and toward the loop by tenuous sheets from the two sheath cells. Laser ablation experiments have demonstrated that the distal tip cells exert a mitogenic influence on the adjacent germ cells (Kimble and White 1981): If the distal tip cells are destroyed, the germ cells stop dividing and enter meiosis. Normally, germ cells arise in the distal 25% of the ovary, presumably in the region of influence of the distal tip cell. Beyond this region, germ nuclei enter into meiosis, as they move toward the loop, first attaining the pachytene stage of meiosis 1, then progressing to diplotene as they reach the loop, and ultimately to diakinesis in the oviduct prior to fertilization.

The germ cells in the ovary are syncytial, each having cytoplasmic bridges to a central core, the rachis. At any given point, there are generally about 12 nuclei surrounding the rachis (Fig. 18), which extends nearly to the distal extremity of the ovary. Germ cells are probably interconnected throughout development; a cytoplasmic bridge is already present between the two germ-line cells in a 450-minute-old embryo (Sulston et al. 1983). The rachis has been shown to contain RNA synthesized from the pachytene nuclei (Gibert et al. 1984). The cytoplasm in the more proximal regions of the rachis contains lipid droplets and other cytoplasmic components typical of ooplasm. It seems likely, therefore, that oocyte cytoplasm is assembled in the rachis and that the approximately 1300 germ nuclei in the distal ovary export the products of their transcription into this structure. This contribution by all the germ cell nuclei to a common pool of ooplasm allows efficient production of oocytes with a large cytoplasmic to nuclear ratio.

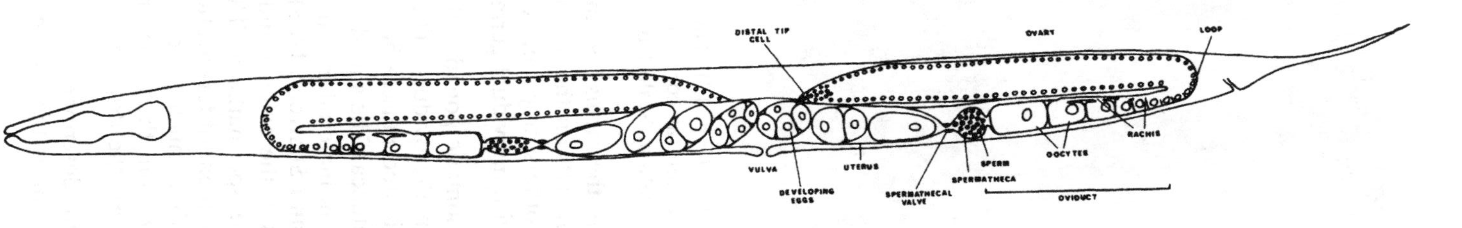

Figure 17 Organization of the hermaphrodite gonad.

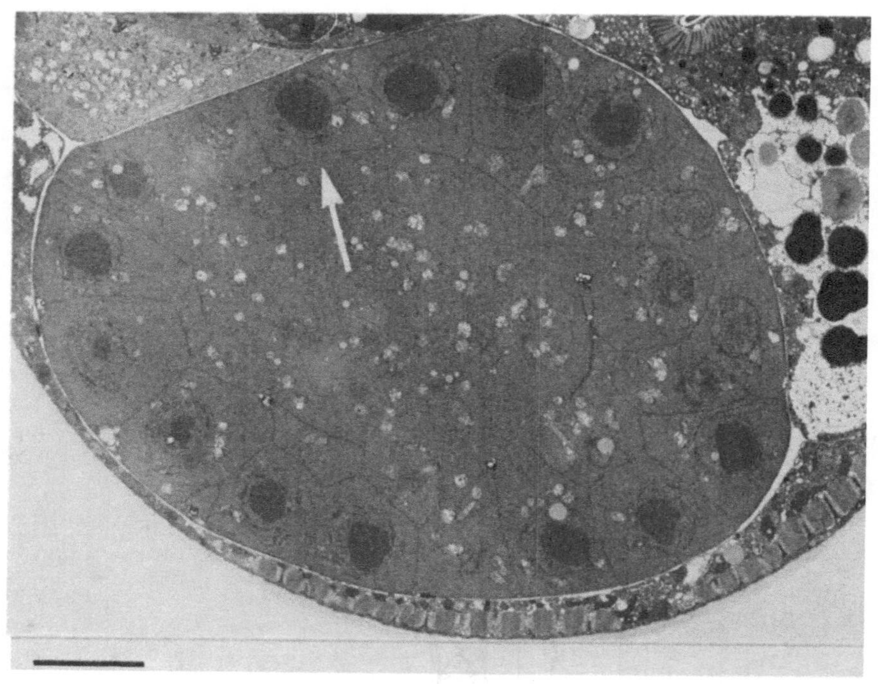

Figure 18 Section through the ovary (Fig. 17). The germ cells are incompletely cellularized with connections to a central core, the rachis. These connections (arrow) are maintained until fertilization. Scale bar represents 5 μm.

The Loop. In the loop region, the rachis becomes confined to the inside of the loop. Some of the germ cells on the outside rapidly increase their cytoplasmic volume, whereas others may die. As a result, ooplasm from the rachis becomes parceled with individual germ nuclei to form oocytes.

The Oviduct. Four pairs of somatic gonad cells with ventrally situated nuclei ensheath the oocytes in the oviduct. These myoepithelial sheath cells contain longitudinally oriented muscle filaments, which act to push the oocytes toward the spermatheca. The sheath cells probably also play a role in the transportation of yolk material into the oviduct from its site of synthesis in the gut (Kimble and Sharrock 1983). The oocytes maintain their connections to the rachis in the oviduct and continue to increase in size as they move toward the spermatheca. They seem to break this connection at or close to the moment of fertilization. During the transit of the oocytes in the oviduct, their nuclei proceed from diplotene into diakinesis. Just prior to fertilization, the oocyte nucleus moves from the center of the cell position toward the distal end.

The Spermatheca. Fertilization (described in Chapter 7) occurs either at the proximal end of the oviduct or as the oocyte enters the spermatheca.

The spermatheca is made up of 22 rather small endothelial cells, connected together by an elaborate network of desmosomes. The sperm become located within the spermatheca adjacent to its inner surface. To enter the uterus, eggs must pass through an elaborate valve at the proximal end of the spermatheca (Fig. 19). This valve consists of a single cell with four nuclei. There is an annulus of filamentous material in the cytoplasm of this cell, which has elaborate folds of plasma membrane on the inside surface of the annulus. The nuclei of the valve are situated in an out-pocketing of the plasma membrane, connected through an isthmus to the main body of the cell. The valve opening must expand to several times its resting diameter when an egg passes through. In the process, a few sperm are often swept along with the egg, but these usually migrate back through the valve and return to the spermatheca.

The Uterus. This structure comprises 50 cells, most of which are endothelial cells forming the wall. A few are specialized cells that seem to mediate the attachment of the lateral uterine walls to the seam cells in the vicinity of the vulva. The actual attachments are to the gonad's bounding

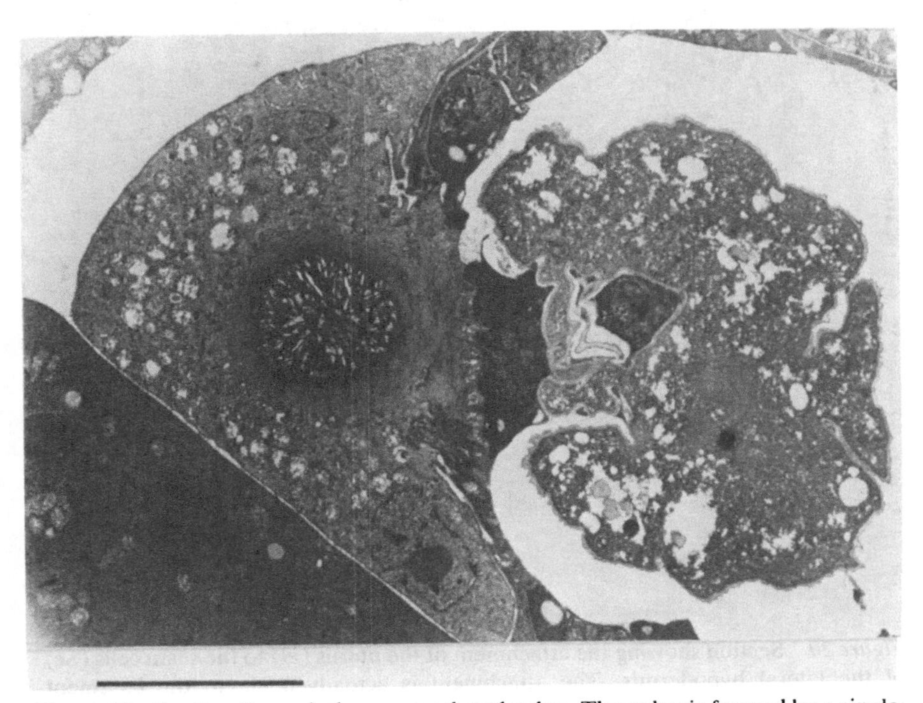

Figure 19 Section through the spermathecal valve. The valve is formed by a single cell with four nuclei and is situated at the junction of the uterus and spermatheca (Fig. 17). Scale bar represents 1 μm.

basement membrane, which is considerably thicker at these points (Fig. 20).

Eight muscle cells are associated with the uterus. In each half-uterus, there is one set of two muscles distal to the vulva that wrap completely around the uterus, and a second set of two muscles proximal to the vulva

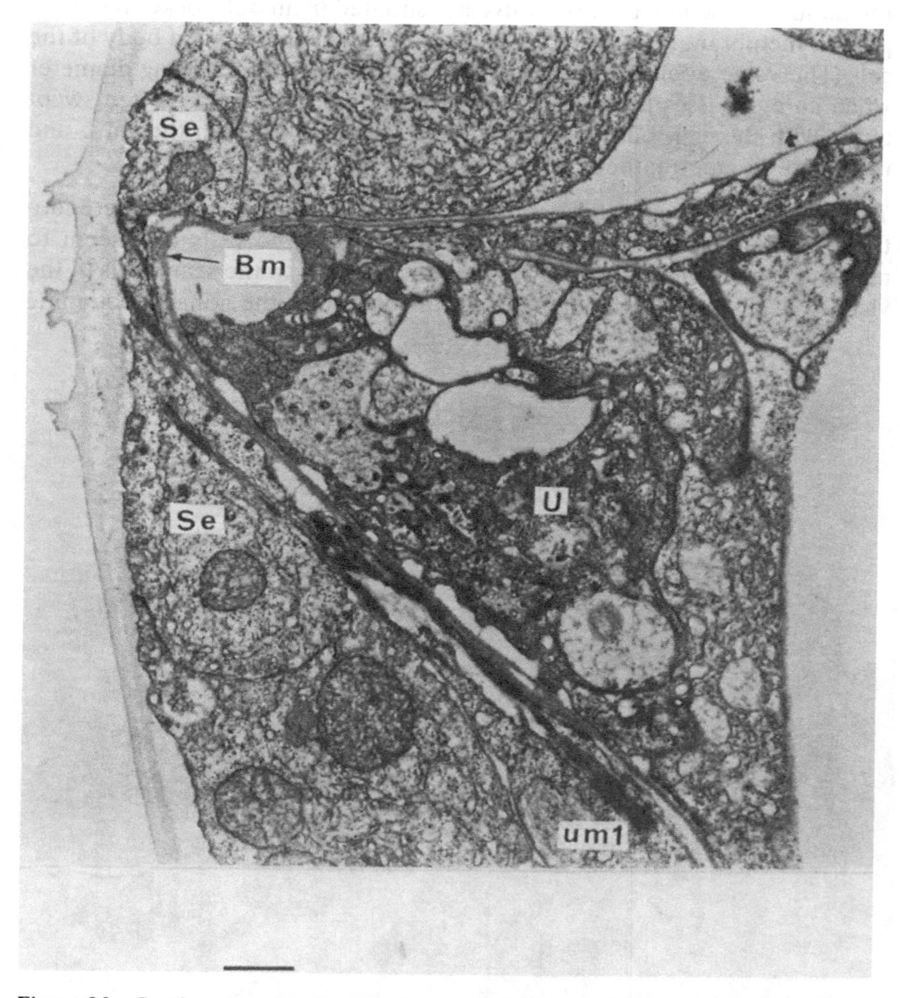

Figure 20 Section showing the attachment of the uterus (U) to the seam cells (Se) of the lateral hypodermis. The attachment is actually made to the basement membrane (Bm), which is markedly thicker in the region of contact. The um1 uterine muscles also attach to the body wall at this point. Scale bar represents 1 μm.

that attach to the lateral body wall (Fig. 16). The uterine muscles are coupled together and are also coupled to the vulval muscles by gap junctions. In operation, the uterine muscles presumably squeeze the eggs out of the uterus, and the vulval muscles open the vulval orifice, allowing eggs to be expelled.

During larval development, a specific cell in the developing uterus, termed the anchor cell, stimulates the adjacent cells of the ventral hypodermis to divide and form the cells of the vulva (Kimble 1981a). The anchor cell makes the initial connection between the uterus and the vulva. It subsequently disappears, opening a passageway between the uterus and the vulva (see Chapter 5).

G. Sex-specific Structures—Male

In the adult male, the tail is specialized into an elaborate copulatory apparatus, the copulatory bursa. This structure has been described in detail by Sulston et al. (1980). The male tail lacks the rectum and the long tail spike of the adult hermaphrodite, but instead has a cloaca, which exists at the center of a broad fan (Fig. 21). The fan is richly endowed with sensory receptors (the rays), and there are additional receptors, anterior and posterior to the cloaca (the hook sensillum and postcloacal sensilla, respectively). A prominent structure at the junction of the alimentary and genital tracts, the proctodeum, houses two rectractable spicules that are deployed during copulation.

On encountering a hermaphrodite, a male will run its copulatory bursa backwards along the hermaphrodite's body, turning at the ends by means of characteristic ventral arching movements of its tail. These movements continue until the vulva has been located. The male then locks onto the hermaphrodite by inserting its two copulatory spicules into the vulva. It eventually ejaculates, passing sperm through its vas deferens and the hermaphrodite vulva into the uterus of the hermaphrodite. Laser ablation studies have shown that the sensory rays are necessary for the backing and turning movements, and the hook sensillum, for the location of the vulva (J. Sulston, pers. comm.).

The ventral arching of the male tail during copulation and the extension of the spicules are mediated by a set of male-specific sex muscles that develop during the L3 and L4 larval stages. Most of the male-specific nervous system also develops during this period, in addition to the neurons that are common to the nervous systems of both sexes in the lumbar ganglia, the preanal ganglion, and the dorsorectal ganglion. One additional ganglion, the cloacal ganglion, has no counterpart in the hermaphrodite. Most of the neurons in these ganglia project axons into the neuropil of the preanal ganglion.

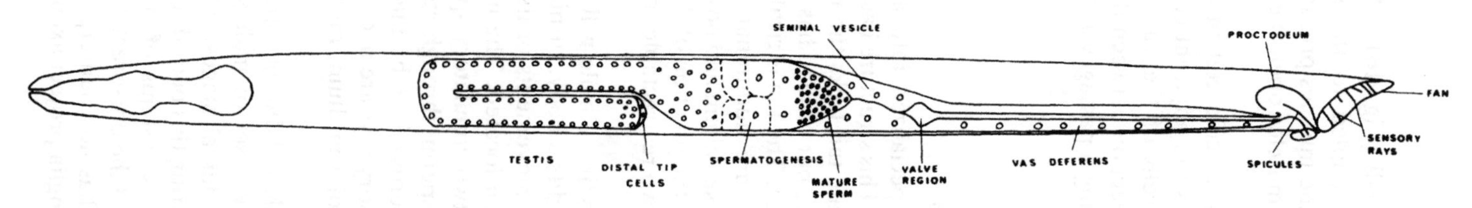

Figure 21 Organization of the male gonad.

1. The Fan

This acellular structure is formed from the adult cuticle matrix (Sulston et al. 1980). A set of 18 sensilla, with their sensory endings in the fan, are known as the rays. The rays are unusual in that they only have one structural cell associated with them, unlike other sensilla, which have an associated sheath cell and socket cell. Each ray has two associated neurons, RNA and RNB, which project into the neuropil of the preanal ganglion. The rays are not identical, however. Rays 1, 5, and 7 open out on the dorsal side of the fan; rays 3 and 9 on the edge; and rays 2, 4, 6, and 8 on the ventral side. Moreover, the neurotransmitter dopamine is present only in the R5A, R7A, and R9A neurons; and ray 6 has a different appearance at its tip from the others.

The lateral hypodermal (seam) cells in the male tail differ from their more anterior counterparts. The posterior five seam cells of the adult male (referred to as set cells) fuse together but do not fuse with the anterior seam cells, which by this time are also fused. They have many of the properties of the anterior seam cells, such as the presence of large Golgi bodies and the lateral shrinking during lethargus, but they do not produce alae. The male alae terminate at the set/seam junction. Though it is likely that the set cells normally produce much of the cuticular material of the fan, they are not essential, and their removal results in only a slightly altered fan morphology (Sulston et al. 1980).

2. The Cloaca

There are three sensilla around the cloaca: an anterior sensillum associated with a sclerotic hook (the hook sensillum) and two posterior postcloacal sensilla. Unlike the ray sensilla, each of the cloacal sensilla has both a sheath cell and a socket cell. Two neurons are associated with the hook sensillum: One (HOA) extends to the outside through a hole in the socket cell, and the other (HOB) ends near the sheath cell/socket cell junction and has a striated rootlet. Three neurons are associated with each of the two postcloacal sensilla; one extends to the outside through a hole in the socket cell and has a short striated rootlet (PCA), whereas the other two (PCB and PCC) end in the sheath cell. PCC has a long striated rootlet.

3. The Proctodeum

The proctodeum contains the two spicules, as well as the apparatus used for their extension and retraction. The spicules consist of spikes, U-shaped in cross section, covered with sclerotized cuticle. Each contains two neurons (SPV and SPD), two sheath cells, and four socketlike structural cells, which probably secrete the cuticular material of the spicules. Each spicule is controlled by two extensor and two retractor muscles. When the spicules extend, they are guided out of the cloaca by the thick sclerotic lining on the roof of the proctodeum (the gubernaculum).

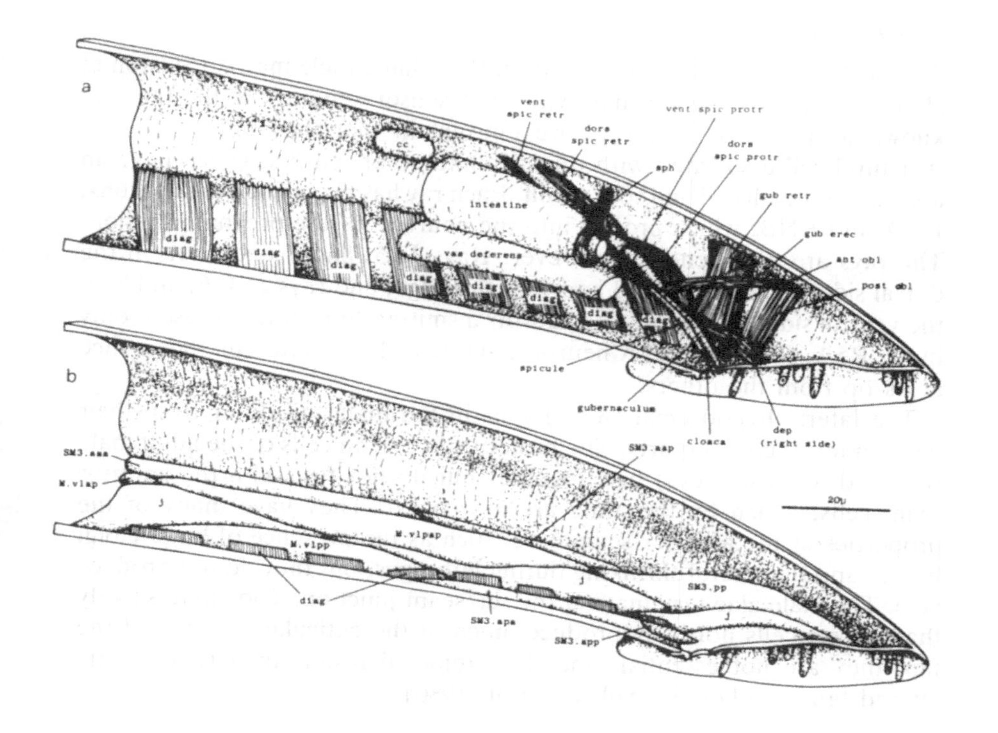

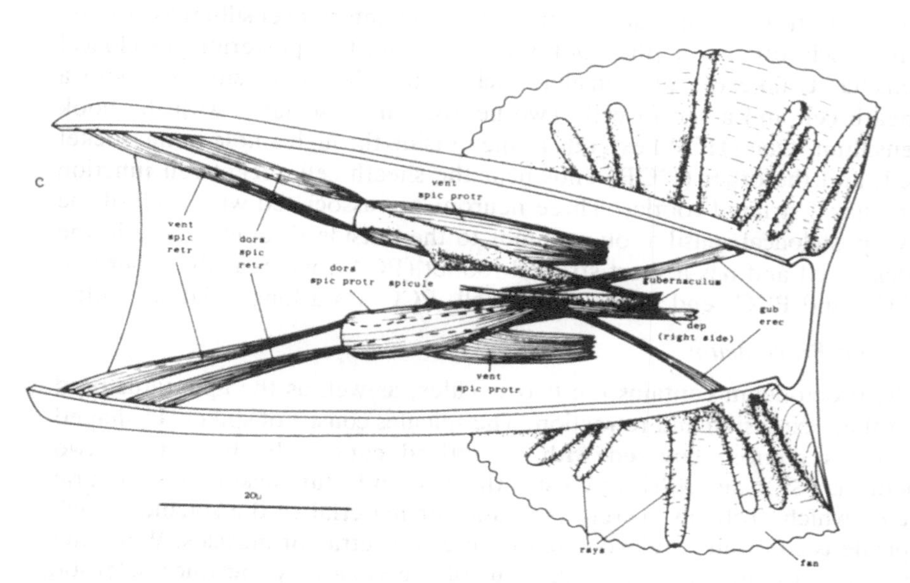

Figure 22 (*See facing page for legend.*)

4. Sex-specific Muscles

A total of 41 sex muscles are added to the male tail during the L4 stage; Figure 22 shows their arrangement in the adult. The characteristic ventral arching of the male tail is facilitated by a set of 15 diagonal muscles distributed along the posterior body and possibly also by 4 posteriorly located oblique muscles. The diagonal muscles have focal attachments at each end, in contrast to the body muscles, which have distributed attachments. They are attached to the body wall ventrally and laterally. Each spicule has two associated protractor muscles and two retractor muscles attached to its proximal end. The retractors are also attached to the body wall in front of the cloaca, whereas the protractors are attached to the body wall near its posterior extremity. The gubernaculum is also controlled by two sets of muscles, two erectors and two retractors.

Some of the preexisting tail muscles that are common to the hermaphrodite are modified during development of the male tail. The sphincter muscle at the end of the intestine is considerably enlarged in the male, probably because the male has no rectum, which leaves the intestinal sphincter as the only structure that can close off the gut. The anal depressor muscle of the male reorganizes its attachments in the adult and appears to act as an accessory to the spicule protractors (Sulston et al. 1980).

5. The Nervous System

The male nervous system has not been reconstructed in as much detail as the hermaphrodite nervous system (most of which is also present in the male); however, all the constituent neurons have been identified and named (Sulston et al. 1980). All of the male-specific nervous system develops postembryonically with the exception of four neurons in the head (CEM), which are components of the cephalic sensilla and are accessible to the outside through holes in the socket cells.

Two classes of male-specific motor neurons, CAn and CPn, are distributed along the length of the ventral cord. Each of these classes has eight members. CPn neurons innervate the diagonal muscles ventrally, and CAn

Figure 22 Muscles in the male tail. Contractile elements are represented, rather than the cell outlines of the internal muscles. (*a*,*b*) Left-hand view of right half of tail. (*a*) Internal muscles; (*b*) ventral wall body muscles and ventral attachments of diagonal muscles; (*c*) dorsal view of spicule muscles (right-hand dorsal spicule protractor and arm of reorganized depressor muscle have been removed to show spicule and ventral spicule protractor). The posterior ends of the ventral spicule protractors attach to the ventral body wall. (ant) Anterior; (post) posterior; (dors) dorsal; (vent) ventral; (cc) coelomocyte; (dep) anal depressor; (diag) diagonal; (erec) erector; (gub) gubernacular; (obl) oblique; (protr) protractor; (retr) retractor; (sph) sphincter muscle; (spic) spicule; (j) body-wall muscle present at hatching. (Reprinted, with permission, from Sulston et al. 1980.)

neurons send commissures up to the dorsal cord. Most of the remainder of the male-specific nervous system is made up of neurons from the various sensory sensilla of the copulatory bursa. There are relatively few male-specific interneurons and motor neurons.

6. *The Male Gonad*

The Testis. The structure of the male gonad has been described by Hirsh et al. (1976), Klass et al. (1976), and Kimble and Hirsh (1979). Its distal extremity is situated near the center of the body (Fig. 21). There are two distal tip cells, which, like the single distal tip cells at the extremities of the hermaphrodite ovaries, exert a mitogenic influence on the germ cells in their immediate proximity (Kimble and White 1981). Germ cells farther away from this region enter meiotic prophase. The testis runs anteriorly for a short distance and then reflexes back and runs posteriorly until it connects with the seminal vesicle. The germ cells of the testis are ensheathed by a basement membrane and are arranged around a central core (the rachis) with which they are all connected, like the germ cells in the hermaphrodite ovary.

The Seminal Vesicle. This structure consists of an inner tube of 20 apparently secretory cells, surrounded by a thin sheet made up of the cytoplasmic processes of three large cells. The vesicle has a larger diameter than either the adjacent testis or the vas deferens. The spermatocytes mature in the seminal vesicle, undergoing two meiotic divisions there to form spermatids (Wolf et al. 1978; Ward et al. 1981). The spermatids are stored in the seminal vesicle until they are ejaculated. Details of spematogenesis are described in Chapter 7.

The Vas Deferens. The vas deferens is a complex structure that has not been completely characterized. It consists of 30 cells, probably including three or more cell types (N. Wolf et al., pers. comm.). Many of the cells contain secretory granules. Some may also be motile; the anterior extremity of the vas deferens seems to act as a valve that releases sperm when the animal ejaculates. The vas deferens is connected to the U cells of the proctodeum (Sulston and Horvitz 1977) via a specialized linker cell situated at the growing tip of the developing gonad. This cell is killed by one of the U-cell descendants when it reaches the proctodeum, thereby opening a passageway between the lumen of the vas deferens and the cloaca (Sulston et al. 1980).

5
Cell Lineage

John Sulston
MRC Laboratory of Molecular Biology
Cambridge CB2 2QH, England

The Nematode Caenorhabditis elegans. © 1988 Cold Spring Harbor Laboratory. 0-87969-307-X/88 $1 + .00.

I. INTRODUCTION

The cell lineage of an organism is a description of the route by which all the cells of the adult are derived from the zygote. We use the term to imply a precise knowledge of both division pattern and cell fate. The somatic cell lineage of *Caenorhabditis elegans* is largely invariant and is known in its entirety, as is the postembryonic somatic cell lineage of the related nematode *Panagrellus redivivus*. The postembryonic germ cell lineage is a general proliferation under the control of the somatic cells of the gonad (see Chapter 7) and is probably indeterminate.

What is the point of knowing the cell lineage? First, it provides us with a concise description of the development of the wild-type animal, in which both the detailed pattern of cell fates and the ancestral relationships between tissues are precisely defined. Second, it is a basis for the analysis of development in individuals made defective by experimental intervention or mutation and for comparison with other species.

Because of its relative invariance, cell lineage occupies a special place in *C. elegans* research. It is similarly important in the study of various other organisms and has been extensively explored in the leech (Stent and Weisblat 1982). Cell lineage probably plays a major role in the development of insect nervous systems (Taghert et al. 1984), though apparently not in the determination of photoreceptors (Lawrence and Green 1979). On the other hand, in the development of many larger organisms, the behavior of cell groups seems to be more significant than that of individual cells. In the insect epidermis, these groups are particularly well defined spatially and are known as compartments. At least part of vertebrate development involves a succession of cell groups that become progressively subdivided, each subgroup giving rise to a limited subset of the tissues of the mature animal. Speaking generally, one can say that development of any living organism proceeds via a lineage of developmental units, but in the nematode these units happen, for the most part, to be directly observable as individual cells.

II. METHODS

The invention that has made it possible to follow the entire cell lineage of *C. elegans* is differential interference contrast microscopy (Nomarski 1955). This technique visualizes differences in refractivity in a shallow focal plane, at very high resolution and in relatively thick specimens. Nuclei stand out against the more refractive cytoplasm in which they lie and are used as cell markers (Fig. 1). In the intact, wild-type animal, cell boundaries are visible only where cells are not in contact with one another or have a distinct cortical layer, although they can often be revealed under special circumstances (e.g., by laser damage, starvation, or osmotic shock,

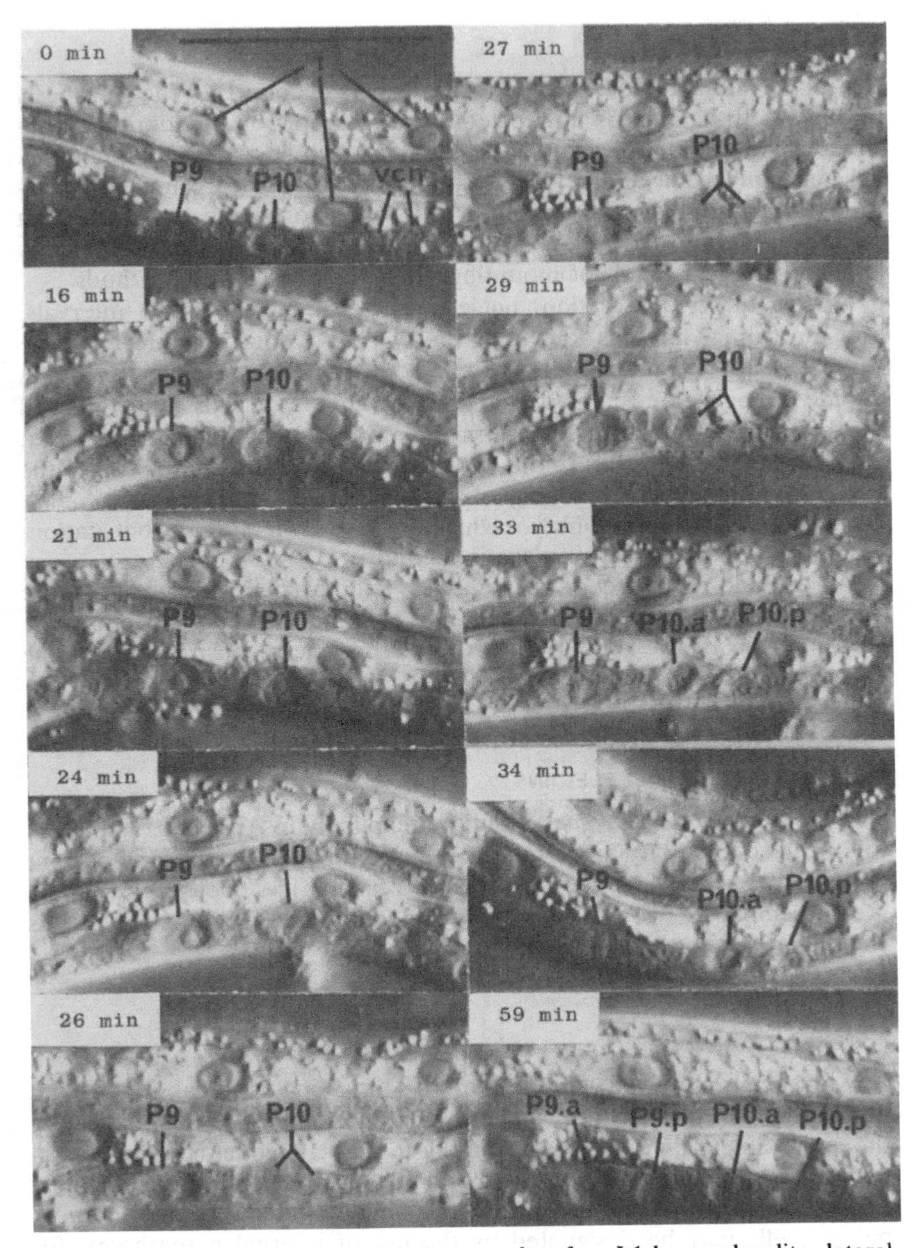

Figure 1 Cell division. Sequential photographs of an L1 hermaphrodite, lateral view (Nomarski optics). (vcn) Ventral cord neurons. 0 min, interphase; 16 and 21 min, P10 prophase; 24 min, P10 metaphase; 26 min, P10 anaphase; 27 min, P10 telophase; 29 min, P9 prophase; 33 and 34 min, P9 metaphase. Bar represents 20 μm. (Reprinted, with permission, from Sulston and Horvitz 1977.)

or in the mutant *clr-1* [E. Hedgecock, pers. comm.]). When properly mounted, the animal is unaffected by microscopic observation and continues to feed and grow, so that development can be watched in real time. The late embryonic and larval lineages must be followed by direct observation, because of the small size of the cells, on the one hand, and movement of the animal, on the other; but video recording is a useful aid for early embryogenesis. Details of mounting methods are given in Methods

During direct observation, nuclei are sketched freehand at intervals, ranging from 2 minutes for young embryos to 1 hour or more for some larval tissues. Depth can be recorded by a color code. To avoid errors in rapidly changing areas, one should limit the number of cells being followed to the capacity of one's short-term memory; in slowly changing areas, many cells, or even several animals, can be watched in rotation. To key the drawings to the animal unambiguously, suitable landmarks must be found. In the case of the early embryo, which is relatively featureless but stationary, a cross hair in the eyepiece is helpful.

At the end of a lineage that has been followed in this way, a more extensive set of nuclei and landmarks is sketched, so as to allow the identification of the cells in electron microscopic reconstructions of the same or different specimens.

III. CELL LINEAGE OF *C. ELEGANS*

The cell lineage of *C. elegans* is laid out explicitly in Appendix 3, and the following commentary is intended merely to highlight its more important features. Because the embryo is discussed further in Chapter 8, relatively greater emphasis will be placed on postembryonic development in this chapter.

A. Nomenclature

Key blast cells are given arbitrary names comprising uppercase letters and numbers. Their progeny are named by adding lowercase letters, indicating the approximate division axis according to an orthogonal coordinate system (a, anterior; p, posterior; l, left; r, right; d, dorsal; v, ventral); the next generation of cells is named by appending additional letters, and so on. A pair of cells may be designated by the use of internal parentheses, for example, M.v$({}^{l}_{r})$paa means both M.vlpaa and M.vrpaa. In the lineage drawings, a, d, and l are represented by left-hand branches, and p, v, and r by right-hand branches; branches are understood to be a/p unless otherwise labeled.

Times quoted are for development at 20°C.

B. Founder Cells

Development of the embryo begins with a series of markedly unequal divisions that give rise to the six founder cells: AB, MS, E, C, D, and P_4 (Fig. 2). These cells, described and named in the classical literature, have distinctive properties as defined both by their division rates (Deppe et al. 1978) and by the general nature of their progeny. They correspond roughly, but not precisely, to the germ layers recognized in higher animals (see Section VIII).

The clone of cells derived from each founder cell behaves in a characteristic way, as follows:

AB. The progeny initially spread over the surface, though some of them lie inside the head temporarily at an early stage. Toward the end of gastrulation, the AB pharyngeal precursors enter the interior through the ventral side of the head; later, the neuroblasts, four muscles, and the rectal cells sink inward and become covered by the hypodermal cells.

MS. Eight cells are formed on the ventral surface. In the course of the next two rounds of division, the pharyngeal precursors enter the interior and form two rows in the head; meanwhile, the body muscle, coelomocyte, and somatic gonad precursors insinuate themselves between the intestine and the surface layer of AB cells.

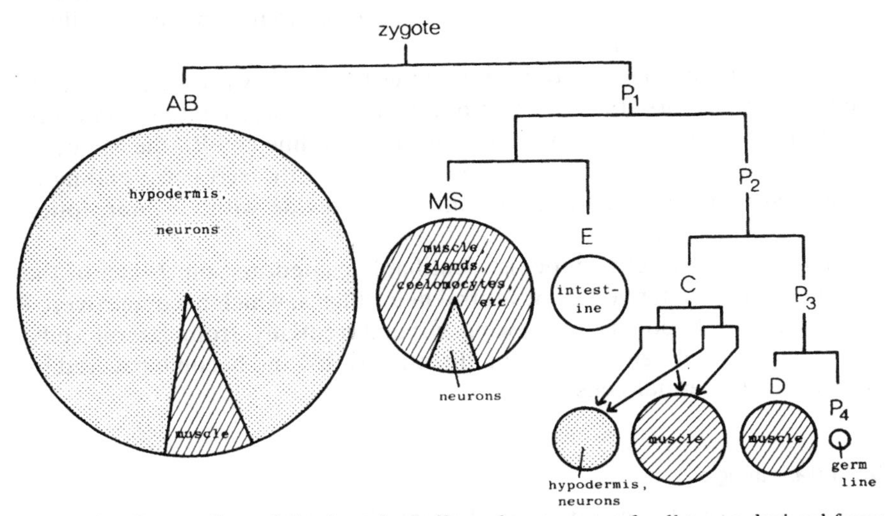

Figure 2 Generation of the founder cells and summary of cell types derived from them. Areas of circles and sectors are proportional to numbers of cells. Stippling indicates typically ectodermal tissue; striping indicates typically mesodermal tissue. (Reprinted, with permission, from Sulston et al. 1983.)

E. All progeny are intestinal. E divides on the ventral surface into two cells, which are the first to enter the interior in the course of gastrulation. Further division leads to a cylinder of cells with distinctively granular cytoplasm lying along the body axis.

C. Starting from the posterior dorsal side of the embryo, the cells spread anteriorly and ventrally. The more posterior ones are body-muscle precursors, which travel around to the ventral side and enter the interior immediately after D. Most of the remaining cells are employed in forming the dorsal hypodermis over the posterior two thirds of the body; their nuclei migrate contralaterally (see Section VII).

D. All progeny are body muscles. The clone enters the interior after two rounds of division, at the same time as $MS(^{a}_{p})p$.

P_4. All progeny are germ line. P_4 enters the interior after E and divides into two cells, which extend lobes into the intestine.

C. Early Cleavage and Gastrulation

At first, all the founder cells divide fairly symmetrically, producing daughters of approximately equal size. Pairs of daughters that will give rise to bilaterally symmetrical cell groups are generated in the first-round divisions of MS, C, D, and P_4 and in the second-round divisions of AB and E. However, at this stage the axes of the embryo, although implicit in cellular terms, are very distorted.

Gastrulation begins at about 100 minutes (28 cells), with the entry of the intestinal precursors (Ea and Ep) from the ventral surface into the interior. It continues progressively with entry of the germ-line cell P_4, the MS cells, some of the C and D cells and, finally, some of the AB cells, ending at about 300 minutes. All the cells continue to divide without interruption during these movements.

There is no distinctive process identifiable as neurulation. Much of the nervous system is derived from ventral neuroblasts, which become covered by a sheet of hypodermis growing circumferentially from lateral cells. Dorsal and lateral neuroblasts sink directly inward and become covered by adjacent hypodermal cells.

D. Mid-cleavage

As time goes on, the cell divisions become less synchronous and, in some cases, less equal. At about 200 minutes (200 cells), a series of highly unequal divisions is seen, followed by the deaths (14 in number) of the smaller daughters so generated. This is the first clear sign of differentiation after the separation of the founder cells. The principal axes of the organism now become fully apparent, as similar sublineages (see Section X) unfold on either side. These events follow closely after the beginning of bulk RNA transcription (Hecht et al. 1981a). Therefore, this stage corresponds, in some respects, to the mid-blastula transition described for amphibian embryos by Newport and Kirschner (1982).

E. Morphogenesis

The cell divisions succeed one another ever less rapidly, and few occur at all after 400 minutes. At this point, the embryo begins to elongate and move, and within 2 hours is some three times the length of the egg. The tissues, already laid down in rudimentary form, take on their mature appearance and neuronal processes grow out. The embryo continues to writhe within the egg for a further 4 hours before it hatches at 800 minutes; during this time, just two cells divide and three die.

F. L1 Larva

Hatching is another major transition, because now the embryo begins to feed and grow. Many embryonic gene functions become necessary at this time, presumably as a consequence of the dilution of maternal gene products and the requirement for new ones (for examples, see Horvitz and Sulston 1980).

After 3 hours, cell division starts again—not involving the majority of cells, as in the embryo, but only the selective division of specific blast cells. Many of the blast cells are hypodermal and have already served to synthesize parts of the L1 cuticle (as judged by the presence of secretory bodies within them). For ease of reference, these blast cells are given short names; they are designated on the embryonic lineage charts by arrowheads and can be located in the drawings of the L1 anatomy (Appendix 3).

First to divide are the two Q cells; they are neuroblasts lying laterally in the posterior half of the body. The behavior of their progeny is asymmetric: those on the left side travel backward, and those on the right travel forward. A little later, the nuclei of the Pn cells move from their lateral positions through the thin layer of cytoplasm overlying the body muscle and into the ventral hypodermal ridge. This migration must be an active process, for it is accurately timed and requires the nuclei to flatten remarkably as they squeeze past the muscle (Sulston and Horvitz 1977). In the ventral ridge the Pn nuclei divide to yield hypodermal cells and neurons. Most of the latter are ventral motor neurons, and their appearance coincides with a reworking of existing neuronal connections (White et al. 1976; see Chapter 11). Other neuroblast divisions occur in the tail.

The hypodermal blast cells divide in a characteristic stem-cell pattern, which is repeated at the beginning of each larval stage. Most of the non-stem-cell daughters that result from this series of divisions fuse with the large hypodermal syncytium (hyp 7), having first become tetraploid by endoreduplication (Hedgecock and White 1985).

The postembryonic mesoblast, M, divides to give additional body muscles, a set of sex myoblasts and, in the hermaphrodite, two coelomocytes.

During L1 lethargus, some of the intestinal nuclei divide, without any accompanying cytokinesis. At about this time, and subsequently at the end of each larval stage, all the intestinal nuclei endoreduplicate; their final ploidy in the adult is thus 32C (Hedgecock and White 1985).

In each sex, the blast cells Z1 and Z4 follow homologous lineage patterns, generating three regulatory cells and the precursor cells of the gonadal somatic structures. These regulatory cells comprise two distal tip cells (which control germ-line proliferation in both sexes and also gonad morphogenesis in hermaphrodites), the anchor cell in hermaphrodites and the linker cell in males. The linker cell is selected from a pair of equivalent cells in the late L1; the anchor cell is similarly selected in the late L2 (see Figs. 2 and 3 in Kimble 1981a). In the hermaphrodite the gonad retains twofold rotational symmetry throughout development, but in the male the cells rearrange into an asymmetric pattern.

At hatching, males can be readily distinguished from hermaphrodites by the following criteria: In males, (1) one of the left coelomocytes lies posterior to the gonad primordium; (2) the hermaphrodite-specific neurons are absent; (3) the nuclei of rectal cells B and Y are enlarged (see Fig. 5 in Sulston and Horvitz 1977). A fourth anatomical distinction, the presence of the four CEM neurons in the head of the male, is not diagnostically useful at this age.

G. L2 Larva

L2 is a time of growth with rather little cell division or obvious elaboration of tissues. In the young L2, the hypodermal Vn.p cells undergo one proliferative round of division; the daughters, together with the H-derived blast cells, then divide once in their stem-cell pattern. V5.pa generates the postdeirid ganglion, and neuroblast G2 adds two neurons to the ventral ganglion. There are rectal blast cell divisions in the male. The mesoderm and the somatic gonad are largely amitotic, but certain precursors for the next phase of development migrate into position: For example, the hermaphrodite sex myoblasts M.v$(^{l}_{r})$paa move anteriorly to lie on either side of the gonad.

The gonad continues to increase in size, as a result of the continued division of the germ cells. It elongates because of the outward movement of cells with "leader" function (Fig. 3). The hermaphrodite gonad extends two arms, one anteriorly and one posteriorly, each led by a distal tip cell. The male gonad extends a single arm, initially anteriorly but soon reflexing

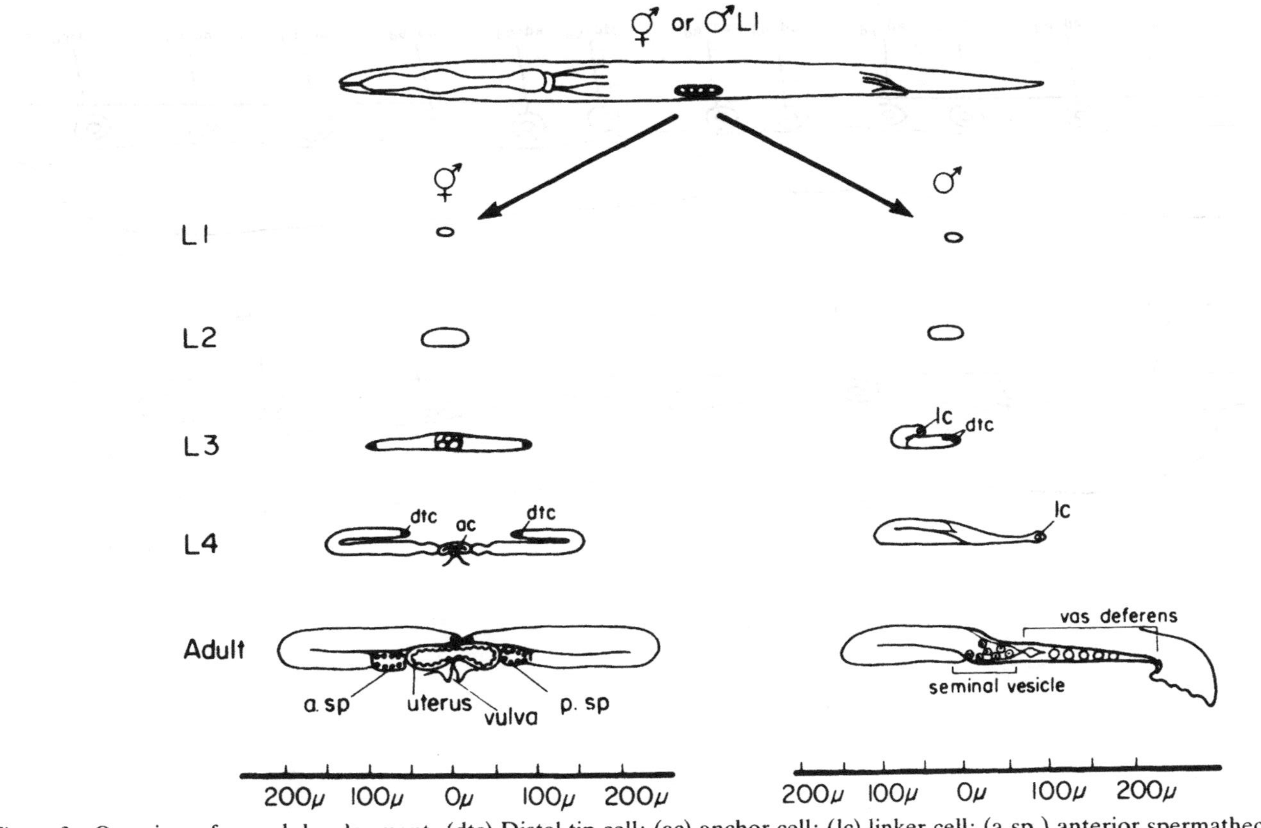

Figure 3 Overview of gonad development. (dtc) Distal tip cell; (ac) anchor cell; (lc) linker cell; (a.sp.) anterior spermatheca; (p.sp.) posterior spermatheca. (Reprinted, with permission, from Kimble and Hirsh 1979.)

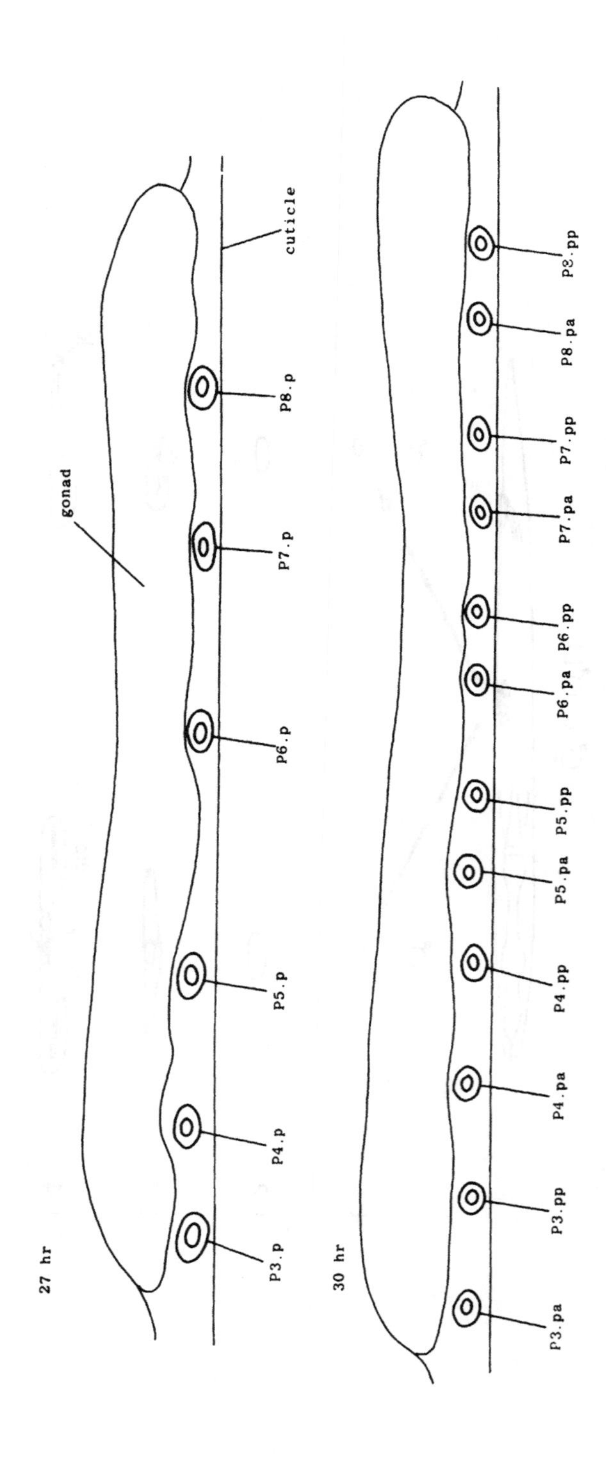
27 hr
gonad
cuticle
P8.p
P7.p
P6.p
P5.p
P4.p
P3.p
30 hr
P8.pp
P8.pa
P7.pp
P7.pa
P6.pp
P6.pa
P5.pp
P5.pa
P4.pp
P4.pa
P3.pp
P3.pa

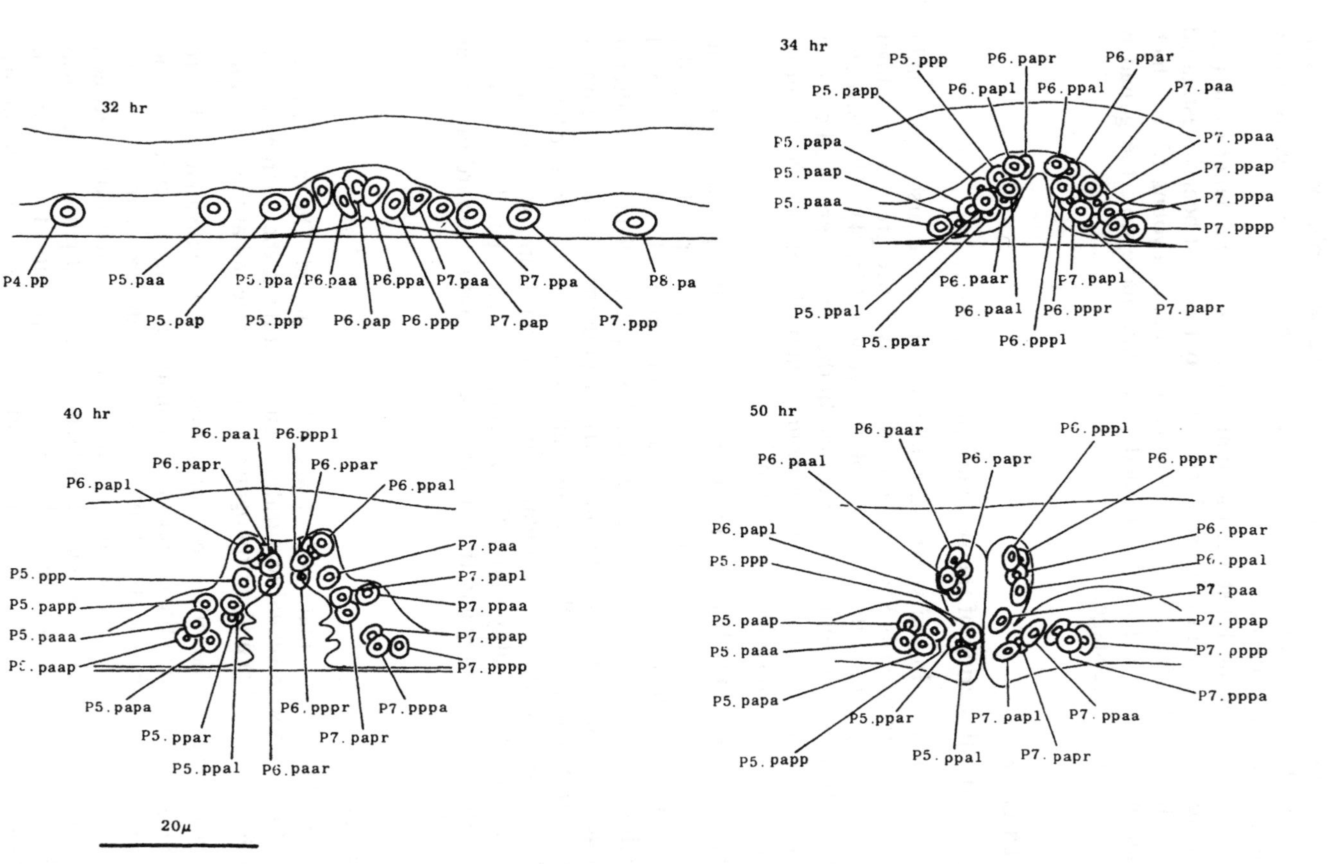

Figure 4 Vulval development, left lateral view. (Reprinted, with permission, from Sulston and Horvitz 1977.)

to grow posteriorly, led by the linker cell. In the hermaphrodite gonad, one of two alternative configurations of the four central cells is selected (see Fig. 2 in Kimble 1981a).

At the end of the L2 stage, the animal may enter the dauer cycle (see Chapter 12). The cell lineages of this pathway have not been investigated carefully, but the dauer larva seems to correspond to a young L3, in which some hypodermal cells have divided and some have not; the gonad is arrested at the L2 stage.

H. L3 Larva

L3 sees the flowering of the sexually specific lineages.

Hermaphrodite. In the ventral hypodermis, three precursors (normally P5.p–P7.p), previously selected from the equivalence group (see below) P3.p–P8.p, generate the 22 cells of the vulva (Fig. 4). The two sex myoblasts divide to yield 16 sex muscles. In the gonad, germ–cell division continues and somatic cell division begins again; the arms reflex to grow back toward the middle of the animal.

Male. In the ventral hypodermis, two precursors (P10.p and P11.p in the wild type), selected from the equivalence group P9.p–P11.p (Fig. 8), divide to add 16 neuronal, glial, and hypodermal cells to the preanal ganglion. In the ventral nerve cord, the Pn.aap cells divide once after a characteristic enlargement. In the lateral hypodermis, 18 ray precursors are generated and enter their terminal sublineages. The male-specific blast cells continue to divide and enlarge around the rectum, so that the tail becomes swollen visibly under the dissecting microscope. The sex myoblasts divide while they and their progeny are migrating toward their final positions.

I. L4 Larva

In the early L4, the somatic cell lineages come to an end. The remainder of L4 is taken up with sexual maturation, culminating in the opening of the hermaphrodite vulva and morphogenesis of the male tail at the final molt.

Hermaphrodite. The vulval cells move together and create an infolding of the body wall to which the center of the gonad is attached by means of the anchor cell. The somatic structures of the gonad differentiate visibly into uterus, spermatheca, and oviduct, with characteristic junctions between the three regions. The sex muscles attach in a defined pattern; they will later be used in egg laying, eight of them being responsible for opening the vulva and its attachment to the gonad, and eight for squeezing the uterus. At the L4 molt, the vulva everts, creating a free passage to the exterior.

Male. The gonad continues to grow posteriorly until the linker cell contacts derivatives of U and attaches to them. The vas deferens, the seminal vesicle, and the valve between them differentiate. The various sensilla (rays, hook, postcloacal sensilla) attach to the body wall and form characteristic cuticular specializations. The cells around the rectum form the proctodeum, which contains the copulatory spicules and is invested with numerous sex muscles. Other sex muscles lie ventrally, some lie longitudinally, and others extend obliquely between ventral and lateral points in the body wall.

During the 2 hours of L4 lethargus, a major reorganization of the male tail generates the copulatory bursa (Fig. 5). The hypodermis moves anteriorly, carrying ray cells and other neurons embedded within it (see Figs. 2 and 3 in Sulston et al. 1980). The latter part of this movement occurs after the outer layer of adult cuticle has been laid down and results in the spinning out of the ray cell processes; the resulting balloon of cuticle then collapses onto itself in a precise fashion, trapping the rays and forming the fan (see Fig. 4 in Sulston et al. 1980). The spicules are lengthened at the same time, by virtue of the attachment of their retractor muscles to the moving hypodermis. One muscle, the anal depressor, becomes completely reorganized: its original vertical filaments disappear and are replaced by horizontal ones. The linker cell is phagocytosed by the progeny of U, so that the vas deferens, now attached to the proctodeum by K and K′, is then open to the exterior.

J. Adult

The adult continues to grow after the final molt, but no further somatic cell divisions have been seen. The germ line continues to generate gametes for several days.

IV. CELL LINEAGE OF *PANAGRELLUS REDIVIVUS*

Numerous classical publications attest to the similarity in overall plan of early embryonic lineages throughout the class Secernentea. These studies have been somewhat extended by Sulston et al. (1983), but late embryogenesis has not been followed in detail in nematodes other than *C. elegans*.

Sternberg and Horvitz (1981, 1982) have followed the postembryonic lineages of *P. redivivus* in their entirety. With just one exception, the set of blast cells in this nematode is identical with that in *C. elegans*. However, the subsequent lineages, although similar in some respects, offer a variety of intriguing variations upon the theme of *C. elegans*. It is important to note that conclusions about cell fates in *P. redivivus* are based solely on light microscopy, not on ultrastructure.

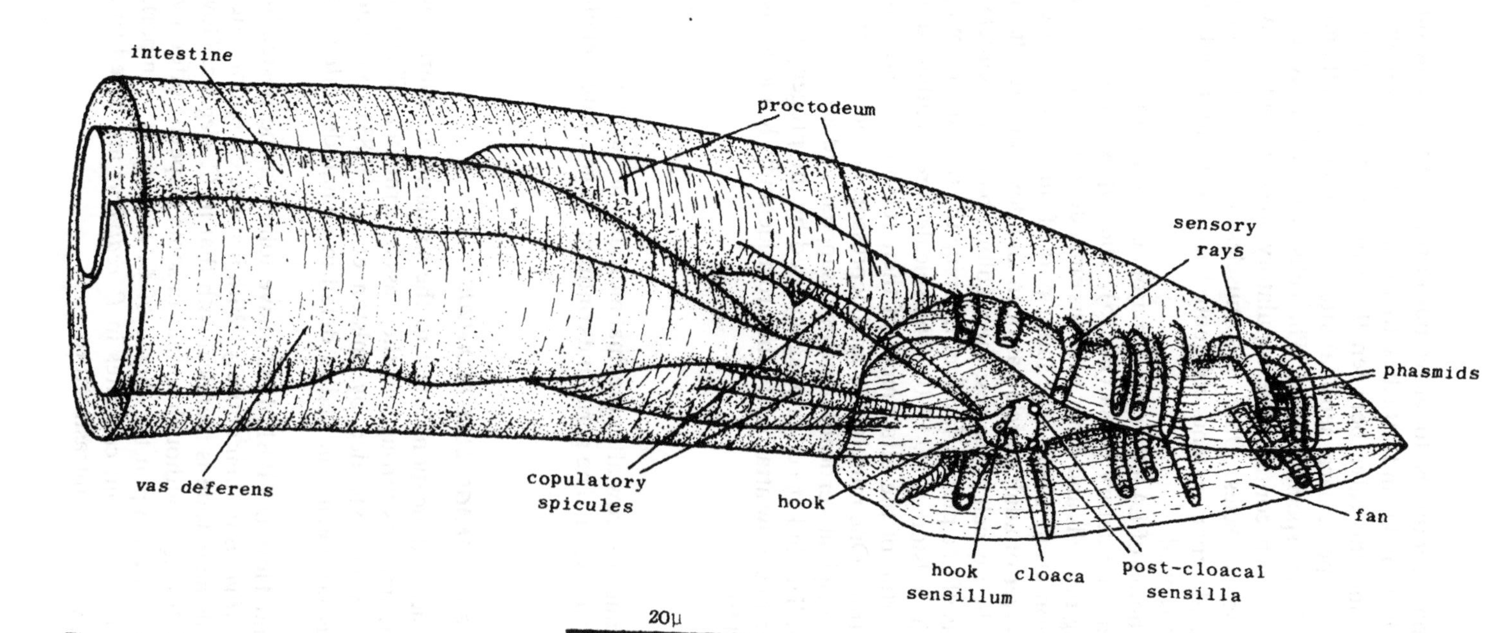

Figure 5 Young adult male tail, left subventral view. (Reprinted, with permission, from Sulston et al. 1980.)

A. Size

P. redivivus is larger than *C. elegans* and might be expected to contain more cells. Indeed, the hypodermal lineages are more extensive, though there are actually fewer body muscles. The hypodermal lineages are more variable, which may also be a correlate of greater size—perhaps a larger-scale version of the weak proliferative regulation (see below) seen in *C. elegans*.

B. Lineage Transformations

Sternberg and Horvitz (1981, 1982) classified the cell lineage differences between the two species into four types.

1. Change in number of rounds of division, leading to the presence of more copies of particular cells in one species than in the other.
2. Change in the terminal fate of one cell to the terminal fate associated with another cell.
3. Polarity reversal. The fates of the anterior and posterior daughters of a given cell are exchanged.
4. Altered segregation. Part of the developmental potential associated with a given cell in one species is associated with its sister in the other.

C. Sublineages

Cell lineages of *P. redivivus*, like those of *C. elegans*, can be analyzed usefully in terms of sublineages (see below). A striking example is a hypodermal sublineage that is repeated no fewer than 70 times in *P. redivivus* but is represented by a single division in *C. elegans* (see Fig. 24 in Sternberg and Horvitz 1982). Some sublineages are identical in the two species but are invoked slightly differently; for example, the ray sublineage (Fig. 6) is repeated 14 times in the *P. redivivus* male (by progeny of V6 and T) and 18 times in the *C. elegans* male (by progeny of V5, V6, and T).

D. Gonad Morphology

Among the most obvious and most easily interpretable differences between the two species are those associated with monodelphy (i.e., the formation of a single-armed gonad) in the female of *P. redivivus*. The death of Z4.pp, the putative posterior distal tip cell, is, by itself, sufficient to account for the absence of germ cells from the posterior arm (cf. Chapter 7; Kimble and White 1981). The accompanying reduction of the soma of the posterior arm to a postvulval sac can be ascribed to reduced division of the relevant precursors (PR, PL) and to the deaths of their posterior daughters. Observations on the mutant *ced-3* (H. Ellis, pers.

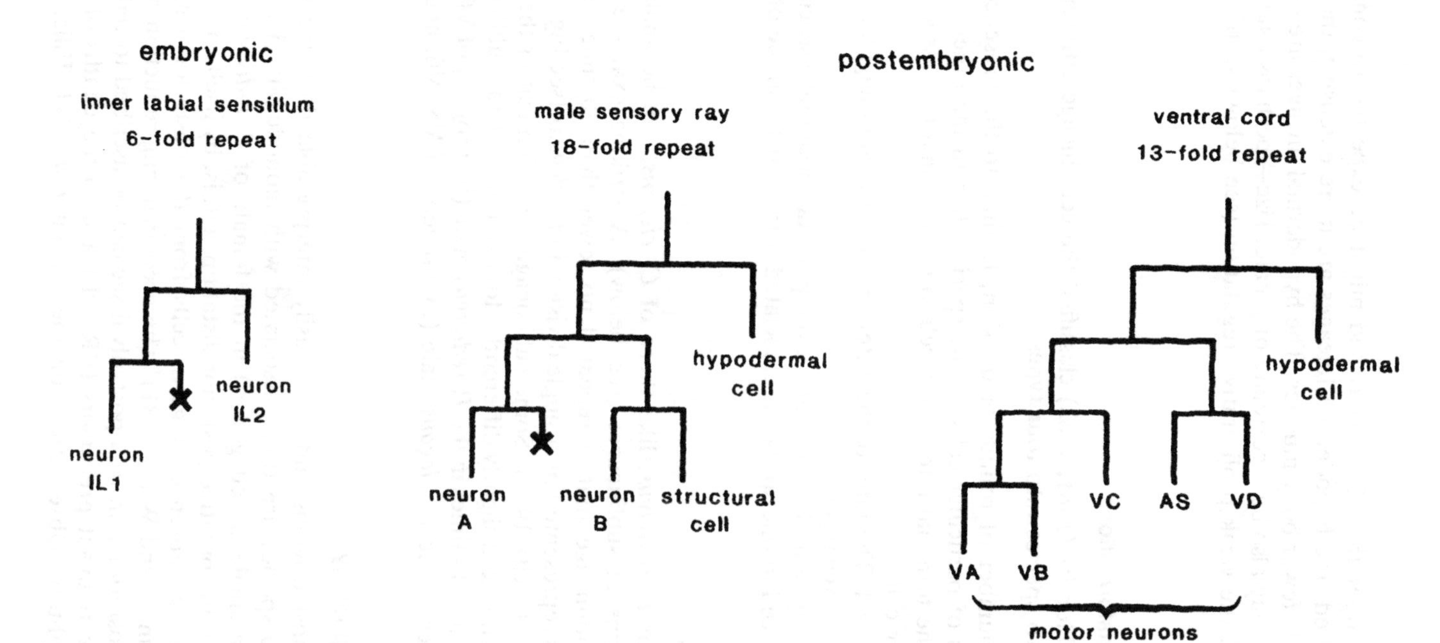

Figure 6 Repetitious sublineages (see Appendix 3 for full contexts of these fragments). Inner labial sublineage is invariant; the other two display local and sexual variation. (Reprinted, with permission, from Sulston 1983.)

comm.) suggest, however, that cell death has no essential function in *C. elegans*; by analogy, it is likely that cell death is not necessary to prevent proliferation of PRp and PLp in *P. redivivus*.

PRp and PLp do sometimes divide once, in which case both daughters die. Sternberg and Horvitz (1981) suggested that this unusual and obviously imperfect arrangement may reflect a recently evolved pattern.

E. Morphogenesis

Notwithstanding the association of many differences between *P. redivivus* and *C. elegans* with cell lineage changes, some major alterations simply involve morphogenesis. The best example is the maturation of the male tail (see description above). In *P. redivivus* the hypodermal movement is much more limited than it is in *C. elegans*; consequently, no bursa is formed and the sensory processes form papillae (i.e., sensilla that protrude only slightly from the hypodermal surface) rather than rays. Analogously, papillae arise in *C. elegans* individuals that produce ray cell groups outside the zone of hypodermal withdrawal (Sulston and White 1980). A similar argument applies to the spicules (Sternberg and Horvitz 1982).

V. CELL DETERMINATION

The cell lineage of *C. elegans* is largely invariant. With certain well-defined exceptions, the patterns of cell division, death, and differentiation are constant from individual to individual, and no great differences are seen in timing.

The few variable aspects of the lineage are of obvious interest and are discussed below. How is the reproducibility of the remainder achieved? One possibility is that most cells are internally programmed as a result of their ancestry and are thus limited to a single fate. We must remember, however, that at any given moment in development, not only does each blast cell have a predictable future but also a reproducible position and a defined set of neighbors. An equally plausible possibility, therefore, is that cells are programmed externally, but the cues are so reproducible from one individual to another that changes in fate are seldom seen. In favorable cases, we can distinguish between these two possibilities by perturbing the system experimentally.

A. Experimental Methods

Two general approaches have been useful. The more direct one is to separate cells from their neighbors physically. So far, this has been done only for the early egg (Laufer et al. 1980), and the results are described in Chapter 8. The second approach is to kill cells in situ. Classically, experimenters have used ultraviolet light, crushing, or ligation for this purpose, but these methods have now been largely superseded by irradiation with a pulsed dye laser (Sulston and White 1980), which provides great precision, convenience of operation, rapid killing, and apparently complete removal of the dead cells. Usually a cell can be made to disappear entirely within a few hours, probably by autolysis and subsequent phagocytosis by neighboring cells. In all the experiments discussed below, the regulative effects stated are only obtained if the killing is done early enough. At sufficiently late times, there is no effect at all, and at intermediate times, the outcome is, not surprisingly, variable and difficult to interpret.

Physical transfer of cells within an animal or between animals (i.e., grafting) would be desirable but has not yet been achieved.

B. Cell Autonomy

The majority of cell-killing experiments do not lead to any qualitative change in the fates of the remaining cells, though a certain amount of regulation in shape is customary. When the fate of a cell is unaffected by the ablation of any of its neighbors, we can reasonably hypothesize that it is determined autonomously. However, other than in the very early embryo (see Chapter 8), the proposal of cell autonomy is based on purely negative evidence and, to that extent, lacks conviction. There are three important ways in which negative results from ablation experiments may be misleading: (1) Parts of the target cell may survive for a time and continue to signal their presence to their neighbors. However, laser surgery appears to trigger death of the whole cell: Persistence of cortical fragments cannot be ruled out, but dynamic exchange of information involving transcription and translation can be effectively dismissed. (2) Adjacent cells may continue to develop normally because their overall positions in the animal are unchanged, or because of redundancy among the signaling cells. In some areas, more extensive ablations have been carried out, so as to shift the intact cells and disrupt their normal interactions as much as possible. (3) There may be insufficient time for the intact cells to respond to the loss of a neighbor, even though they may be capable of doing so, in principle. This limitation is most likely to prevail in terminal sublineages and in the embryo—the very areas in which least regulation is seen.

Bearing in mind these caveats, we can say that many cells appear to develop autonomously with regard to local interactions, but as yet little is known about possible global signals. Most of the cell interactions that have been described take place during early larval development. However, the early divisions of the founder cells are relatively unexplored and recent findings indicate some interactions at this stage as well (Priess and Thomson 1987; E. Schierenberg, pers. comm.; see Chapter 8).

C. Interaction between Cells of Different Potential

The development of certain tissues is dependent upon the presence of cells that do not, themselves, contribute to them. The controlling and dependent cells are neither analogous (similar in fate) nor homologous (similar in ancestry).

1. The development of the vulva is induced by the anchor cell of the gonad (Kimble 1981a). If the anchor cell is killed, all the vulval precursors adopt their tertiary fate (see below), dividing once and then joining the hypodermis. If all the gonadal cells are killed except for the anchor cell, the vulva develops normally (Fig. 7).
2. The continuous division of the germ-line cells requires the presence of the distal tip cells (Kimble and White 1981). If the distal tip cells are killed, mitosis ceases and all the germ cells enter meiosis (Chapter 7).
3. There is a complex relationship between the anchor cell pair and the neighboring "nonanchor" cell pair in the gonad. The nonanchor cells cannot substitute for the anchor cell as its own homolog can (Table 1), but they do influence and are influenced by the anchor cell (see Fig. 16 in Kimble 1981a).
4. The daughters of AB are initially equivalent. The distinction between them seems to be caused by interaction with P_1 and its derivatives. The presence of P_1-derived cells is necessary for AB to generate muscle and other pharyngeal cells (Priess and Thomson 1987).

D. Replacement Regulation and Equivalence Groups

Replacement regulation in the nematode is defined as the generation of structures from alternative progenitors after destruction of their normal progenitors. The groups of cells that can regulate in this way are shown in Table 1, and a particular example is shown in Figure 8; they are known as equivalence groups, to denote the fact that their component cells are more or less equivalent in developmental potential (the extent of the evidence varies from group to group). None of the equivalence groups are exclusive clones, but the members of a group often share substantial homology.

There are two or more fates open to the members of a given equivalence group, and their order of priority can be investigated by laser ablation experiments. The fate of highest priority is called primary and is adopted by a single surviving cell. The fate of next lower priority is called secondary, and the next, if present, tertiary (e.g., Fig. 8; P11.p, P10.p, and P9.p fates are primary, secondary, and tertiary, respectively).

A number of these equivalence groups are apparent in intact animals, because of natural variation in selection of cells for the various fates (Table 1). These are cases in which a pair of equivalent cells straddles the midline (or center of symmetry in the case of the gonad) so that either cell has a

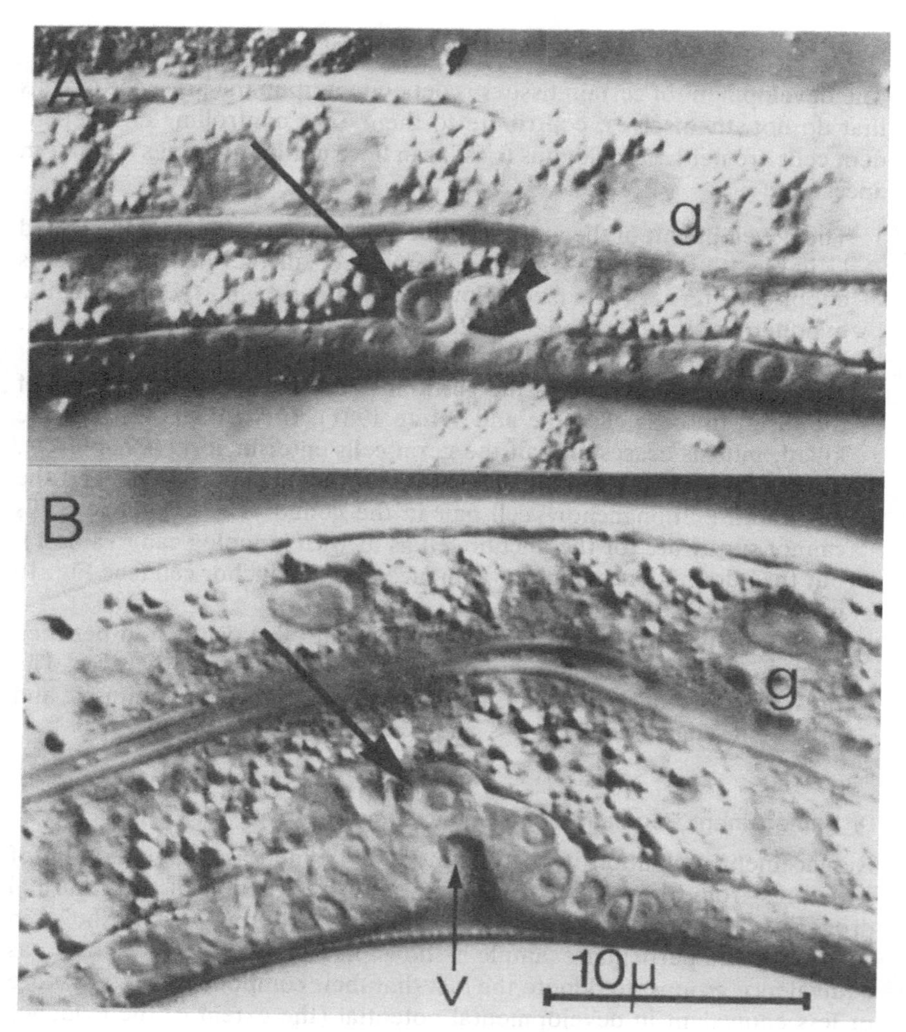

Figure 7 Isolation of an anchor cell precursor (Nomarski optics). (*A*) Anchor cell (arrow) is the only remaining cell in the gonad. The debris left by the ablation of six blast cells is minimal (arrowhead). (*B*) Isolated anchor cell (large arrow) induces the vulva (v). (g) Gut. (Reprinted, with permission, from Kimble 1981a.)

chance (though not necessarily an equal one) of recruitment to the primary fate.

Equivalence groups bear a superficial resemblance to the compartments described for insects (e.g., Lawrence 1981), in that they comprise blast cells of identical developmental potential that do not arise clonally. They differ markedly, however, in that the nematode blast cells have a strictly

Table 1 Known equivalence groups

	Tissue	Group	Members	1°	2°	3°	Natural variation observed	*Panagrellus* (where known)
Embryo	Vent hyp	ABp(l/r)paaaapa	2	duct	G1[a]		–	
Embryo	Vent hyp	ABp(l/r)apaapa	2	G2[a]	W[a]		–	
L1	Vent hyp	P1/2(L/R)	2	P1[a]	P2[a]		+	
L1	Vent hyp	P3/4(L/R)	2	[b]			+	+
L1	Vent hyp	P5/6(L/R)	2	[b]			+	+
L1	Vent hyp	P7/8(L/R)	2	[b]			+	+
L1	Vent hyp	P9/10(L/R)	2	[b]			+	+
L1	Vent hyp	P11/12(L/R)	2	P12[a]	P11[a]		– ?	
L2–L3	Vent hyp herm	P3.p–P8.p	6	P6.p[a]	P5.p[a], P7.p[a]	P3.p[a], P4.p[a], P8.p[a]	–	P4.p–P7.p
L2–L3	Vent hyp male	P9.p–P11.p	3	P11.p[a]	P10.p[a]	hyp	–	+
L2–L3	Lat hyp male	V?–V6	3+	V6[a]	V5[a]	V4[a]	–	
L2–L3	Male proct	Ba(l/r)aa	2	Bα[a]	Bβ[a]		+	
		Ba(l/r)pp	2	no preference			+	
		Ba(l/r)apaav	2	hyp	death		+	
		Bγa(l/r)d	2	no preference			+	
L2–L3	Gonad herm	Z1.ppp/Z4.aaa	2[c]	ac	vu[a]		+	–
L2–L3	Gonad male	Z1.paa/Z4.aaa	2	1c	vd[a]		+	+

Abbreviations: (hyp) Hypodermis; (herm) hermaphrodite; (proct) proctodeum; (ac) anchor; (vu) ventral uterus; (lc) linker cell; (vd) vas deferens.
[a]Precursor cell, giving rise to sublineage.
[b]See next group.
[c]Second interacting pair present but no other replacement.

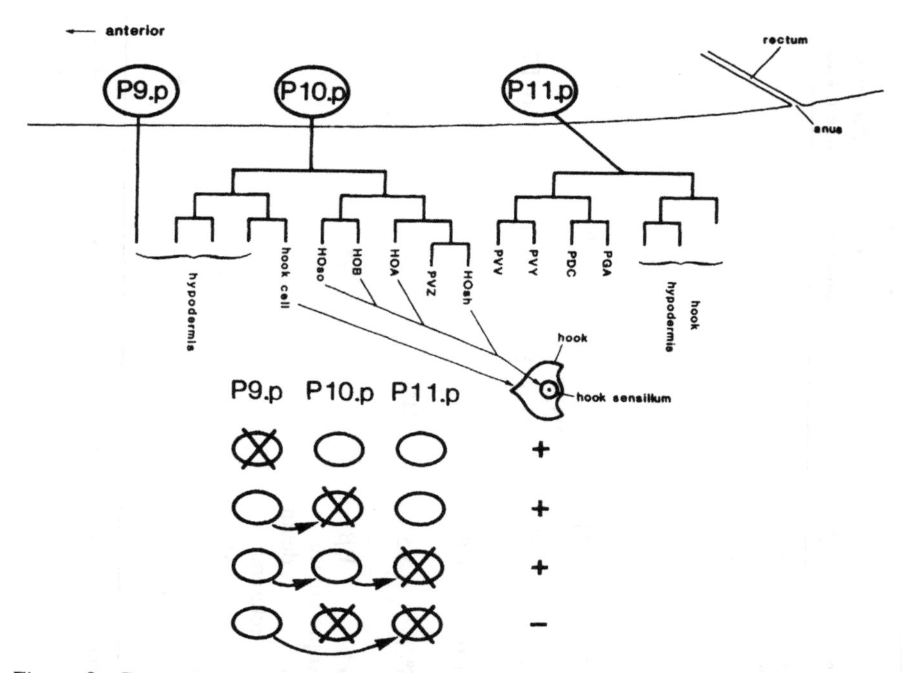

Figure 8 Preanal equivalence group in the male. Schematic lateral view showing the three members of the group (P9.p, P10.p, P11.p) and the lineages to which they give rise in the intact animal. The hook and its sensillum, characteristic of the P10.p lineage, are depicted in ventral view. The table beneath summarizes the replacement regulation observed after four different cell ablation experiments, e.g., after the ablation of P10.p, P9.p moves into its place and generates a typical P10.p lineage. (Reprinted, with permission, from Sulston 1983.)

limited repertoire of lineages and never proliferate to compensate completely for a missing member. A closer analogy to the insect compartment is seen in those nematode tissues that display space-filling regulation (see Sections E and F, below).

E. Form Regulation

After adopting their qualitative fates, which are subject to the constraints of ancestry and cell interaction, as described above, cells may still display plasticity of form. In larvae, a number of tissues are constructed in a space-filling manner, such that several similar cells grow until they meet their neighbors. The best examples are found in the intestine and the hypodermis (Sulston and White 1980). Sometimes, form regulation is more specific: For example, if one of the paired spicule retractor muscles of the male is lost, its fellow forks and attaches to the body wall at two points.

F. Proliferative Regulation

Although the hypodermal blast cells do not show precise replacement regulation (as do the ray precursors, for instance), removal of several contiguous blast cells can cause occasional extra divisions of the remainder (Sulston and White 1980). Regulation of this sort is seen also in the gonad (Kimble 1981a). These responses, weak as they are, differ from replacement regulation and are more analogous to the space-filling divisions of cells within compartments of the insect epidermis. It is significant that the nematode, though more disposed toward form regulation than proliferative regulation, does possess both mechanisms.

G. Axes, Vectorial Regulation, and Polarity Reversal

As a minimal requirement for the execution of the observed lineage, most cells must be aware of the anterior/posterior axis, in some way, for nearly all determinative divisions (i.e., those in which one daughter differs from the other) are anterior/posterior. Under normal circumstances, the cells may well remain aligned in a head-to-tail fashion from the time of their birth, but the existence of a gradient or other extracellular signal is by no means ruled out as an alternative or additional source of anterior/posterior information. In addition, particular cells probably need to know whether they lie dorsal or ventral, or on the left or right side of the animal.

In some circumstances, a blast cell reacts to the killing of a neighbor by reversing the polarity of its lineage. Two examples of compensation in this way have been seen in the gonad (Kimble 1981a) and have been termed vectorial regulation because the respective outcomes are predictable. Extreme isolation experiments, in which many of the neighbors of a given blast cell are removed, can cause unpredictable polarity reversals: Examples have been seen in the gonad (Kimble 1981a) and among the male ray precursors (Sulston and White 1980).

H. Limits to Proliferation

In addition to the cell constancy that is imposed upon the animal by the precise pattern of cell division, differentiation, and death, there are a number of less well-defined limits to cell proliferation. These limits do not serve any obvious function and are, perhaps, vestiges of more primitive controls. Some examples follow: (1) Isolated blastomeres are disinclined to produce more cells than they do in the intact embryo, and cell lines have not yet emerged from cultures of nematode embryos. (2) In the mutant *lin-5*, whose cells fail to undergo cytokinesis, various precursor nuclei generate about as much DNA as their combined progeny would have done in the wild type (Albertson et al. 1978). (3) The repetitive sublineages created (or revealed) by the mutant *unc-86* (Chapter 6) do not continue for long (Chalfie et al. 1981). However, these limits can be circumvented by certain mutations, notably in heterochronic loci (Chapter 6).

VI. PROGRAMMED CELL DEATH

Cell death is a prominent feature of the lineage, accounting for one in eight of all somatic cells produced. The topic has been specifically reviewed by Horvitz et al. (1982a). Cell death is observed directly by the transiently increased refractivity of dying cells (Fig. 9) and can be scored more or less reliably post hoc in mutants that block various stages of phagocytosis (*ced-1*, *ced-2* [Hedgecock et al. 1983]; *nuc-1* [Sulston 1976]). It does not occur at random but, with few exceptions, is a predictable fate of particular cells (like any other terminal differentiation) and follows a well-defined course.

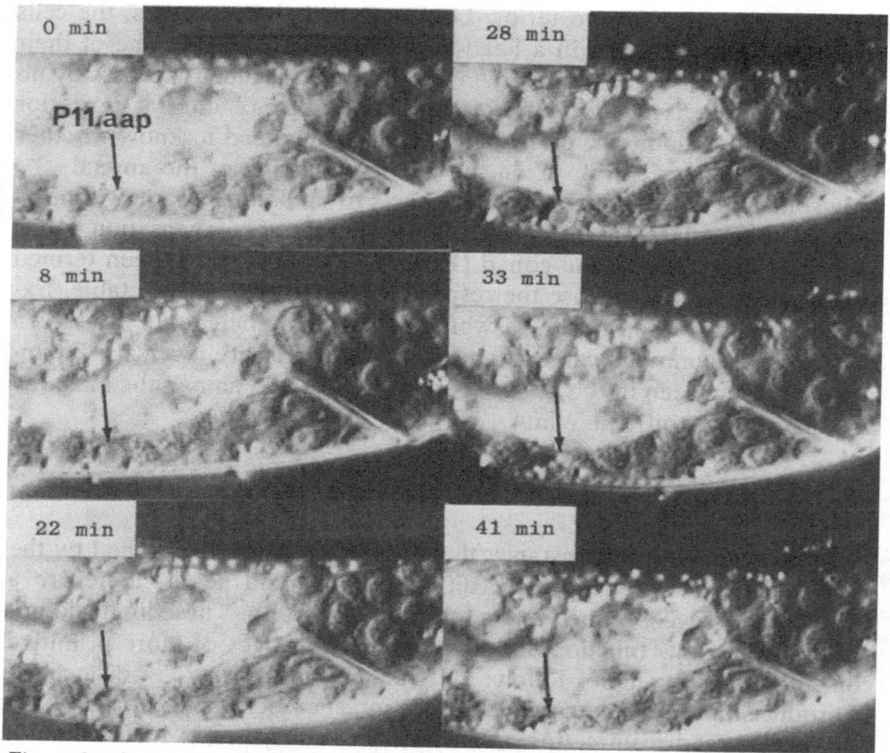

Figure 9 Cell death. Sequential photographs of an L1 hermaphrodite, lateral view (Nomarski optics). Arrow indicates dying cell, P11.aap. Bar represents 20 μm. (Reprinted, with permission, from Sulston and Horvitz 1977.)

Most of the cell deaths do not appear to result from competition for targets, which is an important phenomenon in vertebrates (reviewed by Oppenheim 1981). Evidence comes from the fact that many doomed cells die very soon after their births and from the detailed studies of Robertson and Thomson (1982), who showed that processes did not grow from a sample of doomed neuronlike cells prior to their deaths. In the male, however, there are two pairs of somatic cells that apparently do compete for targets (B.a($^{l}_{r}$)apaav and B.a($^{l}_{r}$)d; see Table 1). There is also a high incidence of death in the germ line of aging or starved animals, which may well be a competitive affair.

A. Tissue Specificity

The majority of deaths are ectodermal, and, in particular, many of the doomed cells are closely related to neuroblasts. There are mesodermal cell deaths in the embryo of *C. elegans* (and in the larva as well in the case of *P. redivivus*). There are no endodermal cell deaths in *C. elegans*, but they have been seen in the embryos of *P. redivivus* and *Turbatrix aceti* (Sulston et al. 1983). In each sex of *C. elegans*, a single cell (the anchor cell of the hermaphrodite and the linker cell of the male) dies in the somatic gonad, having first been responsible for joining the gonad to the appropriate point on the body wall. *P. redivivus* has additional deaths in the somatic gonad, related to its monodelphic form, as discussed in Section IV.D. In both nematodes, some germ cells die (see above).

B. Why Discard Cells?

The loss of cells that have already performed some function is easy to understand: Biology is full of disposable structures ranging from individual cells to entire organisms. *C. elegans* is no exception: The tail-spike cells die in the embryo after generating a bundle of filaments that appear to scaffold the finely tapering tip, and the male linker cell dies at the final molt after leading the vas deferens to the cloaca and mediating its attachment thereto.

Cells that are born only to die need some other explanation. One can postulate that they all function transiently, but both time and opportunity seem to be lacking in many cases (e.g., Robertson and Thomson 1982). The most attractive hypothesis at present is that the deaths are a means of discarding unwanted information (either nuclear or cytoplasmic) in much the same way that sets of chromosomes are discarded in polar bodies at meiosis. This concept is part of the sublineage hypothesis (see Section X).

In some cases, there is a marked delay between the birth and the death of a cell, even though it appears to serve no function in the intervening period. The sexually dimorphic deaths in the embryo are of this sort. Such a delay in the initiation of the editing process suggests, though it does not prove, the existence of an extrinsic signal.

C. Why Kill Cells?

A question that is sometimes raised is how far the organism is advantaged by killing and scavenging unwanted cells, which, although forming a substantial proportion of the total cell number, account for less than 1% of the adult biomass. The existence of mutations in the genes *ced-3* and *ced-4* (Chapter 6; Ellis and Horvitz 1986) has invalidated the suggestion that deaths are strictly necessary to prevent unwanted cells dividing or otherwise interfering with development (though lesser effects cannot be ruled out, and death of the linker cell, which does not require wild-type *ced-3* or *ced-4* activity, may be desirable for male fertility). In truth, a combination of factors may have been responsible for selection of the cell death program, including marginally deleterious effects of some superfluous cells and a greater significance of biomass recovery at certain stages (e.g., in the embryo, and in starved animals, where germ cell deaths assume greater significance). Once the program was available, it would plausibly have come to be used in all relevant circumstances.

D. Mechanism

The majority of cell deaths are probably suicides (Horvitz et al. 1982a). They are prevented neither by mutations that block phagocytosis (Hedgecock et al. 1983) nor by the previous killing of neighboring cells, and most of them occur only 30–90 minutes after the doomed cells are born. In some cases, the deaths are clearly anticipated by the mother cells: For example, early cell deaths in the embryo are presaged by grossly unequal cell divisions, from which the doomed cells emerge with only a tiny fraction of the maternal cytoplasm.

There are two deaths, however (and there may well be others yet to be discovered), that are known to be murders, because they can be prevented by the prior ablation of other cells. The first is that of the male linker cell, which is necessary for leading the growth of the male gonad and for attaching it to the cloaca. This attachment requires a second specific cell type derived from U (Sulston et al. 1980). Following attachment, one of the U derivatives engulfs and destroys the linker cell. If U is ablated at hatching, neither attachment nor destruction of the linker is seen, and the animal is infertile. The second murder is the death of B.alapaav or B.arapaav, one of which is engulfed by P12.pa while the other survives as a component of the cloaca; again, in the absence of P12.pa, both cells survive.

E. Sexual Dimorphism

The embryo and the larva generate sexual dimorphism in rather different ways. The former generates the same set of cells in both sexes and then (after a delay) kills those not needed; the latter takes the same set of blast cells but creates different patterns of cell lineage and fate in the two sexes. The sexual differences in the embryo are slight, so the throwaway procedure makes greater economic sense at that stage. It is interesting that in embryos of both sexes, the unwanted cells begin to differentiate before their deaths: the four male-specific cells by sending processes into the cephalic sensilla, and the two hermaphrodite-specific cells by migrating appropriately.

VII. MIGRATIONS

Three sorts of migration can be distinguished:

1. Concerted movements of cells. These occur during gastrulation, in the generation of the anterior sensilla at the comma stage, and in the maturation of the male tail at L4 lethargus.
2. Individual movements of cells. Individual cells frequently move short distances past their neighbors, but only a few undertake long-range migrations. In other words, the cell lineage places most cells in, more or less, their correct relative positions. The most notable exceptions are
   a. In the embryo: the postembryonic mesoblast M, its contralateral homolog, the somatic gonad precursors Z1 and Z4, and some lateral neurons.
   b. In developing larvae: progeny of $Q(^{\mathrm{L}}_{\mathrm{R}})$, and the hermaphrodite and male sex myoblasts.
3. Nuclear migrations. Two cell classes display reproducible and accurately timed nuclear migrations: namely, the dorsal hypodermal cells in the embryo and the ventral neuroblasts in the L1. Mutations at two loci (*unc-83* and *unc-84*) interfere with both processes (Chapter 6).

VIII. EMBRYONIC GERM LAYERS

Simple as it is, the nematode is clearly triploblastic, as judged by the behavior of its cells at gastrulation. The embryonic endoderm is generated by founder cell E; the mesoderm by MS, Cap, Cpp, and D; and the ectoderm by AB, Caa, and Cpa. However, the part of AB that contributes to the pharynx enters the interior at the end of gastrulation and could be classified either as embryonic ectoderm or embryonic mesoderm.

By and large, the terminal progeny of these precursors differentiate in a manner characteristic of their own germ layers, for example, the nervous system is principally formed from embryonic ectoderm, and the musculature from embryonic mesoderm. A few cells, however, transgress these supposed developmental boundaries: Some of the pharyngeal muscles, the rectal muscles, and one body muscle are formed from AB, whereas some of the pharyngeal neurons are formed from MS (Fig. 2).

IX. SEGMENTATION?

During morphogenesis of the embryo, a periodic repeat in the hypodermis of the body becomes apparent. Although the precursors (Vn, Pn) involved in this repeat give rise to repetitive sublineages, no trace of a parallel periodicity is seen in the embryonic cell lineage, and the wild-type nematode retains little periodicity at maturity. Nevertheless, the hypodermal repeat is an obvious unit structure, and programming of the number of repeats and their subsequent specialization could have been an important mechanism in evolution. It will be interesting to discover whether there is any relationship at all between the limited "segmentation" of the nematode and the true segmentation seen in higher animals.

X. SUBLINEAGES

If the observed indications of cell autonomy are not misleading, it follows that much terminal development, both in the embryo and in the larva, takes place via intrinsically determined sublineages, often referred to simply as sublineages. This means that many precursor cells, whether designated by cell interaction or by ancestry, undergo a fixed pattern of cell divisions without further reference to their neighbors, and give rise to a more or less fixed array of progeny cell types.

That intrinsic programming of precursors can occur is suggested by the repeated appearance of the same sublineage at different times and in different places—for example, the ray sublineage, the ventral cord sublineage, and the hypodermal sublineage of *P. redivivus*. That it does occur has been confirmed in particular cases by extensive laser ablation studies—for example, in the rays, the hook, and the gonad.

Purely intrinsic programming is a particular extreme, which real lineages may approach more or less closely. Of course, at any given stage, the growth and division of cells depend upon an appropriate nutrient environment and global controls (see above). The point is that many blast cells appear to have but one possible fate within the context of the nematode.

Implicit in the notion of sublineage is the possibility of diagnosing the commitment of precursor cells by observation of the lineages to which they give rise. At present, this approach is very important. It has been used, for example, to identify ray precursors, the two kinds of vulval precursors, the various sorts of lateral neuroblasts, and so on. Without the use of sublineages as cell markers, much less progress would have been made in analyzing regulation, homoeotic mutants, and heterochronic mutants (Chapter 6). In due course, however, this technique will be supplemented, and to some extent supplanted, by more direct methods of cell identification such as antibody recognition. This is particularly true of the embryo, where division patterns are more difficult to observe and are less clearly diagnostic than they are in the larva.

XI. HOMOLOGY, ANALOGY, AND SYMMETRY

An important theme of the cell lineage is the interplay between homology and analogy—that is, between similarity of ancestry and similarity of fate. For the purposes of this discussion, we extend the traditional meaning of analogy, which is usually restricted to similar structures of dissimilar origin, to include similar structures of similar origin. Fate is inferred from form or behavior or both; for example, two cells may be classed as analogous because they make similar connections in the nervous system or because they give rise to similar sublineages.

For the most part, cells that are homologous are also analogous; such behavior is readily comprehensible in terms of either common inheritance or similar interactions with neighbors. Bilaterally symmetrical cells usually adopt similar fates, as do cells derived from corresponding positions in identical sublineages. More interesting are the places where decisions are not correlated with ancestry; some examples follow:

1. The fundamental plan of the animal is bilaterally symmetrical. Much of the bilateral symmetry can be traced back to a series of equational divisions of early blastomeres, whose daughters give rise, via symmetrical sublineages, to similar structures on the left and right sides. Such pairs of sisters are homologous analogs. The generation of a unique cell necessarily requires that symmetry be broken at some point (Fig. 10). Sometimes, however, additional and seemingly redundant symmetry breakages are attached to the functional event (see Fig. 20 in Sulston et al. 1983). In other places, especially in the head of the embryo, symmetrical sublineages give rise to the left and right sides, but their precursors do not have similar ancestries: They are nonhomologous analogs. At these points, symmetry is broken and then remade with differently assorted components (Fig. 11).

   During late embryonic and early larval development, symmetry breakage often involves confrontation of homologs across the midline of the animal, and the outcome (as to which homolog adopts which fate) is decided by cell interaction (see above). At other points, notably in the young embryo, the homologs are not in contact and the outcome does

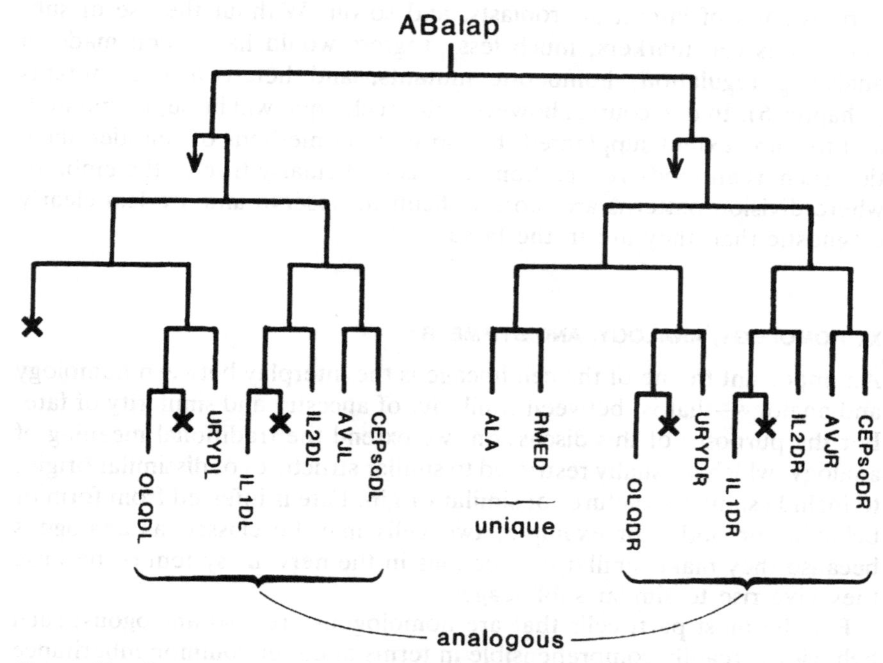

Figure 10 Generation of unique cells by symmetry breakage. (Reprinted, with permission, from Sulston 1983.)

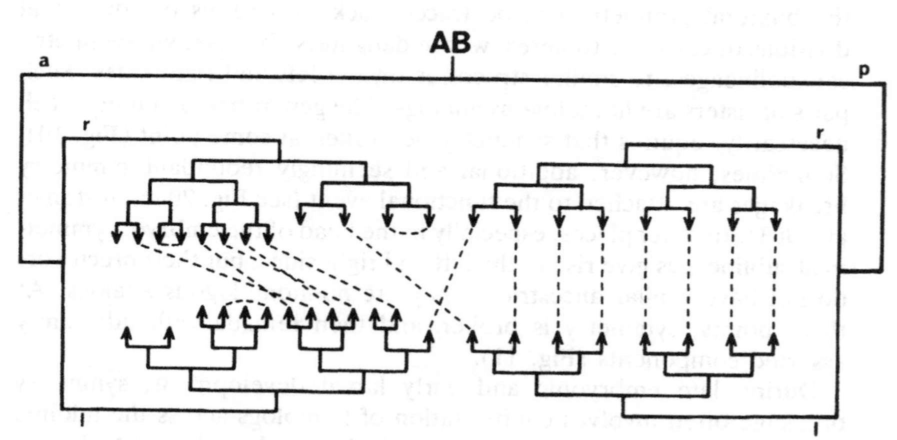

Figure 11 Origin of bilateral symmetry in the embryo. Pairs of precursors that are analogous are linked by dotted lines. ABp gives rise largely to homologous analogs; ABa to nonhomologous analogs. (Reprinted, with permission, from Sulston 1983.)

not seem to involve cell interaction at the time of determination. Thus, it appears that the accurate bilateral symmetry of the mature animal arises in part from a programmed series of separate decisions.

2. The generation of rotational symmetry also requires prior breakage of bilateral symmetry. The vas deferens and the pharynx both have threefold rotational axes of symmetry, but they are constructed according to entirely different rules. In the vas deferens, three equivalent precursors generate identical sublineages. In the pharynx, part of the structure is formed by bilaterally symmetrical lineages, and the rest arises in a piecemeal manner.
3. Some sublineages, instead of (or as well as) being repeated in a symmetrical fashion, appear at several points along the length of the animal (see above). Once again, the precursors of a given repeated sublineage are nonhomologous analogs.

XII. LINEAGE PATTERNS

A danger in discussing lineage patterns is that it is too easy to give the impression that cell divisions per se are of primary importance. It is incorrect, for example, to say that division pattern determines cell fate. A more accurate statement is that ancestry, in a number of cases, determines fate and that its influence is expressed via particular sequences of cell divisions. The lineage shows, however, particularly in the embryo, that a given cell fate can be reached by more than one ancestral route. Indeed, the variety of division patterns that are observed (1) for homologous cells, (2) in the formation of analogous cells, and (3) in the generation of similar structures (e.g., sensilla) suggest that there simply are no overall rules and that perhaps any combination of division patterns and fates is possible in principle.

Why, then, should we think about lineage patterns at all? First, we use sublineages to characterize precursors, and we use differences in them to characterize mutants. Second, however elaborate the patterns are, they must be controlled by distinct instructions. Third, although any pattern may be possible, presumably some are more probable than others, because they either require less complex instructions or, for some other reason, can arise more easily in the course of evolution. It is, therefore, worth looking for patterns that arise frequently and for simple transformations between them. Kimble (1981b) has reviewed this subject, which has also been discussed extensively by Sternberg and Horvitz (1981, 1982) in connection with *P. redivivus*. Here, we mention only the three most important patterns:

1. In the simplest pattern, the clone, a given precursor generates cells of a single type. In an exclusive clone, all the progeny of a single precursor are used to generate a single tissue that has no other source of cells.

Two important examples are the germ line and the intestine, both of which are exclusive clones derived from the early embryonic blast cells P_4 and E, respectively. These are exceptional cases, for the majority of tissues are polyclonal in origin. However, there are many examples of partial clonal derivation, in which either the progeny are not all of the same sort or do not generate an entire tissue—for example, neuroblasts that yield principally one class of motor neuron, and myoblasts that yield clones of body muscle.

2. In the stem cell pattern, a precursor divides to yield one daughter that is like itself and one that is different. This pattern is apparent in the segregation of the germ line during the generation of the founder cells (Fig. 2), in the lateral hypodermis, and in the male seminal vesicle, but as an underlying mechanism it is thought to be far more widespread. It came to assume additional significance with its revelation in certain neuroblast sublineages after mutation of the gene *unc-86* (Chapter 6; Chalfie et al. 1981). This observation raised the possibility that simple division patterns, such as the stem-cell pattern, might be modified repeatedly to yield ultimately the entire complex lineage of the wild-type animal.
3. The third pattern, and perhaps that most characteristic of the nematode, is the repeated sublineage (see Section X). It can arise either as a duplication (i.e., division of a single precursor whose daughters then generate the same sublineage) or by the similar behavior of unrelated cells (nonhomologous analogs).

These patterns, and others derived from them, can be formally interconverted by simple transformations. A major aim of comparative linealogy and studies of lineage mutants is to discover which, if any, of these plausible operations have a basis in fact, and how they are defined biochemically.

XIII. EVOLUTION

In conclusion, it is worth placing the cell lineage of *C. elegans* in an evolutionary perspective. We imagine that early predecessors of the nematode would have used relatively simple lineage patterns. As described above and in Chapter 6, studies of mutants and other species suggest ways in which these patterns may have been modified in the course of natural selection. Some modifications are relatively minor, such as the death of an unwanted neuron; others involve numbers of cells, such as the generation of an additional copy of an existing sublineage by a novel precursor, or a change involving a regulatory cell. Modifications may initially have been accomplished under regulatory control, but if so, the subsequent loss of many of the regulatory elements has led to a less flexible, but perhaps more efficient, system. Both the complexity of the lineage and the limited extent of cell migration can be explained by supposing that new precursors were selected more for their positions than for their intrinsic properties.

ACKNOWLEDGMENTS

The primary consultants for this chapter are Bob Horvitz, Judith Kimble, Einhard Schierenberg, and Paul Sternberg. Many others have contributed valuable comments—particularly Donna Albertson, Marty Chalfie, Ed Hedgecock, Jonathan Hodgkin, Cynthia Kenyon, Jim Priess, Bob Waterston, John White, and Bill Wood. J.S., however, is answerable for all errors, omissions, and inanities.

6

Genetics of Cell Lineage

H. Robert Horvitz
Department of Biology
Massachusetts Institute of Technology
Cambridge, Massachusetts 02139

I. INTRODUCTION

Why isolate *Caenorhabditis elegans* mutants that are abnormal in cell lineage? A biochemist might answer, "To identify molecules involved in generating particular patterns of cell divisions or in specifying particular developmental fates." A physiologist might answer, "To obtain individuals lacking particular cells, so as to help assign functional roles to those cells" (or, conversely, "to obtain individuals with particular functional—or developmental—defects, so as to identify those cells normally needed for those functions"). And a geneticist might answer, "To define the logic of the network of genetic interactions and to examine in vivo the effects of perturbing individual molecular elements of that network." Of course, mutants offer approaches toward all of these goals.

Many cell-lineage mutants of *C. elegans* have been isolated and characterized in detail both developmentally and genetically. The complete knowledge of the cellular anatomy and cell lineage of *C. elegans* (see Chapters 4 and 5 and Appendixes 1–3) has allowed the definition of developmental defects at the level of resolution of single cells. The small size of this animal, as well as its rapid life cycle, self-fertilizing nature, and ease of handling have made sophisticated genetic manipulations possible (see Chapter 2).

In this chapter, we describe some of the methods underlying and conclusions derived from studies of the genetics of *C. elegans* cell lineage. Rather than presenting a comprehensive catalog of the approximately 50 known genes with characterized effects on cell lineage (such information is available from the references cited in Appendix 4), we will instead focus on specific illustrative examples and attempt to provide an overview of what has been learned. A number of reviews are available that describe specific cell-lineage mutants and discuss other aspects of how genes control nematode cell lineage, including the genetic implications both of the wild-type cell lineage and of interspecies differences in cell lineage observed between *C. elegans* and another nematode species, *Panagrellus redivivus* (Horvitz et al. 1983; Sternberg and Horvitz 1984; Fixsen et al. 1985; Greenwald 1985a; Hedgecock 1985; Kenyon 1985b; see also Chapter 5).

II. ISOLATION OF CELL-LINEAGE MUTANTS

Mutants of *C. elegans* altered in cell lineage have been obtained in a variety of ways. Initially, cell-lineage mutants were sought using the dissecting microscope to identify mutants abnormal in specific morphological structures and/or in specific behaviors. Such mutants were then examined for defects in the cell lineages that generate the cells of those structures or that control those behaviors (Horvitz and Sulston 1980). For example, postembryonic cell divisions produce many components of the egg-laying system of the hermaphrodite (the vulva, the vulval and uterine muscles, and some of the neurons that innervate those muscles) (Sulston and Horvitz 1977). Because egg laying is not essential for propagation (eggs can hatch in utero, and larvae can later crawl free from their parent) (Horvitz and Sulston 1980), viable egg-laying-defective mutants could be isolated and examined for cell-lineage defects. In this way, mutants abnormal in the development of all known essential components of the egg-laying system have been obtained (Sulston and Horvitz 1981; Trent et al. 1983). Similarly, searches for mating-defective males (Hodgkin 1983a; see Chapter 9) or touch-insensitive animals (Chalfie and Sulston 1981; see Chapter 11) have also identified cell-lineage mutants.

Several types of direct screens for abnormalities in cell numbers and/or types have been used for the isolation of cell-lineage mutants. The examination, with Nomarski optics, of living animals descended from mutagenized ancestors has led to the identification of mutants with defects in cell-division patterns, cell deaths, or cellular morphology (Hedgecock and Thomson 1982; Hedgecock et al. 1983; Ellis and Horvitz 1986; Kenyon

1986; E. Hedgecock, pers. comm.). The examination of populations of fixed F2 or F3 progeny of mutagenized hermaphrodites stained with Feulgen, Hoechst 33258, or diamidinophenylindole (DAPI) DNA stains has revealed several mutants abnormal in cell number; strains of such mutants have been established from the living siblings, progeny, or parents of the fixed animals (Horvitz and Sulston 1980; W. Fixsen and R. Horvitz, in prep.).

Using these approaches, hundreds of mutants abnormal in postembryonic cell lineages have been isolated. Many additional cell-lineage mutants have been obtained by isolating mutations that interact with previously identified cell-lineage mutations. Intragenic suppressors (Greenwald et al. 1983; Ambros and Horvitz 1984), as well as extragenic suppressors (Ambros and Horvitz 1984; J. Thomas et al., pers. comm.) and enhancers (E. Ferguson and R. Horvitz, in prep.), have been characterized. Table 1 lists those genes known to affect *C. elegans* postembryonic cell lineages, as well as genes known to affect two other aspects of development, namely cell migration and programmed cell death.

In addition to mutants characterized as having specific abnormalities in cell lineage, many mutants have been identified that are grossly defective in development, that is, that fail in embryogenesis or larval growth or that develop into sterile adults (Hirsh and Vanderslice 1976; Meneely and Herman 1979; Miwa et al. 1980; Cassada et al. 1981a). Some of these mutants are known to be abnormal in patterns of early embryonic cell division (Schierenberg et al. 1980; Wood et al. 1980; Denich et al. 1984), and others are cell-lineage mutants as well. In general, these mutants probably define genes that are necessary for embryogenesis but that are not involved in specific aspects of cell lineage; for example, genes needed for basic metabolic functions could well be of this class. In contrast, a few of the genes identified on the basis of defects in embryogenesis (such as those that affect the asymmetry of the first embryonic cell division [Hirsh et al. 1985a; K. Kemphues and J. Priess, pers. comm.; see Chapter 8]) could well have functions that are primarily developmental in nature.

Most genetic studies of cell lineage have focused on the postembryonic cell lineages, in part, because the complete postembryonic lineages (Sulston and Horvitz 1977; Kimble and Hirsh 1979) were elucidated earlier than the complete embryonic lineages were (Sulston et al. 1983). In addition, the postembryonic lineages are easier to study: First, the fates of cells generated by the postembryonic lineages can be readily identified (complete cell-division patterns and subsequent terminal differentiation can be observed without having to contend with the technical difficulties of following cells through late embryogenesis); second, and very importantly, viable and fertile mutant strains abnormal in many and perhaps all of the nongonadal postembryonic lineages can be established.

Table 1 Postembryonic genes that affect cell lineages, cell deaths, and cell migrations

Homeotic, spatial: *ces-3**, *let-23*, *lin-1*, *lin-2*, *lin-3*, *lin-7*, *lin-8*, *lin-9*, *lin-10*, *lin-11*, *lin-12**, *lin-13**, *lin-15**, *lin-16*, *lin-17*, *lin-18*, *lin-20*, *lin-22**, *lin-24*, *lin-25*, *lin-26*, *lin-31*, *lin-32*, *lin-33*, *lin-34*, *lin-35*, *lin-36*, *lin-37*, *lin-38*, *lin-39**, *lin(n300)*, *mab-3*, *mab-5**

Homeotic, temporal (heterochronic): *lin-4*, *lin-14*, *lin-28**, *lin-29*, *unc-86*

Homeotic, sexual: *her-1*, *fem-1*, *fem-2*, *fem-3*, *tra-1*, *tra-2*, *tra-3*, *sdc-1*, *egl-41*, *egl-1**, *lin-8**, *lin-9**, *lin-12**, *lin-13**, *lin-15**, *lin-22**, *lin-28**, *lin-35**, *lin-36**, *lin-37**, *lin-38**, *mab-5**

Blocks: *lin-5*, *lin-6*, *lin-30*, *unc-59*, *unc-85*

Programmed cell deaths: *ced-1*, *ced-2*, *ced-3*, *ced-4*, *ces-3**, *nuc-1*, *egl-1**, *lin-39**

Cell migrations: *egl-5*, *egl-15*, *egl-17*, *egl-18*, *egl-20*, *egl-27*, *egl-43*, *egl(n1332)*, *egl(n1392)*, *emb(e1933)*, *hch-1*, *lin-21*, *lin-32*, *mab-5**, *mig-1*, *mig-2*, *mig-3*, *mig-4*, *mig-5*, *mig(ct41)*, *mig(ct78)*, *lin-21*, *hch-1*, *unc-5*, *unc-6*, *unc-11*, *unc-39*, *unc-40*, *unc-73*, *unc-83*, *unc-84*, *vab-8*

Other or insufficiently characterized to classify: *lin-19*, *lin-23*, *lin-27*, *mab-2*, *mab-9*

Genes indicated with an asterisk (*) are listed in multiple categories. For example, *lin-28* mutations cause heterochronic development but also lead to the formation of normally male-specific rays in the hermaphrodite; *lin-28* is shown both as homeotic, temporal and as homeotic, sexual. Similarly, the *lin-39(n709)* mutation causes those cells that would normally differentiate into VC neurons to express the fate of their spatial homologs and undergo programmed cell death instead; thus, *lin-39* is shown in both the homeotic, spatial and the programmed cell death categories.

III. CHARACTERIZATION OF CELL-LINEAGE MUTANTS

A mutant identified on the basis of a particular cell-lineage defect must be characterized both phenotypically (to establish the nature and extent of its developmental abnormalities) and genotypically (to reveal both how that mutation perturbs normal gene action and how the gene defined by that mutation normally functions). The approaches taken to characterize phenotypically an initial set of 24 cell-lineage mutants are indicative of the general strategies that have been used (Horvitz and Sulston 1980; Sulston and Horvitz 1981). First, aspects of gross anatomy were examined: general morphology, the presence of specific anatomical structures (such as the vulva), and the presence of specific, easily scored cells (the ventral motor neurons and certain peripheral dopaminergic neurons). Second, simple behaviors were observed: locomotion, egg laying, and touch sensitivity.

Third, and to some extent directed by the findings of the studies already completed, anatomy and cell lineages were examined using Nomarski optics. In the course of this work, it was observed that all of these mutants were variable in their phenotypes, that is, the mutations in these mutant strains were of incomplete penetrance (not all individuals displayed a mutant phenotype) and/or variable expressivity (the degree of the mutant phenotype expressed varied among individuals). This variability in phenotype, which is observed even in "null" (complete loss of function) mutants, has continued to characterize the many cell-lineage mutants that have been isolated since. The contrast between the variable lineages of these mutants and the essentially invariant lineage of the wild-type animal (as well as the relatively invariant phenotypes of other *C. elegans* mutants) remains unexplained and intriguing.

The genetic characterization of cell-lineage mutants involves the same types of experiments utilized to characterize any mutant (see Chapter 2). First, mutant phenotypes are identified as dominant or recessive, tested for temperature sensitivity, and examined for single gene Mendelian inheritance. Next, the mutation responsible for the phenotype is mapped and tested for its ability to complement mutations in closely linked genes. The nature of the mutant allele must then be established (does it result in a loss, an increase, or an alteration in gene function?) and the null (complete loss of function) phenotype determined. The time of gene action can be investigated by performing temperature-shift experiments using temperature-sensitive alleles, and the site of gene action can be examined by constructing genetic mosaics or by using a laser microbeam (see Chapter 5) to see whether the expression of the mutant phenotype can be eliminated by the ablation of a cell or set of cells other than those that display the abnormality (two examples of this type of experiment are described below). Finally, interacting genes can be identified by obtaining suppressor and enhancer mutations. Examples of these types of genetic analyses of cell-lineage mutants can be found in several published manuscripts (Horvitz and Sulston 1980; Greenwald et al. 1983; Ferguson and Horvitz 1985; Kenyon 1986; Ferguson et al. 1987).

IV. OVERVIEW OF CELL-LINEAGE MUTANTS

Perhaps the most striking characteristic of *C. elegans* mutants abnormal in postembryonic cell lineages is their existence. A priori, one might have imagined that any mutation that altered patterns of cell divisions and cell fates might so generally perturb development as to result in embryonic lethality. The large number of viable and fertile mutant strains abnormal in cell lineage clearly refutes this expectation. Alternatively, it might be thought that the lineage specificity seen in these mutants reflects a differential, rather than an absolute, need for particular gene functions in different cell lineages; in other words, perhaps the existing mutations have reduced, but not eliminated, the functions of particular genes, and true null mutations would indeed result in a failure in embryogenesis. Again, available data argue against this possibility as a general explanation for the specificity observed: Genetic analyses of many of the genes defined by cell-lineage mutants have established the null phenotypes of these genes (for a discussion of the criteria employed to identify null alleles, see Chapter 2), and many appear not to have embryonic lethal null phenotypes (Horvitz and Sulston 1980; Greenwald et al. 1983; Ambros and Horvitz 1984; Ferguson and Horvitz 1985; Kenyon 1986). Thus, we are left with the conclusion that different genes function in different cell lineages.

It is important to note that the prevalence of viable and fertile cell-lineage mutants should not be interpreted as evidence that most genes involved in cell lineage are so specific in their actions. In general, cell-lineage mutants have been sought by isolating individuals that generate viable and fertile strains. Thus, only mutations that do not affect most cell divisions have been selected for study.

A second interesting attribute of those cell-lineage mutants analyzed so far is that most of them can be considered to be "homeotic." Specifically, most are missing cells of certain types and have extra copies of cells of certain other types. On closer examination, these mutant phenotypes have proved to result from transformations in cell fates: A particular cell, A, adopts the fate of another cell, B, resulting in the loss of A (or of the cells normally generated by A) and the duplication of B (or of the cells normally generated by B). We consider such transformations to be homeotic by analogy with insect homeotic mutants, in which one body part is replaced by another part normally found elsewhere in the animal (see, e.g., Ouweneel 1976; Morata and Lawrence 1977).

Some *C. elegans* mutants are obviously homeotic at the level of gross morphology, for example, "multivulva" mutants have extra vulvalike structures (Fig. 1a,b) (Horvitz and Sulston 1980; Ferguson and Horvitz 1985), and the mutant *lin-22(n372)* has extra copies of specific sensory structures—the postdeirids and, in males, the rays (Fig. 2) (Horvitz et al. 1983; W. Fixsen and R. Horvitz, in prep.). However, many nematode mutants can be recognized as homeotic only when studied at the level of single cells. For example, an animal homozygous for a loss-of-function mutation in *lin-12* appears in the dissecting microscope to be scrawny, with a large protrusion at the normal position of the vulva (Fig. 1c); when examined developmentally, using Nomarski optics, animals of this genotype can be seen to express homeotic transformations in the fates of at least ten sets of cells (Greenwald et al. 1983). In general, genes with homeotic effects seem likely to act in the specification of cell fates; for each of these genes, the wild-type allele results in one fate, and a mutant allele results in an alternative fate.

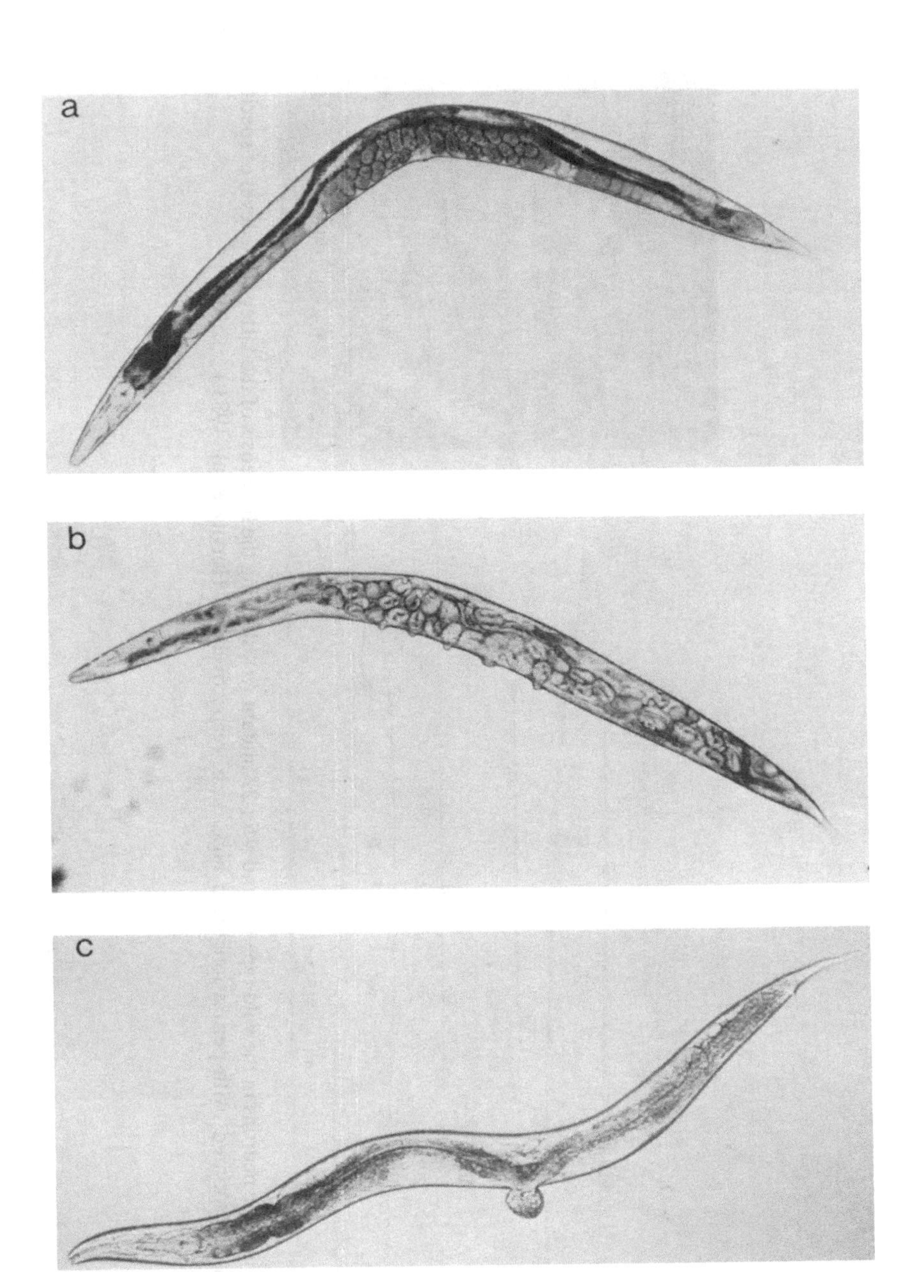

Figure 1 Bright-field photomicrographs comparing the wild type (*a*) to a multivulva mutant, *lin-15(n309)* (*b*), which shows an obvious homeosis at the level of gross morphology, and to a *lin-12(null)* mutant, *lin-12(n137 n720)* (*c*), which has a homeosis visible only at the level of individual cells. (*a,b*) Reprinted, with permission, from Figs. 1a and 3f, respectively, in Ferguson and Horvitz (1985). (*c*) Reprinted, with permission, from Fig. 1c in Greenwald et al. (1983).

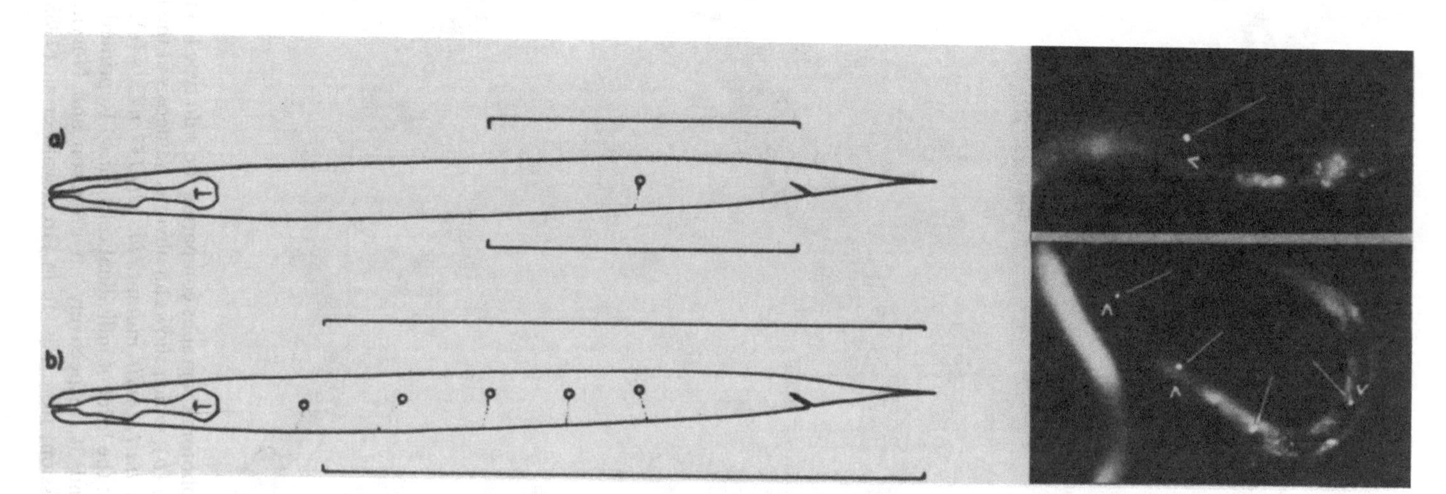

Figure 2 Dopaminergic neurons in the wild type (*a*) and a *lin-22* mutant (*b*), showing the homeosis of the latter at the level of specific identified neurons. (Reprinted, with permission, from Figs. 3a,b, respectively, in Horvitz et al. 1983.)

At this point, we should discuss our use of the word fate. All cells have fates. The fate of a cell produced by a terminal division is either to differentiate into a cell of a specific type or to undergo programmed cell death (Sulston and Horvitz 1977; Horvitz et al. 1982a). The fate of a cell produced at an intermediate point in a cell lineage is to generate a specific pattern of cell divisions and a specific set of descendant cells. It is worth noting that our use of the word fate has derived from the observation that many cell-lineage mutations cause certain blast cells to behave like certain other blast cells. Thus, as discussed above, we interpret such mutants as displaying homeotic transformations in the fates of specific cells. This definition of fate also allows us to consider the *C. elegans* cell lineage as a map of the decisions that occur during development: Every division involves a cell with one fate dividing to produce two cells that, in general, differ both from each other and from their parent cell in their fates (Fig. 3a). Thus, a major step toward understanding the genetic basis of the *C. elegans* cell lineage would be the identification and characterization of genes involved in causing sister cells or mother and daughter cells to have different fates. Mutations in such genes would presumably cause sister cells or mother and daughter cells to have the same fates (Fig. 3b,c). As noted below, a number of mutations of these types have been identified (e.g., *lin-11*, *lin-17*, *lin-18*, *lin-26*, and *unc-86*).

A third striking feature of cell-lineage mutants is the nature of their pleiotropies. Most genes analyzed so far affect multiple cell lineages. Nonetheless, mutations in only a few of these genes (*lin-5*, *lin-6*, and perhaps *unc-59* and *unc-85*) appear broadly pleiotropic and are likely to affect general cellular processes such as DNA replication and cytokinesis (Albertson et al. 1978; Sulston and Horvitz 1981; White et al. 1982). Many

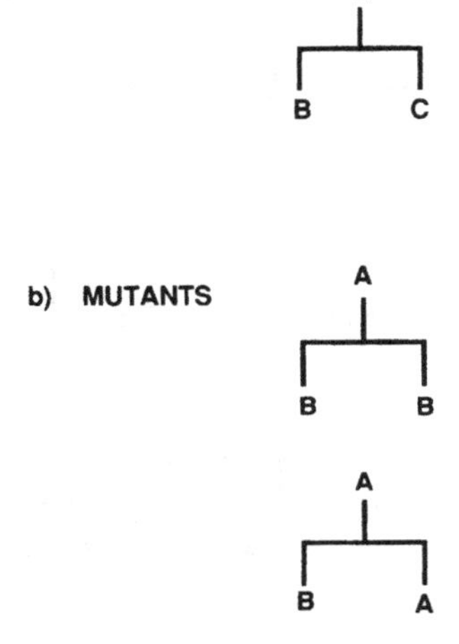

Figure 3 (*a*) Any cell division in the *C. elegans* cell lineage can be considered as involving a mother cell with one fate (A) dividing to produce daughter cells with fates B and C. (*b*) A mutant in which A divided to produce two Bs would be a candidate for defining a gene that normally functions to make the B and C sisters differ in their fates. A mutant in which A divided to produce one B and one A would be a candidate for defining a gene that normally functions to make the C daughter differ from its mother in its fate. (Adapted from Fig. 1 in Hedgecock 1985.)

other genes of this class clearly exist (M. Chalfie et al., pers. comm.), but attention so far has focused on genes that appear to be more specific in their actions.

In most cell-lineage mutants that have been studied in detail, the sets of cells affected and/or the nature of the defects expressed by those cells affected can be seen to be related (although often with some seemingly exceptional cases). For example, *unc-83* and *unc-84* mutations affect two sets of cells that undergo a particular type of migration (Sulston and Horvitz 1981; see below). *unc-86* mutations cause a specific set of neuroblasts to express the fates of their own mother cells (Fig. 4) (Chalfie et al. 1981). *lin-17* mutations cause specific pairs of sister cells that would normally express different fates to express the same fate (Fig. 5) (Sternberg and Horvitz 1984 and in prep.; Ferguson et al. 1987). The *lin-22(n372)* mutation affects four pairs of morphologically similar lateral ectoblasts that normally undergo the same cell lineage (Fig. 6) (Horvitz et al. 1983; W. Fixsen and R. Horvitz, in prep.). *mab-5* mutations cause a variety of posterior-specific cells to express normally anterior-specific fates (Kenyon 1986). The *lin-32(e1926)* mutation causes a variety of neuroblasts to act as nonneural ectoblasts (C. Kenyon and E. Hedgecock, pers. comm.). *lin-12* mutations affect a number of sets of cells with fates specified by cell–cell interactions (Greenwald et al. 1983; Horvitz et al. 1983). *lin-14* and *lin-28* mutations affect a variety of cell lineages, in each case apparently causing a set of stage-specific developmental events to occur precociously (Ambros and Horvitz 1984; see below). In many of these cases, the nature of the pleiotropies has provided the basis for proposing how these genes might function.

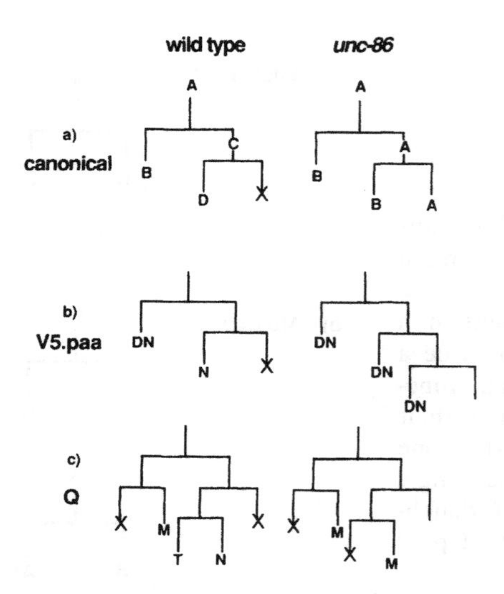

Figure 4 *unc-86* mutations cause certain daughter cells to express the fate of their own mother cells. In the wild type (*left*), cell A divides to produce an anterior daughter B and a posterior daughter C. In *unc-86* animals (*right*), cell A divides to produce a posterior daughter like itself, which then divides to produce an anterior B cell and another posterior A cell. (*a*) Canonical lineage; (*b*) V5.paa lineage; (*c*) Q lineage. (X) Programmed cell death; (DN) dopaminergic neuron; (N) nondopaminergic neuron in *b*; (M) migrating neuron; (T) "touch" neuron; (N) other neuron in *c*. (Adapted from Fig. 4 in Horvitz et al. 1983.)

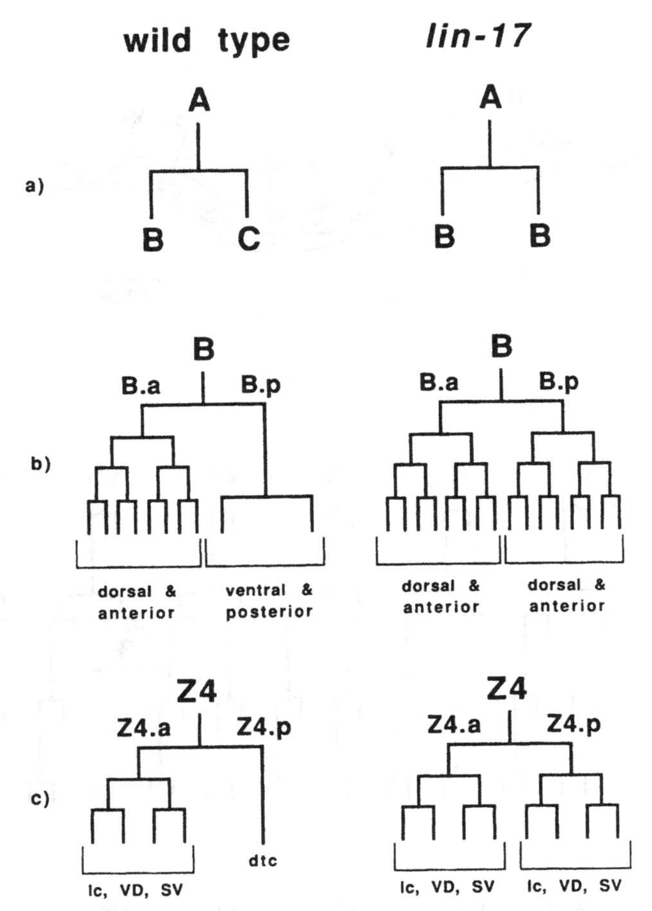

Figure 5 *lin-17* mutations cause certain sister cells to express the same fates. In the wild type (*left*), cell A divides to produce an anterior daughter B and a posterior daughter C. In *lin-17* animals (*right*), cell A divides to produce two identical daughter cells. (*a*) Canonical lineage. (*b*) Male B lineage. In wild-type males, B divides asymmetrically to produce daughter cells of different sizes, B.a and B.p. By the L2 molt, B.a produces eight progeny that lie dorsally and anteriorly within the developing proctodeum, whereas B.p produces two progeny that lie ventrally and posteriorly. All ten progeny undergo several rounds of division during the L3 stage (not shown). In *lin-17* males, B divides to produce daughter cells of the same size, each of which produces eight progeny that lie dorsally and anteriorly by the L2 molt. (*c*) Male Z4 lineage. In wild-type males during the first two larval stages, Z4 produces one cell that becomes a distal tip cell (dtc) and another that generates four cells, including two seminal vesicle cells (SV), a vas deferens precursor cell (VD), and a cell that becomes either a linker cell (lc) or a VD cell. In *lin-17* males, Z4 can generate eight progeny during the first two larval stages. Generally, these animals have no dtc and extra lc, VD, and SV cells. (Adapted from Fig. 1 in P. Sternberg and R. Horvitz, in prep.)

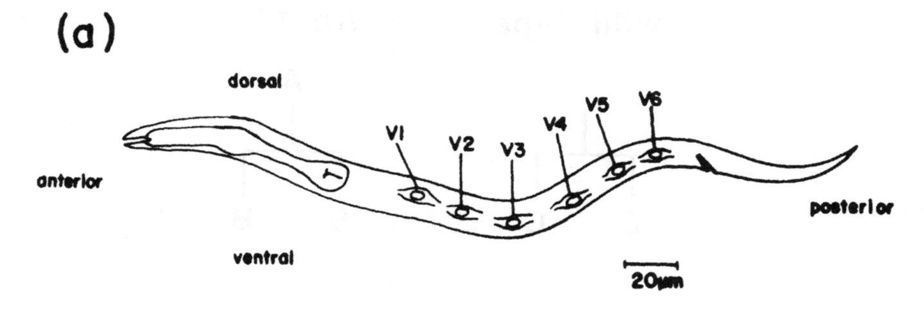

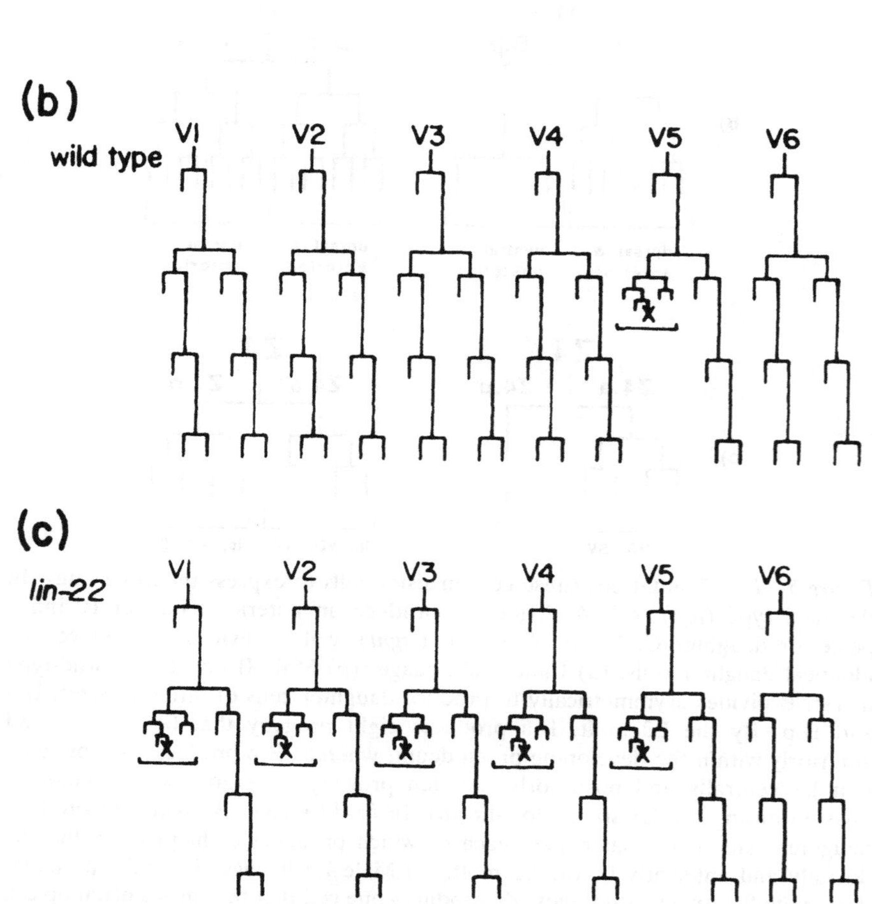

Figure 6 In a *lin-22* mutant, the fates of lateral ectoblasts V1–V4, but not of V6, are transformed to that of V5. (*a*) Schematic left lateral view of a first-stage larva showing the positions of V1–V6. (*b*) V1–V6 lineages in the wild type. (*c*) V1–V6 lineages in a *lin-22* mutant. (X) Programmed cell death. (Adapted from Fig. 2 in Horvitz et al. 1983.)

The homeotic transformations observed in *C. elegans* cell-lineage mutants include spatial, temporal, and sexual transformations in cell fates. In a spatial transformation, a cell adopts the fate of a cell elsewhere in the animal. Such spatial transformations can involve pairs of cells that are morphologically similar (*lin-20*, *lin-22*; Hedgecock 1985; W. Fixsen and R. Horvitz, in prep.), developmentally homologous (*lin-12*; Greenwald et al. 1983), and/or lineally related in a specific way, for example, sisters (*lin-11*, *lin-17*, *lin-18*, *lin-26*; Fixsen et al. 1985; Ferguson et al. 1987; P. Sternberg and R. Horvitz, in prep.). In many cell-lineage mutants, such as most of those that affect vulval development (see below), the cells involved in spatial transformations are both morphologically similar and developmentally homologous. In a temporal transformation, a cell adopts the fate of a cell present at another time in development. Mutants expressing such temporal transformations are said to be "heterochronic" (Ambros and Horvitz 1984), by analogy with the evolutionary concept of heterochrony, which involves interspecies differences in the relative timing of developmental events (Gould 1977). Heterochronic mutants can be either "precocious" or "retarded." In precocious mutants (*lin-28* and recessive *lin-14* mutants), certain stage-specific events occur during earlier stages than in the wild type (and certain normally early events do not occur at all), whereas in retarded mutants (*lin-4*, *lin-29*, and dominant *lin-14* mutants), certain stage-specific events occur during later stages than in the wild type (and certain normally later events do not occur). In a sexual transformation, a cell in one sex adopts the fate of a cell in the other sex. There exist a variety of mutants altered in some or all of the sexually specific cell lineages. Most of these mutants (e.g., *sdc-1*, *her-1*, *fem-1*, *fem-2*, *fem-3*, *tra-3*, *tra-2*, *tra-1*, and *egl-41* mutants) are discussed in detail in Chapter 9; for this reason, we will not consider them in this chapter. Similarly, a number of mutants isolated on the basis of defects in male-specific mating behavior and found to be abnormal in cell lineage (Hodgkin 1983a) are described in Chapter 9, rather than in this chapter.

V. DEVELOPMENTAL CONTROL GENES

That a mutation causes a homeotic transformation does not necessarily imply that the gene defined by that mutation normally has as its primary function the control of developmental decisions. A homeotic gene could simply define a necessary step in a pathway that specifies one particular developmental outcome as opposed to an alternative, default outcome. For example, some of the homeotic genes that affect the development of the vulva of the *C. elegans* hermaphrodite (Sulston and Horvitz 1981; Ferguson and Horvitz 1985; Ferguson et al. 1987) could be of this class. Vulval development is normally induced by the gonadal anchor cell (Sulston and White 1980; Kimble 1981a). If the anchor cell is ablated using a laser microbeam, the vulval precursor cells fail to express their normal fates (vulval cell lineages) and, instead, express another fate (a particular nonvulval cell lineage). In other words, the absence of an anchor cell leads to a homeotic transformation in the fates of the vulval precursor cells. Mutations that transform vulval precursor cells from their normal fates to their alternative potential fate could act by disrupting the activity of any gene required for the generation and/or functioning of the anchor cell. Clearly, not all such genes need function primarily in the control of developmental decisions.

How then can one identify genes that control development? One plausible criterion is the existence of genetically opposite mutant states, for example, null alleles, on the one hand, and overproduction or constitutive-like alleles, on the other (see Chapter 2), that result in opposite mutant phenotypes. Genes defined in this way could not simply provide functions that are necessary prerequisites for particular developmental decisions. In the example above, the constitutive expression of a gene required to generate a functional anchor cell would not generally cause nonvulval cells to express vulval cell lineages, the transformation opposite to that caused by an absence of gene activity. We discuss below two genes that affect cell lineage and fulfill this criterion, *lin-12* and *lin-14*. Other genes defined by homeotic mutations may also function as developmental control genes, but, as discussed above, alternative roles for such genes cannot be excluded.

A. *lin-12*, a Gene That Specifies the Fates of Cells with Fates Controlled by Cell Interactions

The *lin-12* locus was defined by mutations that affect the vulval cell lineages (Ferguson and Horvitz 1985). The genetic analysis of this locus (Greenwald et al. 1983) followed the general strategy outlined in Section III. First, because the dominant effects of the original *lin-12* alleles suggested that they were not null alleles, recessive alleles were sought and obtained as intragenic suppressors of the dominant phenotype. Second, many of these recessive alleles were shown to be null alleles. Third, the dominant phenotype caused by the original alleles was shown to be enhanced when in *trans* to a wild-type allele, as compared with a null allele, suggesting that the original alleles resulted in an increase of an essentially normal gene activity. Thus, these genetic experiments identified opposite classes of *lin-12* alleles, those that eliminate *lin-12* function (*lin-12[0]*) and those that cause an overproduction or constitutivelike expression of *lin-12* function (*lin-12[d]*).

These two classes of *lin-12* alleles result in opposite homeotic transformations in cell fates (Greenwald et al. 1983; Horvitz et al. 1983). For example, if we consider certain pairs of cells, represented as cell a and cell b, that normally adopt distinct fates, A and B, the fates of these cells in the wild type and in *lin-12* mutants are

	Cell a	Cell b
lin-12(d)	A	A
lin-12(+)	A	B
lin-12(0)	B	B

Considered in conjunction with the genetic experiments described above, these observations indicate that for those cells affected by *lin-12*, not only is *lin-12* activity necessary for the expression of the A fate (the absence of *lin-12* activity causes a cell that normally expresses the A fate to express the B fate instead), but it is also sufficient for the expression of the A fate (an increase in *lin-12* activity causes a cell that normally expresses the B fate to express the A fate instead). Thus, the level of *lin-12* activity specifies cell fates: A high level specifies the A fate, and a low level specifies the B fate.

There are at least ten sets of cells that are affected by *lin-12* mutations in this way. One example is provided by two embryonic blast cells, AB-plapaapa and ABprapaapa (for cell nomenclature, see Chapter 5 and Appendix 1). One of these cells normally becomes an ectoblast (G2), and the other normally becomes a neuroblast (W). In *lin-12(d)* mutants, both cells express the G2 fate, and in *lin-12(0)* mutants, both express the W fate (Fig. 7). G2 and W are members of an equivalence group (see Chapter 5); specifically, if a precursor to ABplapaapa (which normally becomes G2) is ablated using a laser microbeam, the other cell (which normally becomes W) can express the G2 fate. This result indicates that these two cells share at least one developmental potential and that the fate expressed by the normal W precursor depends on an interaction with the normal G2 precursor.

Of the sets of cells affected by *lin-12*, many have been demonstrated in this way to have fates that are controlled by cell interactions. Thus, *lin-12* may function in intercellular communication. A molecular analysis of the *lin-12* gene (Greenwald 1985b) has strongly supported this hypothesis. Based on the DNA sequence of the *lin-12* gene, the *lin-12* protein product appears to be structurally related to a family of mammalian proteins that includes epidermal growth factor (EGF) and the low-density lipoprotein (LDL) receptor. Given the current understanding of the localization and activities of the members of this protein family, it seems likely that at least one domain of the *lin-12* product is located extracellularly as a part of a secreted or integral membrane protein. Such a protein would be an excellent candidate for mediating intercellular communications.

The time of *lin-12* action has been examined by performing temperature-shift experiments using a temperature-sensitive allele (Greenwald et al. 1983). The temperature-sensitive periods for the effects of *lin-12* on two

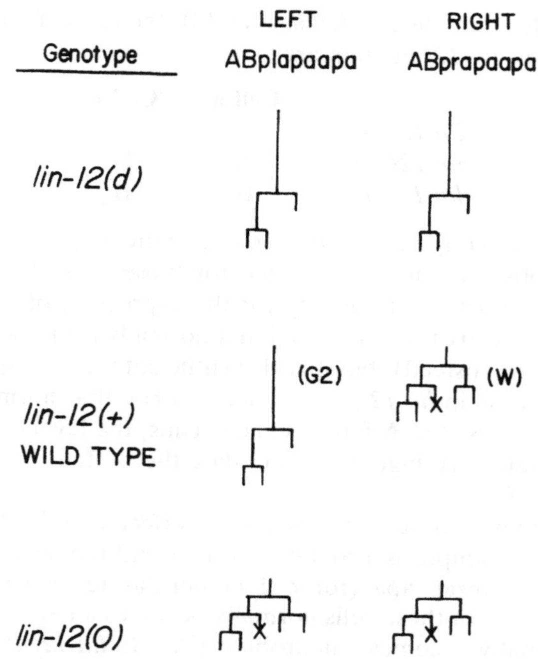

Figure 7 The fates of certain pairs of contralateral homologs are transformed in opposite directions by *lin-12(d)* and *lin-12(0)* mutations. The *lin-12(d)* allele was *n137*, and the *lin-12(0)* allele was *n137 n720*. (W) A neuroblast that divides during the first larval stage; (G2) an ectoblast that divides during the second larval stage. (Adapted from Fig. 6b in Horvitz et al. 1983.)

different sets of cells have been found to correspond to the times of the determination of those cells as ascertained by cell ablation experiments. (The time of irreversible determination must be within the interval between the latest time at which the fate of one cell can be altered by the ablation of another cell and the time at which the fate of the first cell is expressed.) These results are consistent with the hypothesis that *lin-12* functions in the determination of cell fates.

B. *lin-14*, a Gene That Specifies the Stage-specific Expression of Cell Fates

Opposite allelic states of the *lin-14* locus induce opposite homeotic transformations in cell fates (Ambros and Horvitz 1984, 1987). These transformations are temporal; that is, certain cells express fates that ordinarily would be expressed at either earlier or later larval stages of development. Such heterochronic mutations affect some, but not all, stage-specific developmental events and thereby alter the relative timing of different developmental events.

Dominant *lin-14* mutations (*lin-14[d]*) cause an elevated or constitutive-like expression of gene function, as shown by the enhancement of the dominant phenotype when the mutant allele is in *trans* to a wild-type, as opposed to a null (*lin-14[0]*), allele. *lin-14(d)* mutants display retarded development, that is, certain stage-specific events occur at abnormally late developmental stages. For example, *lin-14(d)* mutants undergo supernumerary larval molts, during which larva-specific cuticle is synthesized. This reiterated expression of earlier developmental events can lead to the retardation (and sometimes to the elimination) of succeeding events, such as the synthesis of adult-specific cuticle. Thus, although these animals become reproductively mature at the normal time, a true adult never develops. Other retarded events include stage-specific patterns of cell division and differentiation. An example involving the lateral hypodermal T lineage is illustrated in Figure 8, a and b. A cell-division pattern that occurs during the first larval stage of wild-type animals instead occurs during both the first and the second larval stages of *lin-14(d)* animals.

Null *lin-14* mutations (*lin-14[0]*) are recessive and result in a phenotype that is opposite to that caused by *lin-14(d)* mutations. Specifically, *lin-14(0)* mutations cause precocious development, in which certain developmental events occur earlier than in wild-type animals. For example, *lin-14(0)* mutants produce an adult cuticle one larval stage early. Stage-specific patterns of cell division are similarly affected. Thus, during the first larval stage in *lin-14(0)* animals, the lateral hypodermal T lineage generates a cell-division pattern expressed by wild-type animals during the second larval stage (Fig. 8c).

These observations indicate that *lin-14* acts to determine the timing of certain cell fates. Specifically, mutations that elevate *lin-14* activity cause cells at later stages to express fates normally expressed only by cells at earlier stages, and mutations that eliminate *lin-14* activity cause cells at earlier stages to express fates normally expressed only by cells at later stages. In other words, high levels of *lin-14* activity result in earlier fates, and low levels of *lin-14* activity result in later fates (Fig. 9a). These observations suggest that during normal development, *lin-14* activity may be high at early times and low at later times, thereby resulting in the expression of early fates at early times and late fates at late times (Fig. 9b). Consistent with this hypothesis, molecular studies of *lin-14* gene expression have shown that *lin-14* mRNA levels decrease with time in wild-type animals (G. Ruvkun et al., pers. comm.).

This view of *lin-14* action is somewhat oversimplified. In addition to *lin-14(0)* and *lin-14(d)* alleles, three other classes of *lin-14* alleles have been identified (Ambros and Horvitz 1987). The five classes of *lin-14* mutant alleles lead to five distinct heterochronic phenotypes. Complementation,

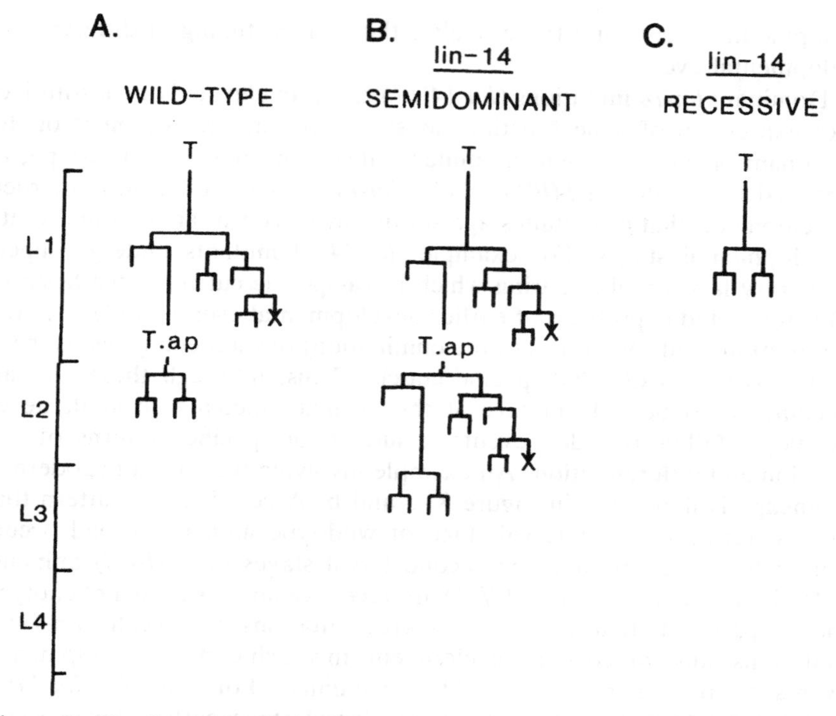

Figure 8 The fates of cells in the lateral hypodermal T-cell lineages are transformed in opposite directions by *lin-14(d)* and *lin-14(0)* mutations. The *lin-14(d)* allele was *n536*, and the *lin-14(0)* allele was *n536 n540*. (*a*) Wild type. (*b*) A *lin-14(d)* mutant displays retarded development, expressing normally first larval stage-specific division patterns in the second larval stage. (*c*) A *lin-14(0)* mutant displays precocious development, expressing normally second larval stage-specific division patterns in the first larval stage. (Reprinted, with permission, from Ambros and Horvitz 1984. Copyright 1984 by the AAAS.)

temperature-shift, and gene dosage experiments have revealed the genetic basis of these different phenotypes: The *lin-14* locus controls two distinct activities, which act at different times during development and are separately mutable to either reduced or elevated levels of gene function.

VI. GENETIC INTERACTIONS AND PATHWAYS: VULVAL DEVELOPMENT

The characterization of mutants abnormal in cell lineage has identified *lin-12* and *lin-14* as two genes that appear likely to function in controlling developmental decisions. The elucidation of how these genes act could provide an important step toward an understanding of the mechanisms involved in specifying cell lineage. However, neither of these genes func-

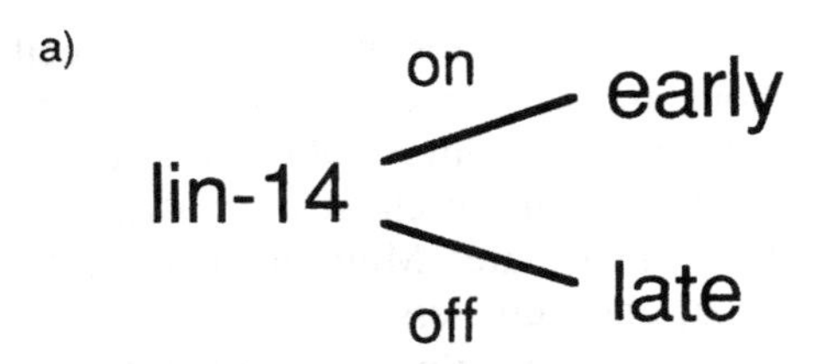

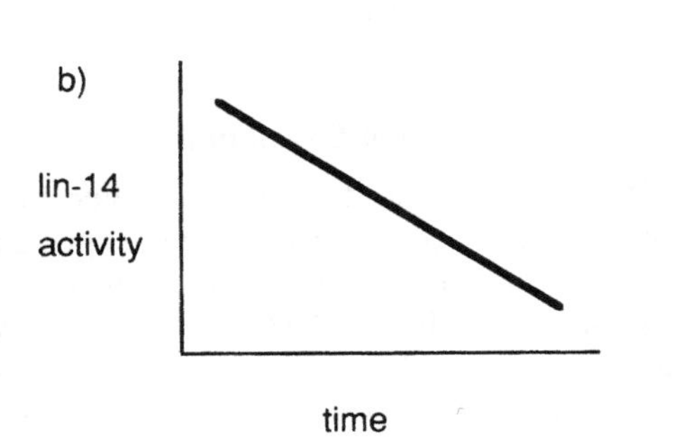

Figure 9 (*a*) The level of *lin-14* gene activity specifies whether certain cells express early or late developmental fates. (*b*) A model for the level of *lin-14* activity during development. Early in development, *lin-14* activity is high, resulting in the expression of early fates. Later in development, *lin-14* activity is low, resulting in the expression of later fates.

tions in isolation. Each must be interacting within a pathway or network of other genes. To understand the genetic and molecular control of cell lineage, it will be necessary to identify and characterize such sets of interacting genes. Such studies should help reveal both the complexity and the logic of the developmental pathways involved. For example, the utilization of macromolecular assemblies as modules in phage morphogenesis (Edgar and Lielausis 1968) and the dependent and independent pathways of the yeast cell cycle (Hartwell et al. 1974) were both discovered by studying sets of interacting genes.

The *C. elegans* cell lineages for which the greatest number of genes and gene interactions have been examined are those responsible for the generation of the hermaphrodite vulva. More than 200 mutations that affect vulval cell lineages have been described (Horvitz and Sulston 1980; Greenwald et al. 1983; Ambros and Horvitz 1984, 1987; Ferguson and Horvitz 1985; Ferguson et al. 1987; W. Fixsen and R. Horvitz, in prep.). The interactions among many of these mutations have been determined and used to help define a genetic pathway for vulval development (Ferguson et al. 1987; P. Sternberg and R. Horvitz, in prep.). In this section, we discuss these vulval cell-lineage mutants, both to illustrate the general approaches and findings of genetic studies of cell lineage and also to provide examples of the ways in which genetic pathways can be determined.

The vulval cell lineages were chosen for detailed genetic analysis for three reasons. First, these lineages are technically easy to study: They involve relatively few, highly visible precursor cells that undergo only three rounds of divisions over a time interval of only about 5 hours (Sulston and

Horvitz 1977). Second, the vulval cell lineages involve both cell autonomous and cell nonautonomous determination (Sulston and Horvitz 1977; Sulston and White 1980; Kimble 1981a; Sternberg and Horvitz 1986). Third, two major classes of vulval cell-lineage mutants are easy to recognize and isolate: Multivulva mutants generate multiple vulvalike structures, and vulvaless mutants fail to generate a vulva; furthermore, these mutants are viable as homozygotes (Horvitz and Sulston 1980; Ferguson and Horvitz 1985).

A. Wild-type Vulval Development

Vulval development in the wild-type hermaphrodite involves a set of six hypodermal precursor cells that express three distinct fates (called 1°, 2°, and 3°) in a precise anterior–posterior spatial pattern (3°-3°-2°-1°-2°-3°) (Sulston and Horvitz 1977; Sulston and White 1980). Each of these fates is a particular cell lineage. The progeny from the 1° and 2° lineages form the vulva, and the progeny from the 3° lineages join the syncytial hypoderm that envelops the animal. A model for the roles of cell autonomous and cell nonautonomous factors in vulval development has been proposed (Fig. 10) (Sternberg and Horvitz 1986). Specifically, the six hypodermal precursor cells are equivalent in developmental potential, each capable of expressing the 1°, 2°, and 3° fates. According to this model, which of these fates each cell expresses is specified by a graded signal from a single cell in the gonad, the anchor cell, which induces vulval formation. A strong signal from the anchor cell induces the expression of the 1° fate, a weak signal induces the expression of the 2° fate, and the absence of an anchor cell signal results in the expression of the 3° fate. Thus, the hypodermal cell closest to the anchor cell expresses the 1° fate, the two next closest hypodermal cells express the 2° fate, and the three most distal hypodermal cells express the 3° fate. Once the anchor cell signal has acted on a hypodermal cell to specify its fate, the subsequent expression of that fate (which involves a series of cell divisions and the generation of a number of cell types) occurs in a cell autonomous manner.

Aspects of this model for vulval development have been derived from studies of mutants. For example, the equivalence in developmental potential of the six hypodermal cells was established, in part, based on the observation that a number of cell-lineage mutations affect these six cells specifically and equivalently (Kimble 1981a; Sulston and Horvitz 1981). Mutant studies have also revealed that the anchor cell signal acts at a distance rather than via direct cell–cell contact: In *n1321* and *rh51* animals, in which the gonad is displaced dorsally so that it cannot be in direct contact with the vulval precursor cells, the anchor cell can still induce both the 1° and 2° fates (J. Thomas and R. Horvitz, in prep.; E. Hedgecock, pers. comm.). Other mutant studies have provided the basis for suggesting

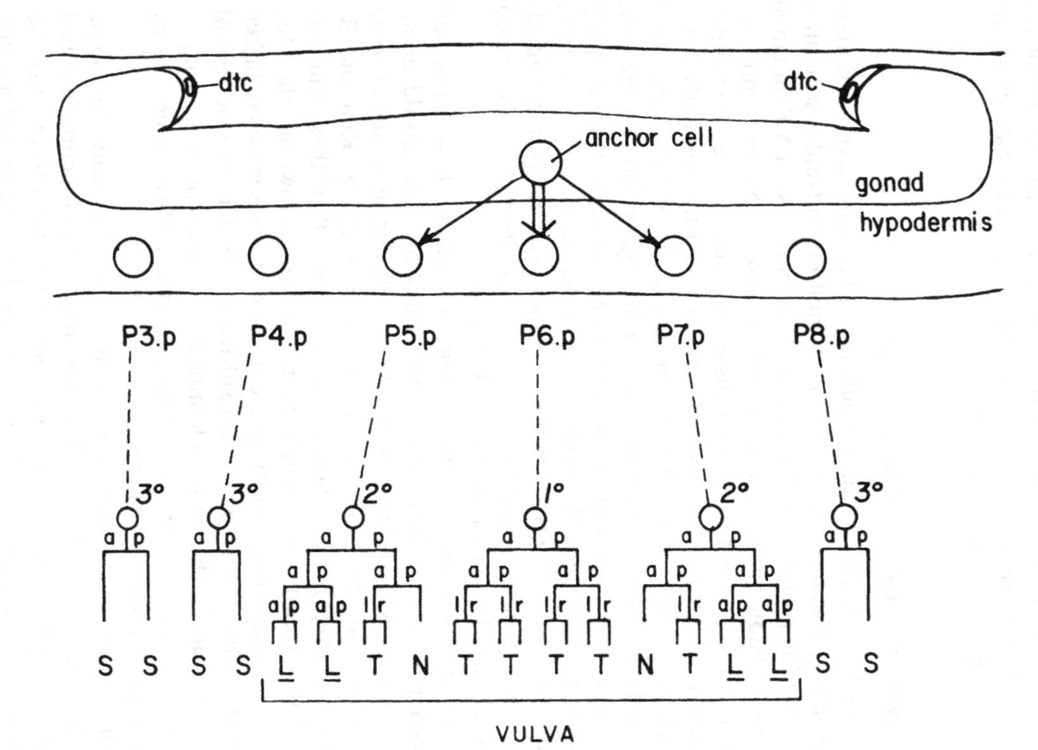

Figure 10 A model for vulval development. A graded signal from the gonadal anchor cell induces the expression of vulval cell lineages (1° and 2°) by a set of six equipotent potential vulval precursor cells. Noninduced precursor cells express a nonvulval (3°) lineage. The vulva is formed from the 22 descendants of the 2° and 1° lineages. (Adapted from Fig. 2a in Sternberg and Horvitz 1984.)

that the anchor cell acts via a graded signal, rather than by initiating a cascade of sequential cell–cell interactions. Specifically, in *unc-84* animals, a vulval precursor cell can express the 2° fate in the absence of any cell expressing the 1° fate (Sternberg and Horvitz 1986). The simplest interpretation of this observation argues against a model in which the anchor cell stimulates one hypodermal cell to express the 1° fate and that cell stimulates its neighbors to express the 2° fate.

B. Genes Involved in the Vulval Cell Lineages

Genes involved in the control of the vulval cell lineages have been identified by isolating mutants that either have multiple vulvalike structures (multivulva or Muv mutants) or that lack a vulva (vulvaless or Vul mutants) (Horvitz and Sulston 1980; Ferguson and Horvitz 1985). Muv mutants have been identified directly, using a dissecting microscope to examine vulval structure. Although Vul mutants can be identified similarly, such mutants generally have been obtained on the basis of their functional defects in egg laying: The vulva is the opening through which eggs are laid, and therefore Vul hermaphrodites cannot lay eggs; because their progeny hatch in utero and consume the body (but not the cuticle) of the parent, Vul animals turn into "bags of worms," which are easily recognized. Strains of such egg-laying-defective mutants have then been examined microscopically to identify those that lack a vulva.

Many Muv and Vul mutants have been isolated in these ways and characterized genetically (Horvitz and Sulston 1980; Greenwald et al. 1983; Trent et al. 1983; Ambros and Horvitz 1984; Ferguson and Horvitz 1985; Ferguson et al. 1987; W. Fixsen and R. Horvitz; S. Kim and R. Horvitz; both in prep.). The phenotypes of almost all of these mutants have been shown to result from single-gene mutations. However, the Muv phenotype of one strain requires the presence of mutations in two unlinked genes, *lin-8 II* and *lin-9 III* (Horvitz and Sulston 1980). The mutagenesis of *lin-8* or *lin-9* animals has led to the identification of mutations defining an additional four genes that can mutate to generate a Muv phenotype by interaction with *lin-8* or *lin-9* (Ferguson et al. 1987).

Twenty-eight genes that can mutate to generate a Vul or Muv phenotype have been identified so far. To elucidate how these genes function in vulval development, genetic analyses have been performed both to identify the null phenotypes of these genes and to determine how the mutations that define these genes affect gene function (Greenwald et al. 1983; Ferguson and Horvitz 1985; Ambros and Horvitz 1987; W. Fixsen and R. Horvitz, in prep.). Eight of the 28 genes appear to have Muv or Vul null phenotypes; thus, these genes normally function in vulval development and are not required for either viability or fertility. Nine genes have lethal or sterile null phenotypes, with their Muv or Vul phenotypes resulting from a partial loss of gene function in most cases; thus, these genes normally function during vulval development but, in addition, are required for growth and maturation. Consistent with this observation, four of these nine genes (*lin-12*, *lin-14*, *lin-15*, and *let-23*) have been found to be involved in the control of a variety of cell lineages besides those of vulval development (Greenwald et al. 1983; Ambros and Horvitz 1984; Fixsen et al. 1985). Three genes have been identified only by dominant mutations and have unknown null phenotypes; although these genes may normally function in vulval development, it is possible that they affect vulval lineages only as a consequence of mutations that cause ectopic gene expression. The other nine genes have unknown null phenotypes and are defined by recessive mutations, which generally cause a reduction or loss of gene function; thus, these genes probably normally function in vulval development, although whether they are required for viability and fertility remains to be determined.

One goal of the genetic analysis of the vulval cell lineages is to define all genes involved relatively specifically in these lineages. (Although there must be many genes that function not only in the cells responsible for vulval development but also in many and perhaps all other cells, the approach of isolating viable mutants abnormal in vulval cell lineages should avoid most such genes.) Multiple alleles of 23 of the 28 genes have been obtained, and it seems likely that most genes that can mutate to a fertile Vul or a fertile Muv phenotype have been identified (Ferguson and Horvitz 1985). Nonetheless, additional genes involved in the vulval cell lineages no doubt remain to be identified. For example, genes that can mutate to generate sterile Muv or sterile Vul animals, as well as genes with Muv or Vul alleles that display maternal effects, would not necessarily have been discovered in the studies done so far. Also, it may be possible to identify genes with redundant functions only as a result of rare multiple mutations or, in some cases, as a result of rare dominant mutations (see Greenwald and Horvitz 1980; Park and Horvitz 1986a). Finally, there may be genes involved in the control of the vulval lineages that cannot mutate to give a Muv or Vul phenotype. Such genes must be identified in other ways, for example, on the basis of other phenotypic abnormalities in vulval structure or function or as suppressors or enhancers of known genes. Thus, known extragenic suppressors of *lin-12*, *lin-1*, and *lin-10* (J. Thomas et al., pers. comm.) might define additional genes involved in the vulval cell lineages.

C. A Genetic Pathway for Vulval Development

The analysis of 28 genes that affect the vulval cell lineages has made it possible to define a detailed genetic pathway for the specification of these lineages (Ferguson et al. 1987). One straightforward approach that has been used in establishing the order of gene action in vulval development involves the direct observation of the vulval cell lineages in mutant animals. These studies have precisely identified the developmental bases of the Muv and Vul mutant phenotypes. Because most Muv and Vul mutations act by reducing or eliminating gene activity, the functions of particular genes can be inferred from the corresponding mutant phenotypes. Specifically, (1) mutations that result in defects manifested before the onset of the vulval lineages can define genes needed for the *formation* of a functional anchor cell or of functional vulval precursor cells; (2) mutations that cause some precursor cells to adopt fates characteristic of other precursor cells can define genes needed for the normal *determination* of precursor cell fates; and (3) mutations that affect the lineage generated by those cells determined to express a particular vulval fate can define genes needed for the *expression* of that fate.

Thus, phenotypic defects have been used to assign genes provisionally to one of three general steps in the pathway of vulval development. These assignments have been tested in two ways. First, double mutants have been constructed to examine gene interactions; mutations that prevent an early step in the pathway generally should be epistatic to mutations that prevent a later step. Second, periods of temperature sensitivity have been defined for temperature-sensitive mutants; a temperature-sensitive period must be prior to or concurrent with all steps in the pathway for which a gene is needed. Using the criteria of mutant phenotypes, epistatic interactions, and temperature-sensitive periods, 23 of the 28 genes have been assigned to particular steps in the pathway for vulval cell lineages.

In defining the genetic pathway of vulval development, one basic question involves the site of action of each gene. Specifically, because the determination of precursor cell fates requires activities both in the gonad (to produce the inducing signal of the anchor cell) and in the hypodermis (to respond to this signal), any gene affecting the determination of precursor cell fates could, in principle, act in either the gonad or the hypodermis. Muv mutants cause precursor cells that normally generate the nonvulval (3°) fate (because, according to the current model, these cells do not receive the inductive signal) to generate a vulval (1° or 2°) fate instead. This Muv phenotype could be caused either by a gonadal defect, resulting in an elevated level of inductive signal, or by a hypodermal defect, resulting in a hypersensitivity to or an independence from the inductive signal. To distinguish among these possibilities, a laser microbeam has been used to ablate the gonad: If a Muv mutant phenotype were caused by an elevated level of inductive signal, that Muv mutant phenotype should be dependent on the presence of the gonad. All Muv mutants examined in this way displayed a Muv phenotype in the absence of a gonad, suggesting that each of the genes defined by these Muv mutations has a site of action within the hypodermis.

Similarly, Vul mutants abnormal in the determination of hypodermal precursor cell fates could be defective in either the gonad or the hypodermis. In Vul mutants of this class, precursor cells that normally generate the vulval fates (1° or 2°) generate the nonvulval (3°) fate instead. This Vul phenotype could be caused either by a gonadal defect, resulting in a failure to produce the inductive signal, or in a hypodermal defect, resulting in a failure to respond to the inductive signal. To distinguish between these possibilities, double mutants have been constructed that carry a Vul mutation and a "signal-independent" Muv mutation of the class described above: Just as the physical ablation of the gonad does not affect the Muv phenotype of these mutants, neither should the genetic ablation of the gonadal signal by a Vul mutation affect this phenotype. Thus, if a Vul mutation affects the phenotype caused by any signal-independent Muv mutation, the Vul mutation must act outside the gonad, presumably within the hypodermis. On the other hand, if all signal-independent Muv mutations are epistatic to a Vul mutation, the site of action of the Vul mutation could be either within the gonad or within the hypodermis. Based on this criterion, all Vul genes involved in the determination of precursor cell fates appear to have a site of action within the hypodermis.

Genetic interactions have also been used to identify genes involved in the expression, as opposed to the determination, of the vulval cell fates. Such genes can be identified because they affect a particular vulval cell fate (1°, 2°, 3°) independently of which precursor cells express that fate. For example, mutations in *lin-11* cause the two cells that express a 2° fate in the wild type to express an abnormal 2°-like fate instead. In a double mutant carrying a *lin-12(d)* mutation, which causes all six potential vulval precursor cells to express a 2° fate, a *lin-11* mutation causes all six precursor cells to express the abnormal 2°-like fate. In a double mutant carrying a *lin-12(0)* mutation, which causes all six precursor cells to express non-2° fates, a *lin-11* mutation does not cause any of the six precursor cells to express the abnormal 2°-like fate. Thus, *lin-11* mutations alter the fates of all and of only those cells determined to express a 2° fate, indicating that *lin-11* normally functions in the expression of the 2° fate.

The studies summarized in this section have allowed Ferguson et al. (1987) to propose a genetic pathway for vulval development (Fig. 11). The first five genes in this pathway function in the generation of cells with the potential to express normal vulval lineages. Two of these genes are necessary for the migrations of the parents of the vulval precursor cells, two are necessary for the cells that normally become the precursor cells to adopt their normal fates (instead of fates characteristic of other cells), and one specifies the time at which the precursor cells are able to express vulval lineages. Once potential vulval precursor cells are generated, 15 genes function in the determination of precursor cell fates. As described above, all 15 of these genes act within the vulval precursor cells and thus define a

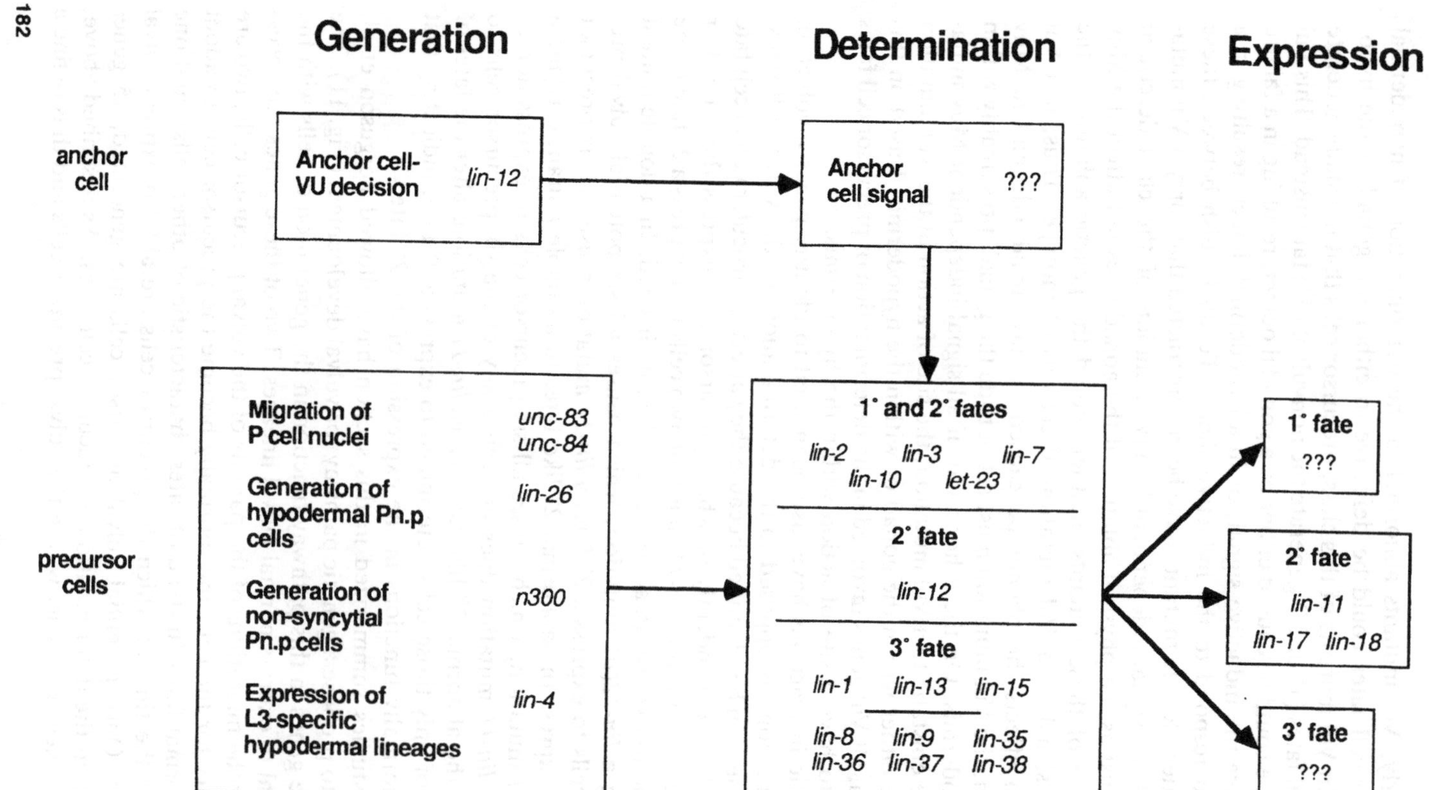

Figure 11 A genetic pathway for the specification of the vulval cell lineages. (Adapted from Fig. 3 in Ferguson et al. 1987.)

system involved in the intracellular response to the extracellular signal that induces vulva formation. This system might include membrane or cytoplasmic proteins that function in "second messenger" systems, as well as nuclear proteins that function to control gene expression. After the fates of particular precursor cells are determined, three other genes function in the expression of the 2° vulval cell fate.

The actions of and interactions among some of the genes involved in the determination of precursor cell fates have been analyzed further (P. Sternberg and R. Horvitz, in prep.). For example, certain double mutant combinations have been found to result in the expression of a particular fate by all six cells, whether or not the anchor cell signal is present. Thus, in a *lin-12(0)* double mutant with a Muv mutation, all six potential vulval precursor cells express the 1° fate; in a *lin-12(d)* double mutant with a Vul mutation, all six express the 2° fate; and in a *lin-12(0)* double mutant with a Vul mutation, all six express the 3° fate. These observations suggest that the states of genes previously identified are sufficient to specify completely the fates of the vulval precursor cells independently of the anchor cell signal.

VII. PROGRAMMED CELL DEATH

Of the 1090 somatic nuclei generated by the cell lineage of the *C. elegans* hermaphrodite, 131 undergo naturally occurring or programmed death (for review, see Horvitz et al. 1982a). Because these cell deaths all involve the same sequence of morphological changes (Sulston and Horvitz 1977; Sulston et al. 1983) and require the activities of the same genes (Hedgecock et al. 1983; Ellis and Horvitz 1986), undergoing programmed cell death can be regarded as expressing a specific cell fate, much like differentiating into an intestinal cell or a dopaminergic neuron. One can imagine two classes of genes that might be involved in programmed cell deaths: (1) "Determination genes," which decide which cells are to die, and (2) "differentiation genes," which are necessary to express the fate of programmed cell death.

Cell death determination genes might be recognized on the basis of mutations that alter patterns of cell death without affecting the machinery necessary for cell death per se. Such mutations would cause particular cells that normally survive to die, or particular cells that normally die to survive. One mutation of this second class has been identified: The mutation *ces-1(n703)* prevents some, but not all, programmed cell deaths; for example, the sisters of the pharyngeal NSM neurons survive in *ces-1(n703)* animals (R. Ellis et al., pers. comm.). The opposite class of transformation is caused by mutations in the genes *egl-1* and *lin-39*. Thus, *egl-1* mutations cause the cells that would normally become the HSN motor neurons to undergo programmed cell death (Trent et al. 1983; Ellis and Horvitz

1986). Similarly, the *lin-39(n709)* mutation causes the cells that would normally become the VC motor neurons to die (Horvitz et al. 1982a; Fixsen et al. 1985). The supernumerary cell deaths of both *egl-1* and *lin-39* animals are morphologically similar to and involve the same genes as (see below) the cell deaths that occur during wild-type development. Thus, these mutants appear to cause the ectopic expression of the normal fate of programmed cell death.

In contrast, there are a number of mutations that cause particular cells to die but probably do not define programmed cell death determination genes. Mutations in the genes *mec-4*, *deg-1*, *lin-24*, and *lin-33* result in the degeneration of certain cells in ways that are morphologically distinct from normal programmed cell deaths and that do not depend on (at least some of) the genes required for normal programmed cell deaths (Chalfie and Sulston 1981; Chalfie 1984b and pers. comm.; Ferguson et al. 1987). These mutations all cause dominant cell-death phenotypes. It seems likely that the cytotoxic effect of each of these mutations results from the expression of too much of or an altered form of a gene product normally expressed in the cells that die or from the ectopic expression of a product normally not expressed in those cells.

Five cell death differentiation genes have been identified. These genes have defined a genetic pathway for programmed cell death. Two genes, *ced-3* and *ced-4*, are necessary for the onset of programmed cell death (Ellis and Horvitz 1986). In *ced-3* or *ced-4* mutants, cells that would normally die survive, differentiate, and, in at least some cases, function (Ellis and Horvitz 1986; L. Avery and R. Horvitz, in prep.). *ced-3* and *ced-4* animals do not display any gross phenotypic abnormalities, which indicates that programmed cell death is not an essential aspect of *C. elegans* development. It seems likely that the *ced-3* and *ced-4* genes act in response to cell death determination genes like those described above.

Two genes, *ced-1* and *ced-2*, are required for the engulfment of dying cells by their neighbors; in *ced-1* or *ced-2* mutants, cells that initiate cell death undergo the early morphological changes associated with programmed cell death, but the phagocytosis of these cells by their neighbors does not occur (Hedgecock et al. 1983). The gene *nuc-1* controls a nuclease that is necessary for the degradation of the DNAs of cells that undergo programmed cell death; in *nuc-1* mutants, the remains of cells that have died can be visualized as pycnotic nuclei containing undegraded DNA (Sulston 1976; Hedgecock et al. 1983). Gene interaction experiments have established that *ced-1* and *ced-2* are epistatic to *nuc-1* and that *ced-3* and *ced-4* are epistatic to all three other cell death genes, indicating an order of gene action of (*ced-3*, *ced-4*), (*ced-1*, *ced-2*), *nuc-1* (Fig. 12) (Ellis and Horvitz 1986).

Like other cell fates, programmed cell death could be specified by factors either intrinsic or extrinsic to the cell expressing the fate. In other

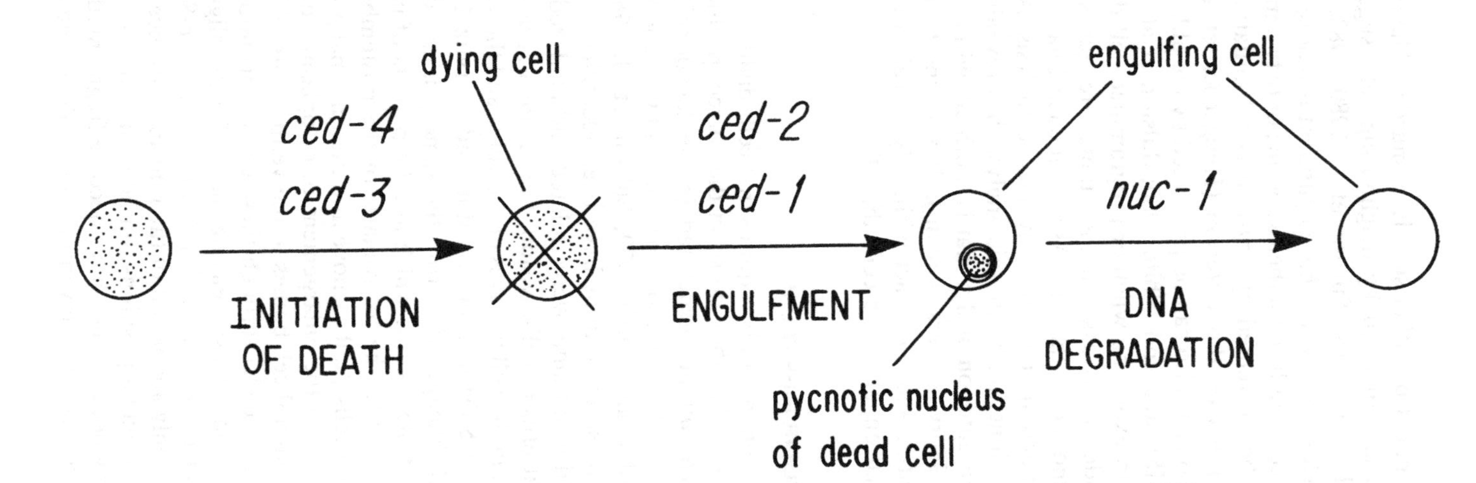

Figure 12 A genetic pathway for programmed cell death. (Adapted from Fig. 10 in Ellis and Horvitz 1986.)

words, cell death might be either suicide or murder. The observation that many dying cells are engulfed by neighboring cells suggested that cell deaths could well be murders (Sulston et al. 1980, 1983; Robertson and Thomson 1982). However, when this engulfment is blocked by a mutation in *ced-1* or *ced-2*, cell deaths nonetheless occur; only two male-specific cell deaths appear to be prevented by *ced-1* or *ced-2* mutations and, hence, appear to be murders by this criterion (Hedgecock et al. 1983). (The deaths of these two cells also can be prevented by the ablation of the cells that engulf them [Sulston and White 1980; Sulston et al. 1980].)

One approach to exploring whether programmed cell deaths might be regarded as suicides is to ask whether there are genes that must be expressed within those cells that die. As described in Chapter 2, one way to identify the site of action of a gene is by mosaic analysis. An examination of *ced-3* and *ced-4* function by mosaic analysis has revealed that both of these genes act cell autonomously; that is, both act within the dying cells (J. Yuan and R. Horvitz, pers. comm.). It remains to be determined whether *ced-1*, *ced-2*, and *nuc-1* act within dying cells or, alternatively, within the cells that engulf those cells that die.

VIII. CELL AND NUCLEAR MIGRATIONS

C. elegans development not only requires that cells divide and differentiate appropriately but also that they do so in the correct positions. Although most cells in *C. elegans* are generated in the locations in which they ultimately function, some cells undergo significant migrations (Sulston and Horvitz 1977; Kimble and Hirsh 1979; Sulston et al. 1983). Mutations affecting many of these migrations have been identified.

Some cell-migration mutants were isolated and analyzed because they are abnormal in patterns of cell division. For example, *unc-83* and *unc-84* mutants are altered in the cell lineages of the vulva and the ventral nervous system (Horvitz and Sulston 1980; Sulston and Horvitz 1981). Both of these defects in cell lineage arise from the same primary developmental lesion: The nuclei of the P cells, which are the parents of both the vulval precursor (Pn.p) cells and the neuroblasts of the postembryonic ventral nervous system (Pn.a cells), fail to move into their normal positions. These P cell migrations involve the enlargement of cytoplasmic extensions that run from the ventrolateral cell bodies to the ventral cord, followed by the migration of the P cell nuclei through these extensions into the cord; the residual cytoplasm soon follows, leaving the 12 P cells aligned along the ventral cord (Sulston and Horvitz 1977). In *unc-83* and *unc-84* animals, the P cell nuclei migrate only about half of their normal distance. These nuclei then return to their original positions, initiate karyokinesis, and (usually) degenerate, probably because they have become fragmented, leaving portions of migrated cytoplasm in the ventral cord (Sulston and Horvitz 1981;

W. Fixsen and R. Horvitz, in prep.). The degeneration of these cells precludes the formation of both the vulva and the adult ventral nervous system. These observations suggest that there are both cellular and nuclear components to the P cell migrations and that only the nuclear component is affected by mutations in *unc-83* and *unc-84*.

Mutations in the *unc-83* and *unc-84* genes lead to one other defect: The nuclei of the major hypodermal syncytium are displaced from their normal dorsolateral positions into the dorsal cord (Sulston and Horvitz 1981; W. Fixsen and R. Horvitz, in prep.). The identification of this pleiotropy led to the discovery that during embryogenesis there is a migration that appears similar to that of the postembryonic P cells (Sulston and Horvitz 1981). Specifically, the cells that will form the major hypodermal syncytium are generated in dorsolateral positions, from which they extend cytoplasmic processes that run dorsally to the contralateral side; each nucleus moves along its process, crosses the dorsal midline, and comes to lie in its final, lateral position. In *unc-84* and most *unc-83* mutants, these nuclei move only halfway and stop at the dorsal midline. This abnormality does not have any gross effect on phenotype.

These two migrations, which involve the development of cytoplasmic extensions followed by nuclear movement along these extensions, are distinctive; no other migrations of this type are observed during *C. elegans* development (Sulston and Horvitz 1977; Kimble and Hirsh 1979; Sulston et al. 1983). The defects of *unc-83* and *unc-84* mutants suggest that the products of these two genes function specifically in nuclear movements that occur during this particular type of migration.

Genes involved in a number of specific cell migrations have been sought by isolating mutants defective in those migrations. For example, as discussed below, the migrations of the HSN motor neurons, the SM sex myoblasts, and the Q neuroblasts have been studied genetically in this way. Many of the mutants obtained appear to be defective only in particular cell migrations, suggesting that there are genes that function only in specific migrations. However, at least some of these mutants display low penetrance effects on additional cell migrations; furthermore, the null phenotypes of these genes are generally unknown. Thus, the specificity of gene function in particular cell migrations remains to be established. The genetics of cell migration has been reviewed recently by Hedgecock et al. (1987).

The HSN motor neurons, which innervate the vulval muscles and drive egg laying, undergo a long-range migration from the postanal region of the tail to the middle of the body (Sulston et al. 1983). The screening of a large number of egg-laying-defective mutants has identified a number that are abnormal in the HSN migrations (Trent et al. 1983; G. Garriga et al., pers. comm.). These migration mutants have defined eight genes, *egl-5*, *egl-18*, *egl-20*, *egl-27*, *egl-43*, *mig-1*, *n1332*, and *n1393*. At least some of these mutants are abnormal in additional aspects of HSN development, which suggests that these genes may function not in HSN migration per se but rather in some earlier or more general step(s) of HSN development, such as the determination of the HSN fate.

The hermaphrodite SM sex myoblasts, which generate the vulval and uterine muscles necessary for egg laying (Sulston and Horvitz 1977), also undergo long-range migrations. The final location of the SM cells appears to be controlled by an interaction with the gonad. Specifically, in the mutant *n1321*, in which the gonadal primordium can be displaced either anteriorly or dorsally, the SM cells migrate to the position of the displaced gonad (J. Thomas and R. Horvitz, in prep.). Furthermore, in wild-type animals in which the gonadal primordium has been ablated, the SM cells initiate their migrations normally but stop at variable positions some distance either before or beyond their normal termination points (J. Thomas and R. Horvitz, in prep.). The examination of egg-laying-defective mutants for the presence of vulval muscles has led to the identification of two genes involved in SM cell migrations, *egl-15* and *egl-17* (Trent et al. 1983; M. Stern et al., pers. comm.). In *egl-15* and *egl-17* mutants, the SM cells stop migrating prematurely. This migration defect is caused by an inhibitory signal from the gonad, because when the gonadal primordium in *egl-15* and *egl-17* animals is ablated with a laser microbeam, the SM cells migrate further anteriorly. It seems likely that either the gonadal signal or the SM cell response to this signal is abnormal in *egl-15* and *egl-17* animals.

The two Q neuroblasts, QL and QR, are left–right homologs and have nearly equivalent fates, each generating three neurons and two cells that undergo programmed cell death (Sulston and Horvitz 1977; Sulston et al. 1983). Each of the Q cells, as well as many of their descendants, undergoes long-range migrations. However, whereas QL and its descendants migrate posteriorly, QR and its descendants migrate anteriorly. A number of mutations have been isolated that reverse the direction of migration of either QL or QR descendants. Mutations in *mab-5* and at least six other genes cause QL descendants to migrate anteriorly (Chalfie et al. 1983; Hedgecock et al. 1987). In *mab-5* animals, QL migrates posteriorly as usual, but after it divides, its daughters reverse direction and migrate anteriorly. This observation suggests that the migrations of the various Q intermediates may be under somewhat different controls. The only mutation known to reverse the direction of the QR descendants, *e1751*, is semidominant and thought to result in altered gene function. Taken together, the phenotypes of these mutants suggest that the normal asymmetry of Q cell migrations is achieved by modifying QL behavior from a "ground state" in which both QL and QR descendants migrate anteriorly. Based on the effects of *mab-5* mutations on both cell migrations and cell lineage, it has been proposed that this gene functions as a component of a system of anterior–posterior positional information (Kenyon 1986).

A few genes have been identified that have obvious effects on multiple cell migrations. For example, the mutant *mig-2(rh17)* is abnormal in the migrations of the HSN neurons, the Q descendants, the CAN cells, and the mesodermally derived coelomocytes (Hedgecock et al. 1987). Similarly, *mig(ct41)* animals are abnormal in the migrations of the HSN neurons, the ALM neurons, and the CAN cells (J. Manser and W. Wood, pers. comm.). Because many migrations are affected in these mutants, and generally the migrations affected are reduced in extent but not abolished, it is possible that the null phenotypes of these genes are more severe and that some of these genes function in many or all long-range cell migrations.

IX. PROSPECTS AND PERSPECTIVES

The studies described in this chapter have been possible because of the advantages offered by *C. elegans* as an experimental organism both for the examination of cell lineage at the level of resolution of single cells and for detailed and sophisticated genetic analyses. Those cell-lineage mutants isolated and characterized so far have begun to allow nematode biologists to attain the goals outlined at the beginning of this chapter. In addition, these mutants have helped identify new avenues of research.

For example, it is likely that many more genes involved in *C. elegans* cell lineage remain to be discovered. Few of the possible types of mutant searches have been attempted, and there are many lineages for which mutants have not been sought. The phenotypes of existing mutants suggest that genes specific for unstudied lineages are likely to exist. Furthermore, as discussed above, even those lineages that have been examined most thoroughly (such as the vulval cell lineages) have not been saturated genetically. The isolation of mutations that either generate new phenotypes or interact with existing mutations should lead to the definition of new genes involved in cell lineage. The continued characterization of such genes and gene interactions should reveal the logical bases of the genetic specification of cell lineage.

One important direction for the further study of existing cell-lineage mutants is mosaic analysis (Herman 1984; Kenyon 1986; also see Chapter 2). Determining the sites of action of genes such as *lin-12* (involved in intercellular signaling), *lin-14* (involved in specifying developmental stage), and those genes involved in long-range cell migrations should contribute significantly to the understanding of the developmental biology of *C. elegans*. The resolution of mosaic analysis should be enhanced by the findings that the genes *ced-3* and *ced-4* (which are required for the onset of all 131 programmed cell deaths in the hermaphrodite) and *ncl-1* (which, when mutant, results in the enlargement of the nucleoli of all cells) act cell autonomously (see Section VII; also, E. Hedgecock; J. Kimble; both pers.

comm.). Using these genes as markers may allow the mapping of gene function at the level of resolution of single cells.

Perhaps the most exciting prospect for the future analyses of cell-lineage genes involves molecular biology. Many of the genes described in this chapter are likely to play interesting roles during *C. elegans* development. The molecular cloning of these genes should make it possible to determine biochemically where, when, and how these genes function. A variety of techniques have been developed for the cloning of *C. elegans* genes defined only by mutations (see Chapter 3). So far, four cell-lineage genes have been cloned: *lin-12* (Greenwald 1985b), which appears to be involved in intercellular signaling; *lin-14* (G. Ruvkun et al., pers. comm.), which specifies developmental stage; *lin-10* (S. Kim and R. Horvitz, in prep.), one of the Vul genes involved in the intracellular system that specifies vulval precursor cell fates in response to the extracellular inducing signal from the anchor cell; and *unc-86* (M. Finney and R. Horvitz, in prep.), which functions in making certain daughter cells different from their mother cells.

Molecular studies using cloned genes may reveal interesting DNA sequences (as in the case of *lin-12*) and interesting times, sites, and/or levels of expression (as in the case of *lin-14*). In addition, DNA clones offer the possibility of analyzing gene interactions at the level of molecules. For example, an examination of the expression of *lin-10* and other genes that function at a similar stage of vulval development in both wild-type and mutant animals could reveal the specificity of the expression of each gene for the vulval precursor cells, as well as reveal which genes precede and which genes follow each other functionally, thereby helping to define a molecular genetic pathway for the vulval cell lineages.

ACKNOWLEDGMENTS

I am grateful to the members of our laboratory and to the many other colleagues in the *C. elegans* community who have shared so generously their data, their ideas, and their friendship over the years. Our studies have been supported by U.S. Public Health Service research grants and a research career development award.

7

Germ-line Development and Fertilization

Judith Kimble
Department of Biochemistry
and Laboratory of Molecular Biology
University of Wisconsin–Madison
Madison, Wisconsin 53706

Samuel Ward
Department of Embryology
Carnegie Institution of Washington
Baltimore, Maryland 21210

I. INTRODUCTION

Classically, nematode germ cells have provided an excellent source of material for studies of fundamental questions of cell biology. In 1883, Van Beneden used the germ cells of the parasitic nematode, *Parascaris equuorum* to show that germ cells contain only a haploid number of

The Nematode Caenorhabditis elegans. 0-87969-307-X/88 $1 + .00.

chromosomes, compared with the diploid number found in somatic cells. This study led Boveri and others to propose that chromosomes are the cellular component that provides genetic continuity from one generation to the next (Wilson 1896). Subsequently, several workers exploited the simplicity and clarity of the cellular architecture of nematode germ cells to explore chromosome structure and the role of the centrosome in cell division.

Today, nematode germ cells continue to provide unique advantages for the study of basic questions of cell biology and development. How is a germ cell signaled to leave the mitotic cell cycle and to enter meiosis? How is a germ cell specified to differentiate as a sperm or as an oocyte? And how is this differentiation carried out? What is the molecular mechanism of ameboid motion? The answers to these questions are beginning to emerge from studies of *Caenorhabditis elegans* germ cells, using a combination of morphological, genetic, and biochemical approaches.

In this chapter, we cover those aspects of *C. elegans* germ-line development that are unique to germ-line tissue. These include the processes of spermatogenesis, oogenesis, and fertilization, as well as a discussion of the control of germ-line proliferation and the onset of meiosis. In addition, we summarize our current knowledge of the cell biology of *C. elegans* germ cells. Here, we know much more about sperm than we know about oocytes—in part, because sperm and spermatocytes can be isolated in quantity and in part because spermatogenesis can be studied in vitro. Embryonic determination of the germ lineage is covered in Chapters 5 and 8, and the decision between spermatogenesis and oogenesis is discussed in Chapter 9.

II. OVERVIEW OF GONADAL ANATOMY AND DEVELOPMENT

The main features of the anatomy and development of the *C. elegans* reproductive system are summarized below and diagramed in Figure 1. More thorough discussions of these subjects can be found in Chapters 4, 5, and 9 and in the original papers (Hirsh et al. 1976; Klass et al. 1976; Kimble and Hirsh 1979; Kimble and White 1981).

A. The Gonadal Primordium

In both sexes, an L1 worm hatches with a gonadal primordium consisting of four cells: two precursors of the somatic gonad, Z1 and Z4, and two precursors of the germ-line tissue, Z2 and Z3. Z1 and Z4 descend from the embryonic precursor cell, EMS; Z2 and Z3 arise as daughters of the embryonic stem cell, P_4. The gonadal primordia of the two sexes are morphologically identical; however, the adult gonads of the two sexes differ dramatically. The adult hermaphrodite gonad retains the twofold

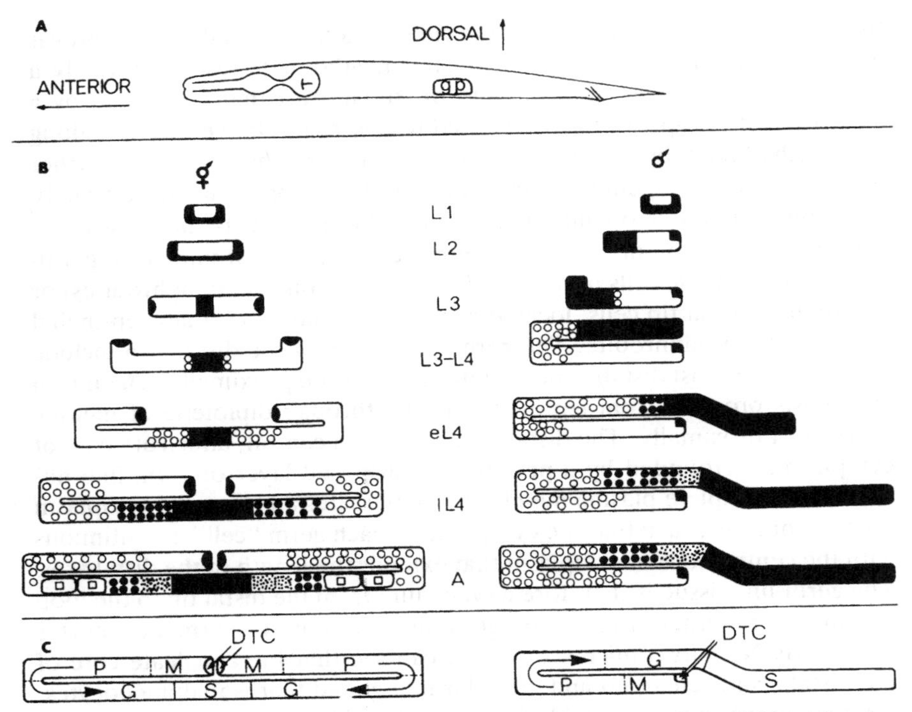

Figure 1 Gonadogenesis and adult anatomy in hermaphrodites (*left*) and males (*right*). (*A*) The mid-ventral position of the gonadal primordium (gp) is the same in both sexes. (*B*) Morphology of the gonad and distribution of cell types within the gonad at consecutive stages of postembryonic development. Somatic tissue is black; germ-line tissue is clear. Mitotic regions of germ-line tissue are blank. (○) Early meiotic (leptotene to pachytene) nuclei; (●) primary spermatocyte nuclei, (▨) sperm; (□) oocyte nuclei. (*C*) Spatial organization of adult gonad. The mitotic (M), pachytene (P), and gamete-forming (G) regions are demarcated by dashed lines. (S) Somatic tissue; (DTC) distal tip cells. Arrows indicate the polarity of maturation of the germ-line tissue, and point proximally. (Reprinted, with permission, from Kimble and White 1981.)

rotational symmetry of the four-cell primordium, whereas the adult male gonad is asymmetrical. This sexual dimorphism results from differences in development of the somatic gonad in the two sexes.

B. Anatomy of the Adult Gonad

The hermaphrodite adult reproductive system consists of two tubular ovotestes, one anterior and one posterior. The ovotestes are joined centrally by two spermathecae and a uterus; the uterus opens mid-ventrally to the exterior via the vulva. The male reproductive system consists of a single tubular testis that is connected to the cloaca via the seminal vesicle and vas deferens. Each hermaphrodite ovotestis and the male testis is actually a U-shaped tube possessing distal and proximal arms that are joined by a loop. (The distal–proximal axis is used to describe relative position along the length of the tubular gonad; a distal structure is defined as being further from the gonadal opening to the exterior than a proximal structure. In hermaphrodites, this opening is the vulva; in males, it is the cloaca.)

In both sexes, the distal arm is composed primarily of immature germ-line tissue. Somatic cells in the distal arm include one (hermaphrodites) or two (males) distal tip cells, located at the apex, and two somatic epithelial cells in the hermaphrodite. The germ-line nuclei of the distal arm include mitotic nuclei most distally and meiotic nuclei more proximally. The nuclei in meiosis progress from leptotene distally through diplotene of meiotic prophase I proximally. The distal arm contains a central, anucleate core of cytoplasm, surrounded by a peripherally disposed layer of germ-line nuclei. An incomplete plasma membrane demarcates an alcove of cytoplasm for each nucleus, such that the cytoplasm of each germ "cell" is continuous with the central core of cytoplasm that extends the length of the distal arm. The germ-line tissue is therefore a syncytium from the distal tip to the loop region where differentiation of germ-line cells into sperm or oocytes begins. By Nomarski microscopy, one can see that the anucleate core of the distal arm is enlarged and granular in a gonad that is making oocytes, whereas it is barely detectable in a gonad making sperm.

In both sexes, the proximal arm is the site of gametogenesis. In hermaphrodites, the germ line of this arm is encapsulated by a somatic contractile epithelial sheath, or oviduct. In males, the germ line is only partly ensheathed by the distal cells of the seminal vesicle.

Sperm are produced continuously in males, whereas they are made only transiently in hermaphrodites. Sperm maturation culminates at the proximal edge of the proximal arm in the two meiotic divisions that generate four haploid sperm from each tetraploid primary spermatocyte. After about 150 sperm are made in each ovotestis, germ cell differentiation switches to oogenesis, and thereafter only oocytes are produced. Developing oocytes are arranged in single file along the hermaphrodite proximal arm. Oocytes become arrested at diakinesis in meiotic prophase I. When an oocyte is fertilized, the zygote moves through the spermatheca to the uterus where meiosis of the oocyte nucleus is completed. Two polar bodies are extruded, and the two pronuclei become apposed after a complex series of movements. The pronuclear membranes then break down, and embryonic divisions begin (Albertson 1984a). No zygote nucleus is formed.

C. Postembryonic Development of the Gonad

Soon after the first divisions of Z1 and Z4, the hermaphrodite gonad can be easily distinguished from the male gonad. In hermaphrodites, Z1 and Z4 together generate 12 cells during L1. During L2, ten descendants of Z1 and Z4 coalesce toward the center of the gonad. This rearrangement establishes two separate regions of germ-line tissue—one anterior and one posterior to the central cluster of somatic cells. These two regions of germ-line tissue develop into the anterior and posterior ovotestes. As larval development proceeds, the somatic cells in the center of the gonad continue to divide according to an invariant pattern of divisions to generate the uterus, spermathecae, and oviducts of the adult hermaphrodite gonad. The two growing ovotestes elongate, one anteriorly and the other posteriorly, during L3. Then, early in L4, each ovotestis turns 180° and continues to elongate in the opposite direction. This directed elongation forms the U-shaped ovotestis. A somatic *distal tip cell* is positioned at the tip of each elongating ovotestis. These cells do not divide and remain in place throughout adulthood.

In males, Z1 and Z4 together generate ten cells during L1 and L2. The original twofold rotational symmetry of the gonadal primordium is abandoned during L1, as eight somatic cells move to the anterior end and two somatic cells, the distal tip cells, become located at the posterior end of the developing gonad. The testis elongates first in an anterior direction and then reflexes and elongates posteriorly towards the cloaca to make the U shape of the single testis. During L3 and L4, the somatic cells at the elongating end of the gonad generate the vas deferens and seminal vesicle of the male somatic gonad. As in hermaphrodites, the distal tip cells remain at the distal end with no further division.

D. Postembryonic Development of the Germ Line

The postembryonic divisions of the germ-line precursor cells, Z2, Z3, and their descendants are variable. Germ-line divisions are unpredictable with respect to time and plane of division. The number of germ-line nuclei increases during larval growth from 2 to about 1000 in hermaphrodites and to about 500 in males. This number continues to increase during adulthood, because the most distal germ-line descendants continue mitotic division throughout adulthood.

Although embryonic determination of the germ lineage appears to rely on cell ancestry (Chapter 5), postembryonic determination of the germ-line descendants as mitotic or meiotic nuclei and as sperm or oocytes has no apparent dependence on cell ancestry. As discussed below, determination of germ-line nuclei as either mitotic or meiotic relies on their interactions with a somatic cell—the distal tip cell. Determination of germ cells as sperm or oocytes is discussed in Chapter 9.

Germ-line cells first leave the mitotic cell cycle to enter meiosis at the proximal edge of the developing ovotestis or testis. Because the earliest stages of meiotic prophase are difficult to detect, the appearance of pachytene nuclei has been used to indicate entry into meiosis. In hermaphrodites, pachytene nuclei are first detected during L3 lethargus (33–34 hr, 20°C, after hatching). In males, they are first seen in mid-L3 (29–32 hr, 20°C, after hatching). As the gonads elongate, more and more distal nuclei enter meiosis. However, a zone of nuclei at the distal end of the gonad always remains mitotic.

Gamete differentiation first occurs at the proximal edge of the developing ovotestis or testis. Primary spermatocytes are first detected at mid-L4 in males and at late L4 in hermaphrodites. More distal nuclei begin to undergo gametogenesis subsequently, so that a progression of stages of gamete maturation can be observed in the proximal arm, with more immature gametes found near the loop region and more mature ones found near the spermatheca. Oocytes begin to form soon after the hermaphrodite molts to adulthood.

III. CONTROL OF PROLIFERATION AND ENTRY INTO MEIOSIS

A. Description of Distal Tip Cells

Two somatic cells, the distal tip cells, regulate germ-line proliferation and entry into meiosis (Kimble and White 1981). These two regulatory cells arise in homologous positions of the lineages of Z1 and Z4 (see Chapter 5). In hermaphrodites, Z1.aa and Z4.pp occupy the anterior and posterior distal tips, respectively; in males, Z1.a and Z4.p both occupy the single distal tip. In hermaphrodites, Z1.aa and Z4.pp both perform *two* functions essential to gonadal development. One is regulation of germ-line proliferation; the other is control of morphogenesis of the U-shaped germ-line tube. To accomplish the morphogenetic or "leader" function, the cell with leader activity guides the developing germ-line tube as it elongates in one direction, reflexes, and elongates in the other direction to complete the U shape. In males, the two distal tip cells perform a single function: regulation of germ-line proliferation. Another cell in the male gonad, the linker cell, carries out the leader function.

Because the only known function of the male distal tip cells is regulation of germ-line proliferation, their ultrastructure might provide clues about the cellular basis of that regulation. Both male distal tip cells are flat, small cells with a cytoplasm containing free ribosomes and some internal membranes (Golgi and endoplasmic reticulum) (Kimble 1978). The distal tip cells are located within the basal lamina that borders the entire gonad. The cellular membrane of the distal tip cells is closely apposed to the outer membrane of the germ-line syncytium, but no specialized junctions have been observed between them. The ultrastructure of the hermaphrodite distal tip cells is similar to that of the male linker cell (Kimble 1978). Both cells have extensive rough endoplasmic reticulum and Golgi cisternae. This ultrastructure suggests that leader function may require secretory activity.

B. Evidence for a Regulatory Role of the Distal Tip Cell

The regulation of proliferation of the germ-line cells by the somatic distal tip cells was discovered in a series of laser ablation experiments (Kimble and White 1981). As described in Chapter 5, individual cells can be killed with a laser microbeam. Laser ablation of the distal tip cells in either sex and at any time during gonadal development leads to arrest of mitosis and initiation of meiosis in all descendants of Z2 and Z3 (Figs. 2 and 3). Thus, if the distal tip cells are killed during L1, soon after their formation, the few germ cells that have been generated stop dividing after a few more mitoses, enter meiosis, and differentiate as sperm. If the distal tip cells are killed later, again all cells of the germ lineage stop dividing, enter meiosis, and differentiate as sperm or oocytes (depending on the sex and stage of development.) In males, both distal tip cells must be destroyed to obtain this effect. In hermaphrodites, if one distal tip cell is killed, only germ cells occupying the ovotestis of the ablated cell enter meiosis. Ablation of the immediate precursors of the distal tip cells in hermaphrodites (Z1.a and Z4.p) mimics the effect of killing the distal tip cells. However, if the two precursor cells of the somatic gonad, Z1 and Z4, are killed, the germ cells do not divide mitotically and do not enter meiosis. Indeed, if the precursors to Z1 and Z4 are killed in the embryo, the two germ cells die during L1.

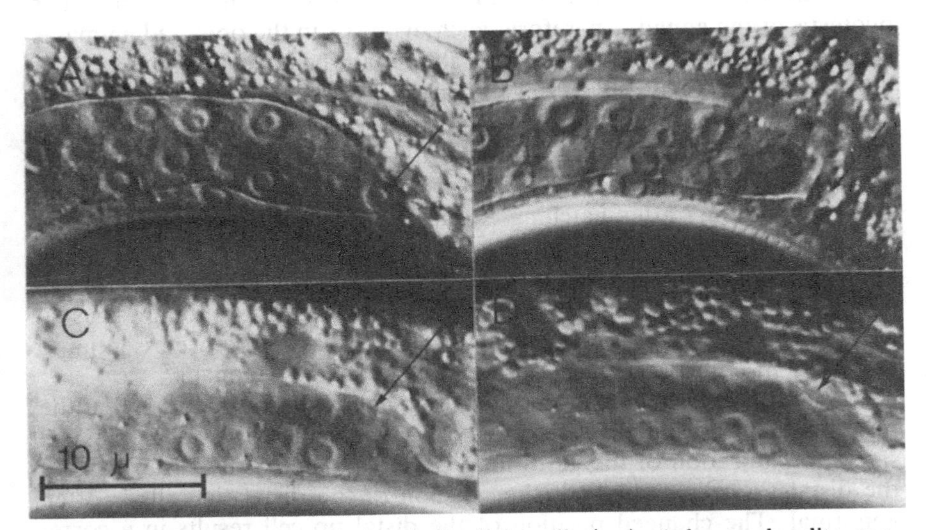

Figure 2 Nomarski micrographs of distal tip cells (→). A hermaphrodite posterior distal tip cell is shown in an L3 before (*A*) and after (*B*) laser ablation. A male distal tip cell is shown in an L3 before (*C*) and after (*D*) laser ablation. (Reprinted, with permission, from Kimble and White 1981.)

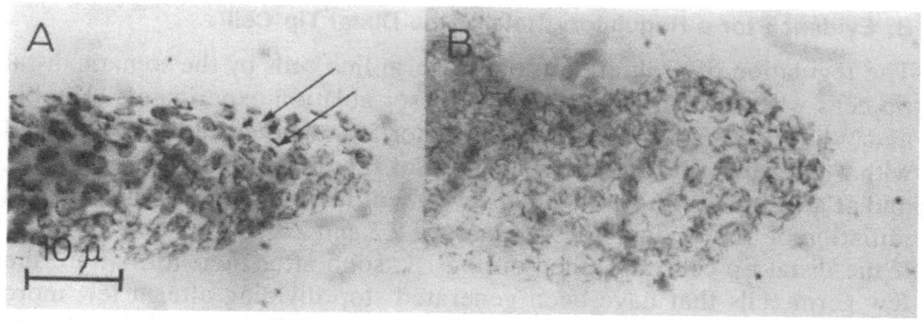

Figure 3 Chromosome morphology of nuclei before and after distal tip ablation observed in Feulgen-stained preparations of dissected gonads. (*A*) Nuclei in the distal end of an unoperated hermaphrodite gonad are not meiotic. Most are in mitotic interphase and two are dividing (→). (*B*) Nuclei in the distal end of a hermaphrodite gonad, about 24 hr after killing the distal tip cell, are all in pachytene. No mitotic figures are seen. (Reprinted, with permission, from Kimble and White 1981.)

An alteration in the distance between the distal tip cell and the most proximal germ-line nuclei shows that the influence of the distal tip cells over neighboring germ cells acts over a distance. If one of the two germ-line precursor cells, Z2 or Z3, is destroyed by laser ablation in a hermaphrodite larva that has just hatched, the remaining precursor cell normally contributes half its descendants to one ovotestis and half to the other ovotestis of the gonad. Therefore, each ovotestis in the operated animal is smaller than normal, containing only about half the normal complement of germ-line nuclei at any given stage of gonadal development. In unoperated hermaphrodites, the most proximal germ cells enter meiosis 33–34 hours after hatching. However, in animals with the gonad reduced in size by ablation of Z2, the most proximal germ-line nuclei do not enter pachytene until 38–44 hours after hatching. Control experiments demonstrate that this delay is not simply an artifact of secondary damage caused by the ablation. For example, if the distal tip cells are destroyed in addition to Z2, nuclei of the shorter gonadal arm display a pachytene morphology as usual at 33–34 hours. The presence of the distal tip cell is therefore essential to the delay in entry to meiosis observed in the smaller gonads.

An alteration in the position of distal tip cells confirms the idea that the position of the distal tip cell activity is responsible for establishing the spatial organization of the germ-line tissue. In males, if the sister cells of the distal tip cells are killed, the position of the distal tip cells is often abnormal. The changed position of the distal tip cell results in a corresponding shift in the axial polarity of the germ-line tissue. For example, if the distal tip cells become located at the anterior end of the gonad, the germ cells at the anterior end remain mitotic and the germ cells at the posterior end enter meiosis and begin spermatogenesis. The polarity of the germ-line tissue is therefore reversed by a change in the position of the distal tip cells.

C. A Gene Central to Control of Germ-line Proliferation, *glp-1*

One gene has been identified that appears to affect the distal tip cell control over germ-line proliferation. Many mutant alleles of this gene, *glp-1 III*, have been isolated (Austin and Kimble 1987; Priess et al. 1987). The germ line of animals of either sex, when homozygous for any of the *glp-1* alleles, is defective in the decision between mitosis and meiosis. Thus, the germ-line cells that are present at hatching divide mitotically only a few times. The descendant germ cells all enter meiosis and differentiate as sperm earlier than normal. This mutant phenotype is essentially the same as the effect observed after ablating both distal tip cells during L1. In the mutant, however, the distal tip cells are still physically present: Z1.aa and Z4.pp in hermaphrodites and Z1.a and Z4.p in males assume their normal morphology and position. In *glp-1* mutants, the leader function of the hermaphrodite distal tip cell is not affected. Therefore, an ovotestis in a *glp-1* homozygous animal assumes its normal U shape but is filled throughout with only a small number of sperm. The function of *glp-1* is unknown. It is plausible that the gene encodes a signal or a receptor in the distal tip cell–germ-line interaction.

Animals homozygous for weak alleles of *glp-1* make sufficient germ-line cells to produce some embryos; however, these embryos die. This embryonic lethal phenotype of *glp-1* shows a strict maternal effect, so that a homozygous *glp-1* hermaphrodite mated to a wild-type male produces only dead embryos, whether or not they are homozygous or heterozygous for *glp-1*. A temperature-sensitive allele of *glp-1* suggests that activity of the *glp-1*-gene product is required only early in embryogenesis for embryonic viability but that it is required throughout larval development for growth of the germ line. Thus, *glp-1* encodes a product that is required for the proper growth and development of the germ line but also for development of the embryo (also see Chapter 8).

IV. SPERMATOZOA AND SPERMATOGENESIS

A. Morphology of Spermatozoa

C. elegans spermatozoa, like those of other nematodes and certain crustaceans, are nonflagellated crawling cells. They are approximately 4.5 × 7 μm, with a single knobby pseudopod protruding from one side of a hemispherical cell body (Fig. 4a) (Wolf et al. 1978; Nelson and Ward 1980; Nelson et al. 1982). When spermatozoa are conventionally fixed, sec-

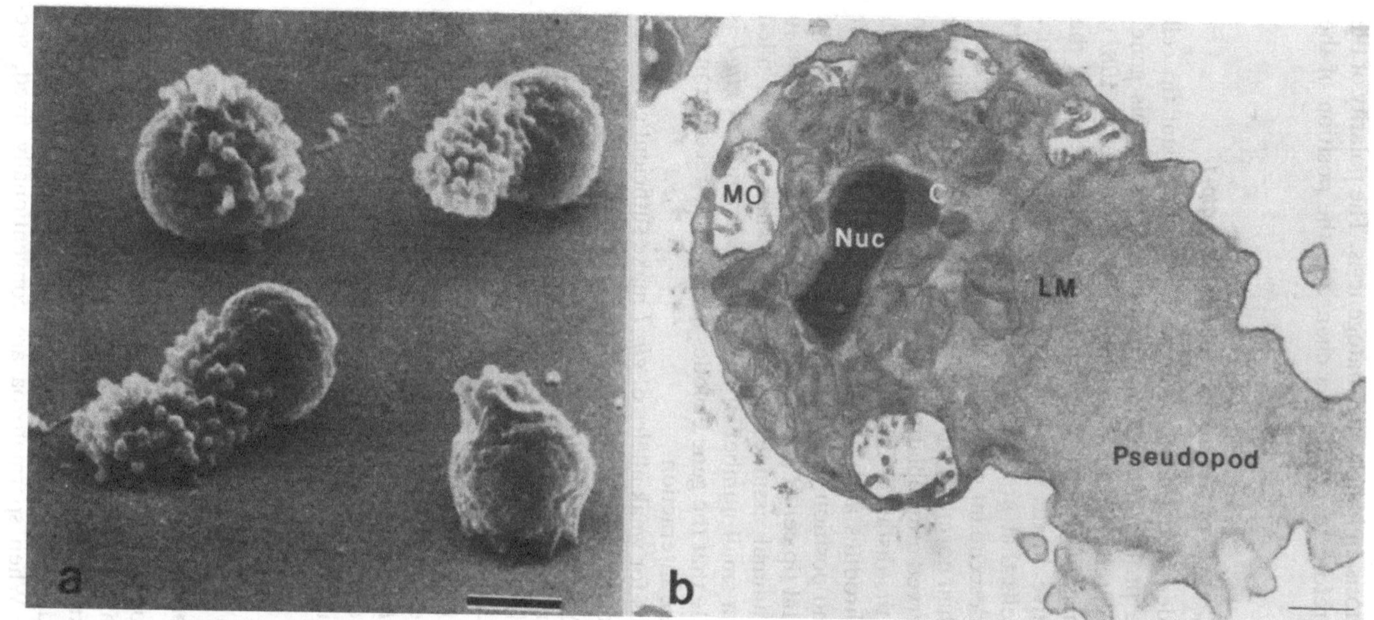

Figure 4 (*a*) Scanning electron micrograph of spermatozoa. The cell on the lower right was crawling over the substrate when fixed; the other cells were not properly attached by their pseudopods to crawl, but they were moving their pseudopods. (*b*) Transmission electron micrograph. The organelles concentrated in the rounded cell body include the nucleus (Nuc), which is surrounded by a less electron-dense halo containing the centrioles (C), mitochondria, and the MOs that have fused with the plasma membrane, forming a stable pore and releasing their fibrous glycoprotein contents. A thin laminar membrane (LM) separates the organelles from the amorphous contents of the pseudopod. No microtubules or microfilaments are found in the spermatozoa, but with other fixation conditions, 2-nm-diameter. filaments are abundant in the pseudopod cytoplasm (Roberts 1983). (Reprinted, with permission and slight modification, from Ward et al. 1981 [copyright, permission of the Rockefeller University Press].)

tioned, and examined by transmission electron microscopy, the pseudopod appears granular, without organelles, microfilaments, or microtubules (Fig. 4b). When fixed more stringently and examined carefully in sections or as whole mounts by high voltage electron microscopy, the pseudopod is found to contain numerous thin filaments, 2–3 nm in diameter (Roberts 1983). These filaments have recently been shown to be composed of the major sperm protein (see Section IV.B; T. Roberts, pers. comm.).

The pseudopod is separated from the spermatozoan cell body by several rows of thin laminar membranes, which extend throughout the cell-body cytoplasm (Fig. 4b). In addition to these membranes, the cell-body cytoplasm contains numerous mitochondria and distinctive membranous organelles (MOs). These are complex, Golgi-derived vesicles found in the sperm of most nematodes. The MOs have a spherical head portion with an electron-dense collar and a highly invaginated membrane body. Most of the MOs in the spermatozoa have fused with the plasma membrane, forming stable pores on the cell body and releasing a fibrous glycoprotein, which adheres to the outside of the cell body. The nucleus consists of condensed chromatin surrounded by an RNA-containing halo in which a paired centriole is embedded. There is no nuclear membrane. Spermatozoa lack many common cytoplasmic structures and organelles such as microtubules, microfilaments, ribosomes, Golgi apparatus, or lysosomes. These structures and organelles are segregated out of the developing sperm, as described below.

B. Protein Composition of Sperm

Sufficient sperm for biochemical analysis can be obtained by growing mutant strains that produce males at high frequency (*him* mutants) in liquid culture. Males are separated from hermaphrodites and larvae by filtration through nylon mesh filters. Sperm are obtained by squashing males between two glass plates and isolating the released sperm by filtration and sedimentation (Klass and Hirsh 1981; Nelson et al. 1982). A 2-liter culture can yield 5×10^8 cells, which contain about 8 mg of total protein.

When purified sperm are disrupted by SDS treatment and the proteins are fractionated on two-dimensional gels, more than 500 polypeptides are found (Fig. 5). Tubulin and actin are minor spots, consistent with the observed absence of microfilaments and microtubules in the sperm. Some of the tubulin presumably represents subunits of the centriole, but contaminating spermatocytes may also contribute and could account for all of the actin found in sperm preparations. The absence of actin in an ameboid crawling cell is astonishing and has led to the discovery of a new mechanism for cellular locomotion (see Section IV.F).

A prominent spot on two-dimensional gels of sperm proteins is made up

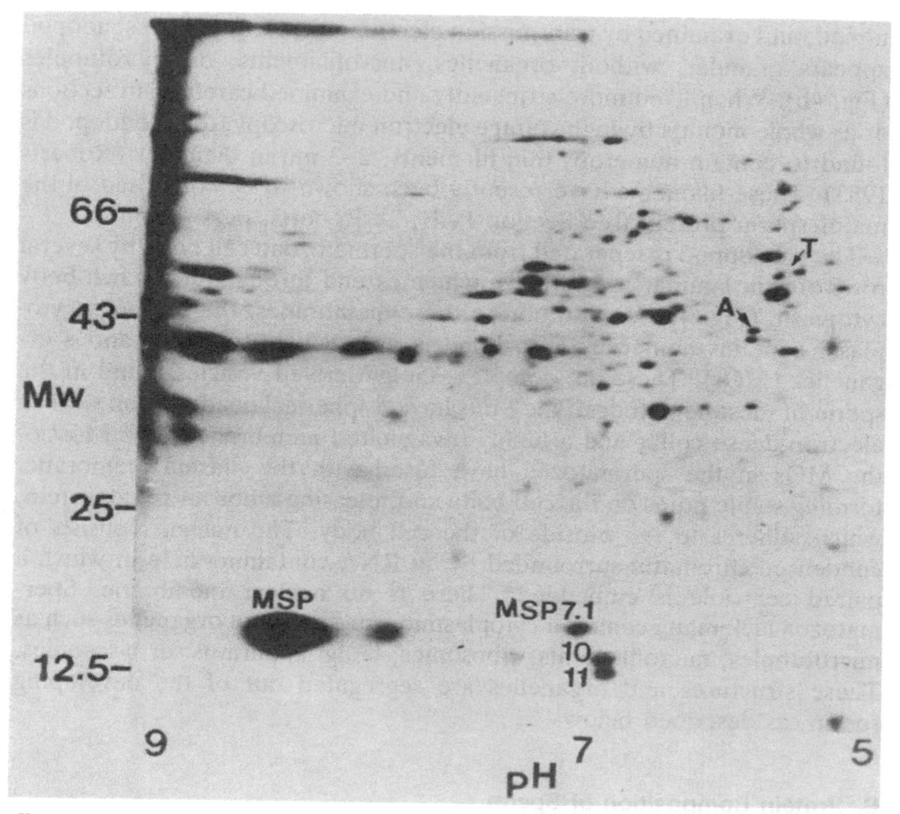

Figure 5 Two-dimensional gel of sperm proteins. ^{35}S-labeled sperm were dissociated in SDS and fractionated by isoelectric focusing from left to right (pH gradient indicated on the bottom) and then by SDS-PAGE from top to bottom (positions of molecular-weight standards are indicated on the left). Proteins were visualized by fluorography. Actin (A) was identified by coelectrophoresis with worm actin; α- and β-tubulins (T) were identified by coelectrophoresis and immunoblotting with anti-α-tubulin. MSPs and an isoelectric variant, MSP7.1, are indicated. (10,11) Sperm-specific proteins that were cut out of gels and used to prepare antibodies. (Reprinted, with permission, from Ward 1986 [copyright, permission of the Rockefeller University Press].)

of at least three and probably more members of a family of similar proteins called the major sperm proteins (MSPs). All three classes of MSPs are detected among proteins synthesized in vitro from male mRNA, so each must represent one or more independently synthesized polypeptide chains. MSPs are not detected by immunofluorescent staining or immunotransfer labeling in any worm tissue except spermatocytes and sperm, nor can their mRNA be detected (see Section IV.C); therefore, the MSPs are sperm-specific proteins. In sperm, they are present in the cytoplasm of both the cell body and the pseudopod where they polymerize into 2–3-nm filaments (Ward and Klass 1982; Roberts et al. 1986).

In addition to the MSPs, two-dimensional gels resolve a number of other polypeptides that appear to be sperm-specific because they are not found on gels of other worm tissue. Some of these polypeptides are numbered in Figure 5. Antibodies to the polypeptides p10, p11, and others have been prepared and used to confirm their sperm specificity. The functions of these proteins are unknown, but like the MSPs, p10 and p11 are both found in the cytoplasm of the pseudopod (Ward 1986).

In addition to these cytoplasmic proteins, sperm-specific membrane proteins have been identified by monoclonal antibodies prepared against whole sperm. Three different antibodies have been found to react with a polypeptide antigen shared by at least eight different minor sperm-specific proteins (Ward et al. 1986). A gold conjugate of one of these monoclonal antibodies detects the antigens in the plasma membrane, the membrane of the body of the MO, and the contents of the MO (Roberts et al. 1986). The appearance and localization of this antigen during sperm development will be described below.

C. MSP Genes

Genes encoding members of the MSP family have been cloned independently in two laboratories (Klass et al. 1982, 1984; Burke and Ward 1983). Because the messages for these proteins are the most abundant mRNAs in whole males, the genes were easily selected by differential screening of genomic and cDNA libraries with male versus hermaphrodite RNA. The identification of the genes was confirmed by hybrid-selected in vitro translation and by comparing the protein sequence predicted from DNA sequencing to that obtained by partial microprotein sequencing.

When genomic Southern blots are probed with cDNA clones of MSP genes, more than 30 bands are found to hybridize. Thus, the MSPs are encoded by a large multigene family with many more members than are indicated by the three protein components separable by electrophoresis (Burke and Ward 1983; Klass et al. 1984). Fourteen cDNA clones have been sequenced and 13 different sequences found, showing that many of the different MSP genes are transcribed into RNA (Klass et al. 1988; Ward et al. 1988). The protein-coding regions of these genes are 87–90% similar, with nearly all substitutions occurring in the third position of codons, so that the encoded MSP proteins are 96–100% conserved.

Unlike the coding-region sequences, the 3′-untranslated sequences of MSP cDNA clones are not conserved. Synthetic nucleotide probes corresponding to these sequences have been used to probe genomic Southern blots. Each hybridizes to only one or a small subset of the MSP genes

(Ward et al. 1988). Many of the transcribed MSP genes have been isolated from a genomic library using these synthetic probes.

To study the location and organization of transcribed MSP genes, cosmid clones containing larger regions of DNA that include some of these sequences have been identified by the DNA fingerprinting method (Coulson et al. 1986). A total of 41 MSP genes have been identified on cosmid clones out of an estimated total of 60 MSP genes in this multigene family. From DNA sequencing and hybridization of DNA with probes from different parts of the MSP coding sequence, it is estimated that about half of the MSP genes are pseudogenes.

Thirty-nine of the MSP genes, including all those known to be transcribed, are organized into 6 clusters composed of 3–13 genes each (Ward et al. 1988). Within each cluster, the genes are usually separated by several kilobase pairs of DNA. Pseudogenes are interspersed among functional genes.

These six clusters of MSP genes have been mapped, by overlap with known genes or by in situ hybridization, to only three chromosomal locations, the left arm of chromosome II and the left and center of chromosome IV. Among the MSP genes on chromosome IV are additional sperm-specific genes, including the gene for p10 (Ward et al. 1988).

D. Spermatogenesis

The determination of cells to become sperm is discussed in Chapter 9. This section will describe the morphological changes leading from spermatocyte to spermatozoan, based on electron microscopy of the male testis and of spermatocytes and spermatids developing in vitro (Klass et al. 1976; Wolf et al. 1978; Nelson and Ward 1980; Ward et al. 1981, 1983). Development proceeds linearly from the spermatogonial cells in the distal tip of the male testis to the spermatozoa adjacent to the *vas deferens*, making it possible to follow all stages in longitudinal sections or serial cross sections. In addition, a medium that supports spermatozoan motility also allows some primary spermatocytes to develop in vitro to form spermatids, so these steps can be carefully studied by time-lapse video recording. All the intermediates described below for male spermatogenesis have also been seen by electron microscopy in hermaphrodites, so spermatogenesis in males and hermaphrodites is likely to be identical. Male and hermaphrodite spermatozoa are indistinguishable by electron microscopy and by their motility in vitro, but they can be distinguished functionally (see Section VI).

About 90 minutes are required, either in vivo or in vitro, for the tetraploid primary spermatocyte to complete meiosis and form four haploid spermatids. The morphological changes are summarized in Figure 6.

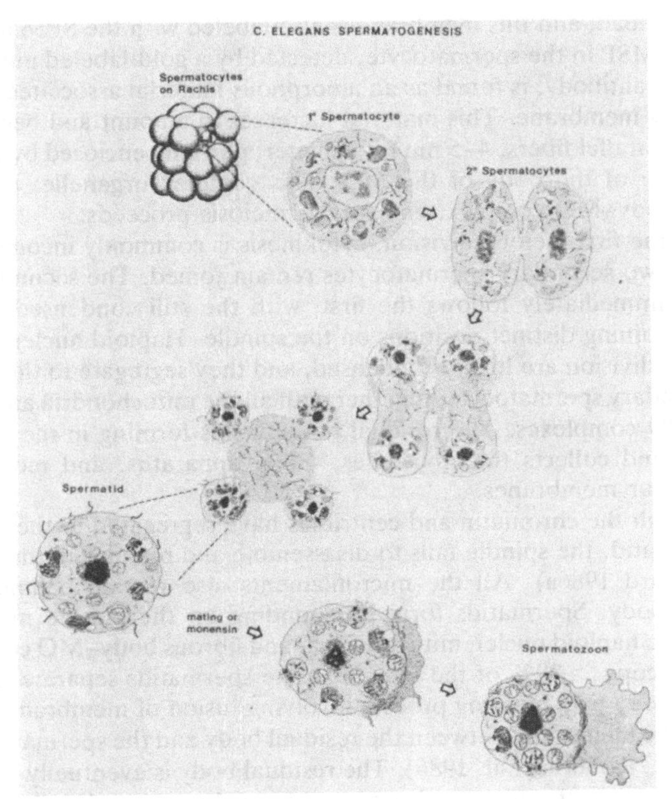

Figure 6 Summary of sperm development. Spermatocytes enter meiosis and proceed through pachytene, attached to a central core of cytoplasm called the rachis. They bud from this core and complete meiosis, forming four haploid nuclei that collect organelles and form spermatids. These bud off a central region of cytoplasm, which is left behind as a residual body. Spermatids are triggered to mature to spermatozoa by mating, or in vitro by monensin, weak bases, or proteases. This maturation includes fusion of the MOs, extension of the pseudopod, and initiation of motility. (Reprinted, with permission, from Ward et al. 1981 [copyright, permission of the Rockefeller University Press].)

The development of the peculiar sperm-specific MO and the association of MSPs with this developing organelle have been studied in detail using both fluorescent and gold-labeled antibodies to follow the localization of sperm-specific antigens (Ward and Klass 1982; Roberts et al. 1986). The MOs first form from the Golgi apparatus in the primary spermatocyte and are recognized by the characteristic electron-dense collar and head. The contents of the head are labeled by a sperm-specific monoclonal antibody (SP56). Subsequently, the membrane of the body of the organelle extends from the head, and this membrane is also labeled with the SP56 antibody. The first MSP in the spermatocyte, detected by a gold-labeled monoclonal anti-MSP antibody, is found as an amorphous material associated with the MO body membrane. This material increases in amount and becomes an array of parallel fibers, 4–5 nm in diameter, partially enclosed by a double membrane of the body of the MO. This complex organelle, called the fibrous body–MO complex, enlarges as meiosis proceeds.

After the first meiotic division, cytokinesis is commonly incomplete, so that the two secondary spermatocytes remain joined. The second meiotic division immediately follows the first, with the still condensed chromosomes retaining distinct positions on the spindle. Haploid nuclei resulting from this division are highly condensed, and they segregate to the poles of the secondary spermatocyte, together with all the mitochondria and fibrous body–MO complexes. The residual body begins forming in the center of the cell and collects the ribosomes, Golgi apparatus, and most of the intracellular membranes.

Although the chromatin and centrioles have segregated to the developing spermatid, the spindle fails to disassemble and remains in the residual body (Ward 1986a). All the microfilaments also remain behind in the residual body. Spermatids form by rounding up the plasma membrane around the haploid nuclei, mitochondria, and fibrous body–MO complexes (which occupy >30% of the volume). The spermatids separate from the residual body by a budding process involving fusion of membrane vesicles to form the membrane between the residual body and the spermatid (Ward et al. 1981; Roberts et al. 1986). The residual body is eventually degraded and resorbed in the male and presumably in the hermaphrodite as well.

From the antibody-labeling experiments, the fibrous body–MO complex is a transient organelle used to transport both cytoplasmic and membranous sperm-specific proteins to the spermatid. Conversely, the proteins actin and tubulin are prevented from appearing in the spermatid by remaining assembled into fibers after meiosis so that they are left behind in the residual body.

After the spermatid buds off the residual body, the fibrous body disassembles and distributes the MSPs throughout the cytoplasm (Ward and Klass 1982; Roberts et al. 1986). The MO body becomes more compact and moves to the periphery of the spermatid. In males that do not copulate, development arrests with accumulation of more than 3000 spermatids. A single male, mated with several hermaphrodites, is capable of siring more than 2500 progeny (Hodgkin 1983a).

Spermatids are spherically symmetric sessile cells. Those produced in males complete maturation to spermatozoa (spermiogenesis) following copulation. Spermatids transferred to hermaphrodites by copulation form pseudopods and begin movement in the uterus within 5 minutes of transfer. Some spermatids left behind in the male also mature to spermatozoa and can be observed crawling in the vas deferens and seminal vesicle. By analogy to *Ascaris,* where an unidentified substance from the vas deferens triggers spermatid maturation (Burghardt and Foor 1978), *Caenorhabditis* spermiogenesis may also be induced by a secretion from the vas deferens, but this has not been demonstrated. Spermiogenesis can be initiated in vitro with high efficiency and reproducibility by three different treatments: the ionophore monesin, weak bases such as triethanolamine, and proteases (Nelson and Ward 1980; Ward et al. 1983). The first two act by increasing the intracellular pH of the spermatid from its resting value of 7.1 to 7.7 or more. Intracellular pH changes trigger many developmental events and have many potential sites of action (reviewed in Nuccitelli and Deamer 1982). Within 2 minutes of this pH increase the spermatid initiates surface movements of its membrane, rearranges its surface projections, fuses its MOs with the posterior end of the cell, and extends its pseudopod anteriorly. This process is completed within 5 minutes, both in vivo and in vitro. Spermiogenesis is an energy-dependent process, which is blocked by metabolic energy inhibitors. Protease treatment triggers the same maturation but without increasing the intracellular pH.

Motility of the pseudopod is not necessary for its formation, because certain concentrations of the respiratory inhibitor azide block motility but not pseudopod formation, and because some mutants form nonmotile pseudopods (see Section IV.F).

E. Mutants with Altered Sperm

One of the reasons for studying spermatogenesis is that mutants affecting sperm development are easily obtained. Sperm-defective mutants are sterile hermaphrodites that can produce progeny if mated to males. If such a homozygous mutant strain produces males that behave normally and copulate with hermaphrodites but yield no outcross progeny, the mutant is most likely sperm defective. The class of sperm-defective mutants most intensively studied are those that make infertile sperm in normal number. Such mutants were originally designated fertilization-defective or *fer* mutants (Ward and Miwa 1978), but all subsequently isolated sperm-defective mutants are named *spe*, irrespective of where in spermatogenesis they are blocked. More than 60 mutations affecting spermatogenesis have been obtained and assigned to more than 40 different genes (Ward and Miwa 1978; Argon and Ward 1980; Ward et al. 1981; S. L'Hernault et al., in prep.). Some of these are summarized in Appendix 4B. Mutations blocking at every stage of sperm development have been obtained. Some mutants accumulate apparently normal intermediates, such as spermatocytes or spermatids that do not mature to spermatozoa. Other mutants accumulate aberrant cells, such as spermatids with crystalline inclusions or spermatozoa that have failed to fuse their MOs with the plasma membrane (Ward et al. 1981). Still other mutants make normal looking motile spermatozoa that fail to fertilize eggs.

F. Sperm Motility

Hermaphrodite spermatozoa must crawl to locate themselves properly in the spermatheca and to avoid being swept away by fertilized eggs passing into the uterus and out of the vulva. Male spermatozoa must be able to crawl from the region in the uterus where they are deposited during copulation to the site of fertilization, the spermatheca, a distance of more than 200 μm. These movements can be observed easily in the transparent hermaphrodite using Nomarski microscopy and time-lapse video recording and can be analyzed in vitro (Nelson et al. 1982). Spermatozoa crawl at about 20 μm/minute, but they can move as fast as 43 μm/minute for short intervals. The pseudopod attaches to the substrate and pulls the cell forward, with the cell body either dragged passively behind or carried atop the rear part of the motile pseudopod. As the cell crawls, new substrate attachments are found continuously under the front of the pseudopod and, in extreme cases, contact under the leading edge of the pseudopod alone provides sufficient traction for locomotion (Roberts and Streitmatter 1984). Cells remain motile for several hours in vitro.

The mechanism of motility cannot be an actin–myosin-based contractile system, because the spermatozoa contain almost no actin and no detectable myosin, and their movements are unaffected by drugs that interfere with microfilaments. Instead, the spermatozoa propel themselves by tip to base flow of membrane over the pseudopod surface (Roberts and Ward 1982a,c). Spermatozoa insert new membrane components at the tips of their pseudopods, and these components flow backwards to be taken up at the base of the pseudopod. This propels the cell forward, because the pseudopod membrane is being continuously rebuilt at the tip, creating new attachment sites under the leading edge of the advancing cell.

Some of the evidence establishing this membrane flow is the observation that labeled lectins, antibodies, and polystyrene beads attached to the pseudopod all move from the tip to the base, as do fluorescently labeled lipids inserted into the membrane. It is not known what provides the driving force for this membrane flow, what establishes its polarity, or how membrane components are inserted and removed.

V. OOGENESIS

A. Description of Oogenesis

Oogenesis begins at the loop between the distal and proximal arms of the hermaphrodite ovotestis. Nuclei at the loop are present only at the outer side of the bend, and, compared with the distal arm, there is a marked increase in volume of cytoplasm surrounding each nucleus in this region. Oocytes appear to result from the packaging of a single peripheral nucleus with cytoplasm from the central core of the distal arm. Generally, the first oocytes produced are observed just after the molt to adulthood.

The proximal arm of adult ovotestis contains a single file of enlarging oocytes. Six bivalents can be distinguished in each nucleus. Mature oocytes are arrested in meiosis at diakinesis of meiotic prophase I. Both meiotic divisions occur after fertilization. The cytoplasm of maturing oocytes contains a large number of osmiophilic granules. Comparison of the ultrastructure of oocytes made in a wild-type animal to that of oocytes made in a mutant that does not contain yolk suggests that these granules contain yolk (Doniach and Hodgkin 1984).

B. Yolk Proteins and Their Synthesis

The yolk proteins of *C. elegans* were first identified as four abundant proteins specific to adult hermaphrodites (Klass et al. 1979). They include two related proteins (yp170A and yp170B) and two unrelated proteins (yp115 and yp88). All four are glycoproteins (Sharrock 1983). Only three polypeptides specific to adult hermaphrodites are readily observed among the products of RNA translated in vitro (Sharrock 1984). Two of these correspond, with little or no modification, to the yp170A/B doublet. The third is large, with a M_r of about 180,000, and binds antibodies specific for both yp115 and yp88 (Sharrock 1984). These and other observations suggest that the 180,000-M_r protein is a precursor, which is cleaved and modified to yield both yp115 and yp88.

A comparison of proteins synthesized by dissected tissues (intestines, gonads, and body wall) suggests that the intestine is the primary site of vitellogenin synthesis (Kimble and Sharrock 1983). Although the yp115/yp88 precursor is secreted from the intestine (Sharrock 1984), its cleavage products yp115 and yp88 accumulate in animals possessing no gonadal tissue. Therefore, cleavage of vitellogenin most likely occurs in the body cavity of the nematode before uptake by the ovary.

The yolk proteins of *C. elegans* can be visualized in situ by indirect immunofluorescent staining (Sharrock 1983). In the embryo, staining with antibodies against any of the yolk proteins yields a punctate pattern. In the oocyte, yolk is localized in osmiophilic cytoplasmic granules, which accumulate during maturation (Doniach and Hodgkin 1984). Generally, only the two or three oocytes near the spermatheca exhibit the high density of staining characteristic of fertilized eggs.

Yolk protein uptake may be facilitated by the epithelial sheath of the oviduct that surrounds maturing oocytes. Electron micrographs show osmiophilic granules in the cytoplasm of sheath cells and coated pits on both the membranes facing the body cavity and those facing the oocyte. However, oocytes contain these granules even when the epithelial sheath has been eliminated by laser ablation. Therefore, it seems likely that yolk can be taken up by oocytes either directly from the body cavity or after transport through the sheath.

C. Vitellogenin Genes

The vitellogenins are encoded by six genes. A family of five genes (*vit-1*, *vit-2*, *vit-3*, *vit-4*, and *vit-5*) encodes the yp170A and yp170B proteins, and one gene (*vit-6*) encodes the precursor protein for yp115 and yp88 (Blumenthal et al. 1984; Spieth and Blumenthal 1985). Cloning of these genes was facilitated by the abundance of their mRNAs in a poly(A)-containing fraction of *C. elegans* high-molecular-weight RNA. Hybrid–arrest translation experiments show that *vit-1* and *vit-2* correspond to yp170B, whereas *vit-3*, *vit-4*, and *vit-5* correspond to yp170A. However, *vit-1* has a stop codon in frame and is therefore a pseudogene (Spieth et al. 1985a). Moreover, cDNA clones have been obtained only from *vit-2* and *vit-5*, so that *vit-3* and *vit-4* may also be pseudogenes.

The vitellogenin mRNAs for yp170A and yp170B and for the yp115/yp88 precursor are about 5 kb in length (Blumenthal et al. 1984; Spieth and Blumenthal 1985). They are abundant in adult hermaphrodites but are missing from larvae and males. Furthermore, they are highly enriched in RNA from intestines but are not detectable in RNA isolated from gonads or body walls.

The coding regions of the *vit* genes are highly conserved (Spieth et al. 1985b). In the region coding for the first 100 amino acids, the *vit-1*/*vit-2* subfamily is 70% homologous to the *vit-3*/*vit-4*/*vit-5* subfamily, and *vit-6* is about 50% homologous to the *vit-1*–*vit-5* family. All six genes begin with a highly conserved signal sequence, and all have a TATA box at about position −30 from the start site of transcription. The nucleotide sequence of *vit-5* has been completed (Spieth et al. 1985a). The *vit-5* message is 4869 nucleotides long, including untranslated regions of 9 bases at the 5′ end and 51 bases at the 3′ end. *vit-5* contains four short introns totaling 218 bp. The predicted vitellogenin, yp170A, has a molecular weight of 186,430.

Two highly conserved heptameric sequences are found in the 5′-flanking regions of all six *vit* genes from *C. elegans* (Spieth et al. 1985b) and in the same regions of an additional five *vit* genes from *Caenorhabditis briggsae* (T. Blumenthal, pers. comm.). These heptameric sequences have not been observed in any other gene cloned from *C. elegans*. The first heptamer, called box 1 (TGTCAAT), appears in both orientations and repeats between four and six times per promoter region, allowing a 1-bp mismatch. The second heptamer, called box 2 (CTGATAA), is present in only one or two copies per promoter region.

With the exception of *vit-3* and *vit-4*, whose coding regions are separated by only about 3 kb, the vitellogenin genes of *C. elegans* do not appear to be clustered. Approximately 50 kb of genomic DNA has been cloned from around each of the four loci of the *vit-1–vit-5* family without revealing additional linkage between the genes (Heine and Blumenthal 1986). Mapping of restriction-fragment-length polymorphisms has localized *vit-1* and the *vit-3*/*vit-4* pair to the X chromosome. In addition, in situ hybridization experiments indicate that all members of the *vit-1–vit-5* family map to the X chromosome (D.G. Albertson, pers. comm. and Appendix 4D).

D. Oogenesis Mutants

No broad search for mutants specifically defective in oogenesis has been carried out to date. A number of embryonic lethal mutants have been obtained (called *zyg* or *emb*) in which a gene product contributed by the mother (presumably via the oocyte) is defective (Hirsh and Vanderslice 1976; Schierenberg et al. 1980). These mutants are discussed in more detail in Chapter 8. More recently, recessive mutants, in which hermaphrodites produce sperm continuously and make no oocytes, have been isolated (J. Kimble, unpubl.). In addition, sterile mutants that are defective in oogenesis have been isolated during an intensive screen for mutants of one particular region (Sigurdson et al. 1984). These have been named *ooc* mutants, but little is known about their defects.

VI. FERTILIZATION

A. Hermaphrodite Self-fertilization

The process of fertilization and the oocyte's response to sperm penetration were described many years ago (reviewed in Nigon et al. 1960; Nigon 1965). These events have been studied more recently by direct observation with Nomarski microscopy and video recording of fertilization in hermaphrodites (Hirsh et al. 1976; Ward and Carrel 1979). Sperm mature in the hermaphrodite as they pass from the testis into the spermatheca, where they can be seen crawling actively, pushing their pseudopods against the spermathecal walls. Electron micrographs show that the pseudopods protrude into invaginations in the spermathecal cells, presumably to anchor the spermatozoa in place.

Development of the spermatheca is described in Chapter 5. It is a folded tube about 100 μm long, with a constriction at the distal end connecting to the oviduct and a complex valve at the proximal end connecting to the uterus. When an oocyte matures in the oviduct, it is pushed up against the constriction by contractions of the oviduct sheath. The oocyte nucleus moves toward the distal end of the cell, and the nuclear membrane breaks down about 2 minutes before the ripe oocyte is pushed through the constriction by stronger contractions of the oviduct sheath. Multiple sperm contact the oocyte as it enters the spermatheca, but only one sperm penetrates and fertilizes the egg. The mechanism for preventing polyspermy is not known. From observations by light microscopy, it appears that the sperm is enveloped by the egg, probably pseudopod first as in *Ascaris* (Foor 1970), but electron micrographs of sperm penetration have not been obtained. Fertilization of the egg is recognized by a sudden increase in granule movement in the egg cytoplasm, quickly followed by eggshell formation. The fertilized egg remains in the spermatheca for 3–10 minutes. Contractions of the spermatheca then push the egg against the spermathecal valve, which dilates to allow passage of the egg into the uterus.

The egg usually carries several sperm with it into the uterus. These sperm crawl back through the spermathecal valve to regain their positions on the walls of the spermatheca, so that nearly every sperm eventually fertilizes an oocyte. Oocytes continue to mature after sperm depletion so that in contrast to the situation in almost all other animals, the number of progeny in *C. elegans* is limited by the number of sperm rather than the number of eggs (Ward and Carrel 1979).

The oocytes that are made in old hermaphrodites depleted of sperm or in mutants with defective sperm undergo partial maturation as they pass through the spermatheca. Their nuclear membrane disappears, and they resume meiosis. No eggshell forms, however, and granule motions of the egg cytoplasm are greatly reduced or absent, suggesting that these two events are triggered by sperm contact. After entering the uterus, unfertilized oocytes undergo one nuclear division, and the oocyte nucleus returns to the center of the cell. The DNA and chromosomes replicate, with nuclear membrane breakdown timed roughly as expected for the AB cell lineage but without cytokinesis or karyokinesis. This produces highly polyploid unfertilized oocytes, which are expelled through the vulva. Observation of such oocytes produced by young hermaphrodites is the simplest way to recognize a sperm-defective mutant with the dissecting microscope.

B. Male Cross-fertilization

When males copulate with hermaphrodites, their sperm are deposited through the vulva among the fertilized eggs in the uterus. The sperm then crawl among the eggs to reach the spermatheca. Surprisingly, when they arrive at the spermatheca of a young hermaphrodite, which already has hundreds of her own sperm, male sperm outcompete hermaphrodite sperm and preferentially fertilize oocytes (Ward and Carrel 1979). This appears to occur by displacement of the hermaphrodite sperm from their positions on the spermatheca. A hermaphrodite that has been multiply mated will lose her own sperm so that only outcross progeny will be produced after mating. More commonly, predominantly outcross progeny are produced until male sperm are depleted, and then self progeny reappear. This preferential sperm utilization appears to reflect a difference between male and hermaphrodite sperm, because if a spermless hermaphrodite is successively mated by males of two different genotypes, the second male has no advantage at fertilization.

In addition, mating has the effect of stimulating oogenesis by the hermaphrodite. An old hermaphrodite that has ceased oogenesis will reinitiate oocyte maturation after mating, and a multiply mated young hermaphrodite will produce as many as 1400 progeny, four times the normal number. Both preferential male sperm utilization and stimulation of oogenesis by mating are mechanisms to guarantee utilization of the male sperm in a population of predominantly self-fertilizing hermaphrodites, ensuring that if mating occurs, it will result in production of outcross progeny.

ACKNOWLEDGMENTS

We gratefully acknowledge the assistance of William Sharrock, Tom Blumenthal, and Michael Klass in preparing this chapter. We also thank Jonathan Hodgkin, Donna Albertson, John Sulston, and John White for excellent editorial advice. Research from the Kimble lab was supported by National Institutes of Health grant GM-31816 and a Basil O'Connor Starter Research grant 5-514. Research from the Ward lab was supported by National Institutes of Health grant GM-25243 and the Carnegie Institution of Washington.

8

Embryology

William B. Wood
Department of Molecular, Cellular, and Developmental Biology
University of Colorado at Boulder
Boulder, Colorado 80309-0347

I. INTRODUCTION

Formal aspects of the embryonic cell lineage and some discussion of embryonic cell determination have been presented in Chapter 5. This chapter provides a more complete description of embryogenesis per se, summarizes the embryonic origins of major organs and tissues, discusses further evidence for early cell determination by intrinsic, cell-autonomous determinants versus extrinsic cues, and reviews what has been learned about genetic control of embryogenesis by analysis of embryonic lethal mutants. The chapter concludes with a brief discussion of possible determination mechanisms in the early embryo.

Embryogenesis in *Caenorhabditis elegans*, from fertilization to hatching, takes about 14 hours at 22°C. The process can be conveniently considered

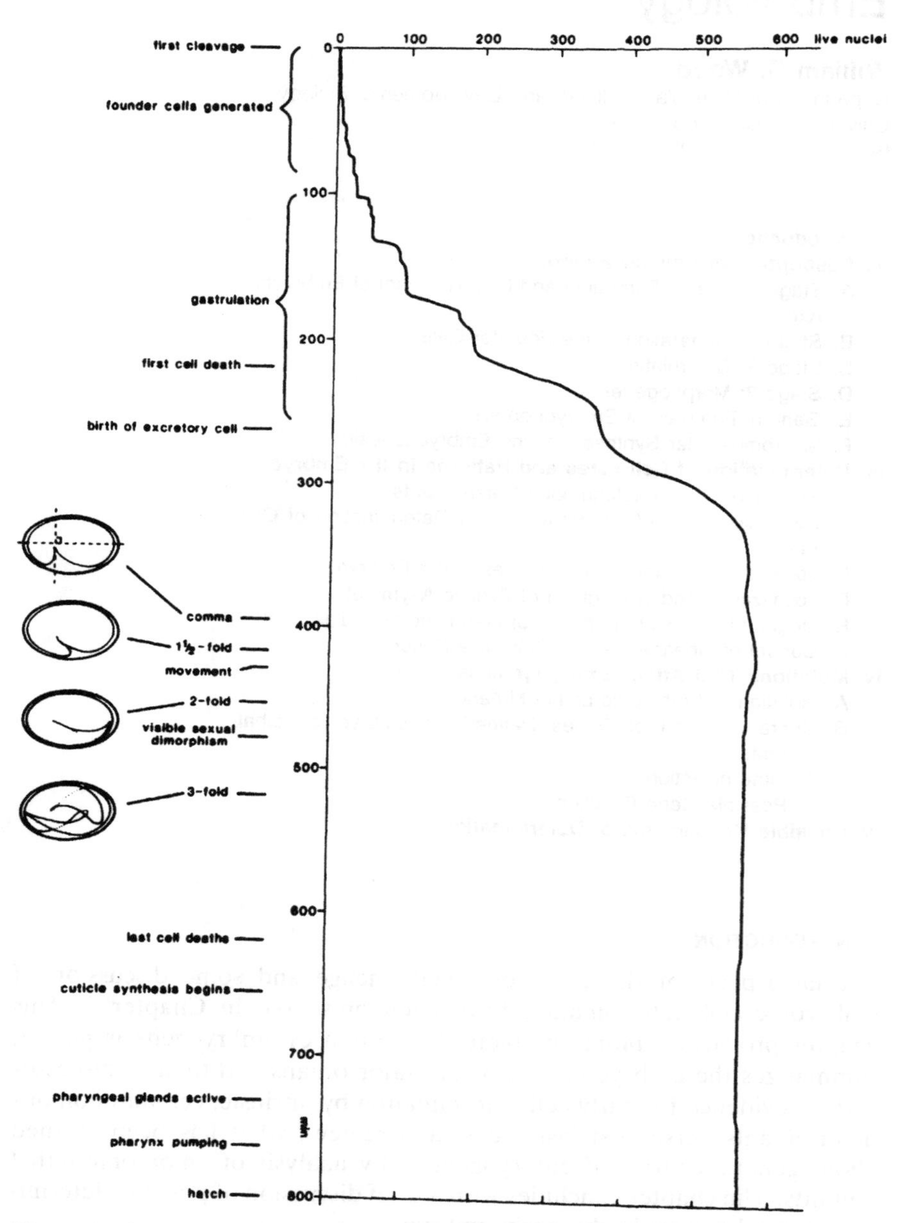

Figure 1 Stages, marker events, and number of nuclei during embryogenesis at 22°C. Fertilization normally occurs at about −40 min. (Reprinted, with permission, from Sulston et al. 1983.)

in three major stages (Fig. 1). The first, including zygote formation and early cleavage, establishment of the embryonic axes, and determination of the somatic and germ-line founder-cell fates, takes place during the first 2 hours after fertilization. The second, including gastrulation, completion of most cell proliferation, and the beginning of cell differentiation and organogenesis, continues until about halfway through embryogenesis. The third stage, including morphogenesis as well as completion of embryonic cell differentiation and organogenesis, occupies the remainder of embryogenesis and concludes with hatching. Early embryogenesis normally (although not necessarily) takes place in the uterus; eggs in healthy well-fed hermaphrodites are normally laid during gastrulation, at 2–3 hours after fertilization.

II. DESCRIPTION OF CELLULAR EVENTS

A. Stage 1: Zygote Formation and Establishment of Embryonic Axes

Mature oocytes pass from the oviduct into the spermatheca and become fertilized by fusion with sperm as they enter the spermatheca, as described in Chapter 7. At this time, the anterior–posterior polarity of the embryo is established, with the sperm pronucleus at the posterior pole (Albertson 1984a). It is not known whether sperm entry defines the posterior pole or whether a functional anterior–posterior polarity is already present in the oocyte before fertilization.

Figure 2 shows Nomarski photomicrographs of an egg in successive stages between fertilization and the end of first cleavage. After fertilization, the maternal pronucleus, which was arrested in first meiotic prophase, resumes meiosis, and by 20 minutes, the first and second polar bodies have been extruded at the anterior pole. Meanwhile, the egg has formed a tough shell around itself, consisting of an inner vitelline membrane impermeable to most solutes, a middle chitinous layer, and an outer layer consisting of lipids and probably collagenous, cross-linked proteins (Chitwood and Chitwood 1974). Also during this time, turbulent cytoplasmic movements are accompanied by contractions of the anterior cell membrane and a pronounced constriction, termed pseudocleavage, near the equator, involving formation and then regression of a cleavage furrow (Fig. 2a–c). Concurrently with pseudocleavage, the egg pronucleus migrates posteriorly toward the sperm pronucleus, which moves a short distance toward the center (Fig. 2b), until the two meet in the posterior half of the egg (Fig. 2c). On the basis of staining of *C. elegans* eggs with anti-tubulin antibodies, the centripetal movement of the sperm pronucleus appears to be mediated by growth of astral microtubules from the centriolar regions adjacent to the sperm pronucleus (Albertson 1984a). Further growth of the asters moves the two juxtaposed pronuclei to the center of

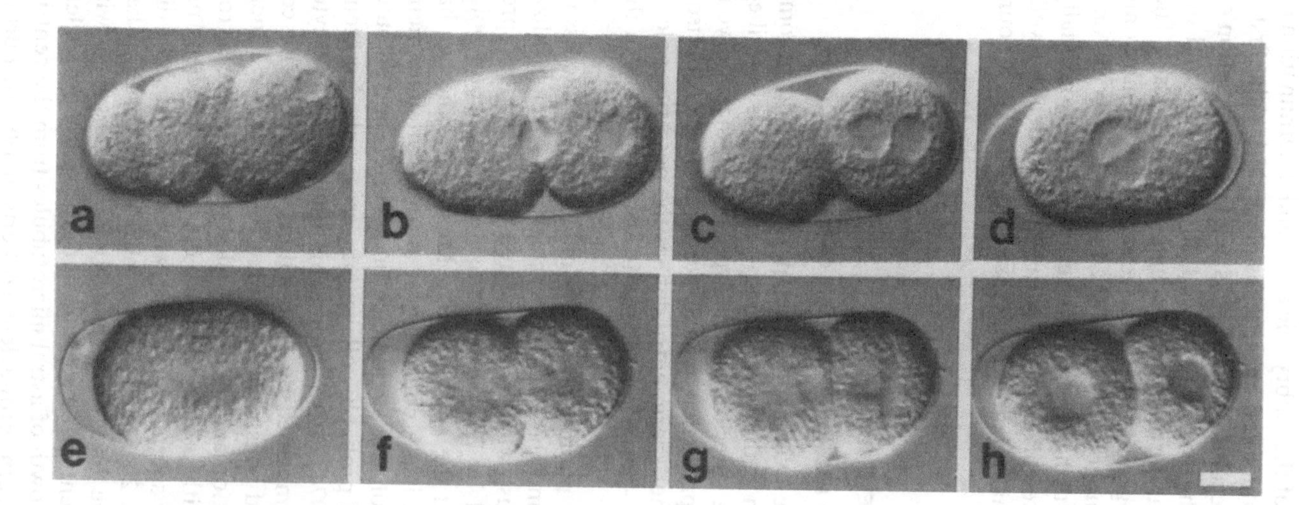

Figure 2 Nomarski photomicrograph of a *C. elegans* embryo at successive stages between fertilization and first cleavage. Embryos in this and subsequent figures are oriented with posterior pole to the right. (*a*) Formation of pronuclei and contractions of anterior membrane (~20 min after fertilization at 25°C). Egg pronucleus at anterior pole is not visible in this focal plane. (*b*) Pseudocleavage and pronuclear migration. (*c*) Meeting of the two pronuclei in posterior half of zygote. (*d*) Movement to center and rotation of pronuclei. (*e*) Formation of the first mitotic spindle. (*f*) Late anaphase and beginning of first cleavage. Centrosomes are visible as round, granule-free regions; note the asymmetric location of the spindle along the anterior–posterior axis. (*g*) Telophase; note disc-shaped centrosome in posterior (P_1) cell, as compared to spherical centrosome in anterior (AB) cell. (*h*) Two-cell embryo. (For additional explanation, see text.) Bar represents 10 μm. (Reprinted, with permission, from Strome and Wood 1983.)

the egg, where they rotate 90° and move to their final position along the anterior–posterior axis, slightly posterior to the center of the egg (Fig. 2d,e). The pronuclear envelopes then break down, and the spindle forms. At anaphase, the posterior centrosome moves backwards, whereas the anterior centrosome remains more or less stationary. At telophase, the posterior centrosome appears disc shaped, compared with the anterior centrosome, which appears spherical. First cleavage begins at about 35 minutes after fertilization (Fig. 2f,g). The first cleavage is asymmetric (unequal), giving rise to the larger anterior ectodermal founder cell AB and the smaller posterior germ-line cell P_1 (Figs. 2h, 4).

At about 10 minutes after completion of the first cleavage, the AB cell begins second cleavage (Fig. 3B), followed 3 minutes later by cleavage of P_1 to produce ethylmethanesulfonate (EMS) and P_2 (Fig. 2i,j). The polari-

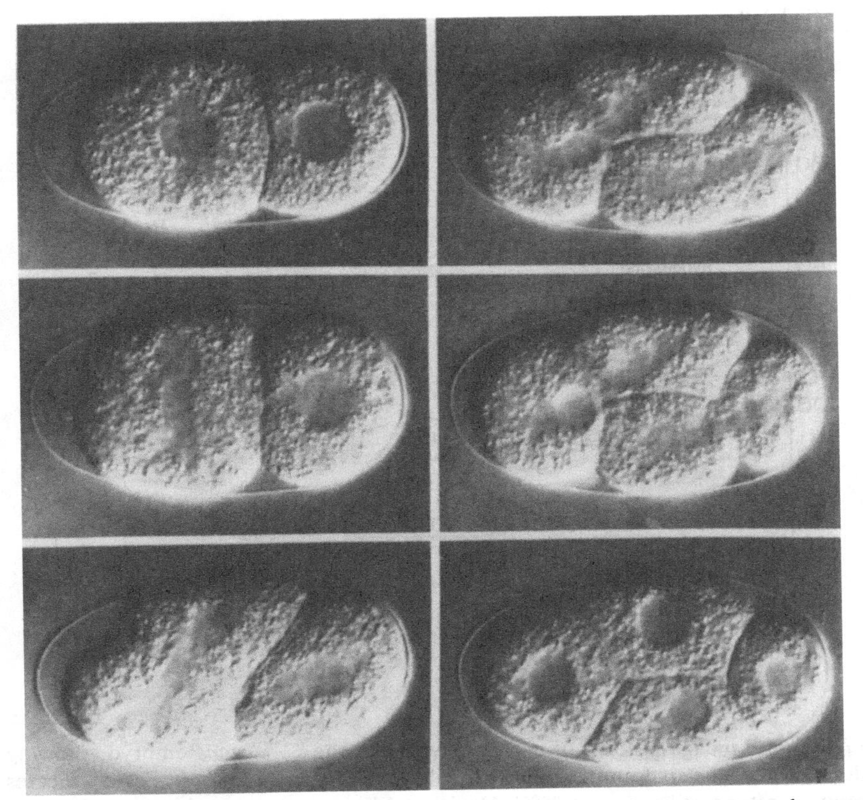

Figure 3 Nomarski photomicrographs of a *C. elegans* embryo between the two-cell and four-cell stages. Note that the larger anterior AB cell begins and completes cleavage somewhat before the smaller P_1 cell. (For further explanation, see text.) (Micrographs courtesy of Ann Cowan.)

ty of the dorsal–ventral axis first becomes apparent during AB cleavage (Fig. 3C), which places the posterior daughter (ABp) dorsal to EMS, leaving ABa at the anterior and P_2 at the posterior pole of the embryo (Fig. 3F).

B. Stage 1: Generation of the Founder Cells

Like first cleavage, the next three P-cell divisions are also unequal, following a stem-cell pattern in which each division gives rise to a larger somatic founder cell and a smaller germ-line (P-cell) daughter (Fig. 4). In each of these divisions, the P-cell centrosome is morphologically distinctive, as in first cleavage. The first somatic daughter is the AB founder cell; the second somatic daughter, EMS, divides unequally to produce the

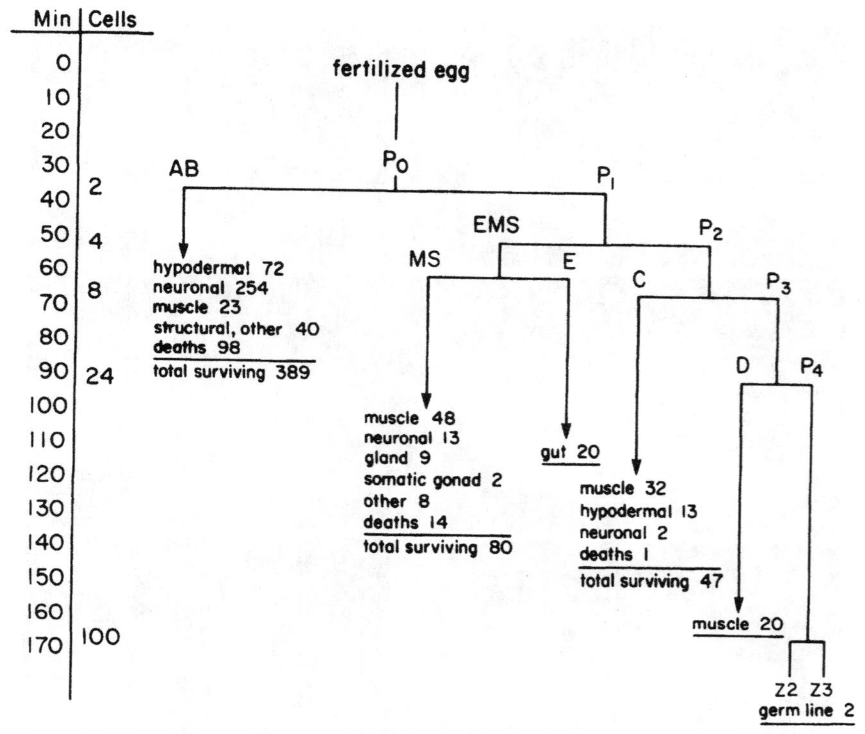

Figure 4 Lineage pattern of early cleavages in the *C. elegans* embryo, showing derivation of the six founder cells AB, MS, E, C, D, and P_4. (*Left*) Scale indicates minutes after fertilization at 25°C. Horizontal lines connecting sister cells indicate approximate times of cleavages. Second scale (*left*) shows number of cells in the embryo with time. Below each founder cell are indicated the type and number of cells (at hatching) that derive from it. (Adapted from Sulston et al. 1983.)

slightly smaller E and the slightly larger MS founder cells. Shortly thereafter, P_2 divides to produce the C founder cell and the germ-line cell P_3, which subsequently divides to produce the last somatic founder cell D and the germ-line founder cell P_4.

The divisions that produce the founder cells are asynchronous as well as unequal (Fig. 5). Following its birth, each founder cell and its progeny exhibit a characteristic rate of division that is roughly proportional to founder-cell size and follows the order of origin, from the fastest, AB, to the slowest, D (P_4 is omitted from the comparison because it divides only once more during embryogenesis; see Fig. 4).

The tissues and cell types that derive from each of the founder cells (Fig. 4) are known from the determination by Sulston et al. (1983) of the complete cell lineage during embryogenesis. The E cell is exclusively an endodermal precursor, giving rise to the entire gut; the D cell is strictly mesodermal, producing only body-wall muscle; and the P_4 cell gives rise only to the germ line. The other three founder cells, AB, MS, and C, produce both ectodermal and mesodermal derivatives, although descendants of AB are primarily ectodermal and those of MS are primarily mesodermal (see Fig. 4; Appendix 3).

C. Stage 2: Gastrulation

Just prior to gastrulation, which begins at about 100 minutes after first cleavage, the 16 AB descendants lie anteriorly and laterally, and the four C descendants lie posteriorly and dorsally. The four MS derivatives lie ventrally near the middle of the embryo, and the two E derivatives, D and P_4, lie ventrally and posteriorly. At the start of gastrulation (Fig. 6a), the two E cells sink inward, followed by P_4 and the MS cells. Later, as the entry zone widens and lengthens (Fig. 6b), most of the remaining myoblasts from the C and D lineages, and then the AB-derived pharyngeal precursors, sink into the interior. The ventral cleft is closed by 290 minutes. As gastrulation proceeds, further divisions of the E cells and the pharyngeal precursors form a central cylinder, and the body myoblasts become positioned between it and the outer cell layer. Cell division and organogenesis continue until about 350 minutes.

D. Stage 3: Morphogenesis

At about 350 minutes after first cleavage, cell proliferation largely ceases and morphogenesis begins (Figs. 1 and 7). The first twitching movements, indicative of muscle function, are observed at about 400 minutes (Fig. 6c). Sexual dimorphism can first be observed at about 470 minutes; a set of four specific neurons undergoes programmed cell death in the hermaphrodite,

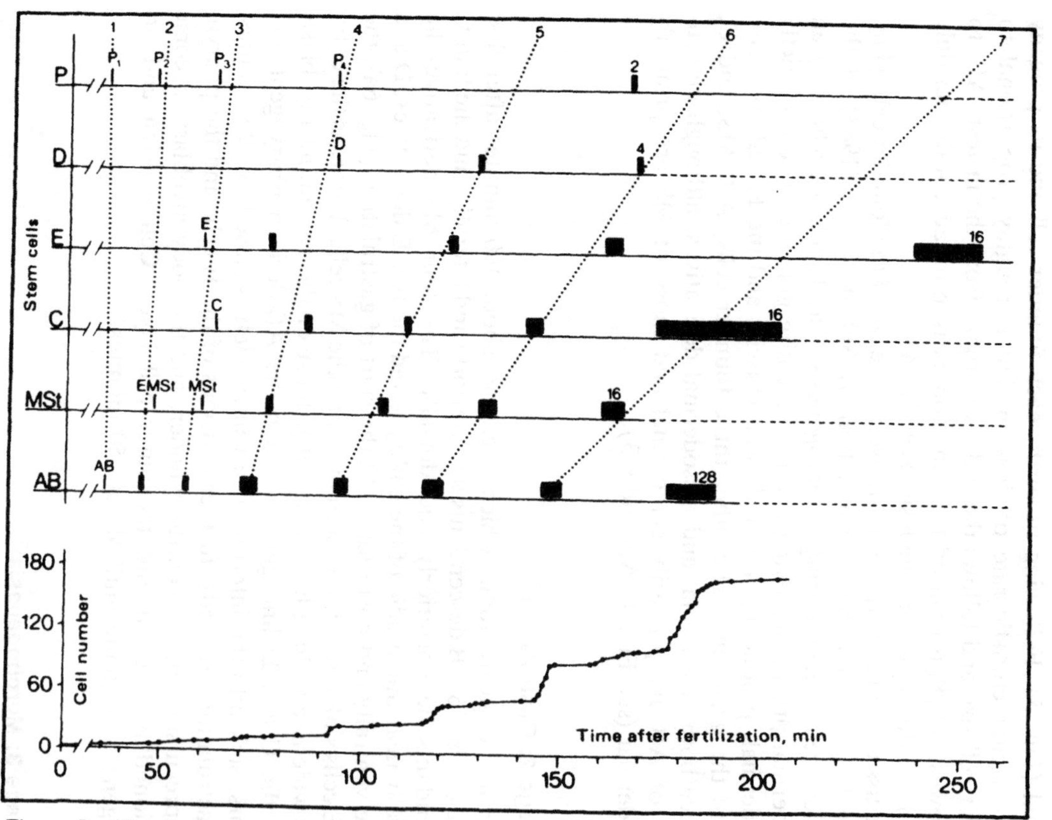

Figure 5 (*See facing page for legend.*)

whereas two other neurons undergo cell death in the male (Sulston et al. 1983).

During the period from about 350 to 650 minutes, the embryo becomes transformed from a spheroid of about 550 cells (Fig. 7) into a cylindrical worm of the same volume and cell number, still inside the eggshell, with an increase in length of about fourfold (Figs. 6d and 8). To accomplish this transformation, cells of the animal change shape coordinately in a process that appears to be controlled by the surface hypodermal cells derived from the AB and C lineages (Priess and Hirsh 1986). These cells, which interdigitate across the dorsal midline and meet along the ventral midline (Fig. 7), are connected by belt desmosomes and contain both actin filaments and microtubules that run circumferentially in alternating bundles around the animal. Lengthening is probably accomplished by tightening of the actin filaments against the internal hydrostatic pressure of the embryo, with the microtubular bundles serving as structural supports to maintain shape. At about 650 minutes, a collagenous cuticle begins to be secreted by the hypodermal cells, and the circumferential bundles begin to disperse. From this point on, shape is maintained by the cuticle, as evidenced by cuticle-defective mutants that elongate normally but then revert to a spheroidal shape late in embryogenesis (Priess and Hirsh 1986). The "pretzel" or threefold-stage embryo (Figs. 6d, 8e), now moving actively, begins pharyngeal pumping at about 760 minutes and hatches at about 800 minutes, probably with the aid of one or more hatching enzymes that catalyze breakdown of the eggshell from within.

E. General Features of Embryogenesis

Contrary to earlier assumptions based on partial lineages (reviewed in Chitwood and Chitwood 1974), there is no strict correlation between founder cells and either germ layers or tissues. Although three of the founder cells, E, D, and P_4, give rise to cells of single tissues, the other

Figure 5 Timing of early cleavages in *C. elegans*, showing differences in cell-cycle times for the six founder-cell lineages. (EMSt and MSt) are designations used previously for EMS and MS, respectively). (*Top*) A horizontal line is marked for each founder cell (stem cell) to show the cell-division events of that lineage. Order from top to bottom is according to length of cell cycle. Time of origin of each founder cell is indicated by a labeled vertical line. Solid black boxes indicate the time, in the indicated lineage, from division of the first cell to division of the last cell to divide in a given round of division. The number over the last box for each lineage indicates the number of cells of that lineage present after that division. Dotted line indicates rounds of cell divisions. Dashed line indicates lineages that were not followed further in these experiments. (*Bottom*) Increase in total nuclei (cell number) with time. (Reprinted, with permission, from Deppe et al. 1978.)

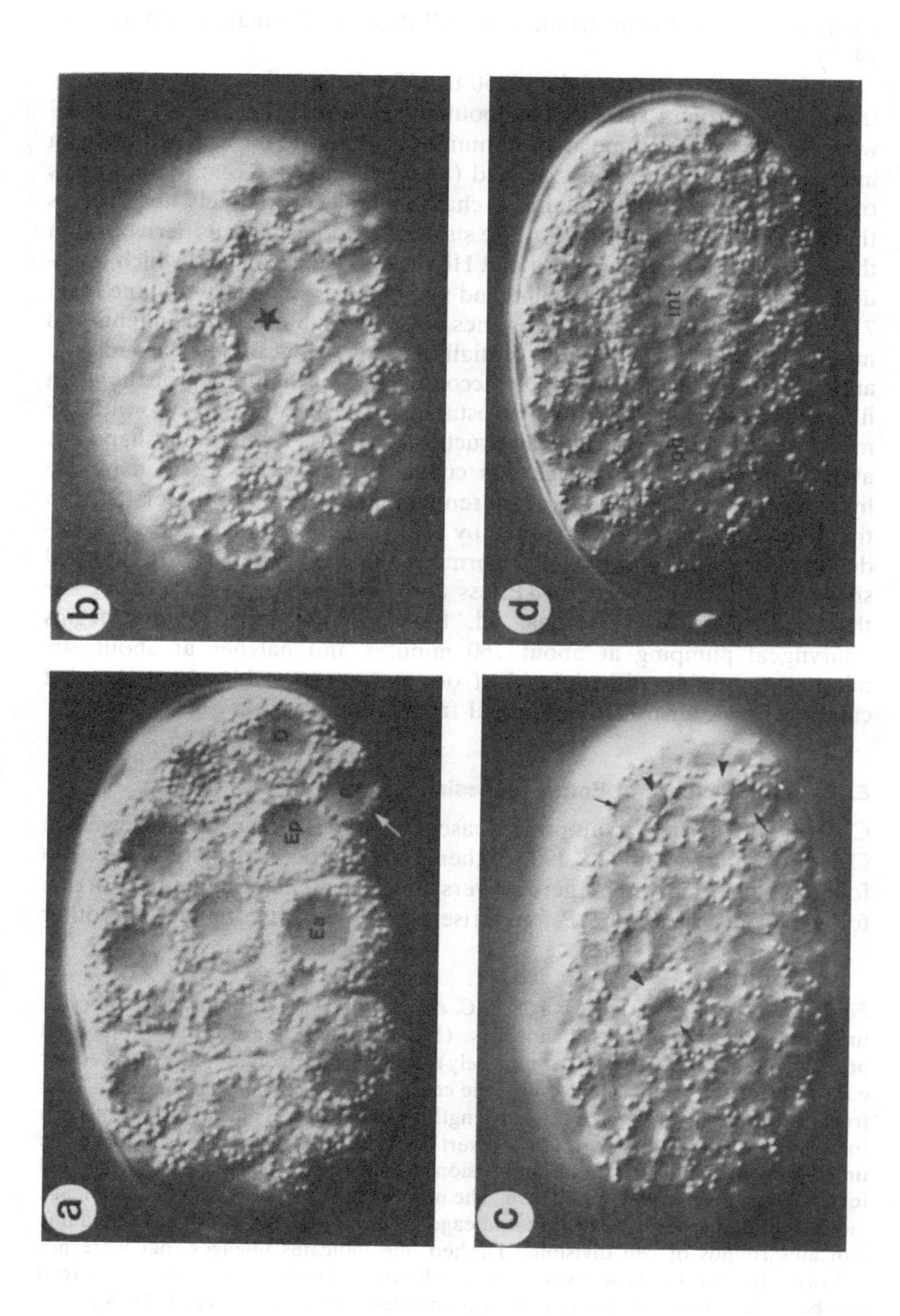
a
Ep
Ea
b
c
d
int
ph

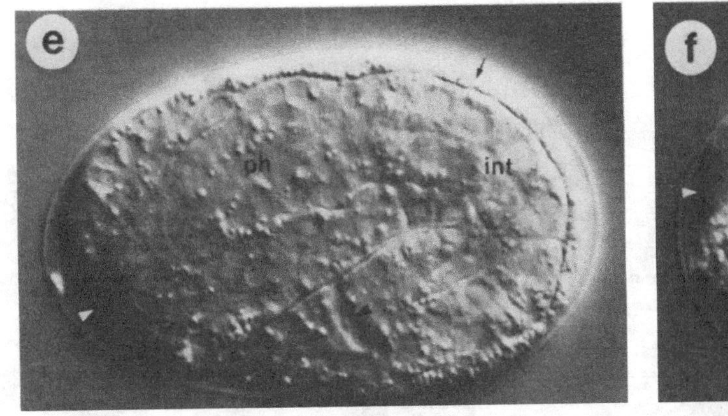

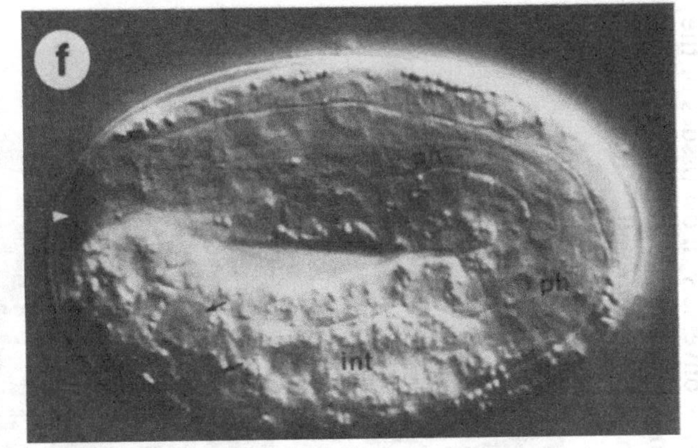

Figure 6 Nomarski photomicrographs of embryos at different stages. (*a*) Beginning of gastrulation (~100 min after first cleavage); left lateral view, focused on mid-plane of embryo. (*b*) Late gastrulation (~210 min); ventral view, focused on ventral surface. (*) Cleft through which the MS cells have just entered. (*c*) Ventral view, focused on ventral surface, which is covered with small neuroblasts (~260 min). Dying cells (arrowheads) are being engulfed by their sisters (arrows). (*d*) Dorsal view, focused on mid-plane (~280 min). (int) Cylinder of intestinal cells (nine nuclei in this focal plane); (ph) cylinder of pharyngeal cells. (*e*) Embryo showing first movement (~430 min). Left lateral view, focused on mid-plane; (int) intestine; (ph) pharynx. (White arrowhead) Anterior sensory depression; (black arrowhead) rectum; (arrows) dorsal hypodermal ridge, heavily loaded with granules. (*f*) Threefold embryo, rolling (~550 min). Only the anterior two thirds of the embryo is in focus. (White arrowhead) Mouth linked by ph to int. (Arrows) Germ cells. (Reprinted, with permission, from Sulston et al. 1983.)

three, AB, MS, and C, each contribute to two germ layers, ectoderm and mesoderm, and to several tissues (Fig. 4). Conversely, most tissues of the adult, with the clear exceptions of the intestine and the germ line, are of mixed lineage, deriving from more than one founder cell.

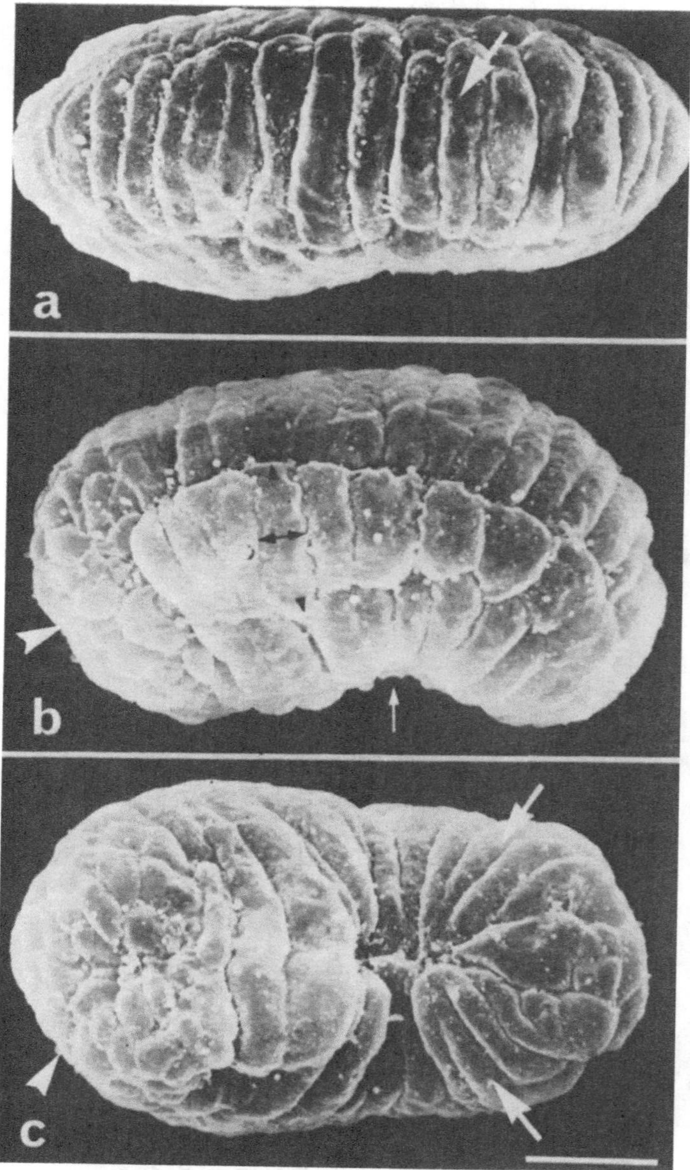

Figure 7 (*See facing page for legend.*)

Patterns of division in the embryo put most cells near their final locations; of the 558 cells present at hatching in the hermaphrodite, only 12 undergo long-range migration during embryogenesis. A common cell fate is programmed cell death; 113 of the 671 cells generated during hermaphrodite embryogenesis die shortly after their birth. Like most other cell fates, these deaths are specific, cell autonomous, and invariant from one embryo to the next. The bilateral symmetry of the animal at the cellular level does not always arise from bilaterally symmetrical lineage patterns during embryogenesis, especially in the nervous system. In fact, the embryo is bilaterally asymmetric from the six-cell stage onward, and the lineage patterns later in embryogenesis appear to have been modified during the course of evolution to produce bilateral symmetry (Sulston et al. 1983). (For further discussion of lineal patterns and embryonic origins of the various tissues, see Chapter 5.)

F. Macromolecular Synthesis during Embryogenesis

Little is known about the details of macromolecular synthesis in the embryo. Polyadenylated mRNAs are first detected in nuclei at about the 100-cell stage (~150 min) by in situ hybridization to squashed embryos using labeled polyuridylate as a probe (Hecht et al. 1981a). This point may correspond, in some respects, to the midblastula transition described in amphibian (Newport and Kirschner 1982) and dipteran embryos (Edgar et al. 1986). However, more sensitive assays using staged early embryos to measure run-on transcription in extracts (I. Schauer and W.B. Wood, unpubl.) or incorporation of radioactive nucleosides into RNA by permeabilized embryos (L.G. Edgar and J.D. McGhee, pers. comm.) indicate some transcription during earlier cleavage stages. Collagen mRNA and an increase over maternal levels of actin mRNA are first detected at about 150 minutes by in situ hybridization using cloned fragments of the corresponding genes as probes (Edwards and Wood 1983). One tissue-specific enzyme, a gut esterase, is first detected at about 200 minutes (Edgar and

Figure 7 Scanning electron micrographs of the apical surfaces of hypodermal cells in embryos prior to elongation. (*a*) Dorsal view. Arrow points to a single dorsal hypodermal cell. The left and right rows of dorsal hypodermal cells have interdigitated and now form a single longitudinal row. (*b*) Lateral view. The longitudinal rows of dorsal (▲), lateral (↔), and ventral (▼) hypodermal cells are visible. The embryo has a small ventral bend (arrow). (*c*) Ventral view. Arrows point to individual ventral hypodermal cells. Right and left pairs of ventral hypodermal cells meet at the ventral midline of the embryo. (*b, c*) The anterior (left) end of the embryo has not yet been enclosed by hypodermal cells, and neural precursors are visible (arrowheads). Bar represents 10 μm. (Reprinted, with permission, from Priess and Hirsh 1986.)

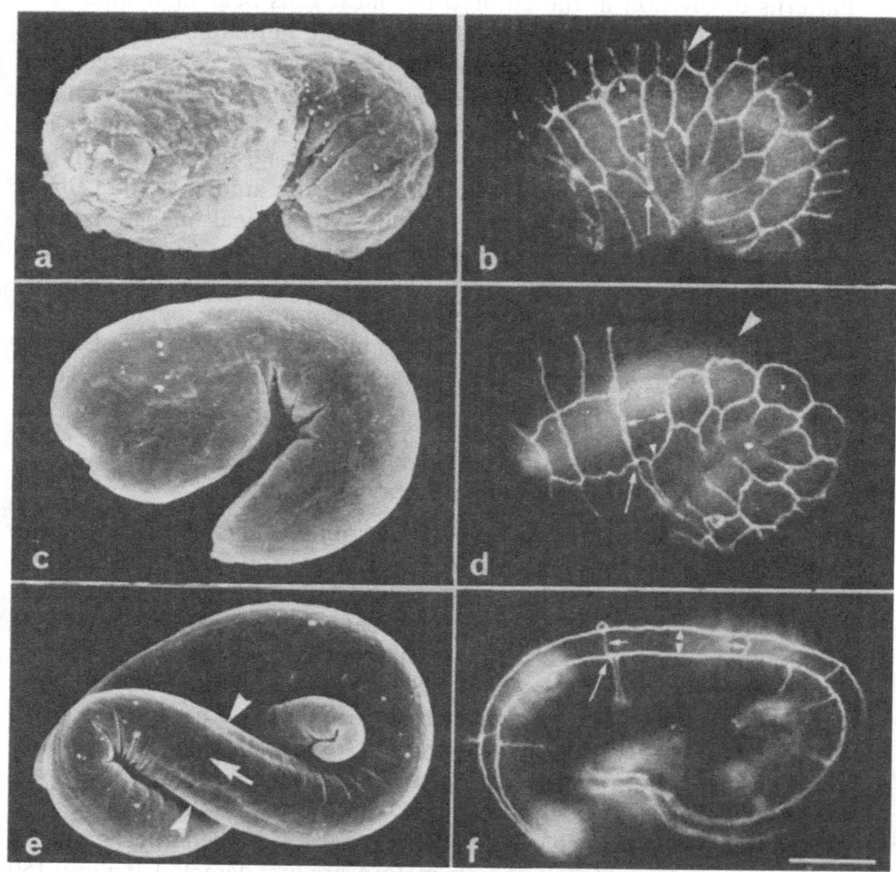

Figure 8 Changes in body and hypodermal cell shape during elongation. The left–right images are scanning electron and MH27–immunofluorescence micrographs, respectively, of comparable embryonic stages before (*a,b*), in the early stages (*c,d*), and in the final stages (*e,f*) of elongation. (*a*) The embryonic sheath covers the embryo during elongation (see also *c, d*) such that individual hypodermal cell boundaries (as seen in Fig. 7b) are no longer visible from the surface. In later stages of elongation (*e*), ridges are visible over the dorsal and ventral hypodermal surface (arrowheads) but not over the lateral hypodermal surfaces (arrow). (*b*) The boundaries of the hypodermal cells are visible in embryos stained with MH27 (an antibody that recognizes belt desmosomes) and viewed in the fluorescence microscope (cf. Fig. 7b). An example of shape change in a single hypodermal cell during elongation is shown in the series *b,d,f*. (Small white arrowheads) Longitudinal margins of a lateral hypodermal cell; (short white arrows) circumferential margins. Note that the relative positions of hypodermal cells do not appear to change during elongation (long white arrow in *b,d,f*). The dorsal hypodermal cells (large white arrowheads in *b,d*) fuse in the early stages of elongation. Bar represents 10 μm. (Reprinted, with permission, from Priess and Hirsh 1986.)

McGhee 1986). Several tissue-specific antigens become detectable by immunofluorescence staining at about this time as well (Priess and Thomson 1987; S. Strome and W.B. Wood, unpubl.).

III. DETERMINATION OF CELL FATES AND PATTERNS IN THE EMBRYO

A. Evidence for Cell-autonomous Determinants

Several lines of evidence indicate that founder-cell fates in the early embryo are lineally, rather than positionally, determined, apparently by asymmetric segregation of cell-autonomous determinants into the appropriate daughter cells at determinative cleavages. As an assay for determination, a convenient tissue-specific differentiation marker is provided by autofluorescent granules containing tryptophan catabolites that accumulate exclusively in gut cells beginning at about 200 minutes (Babu 1974; Laufer et al. 1980). More recently, the gut-specific esterase mentioned in Section II.F above has been used for the same purpose. Permeabilized embryos, treated at early stages with cytochalasin and, in some experiments, colchicine as well to block further cleavage, were shown to express the autofluorescent gut "granules" (Laufer et al. 1980) or gut esterase (Edgar and McGhee 1986) at about the normal time after first cleavage, and only in gut precursor cells present at the time of cleavage arrest. No expression was observed in arrested one-cell embryos. For subsequent stages, expression was observed in only the P_1 cell in arrested two-cell embryos, the EMS cell in four-cell embryos, and the E cell in eight-cell embryos. This result indicates that expression of gut-specific differentiation markers in E cells and their precursors does not require the normal environment of the developing embryo but, rather, appears to be dictated in a cell-autonomous manner by a determinant that segregates internally from P_1 to EMS to E during the first three cleavages.

In similar experiments using additional differentiation markers, cleavage-arrested one-cell embryos were found to express a hypodermal marker, and the suggestion was made that gut and muscle differentiation pathways may be mutually exclusive, based on the finding that individual cleavage-arrested EMS cells appear to express either one but not both (Cowan and McIntosh 1985).

Results with partial embryos and isolated blastomeres, derived by mechanical or enzymatic removal of the eggshell and lysis of some blastomeres, confirm cell-autonomous determination. Partial embryos derived from P_1 alone show twitching, indicative of muscle function, as well as expression of gut markers, and isolated E cells give rise to progeny that express gut markers in the absence of other cells (Laufer et al. 1980; Edgar and McGhee 1986).

Rudimentary cytoplasmic transfer experiments carried out using laser microsurgery are consistent with the view that the presumed gut determinant is cytoplasmic (Wood et al. 1984; Schierenberg 1985). The P_1 nucleus can be removed from a two-cell embryo by extrusion through a small hole burned with a laser microbeam into the posterior pole of the eggshell, which subsequently will often reseal, leaving a P_1 cytoplast adjacent to the intact AB cell. The AB cell will then proliferate through several divisions, but fluorescent gut granules are never seen. However, if the P_1 cytoplast is fused to one of the AB cell progeny by laser-induced disruption of the membrane between them, gut granules are subsequently expressed in about half of the treated embryos.

Ablation experiments using a laser microbeam to kill single embryonic cells appear to support the view that most cells throughout embryogenesis are autonomously determined by internally segregating factors, independent of influences from neighboring cells. Except for a few cases in late embryogenesis, there is no evidence for regulation to compensate for removal of a cell by ablation (Sulston et al. 1983). Often, ablation of an embryonic cell causes the cell types comprising the normal progeny of that cell to be absent from the embryo and has no effect on the fates of neighboring cells or their progeny (for further discussion, see Chapter 5). Interpretation of these findings may be complicated by the fact that the remains of an ablated cell persist, at least for a short time, in the embryo.

B. Evidence for Cell-Cell Interactions in Determination of Cell Fates

More recent experiments have demonstrated that cell-cell interactions determine certain cell fates. Production of pharyngeal muscle cells by the AB lineage appears to require an inductive signal, most likely from cells of the MS lineage (Priess and Thomson 1987). Based on staining with a pharynx-specific monoclonal antibody, the AB-derived pharyngeal muscle cells, normally progeny of the ABa blastomere, do not appear in partial embryos following removal of the P_1 blastomere at the two-cell stage or the EMS blastomere at the four-cell stage. In contrast, the MS-derived pharyngeal muscle cells do appear following removal of the AB blastomere at the two-cell stage.

Observation of the first AB cleavage suggests that ABa and ABp might initially be developmentally equivalent cells. The AB spindle in the two-cell embryo is first oriented perpendicular to the anterior–posterior axis (Fig. 3B). During AB cleavage, the spindle slides one way or the other so that one of the two daughters, designated ABp, becomes located posterior to and to one side of the other, thus defining the dorsal side of the embryo and also establishing the future left-right axis (Fig. 3C,D). The symmetry of this cleavage and the apparently random nature of the change in spindle orientation that defines which daughter will become ABa and which will become ABp suggests that these two cells may be equivalent at the four-cell stage and that their different fates are determined by subsequent cell interactions. Priess and Thomson (1987) obtained further support for this view by micromanipulating embryos in which AB cleavage was not yet complete, but AB spindle orientation was already committed, using a blunt microneedle to reorient the dividing AB cells so that the daughter that would have become ABa became ABp instead. These embryos, in which the initial dorsal–ventral and left-right axes have been switched to the alternative configurations, developed normally to hatching and matured into fertile adults.

Experiments with isolated blastomeres suggest that cell–cell interactions may also be important in maintaining correct polarity of early cleavages, which may, in turn, be important for fate determinations (Laufer et al. 1980; Schierenberg 1987; L. Edgar and T. Hyman, pers. comm.). Additional work is needed to ascertain the extent to which cell-cell interactions contribute to determination of embryonic cell fates.

C. Localization of Germ-line Granules in the Embryo

In attempts to observe localization of possible macromolecular lineage-specific determinants in early cleavage, Strome and Wood (1982, 1983) and, subsequently, S. Strome (pers. comm.) injected mice with homogenates of early embryos, prepared hybridoma cell lines from spleen lymphocytes, and screened them by immunofluorescence microscopy for production of monoclonal antibodies recognizing lineage-specific antigenic components. Only one such component has been identified. It is present in early embryos as cytoplasmic granules, termed P granules, which are observed only in the P cells and, subsequently, in all germ-line cells throughout the life cycle, except mature sperm. Seventeen independently derived hybridomas were found to produce antibody that reacts with P granules, out of a total of about 1500 lines tested. Two apparently similar monoclonal antibodies were found in an independent screen by Yamaguchi et al. (1983).

At fertilization (Fig. 9a), P granules are observed by immunofluorescence microscopy to be uniformly distributed as small particles throughout the ooplasm (Strome and Wood 1983). As the pronuclei approach each other, the P granules become localized around the posterior periphery of the embryo (Fig. 9b), so that after first cleavage they are present exclusively in the P_1 cell (Fig. 9c). Similar prelocalization and segregation occur at each of the subsequent P-cell cleavages, always placing the P granules into the germ-line daughter. The granules appear to increase in size and decrease in number during the first three cell cycles. Prior to the cleavage of P_3, they become associated with the nuclear envelope, where they are found during the remainder of the life cycle. Occasionally, a few granules

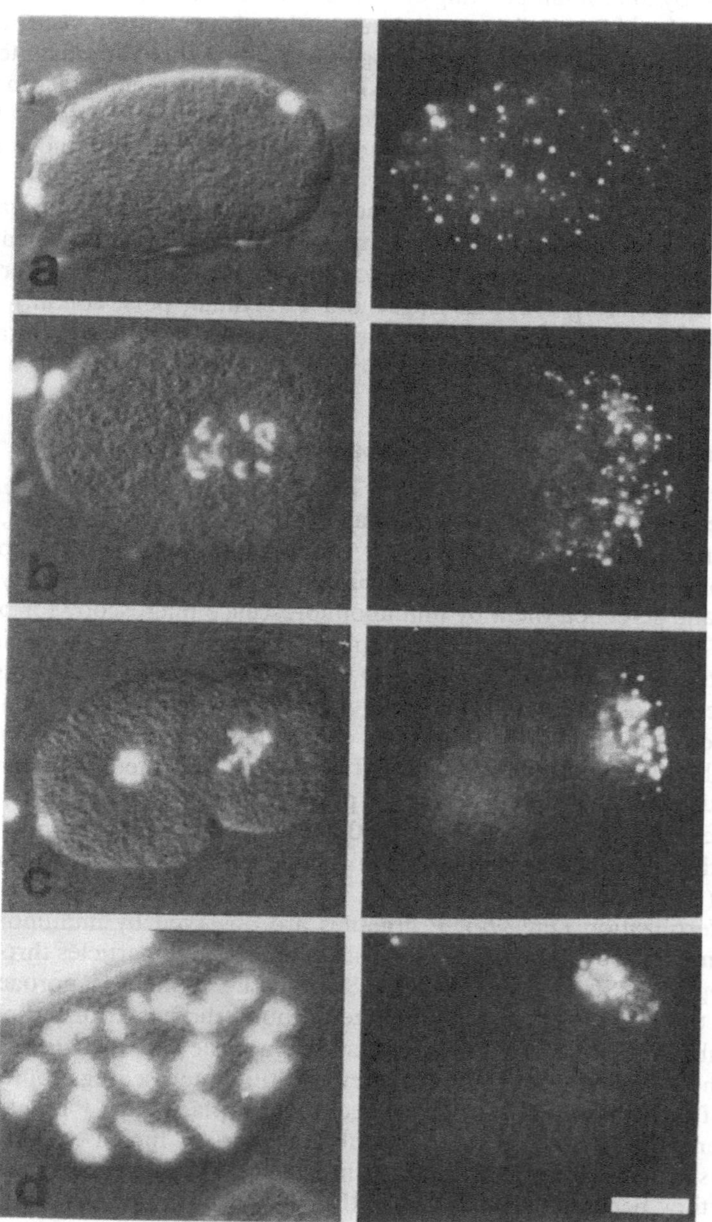

Figure 9 (*See facing page for legend.*)

are found in the D cell immediately following P_3 cleavage (Fig. 9d), but they disappear before the subsequent cleavage of D, as if the granules are unstable in somatic cytoplasm.

D. Localization and Generation of Zygotic Asymmetry

Drug experiments with early embryos indicate that P-granule localization is dependent on the function of an actin-based motility system (Strome and Wood 1983). In permeabilized embryos treated immediately after fertilization with any one of several microtubule inhibitors, such as colcemid, pronuclear migration is blocked, but the P granules nevertheless become localized at the posterior pole. In contrast, in embryos treated with either of the actin microfilament inhibitors cytochalasin B or D, the pronuclei migrate together, but P granules coalesce near the center of the embryo, rather than becoming localized posteriorly. Moreover, none of the other normal manifestations of zygotic anterior–posterior asymmetry are observed, such as the early contractions of the anterior membrane and the characteristic disc-shaped morphology of the posterior centrosome at first cleavage (see Fig. 2). Given this apparent involvement of microfilaments in generating zygotic asymmetry, it is interesting that F actin itself appears to become asymmetrically reorganized in the zygote, localizing in the anterior region (Strome 1986a). These results suggest that at fertilization, two independent systems of cytoskeletal machinery are set in motion: a microtubule-based system necessary for moving the pronuclei and subsequently mediating mitosis, and an actin-based system responsible for posterior localization of the P granules and for generation of several other asymmetries in the zygote.

Figure 9 Localization of P granules after fertilization. Embryos cut out of hermaphrodites were fixed and stained with DAPI (diamidinophenylindole) to visualize chromosomes and with an appropriate monoclonal antibody and fluoresceinated secondary antibody to visualize P granules. (*Left* panels) Nomarski images simultaneously epiilluminated with an appropriate wavelength to show DAPI-stained chromosomes. (*Right* panels) The same set of embryos epiilluminated with a different wavelength to visualize antibodies bound to P granules. (*a*) Zygote after completion of meiosis prior to pronuclear migration. In the Nomarski–DAPI image, the egg pronucleus and both polar bodies can be seen at the anterior pole (*left*), and the sperm pronucleus near the posterior pole. P granules are dispersed throughout the cytoplasm. (*b*) Zygote at the time of pronuclear conjunction, showing localization of P granules around the posterior periphery. (*c*) Two-cell embryo in which the P_1 cell (posterior) is in prophase and P granules are prelocalized in the region of cytoplasm destined for the next P-cell daughter. (*d*) Embryo (26-cells) in which most of the P granules are localized in the P_4 cell, but some are detected in its sister cell, D, below and to the right. Bar represents 10 μm. (Reprinted, with permission, from Strome and Wood 1983 [Copyright M.I.T. Press].)

E. Origin and Possible Function of Germ-line Granules

The P granules observed throughout embryogenesis appear to be maternal in origin, based on lack of an obvious increase in total P-granule immunofluorescence staining. Consistent with this view are experiments with a fertile mutant (Wood et al. 1984) whose P granules do not stain with one of the anti-P granule monoclonal antibodies (K76). When mutant hermaphrodites are mated to wild-type males, the outcross progeny embryos, which are heterozygous for the mutation, show no antigen detectable by immunofluorescence using K76 antibody until the start of gonad proliferation midway through the first larval stage.

Although the segregation behavior of P granules is as expected for a lineage-specific determinant, there is so far no evidence that they are determinative for the germ line, and their function remains unknown. However, they are probably homologous to the granular inclusions seen by electron microscopy in the germ-line cytoplasm of many organisms and are called germinal plasm, nuage, or polar granules (for review, see Eddy 1975). These inclusions have also been observed in *C. elegans*, where their size distribution and localization in the different P cells of early embryos correlate well with the distributions observed for P granules by immunofluorescence (Krieg et al. 1978; Wolf et al. 1983).

F. Control of Lineage-specific Cell-cycle Periods

In the early embryo, the progeny of a given somatic founder cell divide synchronously, but with cell-cycle periods that are different for each of the five major somatic lineages (Deppe et al. 1978; see Fig. 5). The lineage-specific differences in cell-cycle timing appear to be important in establishing the correct spatial patterns of the cleavage-stage blastomeres (Schierenberg et al. 1980).

Using laser microsurgery on early embryos, Schierenberg and Wood (1985) carried out blastomere fusion and cytoplasmic transfer experiments to investigate the basis for the timing differences between lineages. The results indicate that control of the cell-cycle period is cytoplasmic. The period of a cell from one lineage can be altered substantially by introduction of cytoplasm from the cell of another lineage with a different period. Short-term effects of foreign cytoplasm on the timing of the subsequent mitoses differ, depending on the position of the donor cell in the cell cycle. The observations are consistent with the presence of a cytoplasmic oscillator whose period depends upon the concentration of some component that is partitioned to the different founder cells in different concentrations and that regulates the activity of mitosis-inducing or mitosis-inhibiting factors, or both, and thus controls the period of the cell cycle.

IV. MUTATIONS THAT AFFECT EMBRYOGENESIS

A. Isolation of Embryonic Lethal Mutants

Genes essential for embryogenesis were first identified in screens for temperature-sensitive lethal mutations following EMS mutagenesis, using 25°C and 16°C as the restrictive and permissive temperatures, respectively (Hirsh and Vanderslice 1976). In this and other such screens, carried out subsequently in several laboratories (Wood et al. 1980; Miwa et al. 1980; Cassada et al. 1981a), between 10% and 28% of the lethals recovered have shown phenotypes of embryonic lethality; that is, homozygous mutant hermaphrodites shifted to nonpermissive temperature as third- or fourth-stage larvae self-fertilize to produce embryos that fail to hatch.

The screens cited above identified a total of about 55 genes required for embryogenesis; more recent screens for temperature-sensitive mutations have identified additional genes (e.g., Kemphues 1987, 1988; Priess et al. 1987; H. Schnabel and R. Schnabel, pers. comm.). Still more such genes, so far less well characterized, have been defined by nonconditional (absolute) lethals isolated in screens employing partial chromosome duplications or balancer chromosomes that allow convenient propagation of heterozygotes (e.g., Meneely and Herman 1981; Rogalski et al. 1982; Sigurdson et al. 1984). The frequency distribution of multiple mutations within single genes defined by the temperature-sensitive mutants has been used to estimate the total number of such genes at about 200 (Cassada et al. 1981a); however, this is probably an overestimate of the genes required exclusively for embryogenesis, for reasons discussed below (Section B.2).

B. Characterization of Genes Defined by Embryonic Lethal Mutants

1. Time of Action

Information on when these genes act in development has been obtained from parental-effect tests on the temperature-sensitive mutations that define them (Wood et al. 1980; Miwa et al. 1980; Isnenghi et al. 1983; Hirsh et al. 1985a). Among the 47 genes for which these tests are complete, four classes can be distinguished (Table 1). Mutations in the first class (28 genes) show a *strict* parental effect; that is, the homozygous mutant (*m*/*m*) self-progeny of a heterozygous mutant (*m*/+) hermaphrodite escape embryonic lethality, and the *m*/+ outcross progeny of an *m*/*m* hermaphrodite mated to a wild-type (+/+) male die as embryos. Expression of these genes in the maternal parent therefore appears to be both necessary and sufficient for embryonic survival.

Mutations in the second class (11 genes) show a *partial* parental effect: *m*/*m* self progeny of *m*/+ hermaphrodites survive, but so do *m*/+ outcross progeny of *m*/*m* hermaphrodites. In ten of these cases, survival of outcross progeny depends upon introduction of the + allele from the male parent;

Table 1 Functional classes among 47 genes required for embryogenesis, based on parental-effect tests

Gene	References	Gene	References[a]
Class 1			
Strict parental effect; maternal gene function necessary and sufficient (M,M)			
emb-1 III	2	*emb-23 II*	3
emb-3 IV	2	*emb-25 III*	3
emb-4 V	2	*emb-26 IV*	3
emb-5 III	2	*emb-27 II*	3
emb-6 I	2	*emb-28 V*	3
emb-7 = zyg-4 III	2,1	*emb-30 III*	3
emb-8 III	2	*emb-31 IV*	3
emb-11 IV	3	*emb-33 III*	3
emb-12 I	3	*zyg-1 II*	1
emb-16 III	3	*zyg-2 I*	1
emb-18 V	3	*zyg-9 II*	1
emb-19 I	3	*zyg-11 II*	4
emb-20 I	3	*zyg-12 II* Bergerac	1
emb-21 II	3	*par-1 = zyg-14 V*	4
Class 2			
Partial parental effect; parental[b] or embryonic gene function sufficient (M,N)			
emb-2 III	2	*emb-35 IV*	3
emb-9 = zyg-6 III	2,1	*zyg-3 II*	1
emb-15 X	3	*zyg-7 III*	1
emb-17 I	3	*zyg-8 III*[b]	1
emb-24 III	3	*zyg-10 III*	1
emb-32 III	3	*let-2(g30) X*[c]	3
Class 3			
Strict nonparental; embryonic gene function necessary and sufficient (N,N)			
emb-9 III	2	*let-2(b246) X*[c]	1
emb-29 V	3		
Class 4			
Parental and embryonic; maternal *and* embryonic gene function necessary (N,M)			
emb-13 III	3	*emb-34 III*	3
emb-14 I	3	*zyg-5 II*	1
emb-22 V	3		

[a]References: (1) Wood et al. (1980); (2) Miwa et al. (1980); (3) Isnenghi et al. (1983); (4) Hirsh et al. (1985a).

[b]For *zyg-8* mutants, paternal gene function is sufficient; other mutants in this class require maternal gene function.

[c]Note that the *g30* allele of *let-2* shows a partial parental effect; most alleles, like *b246*, show no parental effect.

therefore, expression of these genes in *either* the maternal parent *or* the embryo is sufficient for embryonic survival. In the remaining unusual case, the *m/m* as well as the *m/+* outcross progeny of *m/m* hermaphrodites mated to *m/+* males survive, indicating that expression of this gene in the paternal parent is sufficient for embryonic survival.

Mutations in the third class (three genes) behave in a strictly nonmaternal fashion; that is, *m/m* self progeny of *m/+* hermaphrodites die as embryos, and *m/+* outcross progeny of *m/m* hermaphrodites survive. Expression of these genes in the embryo is therefore both necessary and sufficient for embryonic survival.

For mutations in the fourth class (five genes), neither the *m/m* self-progeny embryos of *m/+* hermaphrodites nor the *m/+* outcross progeny embryos of *m/m* hermaphrodites survive, indicating that expression of this gene is required both maternally and embryonically for embryonic survival.

The preponderance of maternal effects among the mutationally identified genes required for embryonic survival is consistent with the view that rapid rates of cell division in early embryogenesis necessitate maternal synthesis and storage in the oocytes of most of the gene products required during this period.

More detailed analysis of when these genes act has been carried out for the temperature-sensitive mutations by temperature-shift experiments to determine time and duration of the temperature-sensitive period (TSP) leading to embryonic lethality. Interpretation of such experiments is complicated by the uncertainty of whether the TSP defines the time of synthesis or the time of function of the mutant gene product (for discussion, see Wood et al. 1980). Nevertheless, the results are generally consistent with expectations from the parental-effect tests: strict maternal mutants tend to show the earliest TSPs, either before or close to the time of fertilization, and partial maternal mutants show later TSPs, generally during the first half of embryogenesis. Only three strict nonmaternal complementation groups include temperature-sensitive mutants; these mutants have TSPs during the second half of embryogenesis. (For further discussion of this set of genes, see Wilkins 1986 and Kemphues 1987.)

2. *Possible Gene Functions*

An inherent problem with genetic analysis of embryogenesis is the difficulty of distinguishing genes that control pattern formation and cell fates (the most interesting ones developmentally) from genes that control general metabolic and cellular functions also required in embryogenesis. It has been assumed (probably mistakenly; see below) that null mutations resulting in only a specific embryonic defect are more likely to define genes in the interesting category than mutations causing defects at several different developmental stages. For classifying essential genes in this regard, temperature-sensitive mutations have the advantage that their effects on different stages can be tested by shifting homozygous mutant animals to the nonpermissive temperature at different times in the life cycle. They have the disadvantage that they often do not result in a null phenotype, so that inferences of gene function from temperature-sensitive mutant phenotypes, particularly when based on analysis of only one temperature-sensitive mutant allele, are not reliable. Nevertheless, analysis of temperature-sensitive mutants has provided some useful information on the nature of genes required for embryogenesis.

In the set of 55 genes discussed above that are identified by temperature-sensitive embryonic lethal mutations, only about 14 are possible candidates for exclusively embryonic function, and only 5 of these are defined by more than one allele. Among this set of 14, 11 show strict maternal, 3 show partial maternal, and only 1 shows strict nonmaternal requirements for function. Many of the strict maternal mutants show defects in meiosis or first cleavage (Wood et al. 1980; Miwa et al. 1980; Denich et al. 1984; Kemphues et al. 1986), so that subsequent anomalies may be the result of aberrant chromosome distribution rather than lack of specific factors required for determination of cleavage patterns or cell fates. More extensive analysis of mutations defining two of these loci, *zyg-9* and *zyg-11*, have shown that these are "pure" maternal-effect genes, required at no other stage in the life cycle, which function in some manner to ensure proper formation of the nuclei and organization of cytoplasm in the one-cell embryo (Kemphues et al. 1986). Mutants defective in the single nonmaternal gene (*let-2*) arrest near the end of embryogenesis and may have a defect in basement membrane function (J. Kramer and J. Priess, pers. comm.). Therefore, this set of 14 genes may include few, if any, that play controlling roles in embryogenesis.

More recently Kemphues et al. (1988) have isolated potentially interesting strict maternal-effect mutations at five new loci, designated *par-1* (formerly *zyg-14*; Hirsh et al. 1985a) through *par-5* (for *par*titioning-defective), which may be important in the asymmetric partitioning of cytoplasmic components in the early embryo. These mutations show normal meiosis, affect only embryogenesis, and are characterized by synchronous and usually equal, rather than the normal asynchronous and unequal, early cleavages. In these mutants, P granules generally do not show normal asymmetric localization or segregation; the spindles may be oriented abnormally at second cleavage; gut differentiation markers generally do not appear; and morphogenesis does not take place. However, neither these phenotypes nor embryonic lethality are fully penetrant. Surviving embryos develop into sterile adults; that is, these mutations can also result in a grandchildless phenotype.

A set of dominant maternal-effect, temperature-sensitive embryonic lethal mutations has been isolated with the goal of identifying either haplo-insufficient loci whose gene product levels are important for embryonic development or members of multigene families that would not be identified in screens for recessive mutations (P. Mains et al., in prep.). Eight such mutations have been found defining at least six new loci, one of which appears haploinsufficient; mutations at two others have identified a set of interacting genes that include *zyg-9* and may be required for normal meiosis.

Perrimon et al. (1986) have recently shown that in *Drosophila*, maternal-effect embryonic lethality can result from rare hypomorphic alleles of essential genes whose null phenotypes are general lethality rather than pure embryonic defects. Inclusion of such mutations can lead to overestimates of the number of genes mutable to maternal-effect embryonic lethality. Analysis of 29 maternal-effect lethal loci from a saturation screen of a region of chromosome II indicates that the same phenomenon occurs in *C. elegans* and has led to a more precise estimate of the number of pure maternal-effect genes in *C. elegans* as between 25 and 60 (K. Kemphues et al., pers. comm.).

Consistent with these findings, earlier studies of temperature-sensitive embryonic lethals indicated that many embryonically essential genes appear to have functions that are also required at other stages in the life cycle. Among the set of 55 loci discussed above, 41 fall into this category, based on results of temperature-shift experiments. The secondary defective phenotypes resulting from mutations in these genes include slow larval growth, larval arrest, morphological abnormalities, and defective gonadogenesis resulting in sterility. However, multiple phenotypes can also reflect multiple gene functions, as shown by findings that some homoeotic mutations affecting postembryonic development perturb unrelated lineages at different stages (Chapter 6). A particularly interesting case is provided by the gene *glp-1* (for *g*erm-*l*ine *p*roliferation [Austin and Kimble 1987; see Chapter 7]), which is essential for maintenance of germ cell mitotic division by the distal tip cell in the gonad (Kimble and White 1981). This gene is also required embryonically (Priess et al. 1987) in the induction of AB-derived pharyngeal muscle cells by P_1-derived cells (Priess and Thomson 1987). These observations strongly suggest that the same signaling mechanism may be employed for quite different processes at different stages of *C. elegans* development.

V. POSSIBLE MECHANISMS OF DETERMINATION

The mechanism by which the fates of blastomeres are determined in mosaic embryos represents a long-standing problem. In *C. elegans* there is now evidence that fates can be dictated both by internally segregating factors and by cell-cell interactions. There is considerable evidence in several organisms, including *C. elegans*, that the internally segregating factors are cytoplasmic, rather than asymmetrically distributed chromosomal components. DNA strands derived from *C. elegans* sperm have been shown to be randomly distributed to embryonic cells during embryogenesis, thereby ruling out segregation of sperm DNA to specific cells as a determination mechanism (Ito and McGhee 1987). However, a critically timed round of DNA replication may be required for expression of lineage-specific differentiation markers in ascidians (Satoh and Ikegami 1981), as well as in *C. elegans* (Edgar and McGhee 1988).

Although the cytoplasmic location of fate-determining factors seems very likely, their nature remains unclear. Classical observations of cytoplasmic localization and segregation of visible cytoplasmic components, for example, in the ascidian *Styela* (Conklin 1905), led to the notion of maternally derived cytoplasmic determinants, informational macromolecules (we would assume today) that are segregated specifically to different lineages and later somehow promote the appropriate patterns of differential gene expression. An alternative possibility is that maternally derived components are responsible for producing asymmetric distributions in the early embryo of small molecules or ions. These gradients could confer polarity and positional information, somehow dictating early differential synthesis of embryonic gene products that determine cell fates.

There is currently little basis for deciding between these two possibilities, or eliminating both in favor of another, in *C. elegans* or any other mosaic embryo. In *Drosophila*, convincing experimental evidence indicates that the polar plasm of oocytes and early embryos includes determinative factors that dictate functional germ-line development (Illmensee and Mahowald 1974; Niki 1986). Less direct evidence for germ-line determinants based on morphological criteria was obtained classically for the large parasitic nematode *Ascaris,* in which polar plasm appears to prevent chromosome diminution in the germ line (Boveri 1910). Factors in *C. elegans* polar plasm could be involved in preserving P granules (see Section III.C). The P-granule experiments show that *C. elegans* embryos have the capability of segregating a cytoplasmic component to a specific lineage, but there is no evidence that the granules are determinative factors.

For somatic cells, indirect evidence from cytoplasmic transfer experiments in *C. elegans*, described in Section III.A (Wood et al. 1984), as well as ascidians (Whittaker 1982), supports the view that cytoplasmic factors play a role in early determination events. Moreover, although Priess and Thomson (1987) have demonstrated in *C. elegans* that cell-cell interaction determines the fates of the AB-derived pharyngeal muscle cells, this induction also appears to depend on a maternally supplied factor, based on the surprising finding that *glp-1* mutations preventing the induction show a strict maternal effect (Priess et al. 1987).

Further insights into embryonic determination mechanisms should come from continuing analysis of mutations that affect embryogenesis. As reviewed above, the studies of embryonic lethal mutants are just beginning to be helpful in this regard. Analysis of homoeotic mutations that affect postembryonic cell lineages has identified genes that appear to be directly involved in the specification of different daughter cell fates in determinative cell divisions (see Chapter 6). The determinant model, postulating maternally derived informational macromolecules required for specification of founder-cell fates, would predict the possibility of maternal-effect homoeotic mutants that show altered fates in a particular lineage but are otherwise embryonically normal. The *glp-1* mutants may represent such a class, although no other such mutants have yet been identified. The gradient model would predict that, as in *Drosophila* (reviewed in Gergen et al. 1986), genes expressed in the early embryo, defined by nonmaternal-effect mutations, may be important for early patterning and cell fate determination. So far, too few strictly embryonically acting mutations have been analyzed in *C. elegans* to assess this prediction. Therefore, although mutant analysis may provide the most promising approach to elucidating embryonic determination mechanisms, definitive answers must await the results of more extensive screens for embryonic mutants and more detailed characterizations of mutant phenotypes than have been reported so far.

ACKNOWLEDGMENTS

I am grateful to Donna Albertson for help with the final revision and updating of this chapter and to Lois Edgar, David Hirsh, Ken Kemphues, Jim Priess, Einhard Schierenberg, Susan Strome, and John Sulston for comments and suggestions on earlier drafts.

9

Sexual Dimorphism and Sex Determination

Jonathan Hodgkin
MRC Laboratory of Molecular Biology
Cambridge, England

The Nematode Caenorhabditis elegans. © 1988 Cold Spring Harbor Laboratory. 0-87969-307-X/88 $1 + .00.

I. INTRODUCTION

Sexuality is a phenomenon that occurs throughout the animal kingdom and is therefore of general biological interest. From an evolutionary point of view, sexuality raises the questions of how separate sexes evolved at all and why so many different sexual systems have arisen. From the viewpoint of developmental biology, sexuality is important because the different sexes of an animal result from different developmental pathways, which must be separately established and maintained. Understanding how the cells of a whole animal develop as male rather than female is fundamentally the same problem as understanding how the cells in one part of the animal develop into an intestine as opposed to a muscle, or a wing as opposed to a leg. From this standpoint, the study of sexual differentiation is essentially a problem in developmental biology.

It is a special problem for several reasons. First, sexual specialization is possibly the most far-reaching of the determinative decisions that must occur in development, because correct sexual development requires specialized behavior on the part of a great variety of cells in many different parts of the animal. This is particularly true of *Caenorhabditis elegans*, in which at least 30% of the somatic cells and all of the germ cells are sexually specialized. Second, it is possible to identify the primary sex-determining signal (usually chromosomal, as in *C. elegans*, but environmental in some organisms). This defines the first of the events that eventually lead to different sexual phenotypes, and it is therefore possible to work down from this defined primary step in the hope of eventually identifying the whole set of interactions that occur during sexual development. In other pathways of development, the primary signal (e.g., an ooplasmic determinant, or a particular concentration of a diffusible morphogen) is uncertain, hypothetical, or unknown. Third, sex determination is a phenomenon common to all groups of animals, so it is possible that general principles will be easier to identify in studying sex than in studying more specialized characteristics.

II. SEX-DETERMINING SYSTEMS IN *C. ELEGANS* AND OTHER NEMATODES

Most species in the phylum Nematoda are gonochoristic, that is, they have separate male and female sexes, and reproduction occurs exclusively by cross-fertilization. However, there are many nematodes with unusual modes of reproduction, such as parthenogenesis or hermaphrodite self-fertilization (for review, see Nigon 1965). *C. elegans* is one of these, being a species with hermaphrodite and male sexes, but no female sex. Most natural populations are probably composed predominantly of hermaphrodites, because soil samples gathered in many different parts of the world have yielded populations of *C. elegans* that are almost pure hermaphrodite (Maupas 1900; P. Anderson et al., pers. comm.). Rare males are found in these populations and also in stocks of the standard laboratory strain (var. Bristol, strain N2). The males are invariably *XO* and, therefore, must arise as the result of spontaneous X chromosome loss ($XX \rightarrow XO$), probably during meiosis (Hodgkin et al. 1979). The males are capable of cross-fertilizing hermaphrodites to yield equal numbers of hermaphrodite and male cross progeny. Under most circumstances, the higher growth rate of a hermaphrodite population means that the frequency of males in a population remains close to the rate of spontaneous X chromosome loss ($<0.5\%$).

There are close relatives of *C. elegans*, such as *Caenorhabditis remanei*, (Sudhaus 1974), which are gonochoristic. The existence of such species and the prevalence of gonochorism among nematodes in general make it likely that the hermaphrodite sex of *C. elegans* is a secondary specialization. Comparison of *C. elegans* and *C. remanei* suggests that the only difference between the *C. elegans* hermaphrodite and a true female lies in the germ line: Hermaphrodite gonads produce a limited number of sperm during the fourth larval stage before switching over to oogenesis in the adult. Therefore, the hermaphrodite can be regarded as a modified female.

The basis of sex determination in most nematodes is chromosomal: Females (or hermaphrodites) are *XX*, and males are *XO*. *C. elegans* has this standard mechanism, with *XX* hermaphrodites and *XO* males, so one can hope that what is learned about *C. elegans* will prove to be true of nematodes in general. Different mechanisms have been described for some nematodes, such as alternative chromosomal mechanisms (e.g., the multiple X chromosomes of *Ascaris lumbricoides;* for review, see White [1973]) or environmental sex determination. For instance, in several parasitic nematodes (e.g., *Mermis* and *Meloidogyne*), it is believed that male development is favored by stressful conditions such as crowding (Christie 1929; Triantaphyllou 1973).

III. SUMMARY OF DIFFERENCES BETWEEN HERMAPHRODITES AND MALES

There are extensive differences between the two sexes of *C. elegans*, involving most of the tissues and organs of the animal, as well as its overall organization and size (see Figs. 1 and 3). Some structures appear to be identical in adult hermaphrodite and adult male (e.g., the main body muscles, the pharynx, and the excretory system), but a large part of the animal shows sexual specializations of one kind or another. Of the somatic nuclei in the adult (959 in the hermaphrodite and 1031 in the male), no more than about 650 appear to be sexually indifferent; the remainder ($\sim$ 30% of hermaphrodite nuclei and 40% of male nuclei) are sexually specialized. In addition, the germ line develops differently in the two sexes. The various differences between the two sexes are discussed in Section III.A–H. For more detailed descriptions, see Chapters 4, 5, and 7.

A. Germ Line

The physiology and development of the germ line are discussed more extensively in Chapter 7; relevant aspects are summarized here. The germ line is the largest tissue in adults of both hermaphrodite and male, as measured by the number of nuclei. In contrast to the somatic tissues, which undergo a limited and invariant set of cell divisions, the nematode germ line undergoes general proliferative growth during larval development, with no fixed lineage (Kimble and Hirsh 1979).

The hermaphrodite germ line is derived from two L1 precursor cells, Z2 and Z3, and the descendants of these cells colonize the two arms of the somatic gonad. Germ cell proliferation continues in the distal arms, so that a total of at least 1000 nuclei are generated in each ovary during the lifetime of a hermaphrodite. Meiosis begins at the L3/L4 molt, in the proximal arms of the gonad. The first germ cells to differentiate all become sperm; about 40 primary spermatocytes are formed in each arm, giving rise to a total of roughly 320 sperm in all. Subsequent differentiation of germ cells in the adult gives rise exclusively to oocytes, so that the adult hermaphrodite becomes functionally a female. The sperm made previously are stored in the spermathecae and are used to fertilize mature oocytes as they pass through the spermatheca into the uterus. Sperm are utilized with almost 100% efficiency (Ward and Carrel 1979), so each hermaphrodite produces more than 300 self-progeny (mean 329, range 274–374 for 12 animals [Hodgkin 1983c]). After the sperm are used up, oogenesis is arrested but can be stimulated to resume after insemination by a male. A single hermaphrodite can yield over 1400 progeny if it is mated with several males (Hodgkin 1986). Introduction of sperm from a male at any time during adulthood leads to a switch from self-fertilization to cross-fertilization, because the male sperm fertilize oocytes preferentially (Ward and Carrel 1979). It follows that sperm from the hermaphrodite are somehow different from male sperm, though they appear morphologically identical.

The male germ line is also derived from the two precursor cells Z2 and Z3, which undergo a similar mitotic proliferation within the single arm of the male gonad. Meiosis begins earlier than in the hermaphrodite, during the L3 stage. Gametogenesis occurs in the proximal (more posterior) part of the gonad, producing only sperm. Males transfer sperm to hermaphrodites by means of mating, which can occur successfully within 6 hours of the last larval molt. Males will continue to mate with many different hermaphrodites over a period of up to 6 days at 20°C. A single male has been observed to sire as many as 2871 progeny in all (Hodgkin 1983a).

Because the process of sperm transfer is not wholly efficient (Ward and Carrel 1979), it is likely that a single male can produce 3000–4000 functional sperm during his lifetime.

In both hermaphrodite and male, the distal part of the germ line is maintained in a state of mitotic proliferation by an influence exerted by the distal tip cells at the ends of the gonad arms (Kimble and White 1981). It is believed that germ cells enter meiosis as soon as they reach proximal parts of the gonad and escape from the distal tip cell influence. In the male, only sperm are made, but in the hermaphrodite a choice must be made between spermatogenesis and oogenesis. One of the central questions in studying *C. elegans* sex determination is how this choice is controlled. Normally, the first germ cells to differentiate in the hermaphrodite form sperm, and the germ line later switches to oogenesis, but it is possible to affect this switch by means of mutations, so that all hermaphrodite germ cells form oocytes or all form sperm (see Section V).

B. Somatic Gonad: Anatomy and Development

The gonad primordia of both sexes are identical in the L1 stage: The small somatic blast cells Z1 and Z4 flank the large germ-line precursors Z2 and Z3. The behavior of Z2 and Z3 is summarized in Section III.A; the development of Z1 and Z4 in both sexes has been described in detail by Kimble and Hirsh (1979). In the hermaphrodite, Z1 and Z4 give rise to a total of 143 nuclei, which form the following structures: one anchor cell, a centrally located uterus consisting of 28 dorsal and 32 ventral nuclei, and two symmetrical gonad arms (anterior and posterior). Each gonad arm consists of a junction between uterus and spermatheca (6 nuclei), a spermatheca (24 nuclei), a sheath surrounding the germ cells (10 nuclei), and one distal tip cell. The distal tip cells of the hermaphrodite have a dual role, being responsible both for germ cell proliferation, as described above, and for elongation of the gonad arms during larval growth (leader function). The anchor cell also plays a key role, being responsible for the induction of vulval lineages in the three ventral hypodermal cells P5.p–P7.p (Kimble 1981).

The male gonad follows a different lineage, with Z1 and Z4 generating a total of 56 nuclei. These form the following structures: one linker cell, a vas deferens (30 nuclei), a seminal vesicle (23 nuclei), and two distal tip cells. The male germ cells are unsheathed, being surrounded only by a basement membrane. The distal tip cells are required for germ cell proliferation but have no other known function. The linker cell is responsible for gonad elongation toward the tail, and eventually this cell makes contact with two cells of the proctodaeum, U.lp and U.rp. These cells kill the linker cell, thereby opening a passageway between vas deferens and cloaca (Sulston and White 1980).

C. Intestine

The adult intestine consists of 20 cells, 14 of which are binucleate. All intestinal nuclei undergo endoreduplication during larval growth to reach a DNA content of 32n (Hedgecock and White 1985); consequently, the intestine contains about one third of the somatic DNA in the animal. The intestine is anatomically almost identical in both sexes, but it is functionally specialized for yolk protein production in the adult hermaphrodite (Kimble and Sharrock 1983; Wood et al. 1985). Yolk protein synthesis occurs exclusively in the intestinal cells of the adult hermaphrodite; no trace of yolk is made in larvae or in males. The major yolk proteins, which are among the more abundant proteins synthesized in the adult, consist of two related 170-kD proteins and two smaller proteins (115 kD and 88 kD), the latter two being derived from a 180-kD precursor (Sharrock 1983, 1984). These proteins are secreted into the pseudocoelom and are taken up by the gonad to be incorporated into developing oocytes (Kimble and Sharrock 1983). In situ hybridization experiments using probes for *vit* (yolk protein)-gene transcripts indicate that all 20 cells of the intestine participate in making yolk proteins (Wood et al. 1985). Their synthesis is not dependent on the presence of the gonad, because hermaphrodites whose gonads have been removed by laser ablation in the L1 stage still synthesize yolk proteins in abundance (Kimble and Sharrock 1983). Also, oogenesis is not apparently dependent on yolk protein synthesis, because in certain mutants, e.g., a double mutant *tra-1(0);fem-1(0)*, cells that look like oocytes are produced in the absence of yolk protein synthesis by the intestine (Doniach and Hodgkin 1984).

D. Musculature

All sex-specific muscles are derived from the postembryonic blast cell, M (Sulston and Horvitz 1977). In the hermaphrodite, M gives rise to 14 body-muscle nuclei, 2 coelomocytes, and 2 sex mesoblasts. These mesoblasts migrate anteriorly and then give rise to 16 muscle cells specialized for egg laying: 8 uterine muscle cells (4 type 1 and 4 type 2) and 8 vulval muscle cells (4 type 1 and 4 type 2). The uterine muscle cells are disposed circumferentially around the uterus and presumably squeeze eggs toward the vulva; the vulval muscle cells are arranged in a cross around the vulva and contract to expand the uterus and open the vulva, permitting egg laying.

In the male, M gives rise to 14 body-muscle nuclei and 6 sex mesoblasts. These mesoblasts migrate posteriorly, giving rise to 1 coelomocyte and 41 muscles specialized for copulation. The anatomy and ontogeny of these muscles are described in detail by Sulston et al. (1980). Fifteen form the preanal diagonal muscles (8 right and 7 left), and 10 form the preanal longitudinal muscles (5 symmetrical pairs). Both of these muscle sets probably act to cause tail curling during male mating. Eight other cells form spicule protractors (2 dorsal and 2 ventral) and spicule retractors (2 dorsal and 2 ventral); 4 cells form erectors and retractors of the gubernaculum; and 4 form postanal oblique muscles.

In addition to these postembryonic cells, several of the juvenile muscle cells undergo changes during male maturation: The anal depressor muscle (mu anal) and the rectal sphincter muscle (mu sph) both undergo extensive reorganization (Sulston et al. 1980).

E. Hypodermis and Other Structural Elements

The hypodermis of the hermaphrodite contains one sexually specialized structure, the vulva, which is a transverse slit located midventrally. The ventral hypodermal cells P5.p–P7.p divide during the L3 stage, yielding a total of 22 vulval nuclei; P3.p, P4.p, and P8.p also divide once (Sulston and Horvitz 1977). In the male, P3.p–P8.p do not divide.

The male tail contains several complicated sex-specific structures (see Fig. 1). The lateral hypodermal blast cell pairs V5, V6, and T construct the acellular cuticular fan and its nine pairs of supporting rays. The proctodaeum, the junction between alimentary and genital tracts, contains 26 nuclei, mostly derived from blast cell B, with some contributions from U, F, and K. The two cells U.rp and U.lp are responsible for killing the gonadal linker cell, and the connection between proctodaeum and vas deferens is formed by one B-derived cell, together with two cells K.a and K′.

The two copulatory spicules normally lie in channels associated with the proctodaeum; each consists of three neurons and six structural cells (four socket and two sheath), which are extensively sclerotized, giving them structural rigidity. During copulation they are inserted into the vulva of the hermaphrodite by means of protractor muscles. The gubernaculum is a strip of sclerotized cuticle in the roof of the proctodaeum, which probably serves to guide the tips of the spicules during copulation. There are also three male-specific sense organs associated with the cloaca: the paired postcloacal sensilla, derived from blast cells Y and B, and the preanal hook sensillum, derived from P10.p and P11.p.

The hypodermal tissues may also be responsible for the overall size difference between the two sexes: Mature hermaphrodites are longer (1400 μm as opposed to 1000 μm) and fatter than mature males. Part of this difference may be a consequence of the more voluminous gonad of the hermaphrodite, but a considerable difference in size is also seen in gonad-ablated animals, implying that other tissues, such as the hypodermis, contribute to the larger size of the hermaphrodite.

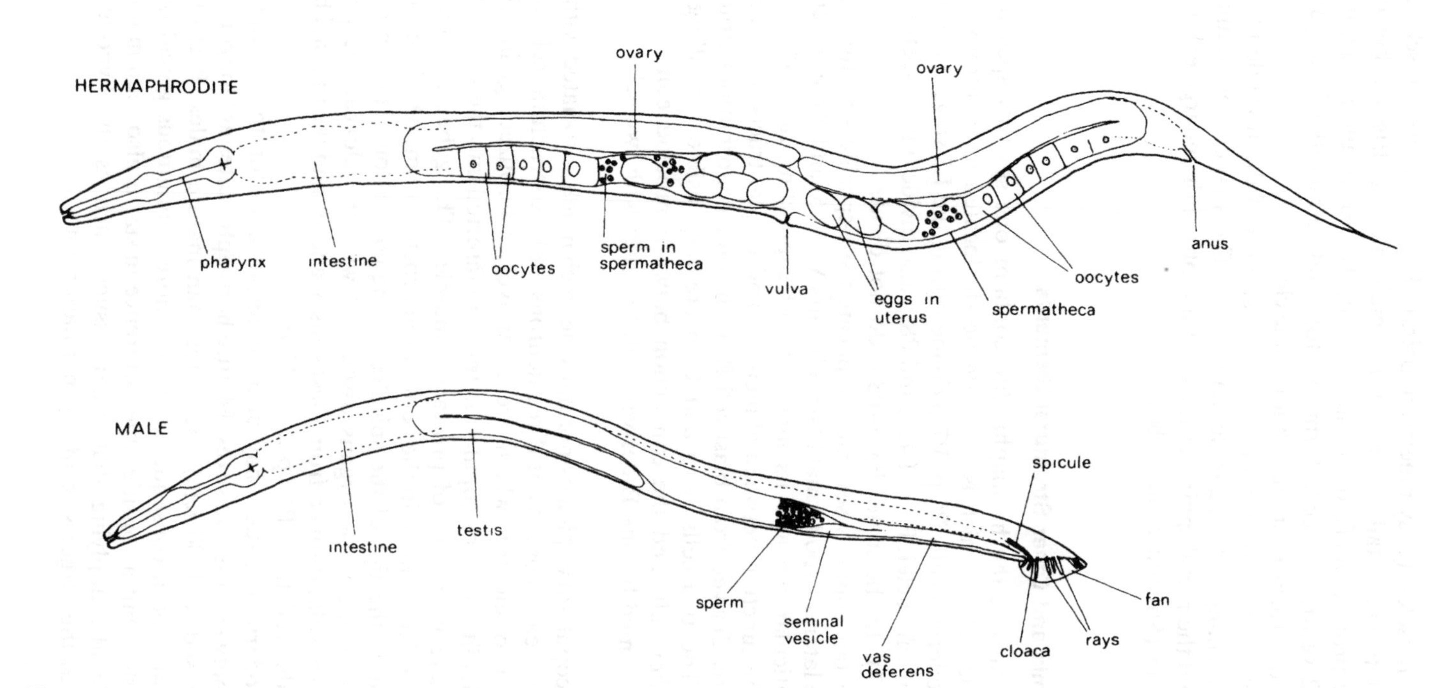

Figure 1 Schematic diagrams of hermaphrodite and male.

F. Nervous System and Behavior

The hermaphrodite nervous system contains 302 neurons, of which at least 12 are sexually specialized. These are the neurons involved in egg laying, which is the only type of behavior displayed by the hermaphrodite that is absent from the male repertoire. The most important components of the egg-laying circuitry are the two hermaphrodite-specific neurons (HSNs), which synapse directly onto the uterine and vulval muscles (White et al. 1986). The six VC neurons, which are also specific to the hermaphrodite, also synapse onto these muscles. Four other neurons common to both sexes synapse onto the HSNs: two mechanosensory neurons (PLML/R) and two interneurons (BDUL/R). Other neurons may also be involved in the egg-laying system, via indirect or humoral inputs. Extensive analyses of egg-laying behavior by means of anatomical, surgical, pharmacological, and genetic methods have been carried out (see e.g., Horvitz et al. 1982b; Trent et al. 1983).

The adult male nervous system contains 381 neurons, of which at least 87 are sexually specialized: These cells are not found in the hermaphrodite. The nervous system of the male is substantially more complicated than that of the hermaphrodite and has been reconstructed only in part, unlike the complete reconstruction of the adult hermaphrodite nervous system (White et al. 1986). In particular, the male nerve ring has not been reconstructed. It is likely that many of the 294 neurons common to both sexes have different circuitry in males and hermaphrodites, as is demonstrably the case for neurons such as PDB and PHC (Sulston et al. 1980).

Most of the additional neurons in the male have cell bodies in the tail; the only exceptions are 1 anterior cord neuron (P2.aap), the 12 male-specific CA and CP cord motor neurons derived from P3–P8, and the 4 cephalic companion cells, CEM, which are embryonic descendants of AB. The CEM cells have sensory endings open to the exterior in the cephalic sensilla (Ward et al. 1975); their anatomy has suggested the hypothesis that they mediate a specific chemotaxis of males toward hermaphrodites, for which some indirect evidence exists (Sulston and Horvitz 1981). However, there is no direct evidence that implicates the CEM cells in this chemotaxis, nor has the putative sexual attractant been identified.

The neurons in the tail include the 18 pairs of sensory neurons innervating the 18 copulatory rays (derived from V5, V6, and T), 2 sets of three sensory neurons innervating the postcloacal sensilla (derived from Y and B), the 2 sensory neurons of the hook sensillum (derived from P10), and the spicule neurons. Each spicule is associated with 2 sensory neurons and 1 motor neuron, all derived from B. In addition, there are 15 interneurons or interneuronlike cells (10 from P9.p–P12.p, 2 from B, 2 from F, and 1 or 2 from U) and 4 motor neurons derived from B, U, and F.

Relatively little is known about the function of the male nervous system, though it has been possible to assign functions to some structures by means of laser ablation experiments (Sulston and White 1980; J.E. Sulston, pers. comm.). For example, the hook sensillum appears to be necessary for efficient location of the vulva by the male tail. Copulation is probably the most complicated behavior exhibited by *C. elegans*, and the underlying anatomy and neural circuitry are also complicated, so a detailed analysis has not yet been carried out. Some puzzles have already been encountered: For example, the mechanism whereby the vas deferens opens to permit ejaculation is obscure, as there appears to be no musculature or neuronal input directly associated with the vas deferens (J.E. Sulston, pers. comm.).

G. Differences Related to X Chromosome Dosage

The dimorphic features of male and hermaphrodite described in Section III are related to sexual differentiation per se. There are additional differences between the two sexes that are a consequence of the different genetic content of hermaphrodite and male.

The most important of these differences is dosage compensation: It is believed that most (but not all) sex-linked genes are transcribed at a higher level in *XO* animals than in *XX* animals, thereby equalizing the expression of genes in the two sexes (Hodgkin 1983c; Wood et al. 1985; Meyer and Casson 1986). Failure to compensate correctly is probably lethal. The analysis of dosage compensation is discussed in Section VII.

A difference directly related to X chromosome dosage is radiation sensitivity: Males are more sensitive to gamma radiation than hermaphrodites (Hartman and Herman 1982a). These investigators showed that the greater sensitivity of males results from their possessing only one X chromosome, because *XO* animals transformed into hermaphrodites by a *her-1* mutation were still sensitive, whereas *XX* animals transformed into males by a *tra-1* mutation were more resistant than *XO* males.

A phenomenon that may have a similar basis is the difference in life span: Johnson and Wood (1982) observed that N2 hermaphrodites have a mean life span of 19.9 days, as opposed to 17.7 days for N2 males. The shorter life span of males might be a consequence of damage to the single X chromosome of the male.

Finally, the fact that the single X chromosome of the male has no pairing partner during meiosis implies that there must be a special mechanism for handling the unpaired X in male meiosis, because there is no loss of X chromosomes during this process: *XO* males produce haplo-X and nullo-X sperm in an exact 1 : 1 ratio (Hodgkin et al. 1979). The mechanism appears to be male specific, because the unpaired X behaves aberrantly in *her-1 XO* hermaphrodites, at both spermatogenic and oogenic meioses. In *XO* hermaphrodites, the unpaired X chromosomes are lost so that about 70%, rather than 50%, of gametes are nullo-X (Hodgkin 1980). Therefore, this is really a difference associated with sexual phenotype and provides a further indication that spermatogenesis in males and spermatogenesis in hermaphrodites are not identical processes.

H. Overview of Sexual Development; Autonomy

Embryonic development is almost identical between the two sexes: The only lineage differences are two sets of programmed cell deaths that occur about two thirds of the way through embryogenesis (Sulston et al. 1983). These cell deaths remove the four CEM cells (in the hermaphrodite) or the two HSN cells (in the male). The only other visible differences in the anatomy of the newly hatched L1 are seen in the nuclear volume of blast cells B and Y and in the placement of one coelomocyte (Sulston and Horvitz 1977).

Postembryonic development is extensively divergent between the two sexes, particularly during late larval development. Much postembryonic development involves addition of sex-specific structures onto an unchanged larval substrate. Some of the dimorphic development results from the division of primordial cells that divide in one sex but not in the other. For example, the three vulval precursor cells (P5.p–P7.p) do not divide in the male but yield 22 nuclei in the hermaphrodite. Conversely, the blast cells B, Y, U, and F survive without dividing in the hermaphrodite, but they undergo complex lineages in the male, yielding a total of 66 nuclei in the adult. Other parts of sexual maturation require different lineage patterns derived from homologous precursors, for example, the different lineages pursued in the two sexes by the gonadal ancestors Z1 and Z4, or the mesoblast M. Overall, then, there are both separate and common primordia for the generation of sexual dimorphism in this organism.

A number of experiments have been carried out to investigate the autonomy of sexual development in *C. elegans*. The results are relevant to the question of whether sex determination in this organism is cell autonomous or whether any diffusible agents such as sex hormones are involved. In general, it has been found that laser ablation of various parts of the animal does not affect the sexual phenotype of the remaining parts. Yolk production by intestinal cells is not affected by ablation of the gonad (Kimble and Sharrock 1983), and ablation experiments on the developing male tail have also indicated a high degree of autonomy (Sulston and White 1980). One exception to the rule of autonomy is observed in the generation of the vulva: The vulval divisions of the ventral hypodermis are dependent on receiving an inductive signal from the anchor cell in the developing hermaphrodite gonad (Kimble 1981a). This signal appears to act only over a short range, however, and is not required for other aspects of sexual maturation.

IV. CHROMOSOMAL BASIS OF SEX DETERMINATION

A. Effect of Altering X Chromosome Dosage

The primary sex-determining signal in *C. elegans* is X chromosome dosage: In a diploid animal (with five pairs of autosomes), the presence of two X chromosomes (*XX*) results in hermaphrodite development, and the presence of one X chromosome (*XO*) results in male development. The critical signal could be the ratio of X chromosomes to autosomes or the absolute number of X chromosomes. These possibilities can be distinguished by varying the X chromosome to autosome ratio (X/A ratio), as summarized in Table 1 (from Nigon 1949a; Hodgkin et al. 1979, 1987a; Madl and Herman 1979). The fact that *4A;2X* tetraploids (i.e., animals containing four sets of five autosomes and two X chromosomes) are male immediately shows that the critical signal is the X/A ratio, not the absolute number of X chromosomes. At an X/A ratio of 0.67 (as in a *3A;2X* triploid) or less, the phenotype is male; at a ratio of 0.75 (as in a *4A;3X* tetraploid) or more, the phenotype is hermaphrodite.

Madl and Herman (1979) investigated the effect of altering X/A ratio in triploid animals by means of duplications of part of the X chromosome (see Fig. 2). A large duplication, *mnDp10* (~25% of the X chromosome), shifted the *3A;2X* phenotype from male to hermaphrodite, but a smaller duplication (*mnDp9*) was only partly effective, tending to cause intersexual development, and an even smaller duplication (*mnDp8*) was ineffective. These three duplications form a nested set, so there must be at least two ratio-affecting sites in the region covered by *mnDp10* but not by *mnDp8*. This region is about one tenth of the X chromosome, which implies that there may be at least 20 sites overall. There must be at least one more site outside *mnDp10* because *AA;XO;mnDp10* is male.

Table 1 Effect of X chromosome/autosome ratio on sexual phenotype

	Diploid	Triploid	Tetraploid
Inviable	*AA;OO*		
Male	*AA;XO*	*AAA;XO*	*AAAA;XX*
		AAA;XX	
Intersex		*AAA;XX;mnDp9*	
Hermaphrodite	*AA;XX*	*AAA;XX;mnDp10*	*AAAA;XXX*
	AA;XXX	*AAA;XXX*	*AAAA;XXXX*
Inviable	*AA;XXXX*		

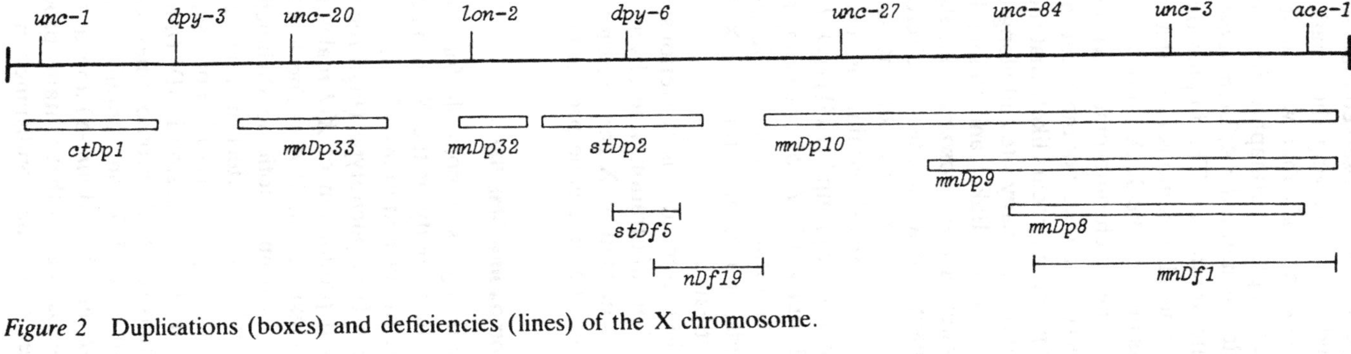

Figure 2 Duplications (boxes) and deficiencies (lines) of the X chromosome.

R.K. Herman (pers. comm.) observed that crossing tetraploid hermaphrodites with diploid males carrying the small duplication *mnDp8* generated *3A;2X;mnDp8* animals that were uniformly male but that the reverse cross, mating diploid hermaphrodites carrying *mnDp8* with tetraploid (*4A;2X*) males, generated *3A;2X,mnDp8* animals that were sometimes hermaphrodite or intersex. This implies that *mnDp8* carries a fourth ratio-affecting site and also demonstrates that there can be a maternal effect on the assessment of the X/A ratio.

Other regions of the X chromosome have not been investigated, although duplications covering most of the X chromosome are now available (see Fig. 2). At this point, it appears likely that the X/A ratio is determined by the cumulative effect of many sites all along the X chromosome, as in *Drosophila* (for review, see Baker and Belote 1983).

It is not known whether the autosomal contribution to the X/A ratio is diffuse, involving many autosomal sites, or discrete, involving only one or a few sites. Variations in autosomal content have not been investigated extensively with regard to sex determination, except that it is known that large duplications of linkage group I (LGI) (*sDp1*, *sDp2*), LGII (*mnDp34*, *mnDp35*), or LGIII (*mnDp37*, *eDp6*) or trisomy for LGIV (see Chapter 2) do not cause masculinization of diploid *XX* animals. Their effects on triploids or tetraploids, or in combination with X chromosome deficiencies, have not been investigated.

How the X/A ratio is computed is mysterious: Most conceivable mechanisms involve some kind of titration, whereby an autosomal product made in limiting quantity is titrated by X chromosome sites (or conceivably by uncompensated X chromosome gene products).

B. Control of X Chromosome Stability

The X chromosome dosage is responsible for sex determination, so that mutations affecting the stability of the X chromosome will alter the sex ratio of populations. Many mutations of this type have been identified, most of which have a Him phenotype (*h*igh *i*ncidence of *m*ales), recognized by an increased production of *XO* males from self-fertilizing *XX* hermaphrodites (Hodgkin et al. 1979). The loss of X chromosomes appears to result primarily from meiotic nondisjunction. Most of the known *him* mutations have only a slight effect, increasing the rate of *XO* production from 0.2% (the wild-type level) to 5% or less, but several cause larger increases. Mutations in three genes on LGIV (*him-3*, *him-6*, and *him-12*) affect the meiotic behavior of all chromosomes, causing generalized nondisjunction (Hodgkin et al. 1979 and unpubl.), whereas three other autosomal genes (*him-1*, *him-5*, and *him-8*) appear to have a large selective effect on the X chromosome, so that mutants produce 20–40% males. In mutants of the latter type, the nondisjunction is associated with, and perhaps caused by, a large decrease in X chromosome recombination, with little or no effect on autosomal recombination (Hodgkin et al. 1979; R.K. Herman, pers. comm.). These genes appear to be specifically involved in X chromosome stability during *XX* meiosis, but they do not affect the handling of the unpaired X in male meiosis. Only one mutation affecting the behavior of the unpaired X has been identified: *him-7(e1480)* causes a small but significant increase in nullo-X male sperm frequency from 50% to 52%, as well as slightly increasing nondisjunction in hermaphrodites (Hodgkin et al. 1979). One mutation with an anti-Him phenotype was isolated by Hartman and Herman (1982b *[rad-4(mn158)]*), which shows reduced X chromosome nondisjunction so that male production drops to 0.03%; it also partly suppresses some *him* mutants.

The existence of these various genes indicates that the meiotic stability of the X chromosome is controlled specifically and suggests that the wild-type rate of nondisjunction may be a selectively advantageous optimum, providing a constant low level of males in natural populations (for a theoretical treatment, see Hedgecock 1976). Wild-type hermaphrodites that have been stressed by brief heat shock produce increased numbers (2–5%) of *XO* male self-progeny (Hodgkin 1983a).

V. MAJOR SEX-DETERMINING GENES

At first sight, the diffuse nature of the primary sex-determining signal, the X/A ratio, suggests that sex is determined polygenically, as the result of a balance between feminizing genes on the X chromosome and masculinizing genes on the autosomes. The existence of a small number of major sex-determining genes shows that this is not the case. Instead, it appears that the X/A ratio provides an initial signal to set the states of these genes, which then act to direct development along a particular sexual pathway. The reason for this belief is that point mutations in the major sex-determining genes can completely transform sexual phenotype, overriding any effect of the X/A ratio.

The effects of mutations in the seven major genes identified so far are summarized in Table 2 and described at greater length in Section V.B. The genes fall into three classes: (1) The three *tra* genes appear to be required in the hermaphrodite but not in the male, because loss-of-function mutations in these genes result in masculinization of *XX* animals but have no obvious effect on *XO* animals. (2) The *her-1* gene appears to be required in the male but not in the hermaphrodite, because *her-1* loss-of-function mutations result in hermaphrodite development in both *XX* and *XO* animals. (3) The three *fem* genes appear to be required for male somatic development and also for spermatogenesis in both males and hermaphrodites, because *fem* loss-of-function mutations result in female development

Table 2 Null and dominant phenotypes of sex-determination genes

	XX phenotype	*XO* phenotype
	A. Null phenotypes	
Wild type	Hermaphrodite	Male
tra-1(0) III	Low-fertility male	Low-fertility male
tra-2(0) II	Male gonad, incomplete male body	Male
tra-3(0) IV	Masculinized gonad and body (maternal rescue)	Male
her-1(0) V	Hermaphrodite	Hermaphrodite
fem-1(0) IV	Female (some maternal rescue)	Female { partial maternal rescue
fem-2(0) III	Female (complete maternal rescue)	
fem-3(0) IV	Female (no maternal rescue)	
	B. Dominant phenotypes (homozygotes)	
tra-1(dom)	Female	Female
tra-2(dom)	Female	Male (variably feminized)
her-1(dom)	Masculinized hermaphrodite	Male
fem-3(dom)	Male germ line, female soma	Male

in both *XX* and *XO* animals. The rest of this section describes the identification and analysis of these genes.

A. Isolation of Sex-determining Mutations

Mutations that cause an animal to develop a sexual phenotype inappropriate to its chromosomal sex (X chromosome to autosome ratio) have been isolated in a variety of ways (for a summary of phenotypes, see Table 2).

Recessive alleles of the three *tra* genes (which cause masculinization) were first isolated as chance segregants from general mutagenesis screens, because the appearance of males in an otherwise self-fertilizing hermaphrodite population is a conspicuous event (Hodgkin and Brenner 1977). Temperature-sensitive alleles of two other genes *fem-1* and *fem-2*, as well as a temperature-sensitive *tra-2* allele, were obtained in the course of general screens for temperature-sensitive sterile mutations (Klass et al. 1976; Nelson et al. 1978; Kimble et al. 1984).

A systematic search for autosomal mutations causing masculinization of *XX* animals was carried out (Hodgkin 1987b), using a strain heterozygous for two sex-linked markers, *dpy-7 +/+ unc-18*. These markers are tightly linked, so almost all *XO* male animals that appeared as a result of nondisjunction were either dumpy or uncoordinated, and could be ignored. The appearance of wild-type males or intersexes indicated the presence of a sex-determining mutation. Approximately 40,000 mutagenized genomes were screened, yielding recessive alleles of the three known *tra* genes, but no new *tra* genes were identified.

A large-scale search for mutations of the opposite type (causing feminization of *XO* animals) was carried out using strains homozygous for *him-5* and *dpy-21* (Hodgkin 1980). These strains segregate approximately 65% dumpy hermaphrodites and 35% nondumpy males, because the *him-5* mutation increases X chromosome loss, and the *dpy-21* mutation is only expressed in *XX* animals (see Section VII.B). Mutations transforming *XO* animals into hermaphrodites were recognized by the appearance of nondumpy hermaphrodites. From an extensive screen, eight mutants of this type were recovered, all of which proved to carry recessive alleles of *her-1*. A dominant feminizing mutation, *tra-1(e1575*sd*)* (originally called *her-2*), was also isolated in this screen. This mutation causes both *XX* and *XO* animals to develop into fertile females (i.e., spermless hermaphrodites). No *fem* mutations were obtained, probably because the maternal effects of the three *fem* genes precluded their isolation (see Section V.B).

A direct screen for dominant *XX*-feminizing mutations was carried out by searching for females in the F1 progeny of mutagenized wild-type (N2) hermaphrodites (Doniach 1986). Dominant mutations of *tra-1* and *tra-2* were obtained, as well as one *fem-3* allele.

Slight masculinization can cause the death of the HSN cells in an *XX* animal, resulting in a fertile but egg-laying defective hermaphrodite. Several weak masculinizing mutations have been isolated in the course of screens for Egl mutants (Trent et al. 1983; C. Desai, pers. comm.). A dominant masculinizing allele of *her-1*, *n695*sd, was obtained, as well as several *tra-2* alleles. Two of the *egl* genes identified in these screens, *egl-16* and *egl-41*, may also prove to be directly involved in sex determination (see below).

Dominant gain-of-function mutations (which cause sexual transformations opposite those caused by recessive alleles) have been obtained in four of the major sex-determining genes: *tra-1* (Hodgkin 1983b), *her-1* (Trent et al. 1983), *tra-2* (Doniach 1986), and *fem-3* (Barton et al. 1987). Dominant mutations of this type can be reverted to a recessive state, using the paradigm of Lifschytz and Falk (1969), and this has been successfully achieved for all four genes. These reversions constitute efficient selections for null or hypomorphic alleles of the genes in question.

Other methods for isolating sex-determination mutants have relied on interactions between the different genes. For example, extragenic suppressors of *her-1* were obtained by searching for males in a *her-1(0)* background; alleles of *tra-1* and *tra-2* were obtained (Hodgkin 1980). An extensive search for suppressors of *tra-3* yielded alleles of *fem-1*, *fem-2*, and *fem-3*, as well as *tra-1(dom)* and *tra-2(dom)* mutations (Hodgkin 1986). Large-scale reversion experiments on temperature-sensitive alleles of *tra-2* (Klass et al. 1976), *fem-1* (Nelson et al. 1978; Barton et al. 1987), *fem-2* (Barton et al. 1987), and *fem-3* (M. Shen, pers. comm.) have been carried out by growing up populations of mutagenized worms at permissive temperature and shifting to restrictive temperature. The *tra-2* reversion yielded only a same-site revertant, but the *fem* reversions have yielded a variety of extragenic suppressors, some of which lie in known genes such as *tra-1*, *tra-2*, and *fem-3*. Interactions between *tra-2* or *tra-3* and the three *fem* genes have been used to set up efficient selections for more alleles of *fem* genes (Doniach and Hodgkin 1984; Hodgkin 1986).

In summary, a variety of screens and selections have been employed to isolate sex-determination mutants. Most of these methods have involved screening populations of 10^5–10^7 or more, so probably most of the major sex-determining genes have been found by now. New genes may well be identified, but it is likely that these will have only minor or tissue-specific effects (e.g., affecting only the germ-line sperm/oocyte choice), and that it will be possible to accommodate them in the scheme set out below (Section V.D). It is worth emphasizing that all of the genes so far identified appear to affect only sexual characters: None has pleiotropic effects on viability or other aspects of the nematode phenotype. Conversely, genes affecting other aspects of the worm, such as musculature or cuticle, do not affect sex determination. Admittedly, lethal mutations with significant effects on sex determination are more difficult to recognize, and some mutations of this type (affecting both sex determination and dosage compensation) are now being studied (see Section VII).

B. Properties of Sex-determining Genes

Each of the seven major genes so far identified has been defined by at least four independent mutations. For most of the seven, either amber alleles or deficiencies or both are known, so that the null phenotypes of the genes are reasonably certain. These phenotypes are set out in Table 2. Dominant gain-of-function mutations have been obtained for four of the genes, and the phenotypes of these alleles are also summarized in Table 2. A brief description of the seven genes follows.

1. tra-1 III

The null phenotype was originally defined by *e1099* (Hodgkin and Brenner 1977; Hodgkin 1983b); this mutation is fully recessive but causes masculinization of homozygous *XX* animals. The nongonadal phenotype of *tra-1(0) XX* animals is indistinguishable from that of wild-type *XO* males: All dimorphic characters examined show complete transformation to a male phenotype (e.g., Fig. 3B). The gonadal morphology is male, but about 30% of the individuals have abnormal stunted gonads. Most of the animals contain sperm, and oocytes are not usually seen. However, the maximum number of sperm made is far below the number made by an *XO* male, and

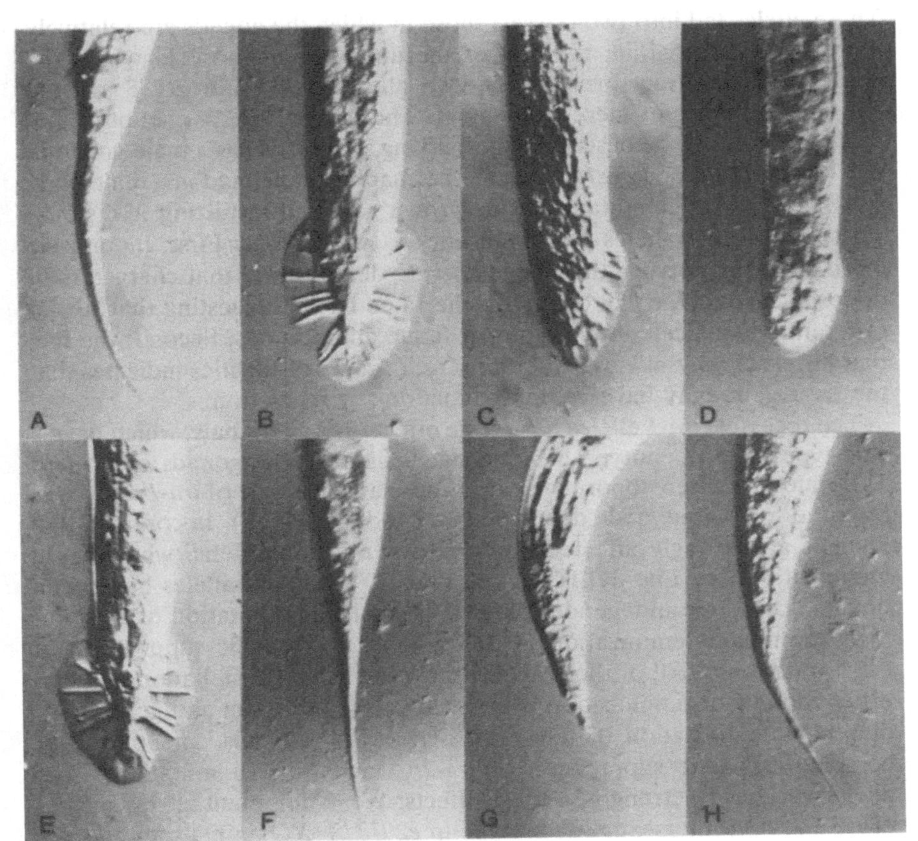

Figure 3 Photographs of tail phenotypes (Nomarski optics). (*A*) Wild-type *XX* (hermaphrodite); (*B*) *tra-1(0) XX* (male); (*C*) *tra-2(0) XX* (incomplete male); (*D*) *tra-1(0) XXX* (abnormal male); (*E*) wild-type *XO* (male); (*F*) *her-1(0) XO* (hermaphrodite); (*G*) *tra-1(dom)/+ XO* (female, incomplete tail spike); (*H*) *tra-1(dom) XO* (female).

only about 30% of *tra-1(0) XX* males are capable of siring progeny. The maximum number of progeny sired by such a male is 262, in contrast to the maximum of 2871 sired by a wild-type *XO* male. An amber allele of *tra-1*, *e1781*, results in a slightly different gonadal phenotype: The gonads are larger and always have male morphology, but apparent oocytes are made in this gonad once the animal reaches maturity (Hodgkin 1987b; T. Schedl, pers. comm.). It is not clear which (if either) of these two alleles results in complete loss of *tra-1* function.

Many weaker recessive mutations of *tra-1* have also been identified, which cause incomplete masculinization of *XX* animals (Hodgkin 1987b).

These alleles fall into two classes: those in which the gonads are relatively more transformed than the rest of the animal, and those in which the gonads are less transformed. The two alleles *e1732* and *e1488* provide extreme examples of these two classes: The respective *XX* phenotypes of these mutants can be described (simplifying somewhat) as a male gonad in a hermaphrodite body (*e1732*) and a hermaphrodite gonad in a male body (*e1488*). The latter mutant can be grown as a self-fertilizing *XX* stock. These weak *tra-1* alleles are significant for two reasons: First, the mosaic intersexual phenotypes that they generate demonstrate that characters of opposite sexual type can coexist in the same body, suggesting that sexual differentiation is cell autonomous or tissue autonomous. Second, the fact that different *tra-1* alleles exhibit different tissue specificities indicates that the *tra-1* gene may have complex functions or regulation.

The recessive *tra-1* alleles have little effect on *XO* animals, which have a wild-type male phenotype in nongonadal tissues. The gonads of *tra-1(0) XO* males are often abnormal, indicating that some part of *tra-1* activity is required for normal male development (Hodgkin 1987b). In contrast, *XO* animals are strongly affected by dominant *tra-1* mutations, which cause feminization (see Fig. 3G,H). More than 20 dominant alleles have been obtained by three independent means (sexual transformation of *XO* animals, dominant feminization of *XX* animals, extragenic suppression of *tra-3* mutations). All of these alleles show the same constellation of properties: dominant feminization of *XO* animals (both soma and germ line), dominant feminization of the germ line of *XX* animals, and partial or complete epistatic suppression of *tra-2(0)* and *tra-3(0)* mutations. The alleles vary in the strength of these effects: Weak dominant alleles such as *e2013* have an intersexual phenotype in *e2013/+ XO* animals and do not fully suppress *tra-2(0)* mutations, whereas strong dominant alleles such as *e1575* cause female development in *e1575/+ XO* animals and do suppress *tra-2(0)* mutations, so that *tra-2(0);tra-1(e1575/+)* is female rather than male. All of these dominant mutations must represent gain-of-function alterations at the *tra-1* locus, because no dominant effects are caused by *tra-1(0)* alleles or by *eDf6*, a deficiency for the *tra-1* region. They are regarded as *tra-1* mutations because all show very tight ($<0.5\%$) linkage to *tra-1(0)* alleles and because the first *tra-1* dominant *e1575* can be reverted by *tra-1(0)* mutations in *cis* but not in *trans* (Hodgkin 1983b, 1987b). Dosage studies on some of the dominant mutations indicate that they cause inappropriate activity of *tra-1*, rather than overproduction. None of them appears to be neomorphic, altering the nature of the *tra-1* product, because none affects the development of the *XX* (female) soma. The simplest explanation of their properties appears to be constitutive *tra-1* activity, but it is unclear whether this represents constitutive transcription or constitutive activity at some post-transcriptional regulatory step.

2. tra-2 II

The null phenotype of *tra-2* is defined by the allele *e1095* and the amber allele *e1425* (Hodgkin and Brenner 1977; Hodgkin 1985b). These mutations have a recessive masculinizing effect on *XX* animals: homozygotes have well-formed male gonads and extensive spermatogenesis, but the male tail anatomy is abnormal, being incompletely masculinized (see Fig. 3C). The *tra-2(0) XX* males never show any sign of mating behavior, suggesting that the nervous system is incompletely masculinized, although most of the male nervous system is present and wild type, as assayed by the criteria of ventral cord neuron counts, presence of CEM neurons, and catecholamine staining (Hodgkin and Brenner 1977). *XO* animals homozygous for a *tra-2(0)* allele are unaffected by the mutation, being wild-type fertile males. In addition to the strong recessive masculinization phenotype, *tra-2(0)* alleles also exhibit weak dominant masculinization of *tra-2/+ XX* animals. These animals have abnormal or missing HSN cells, resulting in a variable Egl (egg-laying defective) phenotype (Trent et al. 1983). The *tra-2* gene therefore appears to be haplo-insufficient.

Many weaker alleles of *tra-2* have been obtained, which cause less complete masculinization. These alleles can be ranked by their transforming effects in an unambiguous allelic series, in contrast to the *tra-1* hypomorphs. Alleles such as *e1209* have an incompletely male tail and an intersexual gonad, often with some induction of vulval divisions (Hodgkin and Brenner 1977). A temperature-sensitive allele, *b202*, has been isolated and characterized in detail (Klass et al. 1976, 1979). The transformer mutation *tri*, described by Beguet and Gibert (1978), has also been shown to be a *tra-2* allele (Hodgkin 1985b): This allele (*f70*) partly transforms *XX* animals so that they are still self-fertile hermaphrodites, but they usually have masculinized tails and are egg-laying defective. Even weaker *tra-2* alleles are known, in which the only sign of masculinization is a recessive Egl phenotype of low penetrance, caused by death or malfunction of HSN cells (Trent et al. 1983). One of these weak alleles, *e1875*, has also been shown to cause an increase in the mean number of self-progeny produced by an *XX* hermaphrodite, suggesting that *tra-2*-gene activity may be involved in limiting the number of sperm (J. Hodgkin, unpubl.).

Additional evidence for the involvement of *tra-2* in controlling hermaphrodite spermatogenesis comes from a series of dominant alleles isolated by virtue of a dominant feminization phenotype in *XX* animals (Doniach 1986). The strongest of these mutations, *e2020* and *e2046*, completely eliminate spermatogenesis from *XX* animals in either homozygotes or heterozygotes but have little effect on *XO* animals, which are fertile males. A low level of oogenesis and yolk production is seen in old adult *e2020 XO* males. Other dominant feminizing *tra-2* alleles such as *e1940* have weaker effects and seem to be loss-of-function alleles in some respects: For example, *e1940/e1095 XX* is intersexual, in contrast to *e1940/+ XX* and *e2020/e1095 XX*, both of which are female. All seven dominant alleles appear to have a much greater effect on germ line than on soma, suggesting that they affect a germ-line-specific control. Three have been shown to be *tra-2* alleles by introducing *tra-2(0)* mutations in *cis*.

3. tra-3 IV

Four alleles of *tra-3* have been obtained, all of which appear to be null because three out of the four are amber alleles and the fourth causes an identical phenotype to the other three. These mutations are fully recessive and have no effect on *XO* animals; in *XX* animals they exhibit a masculinizing effect similar to weak *tra-2* alleles such as *e1209* but only if both mother and zygote are *tra-3(0)*. Homozygous *tra-3 XX* daughters of heterozygous *tra-3/+* mothers are completely wild type, with normal self-progeny brood sizes, but all of their *XX* progeny are masculinized (Hodgkin and Brenner 1977). At low temperature (< 15°C), these *tra-3* homozygotes are occasionally self-fertile intersexes, so homozygous strains can be propagated (Hodgkin 1985b). Introduction of a *tra-3(+)* allele into the zygote by crossing a *tra-3* hermaphrodite with a wild-type or *tra-3/+* male completely rescues normal hermaphrodite development in the *XX*, so either maternal or zygotic (but not paternal) expression of *tra-3(+)* is sufficient for normal development. These observations, together with the fact that *tra-3* amber alleles are suppressed efficiently by all amber suppressors (Kimble et al. 1982; Hodgkin 1985b), is consistent with the belief that the *tra-3*-gene product is required only in very small amounts. The similarity of the *tra-3* phenotype to a weak *tra-2* phenotype, and the fact that the phenotypes of *tra-2* and *tra-2;tra-3* are identical, have been interpreted to mean that the *tra-3* product acts merely as an almost dispensable cofactor for the action of *tra-2* (Hodgkin 1980).

4. her-1 V

More than 18 recessive alleles of *her-1* have been obtained (Hodgkin 1980 and unpubl.; C. Trent et al., in prep.). Most of these alleles (e.g., *e1520*) appear to be null, because the phenotype of *e1520/mDf1 XO* is similar to that of *e1520/e1520 XO* (*mDf1* is a deficiency for the *her-1* region). None of these recessive alleles has any effect on *XX* animals (wild-type hermaphrodites), but all cause complete transformation of *XO* animals into fertile hermaphrodites (Fig. 3F), except for the three weak alleles, which cause only partial transformation. The fertility of these *her-1 XO* hermaphrodites (maximum brood 160 zygotes) is lower than that of *her-1 XX* hermaphrodites. A temperature-sensitive allele, *e1561*, has been isolated: At 15°C, most *e1561 XO* animals are fertile males, and at 25°C, most are fertile hermaphrodites.

A rare dominant masculinizing allele of *her-1*, *n695*, has also been obtained (Trent et al. 1983). Heterozygotes for *n695* have a weak Egl phenotype in *XX* animals, whereas homozygotes have a strong Egl phenotype and exhibit variable (never complete) masculinization of tail structures. The *n695* allele does not affect *XO* animals, and its effects are reverted by *her-1* recessive mutations in *cis*.

5. fem-1 IV

Many alleles of *fem-1* have been obtained (Doniach and Hodgkin 1984), most of which are temperature sensitive like the first allele *hc17* (Nelson et al. 1978). Several stronger, non-temperature-sensitive alleles have been isolated, some of which are probably null (one of these, *e1991*, is amber). These null mutations are recessive and transform both *XX* and *XO* animals into fertile females. The transformation shows a maternal effect: *fem-1 XO* progeny of *fem-1/+ XX* mothers are transformed incompletely into intersexes, whereas *fem-1 XO* progeny of *fem-1 XX* mothers are completely female. Also, *fem-1 XX* progeny of heterozygous mothers are occasionally self-fertile (20% of animals), but *fem-1* progeny of homozygous mothers are never self-fertile.

6. fem-2 III

The first *fem-2* allele, *b245*, is a weak temperature-sensitive feminizing mutation (Kimble et al. 1984). Stronger alleles have been obtained, some of which may be null (e.g., *e2105*; Hodgkin 1986). Their effects are similar to those of *fem-1*, except that there is greater maternal rescue of the *XX* phenotype, so that *fem-2 XX* daughters of *fem-2/+* mothers are hermaphrodites rather than females. Also, *fem-2 XO* animals produced by *fem-2* mothers are female if grown at 25°C but intersexual if grown at lower temperatures. *XX* siblings are female at all temperatures. Several independently isolated alleles show the same temperature sensitivity for the *XO* phenotype, suggesting that the requirement for *fem-2(+)* product is more stringent at 25°C than at 15°C, as with *tra-3(+)*, discussed above.

7. fem-3 IV

More than 20 *fem-3* alleles have been obtained (Hodgkin 1986; Barton et al. 1987). Some appear to be null (e.g., *e1996*), in that they display similar complementation properties to deficiencies in the *fem-3* region (e.g., *eDf18, eDf19*). Their effects are broadly similar to those of *fem-1* and *fem-2*. Both *XX* and *XO fem-3* animals are transformed into fertile females, if derived from a *fem-3* mother. The *fem-3 XX* progeny from a *fem-3/+* mother are also invariably female, but *fem-3 XO* siblings are intersexual. The *fem-3(+)* activity appears to be required in critical amounts, because *fem-3/+ XX* animals are sometimes female rather than hermaphrodite and usually produce fewer self-progeny than wild type. The *fem-3* gene therefore appears to be a haplo-insufficient locus, like *tra-2*, but unlike the other five genes discussed here. Also, *fem-3/+ XO* sons of *fem-3/fem-3* mothers are often partly feminized, but *fem-3/+ XO* sons of *fem-3/+* mothers are never feminized (Hodgkin 1986). This observation indicates that there is an essential maternal contribution of *fem-3*-gene product, in contrast to the maternal contributions of *tra-3*, *fem-1*, and *fem-2*, which are not essential.

Several semidominant masculinizing alleles of *fem-3* have been obtained (Barton et al. 1987). These result in masculinization of the hermaphrodite germ line, but they do not affect hermaphrodite somatic tissues or males. All alleles so far isolated are somewhat temperature sensitive, but they vary in strength. The best characterized allele, *q20*, has been shown to be a *fem-3* allele by intragenic reversion to *fem-3(0)*. At restrictive temperature, all *q20* homozygotes and about half *q20/+* heterozygotes are sterile due to continuous sperm production. Thousands of sperm are produced. Homozygous *q20 XO* animals are fertile males at restrictive temperature.

8. *Other Genes*

Several other loci have been identified that may be involved in sex determination, but little is known about them as yet. Mutations at two of the *egl* loci, *egl-16 X* and *egl-41 V*, cause variable and low-penetrance masculinization of *XX* animals, which is enhanced by other weak masculinizing mutations (Trent et al. 1983; Hodgkin et al. 1985; Villeneuve and Meyer 1987; C. Desai, pers. comm.). Several genes have been identified that appear to be necessary for correct sex determination in the germ line but not in the soma, or vice versa. Some of these, such as *fog-1* or *mab-3*, are discussed in Section VI. The *fog-2* gene is more appropriately mentioned at this point, as it may interact with *tra-2* (T. Schedl, pers. comm.). Recessive loss-of-function mutations in this gene resemble dominant *tra-2* alleles: They have no effect on *XO* animals, which are normal fertile males, but *XX* animals are transformed into fertile females. Mutant *fog-2* strains can therefore be maintained as homozygous male/female stocks.

C. Analysis of Temperature-sensitive Mutations

Temperature-sensitive mutations have been obtained in five of the seven genes described above (*tra-2*, *her-1*, *fem-1*, *fem-2*, and *fem-3*). Extensive analyses by means of temperature-shift experiments have been published for *tra-2* (Klass et al. 1976, 1979) and *fem-1* (Nelson et al. 1978); also some data exist for *her-1* (Hodgkin 1984), *fem-2* (Kimble et al. 1984), and *fem-3* (Hodgkin 1986). Broadly speaking, the temperature-sensitive periods (TSPs) determined by these experiments are consistent with the observed pattern of sexual development, so that the TSPs for all of these genes begin before the earliest sign of sexual dimorphism in the tissues studied. Also, the order and length of the TSPs for the different genes are consistent with the sequence of interactions inferred from epistasis analysis, described in the next section (TSP comparisons are presented by Nelson et al. [1978] and Hodgkin [1984]).

The interpretation of these experiments is complicated by several factors: First, it is clear (by comparison with non-temperature-sensitive alleles of the same genes) that none of the temperature-sensitive mutants is "perfect," in the sense of being fully wild type at permissive temperature and fully mutant at restrictive temperature. Second, it is uncertain whether the temperature sensitivity reflects the time of synthesis of a gene product, its time of action, or some combination of the two. A further complication is encountered with maternal-effect genes such as *fem-1*, *fem-2*, and *fem-3*: A temperature-shift experiment may affect only the zygotic synthesis of a gene product and leave the dowry of maternal gene product unaffected.

Despite these caveats, some important conclusions have been reached. First, TSPs for a given mutation are not identical from tissue to tissue. For example, the TSP for self-progeny production (presumably reflecting germ-line functions) in *tra-2(b202ts) XX* animals extends throughout development, whereas the TSP for hermaphrodite tail development is brief and early, preceding hatching. These differences from tissue to tissue may reflect different thresholds of sensitivity for a given gene product but may also indicate that some tissues become irreversibly committed to one pattern of sexual development while others remain labile. Second, the mosaic intersexual phenotypes that can be generated by temperature shifts at intermediate times show that different sexual phenotypes can be expressed in adjacent tissues within one animal, which argues for cell-autonomous or tissue-autonomous sex determination. Third, the fact that short defined TSPs are observed for some of the genes indicates that their activities become dispensable in many tissues, again suggesting that irreversible commitments occur. The sexual phenotype of some characters (e.g., germ line) can be affected by temperature shifts in late larval or adult life for the genes *tra-2*, *fem-1*, *fem-2*, and probably also *her-1* (P. Schedin and P. Jonas, pers. comm.), suggesting that these gene activities may be required continuously.

Questions of autonomy and commitment would be answered more satisfactorily by means of mosaic analysis (Herman 1984), which has not yet been attempted for any of the sex-determination genes. However, some information on autonomy has come from observations of triploid intersexes, *3A;2X* animals carrying a partial X duplication such that the X/A ratio is in the ambiguous range between 0.67 and 0.75. These animals are mosaics, in which some tissues are hermaphrodite and some are male (Madl and Herman 1979). In a further analysis of such animals, using in situ hybridization with a *vit* gene probe to measure intestinal synthesis of yolk protein, P. Schedin (Wood et al. 1985 and pers. comm.) has shown that in particular, sex determination can occur autonomously in germ line, somatic gonad, intestine, and tail morphogenesis.

D. Epistatic Interactions; Model for Sex Determination

The epistatic interactions between the various sex-determination mutants have been investigated extensively (Hodgkin and Brenner 1977; Nelson et al. 1978; Hodgkin 1980, 1984, 1986; Doniach and Hodgkin 1984). With regard to somatic phenotype, it has been found that the genes can be ranked in a hierarchy of epistasis, in the order *her-1* < *tra-2, tra-3* < *fem-1, fem-2, fem-3* < *tra-1*. Thus, in somatic tissues, recessive (masculinizing) mutations of *tra-1* are epistatic to all feminizing mutations of *her-1*, *fem-1*, and so forth. Also, dominant (feminizing) mutations of *tra-1* are epistatic to all masculinizing mutations of *her-1*, *tra-2*, and so forth.

A slightly different situation is found with regard to germ-line phenotype, for which the hierarchy of epistasis has the order *her-1* < *tra-2, tra-3* < *tra-1* < *fem-1, fem-2, fem-3*. Consequently, *fem(0)* mutations are epistatic to *tra-1(0)* in the germ line but not in the soma, so that the double mutant *tra-1(0);fem-1(0)* has an entirely male body and gonad morphology, but its germ line makes oocytes rather than sperm (Fig. 4).

These epistatic interactions have been interpreted in terms of the model shown in Figure 5, which summarizes the proposed regulatory interactions between the genes. Activation or repression of one gene activity by another is indicated. These interactions need not necessarily take place at the level of transcription, as the activity of a gene can be controlled at many different levels. A model such as this, which has been inferred from genetic data, can provide little information about the molecular nature of the regulatory interactions. In the following explanation of the model, repress (or activate) is used as shorthand for "negatively (or positively) control the activity of." The interactions so far identified provide evidence for three patterns of gene activity: (1) In the *XX* soma: high X/A ratio represses *her-1*; therefore, *tra-2* and *tra-3* are active. These gene activities repress the three *fem* genes; therefore, *tra-1* is derepressed, and female somatic development ensues. (2) In the *XX* germ line, high X/A ratio represses *her-1*; therefore, *tra-2* and *tra-3* should be derepressed. However, *tra-2* is initially and transiently repressed by a germ-line-specific control (probably involving the gene *fog-2* but symbolized by ? in the diagram). In the transient absence of *tra-2* activity, the *fem* genes are derepressed, so their activity promotes a burst of spermatogenesis and represses *tra-1*, thereby permitting an initial phase of spermatogenesis in the hermaphrodite gonad. Subsequently, *tra-2* becomes derepressed; consequently, the *fem* genes are repressed and *tra-1* is activated. In the absence of *fem*-gene activity, the germ line switches to oogenesis, resulting in a functionally female phenotype in the adult hermaphrodite. The *tra-1*

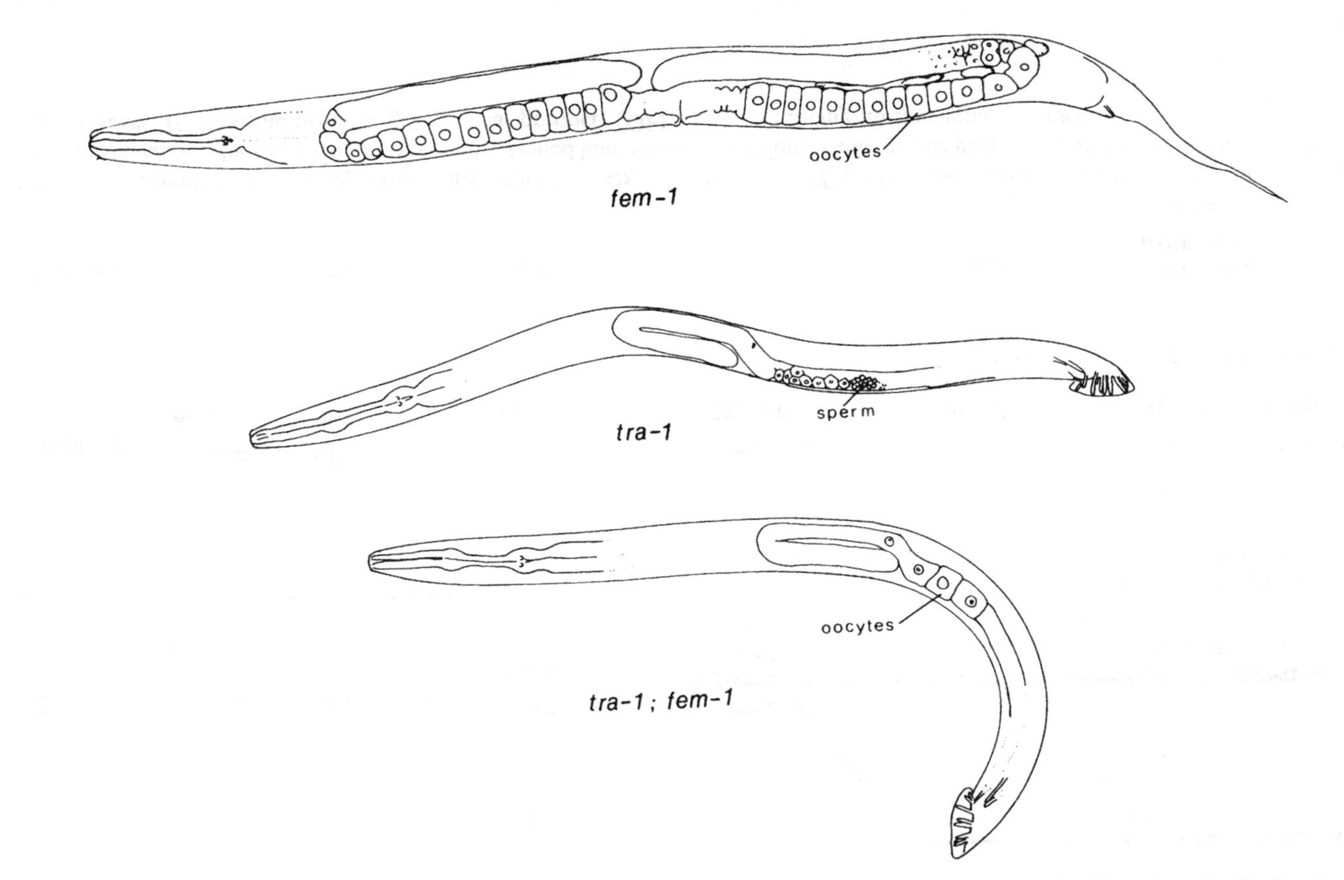

 Figure 4 Interaction of *fem-1* and *tra-1* null alleles, *XX* phenotypes (*XO* phenotypes are identical). (Drawings by T. Doniach.)

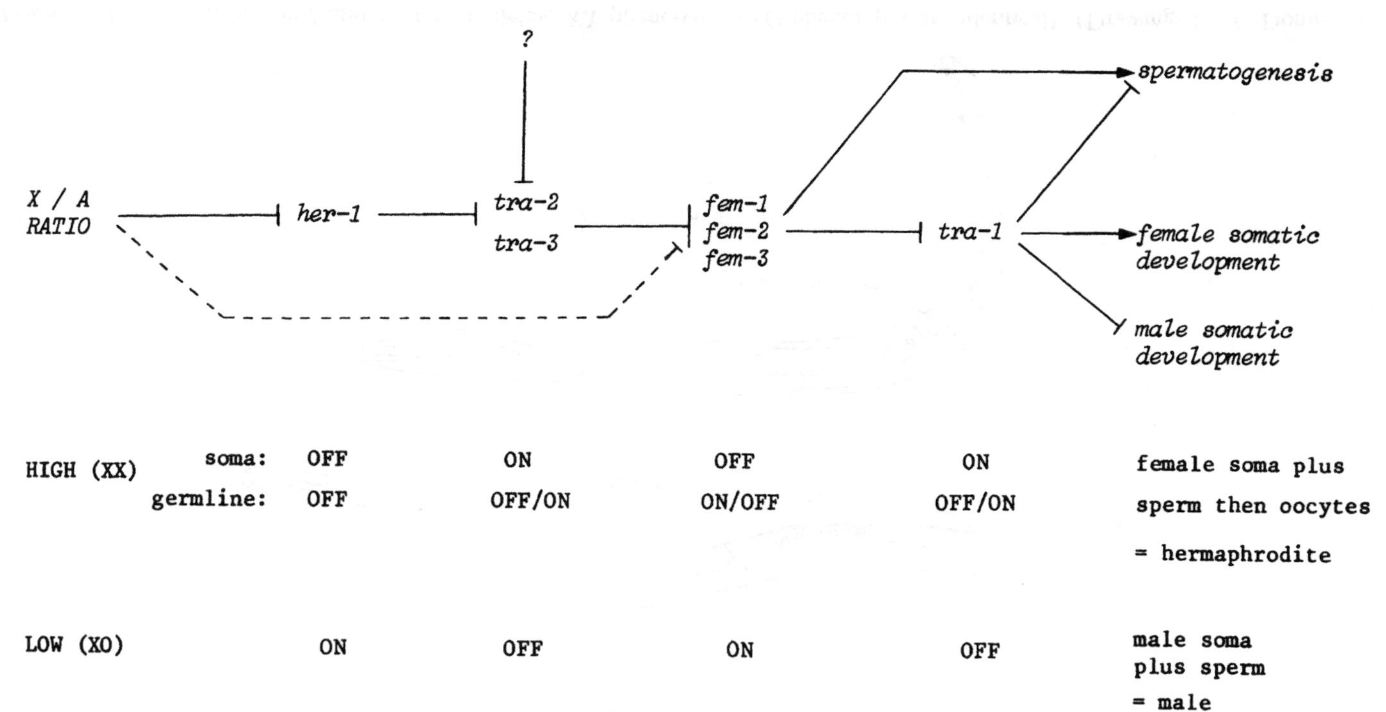

Figure 5 Current model for regulatory interactions in sex determination. (*Top*) Proposed interactions between genes. (Pointed arrows) Activation; (barred arrows) repression; (dashed line) weak interaction. (*Bottom*) The proposed activities of the various genes in the two sexes. (ON) Gene producing functional product; (OFF) gene producing no functional product.

activity may assist the switch by inhibiting spermatogenesis. (3) In the *XO* soma and germ line, low X/A ratio activates *her-1*; therefore *tra-2* and *tra-3* are repressed. This results in full derepression of the *fem* genes that promote spermatogenesis and repress *tra-1*; in the absence of *tra-1* gene activity, male somatic development ensues.

These three patterns of gene activity are summarized in Figure 5 as a set of on/off states, symbolizing presence or absence of each gene activity. It is possible that the real physiological states are high and low, rather than on and off.

Various minor interactions have been identified, in addition to the major interactions summarized above. The most important is a weak direct effect of X/A ratio on *tra-1* activity. The main reason for postulating this interaction is that *tra-2(0),* and *tra-2(0);tra-3(0), XX* animals are masculinized incompletely in the soma, unlike *tra-1(0) XX* animals, which are male in the soma. The abnormal phenotype is ascribed to residual low-level *tra-1(+)* activity in the *tra-2(0) XX* animal. The interaction may be unimportant in wild-type sex determination, as would be the case if the interaction involves direct weak *activation* of *tra-1* by high X/A ratio. Alternatively, the interaction may involve direct weak *repression* of *tra-1* by low X/A ratio, in which case the interaction would be essential for normal *XO* male development. At present, these two possibilities cannot be distinguished, nor is it certain whether the weak effect acts directly on *tra-1* or via the *fem* genes.

The model as it stands is consistent with all double, triple, and quadruple mutant constructions that have been tested, using mutations of the seven major genes described above. Undoubtedly, further work will refine the model, but the basic structure of the cascade has been strongly supported so far.

The postulated interactions explain only how *tra-1* and the *fem* genes are controlled; they do not explain how these gene activities result in particular developmental events. The simplest hypothesis is that *tra-1* acts to control batteries of target genes, repressing those required for male development and activating those required for female development. The *fem* genes also appear to have a dual role, repressing one gene (*tra-1*) and activating other genes required for spermatogenesis. Hypotheses of this type predict the existence of sex-specific target genes, which are the subject of Section VI.

VI. DOWNSTREAM FUNCTIONS: SEX-SPECIFIC GENES

The postembryonic development of hermaphrodite and male is widely divergent: To what extent is this reflected in the underlying genetic instructions? Four categories of gene can be envisaged (excluding those discussed in Section V, which have a determinative role): (1) genes with very similar or identical functions in both sexes; (2) genes required for both male and hermaphrodite development but functioning differently in the two sexes; (3) genes required only for hermaphrodite development; and (4) genes required only for male development. An attempt to delimit these categories was made by comparing hermaphrodite and male phenotypes for mutations in more than 200 genes (Hodgkin 1983a). It was found that the majority had similar phenotypes in both sexes, implying that most genes have similar roles in both hermaphrodite and male. However, there also exist many candidates for hermaphrodite-specific and male-specific genes, as described below.

A. Hermaphrodite-specific Genes

Of genes that might be specifically required for hermaphrodite development, the most obvious candidates are the *lin* genes, all of which have been identified by means of mutations affecting the hermaphrodite postembryonic lineages (see Chapter 6). The *lin* genes form a very heterogeneous set: Out of 39 genes so far defined, 4 (*lin-5*, *lin-6*, *lin-16*, *lin-19*) affect the majority of postembryonic cell divisions and are therefore probably cell-division-cycle genes. Four are heterochronic genes (*lin-4*, *lin-14*, *lin-28*, *lin-29*) that affect the relative timing of events in postembryonic divisions (Ambros and Horvitz 1984) and are required for both hermaphrodite and male development. However, the temporal program of events in the two sexes is not the same, so that these genes are probably utilized in different ways and for different purposes in males and hermaphrodites. The same is true for most of the other *lin* genes (see Chapter 6). For example, some of the Muv (multivulva) mutations that result in extra vulval divisions in the cells P3.p–P8.p of the hermaphrodite also cause extra divisions in P3.p–P8.p of the male and/or P9.p–P11.p of the male (resulting in a "multihook" phenotype [Sulston and Horvitz 1981; Greenwald et al. 1983]). However, many of the Vul (vulvaless) mutations (defining seven genes) have little or no effect on male phenotype or male mating efficiency (Hodgkin 1983a; Ferguson and Horvitz 1985). For several of these genes (e.g., *lin-2*, *lin-7*), it is likely that the mutations are null, so it appears that these genes are not required for male development but are required for a particular part of hermaphrodite development (vulva formation) and are therefore sex specific. It is still possible that these genes are expressed during male development, even though they have no function in males.

Another large category of possible hermaphrodite-specific genes is the set of *egl* genes (Trent et al. 1983). More than 40 *egl* genes have been identified so far, but many of these (at least 13) affect movement in both sexes and therefore appear to have a neuronal or muscular phenotype that is not sex specific. Others (at least three genes) have pleiotropic effects on the development of the male tail. However, mutations in eight genes, and possibly more, do not affect male development, male behavior, or male mating efficiency and, therefore, may be hermaphrodite specific. The gene *egl-1* is particularly interesting, because *egl-1* mutations result in the programmed cell death of the HSN cells, an event that occurs normally in the male. Another *egl* gene with a striking sex-specific effect is *egl-15*, which affects sex-muscle differentiation in the hermaphrodite but not in the male.

In addition to genes such as these that play roles in the development and function of the hermaphrodite soma, there must be a large number of genes that are required for normal oogenesis. The expression of these genes is likely to be confined to the hermaphrodite, although their gene products are required for normal embryonic development in both sexes. These would therefore be hermaphrodite-specific genes. Several such genes have been identified by embryonic lethal mutations that show a strict maternal effect but do not appear to influence larval development or male fertility (see Chapter 8). The yolk protein genes also fall into this category: As expected, they are transcribed only in adult hermaphrodites (Blumenthal et al. 1984).

B. Male-specific Genes

Almost all *C. elegans* genetics has entailed studying mutations expressed in the hermaphrodite, so one would not expect many male-specific genes to have been identified. A search for *mab* (*m*ale *ab*normal) mutations was carried out by screening mutagenized *him* stocks for the segregation of males that were unable to mate (Hodgkin 1983a). This preliminary search led to the identification of ten *mab* genes. It was found that mutations in seven of these genes had both male and hermaphrodite phenotypes, but the male phenotype was much more conspicuous. For example, the *mab-2* mutation causes division failures in the lateral hypodermal V and T lineages of both sexes. These defects cause no gross change in the anatomy of the hermaphrodite, but in the male they cause extensive disruption of the bursal rays and fan. Also, mutations in *mab-1*, *mab-4*, *mab-8* and *mab-10* do not affect male lineages but do affect male tail morphogenesis, so that the bursal region is swollen and abnormal. In hermaphrodites, the tail is unaffected but the vulva is swollen, so these genes seem to have different but related functions in the two sexes. Three *mab* genes may be truly male specific, because mutant hermaphrodites display no obvious phenotype: *mab-3*, *mab-6*,and *mab-9*. Several independently isolated *mab-3* mutants exhibit the same complex phenotype: The male tail develops abnormally, adult males frequently carry a hermaphrodite tail spike, and the adult male intestinal cells synthesize large amounts of yolk protein (Hodgkin et al. 1985; M. Shen, pers. comm.). The *mab-6* mutant seems to be defective only in tail morphogenesis, like *mab-1*, but the *mab-9* mutant is more severely abnormal. Mutant *mab-9* males exhibit disruptions of the B lineage, that is, a blast cell that divides only in the male.

Sperm are made in both hermaphrodite and male, so the set of genes required for spermatogenesis cannot be regarded as strictly sex specific. However, spermatogenesis appears to be directly controlled by the sex-determination genes (positively by the *fem* genes and negatively by *tra-1*), so at least some of the spermatogenesis genes may be target genes. A gene has been identified that may act at an intermediate level between the sex-determination genes and the spermatogenesis genes: *fog-1 I* (for *f*eminization *o*f *g*erm line [T. Doniach, pers. comm.]). The *fog-1* allele *e1959* causes semidominant partial feminization of *XO* gonads (which form both sperm and oocytes) and recessive feminization of *XX* gonads (only oocytes are made) but has no obvious effect on somatic development. Recessive *mog* mutations with the opposite effect (*m*asculinization *o*f *g*erm line) have also been identified (K. Barton and T. Schedl, pers. comm.).

Relatively few genes specifically required for spermatogenesis have been identified by mutation so far: 9 *fer* genes and 12 *spe* genes (Argon and Ward 1980; Sigurdson et al. 1984; S. Ward et al., pers. comm.). All affect sperm in both male and hermaphrodite. In addition, a surprisingly large family of sperm-specific genes—those encoding the 15-kD major sperm proteins (see Chapter 7)—has been identified by cDNA cloning (Burke and Ward 1983; Klass et al. 1984).

VII. DOSAGE COMPENSATION

A. Evidence for Dosage Compensation

Earlier indirect arguments for X chromosome dosage compensation in *C. elegans* (Hodgkin 1983c) have been supported by more recent direct genetic and molecular evidence (Wood et al. 1985; Meyer and Casson 1986; P.M. Meneely and W.B. Wood, in prep.). It is very unlikely that *C. elegans* achieves compensation by means of *X* inactivation (as occurs in mammals), because heterozygotes for recessive sex-linked mutations never express a mutant phenotype. More probably, compensation is achieved as in *Drosophila*, by increasing transcription from the single X chromosome of the *XO* male, or by decreasing transcription from the two X chromosomes of the *XX* hermaphrodite. There are some indications that this is the case from the phenotypes of hypomorphic alleles of various sex-linked genes, such as *let-2(mn102)* and *lin-15(n767)*, which appear to have a more mutant phenotype in *XX* heterozygotes over a deficiency (i.e., *XX* hemizygotes) than in *XO* hemizygotes (Wood et al. 1985). However, other genes behave as if uncompensated: The sex-linked amber suppressors *sup-7* and *sup-21* are expressed at a higher level in *XX* animals than in *XO* animals (Hodgkin 1985b), and the unusual mutation *let-7(mn112) X* shows a reversed pattern of expression. *let-7 XX* (and *let-7/mnDf5 XX*) animals die at L3, whereas *let-7 XO* animals die at late L4 (Meneely and Herman 1979).

Experiments measuring the levels of two enzymes, products of the sex-linked genes *ace-1* and *nuc-1*, have indicated that dosage compensation occurs (J.G. Duckett, quoted in Bull [1984]; W.B. Wood, pers. comm.). More direct experiments on the expression of sex-linked genes have been carried out by measuring mRNA levels for several different sex-linked genes (Meyer and Casson 1986; L. Donahue et al., in prep.). These experiments have shown that most sex-linked transcripts have similar levels in *XO* and *XX* animals, rather than being in a ratio of 1 : 2, which argues strongly for dosage compensation at the transcriptional level.

B. Dosage Compensation Control Genes

Four autosomal genes have been identified that may be involved in controlling X chromosome expression, because mutations in these genes exhibit phenotypes dependent on X chromosome dose. All four genes have similar properties; two (*dpy-21 V* and *dpy-26 IV*) have been examined in detail (Hodgkin 1983c; Meneely and Wood 1984). Two alleles are known for each of these genes; those of *dpy-26* are probably null or close to null. Homozygous *dpy-26 XX* progeny of *dpy-26/+* mothers are viable and slightly dumpy but produce self-progeny consisting mainly of inviable *XX* animals that die as embryos or young larvae, a few very abnormal, severely dumpy *XX* adult hermaphrodites, and a few fully viable, essentially wild-type *XO* males. Therefore, this gene is a maternal-effect *XX* lethal. The lethality is known to be associated with X chromosome dosage rather than sexual phenotype because *dpy-26 XO* hermaphrodites or females (generated by *her-1*, *fem-1*, or *tra-1(dom)* mutations) are viable and non-dumpy, whereas *dpy-26 XX* males (generated by *tra-1* mutations) are not viable. The production of *XO* males by *XX dpy-26* hermaphrodites is the result of an additional meiotic nondisjunction phenotype.

Mutations in two other genes, *dpy-27(rh18) III* and *dpy-28(y1) III*, appear to have similar effects to those of *dpy-26(n199)*, except that the level of *XX* lethality is slightly lower, and there is no meiotic phenotype (Meyer and Casson 1986; Hodgkin 1987a).

Two mutant alleles of *dpy-21*, *e428* and *e459*, do not cause lethality in *XX* animals, but they result in a Dumpy phenotype that is not observed in *XO* males, *XO* hermaphrodites, or *XO* females. In the presence of three X chromsomes, *dpy-21(e428)* is lethal, and it is also lethal in combination with homozygous large duplications of the X chromosome (*mnDp10*, *mnDp25*), which shift the X/A ratio to more than about 1.25 (Meneely and Wood 1984). These investigators showed that the Dumpy phenotype of *e428* is expressed at X/A ratios of more than about 0.63 in *XO* animals carrying partial X chromosome duplications. Duplications from at least five different regions of the X chromosome (see Fig. 2) were observed to interact additively, which shows that the *dpy-21* phenotype depends on multiple X chromosome sites, that is, a cumulative X/A ratio, as with sex determination. However, the thresholds for Dumpy phenotype and hermaphrodite development are not identical, because at ratios between 0.65 and 0.68, animals are dumpy but superficially male or intersexual; also *dpy-21 4A;3X* animals (ratio 0.75) are nondumpy hermaphrodites (Hodgkin 1983c). Similarly, *dpy-27 4A;3X* animals are viable nondumpy hermaphrodites (Hodgkin 1987a).

The properties of these genes suggest that they may be necessary to reduce expression of the X chromosome in *XX* animals. Some evidence from measurements of X chromosome transcript levels supports this notion: Higher levels are observed in *dpy-21*, *dpy-27*, and *dpy-28 XX* animals than in wild-type *XX* animals (Meyer and Casson 1986; L. Donahue et al., in prep.).

Mutations of the opposite type (i.e., *XX* viable, *XO* inviable) are less well characterized. Mutations in two genes (*dpy-22 X, dpy-23 X*) appear to have more severe effects on *XO* animals than on *XX* animals (Hodgkin and Brenner 1977) and may also affect the expression of sex-linked hypomorphic alleles (Wood et al. 1985). Other candidates for *XO*-specific lethal mutations have recently been identified. One of these, an X-linked mutation *y9*, leads to decreased X chromosome expression and also acts as a suppressor of *XX* lethality in *dpy-28* animals (L. Miller and B.J. Meyer, pers. comm.).

C. Relationship between Dosage Compensation and Sex Determination

Various interactions between dosage compensation and sex determination have been observed (Hodgkin 1983c; Meneely and Wood 1984), the significance of which is uncertain at present. Meneely and Wood (1984) observed that an apparent *dpy-21* allele, *ct16*, caused feminization of *XO* animals, but it is more probable that this phenotype is, in fact, abnormal male development caused by a *mab-3* mutation present in this strain (Hodgkin 1987a; M. Shen, pers. comm.). These investigators also showed that *dpy-21(e428)* caused *XO;mnDp10* animals to develop into dumpy intersexes, as compared with *dpy-21(+);XO;mnDp10* animals, which are male. A similar but lesser effect is seen with the two duplications *mnDp9* and *mnDp25* but not with any smaller duplications. Thus, it appears that *dpy-21(e428)* can feminize when the X/A ratio is close to the critical value of about 0.68. If the *dpy-21* mutation increases X chromosome transcription, as seems to be the case (Meyer and Casson 1986; L. Donahue et al., in prep.), such an effect might be expected, if the X/A ratio assessment for sex determination takes place at the level of RNA rather than DNA. Hodgkin (1987a) observed a more extreme effect with *dpy-27*: Normal triploid *3A;2X* animals are males (Table 1), but *dpy-27 3A;2X* animals are hermaphrodites.

An interaction between *dpy-21* and *tra-1* has been observed: *XX* animals homozygous for *dpy-21* and *tra-1(0)* alleles, such as *e1099*, have an abnormal male tail and are unable to mate. This phenotype is similar to that of *tra-1(e1099) XXX* animals, which are also abnormal dumpy males (Fig. 3D). Meneely and Wood (1984) reported partial suppression of a weak *tra-1* allele, *e1076,* by *dpy-21(e428)*, but this allele is variable in phenotype, and the apparent suppression may not be significant. In *tra-1(e1099);dpy-21 XX* males and in *tra-1(e1099) XXX* males, the gonads are abnormal but do not contain oocytes or anchor cells, so it is not clear whether the phenotype of these animals represents true feminization or merely an abnormality resulting from genetic imbalances (i.e., excessive X chromosome transcription).

The clearest indication of a direct link between dosage compensation and sex determination comes from recent evidence that the *egl-16* gene (now renamed *sdc-1*, for *s*ex determination and *d*osage *c*ompensation) is involved in control of both processes (Villeneuve and Meyer 1987). Recessive mutations in *sdc-1* have no apparent effect on *XO* animals but cause apparently elevated levels of *X* gene expression, as well as partial masculinization, in *XX* animals. The masculinization occurs through effects on the sex-determination pathway that appear to be upstream of the genes described previously. The *sdc-1* gene could therefore be equivalent to the *Drosophila Sxl* gene, which also controls both dosage compensation and sex determination (Cline 1984).

VIII. CONCLUSION

A. Comparisons with *Drosophila* and Other Animals

The only other organism in which the genetics of sex determination has been thoroughly investigated is *Drosophila melanogaster* (for review, see Baker and Belote 1983). It is interesting to compare sex determination in these two animals, which are phylogenetically very remote from each other. Some broad similarities emerge, suggestive of convergent solutions at the level of overall organization, but the systems differ in detail. In both animals, the key sex-determining signal is the X/A ratio in the zygote, which is responsible for initiating correct development in three largely independent processes: dosage compensation, somatic sexual phenotype, and germ-line sexual phenotype. As noted above, there appears to be one master control gene in *Drosophila*, *Sxl*, which regulates at least two processes: dosage compensation and somatic sexual phenotype. The *sdc-1* gene may serve similar control functions in *C. elegans*. A set of genes specific to fly dosage compensation (*msl*, *mle*, etc.) has been identified; mutations in these genes are *XY* lethal and *XX* viable, in contrast to most mutations in the putative dosage compensation genes of *C. elegans* (see Section VII.B), which are *XX* lethal and *XO* viable. It may be that the mechanism of transcriptional dosage compensation is rather different in these two organisms.

There is also a set of genes that act specifically in somatic sex determination of *Drosophila*: *tra*, *tra-2*, *ix*, and *dsx*. These genes are believed to form a cascade somewhat analogous to the cascade inferred for *C. elegans*, with mutations in the last two genes, *dsx* and *ix*, being epistatic to those in the other genes. However, the *dsx* gene appears to be bifunctional, producing transcripts essential for both male and female development, in contrast to the nematode *tra-1* gene, which appears to be essential only for hermaphrodite development and almost dispensable for male development. Also, most of these genes do not affect germ-line sexual phenotype, with the exception of the *Drosophila tra-2* gene, which is required for normal spermatogenesis. This role is unexpected, because the *Drosophila tra-2* gene is otherwise not required in the male but is required in the female soma. The control of germ-line sexual phenotype in *Drosophila* is still largely mysterious, in contrast to the situation in *C. elegans*.

Although the same genes affect both germ-line phenotype and somatic phenotype in *C. elegans*, the postulated gene interactions are not the same in these two divisions of the animal. Also, it is harder to achieve perfect sexual transformation of the germ line than of the soma: *tra-1 XX* males, *her-1 XO* hermaphrodites, and *fem-1 XO* females are less fertile than their wild-type counterparts. It appears that the control of germ-line sexual phenotype is more complicated than that of the soma, in both *Drosophila* and *C. elegans*.

One incidental consequence of the work on sex determination in *C. elegans* has been the creation of strains with formally quite different sex-determining mechanisms. For example, the X–IV translocation *mnT12*, which joins the whole X chromosome to LGIV, can be used to construct an *XX* hermaphrodite/*XY* male system, in which the *mnT12* double chromosome is a neo-X, and chromosome IV is a neo-Y (Sigurdson et al. 1986). Second, a variety of strains with male/female sex determination have been created, in some cases by means of mutations at single loci, such as *tra-2 II*, *tra-1 III*, or *fog-2 V*. Artificial WZ female/ZZ male strains have been created by combining null and dominant (constitutive) mutations of *tra-1* (Hodgkin 1983b); in these strains the X chromosome has become irrelevant to sex determination, and instead chromosome III is a sex chromosome, either Z, carrying *tra-1(0)* or W, carrying *tra-1(dom)*. Finally, temperature-sensitive mutations in the various major autosomal sex-determining genes (see Section V.C) permit environmental, as opposed to chromosomal, sex determination. It is easy to see how switch genes like these could evolve to become sensitive to an environmental variable such as temperature or nutrient availability and thereby permit evolution of environmental sex determination. These examples illustrate how readily one sex-determining mechanism can be changed into another and may be illuminating in the wider context of the evolution of sex determination (for review, see Bull 1984).

B. Unanswered Questions

The present model for sex determination in *C. elegans* leaves many questions unanswered, some of which will probably be resolved as molecular aspects of the system are investigated. In mechanistic terms, little is known about the molecular details of the regulatory interactions or how the genes are maintained in their various states of activity. The first step in the postulated cascade, the assessment of X/A ratio, is a particularly challenging problem at the molecular level.

Other problems may be resolved by further genetic or mosaic analysis, such as how the switch from hermaphrodite spermatogenesis to hermaphrodite oogenesis is achieved, and whether this switch involves events solely within the germ-line cells. Further mutant hunts and mutant characterization might also provide more information about the downstream sex-specific target genes. Sex-specific genes might also be sought by biochemical methods, such as screening for sex-specific antigens or sex-specific transcripts.

There are also general questions of organization. What is the purpose of the cascade of interactions? Why are there so many genes? A partial answer may be that the cascade is necessary in order to permit germ-line-specific modulation, which allows the production of sperm in a female body. This does not explain the details, such as why there should be three *fem* genes or why each of the three shows a strong maternal effect. As with many phenomena in animal development, it may be that some of the organization is merely an accident of evolutionary history and that teleological explanations should not be sought too deeply. It is also true that the cascade is no more complicated than some of the regulatory interactions that have been studied in prokaryotes or in yeast and that other developmental gene control circuits may turn out to be at least as complicated, once they are studied in comparable detail.

ACKNOWLEDGMENTS

Valuable criticisms of this chapter were provided by Bob Herman, Bob Horvitz, Judith Kimble, Michael Shen, Bill Wood, and others, to whom I am very grateful.

10
Muscle

Robert H. Waterston
Department of Genetics
Washington University School of Medicine
St. Louis, Missouri 63110

I. INTRODUCTION

Studies of *Caenorhabditis elegans* muscle have sought to determine the role of specific components, both in the assembly of the myofilament lattice during development and in the contractile process itself. The proper assembly of the muscle structure is likely to involve principles that are generally important in the spatial differentiation of the organism, and muscle, with many of its major components known and with its well-defined structure, has advantages for discovering these principles. The propensity of the major components for self-assembly in vitro is well documented, but how the assembly is controlled and modulated in vivo to yield a well-defined lattice structure is unclear. The variety of forms that the major muscle filaments assume—from their less ordered cytoplasmic forms, to smooth muscle, to the almost crystalline arrangement in insect flight muscle—attests to the importance of this control. The fundamental mechanism of force generation in all these forms is known to involve the sliding of myosin-containing thick filaments past thin filaments. Despite the universal importance of the actomyosin mechanism, the details of force generation—how the chemical energy of ATP is transduced to the mechanical work of movement—are only poorly understood.

A combination of genetic, biochemical, and morphological approaches has been applied to the study of *C. elegans* muscle assembly and function. This combination of approaches has been very powerful in understanding analogous processes in prokaryotes and is proving successful in the worm. The general advantages that *C. elegans* has for studying mutations affecting movement apply as well to the subset of those mutations affecting muscle (Brenner 1974). Animals with extremely limited function of the body-wall musculature are viable and fertile; growth on a lawn of *Escherichia coli* requires little movement for feeding, and the animals self-fertilize internally so that mating is not required for propagation. The conservation of muscle proteins and their relative abundance has made identification of the principal components of the body-wall lattice straightforward and has facilitated work on specific proteins such as myosin. Finally, the precise, periodic structure of the myofilament lattice in the wild type is easily monitored using light microscopy. The known correlation of structures visible in the light microscope with those identified in the electron microscope facilitates the location of specific muscle proteins to defined structures of the myofilament lattice. Polarized light microscopy can be used to detect abnormalities of muscle structure in the living animal so that mutants affecting this structure can be readily distinguished from other mutants affecting movement. This criterion for identifying muscle mutants undoubtedly excludes some mutants that affect components of the muscle cell without altering assembly and may include some mutants with primary defects in other tissues that only induce changes in muscle structure secondarily; nonetheless, polarized light microscopy has provided a useful point of departure.

This chapter will review the progress made by application of these varied approaches to the problem of muscle assembly and contraction. Muscle mutants and muscle genes have figured extensively in other studies that do not bear directly on questions of muscle formation and function. For example, early cloning methodologies were developed in studies of muscle genes. Muscle genes have been of primary importance in studies of transposons in *C. elegans* (Moerman and Waterston 1984; Eide and Anderson 1985b; Moerman et al. 1986; see Chapter 3). Muscle mutants have also been of prime importance in genetic studies, such as the development of fine-structure genetic maps and the identification and study of tRNA nonsense suppressors (see Chapter 2). These topics will be dealt with here only as they relate to problems of muscle formation and function.

This chapter begins with a description of the morphology of wild-type muscles, followed by a summary of experiments to define the molecular composition of muscle. It then summarizes genetic studies and briefly describes the catalog of existing mutants. Most of the remainder of the chapter is devoted to specific identified genes and their protein products, myosin heavy chain, actins, paramyosin, and *unc-22*. The final section discusses some of the likely steps in muscle assembly.

II. MUSCLE STRUCTURE IN THE WILD TYPE

Of the several sets of muscles in *C. elegans*, the body-wall musculature is most prominent, both in terms of cell number and total mass (Figs. 1 and 2). It is arranged in four strips, lying left and right subdorsally and left and right subventrally (Fig. 1). Generally, the dorsal quadrants oppose the ventral quadrants, and vice versa. The backward propagation of a contractile wave along the dorsal side accompanied by a similar antiphase wave along the ventral side produces the sinusoidal forward movement of the animal (see Chapter 11). Each quadrant contains 24 mononucleate, diploid cells, except for the ventral left quadrant, which contains 23 (Sulston and Horvitz 1977). The elongate, spindle-shaped cells are arranged in overlapping pairs in the anterior portion of the animal, with progressively less overlap and less pairing toward the posterior end (Fig. 1). As a result, almost two thirds of the muscle cells are located anterior to the vulva. As in other nematodes, three parts of each muscle cell can be distinguished: (1) the cell body, containing the nucleus and cytoplasmic organelles, (2) the arm, a process extending from the cell body to either the dorsal or ventral nerve cord to receive synaptic input from the motor neurons, and (3) the spindle, the region that contains the contractile myofilament lattice itself. In *C. elegans*, the myofilament lattice is limited to a single 1–2-μm deep zone, lying just under and parallel to the hypodermis.

The other muscles of *C. elegans* have not been studied as intensively, but their general structure is understood (Fig. 2). The single sarcomere arrangement of most of these muscles contrasts with the more complicated structure of the body-wall myofilament lattice, described below. Consequently, they have been useful in localizing specific components in the lattice structure. These muscles have also provided interesting examples of differential gene expression within multigene families.

The myofilament lattice of the body-wall musculature forms an obliquely striated array (Rosenbluth 1965; Hirumi et al. 1971), the structure of which is most easily understood by beginning with the basic unit from which the overall structure can be built (Fig. 3a). This structural unit is analogous to the sarcomere of vertebrate muscle. The myosin-containing thick filaments are centrally placed and overlap with two sets of actin-containing thin filaments, one set extending in from either end of the unit.

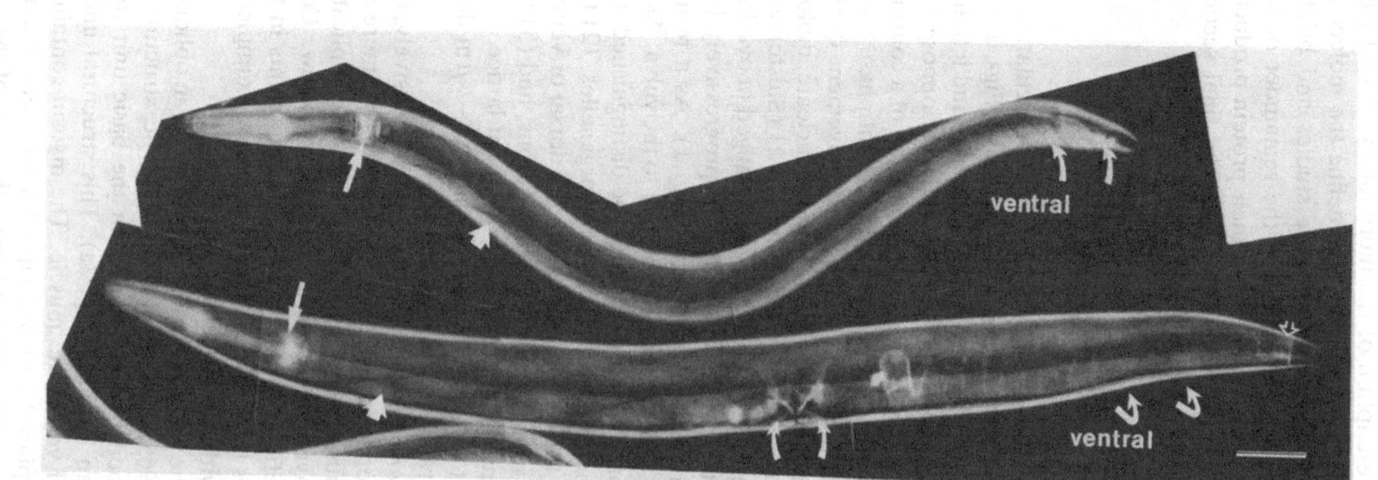

Figure 1 The muscles of *C. elegans I*. A hermaphrodite (*bottom*) and a male (*top*) adult are shown after staining with fluorescently labeled phalloidin, a fungal toxin that binds to filamentous (F)-actin and thus reveals the various muscles of the worm. The male lies on its side, with the ventral surface nearest the hermaphrodite, whereas the hermaphrodite has been rotated to partially reveal the ventral surface. The bulk of the signal comes from the body-wall muscles that extend from the tip of the head to the tail, in each of the four quadrants (short arrow). The pharyngeal muscles are located anteriorly (long arrow). The vulva muscles of the hermaphrodite are ventrally positioned midway down the animal (curved arrows), and the male sex muscles are located in the tail (curved arrows). The anal depressor muscle is also visible (open arrow), but the intestinal muscles, just anterior to the anus, are not well shown. Nonmuscle tissues containing F-actin also stain, including the microvillus border of the intestine (seen in the male), the spermathecal cage of the hermaphrodite, and the cortex of the maturing oocytes. Bar represents 50 μm. (Photograph by G.R. Francis and R.H. Waterston.)

In the nematode, the thin filaments are attached to dense bodies, rather than Z lines as in vertebrate muscle, and each dense body originates at the cell membrane adjacent to the hypodermis. The thick filaments are stacked in columns extending into the cell on either side of an amorphous electron-dense material analogous to the M line of vertebrate striated muscles (Francis and Waterston 1985). Like the dense body, the M-line analog also appears to be anchored in the membrane; it probably functions to establish and maintain the alignment of the thick filaments.

Despite these basic similarities, the body-wall muscles differ significantly from vertebrate muscles in three respects. First, nematode muscle is obliquely striated rather than cross striated as in vertebrates (Rosenbluth 1965), as illustrated in Figure 4. The filaments themselves are longitudinally oriented, as determined by electron microscopy, but adjacent structural units are offset relative to one another by more than a micron, rather than aligned as in cross-striated muscle. Thus, the observed striations are at an angle of only 5–7° to the longitudinal axes of both the filaments and the animal (Mackenzie and Epstein 1980). As a result, transverse sections sample adjacent contractile units at different levels (Fig. 5). However, all the contractile units are held in register from the cell membrane into the interior along the length of the animal by the dense bodies and M lines. This organization results in the simple pattern of striations, superficially resembling that of vertebrate muscle, which is seen when the animal is viewed from the surface by light microscopy.

Second, the thick and thin filaments differ in size and composition from those of vertebrates. In the *C. elegans* adult, the thick filaments are estimated to be generally about 10 μm in length and taper in diameter from 33.4 nm centrally to 14.0 nm distally (Fig. 5; Mackenzie and Epstein 1980; Epstein et al. 1985). Vertebrate thick filaments in striated muscle are 1.6 μm in length and 12.0–14.0 nm in diameter (for review, see Harrington 1979). Also, in contrast to vertebrates, the thick filaments of *C. elegans*, like those of many other invertebrates, contain paramyosin as well as myosin (Fig. 3; Epstein et al. 1974; Waterston et al. 1974). The thin filaments are also longer in *C. elegans* (~6 μm) than in vertebrates (1 μm) but are similar in diameter. They are composed of actin, tropomyosin, and probably troponin, although the subunit composition of the latter has not been defined.

Third, in addition to these differences in lattice structure, *C. elegans* muscle differs from vertebrate muscles in its attachment. As in vertebrates, the ends of the cells contain attachment plaques, to which the longitudinally oriented thin filaments attach from only one direction. These structures are thus analogous to the attachment plaques found in half-I-bands at the myotendinous junctions of skeletal muscle and the intercalated disks of cardiac muscle. However, although some tension is undoubtedly transmitted between cells via attachment plaques, most seems to be transferred

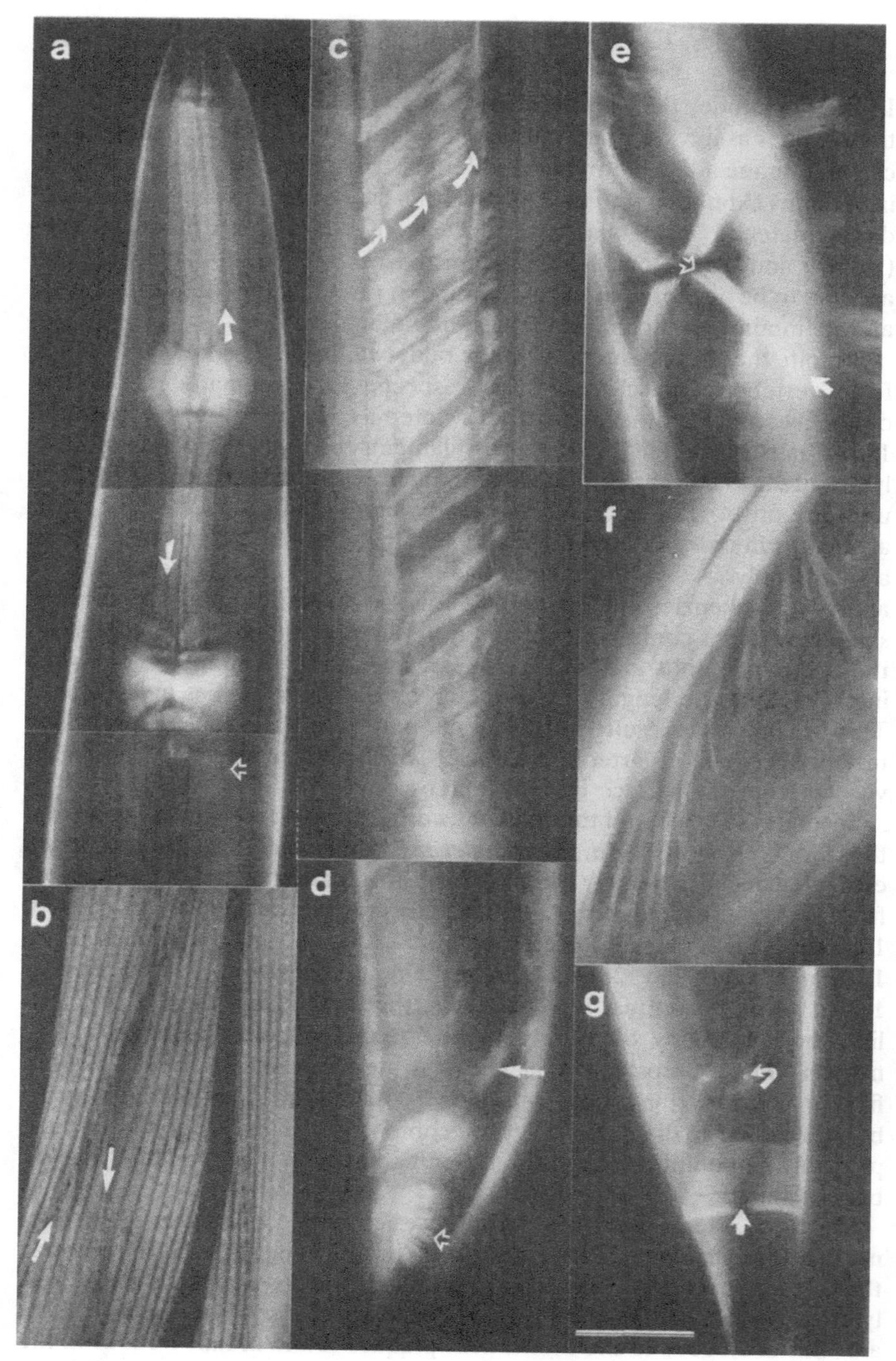

Figure 2 (*See facing page for legend.*)

directly to the cuticle through a series of lateral attachments (Fig. 6). As mentioned earlier, the dense body and M-line analog appear to be anchored in the cell membranes, and small projections extend from these areas of the plasma membrane into the basement membrane (Francis and Waterston 1985). The overlying hypodermal syncytium contains an extensive array of half-desmosomes with associated filaments on each cell face.

Figure 2 The muscles of *C. elegans II*: The organization of the various muscle groups of the worm is shown in more detail in these higher magnification micrographs of animals stained with phalloidin, as in Fig. 1. In all cases, the animals are oriented with anterior toward the top and ventral to the left. (*b*,*c*) The animals have been rotated slightly to bring the dorsal (*b*) or ventral (*e*) surface into view. The magnification is the same throughout. (*a*) The pharynx has radially oriented myofilaments running between the triangular lumen and the outer membrane. Contraction of these muscles serves to open the lumen and move the grinder of the posterior bulb. Arrows point to the gaps between half-I-bands and mark the center of the single sarcomere. Behind the pharynx, the microvillus border of the intestine can be seen (open arrow). (For details, see Albertson and Thomson 1976.) (*b*) The body-wall muscles contain alternating light and dark bands that reflect the underlying alternation of thin and thick filaments (for more detailed descriptions, see later figures). Cell boundaries result in discontinuities in the regular banding pattern (arrows). (*c*) The diagonal muscles of the male tail consist of seven or eight muscle cells on each side of the animal and function during copulation to bend the tail ventrally toward the hermaphrodite. The filaments are organized into three sarcomeres, as revealed by the gaps (arrows) between I bands. (*d*) In addition to the 15 diagonal muscles, the male tail contains 26 other muscle cells that function in copulation (for details, see Sulston et al. 1980). The image shows portions of the anterior and posterior oblique muscles (open arrow) and the spicule protractor (solid arrow) and retractor muscles. A portion of the reoriented anal depressor muscle can also be seen. (*e*) The 16 sex muscle cells of the hermaphrodite consist of 8 associated with the uterus and 8 others with the vulva. Four of the vulval cells are shown in this focal plane, with insertions of one cell marked near the vulva opening (open arrow) and the lateral hypodermis (solid arrow). Filaments from other cells can also be seen, and the out-of-focus body-wall muscle cells passing around the vulva are also apparent. (*f*) The two intestinal muscles, one on each side, are thin sheetlike cells that pass from the surface of the intestinal cells to the ventral hypodermis. They contain relatively few filaments, seen here as wisps of filament bundles. (*g*) The anal muscles consist of the anal sphincter muscle and the H-shaped anal depressor muscle. The anal sphincter, seen in optical cross section (curved arrow points to the dorsal half of the muscle), circles the intestine at its junction with the rectum and serves to close off the former from the latter during defecation. The left half of the anal depressor is seen here inserting ventrally on the dorsal surface of the rectum and anus and dorsally on the subdorsal hypodermis. The contractile apparatus is organized as a sheet of filaments forming a single sarcomere, with the center marked by an obvious gap (large arrow) separating half-I-bands. Bar represents 20 μm. (Photographs by G.R. Francis and R.H. Waterston.)

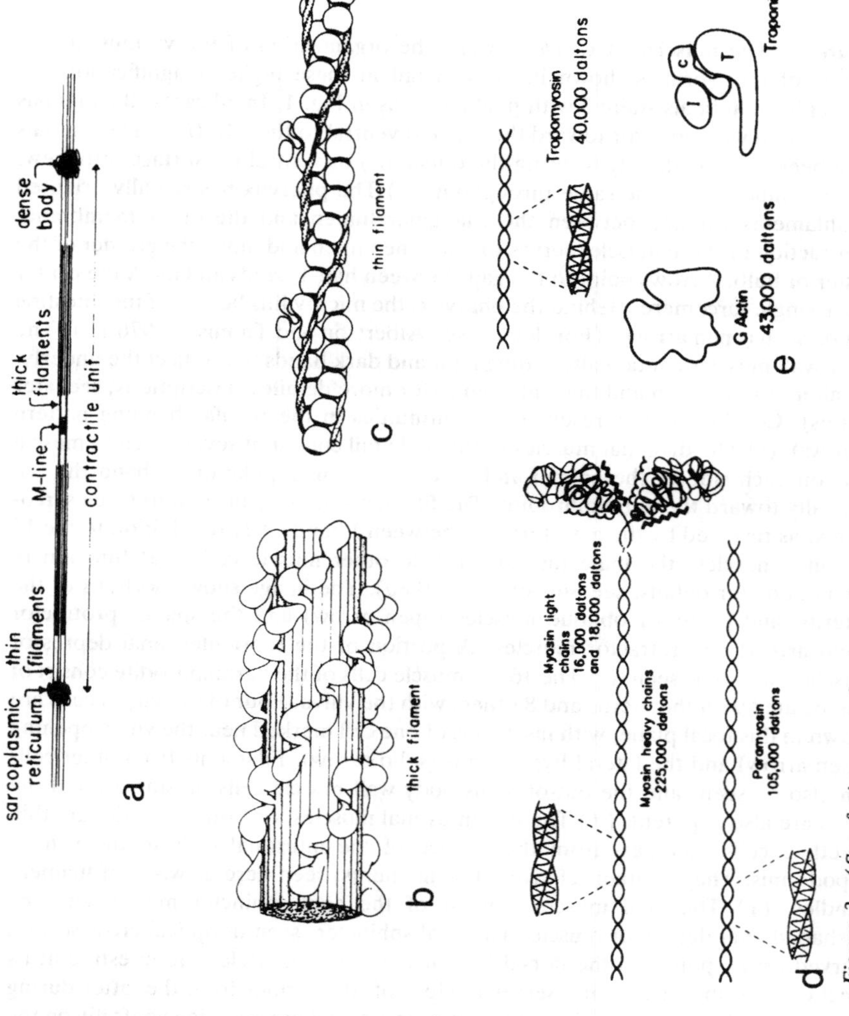

Figure 3 (*See facing page for legend.*)

These desmosomes are generally limited to the regions where the hypodermis contacts muscle cells and are arranged in bands that line up roughly with the dense bodies and M-line analogs on one side and with the annuli of the cuticle on the other (see below). Local contraction of the lattice brings together dense bodies on either end of a contractile unit. As a result of the attachment of the dense bodies through the hypodermal desmosomes to the cuticle, the overlying cuticle is shortened. When coordinated by the nervous system, this shortening can produce a bend in the animal.

Just how tightly the dense body is coupled to the cuticle is unclear. Some longitudinal transfer of developed tension from one contractile unit to the next in series undoubtedly occurs via the bipolar dense bodies and the attachment plaques; combined with some elasticity in the coupling of dense bodies to the cuticle, this transfer probably allows the developed tension to

Figure 3 (*a*) The contractile unit of *C. elegans* muscle is analogous to the sarcomere of vertebrate muscles. Thin filaments are anchored at one end to dense bodies (analogs of the vertebrate Z line) and overlap with bipolar thick filaments, whose alignment is maintained by the M-line components. As the thin filaments are drawn in along the thick filaments, the dense bodies are pulled closer to one another and shortening occurs. (*b*) A schematic representation of a segment of the thick filament, showing the myosin heads on the surface of a filament composed of myosin rods and other components. The four-stranded symmetry depicted here is based on the filament structure determined for other invertebrate thick filaments of similar diameter. (Redrawn from Crowther et al. 1985 and Kensler et al. 1985.) (*c*) A schematic representation of a thin filament, based on the structure determined for vertebrate and invertebrate muscles. The two-stranded actin helix contains tropomyosin in the groove between the two actin strands. Troponin lies along the tropomyosin. (Redrawn from Cohen 1975.) (*d*) The principal known components of the invertebrate thick filament are myosin and paramyosin. Myosin has six subunits, comprising two identical heavy chains and two pairs of light chains. Paramyosin has two identical subunits of 105,000 daltons. Almost half the myosin heavy chain and virtually all the paramyosin molecule consist of an α-helical coiled-coil rod. In addition, myosin has two identical globular heads which, together with the light chains, contain the majority of the mass of the molecule. The head possesses actin-binding and ATPase activities and is undoubtedly the site where the conformational changes occur that underlie muscle contraction. (*e*) The components of the thin filament are sketched here. Tropomyosin has two subunits of about 40,000 daltons arranged in a rodlike α-helical coiled coil. The globular, or G-actin, monomer, drawn here on approximately the same scale as the coiled-coil inset of tropomyosin, has a large and small domain as determined by X-ray crystallography of mammalian actins. Troponin, on the same scale as actin, has three subunits in mammalian muscle. One subunit, troponin-T, binds to tropomyosin; another, troponin-C, binds to Ca^{2+}; and the third, troponin-I, binds actin. Together these subunits, as mediated by tropomyosin, serve to regulate contraction in response to Ca^{2+} levels in mammalian muscle. The subunit composition of troponin in *C. elegans* has not been defined.

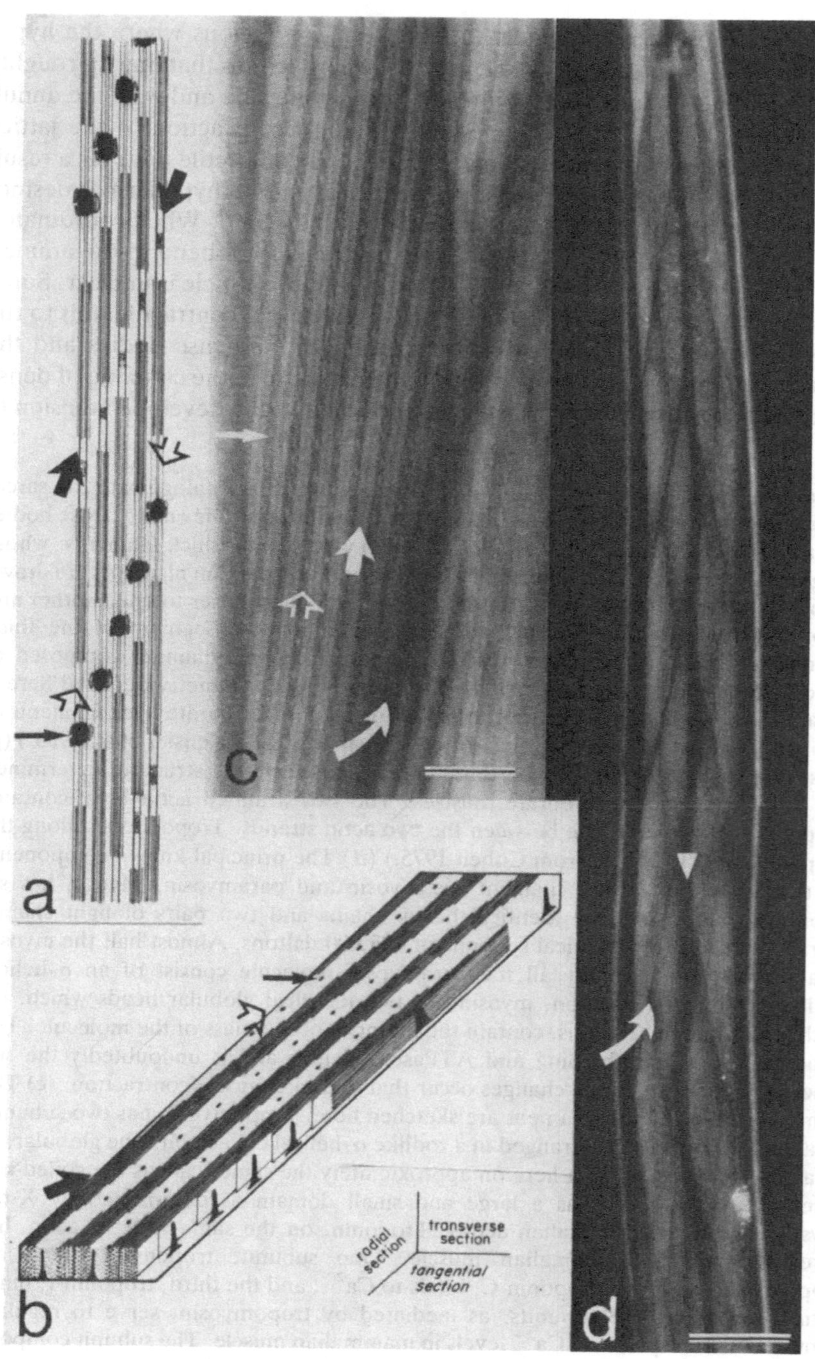

Figure 4 (*See facing page for legend.*)

be distributed more uniformly over a local area. The Pn hypodermal cells can migrate through the network of desmosomes during the L1 stage without apparent impairment of movement. This migration, as well as the molting process, implies that these attachments can be broken down and regenerated as required.

The control of the contractile process via signals received from the nervous system and from other muscle cells is discussed in Chapter 11. The more proximate control appears to be via a sarcoplasmic reticulum, which has received little attention. Sheet-like membrane sacs, which stain with Fe^{2+}, lie against the plasma membrane between M-line analogs and dense bodies and extend into the lattice along the sides of the dense bodies (Fig.

Figure 4 The structure of the body-wall musculature, as viewed by light microscopy. (*a,b*) Illustrations of the relationship of the contractile units to the pattern of muscle organization seen in the light microscope, in high- (*c*) and low- (*d*) magnification polarized light micrographs of an unfixed young adult. Anterior is at the top of each panel. (*a*) A series of offset contractile units to illustrate the arrangement of the myofilaments from a surface view. The filaments lie parallel to the longitudinal axis of the animal, but rather than being held in lateral register as in vertebrate striated muscle, adjacent units are staggered. This stagger gives rise to striations that are at an oblique angle to the long axis of the filaments, as indicated by the large open and solid arrows. The angle used in the diagram, for ease of illustration, is about 20°, but in the animal the angle is only about 6°. The small solid arrow points to a dense body. (*b*) Lower-magnification sketch in three dimensions shows how the surface view relates to overall muscle structure. On the surface, the large solid arrow again points to a band of thick filaments (A band), and the large open arrow points to a thin-filament region (I band). (The thin filaments have been omitted from the diagram for clarity.) In contrast to the lateral stagger, the thick and thin filaments are maintained in register in the radial direction by the M line and dense bodies, respectively. The dense bodies appear as points in the thin-filament band in surface view (small arrow) but as fingerlike projections into the cell on cross section or longitudinal section. (*c*) Most of the myofilament lattice of one muscle cell and portions of two others, out of the plane of focus. The cell boundaries are apparent as a lack of signal disrupting the regular pattern (curved arrow). The bright A bands (large solid arrow) correspond to the regions containing thick filaments, either alone (central dark strip) or overlapping thin filaments (brighter outer portion of the band). The dark I bands represent (large open arrow) areas of thin filaments only, and the row of bright dots therein correspond to the discontinuous dense bodies (small solid arrow at left). Bar represents 7 μm. (*d*) Low-power micrograph showing the dorsal aspect of a young adult, in which the left and right quadrants are separated by a dark strip corresponding to the dorsal hypodermal ridge and nerve cord (arrowhead). The staggered pairs of spindle-shaped cells of each quadrant, as outlined by their darker margins (curved arrow), overlap anterior and posterior pairs. The oblique striations within each cell are just visible at this magnification. Bar represents 50 μm. (Parts of the figure are reprinted, with permission, from Waterston and Francis 1985.)

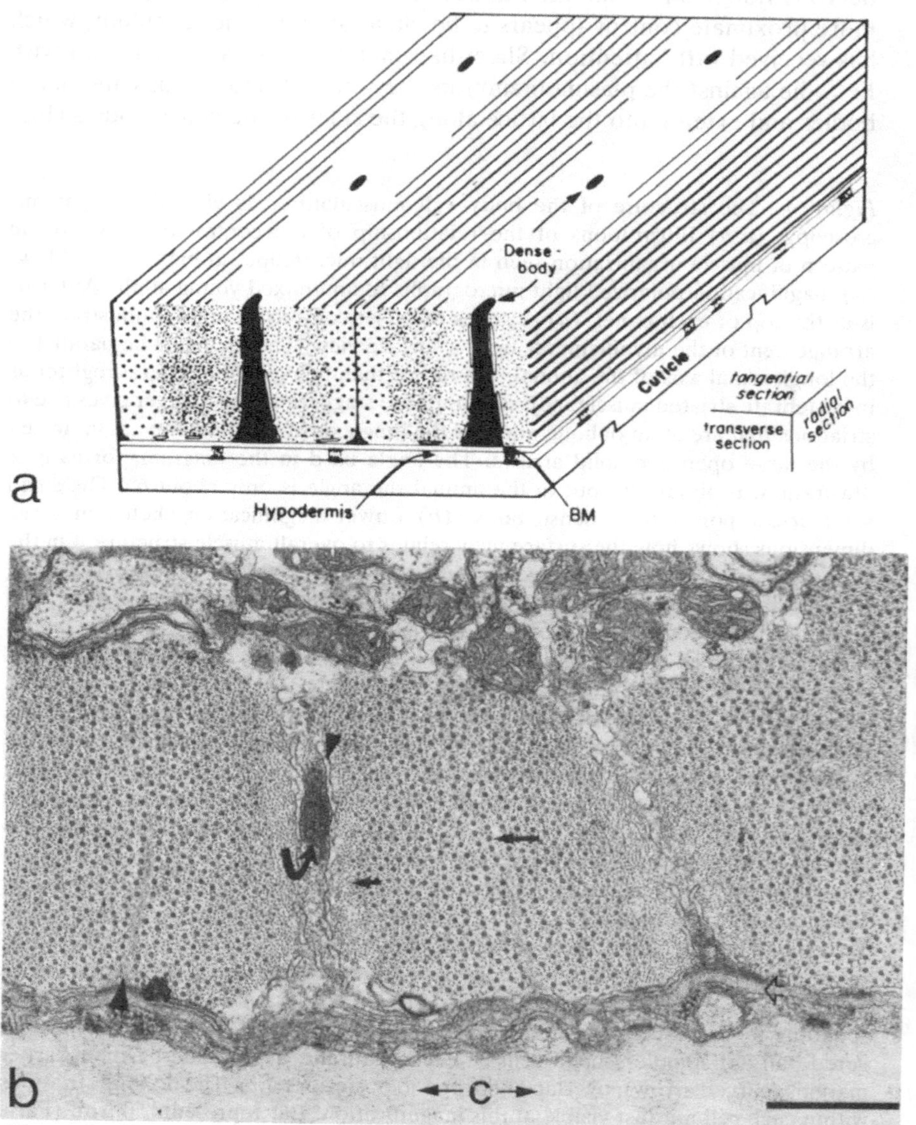

Figure 5 (*See facing page for legend.*)

5; R.H. Waterston, unpubl.). Presumably, electrical signals are transferred from the plasma membrane to the adjacent vesicles and then into the cell, with concomitant release of Ca^{2+}. No equivalent to the T-tubule system is found in *C. elegans*; the need for such a system is apparently obviated by the direct apposition of parts of the sarcoplasmic reticulum to the plasma membrane.

III. COMPOSITION OF *C. ELEGANS* MUSCLE

The major muscle components of *C. elegans* are similar to those of other animals (Fig. 3; for review, see Harrington 1979). Early work demonstrated that *C. elegans* extracts contain myosin with both heavy- and light-chain subunits, actin, paramyosin, tropomyosin, and troponinlike proteins (Epstein et al. 1974; Waterston et al. 1974; Harris et al. 1977; G.R. Francis and R.H. Waterston, unpubl.), which were presumed to be muscle components. Genetic evidence has confirmed that the myosin heavy chain coded by the *unc-54* gene, paramyosin coded by the *unc-15* gene, and actins coded by the *act-1* and *act-3* genes are all body-wall muscle constituents (see Sections VI, VII, and VIII).

The location of these and other proteins has been defined further by using antibodies raised against isolated components to reveal their positions in the intact structure. This approach has been applied both to the

Figure 5 The body-wall muscle structure in transverse section. (*a*) Three-dimensional sketch illustrating the relationship of the transverse section to overall muscle organization and the position of the myofilament lattice relative to the hypodermis and cuticle. The contractile apparatus lies as a sheet of filaments just inside the muscle cell membrane that apposes the hypodermis. A thick basement membrane (BM) separates the two cells. Outside the hypodermis is the cuticle. As the filaments are oriented longitudinally, transverse sections of the worm show the filaments in cross section. Because the contractile units are held approximately in register in the radial dimension, but staggered relative to one another in the tangential plane, a relatively simple pattern of alternating thick- and thin-filament bands results. The thin filaments have been omitted from the tangential view for simplicity. (Adapted, with permission, from Francis and Waterston 1985.) (*b*) Electron micrograph showing many of the features diagramed above. Mitochondria cluster at the boundary of the lattice and the cell body above, and the hypodermis and cuticle (C) lie below, separated from the muscle cell by the basement membrane (open arrow). Portions of the two dense bodies are seen, one cut in glancing section (curved arrow) and the other cut near the base at the right of the image. The membranous sacs (small arrowhead) surrounding the dense body are present throughout the center of the thin-filament (I) band (short arrow) and extend under the thick-filament (A) band along the muscle cell membrane. The M line (large arrowhead) runs through the center of the thick-filament (A) band (long arrow). Bar represents 0.5 μm. (Photograph by J.N. Thompson and R.H. Waterston.)

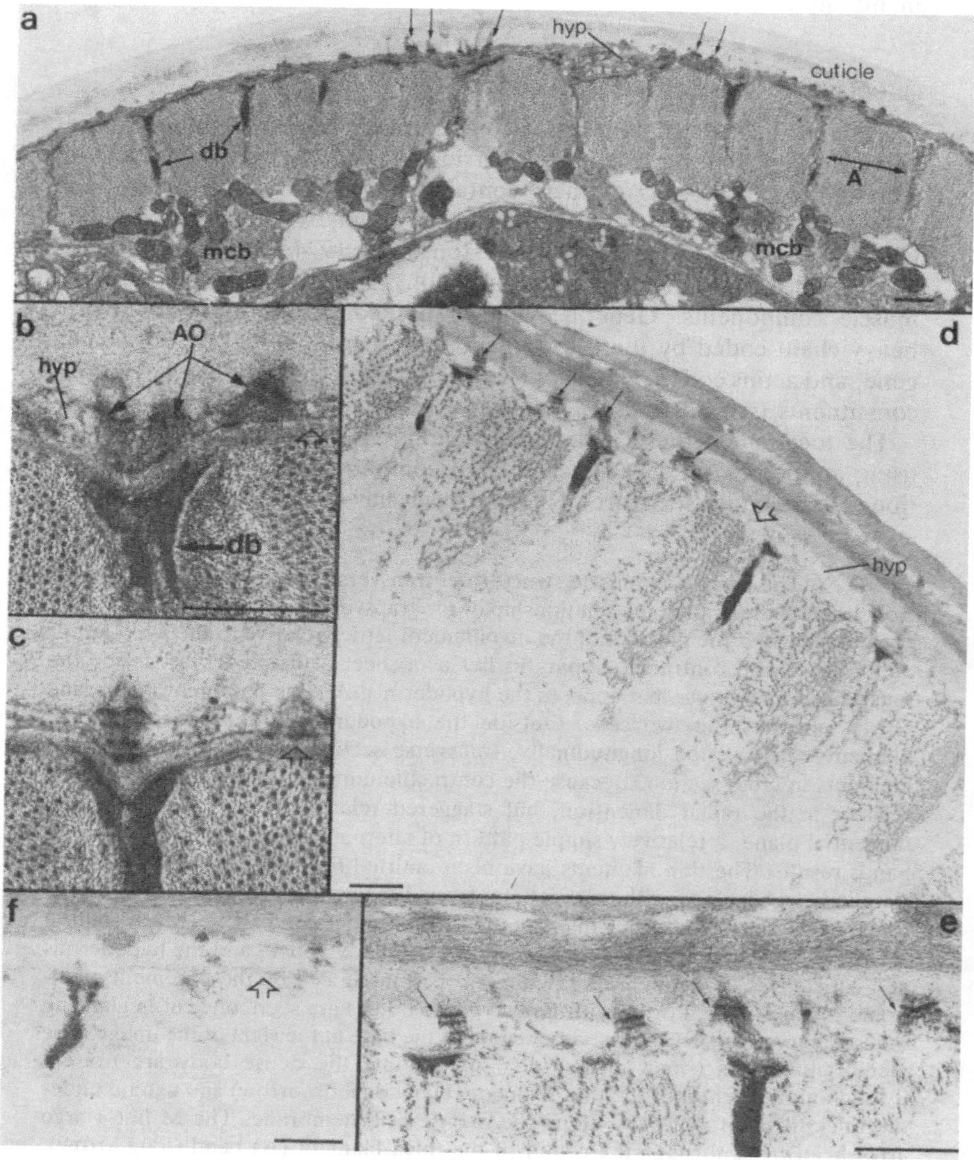

Figure 6 (*See facing page for legend.*)

thick filament and to the accessory structures that presumably align the myofilaments and transmit the tension developed during contraction through the hypodermis to the overlying cuticle. In the case of the accessory structures, the antibodies have also helped to identify new muscle proteins.

A. The Thick Filament

Thick filaments in *C. elegans* are composed principally of myosin and paramyosin (Figs. 3b,d). Paramyosin is found in the body-wall musculature, pharynx, and other minor muscles and is encoded by the single gene *unc-15* (see Section VII). In contrast, the major thick-filament protein, myosin, is represented by four electrophoretically resolvable myosin heavy-chain isoforms in *C. elegans*, termed myoA, myoB, myoC, and myoD (Epstein et al. 1974; Schachat et al. 1977b; Waterston et al. 1982c). Dissection experiments demonstrated that the C and D forms are found in the pharynx, whereas the A and B forms are constituents of the body-wall muscle (Epstein et al. 1974). Genetic and later immunological studies indicate that both A and B forms are expressed in all 95 body-wall muscle cells (Mackenzie et al. 1978a; Miller et al. 1983). myoB is the most abundant, comprising about 70% of the total myosin heavy chain in the adult. myoA accounts for about 20%, and isoforms C and D account for about 5% each. Fractionation of myosins from wild type and from mutants

Figure 6 Relationship of the muscle cells to the hypodermis and cuticle. (*a*) Low-power electron micrograph showing the overall relationship of two muscle cells to the hypodermis (hyp) and cuticle above and the intestine below. The muscle cell bodies (mcb) contain the mitochondria and other cell organelles. An A band (A) is labeled, as are two dense bodies (db). Arrows point to the hypodermal organelles believed to have a role in maintaining attachment of the cuticle to the hypodermis. Bar represents 0.4 μm. (*b,c*) Series of structures between the myofilament lattice and cuticle. The dense body (db) ends immediately adjacent to the cell membrane. Just exterior to the dense body is the basement membrane (open arrow), which often appears at greater intensity next to the dense body. On the membrane surfaces of the hypodermis (hyp), both facing the muscle cell and the cuticle, are half-desmosomes (AO) with fibrous bundles stretching between oppositely facing half-desmosomes. The cuticle in the area above the half-desmosomes is of increased density. (*d–f*) Attachment structures are more apparent in worm fragments in which membranes and cytosolic components have been removed by extraction with salt solutions and nonionic detergents. The open arrow points to the residual basement membrane between muscle cell and hypodermis, and the small solid arrows point to the half-desmosomes. (*f*) Additional extraction with chaotropic agents has removed the desmosomal structures, but the basement membrane is still present. Bar represents 0.5 μm. (Data from G.R. Francis and R.H. Waterston, unpubl.).

lacking myoB or containing an altered myoB indicates that myoA and myoB exist primarily if not exclusively as homodimers, even though produced simultaneously in the same cells (Schachat et al. 1977b, 1978a).

Monoclonal antibodies specific for either myoA or myoB have allowed the locations of these myosins to be determined within the thick filaments of the A band (Miller et al. 1983; see Table 1). Each bipolar thick filament contains both types of myosins, which are confined to distinct regions (Fig. 7). myoA is located in the central region, as inferred from immunofluorescent staining or from direct observation of antibody bound to isolated thick filaments viewed by electron microscopy. myoB is located in the terminal 4.4 μm at each end of the 10-μm filament, giving rise to the model in Figure 7G. The central location of myoA in the region of the bare zone, where myosin is assembled in an antiparallel fashion, and the terminal location of myoB in regions where myosin is assembled exclusively in a parallel fashion, suggest that the heavy chains may be specialized for different assembly functions. In addition, the location of myoA is approximately coincident with the H zone of relaxed muscle, where thin filaments do not overlap the thick filaments. Thus, the different myosins may also have specialized roles in contraction.

To confirm the locations of the two myosins by biochemical procedures, the isolated thick filaments were treated for 30 minutes or longer in various salt solutions (Epstein et al. 1985). The higher the salt concentration is, ranging from 0.10 M to 0.75 M, the greater the solubilization of the filaments. Using myoA antibodies to mark the centers of the filament and myoB antibodies to mark the ends, it was shown that the filaments were solubilized from the ends toward the center, leaving a relatively insoluble region containing myoA. These results confirmed the antibody localization of the A and B isoforms in the thick filament.

Unexpectedly, Epstein and co-workers noted a structure extending beyond the myosin-containing region in partially extracted filaments (Fig. 8; Epstein et al. 1985). The surface of this structure was smooth, and it failed to react to either myoA or myoB antibodies or with antibodies directed against paramyosin. These investigators postulate that this structure represents the core of the thick filament and is composed of an unidentified protein. No protein of abundance comparable to myosin or paramyosin was apparent on gel analysis of the thick-filament preparation used; however, such a core could be composed of several different polypeptides.

B. The Accessory Proteins

In addition to thick and thin filaments, muscle contains dense bodies and M lines, which function to align the filaments into I and A bands, respectively, and to attach the lattice to the cell membrane. There are also likely to be intermediate filaments and possible elastic components by analogy to vertebrate muscles. Francis and Waterston (1985 and unpubl.) took advantage of the relative insolubility of these accessory components and their attachment to the surface of the cell and, in turn, to the hypodermal cells and cuticle to obtain enriched preparations. More than 40 antibodies were recovered against these preparations that reacted with worm fragments, as assayed by immunofluorescence. Some of these antibodies are listed in Table 1, grouped according to the structures they recognize. The specificities of many of these antibodies for electrophoretically separated polypeptides, as assayed on gel blots, are also listed in Table 1.

These antibodies have not only established the composition of several accessory structures but have also revealed new structures not identified previously. For example, the localization of a p400/440 polypeptide to the I band, that is, the portion of the thin filaments not overlapped by the thick filaments, suggests that the I band in *C. elegans* may have components in addition to thin filaments, as has been reported in vertebrate striated muscle (Francis and Waterston 1985). But the most important contribution of these antibodies has been to define some of the components necessary to transmit the force of contraction from the myofilaments to the cuticle.

Three different sets of antibodies reveal the periodic repeat of the dense bodies within the I band (Francis and Waterston 1985). MH35 and MH40 react with the full depth of the dense body and recognize the protein p107a (Fig. 9). Polyclonal antibodies against this polypeptide cross-react with α-actinin of chick gizzard smooth muscle and not other components of smooth muscle. MH23 and MH24 react with the base of the dense body and with the attachment plaques seen where half-I-bands insert along cell margins. These antibodies recognize a family of polypeptides with relative molecular weights of 107,000 and 110,000. Finally, MH25 reacts with the base of the dense bodies and the base of the M-line analog, very near the muscle cell membrane. These results suggest that dense bodies are made up of three domains, as illustrated in Figure 9D and that attachment plaques contain only the two basal components. The M line apparently shares the MH25 antigen as part of its base.

Other antibodies reveal likely components of structures involved in transmitting the tension developed by the muscle cell to the hypodermis and cuticle (G.R. Francis and R.H. Waterston, unpubl.). MH2 and MH3 react with extracellular material secreted by the muscle cell and concentrated under the M line and dense-body membrane attachment sites. Another series of antibodies give a different staining pattern in hypodermal cells (Fig. 10). This latter pattern consists of double bands, designated hypodermal cell bands, which run circumferentially under the cuticle but are confined to the region underlying the muscle quadrants. The hypodermal cell bands actually consist of a series of minute dots, as seen in high-resolution light micrographs. The antibodies defining these bands

Table 1 Antibodies to muscle-associated components

Antibody[a]	Specificity[b]	Localization by indirect immunofluorescence
	Antibodies recognizing myosin isoforms[c]	
5–6	myoA	Center of A band; body-wall muscles and all others except pharynx
5–2, 5–3, 5–13, 12.1, 28.2	myoB	Lateral regions of A band; body-wall muscles and all others except pharynx
9.2.1, 5–1	myoC	Pharyngeal muscles only
25.1, 5–25, 10.2.1	myoB and myoA or myoD	All thick filaments
	Antibodies recognizing other muscle components[d]	
anti-p107	107 kD, pI 6.1, pI 6.6	All dense bodies, full depth
MH35, MH46	107 kD, pI 6.1	All dense bodies except at cell margins
MH31, MH32	ND	All dense bodies except at cell margins
MH23, MH24, MH37	107 kD, pI 6.6	Base of dense bodies and cell margins at thin-filament insertion sites
MH29, MH39, MH44, MH45, MH47	380 kD, 360 kD	Flanking dense bodies in I-band center
MH2, MH3	190 kD, 180 kD	Sheet of stain limited to muscle quadrant with increased intensity adjacent to dense bodies, M lines, and muscle cell margins
MH25	ND	Intramembranous or extracellular, underlying dense bodies, M lines, cell margins where thin filaments terminate
MH9	200 kD, 230 kD	Nuclei, H zone
MH42	ND	H zone
MH1, MH16	Paramyosin	Weak A band
MH7	ND	I band; rabbit muscle: I band
MH15	ND	A band; rabbit muscle: A band; 3T3 cells: actin cables and colcemid-sensitive intermediate filaments

Antibodies recognizing structures with intermediate filaments[c]

anti-p62	62 kD, 64 kD, 68 kD, pI 5.8	Weak staining of region flanking dense bodies in I-band center; intestinal terminal web; excretory cell
MH33	62 kD, 64 kD, 68 kD, pI 5.8	Intestinal terminal web
MH13	62 kD, 64 kD, 68 kD, pI 6.7	Patches in muscles of several mutants; intestine pharyngeal tonofilaments; excretory cell; nerve ring
anti-p55[c]	55 kD, 68 kD, pI 6.7	Weak staining of region flanking dense bodies in I-band center; hypodermis in muscle quadrants only; intestine; pharyngeal tonofilaments
anti-IFA[f]	68 kD, pI 6.2, 70 kD	Hypodermis in muscle quadrants only; pharyngeal tonofilaments; excretory cell; rectum
MH4	68 kD, pI 6.2, 70 kD	
anti-p70	70 kD	
MH20	ND	Like anti-p55; cytokeratin filament system of PtK1 cells
MH5	300 kD	Hypodermis in muscle quadrants only; hemidesmosomes at ends of tonofilament bundles in pharyngeal cells; rectum; uterus
MH8, MH12, MH46	360 kD	Hypodermis in muscle quadrants only; outer surface of pharyngeal cells; rectum; uterus
MH22, MH27	>6 bands, 80–200 kD	Band desmosomes of all epithelial cells

[a]Most antibodies are mouse monoclonal antibodies and are named with MH followed by an integer or simply with integers. Polyclonal antibodies are designated with the prefix "anti."

[b]Antibody specificities were determined by incubating nitrocellulose replicas of SDS-polyacrylamide gels of nematode fractions with appropriate antibodies; ND, not determined.

[c]From Miller et al. (1986).

[d]From Francis and Waterson (1985 and unpubl.).

[e]From G.R. Francis and R.H. Waterston (unpubl.).

[f]The anti-IFA antibody (Pruss et al. 1981) is a mouse monoclonal that reacts with a determinant present on all classes of vertebrate intermediate filaments.

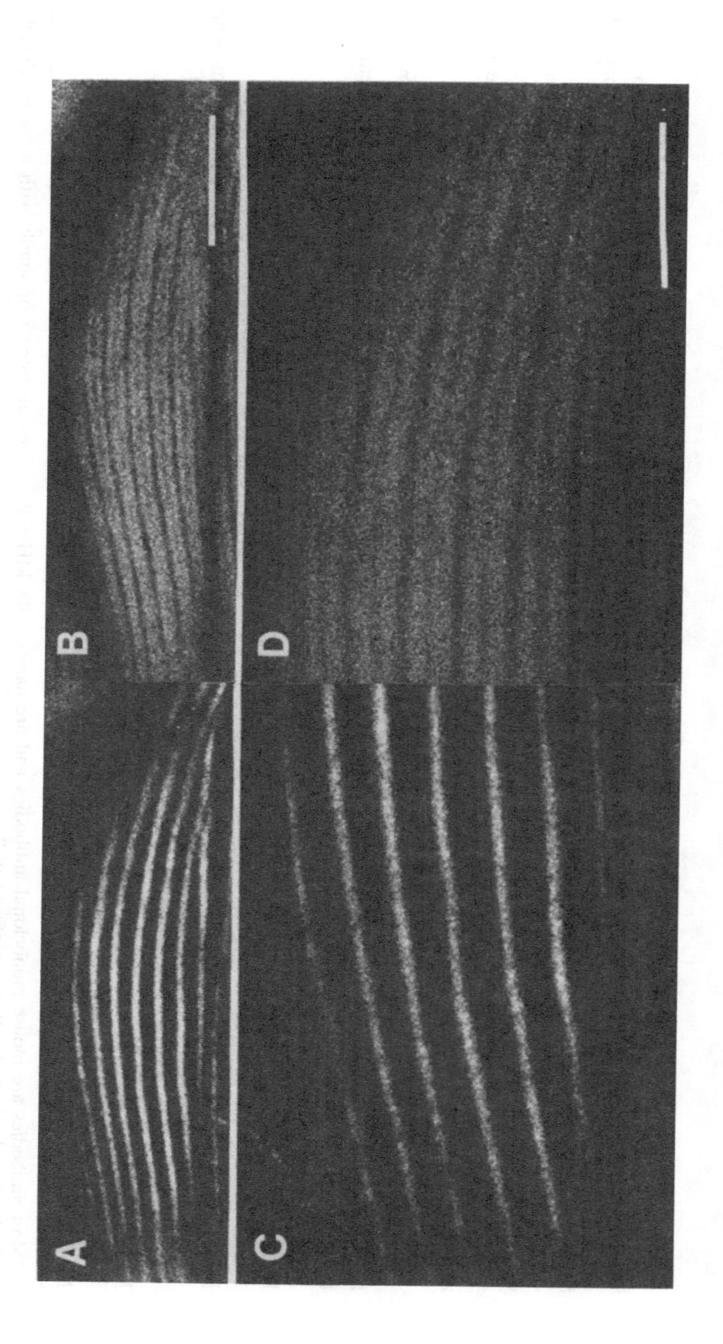
A
B
C
D

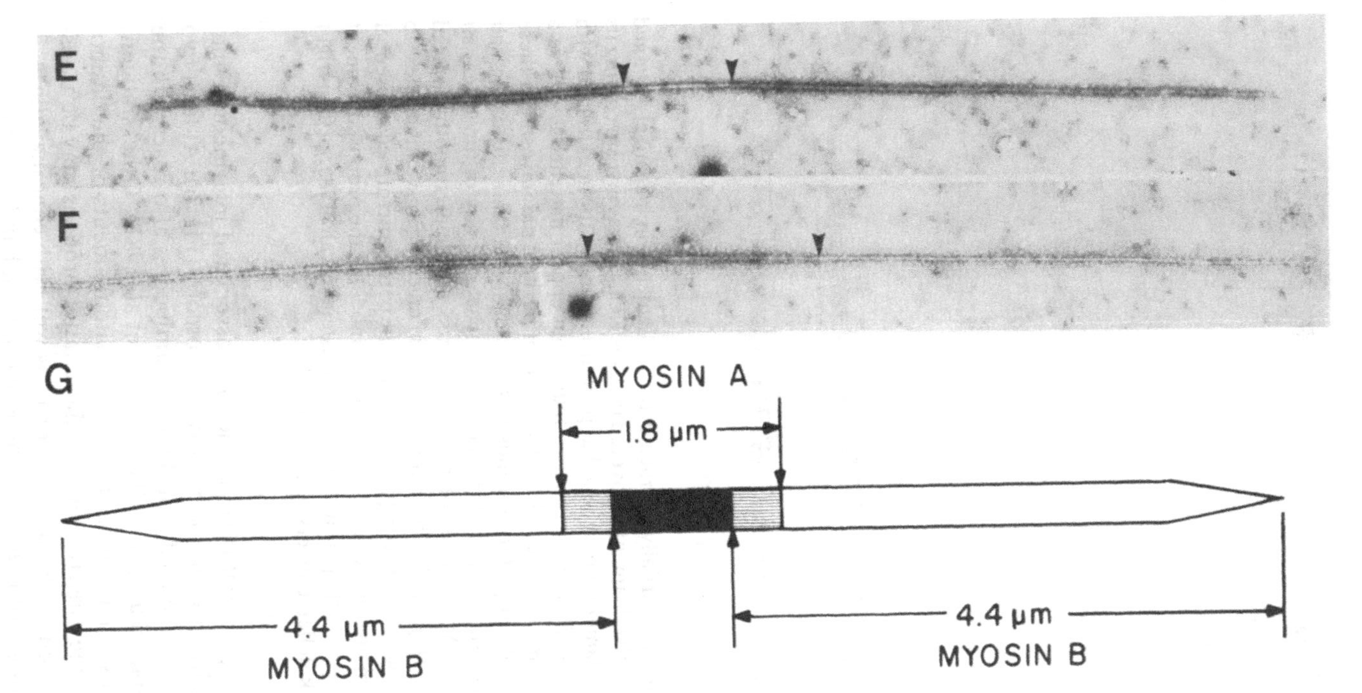

Figure 7 The distributions of myoA (*A, C, F*) and myoB (*B, D, E*) in the thick filament. (*A–D*) Immunofluorescent micrographs of the same cell stained with monoclonal antibodies specific for myoA (*A, C*) and myoB (*B, D*) show that myoA is limited to the central region of the A band, whereas myoB is present only on the outer portions of the A band. (*E, F*) Antibody-decorated purified thick filaments, as seen in these negatively stained preparations, confirm the location of the myosins determined by light microscopy. (*G*) These findings are incorporated into a model of thick-filament structure. (Adapted, with permission, from Miller et al. 1983.) (*A–D*) Bar represents 5 μm.

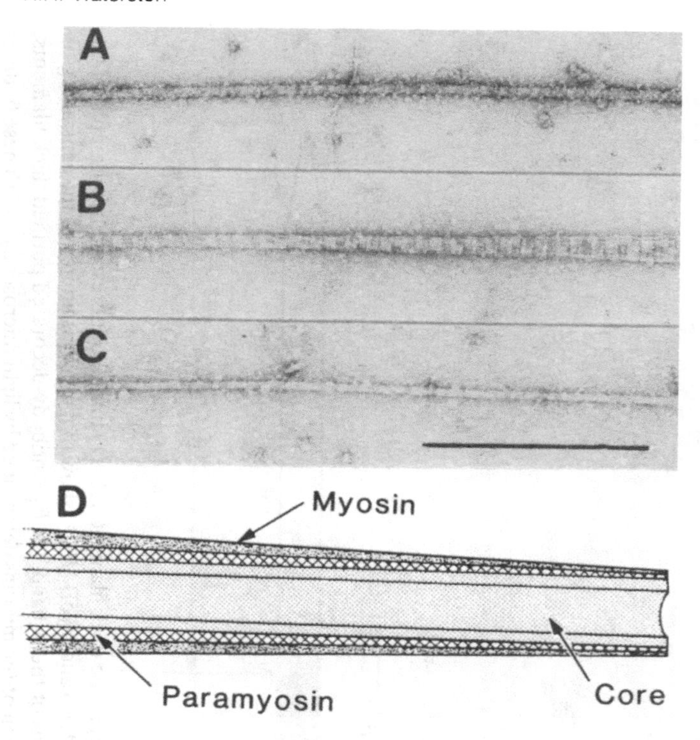

Figure 8 Structure of partially extracted thick filaments in negatively stained preparations. The native filament recovered in 0.1 M KCl (*A*) has myosin coating its surface, giving the filament a roughened appearance. Among those extracted with 0.25 M KCl, filaments (*B*) are occasionally found that appear flattened and have a surface pattern similar to that seen in paracrystals of purified paramyosin. In these filaments, the paramyosin pattern extends to the ends. Most filaments at this salt concentration do not show this pattern, and it is not seen at any of the other salt concentrations used. At the ends of most filaments, a thinner smooth-surfaced structure is found (*C*). This structure does not react against antibodies to paramyosin and is postulated to represent a distinct core structure composed of molecules other than myosin or paramyosin. (*D*) A model of the proposed filament structure. (Reprinted, with permission, from Epstein et al. 1985.) (Copyright permission of Rockefeller University Press.)

probably recognize the half-desmosomes and associated intermediate filaments seen in the hypodermis by electron microscopy (Fig. 6). At least two polypeptides identified on gel blots by these antibodies also react with the general intermediate filament antibody (Pruss et al. 1981). Other antigens may be proteins associated with the intermediate filaments or the desmosomes into which they insert. Some of these latter proteins may be transmembrane components, because they react as well with concanavalin A, a lectin whose binding also reveals the hypodermal cell band pattern.

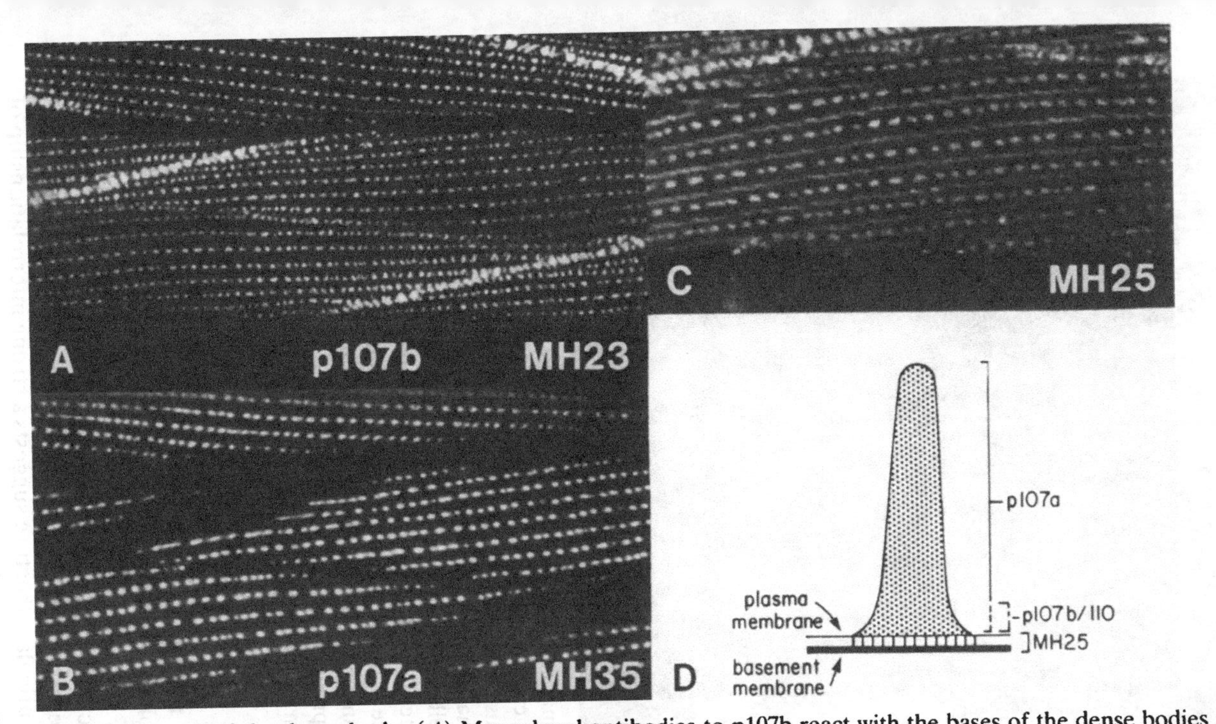

Figure 9 Components of the dense body. (*A*) Monoclonal antibodies to p107b react with the bases of the dense bodies and with the attachment plaques at the cell margins, which appear as two brighter diagonal lines cutting across the rows of dense bodies. (*B*) Monoclonal antibodies to p107a react with the full length of the dense bodies within the I band. The antibodies do not react with regions of the cell containing attachment plaques in half-I-bands, which thus appear as gaps in the pattern. (*C*) Third type of monoclonal antibody, MH25, reacts not only at the sites of dense bodies and attachment plaques, but also at the bases of the M line. Membrane-associated regions of the dense bodies, attachment plaques, and M line share similar structures in electron micrographs (not shown). (*D*) Model of the dense-body structure incorporating these findings. (Reprinted, with permission, from Francis and Waterston 1985.) (Copyright permission of the Rockefeller University Press.)

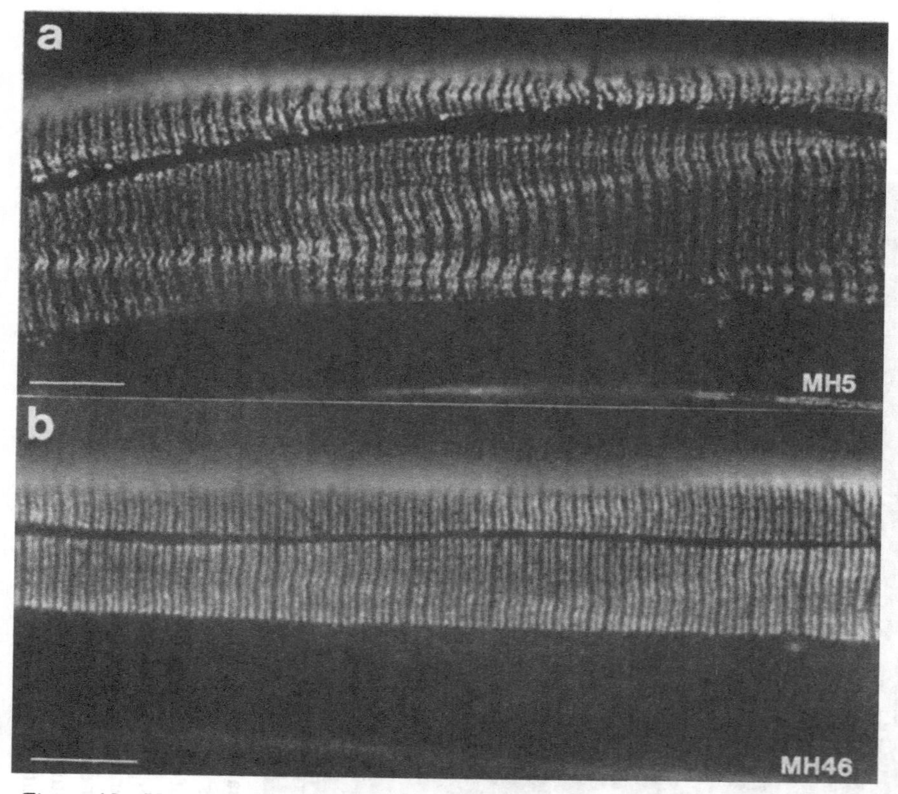

Figure 10 Hypodermal cell antigens. (*a*) The monoclonal antibody MH5 reveals an antigen located in circumferentially oriented bands in the regions of the hypodermis apposed to muscle cells. Each band has a central nonstaining region within it. The signal is enhanced in regions where there are junctions between the underlying muscle cells. (*b*) Other monoclonal antibodies represented here by MH46 produce a pattern similar to that seen in *a*. However, the bands are more uniform in appearance and there is little enhanced staining in regions adjacent to muscle–muscle cell junctions. The longitudinal gap in staining down the center of the quadrant is not reproducibly seen and may represent an artifact of preparation. Bar represents 10 μm. (Data from G.R. Francis and R.H. Waterston, unpubl.)

These antibodies thus define a series of components beginning in the muscle cell at the points of attachment to thin and thick filaments to dense bodies and M lines, continuing through the muscle cell membrane to the intercellular space across the hypodermal cell cytoplasm and membranes to the overlying cuticle. Furthermore, the relationship of these proteins to one another and to cuticular structure suggests a strong interrelationship of muscle and hypodermis in development.

IV. GENETIC ANALYSIS

The ease of genetic analysis is the fundamental reason for using *C. elegans* to study muscle. The isolation of mutant alleles for genes involved in muscle development can provide information about the functions of gene products that might be difficult to obtain in other ways. Examples might include the functions of protein isozymes; modifying enzymes or template proteins that function in myofilament assembly but are not themselves incorporated into the lattice; and new components of the lattice with functions previously unknown. Genetics can also help dissect the functional domains of specific proteins.

Muscle mutants were among the original set of uncoordinated mutants described by Brenner (1974). Since then, a large number of additional mutants representing more than 25 genes have been isolated by several methods (Table 2; Fig. 11). Most of these mutants have been recovered as slow or uncoordinated animals and subsequently identified as having abnormal body-wall muscle by inspection with polarized light microscopy (Waterston et al. 1980). Alternatively, some have been isolated after enriching for slow animals by placing mutagenized worms in a gradient of a chemoattractant such as bacteria: The wild type moves rapidly up the gradient, leaving behind slower and paralyzed animals (Zengel and Epstein 1980b). Screens have been carried out on both the F1 and F2 generations, for dominant and recessive mutations, respectively.

Reversion analysis of mutants representing many muscle genes has been carried out to identify loci that, when mutated, can compensate for the initial defect. Reversion analysis can yield information about the nature of the original allele and its gene, as well as leading to the identification of suppressor mutations in other genes (for a discussion of the potential usefulness of this approach, see Chapter 2). These studies have revealed that the dominant alleles originally recovered in genes *unc-90*, *unc-93*, *unc-105*, and *act-1–act-3*, all revert to wild type at frequencies similar to gene knockout frequencies, suggesting that the original allele in each case is a gain-of-function mutation in a gene whose product is dispensable under laboratory conditions, perhaps because redundant genes exist.

Extragenic suppressors have also been recovered in reversion analysis of muscle mutants. These suppressors include the nonsense suppressor, *sup-5(e1464)*, found as a suppressor of the *unc-15(e1214)* null allele (Waterston and Brenner 1978; see Chapter 2). Other suppressors involve indirect suppression and include new alleles of both known genes and new genes. The mutations *unc-105(n490)* (Besjovec et al. 1985; Park and Horvitz 1986a) and *unc-90(e1463)* (G.R. Francis and R.H. Waterston, unpubl.) produce short, hypercontracted animals that grow poorly. Second-site mutations in other muscle genes, which counteract the hypercontraction induced by *unc-105*, can result in longer animals that often grow better.

Table 2 Genes affecting muscle structure in *C. elegans*

Gene	Linkage group		References
		Gene affecting primarily A-band organization	
unc-15	*I*	Structural gene for paramyosin, normal thick filaments lacking in null alleles, paracrystals present in missense alleles; missense alleles suppressed by *sup-3*	Waterston et al. (1977); Riddle and Brenner (1978)
unc-45	*III*	Sharply reduced thick filament number, myoB particularly disrupted; all five alleles are temperature sensitive	Epstein and Thomson (1974); R.H. Waterston and A.M. Curry (unpubl.)
unc-54	*I*	Structural gene for the major myosin heavy-chain myoB; null alleles have only 30% of wild number of thick filaments; missense alleles altering assembly and subfragment-1 function recovered and representative alleles sequenced; alleles with altered S-1 function act as suppressors of *unc-22*	Epstein et al. (1974); MacLeod et al. (1977a); MacLeod et al. (1981); Karn et al. (1983); Dibb et al. (1985); Moerman et al. (1982)
unc-82	*IV*	Thick-filament number reduced, and larger filamentous aggregates present, composed in part of paramyosin; unusual interactions with *unc-15* mutant alleles	Waterston et al. (1980)
unc-89	*I*	Thick-filament number normal, but filaments not placed in A bands, and no M-line matrix visible in electron micrographs; pharyngeal musculature also abnormal	Waterston et al. (1980)
sup-3	*V*	Improves myoA organization into A bands in *unc-54(0)* and *unc-15* missense backgrounds; overexpression of myoA due in some alleles to duplication of *myo-3* gene	Riddle and Brenner (1978); Waterston et al. (1982b); I. Maruyama and D.M. Miller (pers. comm.).
		Genes affecting primarily I-band organization	
act-I, act-3	*V*	Structural genes for major actin isoforms; putative missense alleles have large aggregates of thin filaments at ends of body-wall muscle cells; putative null alleles have normal structure	Waterston et al. (1984); Landel et al. (1984)
unc-60	*V*	Large aggregates of thin filaments at ends of body-wall muscle cells; A bands relatively intact	Brenner (1974); Waterston et al. (1980)

unc-78	*X*	Large aggregates of thin filaments, with nearly normal A bands; pharyngeal musculature also abnormal	Waterston et al. (1980)
unc-94	*I*	Small clumps of thin filaments at the ends of cells	Zengel and Epstein (1980b)
		Genes affecting attachment of lattice	
unc-23	*V*	Anterior muscle cells detach and degenerate as larvae mature, hypodermal cells swell	Brenner (1974); Waterston et al. (1980)
unc-52	*II*	Growth of lattice retarded in muscle cells behind pharynx, and normal lattice in mutant or suppressed animals may detach from cell membrane	Brenner (1974); Mackenzie et al. (1978b); Zengel and Epstein (1980c); G.R. Francis et al. (pers. comm.)
unc-95	*I*	Clumps of thin filaments containing islands of dense bodies; p107a antigen unattached to cell membrane, p107b and MH25 antigens present in membrane	Zengel and Epstein (1980b); G.R. Francis and R.H. Waterston (unpubl.)
unc-97	*X*	Normal organization of faint shallow sarcomeres that detach under pressure or in older animals	Zengel and Epstein (1980b); Rioux and R.H. Waterston (unpubl.)
		Genes implicated in regulation of contraction	
unc-22	*IV*	Structural gene for a 500-kD component of the A band, thick-filament number about normal, even in *unc-22(0)* animals; resistant to levamisole	Brenner (1974); Moerman and Baillie (1979); Waterston et al. (1980); Moerman et al. (1986)
unc-90	*X*	Irregular A and I bands resembling the muscle structure of animals grown in levamisole; suppressed by *lev-11* and hypercontraction of dominant allele relieved by several muscle mutations	Waterston et al. (1980); G.R. Francis and R.H. Waterston (unpubl.)
unc-93	*III*	Nearly normal muscle structure, but with minor irregularities in A and I structure apparent with polarized light; small irregular densities visible in electron micrographs	Greenwald and Horvitz (1980); Greenwald and Horvitz (1982); I.S. Greenwald (pers. comm); R.H. Waterston (unpubl.)
unc-105	*II*	Resembles *unc-90*; partially suppressed by *unc-22* and *sup-20*, and other muscle mutations relieve hypercontraction of dominant alleles	E.C. Park and H.R. Horvitz (pers. comm.)
sup-10	*X*	Null alleles are suppressors of *unc-93*, and dominant allele resembles *unc-93* dominant allele	Greenwald and Horvitz (1980); Greenwald and Horvitz (1986)

(*Continued on following page.*)

Table 2 (*Continued.*)

Gene	Linkage group		References
		Genes with other effects on muscle structure	
unc-96	*X*	Relatively normal muscle structure; small collections of thin filaments, but with needlelike aggregates of intermediate filament antigens	Zengel and Epstein (1980b); R.H. Waterston and G.R. Francis (unpubl.)
unc-98	*X*	Relatively normal muscle structure; small collections of thin and thick filaments, with needlelike aggregates of intermediate filament	Zengel and Epstein (1980b); G.R. Francis and R.H. Waterston (unpubl.)
unc-87	*I*	Disorganized A and I bands with collection of both thick and thin filaments	Waterston et al. (1980)
unc-27	*X*	Mild disruption of A- and I-band pattern	R.H. Waterston (unpubl.)
		Genes suppressing specific muscle mutants	
sup-9	*II*	Suppressor of *unc-93 (e1500)*, normal muscle structure	Greenwald and Horvitz (1980)
sup-11	*I*	Dominant suppressor of *unc-93(e1500)*	Greenwald and Horvitz (1982)
sup-12	*X*	Recessive suppressor of *unc-60* alleles; inter-dense body region with unusual filament densities and abnormal gonad structure	G.R. Francis (pers. comm.)
sup-13	*III*	Recessive suppressor of *unc-78* alleles, normal muscle structure	G.R. Francis (pers. comm.)
sup-16	*II*	Suppressor of *unc-52(e669)* and at least one other allele	S. Rioux and R.H. Waterston (unpubl.)
sup-20	*X*	Suppressor of *unc-105(n490)*	Park and Horvitz (1986b)

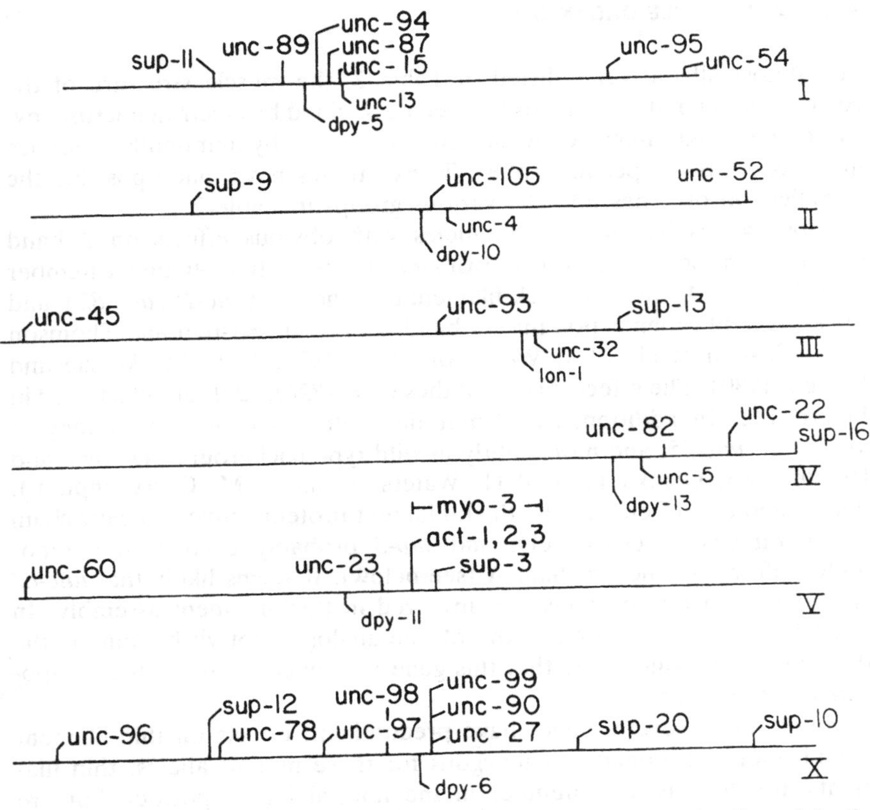

Figure 11 A genetic map showing the position of the genes implicated in the formation and function of normal body-wall muscle. In general, the muscle genes are not clustered but are found throughout the genetic map. Several standard reference genes are shown below the lines.

Using this method, it has been possible to obtain large numbers of *unc-54* and *unc-22* mutants, as well as alleles of other genes. For example, a new gene, *sup-20*, was identified and shown by mosaic analysis to be expressed in muscle cells (Park and Horvitz 1986a,b). The genes *sup-9*, *sup-10*, and *sup-11* were each defined initially by alleles that suppress *unc-93(e1500)* (Greenwald and Horvitz 1980, 1982). Recently, a new allele of *sup-10* has been isolated, which mimics the phenotype of *unc-93(e1500)* animals in a wild-type background (Greenwald and Horvitz 1986). The gene *sup-3* was identified as yielding suppressors of *unc-54* null alleles (Riddle and Brenner 1978; see below). A different type of example is provided by the mutations recovered as suppressors of *unc-22*,which turn out to be a unique class of *unc-54* allele (Moerman et al. 1982). These and other indirect suppressors are included in Table 2.

V. MUTANT MUSCLE ORGANIZATION

For almost all the genes listed in Table 2, the muscle structure of the corresponding mutant animals has been examined by electron microscopy, by polarized light microscopy and, in many cases, by immunofluorescence microscopy using specific probes. These studies have made possible the classification of genes into the various groups in Table 2.

Those genes having mutant alleles with obvious effects on A-band organization include those with primary effects on thick-filament number (*unc-15*, *unc-54*, *unc-45*), thick-filament morphology (*unc-15*, *unc-82*), and thick-filament organization into A bands (*unc-89*) (Epstein and Thomson 1974; Epstein et al. 1974; Waterston et al. 1977, 1980; Mackenzie and Epstein 1980). The effects of one of these, *unc-82(e1220)*, are illustrated in Figure 12a. In addition, *sup-3* mutations alter A-band morphology in *unc-54*, in *unc-15*, and more subtly in wild-type backgrounds (Brown and Riddle 1985; Otsuka 1986; R.H. Waterston and A.M. Curry unpubl.). *unc-54* and *unc-15* encode the thick-filament proteins, myosin heavy chain and paramyosin, respectively, and *sup-3* probably encodes the minor body-wall myosin heavy-chain A (see below). It seems likely that *unc-82* and *unc-45* encode proteins also involved in thick-filament assembly. In *unc-89* animals, the matrix of the M-line analog is not visible among the thick filaments, suggesting that this gene may encode one of the components of the M line.

Mutations in several genes have predominant effects on thin-filament organization. In animals homozygous for these mutant alleles, thin filaments are decreased in number in the normal lattice position but are present in large groups at the ends of cells and in cell processes, as detected both by electron microscopy (Waterston et al. 1980; Zengel and Epstein 1980b) and by fluorescence microscopy (G.R. Francis and R.H. Waterston, unpubl.), using the mushroom toxin phalloidin, which binds specifically to F-actin (Wulf et al. 1979). The *unc-60* (see Fig. 12b), *unc-78* and *act-(st15)*, *act-(st22)*, and *act-(st94)* animals all have very large clumps or patches of phalloidin-positive material, whereas the thick filaments remain largely confined to A bands, and M lines are easily seen. The *act* genes have been shown to encode actin (see below), and it seems likely that the other three genes in this group encode other components required for thin-filament assembly or attachment.

Two genes affect muscle structure primarily in the later larval and adult stages: *unc-23* and *unc-52*. In *unc-23* animals, the muscle cells appear to detach from the cuticle and, subsequently, to degenerate (Waterston et al. 1980). The process begins with the most anterior cells and proceeds caudally, usually being more extensive in one quadrant than others, but never extending beyond the vulva. By electron microscopy, the disruption in the muscle attachment appears to occur in the hypodermal cell. In *unc-52* animals, cells posterior to the pharynx are affected, whereas the more rostral cells remain normal even in older adults (Fig. 12d; Brenner 1974; Waterston et al. 1980). Mackenzie et al. (1978a) noted that sarcomere growth is initially normal but is slowed in later larval stages relative to wild type. Later studies suggested that *unc-54* myosin synthesis is reduced in these mutants relative to the pharyngeal myosins (Zengel and Epstein 1980c). In contrast, studies of *unc-52* mutants with a variety of antibodies suggest that the primary defect affects one of the chain of components anchoring the filaments to the adjacent hypodermis. In animals bearing weak *unc-52* alleles or in animals with partially suppressed amber alleles, muscle growth occurs normally into the adult stage, but as animals age, or with applied pressure, the myofilament lattice detaches from the membrane in occasional cells. The myofilaments slide to the ends of the cell, leaving a central void in the center of the cell. Examination of such cells by immunofluorescence using the antibodies to the dense-body components shows that whereas the MH25 and p107b antigens remain associated with the membrane, the p107a antigen is found with the myofilaments. A defect in one of the attachment components could also explain retarded sarcomere growth, if the attachments also serve as sites of muscle growth.

Other genes possibly involved in muscle attachment are *unc-95* and *unc-97* (Zengel and Epstein 1980b). Although *unc-95* animals do not exhibit the differential effect on head and other muscles, electron micrographs and fluorescent antibody studies indicate that the lattice does detach—the dense body antigen p107a is removed from the muscle cell membrane and is present as islands in the detached filament network (G.R. Francis and R.H. Waterston, unpubl.). The *unc-97* animals have a well-organized myofilament lattice, but the lattice is shallower than that seen in wild type, and its organization can be disrupted with pressure (Zengel and Epstein 1980b), similar to the lattice in partially suppressed *unc-52* animals. In older *unc-97* adults, cells can be found that lack myofilaments centrally and contain myofilamentous aggregates at their ends.

In another group of mutants, the structural defects appear rather minor, compared with the observed changes in motility. This group of mutants might affect primarily the regulation of contraction, with only secondary effects on structure. The genes of this set include *unc-22*, *unc-90*, *unc-93*, and *unc-105*. Both *unc-90* and *unc-105* were identified through the isolation of dominant alleles, shown in later analysis to result in gain of function. Worms homozygous for dominant alleles in either gene are short, stiff, and apparently hypercontracted (Waterston et al. 1980; Park and Horvitz 1986a) and resemble animals reared on the acetylcholine agonist levamisole, both in appearance and in muscle structure. In contrast, most other muscle mutants are flaccidly paralyzed. The *unc-22* mutations confer resistance to levamisole. Mutant animals exhibit an almost continual

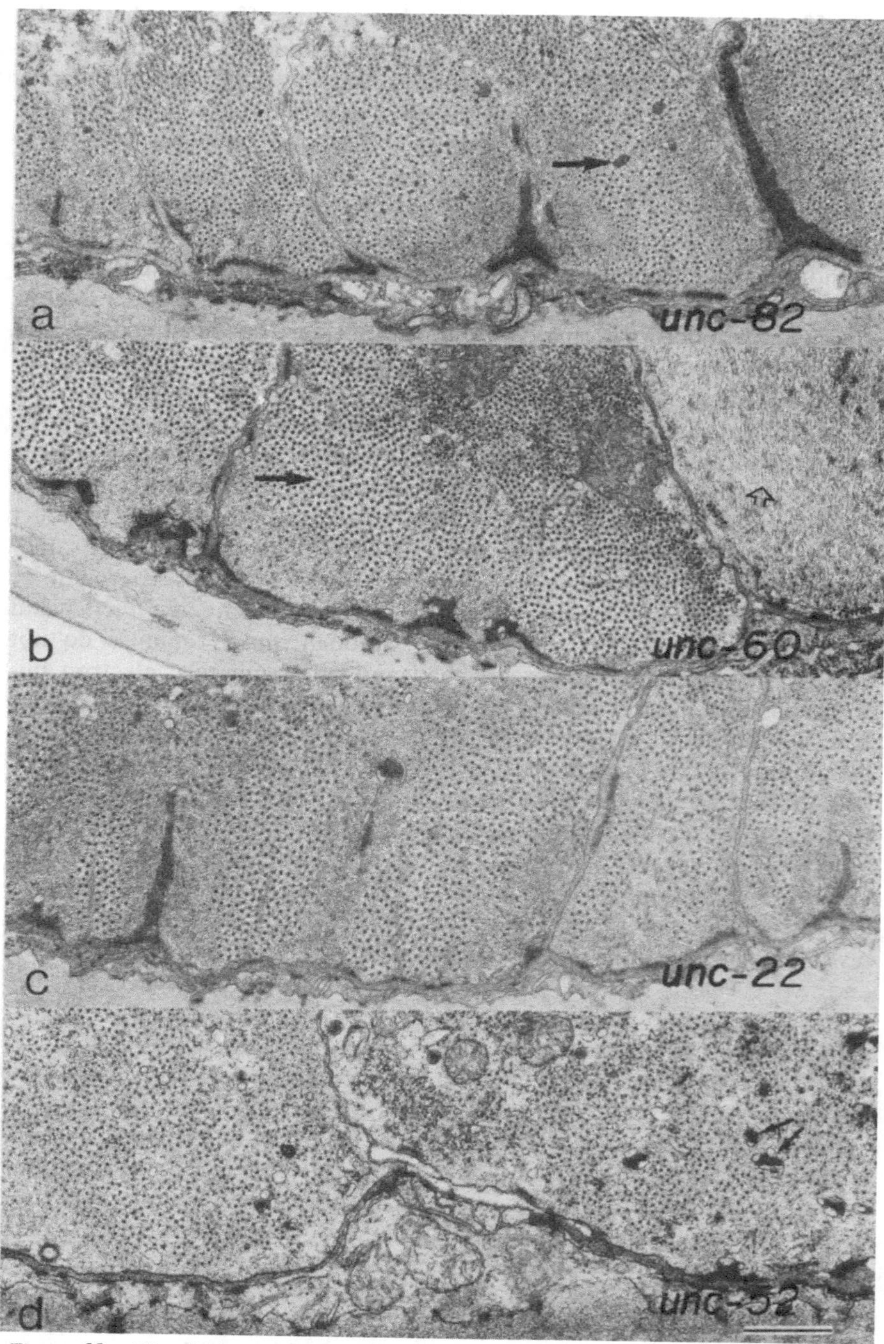

Figure 12 (*See facing page for legend.*)

twitching of the body-wall musculature, enhanced by levamisole (Brenner 1974). The myofilament lattice organization in animals homozygous for weak alleles, such as *s12*, is similar to that of wild-type animals. Animals homozygous for the canonical allele *e66* or the null allele *s32* have disrupted muscle structure, but both thick and thin filaments and the accessory structures are present in apparently normal quantities (Fig. 12c; Waterston et al. 1980; D.G. Moerman and R.H Waterston, unpubl.).

The *unc-93(e1500)* mutation results in an unusual phenotype. When prodded, the mutant animal responds with cycles of first contraction over the length of the animal, followed by relaxation, leading to the phenotypic description "rubber band" (Greenwald and Horvitz 1980). Only minor structural defects have been observed in *unc-93(e1500)* animals. With polarized light, the A and I bands are not defined as sharply as those in the wild type, and in electron micrographs an amorphous electron-dense material is found in both A and I bands (R.H. Waterston, unpubl.). Recently, Greenwald and Horvitz (1986) have isolated an allele of *sup-10* that yields a similar rubber-band phenotype; therefore, the *sup-10* gene can also be included in this group.

The point at which these mutations might act in the regulation of contraction is not known. The *unc-22* protein apparently acts directly in the

Figure 12 Mutant muscle structure. Two mutants are shown, representing classes with primarily altered thick-filament distribution, number, or morphology (*a*) and thin-filament distribution or number (*b*). In addition, two mutants with overall organization defects are shown, in which organization is mildly disrupted (*c*) or severely disrupted (*d*). For examples of wild-type structure, see Figs. 5 and 6. (*a*) *unc-82(e1220)* has unusual accumulations of electron-dense material (arrow) distributed among the thick filaments which, themselves, are only loosely gathered into A bands. By comparison with light and electron micrographs of *unc-15* mutants, these aggregates are likely to be composed in part of paramyosin, a hypothesis confirmed by antibody staining (G.R. Francis and R.H. Waterston, unpubl.). (*b*) *unc-60(e677)* has only a few thin filaments scattered among the thick filaments (closed arrow). Instead, large bundles of thin filaments are found in other regions of muscle cells (open arrow). The thick filaments are not confined to well-delineated A bands, but generally they remain near the muscle cell surface apposing the hypodermis. (*c*) *unc-22(e66)* animals have relatively normal looking muscle with dense bodies and I and A bands all discernible. However, the arrays of thick and thin filaments are of irregular size and shape, and many filaments are seen out of normal position. (*d*) *unc-52(e669)* animals show muscle defects only in later larval stages, and even then anterior muscles are normal. The muscles show a retarded growth of the myofilament lattice, and the filaments that are present become disorganized. A unique feature of this pattern of disorganization is the presence of fragments of dense bodies (arrows) apparently unattached to the plasmalemma. Defects in the basement membrane are also apparent. (Adapted, in part, with permission from Waterston et al. 1980.)

A band (see Section IX), perhaps on myosin itself. The other genes might encode other regulatory proteins of the myofilament lattice, for example, myosin regulatory light chains or the tropomyosin–troponin components. Alternatively, they may act at another step of the excitation–contraction pathway.

The remaining genes—*unc-27*, *unc-87*, *unc-96*, and *unc-98*—are more difficult to classify. Thick- and thin-filament numbers are about normal, but the organization of both A and I bands is mildly to severely disrupted, with *unc-87* animals being the most substantially affected. The L3–L4 stages of the latter are particularly slow, with considerably better movement in older adults (Waterston et al. 1980). This reversal with age contrasts with most muscle mutants, which are most severely affected as adults. The suppression of the *unc-87(e843)* allele by *sup-3* suggests that perhaps the *unc-87* gene encodes a component involved in thick-filament assembly. The *unc-27* animals are also quite paralyzed. The *su115* allele (formerly *unc-99*) results in paracrystallinelike arrays of thin filaments, as identified in electron micrographs (Zengel and Epstein 1980b). No unusual accumulations are detected, however, in phalloidin-stained *unc-27(su142)* animals (G.R. Francis and R.H. Waterston, unpubl.). The *unc-96* and *unc-98* mutant animals move almost as well as wild type, and the myofilament lattice and accessory structures are only mildly disorganized. Interestingly, the *unc-96(su151)* and *unc-98(su130)* alleles result in muscle cells containing needlelike aggregates of material (Zengel and Epstein 1980b) that binds antiintermediate filament monoclonal antibodies but not phalloidin (G.R. Francis and R.H. Waterston, unpubl.).

Many of the identified alleles at suppressor loci have little or no effect on muscle structure in a wild-type background, so that it has not been possible to classify these genes into the groups mentioned above on the basis of mutant morphology. However, it seems likely that many of the suppressors, themselves, encode proteins functioning in the structures altered by the mutations that they suppress. Molecular investigations have confirmed this suggestion for *sup-3* and for *unc-54* suppressors of *unc-22*.

VI. MOLECULAR GENETICS OF MYOSIN HEAVY CHAIN IN *C. ELEGANS*

Myosin has been studied intensively in *C. elegans*. As the major component of the thick filament and the site of conversion of chemical to mechanical energy, myosin is of primary importance in muscle assembly and contraction. The combined molecular and genetic approach possible in *C. elegans* has permitted the study of the myosin heavy-chain genes and their protein products in a manner not possible in most other organisms. A comprehensive review of this work has appeared recently (Karn et al. 1985).

Early biochemical work showed that myosin could be purified by standard procedures from *C. elegans* (Epstein et al. 1974). On denaturing gels, it consists of a heavy chain of 220,000 daltons and two light chains, LC1 of 18,000 daltons and LC2 of 16,000 daltons. The intact *C. elegans* myosin is probably a hexamer similar to other myosins, containing two heavy chains and two of each light chain (Fig. 3). As mentioned earlier (Section III.A), there are a total of four electrophoretically distinct myosin heavy chains, designated myoA, myoB, myoC, and myoD (Schachat et al. 1977b; Waterston et al. 1982b). myoB is most abundant and accounts for about 70% of the total myosin heavy chain in the adult. myoA accounts for about 20%, whereas types C and D are each about 5% of the total. myoC and myoD are exclusively pharyngeal; myoA and myoB are present in the body-wall musculature, vulva muscles, and anal sphincter muscle.

Studies on *C. elegans* myosin were stimulated by the finding that certain *unc-54* mutants contained reduced amounts of myosin and that one allele, *e675*, leads to a shortened myosin chain (Epstein et al. 1974). Mutants at the *unc-54* locus are relatively easy to recover, occurring in approximately 1 in 600 haploid genomes after mutagenesis (Brenner 1974). The adult homozygotes are almost completely paralyzed, are smaller and paler than wild type, and lay no eggs. The animals are fertile, however, so that internally fertilized eggs develop, hatch inside the adult, and then devour it. The larvae do move, although distinctly more slowly than wild type. The generation time of *unc-54* mutants is nearly normal. The easily scored phenotype, combined with good growth of mutants, has made the *unc-54* locus particularly suitable for genetic studies. Methods have been developed that allow the rapid recovery of large numbers of *unc-54* mutant alleles (Anderson and Brenner 1984).

Protein chemical studies proved unambiguously that *unc-54* encodes myoB, the major myosin heavy chain. In animals homozygous for the *unc-54* mutant allele *e190*, total myosin is present at about 30% of wild-type levels and lacks entirely a cyanylation peptide recovered from wild-type myosin (MacLeod et al. 1977b). The *e675* allele results in a shortened myosin heavy chain, as well as a shortened cyanylation peptide (MacLeod et al. 1977a) due to an internal in-frame deletion that removes an internal methionine residue.

A. Molecular Analysis of the Wild-type Myosin Heavy-chain Genes

The genetic and protein chemistry studies provided the necessary background for cloning of the *unc-54* gene (MacLeod et al. 1981). Random cDNA clones prepared from a high-molecular-weight RNA fraction containing myosin message were tested against Southern blots of genomic DNA from wild type and the deletion strain *unc-54(e675)*. One clone hybridized to different-sized restriction fragments from the two strains, and its identity as an *unc-54*-specific sequence was confirmed when the hybridizing fragment from DNA of *unc-54(e190)* was also found to be altered in size. The probe also recognized an mRNA of about 6 kb on Northern transfers, present in wild type but virtually absent in the *e190* allele, which had been shown by in vitro translation experiments to result in a reduced amount of *unc-54* mRNA (MacLeod et al. 1979).

The cDNA clone was used to recover genomic *unc-54* sequences from a λ library. The region of the genomic clones encoding the 6-kb message was determined, and in a remarkably short time the gene and its surrounding regions were sequenced (McLachlan and Karn 1982; Karn et al. 1983). The gene contains eight introns, ranging in length from 38 to 662 bp, as diagramed in Figure 15. Karn et al. (1983) noted that the positions of these introns do not correspond to the sites in the protein susceptible to proteolysis, which have been used traditionally to cleave myosin into smaller units. How the positions of the introns correspond to more rigorously defined domains awaits the crystallographic solution of the three-dimensional structure of the myosin head.

The derived amino acid sequence from *unc-54* provided the first complete sequence of a myosin heavy chain from any organism. The primary sequence and predicted secondary and tertiary structures have been used to construct a model of the likely three-dimensional structure of an α-helical coiled-coil segment of the rod (McLachlan and Karn 1982, 1983). The regular disposition of hydrophobic residues in an alternating 3–4 heptapeptide repeat typical of a coiled-coil structure is superimposed on a longer 28-residue repeat of surface charges. Groups of like charges tend to be clustered, with the groups of positive charges separated by 14 residues from groups of negative charges. At four sites within the rod, a single residue is inserted that shifts the phase of the repeat. No alteration in the stability of the coiled-coil structure is apparent near the light meromyosin–subfragment-2 junction, a region postulated by some to undergo a helix to random coil transition during contraction (Harrington 1979).

Using this model of the myosin rod, a model of myosin interactions in the thick filament has been put forward (McLachlan and Karn 1982, 1983). The amount of stagger between neighboring pairs of myosin molecules giving optimal alignment of the surface positive and negative charges has been estimated to be 98 residues by computer modeling. This amount of stagger of parallel coiled-coil molecules corresponds to 14.6 nm, approximately equal to the 14.3-nm period of myosin heads in thick filaments, as determined by X-ray analysis and other methods (Fig. 13). The slight difference between the calculated optimum and the observed stagger could be accounted for by a slight skewing of the rods about the thick filament.

The conservation of amino acid sequences in the myosin head and, hence, of the corresponding nucleotide sequences in DNA allowed the use of *unc-54* sequences to identify other myosin heavy-chain genes in both *C.*

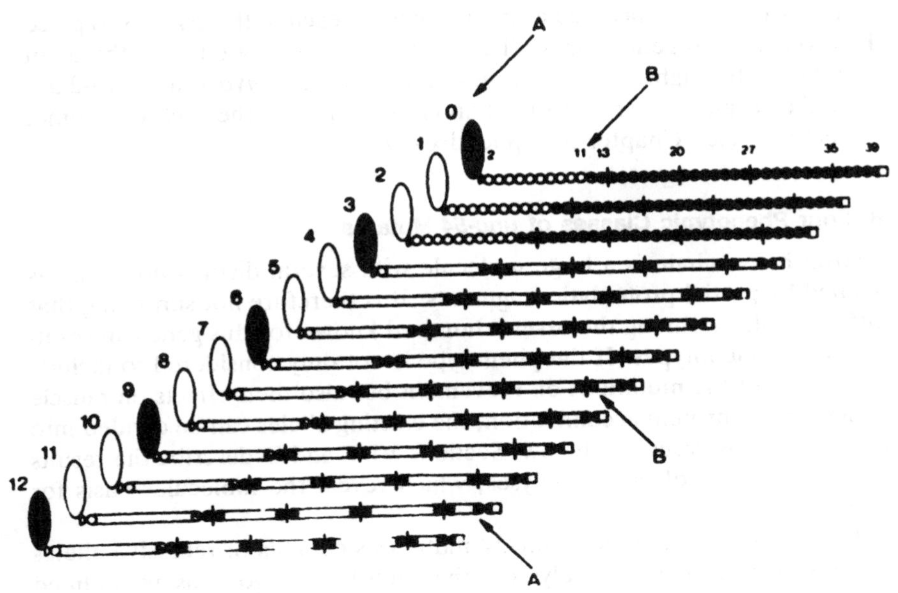

Figure 13 Parallel array of myosin rods drawn to show the relationships between the 98-residue (or 14.3-nm) and 294 (or 43.0-nm) displacements and the periodic structure of the amino acid sequence. In rods numbered *0–2*, open circles represent the 28-residue zones in the short S-2, ending within zone 12; and filled circles represent 28-residue zones in the remainder of the rod. Vertical bars mark the positions of skip residues, and the open square beyond zone 40 represents the nonhelical tailpiece. Rods *3–12* emphasize the half-staggered arrangement of the skip residues. The diagram is a two-dimensional view and should not be taken literally. Arrows (*A, B*) indicate the plane of two different cross sections of arrays, including 7 or 11 rods. (Reprinted, with permission, from Karn et al. 1985.)

elegans and other species. Independent extensive analysis by Karn et al. (1983) and by G.M. Benian and R.H. Waterston (unpubl.), using subclones from the conserved region of the *unc-54* gene, led to identification of three additional myosin heavy-chain genes in *C. elegans*. Karn and co-workers have confirmed by sequence analysis that these three other DNA regions, designated *myo-1*, *myo-2*, and *myo-3*, encode myosin heavy chains; all three genes have now been sequenced.

To determine which gene encodes each of the three electrophoretically defined species of heavy chain, Miller et al. (1986) have used monoclonal antibodies specific for each of the isoforms in combination with expression vectors. For each gene, β-galactosidase fusion proteins were reacted with the panel of monoclonal antibodies. These studies show that *myo-3* encodes myoA, the minor body-wall myosin, and *myo-1* and *myo-2* encode the pharyngeal myosins, myoD and myoC, respectively.

Albertson (1985) has used cosmid clones for each of these genes to place them on the cytogenetic map. The *myo-3* gene is located near the actin complex on the right arm of chromosome V, whereas *myo-1* and *myo-2* are located centrally on chromosome I and right center on the X chromosome, respectively (see Chapter 3; Appendix 4D).

B. Four Phenotypic Classes of *unc-54* Mutants

Myosin heavy chain is a large molecule with several distinct domains, as defined by partial proteolytic fragments. It is therefore not surprising that different alleles among the several hundred known for this gene can result in distinct phenotypes. If the phenotype is not only considered to include the effects of the mutations on movement but also their effects on muscle structure and protein accumulation, the existing alleles can be divided into four classes, as described below. Sections VI.C and D describe the results of genetic and molecular analyses, which reveal the molecular basis for each class.

The predominant class of *unc-54* mutations consists of recessive alleles that result in near total paralysis of the adult homozygote, as exemplified by null alleles such as *e190* and *e1213*. The resulting mutant animals have a disordered array of myofilaments in the body-wall musculature, and the number of thick filaments is reduced to about one-quarter the number found in wild type (Fig. 14a). Little or no banding of the residual filaments is apparent by polarized light. Protein chemistry shows that no *unc-54* myosin heavy chain accumulates in most of these mutants (MacLeod et al. 1977b).

A second large class consists of semidominant alleles that result in slow movement of heterozygous animals. For some of these alleles, homozygotes are viable but often more severely paralyzed than animals homozygous for null alleles (MacLeod et al. 1977b). The majority of alleles of this class, however, are recessive lethals; homozygous animals never move from the pretzel configuration attained at the end of embryogenesis (Dibb et al. 1985; R.H. Waterston, unpubl.). These animals may emerge from the egg, apparently by enzymatic digestion of the egg shell, but they arrest development and die before the second larval stage. A fraction of the alleles of this class are viable when heterozygous with null alleles. Morphological studies of animals homozygous for *e1152* (the canonical allele of this group) show that the number of identifiable thick filaments is reduced, and the thick filaments are not ordered into distinct A bands (Fig. 14c; MacLeod et al. 1977b). Thick filaments are only marginally reduced in number in heterozygotes; more strikingly, however, the filaments are not maintained in organized A bands. Protein chemistry on the homozygous viable alleles indicates that they can contain near normal levels of *unc-54* myosin as, for example, in *e1152* homozygous animals

(MacLeod et al. 1977b), but other alleles of this class apparently can result in a reduced accumulation of the *unc-54* myosin heavy chain (A. Besjovec and P. Anderson, pers. comm.).

The third class contains only two weakly dominant alleles, *e675* and *s291*. Their resulting phenotypes are characterized by near paralysis, accompanied by a fine surface twitch of the homozygote (MacLeod et al. 1977a; Dibb et al. 1985). Thick-filament number is reduced almost to the level seen in null alleles, and the remaining filaments are not organized into distinct A bands (Fig. 14b). In addition, filaments of larger than normal diameter are seen. In each case, the mutant animal contains a shortened *unc-54* heavy chain, the result of an internal in-frame deletion.

Like the first class, the fourth class consists of recessive alleles but with properties that make them quite distinct. The canonical allele of this class, *s74*, when homozygous in a wild-type background, results in slow, stiff animals of normal size. The muscle structure of *s74* homozygotes cannot be reliably distinguished from wild type on an individual basis, but in populations of animals, a slight disruption of the normal A- to I-band repeat is more often apparent in *s74* animals than in wild type. The myoB product of *unc-54(s74)* is present at normal levels, but two-dimensional gel analysis of cyanogen-bromide-cleaved protein compared with wild type reveals a shift in the pI of a 21-kD peptide (Waterston et al. 1982b). Alleles of this class have been recovered as rare dominant suppressors of the *unc-22(s12)* mutation (Moerman et al. 1982) and include some alleles that closely resemble *s74*, such as *s95* and *st134*. Others, such as *s77*, are significantly weaker in their effects on movement and have normal muscle structure despite a change in a highly conserved region of the molecule (Fig. 14d; see Section VI.D). Still others result in abnormal muscle organization in a wild-type background, for example, *unc-54(st130)* and *unc-54(st132)* (D.G. Moerman, pers. comm.).

C. Genetic Analysis of the *unc-54* Gene

The division of the myosin molecule into domains suggested that it might be possible to infer the action of particular mutations by their position within the gene. Furthermore, the localization of the mutation within the gene would limit the sequence analysis necessary to determine the nature of the mutation at the DNA level.

To localize representative mutations within the gene, an intragenic map was constructed (Waterston et al. 1982a). Using flanking unique markers, stable heteroallelic strains were made, and the wild-type animals resulting from recombination or conversion were recovered. The wild-type exceptions were analyzed to determine whether the acquisition of the *unc-54(+)* allele was accompanied by recombination of flanking markers. If so, the

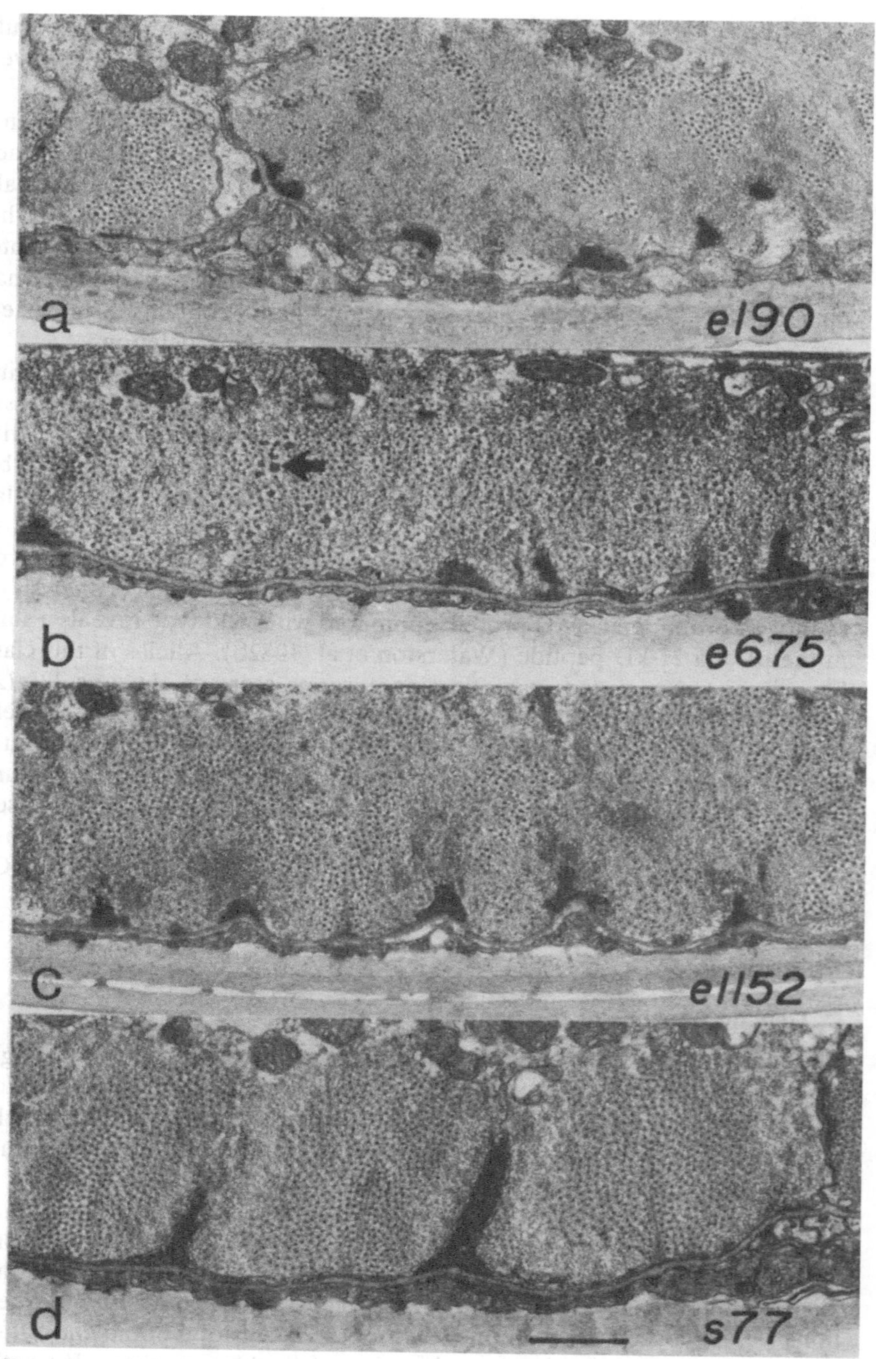

Figure 14 (*See facing page for legend.*)

events were used to order the two *unc-54* alleles by standard four-factor analysis.

A succession of pair-wise combinations were evaluated, resulting in the map shown in Figure 15. The few ambiguities in the results could be accounted for by rare marker loss during strain propagation. In addition, a large proportion of exceptional animals appeared to result from what must be formally interpreted as double-crossover events. This high proportion of apparent double-crossover events, technically termed high negative interference, probably represents gene conversion. The overall genetic size of the *unc-54* gene has been roughly estimated at 0.02–0.04 map units from recombination between alleles at the extremes of the map, consistent with the relatively large physical size of the *unc-54* gene.

D. Sequence Analysis of *unc-54* Alleles

For nearly 20 *unc-54* alleles, the exact nature of the mutation has been determined through DNA sequence analysis. Representative alleles from each of the four phenotypic classes have been included in the analysis, so that the molecular basis for these phenotypes has begun to emerge (Fig. 15; Wills et al. 1983; Dibb et al. 1985; J. Kiff et al., pers. comm.).

The recessive class includes seven nonsense mutations, with both amber and ocher but no opal stop codons identified. In addition, the reading frame is altered by a T : T excision in *e903* and a 450-bp deficiency in *e190* that removes a splice donor site. The two weak twitchers, *e675* and *s291*, as expected from the protein analysis, have deletions in the light meromyosin subfragment of 300 and 1300 bp, respectively, which maintain the reading frame across the deficiencies. In the only dominant allele sequenced,

Figure 14 Muscle structure of the four phenotypic classes of *unc-54* mutants. (*a*) *unc-54(e190)* animals lack myoB entirely, due to an internal deletion in the *unc-54* gene. Thick-filament number is substantially reduced, and the few remaining thick filaments are not organized into A bands. (*b*) *unc-54(e675)* animals make normal amounts of myoB in which a short segment of the rod portion is deleted (MacLeod et al. 1977a; Dibb et al. 1985). Many thick filaments form, but they are not organized into A bands. Also, larger-diameter filaments are present (arrow). (*c*) In *unc-54(e1152)* animals, the myoB contains two amino acid substitutions in the rod portion near its junction with the head. Thick-filament number is intermediate between the null mutants and wild type, but again thick filaments are not organized into ordered arrays. (*d*) The *unc-54(s77)* animals produce a myoB with a substitution in the conserved thiol region of the myosin head (J. Kiff et al., pers. comm.). Muscle structure is generally, but not always (not shown), normal with wild-type numbers of thick filaments organized into A bands. The failure of these animals to move normally thus reflects an intrinsic failure in myosin head function. Bar represents 0.5 μm. (Adapted, in part, from Moerman et al. 1982; J.N. Thomson and R.H. Waterston, unpubl.)

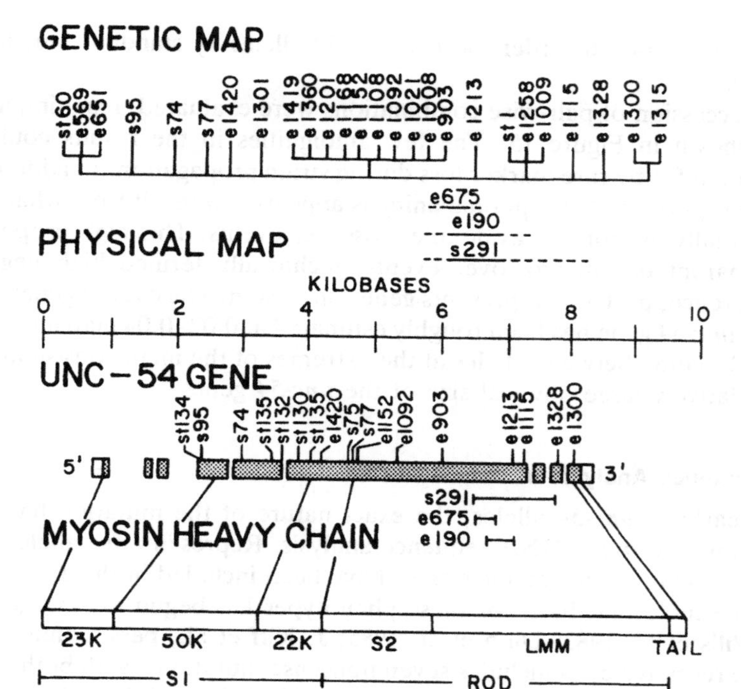

Figure 15 Molecular genetics of *unc-54*. The genetic map of the *unc-54* gene is inverted from its standard map orientation to place the 5′ end of the gene to the left. The physical map of the gene is shown, illustrating the intron–exon structure and the position of the sequenced mutations within it. The structures formed by the myosin heavy chain and the common protelytic digestion fragments are shown, and their relationship to the coding regions of the *unc-54* gene are indicated. (Adapted, with permission, from Dibb et al. 1985; Kiff et al., pers. comm.)

e1152, a Gly–Lys pair is replaced by Arg–Met in the subfragment-2 region just adjacent to its junction with subfragment 1. Finally, seven *unc-22* suppressors in *unc-54* and several intragenic revertants of one of these suppressors have been found to be missense mutations in the myosin head. Three alleles have altered codons for amino acids that flank the predicted myosin ATP-binding site (Walker et al. 1982), and two of these, *st134* and *s95*, occur in adjacent codons. Five intragenic revertants of *s95* all alter the codon affected by the initial *s95* event (J. Kiff et al., pers. comm.). Two other alleles, *s77* and *st135*, on the other hand, alter codons for conserved amino acids near the SH1 thiol residue (J. Kiff et al., pers. comm.), a region also implicated in ATP hydrolysis in some studies. The last two alleles, *st130* and *st132*, alter codons in the central 50-kD fragment and are unusual in this set in that they affect muscle structure, as well as movement phenotypically.

E. Reversion Analysis of *unc-54–sup-3 V*

Reversion analysis has been carried out on a subset of *unc-54* alleles. Several internal revertants have been obtained for nonsense alleles (e.g., *e1300*), for the dominant missense alleles (e.g., *e1152*), and for the *unc-22* suppressor alleles (e.g., *s95* and *st132*) (R.H. Waterston and D.G. Moerman, unpubl.). These latter revertants may prove of interest for defining domains of specific myosin functions.

Reversion analysis of *unc-54* also led to the identification of the indirect suppressor *sup-3 V* (Riddle and Brenner 1978). The *sup-3* suppressors are of variable strength and occur at a forward frequency of approximately 10^{-4}. In cross-suppression tests, *sup-3* suppresses all *unc-54(0)* alleles equally well (where 0 signifies a null allele) but suppresses the dominant missense or in-frame deficiency alleles poorly. *sup-3* also suppresses *unc-15* missense alleles, but not *unc-15(0)* alleles, and suppresses certain alleles of other genes (e.g., *unc-87*).

Stronger *sup-3* suppressors of *unc54(0)* alleles can be recovered from *unc-54(0)*; *sup-3(e1407)* animals by additional rounds of selection for better moving animals, either after mutagenesis or spontaneously (R.H. Waterson, unpubl.). These stronger suppressors, however, are unstable and revert spontaneously at a frequency of 10^{-3} to near basal levels of activity. These genetic properties suggested that tandem duplications of the region might be the basis for the suppressor activity.

These results implied that *sup-3* might either regulate the expression of the gene for the minor body-wall myosin myoA or be the structural gene itself (Riddle and Brenner 1978). Using gel densitometry to measure relative levels of the myosins, it was possible to show that myoA accumulation is increased about twofold in *sup-3(e1407)* animals, either in *unc-54(0)*, *unc-54(e675)*, *unc-15(e73)*, or wild-type backgrounds (Waterston et al. 1982b; Otsuka 1985). These results, together with the demonstration that *myo-3* encodes myoA heavy chain and the positioning of *myo-3* cytogenetically in the region of *sup-3*, all strongly supported the hypothesis.

Experiments have recently been carried out to examine the *myo-3* sequence in *sup-3* mutants (I. Maruyama and D.M. Miller, pers. comm.). The results indicate that *myo-3* copy number is increased both on the *eDf1* deficiency chromosome and in the genome of animals carrying the *sup-3(e1407)* allele. Even higher copy numbers are present in stronger suppressors derived from *e1407*.

Loss-of-function alleles of the *myo-3* gene have also been recovered recently (R.H. Waterston, unpubl.). Animals homozygous for these new alleles, *myo-3(st378)* or *myo-3(st386)*, never change position during embryogenesis and arrest development just after hatching. The paucity of thick filaments in electron micrographs of the muscle of the arrested larvae and the nearly complete lack of movement of the animals, combined with the location of myoA in wild-type animals at the center of normal filaments, suggests that the myoA protein plays a key role in initiating thick-filament formation.

F. Gene Interactions—*unc-54* and Other Muscle Genes

Double mutants carrying an *unc-54* mutation and one of the other mutations affecting muscle generally show simply the additivity of the two phenotypes and in many cases are lethal, for example, *unc-15(e73);unc-54(e190)* or *unc-15(e1214);unc-54(e1301)* (R.H. Waterston, unpubl.). Interesting exceptions to this general finding are *unc-54(0);unc-45(ts)* strains. Either mutation alone, when homozygous, drastically reduces the number of intact thick filaments; curiously, the double mutant at the restrictive temperature shows no additional reduction in thick filament number (Fig. 16; R.H. Waterston, unpubl.). The lack of additivity of the two mutations implies that they both affect the same muscle component, in this case myoB. Indeed, in *unc-45* animals, myoB is not at all associated with A bands, and myoA distribution is no worse than that seen in *unc-54(0)* mutants (Fig. 16; R.H. Waterston and A.M. Curry, unpubl.).

Further evidence that *unc-54* activity modifies myoB to permit its assembly into thick filaments comes from studies with *sup-3*. The strong *unc-54(0)* suppressor *sup-3(e1407;st90 st92)* has little effect on movement or muscle structure of *unc-45* animals (Fig. 16). In contrast, *unc-54(0);unc-45(ts)sup-3* animals are well suppressed and resemble *unc-54(0);sup-3* animals, both in terms of movement and muscle structure (R.H. Waterston and A.M. Curry, unpubl.). These results could be explained if the *unc-45(ts)* mutations result in a myoB that fails to assemble and counters the action of *sup-3* mutations. Inclusion of an *unc-54(0)* mutation in the construct removes the faulty myoB protein and suppression occurs.

VII. PARAMYOSIN

Paramyosin is another major component of the thick filament. An α-helical coiled-coil rodlike protein for most of its length, it probably lies internal to myosin and forms, at least in part, the core of the thick filament (Mackenzie and Epstein 1980). It is present at nearly 1 : 1 molar ratios, with myosin heavy chain in the body-wall musculature; and dissection experiments show it is also present in pharyngeal muscles but at a reduced ratio to myosin (Waterston et al. 1974). Recent evidence indicates that the *unc-15* gene is the only gene for paramyosin in the nematode (H. Kagawa and J. Karn, pers. comm.; S. Rioux and R.H. Waterston, unpubl.).

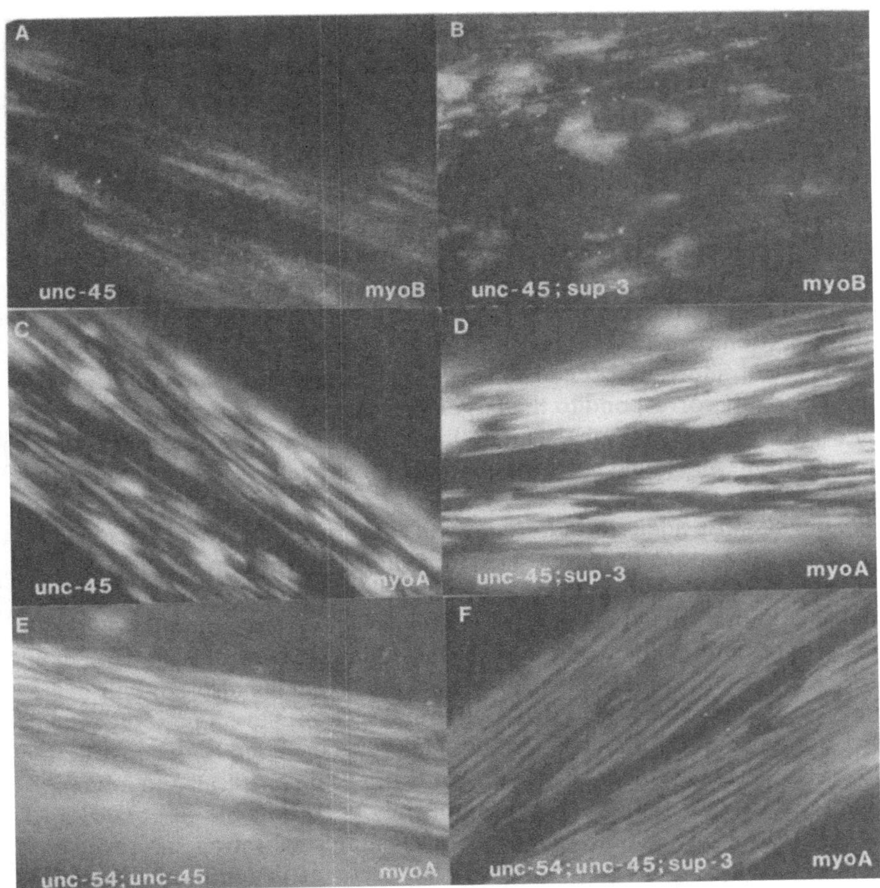

Figure 16 Interactions of *unc-54* with *unc-45(m94)* and *sup-3*. *unc-45* animals grown at restrictive temperature have been reacted with anti-myoB monoclonal antibody 28.3 (*a*) and anti-myoA monoclonal antibody 5.6 (*c*). The myoB staining is faint and shows almost no organization into A bands, but myoA retains some A-band organization. (*b,d*) The *unc-45;sup-3* double-mutant animals are not altered significantly in the distribution of these two proteins. (*e*) The double mutant *unc-54;unc-45* is also not different in its distribution of the myoA polypeptide from *unc-45* alone. However, the triple-mutant animal *unc-54;unc-45;sup-3* (*f*) has significantly improved muscle structure, with *myo-3* staining largely present in A bands. The pattern seen is similar to that of the *unc-54; sup-3* double mutant alone. The central area of reduced staining seen in some of the A bands is unexpected and as yet unexplained. (Data from A.M. Curry and R.H. Waterston, unpubl.)

A. Genetic Analysis of *unc-15*

Mutant alleles of the *unc-15* locus are of two types: apparent null alleles resulting in the absence of a paramyosin accumulation (e.g., *e1214*) and presumed missense alleles affecting assembly but not accumulation (e.g., *e73*) (Waterston et al. 1977). Three alleles of the first type are known; all are poorly viable as homozygotes at 20°C and cannot be grown continuously at 15°C or 25°C (R.H. Waterston, unpubl.). Paramyosin is reduced at least 50-fold in these homozygous mutants. Dissection experiments prove that the pharyngeal paramyosin is also lacking, suggesting that the same gene is expressed in both tissues. Electron micrographs reveal that the *unc-15(0)* mutants are entirely lacking normal thick filaments. Instead, both larger diameter aggregates of electron-dense material and hollow-core filaments are found (Waterston et al. 1977; Mackenzie and Epstein 1980). Attempts to isolate thick filaments from the null mutants result in isolation of short, thickened 1.5-μm-long filaments that contain myosin and probably correspond to the larger aggregates of electron-dense material in electron micrographs. The hollow filaments are also present in these preparations and could correspond to the proposed core structure (Epstein et al. 1985).

The missense alleles of *unc-15* vary in resulting phenotype. These alleles produce alterations in muscle structure similar to those noted for the null alleles, with both hollow core filaments and the larger diameter aggregates, but in addition, they produce an allele-specific, variable number of apparently normal thick filaments, not organized into A bands (Waterston et al. 1977). Animals homozygous for the missense alleles also contain aggregates of paramyosin which, on longitudinal section, have a periodic repeat.

Genetic analysis of *unc-15* mutant alleles has included both construction of a fine-structure genetic map (Rose and Baillie 1981) and reversion analysis. Using *unc-15(e73)*, it has not only been possible to obtain *sup-3* suppressors (see Section VI.E) but also numerous intragenic revertants (Riddle and Brenner 1978). The latter occur at an unusually high frequency, suggesting that many are pseudorevertants, involving second-site mutations partially compensating for the first change. It seems likely that the paramyosin polypeptide chain contains a periodic repeat of charged groups, similar to that found in myosin, and possibly a large number of sites could be changed to compensate for the initial mutation. Curiously, some of the pseudorevertant alleles complement *unc-15(e73)* but cause paralysis or lethality when homozygous themselves (D.L. Riddle, pers. comm.). The latter phenotype could result from simultaneous mutation in a closely linked essential gene; an intriguing alternative hypothesis is that the reversion event produces a paramyosin capable of interacting normally with the original *e73* paramyosin but not with itself.

In addition to these studies, double mutants carrying *unc-15* and other muscle gene mutations have been constructed. The phenotypes of such doubly mutant animals are generally summations of the two single mutant phenotypes. An interesting interaction, however, occurs between *unc-15* and *unc-82* mutations (Waterston et al. 1980). The double-mutant phenotype observed is not predictable and depends on the particular combination of alleles used. For example, the *unc-15(e73);unc-82(e1323)* double mutant moves better than *unc-15(e73)* animals but not as well as *unc-82(e1323)* animals. On the other hand, the *unc-15(e1214);unc-82(e1323)* combination is lethal and the *unc-15(1214);unc-82(e1220)* double mutant resembles *unc-15(e1214)* mutant animals. These results suggest that *unc-82*-gene product may interact with paramyosin, and morphological analysis of the *unc-82* mutants and double mutants supports this hypothesis. The needlelike aggregates found in the *unc-82* animals resemble paramyosin paracrystals; these aggregates are absent in the double mutant with the *unc-15* null allele. The aggregates found in the double mutants with the missense allele *unc-15(e73)* are different from those found in either mutant alone. Perhaps the *unc-82*-gene product modifies paramyosin in the assembly process.

B. Molecular Analysis of the *unc-15* Locus

The *unc-15* gene has recently been cloned (H. Kagawa and J. Karn, pers. comm.). Using a polyclonal antibody directed against paramyosin to screen a bank of random genomic fragments cloned into a plasmid expression vector, a positive clone was obtained that was used to recover larger genomic fragments from a phage library (H. Kagawa, pers. comm.). To confirm the identity of this sequence as one coding for paramyosin, three distinct lines of evidence were developed: (1) Sequence analysis of a small region of the DNA showed that the open reading frame encodes a polypeptide with a sequence typical of α-helical coiled-coil proteins; (2) monoclonal antibodies directed against paramyosin recognized antigens produced by the subcloning of random fragments from the larger λ clones into an expression vector; and (3) a partial sequence of a methionine peptide recognized by a monoclonal antibody matched the predicted amino acid sequence from an expression vector clone recognized by the same antibody.

Confirmation that this sequence corresponds to the *unc-15* genetic locus was obtained directly by showing that mutator-induced alleles of *unc-15* (P. Anderson and A. Besjovec, pers. comm.) contain a Tc1 insertion in the DNA hybridizing to the paramyosin clones (S. Rioux and R.H. Waterston, unpubl.). This insertion is absent in coisogenic revertants of the alleles.

Much of the paramyosin gene has now been sequenced (H. Kagawa, pers. comm.). As predicted, almost all of the conceptually translated polypeptide consists of sequences typical of α-helical coiled-coil proteins. The considerable homology found to the myosin rod portion raises the possibility of a common ancestor for the two genes. The paramyosin sequence, together with the already available sequences for the myosins, may be used to make specific predictions about thick-filament structure.

VIII. ACTIN GENES

Mutations altering the actin sequences in *C. elegans* have been identified using a combination of genetics and recombinant DNA studies. Actin, a universal component of actomyosin-based motility, is extremely conserved in structure throughout the animal and plant kingdoms. It is a globular molecule, with a monomer molecular weight of 42 kD and readily polymerizes to form a two-stranded helical filament (Fig. 3). Its role in contraction is to bind to and activate myosin under appropriate stimuli and to transmit the developed tension to adjoining structures. In muscle, actin interacts with (in addition to myosin) tropomyosin, troponin, α-actinin, and possibly other proteins involved in attaching the thin filament to the Z disk or its equivalent.

The DNA sequences encoding actin in *C. elegans* were initially identified using a heterologous probe from *Dictyostelium* (Files et al. 1983). With the recovery of genomic clones it was apparent that three of the four sequences, *act-1*, *act-2*, and *act-3*, are located in a 12-kb stretch of DNA, whereas the fourth gene is not closely associated with the others (Files et al. 1983; Fig. 17). Each of the genes has been sequenced (M. Wild and D. Hirsh, pers. comm.). *act-1* and *act-3* encode identical polypeptide chains, but *act-2* and *act-4* products differ slightly from these and from each other.

The actin cluster was located on the genetic map, using a closely linked restriction-fragment-length polymorphism (RFLP) detected between the Bristol and Bergerac strains (Files et al. 1983). This first use of the RFLP mapping approach in *C. elegans* placed the cluster between *unc-23* and *sma-1* on linkage group V. Suitable RFLPs could not be identified near *act-4*; this gene was subsequently mapped to the X chromosome cytogenetically (Albertson 1985).

Five semidominant alleles, each affecting thin-filament organization, had been mapped to the same region as the actin cluster (Waterston et al. 1984). Three of these mutations, *st15*, *st22*, and *st94*, are homozygous viable, and all three revert phenotypically to wild type at unusually high frequencies, either spontaneously (2×10^{-6}) or after ethylmethanesulfonate (EMS) mutagenesis (2×10^{-5}). One interpretation put forward to explain these genetic properties was that these mutations affected members of a gene family. In this case, the dominant mutations would represent missense mutations, resulting in a polypeptide of altered function that interferes with the product of other members of the family; the reversion events might represent gene knockout mutations, at least some of which could result from DNA rearrangements. Accordingly, 73 of the revertants were examined by Southern blot analysis for DNA rearrangements in the actin gene cluster (Landel et al. 1984). Four of these revertants exhibited altered restriction fragment sizes, which are most easily interpreted as

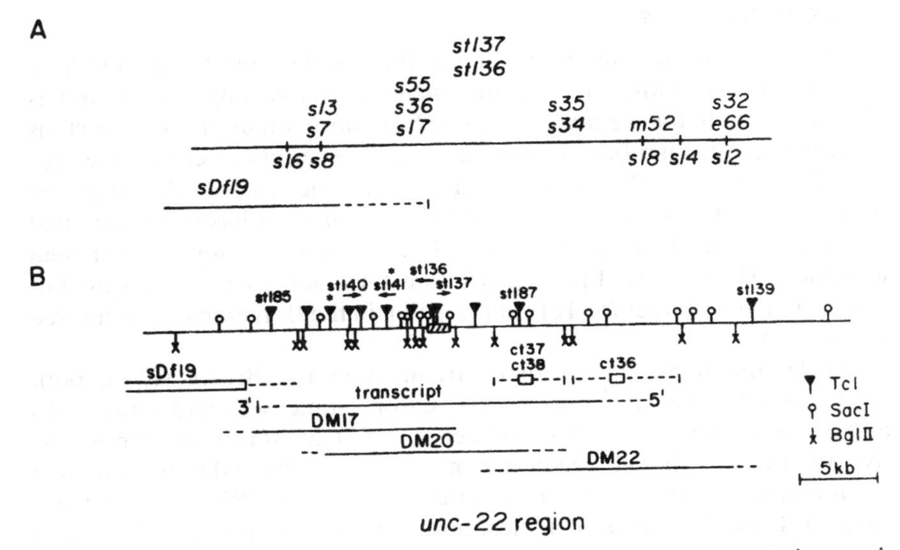

Figure 17 (*A*) Molecular genetics of *unc-22*. The fine-structure genetic map is shown, with the small deficiency *sDf19* (Rogalski et al. 1982) coming in from the left and ending within the gene. (*B*) The cloned region of *unc-22*, with restriction sites for several enzymes shown, along with sites of insertion of the transposable element Tc1, which interrupt *unc-22* function. One insertion, at the extreme right, produces a very weak Unc-22 phenotype. The position of the *sDf19* deficiency and the series of overlapping clones from the region are also shown. (Reprinted, with permission, from Moerman et al. 1988.)

rearrangements or deletions of the cluster. If the rearranged gene indeed represents the gene carrying the initial missense mutation, these results imply that *st15* lies in either *act-1* or *act-3*, *st22* lies in *act-3*, and *st94* lies in *act-1*. At least two of the events could have resulted from unequal pairing between members of the complex and subsequent crossing-over.

The loss of either *act-1* or *act-3* through the rearrangements does not appear to be deleterious (Waterston et al. 1984). The revertant animals move as well as wild type and have normal growth rates and brood sizes. Muscle structure is normal, as viewed by polarized light or electron microscopy, and with phalloidin staining in immunofluorescent microscopy, the I bands are indistinguishable from wild type, both in their organization and staining intensity under the conditions used. With gene-specific probes using Northern analysis, it has been possible to identify each of the specific mRNA products of the four genes (M. Krause and D. Hirsh, pers. comm.). Both *act-1* and *act-3* encode abundant mRNAs; the *act-2* product is less abundant.

IX. THE *UNC-22* LOCUS

The *unc-22* gene was among the first of the muscle genes to be identified: The distinctive "twitcher" mutant phenotype is readily scored and is almost unique to *unc-22* mutants. *unc-22* and *unc-54* mutations interact, as described above, and both forward and reverse selections can be used for *unc-22* mutants. For these reasons, the *unc-22* gene has been the subject of intensive genetic analysis. The *unc-22*-protein product has been described only recently (see below), and its vertebrate homolog, if any, has not been identified. The gene is of particular interest not only as a muscle gene but also as a favored target for Tc1 transposition in the Bergerac BO strain (see Chapter 3).

unc-22 mutations are unusually frequent in the Bristol strain, both spontaneously (6×10^{-7} per haploid chromosome set) and after EMS mutagenesis ($\sim 3 \times 10^{-3}$) (Besjovec et al. 1985). Most isolates carry recessive mutations, although two dominant alleles have been described among the hundreds recovered (D.G. Moerman et al.; D.L. Riddle; both pers. comm.). Even the recessive alleles, however, are conditionally dominant: They exhibit the mutant phenotype in solutions of acetylcholine agonists such as levamisole or 1% nicotine (Moerman and Baillie 1979). This conditional dominance allows detection of new mutations as heterozygotes; reversion can be detected in heterozygotes under standard conditions, where the wild-type allele is dominant.

A. Mutant Phenotype

The phenotype of *unc-22* mutants is unusual among *C. elegans* mutants: All alleles exhibit a pronounced, almost incessant, twitch of the body-wall muscles. The twitch occurs at a subcellular level, independently and simultaneously at multiple sites along the animal (Moerman and Baillie 1979; Waterston et al. 1980). In animals homozygous for weak alleles, such as *e105* and *s12*, the twitch is superimposed on near normal movement. With stronger alleles, including the amber allele *s32*, the adult animals are unable to propel themselves along the surface of the plate. Even the severely paralyzed animals still twitch, however, and this twitching is increased if the animals are prodded. The effects of various drugs on the twitching activity suggest that the twitch may represent a spontaneous relaxation or failure of contraction, rather than spontaneous contraction, as might be inferred from the appellation twitcher. Twitching is exacerbated by choline agonists (Moerman et al. 1982) and is attenuated by γ-aminobutyric acid (J. White, pers. comm.), a likely transmitter in inhibitory neurons of *C. elegans*.

The muscle structure of *unc-22* mutants is variable (Waterston et al.

1980). The *e105* and *s12* alleles result in nearly normal structure, despite the almost incessant twitch. In animals homozygous for the amber allele *s32*, the A bands as visualized by polarized light are severely disorganized, although the overall intensity of the birefringence is not as reduced as it is in thick-filament mutants (e.g., *unc-54*). Electron microscopy confirms that thick filaments are present in near normal numbers but are not restricted to ordered arrays. Other structural elements of the muscle are also present but lack the normal order.

B. Genetics

Genetic investigations of *unc-22* have included both construction of a fine-structure map and reversion analysis of several alleles (see above). The fine-structure map is detailed, positioning 17 alleles in six regions of the gene (Fig. 17; Moerman and Baillie 1979, 1981; Rogalski et al. 1982; Moerman and Waterston 1984). It was constructed in a manner similar to the *unc-54* map, with minor differences. As with *unc-54*, both recombinants and double recombinant events were recovered, and only those events arising by reciprocal recombination were used to construct the map. The distance between extreme alleles in the map is about 0.04 map units, similar to the *unc-54* distance.

C. Molecular Analysis of *unc-22*

The *unc-22* gene has also been used extensively in the genetic analysis of Tc1-related mutator activity found initially in *C. elegans* var. Bergerac (Moerman and Waterston 1984). It was the first gene shown to be a target of that mutator activity; and the high frequency of germ-line insertion and excision of Tc1 with *unc-22* combined with conditional dominance of *unc-22* mutations have made it a gene of choice for genetic analysis of the mutator (see Chapter 3).

The *unc-22* gene was cloned using Tc1 insertions as a tag for the gene. The copy of Tc1 responsible for the Unc-22 phenotype was identified by placing the *unc-22(Tc1)* allele in a low Tc1 copy number background and then comparing the restriction fragments hybridizing in its DNA with those of DNA from coisogenic revertants. The relevant fragment was recovered in a plasmid, and the flanking unique sequence DNA from the clone was used to confirm that the clone derived from the *unc-22* gene.

Genomic λ clones covering more than 30 kb were then recovered and tested against other Tc1-induced alleles, against diepoxybutane-induced alleles, and against *sDf19*, a small deficiency entering *unc-22* from the left and ending within it, as determined by fine-structure mapping (Rogalski et al. 1982; Moerman et al. 1985b; C. Trent and W.B. Wood; G.M. Benian et al.; both pers. comm.). The DNA rearrangements associated with these mutations suggest that the *unc-22* gene spans more than 20 kb, and analysis of the fragments produced by *sDf19* orients the physical map of *unc-22* on the genetic map, as shown in Figure 17.

Cloning of *unc-22* genomic sequences has permitted analysis of the *unc-22*-gene products. From RNA gel blots, *unc-22* DNA hybridizes to an mRNA species of about 14 kb, sufficient to encode a polypeptide of more than 500,000 daltons (G.M. Benian and D.G. Moerman, pers. comm.). The incorporation of Tc1 into this message in Tc1-induced mutants confirms that the mRNA is indeed the *unc-22* transcript (G.M. Benian and D.G. Moerman, pers. comm.). Single-stranded probes indicate the gene is transcribed from right to left on the conventionally drawn genetic map. A restriction fragment from the central region of the gene was placed in an expression vector, and the resulting β-galactosidase–*unc-22* fusion protein, containing more than 60,000 daltons of *unc-22*-encoded polypeptide, was used as an immunogen. The resulting antiserum, after absorption with β-galactosidase, reacts with a polypeptide of more than 500 kD on protein gel transfers (Moerman et al. 1988). The same antiserum, when used for indirect immunofluorescence on worm fragments, locates the *unc-22* protein in the A bands of the body-wall muscle (Moerman et al. 1988). The pattern seen is similar, though not always identical, to the staining observed with a monoclonal antibody to *unc-54* myosin.

These results place the *unc-22* product within the myofilament lattice in the A band. Its function there remains mysterious, but the constant twitch of *unc-22* mutant animals, especially in the presence of choline agonists, suggests that it may help regulate actomyosin interaction. The indirect suppression of the twitching phenotype by only the *unc-54* subfragment-1 mutations raises the possibility that the *unc-22* protein may interact directly with myosin. As yet, however, it is uncertain whether the *unc-22* polypeptide is a component of the thick filament or part of another A-band structure.

The analog of the *unc-22* product in vertebrate muscle, if any, is not known. The size of the *unc-22* protein makes it distinct from thick-filament proteins described previously. Large proteins have been identified in vertebrate muscle, such as titin and nebulin. Titin is estimated to have a monomer molecular weight of 1×10^6 to 2×10^6 and has been localized to the A–I junction and other positions in the sarcomere (Wang et al. 1979). Nebulin is a 400-kD subunit and is found in vertebrate muscle at the N2 line (Wang and Williamson 1980). The functions of these proteins are unknown, but there is speculation that both may be part of a third filament system, in addition to thick and thin filaments, involved in maintaining overall sarcomere architecture and acting perhaps as series elastic components.

X. CONCLUSION

The combination of genetic, morphological, and molecular studies of *C. elegans* muscle has led to the identification of the genetic elements encoding the principal components of the thick and thin filaments. Detailed studies of the altered forms of these proteins produced by the mutant alleles promise to provide insights into the roles of these components in assembly and contraction. The potential of these studies is illustrated by the studies of the *unc-54* gene, where the sequencing of individual mutations has shown that mutations in the thiol region and mutations near the proposed ATP-binding site can result in similar phenotypic changes, suggesting that both regions may be involved in the same function. The specific effects of these mutations on myosin function should be revealed through in vitro studies and will be easier to interpret as the three-dimensional structure of myosin becomes known.

The other muscle genes already identified and those that undoubtedly remain to be discovered may encode other components of the filaments such as myosin light chains or tropomyosin; or they may encode accessory components, such as those identified by monoclonal antibodies, that have important roles in aligning and maintaining the filaments in an ordered array and in transmitting the developed tension from the muscle to the cuticle. Alternatively, some genes may encode proteins that function only in the assembly of the muscle and are not, themselves, incorporated into the lattice. The next few years are likely to see a rapid increase in the number of muscle genes assigned to specific polypeptides. As a result, a detailed picture of the assembly of *C. elegans* muscle should become known.

Already, a general picture of the assembly process is emerging. In both the thick filament and the dense body, the existence of components restricted to specific regions suggests a sequential assembly process, with specialized roles for individual members of gene families. The myoA heavy chain, product of the *myo-3* gene, with paramyosin and/or a distinct core protein, probably nucleates thick-filament formation with subsequent elongation of the filament through the distal addition of myoB, the *unc-54* protein. The idea of a unique role for myoA in assembly is supported indirectly by observations on muscle growth. In young myocytes, just beginning to synthesize muscle-specific proteins (Gossett et al. 1982), myoA becomes localized to A bands before myoB (G.R. Francis and R.H. Waterston, unpubl.). The paucity of thick filaments and the lack of movement in animals homozygous for *myo-3* loss-of-function alleles now provide direct evidence for a unique role of myoA in thick-filament assembly. The converse conclusion seems sound—myoB is not required for initiating filament formation. With the overproduction of myoA resulting from *sup-3* mutations, the thick filaments in the *unc-54* null animals even form relatively organized, albeit somewhat narrow, A bands.

The nature of the switch that limits myoB to the ends of the thick filament is unclear. It cannot simply be at the level of regulation of synthesis, as both myoA and myoB are synthesized in the same cells throughout development at a relatively constant ratio (Schachat et al. 1977b). Perhaps, the observed taper of the filaments leads to a change in conformation that favors one protein over the other. Whether other proteins may be involved in implementing the changeover is unknown. The *unc-45* gene, where mutant alleles affect predominantly myoB assembly, might act at such a step.

The dense body may also be constructed in a stepwise manner with different components used for distinct phases of assembly. The three domains of the dense body are the transmembrane portion, the base of the dense body, and the extension of the dense body into the myofilament lattice (Fig. 9). The assembly of the transmembrane domain could be the initial event in dense-body formation, with p107b added to form the base, followed by p107a to form the bulk of the dense body proper.

Observations of growing muscle indirectly support such a sequence. In embryogenesis, the MH25 antibody, which presumptively recognizes a component of the transmembrane portion, detects its antigen anchored to the hypodermis before either p107b or p107a are detectable. Attachment plaques for half-I-bands are found at the ends of cells in all stages. They are composed only of the transmembrane portion and p107b. As cells grow, these attachment plaques might serve as sites of p107a addition to create a new row of bipolar dense bodies and thus add to the sarcomere number per cell. These observations are only associations of events and do not show cause–effect relationships, but as the genes for these components are identified the ideas presented become testable.

The next few years will undoubtedly see an increasing number of the presently known muscle-affecting genes cloned, either by Tc1 tagging or by walking from existing cloned loci via the physical map being developed by Coulson and Sulston (1986) (see Chapter 3). New genes will also be identified by continuing present approaches and by expanding the mutant searches to recover recessive lethal mutations that either result in complete paralysis such as *myo-3* loss-of-function alleles or alter pharyngeal as well as body-wall muscles. A reasonably complete catalog of the muscle genes will be available in the near future. And ideas about the role these genes play in muscle assembly and function will be formulated as is happening with *unc-22* from a consideration of the mutant phenotypes, from the protein location within the muscle cell, and from DNA sequence analysis. But more definitive tests will be required.

A major challenge to workers in this field will be to develop assays of function and to reconstruct the assembly process in vitro. The ability to express proteins from cloned genes, as is being done now for myosin, will be important. The mutants may provide important intermediates in the assembly process for in vitro studies. The apparent independent stepwise assembly of structures like the dense body and thick filament may simplify the task. Nonetheless, the task is formidable. But already most of the major components of muscle are known, and in *C. elegans* the genes for many of these are already identified. The approaches available for the study of *C. elegans* muscle are as strong as those available anywhere. General principles discovered in the study of muscle are likely to be applicable to other organelles.

ACKNOWLEDGMENTS

I would like to thank the many people who have helped bring this chapter into being: in particular the other contributors to this volume, especially John Sulston, John White, and Jonathan Hodgkin; members of my laboratory, especially Donald Moerman; Henry Epstein; David Baillie; and Jim Priess.

11

The Nervous System

Martin Chalfie
Department of Biological Sciences, Columbia University
New York, New York 10027

John White
MRC Laboratory of Molecular Biology
Cambridge CB2 2QH, England

The Nematode Caenorhabditis elegans. © 1988 Cold Spring Harbor Laboratory. 0-87969-307-X/88 $1 + .00.

I. INTRODUCTION

The nervous system is the most complex organ in *Caenorhabditis elegans*. In hermaphrodites the 302 neurons and 56 glial and associated support cells account for 37% of the somatic nuclei. In males the nervous system is more extensive; the 381 neurons and 92 glial and supporting cells comprise 46% of the somatic nuclei. The nervous system mediates a rich variety of behaviors, despite the small number of component neurons: Animals move forward and backward by propagating sinusoidal waves along their bodies; they perform exploratory movements with their heads as they feed; and they respond to a number of sensory stimuli, including mechanical stimulation, changes in the chemical environment, osmolarity, and temperature. Animals usually respond to changes in their environment with specific movements, but sensory cues are also important in regulating entry and escape from the dauer state (see Chapter 13), egg laying, and feeding. The differences in the number and types of neurons in hermaphrodites and males are reflected in sex-specific behaviors such as egg laying in hermaphrodites and the comparatively complex mating behavior of males.

In this chapter we discuss the structure and function of the *C. elegans* nervous system. Several different approaches have been used to investigate how this system functions. These methods include the use of reconstructions from electron micrographs of serial sections, histochemistry, and studies of the behavioral consequences of pharmacological agents, laser ablations, and genetic lesions.

Because of the small size of the animal, it is at present impossible to study the electrophysiological or biochemical properties of individual neurons. Such studies, however, can be made on the much larger, parasitic nematode *Ascaris suum*. Because many neurons in *Ascaris* and *C. elegans* have analogous morphologies and connectivities, it is likely that they have similar functional properties as well. Indeed, many of the inferences about neuronal function in *C. elegans* are supported by direct electrophysiological studies of *Ascaris* analogs.

II. STRUCTURE OF THE NERVOUS SYSTEM

The nervous system of the *C. elegans* hermaphrodite has been reconstructed in its entirety from electron micrographs of serial sections. A detailed knowledge of the structure (and, hence, connectivity) of a nervous system provides insights both into the functional aspects of the structure and into the developmental mechanisms that may be used in its generation.

The description of the hermaphrodite nervous system that is given in this chapter is based on published reconstructions of the pharynx (Albertson and Thomson 1976), of the sensilla in the head (Ward et al. 1975; Ware et al. 1975), of the tail ganglia (Hall 1977), of the ventral nerve cord (White et al. 1976), and of the complete nervous system (White et al. 1986). A partial reconstruction of the nervous system in the male tail, where most of the sexual differences are seen, has been described (Sulston et al. 1980).

A. General Features of the Nervous System

The *C. elegans* nervous system is made up of two almost independent units. Twenty of the cells are contained within the pharynx (Albertson and Thomson 1976); the remaining neurons are found throughout the body of the animal (Appendix 2). (A pair of cells, the RIP cells, connect the main nervous system to the pharyngeal nervous system.) Most of the cell bodies of these neurons are segregated into ganglia by the arrangement of the basement membrane lining the pseudocoelom. Cell bodies may be physically adjacent and yet be in different ganglia because they are separated by a pair of basement membranes (Chapter 4). Bundles of neuronal processes in *C. elegans* run in longitudinal or circumferential tracts adjacent to the hypodermis (Fig. 1). The major areas of neuropil are the circumpharyngeal nerve ring and the ventral and dorsal nerve cords. In many cases, there is a ridge of hypodermis adjacent to the process bundle (Fig. 2). Neuron cell bodies lie adjacent to the bundle into which their processes project. Anterior to the nerve ring, there are no obvious groupings of cell bodies within the neuropil that encircles the pharynx; this region is loosely referred to as the anterior ganglion. Posterior to the nerve ring, the basement membrane of the pseudocoelom splits the cell bodies adjacent to the nerve ring into four groups (see Fig. 6 in Chapter 4): a small dorsal ganglion, two lateral ganglia, and a ventral ganglion. The excretory duct separates the ventral ganglion from the more posterior retrovesicular ganglion. A row of cell bodies runs down the ventral midline from the retrovesicular ganglion to the preanal ganglion in the tail. There are three other ganglia in the tail: two laterally symmetric lumbar ganglia and a single, small dorsorectal ganglion. Finally, there are two small ganglia, often referred to as the posterior lateral ganglia, situated laterally in the posterior body, and some isolated lateral cells along the length of the body.

The anterior ganglion, the ventral ganglion, and the dorsorectal ganglion are structurally limited. The other ganglia are open in that there are no delimiting boundaries at one end. The retrovesicular ganglion is open and continuous with the ventral cord which, in turn, is open and continuous with the preanal ganglion. Similarly, the lateral ganglia are open and continuous with the lumbar ganglia. Thus, neuronal cell bodies are located in three main compartments, two lateral and one ventral.

There appear to be no functional correlates to the groupings of cells into particular ganglia. Neurons are often more analogous in structure and

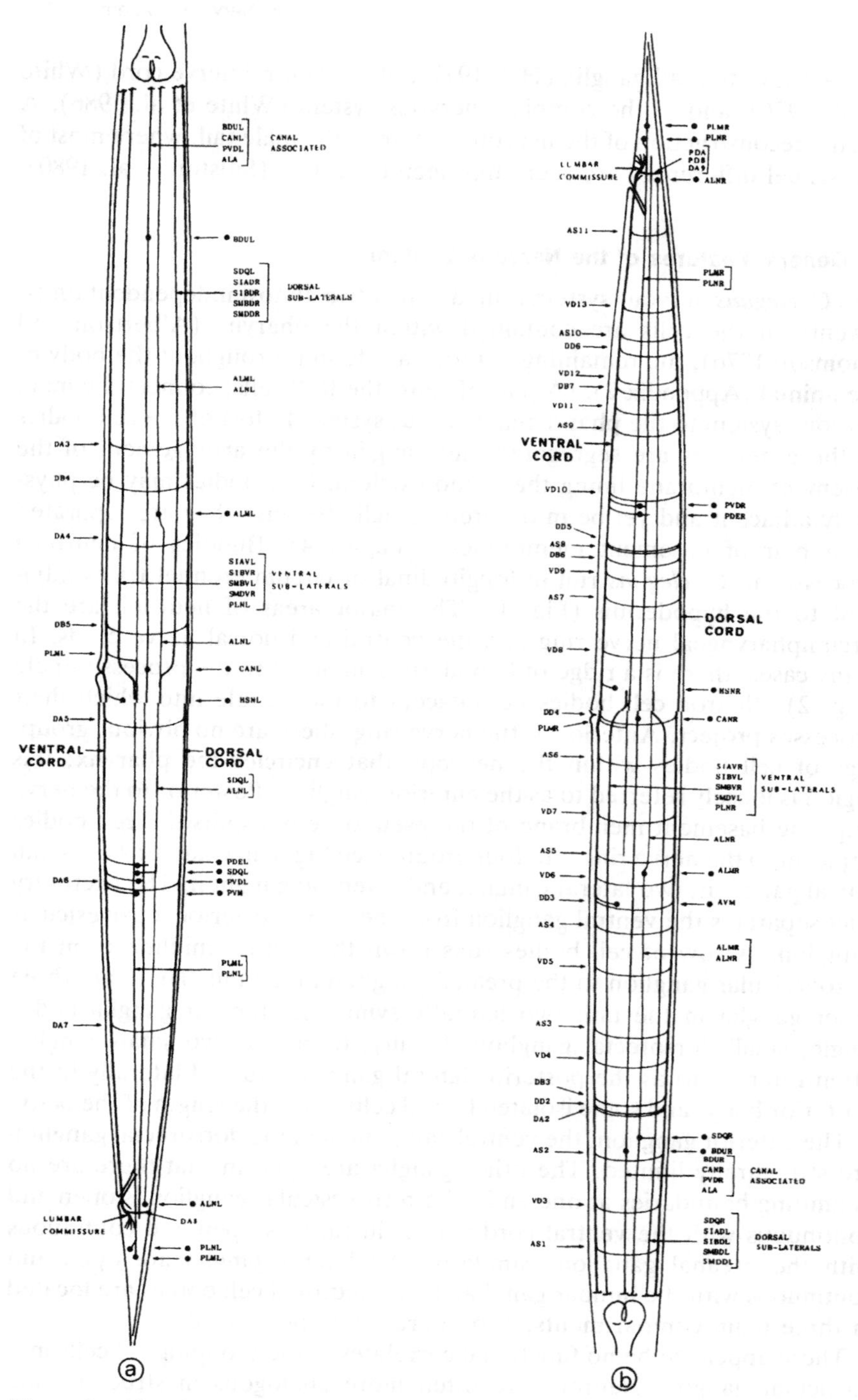

Figure 1 (*See facing page for part c and legend.*)

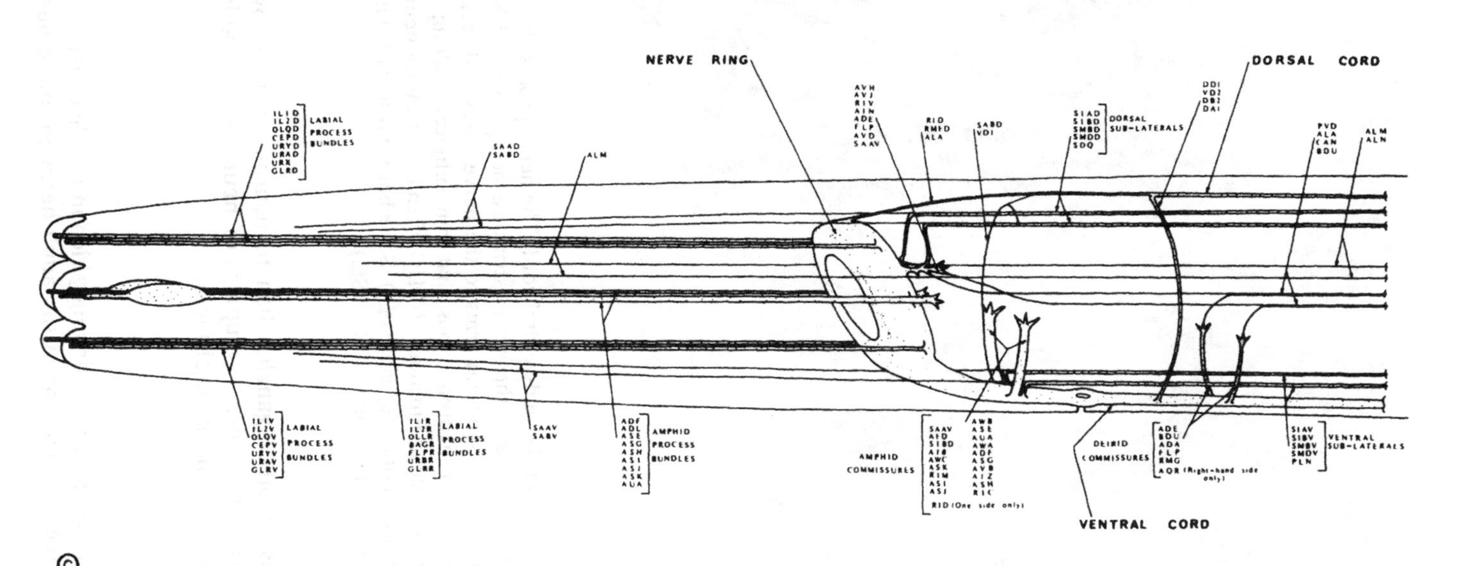

©

Figure 1 Nerve process tracts in the left (*a*) and right (*b*) body and in the head (*c*). Nerve processes generally run in bundles along with the processes of other nerves. These bundles are oriented longitudinally as cords or circumferentially as commissures. The nerve ring is the central region of neuropil in the animal. The ventral cord is the main process bundle that emanates from the nerve ring. It is made up of the processes of interneurons and motor neurons. Most of the processes that run in the dorsal cord are axons of motor neurons that originate in the ventral cord and enter the dorsal cord via commissures. There are four sublateral process bundles which contain the processes of interneurons and motor neurons that originate in the nerve ring. These run anteriorly and posteriorly from the nerve ring. (Reprinted, with permission, from White et al. 1986.)

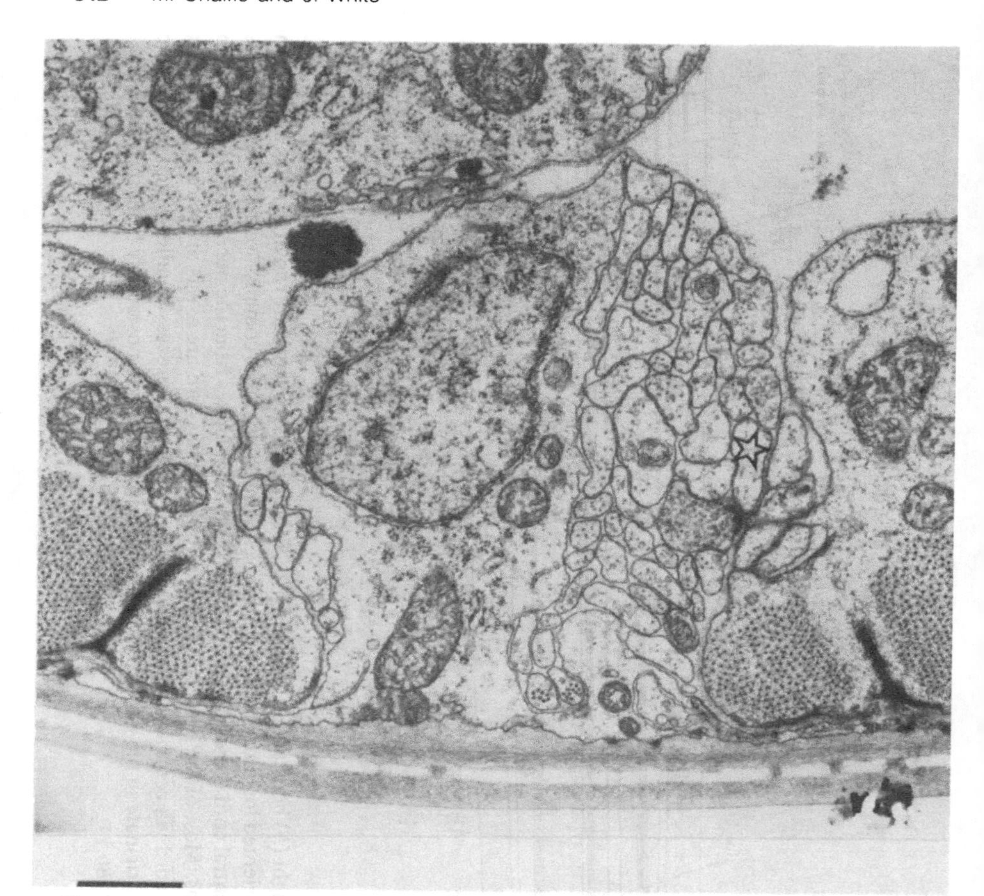

Figure 2 Electron micrograph of the mid-ventral cord. The nerve process bundle runs alongside a ridge of hypodermis, the whole structure being bounded by a basement membrane. Muscle cells send out arms to the nerve cords and receive their synaptic input from motor neuron axons that run subjacent, along the basement membrane. Neuromuscular junctions (NMJs) often have a motor neuron dendrite that dips into the region of the synapse (☆). Scale bar represents 1.0 μm. (Reprinted, with permission, from White et al. 1986.)

connectivity to neurons in other ganglia than to neurons in their own ganglia. Ganglia may simply be local groupings of neuronal cell bodies brought about by extraneous mechanical factors.

B. Neuronal Structure

The nervous system of *C. elegans* is surprisingly rich in the diversity of its neuron types. Each of the 302 neurons in the adult hermaphrodite has a unique combination of properties, such as morphology, connectivity and position, enabling it to be given a unique label (see Appendix 2). By grouping neurons that only differ from each other in position, White et al. (1986) assigned these cells to 118 classes. This may overestimate the number of types of neurons that are intrinsically different, as it is possible that intrinsically similar neurons in different environments have different patterns of connectivity and, hence, have been assigned to separate classes. Nevertheless, there are usually quite striking morphological or ultrastructural differences between classes, suggesting that most of the classes are, in fact, intrinsically different.

The strict classification of neurons into sensory receptors, interneurons, or motor neurons is not generally possible for the neurons of *C. elegans*, because in many cases individual neurons combine two or more of these functions. For example, the inner labial mechanoreceptors (IL1) combine all three functions, making neuromuscular junctions and also synapses onto other neurons (Ward et al. 1975).

1. Neuron Processes

The neurons of *C. elegans* have simple morphologies. Many neurons are monopolar with a single unbranched process; few neurons have more than two processes. Where branches do occur, they are usually very stereotyped and occur in well-defined locations. Neurons with a branched structure generally have the same pattern of branching in different animals. Most variations in branching patterns that have been seen are conservative in that processes end up in the same regions of neuropil (see, e.g., Albertson and Thomson 1976).

When processes of neurons of the same class encounter one another, depending on neuron class, they either run alongside each other until they end or terminate at the point of contact. Those that run alongside each other may or may not join with gap junctions; those that terminate on contact almost invariably join by a gap junction at the point of contact.

2. Sensory Receptor Endings

Sensilla. The head is richly endowed with receptors of various probable sensory modalities. Many are components of specialized sense organs—the sensilla. Each sensillum contains one or more ciliated nerve endings and two nonneuronal cells: a sheath cell and a socket cell (Fig. 3). Socket cells appear to act as interfacial hypodermal cells, connecting the sensillum to the hypodermis. Sheath cells are gliallike cells that envelop the endings of neurons. Although the function of the sheath cells is not known, they could

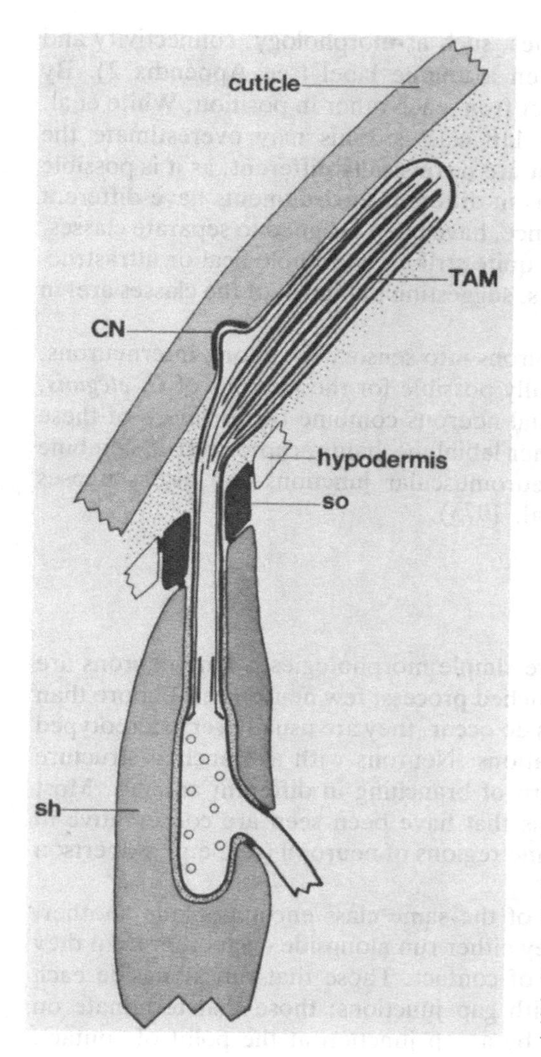

Figure 3 Schematic section through a cephalic sensillum. As in other sensilla, the cilium of the neuron passes through a channel formed by the sheath (sh) and socket (so) cells, thereby penetrating the hypodermis. The neuronal cilium penetrates the sheath cell so that the sheath cell is cylindrical in the region that envelops the cilium. The socket cell, on the other hand, usually has a process that wraps around the base of the sensillum and connects with itself by means of a tight junction. Belt desmosomes attach the neurite to the sheath cell, the sheath cell to the socket cell, and the socket cell to the hypodermis. The distal segment of this cilium is embedded in the subcuticle and contains supernumerary microtubules and darkly staining tubule-associated material (TAM). A small nubbin (CN) extends into the cuticle. Scale bar represents 1.0 μm. (Reprinted, with permission, from Perkins et al. 1986.)

regulate the external chemical environment of the cilia. Their inner surfaces, adjacent to the neural dendrite, are often invaginated extensively, and large numbers of secretorylike vesicles are usually present in the cytoplasm in these regions. Belt desmosomes couple the cilium to the sheath cell, the sheath cell to the socket cell and the socket cell to the hypodermis.

The axonemes of the sensory cilia of nematodes generally have nine doublet microtubules with a variable number of inner singlet microtubules. The A and B subfibers of the doublets have 13 and 11 protofilaments, respectively, as seen in doublets from other organisms. The inner singlet microtubules, however, are unusual in having only 11 protofilaments, a structure that is also found in most of the cytoplasmic microtubules in *C. elegans* (Chalfie and Thomson 1982). Mature sensory cilia do not have basal bodies such as those seen on motile cilia (Perkins et al. 1986). The microtubules of the cilium terminate proximally in a structure that is probably equivalent to the transition zone of motile cilia.

Certain classes of sensilla have a hole in the overlying cuticle, which gives the cilia of the sensillum access to the external environment. Such sensilla are generally considered to be chemosensory, although definitive proof of this is, in most cases, lacking. Among the sensilla of this type are the two amphid sensilla (Fig. 4) and the six inner labial sensilla in the head, and the two phasmid sensilla in the tail.

Some ciliated endings do not penetrate the cuticle and are, therefore, thought to mediate mechanosensation. Several classes of these sensilla have club-shaped specializations that underlie the cuticle distal to the cilium. These each appear to be anchored to the cuticle by means of a small darkly staining nubbin (Fig. 3). Sensilla with these features include the six outer labial sensilla, the four cephalic sensilla, the two deirid sensilla, and the two postdeirid sensilla. There is some ultrastructural variation between the cilia in the sensilla (Perkins et al. 1986), for example, four of the doublets of the axonemes of the quadrant outer labial sensilla (OLQ) extend through the cilium and are joined along their lengths by thick cross bridges to form a square. A small hub runs in the center of the square and is attached to the doublets by fine radial arms. In contrast, the distal regions of the lateral outer labial (OLL) cilium contain darkly staining, amorphous material surrounded by singlet microtubules, most of which are not derived from the cilium. The bases of the cilia are also different; the OLQ cilium has a striated rootlet, whereas the OLL does not. The significance of these structural features is not known.

Touch Receptors. The touch receptors (ALM, PLM, and AVM) are the only receptor neurons with a well-characterized sensory modality (Chalfie and Sulston 1981). The sensory processes are not organized as sensilla but extend along the length of the lateral lines and the ventral midline. Stimulation of the anterior receptors with light touch causes animals to back up, whereas similar stimulation of the posterior receptors provokes animals to move forward. The dendritic receptors lie close to the cuticle. Touch receptor neurons are unciliated but have prominent microtubules in their processes, containing 15 protofilaments instead of the more usual 11 protofilaments (Chalfie and Thomson 1982). These microtubules are discontinuous along the length of the process, having a mean length of approximately 20 μm in the adult (Chalfie and Thomson 1979). The end of the microtubule farthest from the cell body is always adjacent to the plasma membrane.

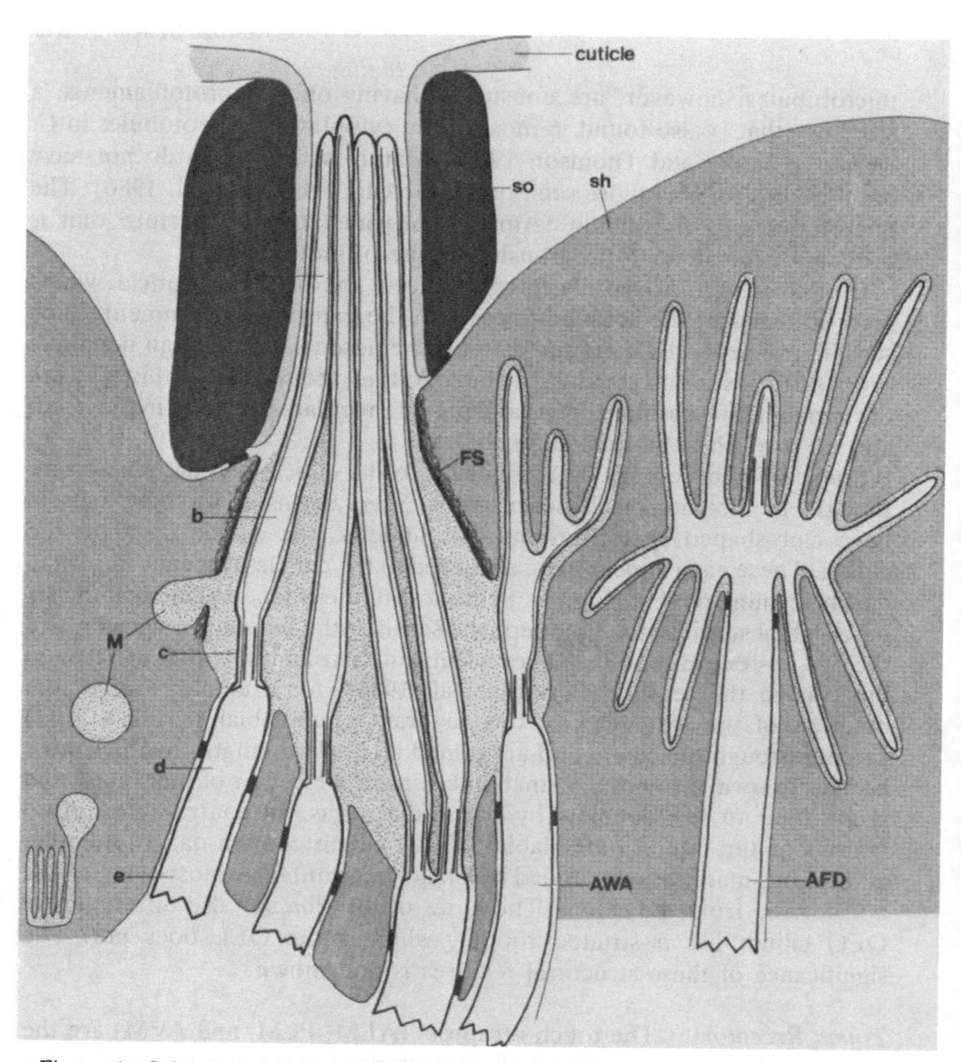

Figure 4 Schematic section through an amphid sensillum. The amphid channel is formed from a socket cell (so) and a sheath cell (sh). Belt desmosomes attach the neurites to the sheath cell, the sheath cell to the socket cell, and the socket cell to the hypodermis. The socket channel is lined with cuticle that is continuous with the external cuticle. The sheath channel is filled with a darkly staining matrix material (M), which appears to be made by the sheath cells. Six singly ciliated and two dually ciliated neurons enter the amphid channel (three are shown). Three neurons (AWA, AWB, and AWC) leave the main process bundle and have endings that are embedded in the sheath cell (only AWA is shown). An additional neuron (AFD) enters the sheath cell separately, where it has an extensively villated ending. The five insets show cross sections of a channel dendrite through the distal segment (*a*), the middle segment (*b*), the transition zone (*c*), the neuron/sheath junction (*d*), and the main dendrite (*e*). Scale bar represents 1.0 μm. (Reprinted, with permission, from Perkins et al. 1986.)

Other Possible Sensory Receptors. There are several other classes of neuron that are strong candidates for sensory receptors. The ciliated neuron AFD invaginates the amphid sheath, although it does not enter the amphid channel (Fig. 4b). It terminates in about 50 villi. AFD may function as a thermoreceptor (Perkins et al. 1986). The neuron classes BAG and FLP both have ciliated endings in the head; BAG also has a striated rootlet. Both classes have unusually shaped endings that are in the form of a bag and a flap, respectively; these endings wrap around extensions of the lateral inner labial socket cells. There are two additional ciliated neurons: (1) AQR, which has its cilium situated just behind the nerve ring; and (2) PQR, which has its cilium situated in the tail. The adult male has several additional classes of ciliated neuron (Chapter 9).

Several other neuron classes have features that suggest that they could be receptors, although they do not have cilia. URA, URB, URX, and URY all have processes that end in the tip of the head; those of URX and URY flatten out into thin sheets in this region. All four classes project into the nerve ring where they are predominantly presynaptic. Other cells, in the pharynx, have free endings that lie adjacent to the cuticle. Albertson and Thomson (1976) hypothesized that such ending may be important for mechanosensation. Several classes of neuron (PHC, PVR, AVG, ALN, and PLN) have long processes that run to the tip of the tail spike, suggesting that this organ could have a sensory function.

Certain classes of motor neuron (ventral cord motor neurons VAn, VBn, DAn, DBn, and ASn, and ring motor neurons SMB and SMD) have long lengths of process distal to the regions where neuromuscular junctions (NMJs) are made. There are few, if any, synaptic contacts made in these regions of process and so their function is not obvious. It has been suggested that in the case of the ventral cord motor neurons, these regions may function as stretch receptors. Such receptors could transduce body posture in one region into motor neuron activity in an adjacent region, thus enabling the propagation of waves of contraction along the body muscles (L. Byerly and R.L. Russell, pers. comm.).

3. Synapses

The pattern of synaptic contacts in wild-type animals is remarkably similar between individuals, enabling a canonical wiring diagram to be made of the connections between classes of neuron (Appendix 2). There is, however, a certain amount of quantitative and qualitative variability. The number of synapses occurring between classes varies, and sometimes synaptic contacts are seen between classes that do not usually synapse to each other. In many cases, the qualitative differences may be explained by the misplacement of a particular process into a new cellular environment (White et al. 1983).

Chemical Synapses. Chemical synapses occur en passant between adjacent processes. The number of synapses between cells varies from 1 to 19 but is typically around 5. The presynaptic process has a vesicle-filled varicosity and a specialized, darkly staining region in the membrane adjacent to the point of contact with the postsynaptic elements. There are generally no visible specializations in the postsynaptic membranes. There is a considerable variation in the physical size of chemical synapses. In some cases, particularly where processes are only adjacent for a short distance, there may be a single large synapse between the processes (Fig. 5a). Conversely, processes that are adjacent for long distances may make many small synapses (Fig. 5b). Usually, one cell is clearly presynaptic and the other is postsynaptic, but some processes synapse onto each other reciprocally.

Several distinct types of synaptic vesicles are seen in presynaptic regions. The most ubiquitous vesicles are spherical, with 35-nm diameters and lightly staining interiors. Some classes of neurons, notably the amphid receptors, have a second type of vesicle coexisting with vesicles of this type. These vesicles are larger (37–53 nm) and have darkly staining cores (Fig. 6). Interestingly, these larger vesicles often seem to be excluded from

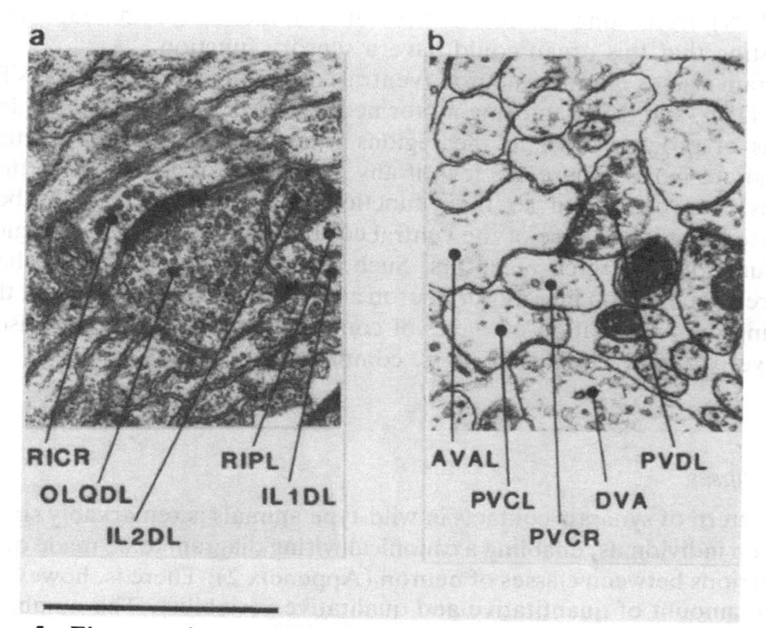

Figure 5 Electron micrographs of chemical synapses. Chemical synapses (*a*, *b*) in *C. elegans* are seen as concentrations of vesicles that are situated adjacent to a darkly staining presynaptic specialization. There is generally no visible postsynaptic specialization. Scale bar represents 1.0 μm. (Adapted from White et al. 1986.)

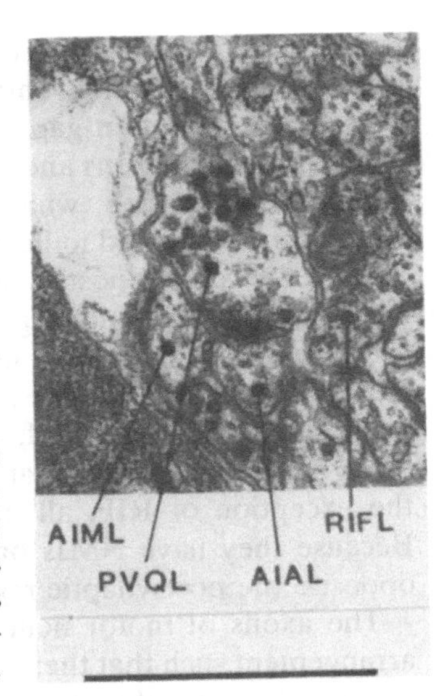

Figure 6 Several neuron classes have synapses that contain more than one type of vesicle. Typically, such neurons have clear vesicles of 36 nm diameter, coexisting with darkly staining vesicles of 37–53 nm. Usually, only the small clear vesicles are situated immediately adjacent to the presynaptic specialization. Scale bar represents 1.0 μm.

the region immediately adjacent to the presynaptic specialization where only small vesicles are present.

Because there are generally no visible postsynaptic specializations on neurons in *C. elegans*, there are often ambiguities in the identification of the postsynaptic partner. In some cases, there is clearly only one postsynaptic partner at a particular synapse, but in other cases there may be two or three, making a dyadic or triadic synapse. It is not always clear in these situations whether all the postsynaptic elements are functional (i.e., have the appropriate receptor) or whether some may simply be passive neighbors. However, there are often multiple instances of particular dyadic or triadic combinations, suggesting that all postsynaptic elements are functional in these cases.

Several classes of neuron have regions of process that are devoid of presynaptic specializations, even though appropriate postsynaptic partners are available. The ventral cord interneurons, AVA, AVB, AVD, and AVE show striking examples of this behavior in the nerve ring, where they have extensive processes that are exclusively postsynaptic. In the ventral cord these interneurons make many synaptic contacts, both to each other and to ventral cord motor neurons. There seems to be no obvious spatial restriction of postsynaptic contacts.

Neuromuscular Junctions. NMJs are special cases of chemical synapses, where at least one of the postsynaptic elements is muscle. Motor neurons do not send out axons that arborize around the target muscles, as in most other organisms. Instead, muscle cells have long neuronlike processes (muscle arms) that run to the process bundles in which motor neuron axons reside (Fig. 2). The body muscles are innervated by motor neurons with axons in the dorsal and ventral cords and on the inside surface of the nerve ring. There are often gap junctions between muscle arms.

Motor neuron axons and muscle arms generally run on either side of the basement membrane, which delimits the process bundle (Fig. 2). Thus, NMJs are constrained to lie in the plane of the basement membrane. There are usually several postsynaptic elements at a NMJ; several muscle arms crowd around and interdigitate at foci corresponding to the presynaptic specializations on a motor neuron axon. Certain classes of neuron (VDn, DDn, RMD, SMD, RME, and RIP) have dendritic processes that are postsynaptic at NMJs and, thus, form diadic synapses. These processes often have short spines that dip into the junctional region (Fig. 2). With the exception of RIP, all of these cells are motor neurons themselves. Because they have NMJs on the side of the animal that is diametrically opposite the postsynaptic regions, they could function as cross inhibitors.

The axons of motor neurons of the nerve ring have a highly ordered arrangement such that there is thus a fairly precise somatotopic mapping of the muscles onto the motor end-plate region (Fig. 7).

Gap Junctions. Gap junctions are seen as regions where the closely apposed membranes of two adjacent cells flatten and become more darkly stained. Gap junctions occur between neurons and between muscle cells. They are rarely, if ever, present between neurons and muscle cells, probably because of the basement membrane that separates the two. The gliallike processes of GLR (which form a sheet overlying the muscle arms on the inside of the nerve ring) are unusual in that they make gap junctions to both muscles and neurons.

Gap junctions frequently occur between adjacent members of a class of neuron if their processes are accessible to each other. Chemical synapses, on the other hand, are almost never seen between adjacent class members. In general, there are fewer gap junctions between cells (usually approximately one to three) than chemical synapses; however, a single gap junction has been demonstrated to be sufficient to mediate functional coupling between neurons (Chalfie et al. 1985).

The muscle arms that interdigitate on the inside of the nerve ring have a striking arrangement of gap junctions, suggesting that the arrangement of gap junctions is regulated (Fig. 8). There are gap junctions between arms from muscles in adjacent quadrants but not between arms from muscles in adjacent rows, even though both sets of arms are equally accessible. Muscles in the same quadrant are connected by gap junctions at their cell bodies, well away from the arms. The presence of a hypodermal ridge between quadrants prevents gap junctions being made between the bodies of muscle cells in different quadrants. Perhaps, when muscle arms grow into the nerve ring, they can discriminate arms from cells to which they are coupled from cells to which they are not coupled.

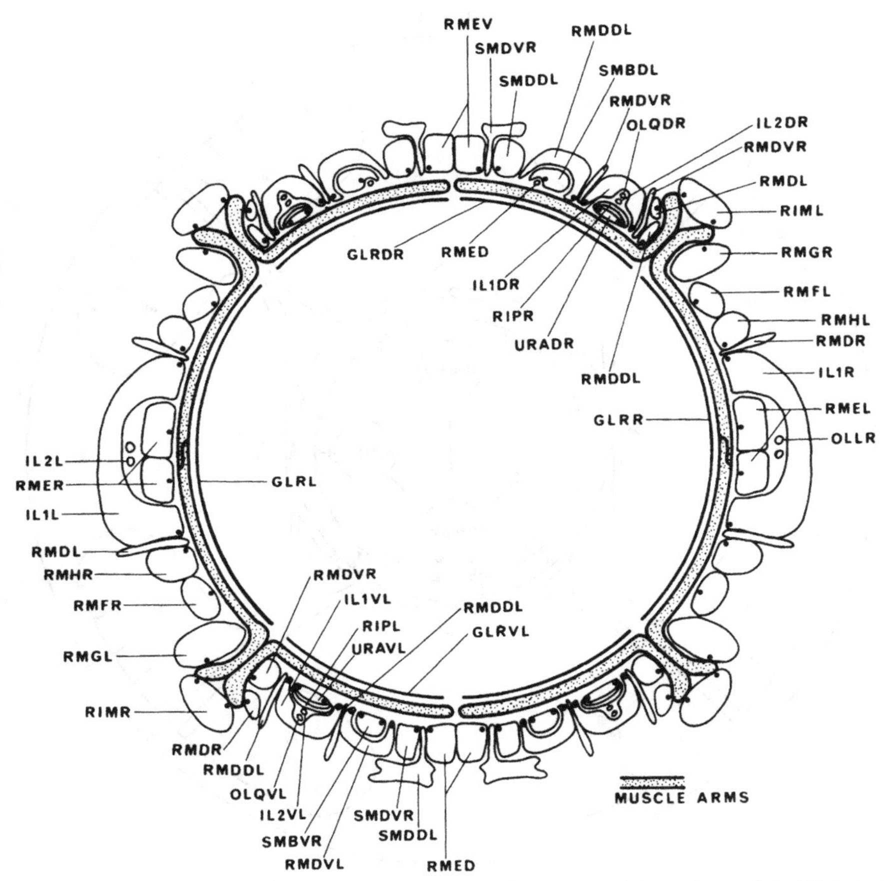

Figure 7 The arrangement of NMJs in the nerve ring. Muscles from the eight rows in the head have arms that project onto the inside surface of the nerve ring, preserving the map of the row positions. The arms are sandwiched between the sheetlike processes of the GLR cells on the inside and the motor neurons of the nerve ring on the outside. Four spurs of muscle arms penetrate into the anterior neuropil of the ring sublaterally where they receive synaptic input from RIM. The other classes of motor neuron form complex but well-defined structures adjacent to the inner surface of the nerve ring. Most NMJs are dyadic, with the dendrites of motor neurons or RIP being the corecipients. The black dots in the processes show the locations of the presynaptic specializations. (Reprinted, with permission, from White et al. 1986.)

C. Organization of Processes within Process Bundles

There is a high degree of spatial order in the bundles of processes that make up the nervous system of *C. elegans*. Some idea of the degree of order may be seen in the cross section of the anterior ventral cord shown in

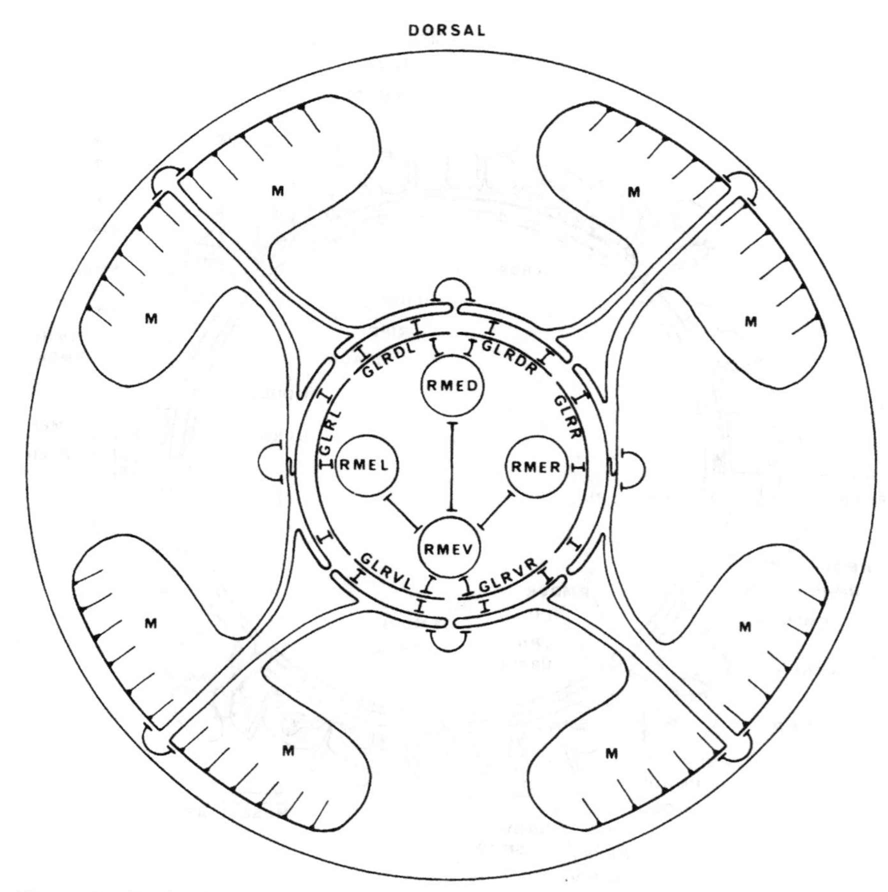

Figure 8 Projection of head muscles onto the nerve ring. The 32 muscles of the head send arms that run to the inside surface of the nerve ring. The arms are highly ordered in this region and map according to the positions of the contractile regions of the muscle cells. Gap junctions couple the arms of muscles to arms of muscles from adjacent quadrants and to GLR cells. RME motor neurons also make gap junctions to the GLR cells. Muscle cells in adjacent rows in the same quadrant are coupled by gap junctions situated on the main body of the cells. There are no gap junctions seen between the arms of these coupled cells, even though they interdigitate extensively. (Reprinted, with permission, from White et al. 1986.)

Figure 9. The cord is fairly bilaterally symmetric in this region; equivalent processes run in similar locations on each side. Much of the same degree of order is seen between comparable regions in different isogenic animals. Most process bundles are made up of processes that grow in more than one direction. Sometimes the direction of growth of a process is opposite to that of all its neighbors, but even in these cases the relative location of the process is maintained over long distances. Processes of neurons that develop postembryonically are often located within groups of embryonically derived processes. Thus, it seems as if processes can insinuate themselves in between preexisting processes within a bundle, so as to run in a specific location.

Laser ablation studies have identified several cases in embryonic development where specific fasciculation to an early "pioneer" neuron is necessary for correct guidance of certain neuronal processes (Durbin 1987). The process of AVG is situated on the right-hand side of the hypodermal ridge and is the first to grow along the length of the ventral cord. If this cell is removed prior to process outgrowth, there is an appreciable disorganization of the later developing processes in the cord. Many more processes than normal run along the left-hand side of the hypodermal ridge in these animals. Processes from PVPR and PVQL show a clear leader/follower relationship. These processes are normally closely associated and run on the left-hand side of the hypodermal ridge. If PVPR is removed, PVQL is misrouted and runs in the right-hand side of the ridge. However, if PVQL is removed, the process of PVPR is unaffected. Similar leader/follower relationships have been shown to occur during postembryonic development. Processes from the BDU cells are required for the appropriate positioning of the branches of a late developing touch receptor (AVM) in the nerve ring neuropil (W.W. Walthall and M. Chalfie, unpubl.).

1. Neighborhoods

The set of processes immediately adjacent to a given process has been defined as the neighborhood of that process (White et al. 1983). A measure of the neighborliness of two processes is the distance over which they are directly adjacent. Processes generally have close associations with one or a few of their neighbors, staying in direct adjacency with them over long distances. Other neighbors may have looser associations and drift in and out of direct adjacency (Table 1). It therefore appears that specific neighbors have a high adhesive affinity for a particular process and may have provided a favored tract along which the growing process selectively fasciculated.

The set of synaptic partners that a neuron may have clearly has to be a subset of its neighbors. The relative size of this subset is usually quite large (Table 1), that is, neurons make synaptic contacts with many of their neighbors (around half on average) and are therefore highly locally connected. There is a certain amount of circumstantial evidence suggesting that neurons are capable of forming synaptic contacts with classes of neuron that are not normally in their neighborhood (Chalfie et al. 1983; White et al. 1983). The confinement of processes into particular restricted neighborhoods is therefore a major determinant of connectivity.

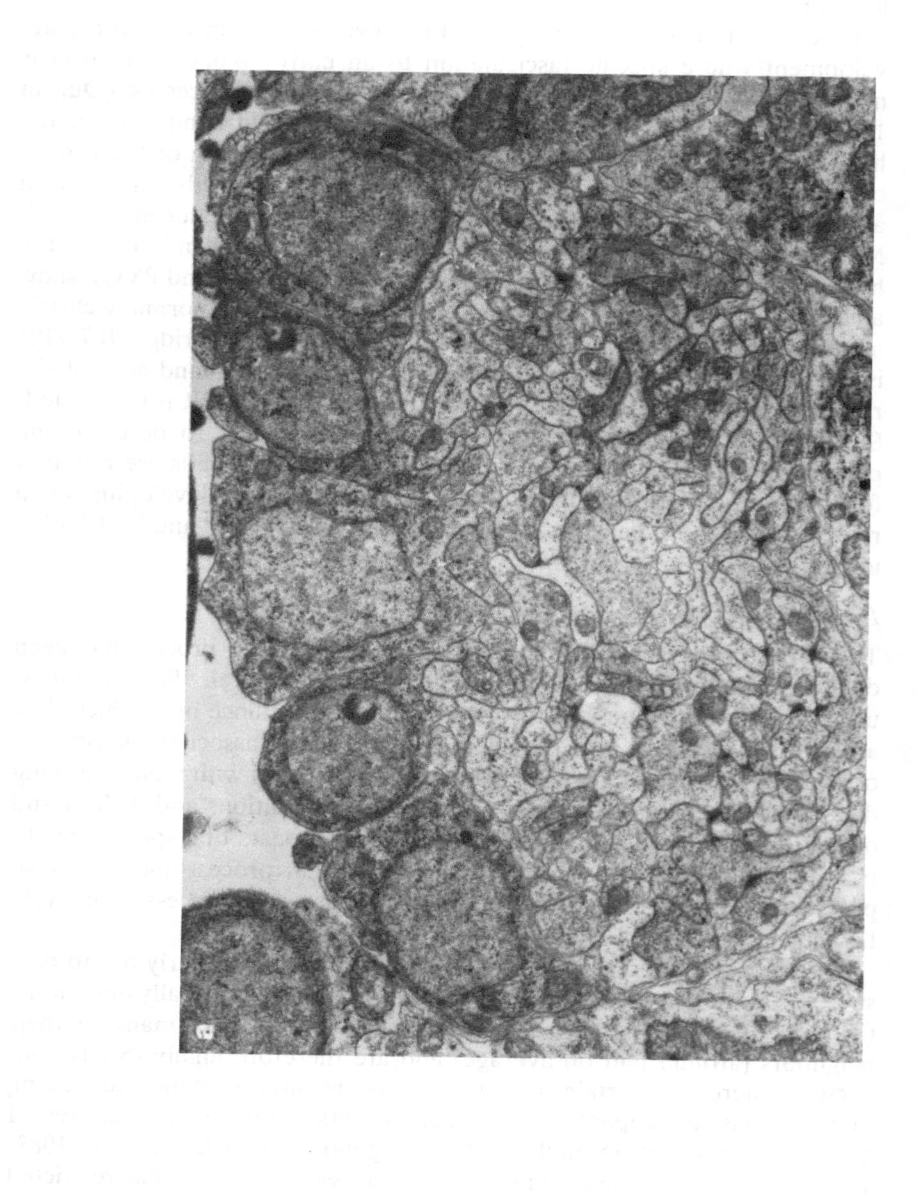

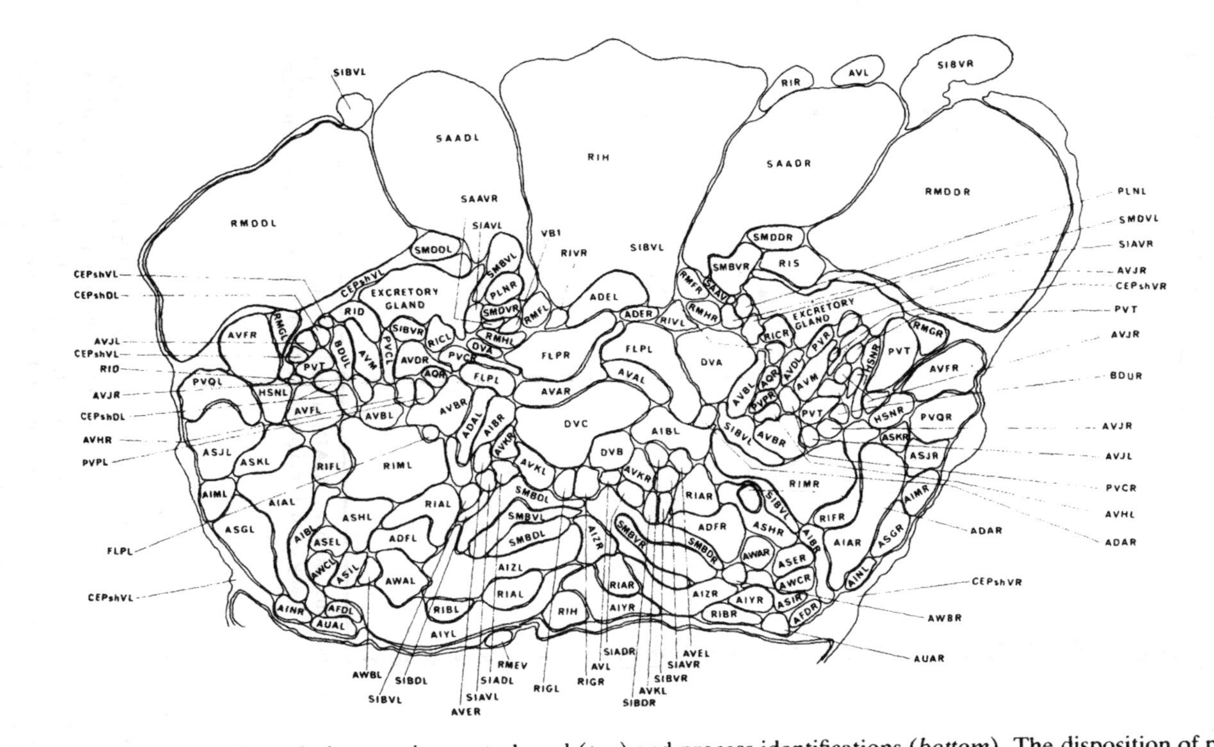

Figure 9 Transverse section through the anterior ventral cord (*top*) and process identifications (*bottom*). The disposition of processes within bundles is highly ordered in *C. elegans*. This can be readily seen by comparing the left- and right-hand sides of the nerve cord, which is fairly bilaterally symmetrical in this region. Processes run alongside a defined set of neighboring processes; this set is maintained over long distances. Some processes take on characteristic morphologies in certain regions. In this region, the processes of SMB flatten and sandwich each other, and AIY takes on a characteristic conformation with its postsynaptic partners, RIB, AIZ, and RIA. (Reprinted, with permission, from White et al. 1986.)

Table 1 Adjacencies and synaptic contacts made by the interneuron AIAR

Neighbor	Relative adjacency	Presynaptic	Postsynaptic	Coupled by gap junctions
AIMR	100		*	
AIBR	98	*		
ASGR	78		*	
ASKR	70		*	
AWCR	62	*	*	
RIFR	47	*		
PVQR	42		*	
ASHR	36	*	*	
ASIR	36		*	*
SMBVR	36			
AIZR	36		*	
ASJR	32			
AIAL	30			*
ASER	23	*	*	
AWCL	23		*	
AIYR	21			
ADLR	13		*	
ASEL	13		*	
ADFR	13			*
SMBDR	13			
SIAVR	13			
VD1	11			
RIS	11			
RMGR	8			
AINR	8			
AWAR	8			*
SIADR	8			
AVL	6			
AINL	4			

The processes of most neurons, such as AIAR, inhabit restricted neighborhoods, yet make synaptic connections with many of their neighbors. (Reprinted, with permission, from White et al. 1983.)

2. *Changes of Neighborhood*

Not all neurons have processes that run in a single neighborhood. Sometimes processes switch reproducibly from one neighborhood to another. This switching increases the number of potential synaptic partners that a neuron may have and also provides a way of linking spatially separated neighborhoods. Changes of neighborhood occur primarily at the transition between the pre- and postsynaptic regions of motor neurons and at mechanical discontinuities such as the junction of two process bundles.

Most motor neurons receive their synaptic input from interneurons that run in the interior of process bundles while their axons run on the surface, underneath the bounding basement membrane. The transition between the two neighborhoods is fairly gradual. This behavior could be a consequence of the presynaptic region of the process (which is spatially restricted), having an adhesive affinity for the basement membrane.

Many processes change neighborhoods abruptly at mechanical discontinuities. This occurs most commonly at the junction of one process bundle with another. For example, the processes of the two AIB cells (interneurons that integrate chemosensory information from the amphid chemoreceptors) pass through three distinct neighborhoods (Fig. 10).

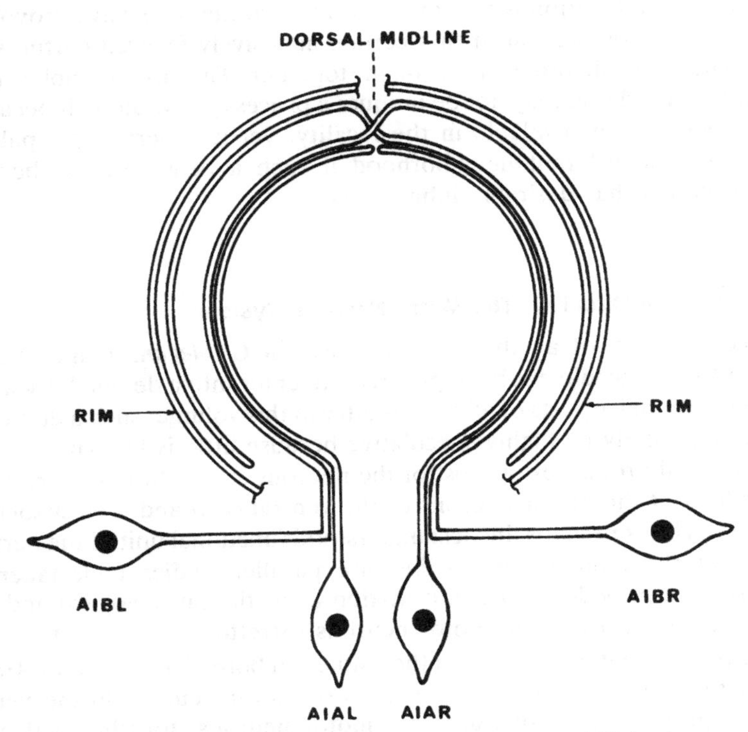

Figure 10 Relative dispositions of processes of AIA and AIB in the nerve ring. The cell bodies of AIA are located in the ventral ganglion, whereas those of AIB are in the lateral ganglia. Processes from the AIB neurons enter the ventral cord where they meet and become closely associated with the processes of AIA. These two sets of processes stay in close association as they turn around the nerve ring until the dorsal midline is reached. At this point, the processes of AIA terminate and the processes of AIB cross the process bundle, making an abrupt change of neighborhood. They then continue running around the ring but in close association with the process of RIM. (Modified from White et al. 1986.)

They first join the amphid commissures that run into the ventral cord. When they reach the cord, they become closely associated with the processes of AIA (another class of amphid interneuron with cell bodies in the ventral ganglion). The processes of AIA and AIB remain in close association as they course into and around the nerve ring to the dorsal midline. At this point, the processes of each AIA neuron meet and terminate. The processes of AIB do not terminate but abruptly turn and run across the process bundle for a short distance, before turning again and continuing around the ring in a new neighborhood, now in close association with the processes of RIM.

Abrupt transitions of neighborhood could be due to changes in adhesive affinity, as has been suggested for the case of motor neuron axons. An alternative explanation is that mechanical discontinuities have provoked a remixing of processes, such that certain adhesively favored partners may be separated while others are brought together. This notion implies that a given neighborhood may not be unique; a process grows along it because it is the best that is available in the locality, yet it is perfectly capable of growing in a different neighborhood if such a neighborhood becomes accessible and has desirable inhabitants.

D. Analysis of Circuitry: The Motor Nervous System

The connectivity of all the neuron classes in *C. elegans* is indicated in Appendix 2. It would not be appropriate to enter into a detailed discussion of the functional aspects of all the circuitry in this volume; such a discussion would necessarily be highly speculative because little is known about the physiological properties of most of the neurons. We will, however, take a closer look at the motor neurons of the ventral cord and their associated circuitry. This is perhaps the best understood functional unit of the nervous system at the moment, partly because parallel studies undertaken on *Ascaris* have provided some information as to the physiological and biochemical properties of the motor neurons (Stretton et al. 1985).

The picture that emerges from the studies on both *C. elegans* and *Ascaris* is that there are several distinct classes of motor neurons in the ventral cord. Separate classes of excitatory motor neurons, together with their associated interneurons, are used for forward and backward locomotion. The inhibitory classes of motor neuron are used for coordinating muscle activation between the dorsal and ventral halves of the animal.

1. Components of the Motor Circuitry

C. elegans normally lies on its side and moves over a surface by propagating dorsal/ventral flexures along its body. It moves forward by propagating backward-directed waves and moves backward by propagating forward-directed waves. The propagating waves are generated by the coordinated activation of the ventral and dorsal body muscles. Ventral body muscles are innervated by motor neurons with cell bodies and axons in the ventral cord (VAn, VBn, VDn, VCn), whereas dorsal muscles are innervated by motor neurons with cell bodies in the ventral cord and axons in the dorsal cord (DAn, DBn, DDn, ASn).

All the VAn and DAn neurons receive the same pattern of synaptic input from cord interneurons (Fig. 11) and both have anteriorly directed axons,

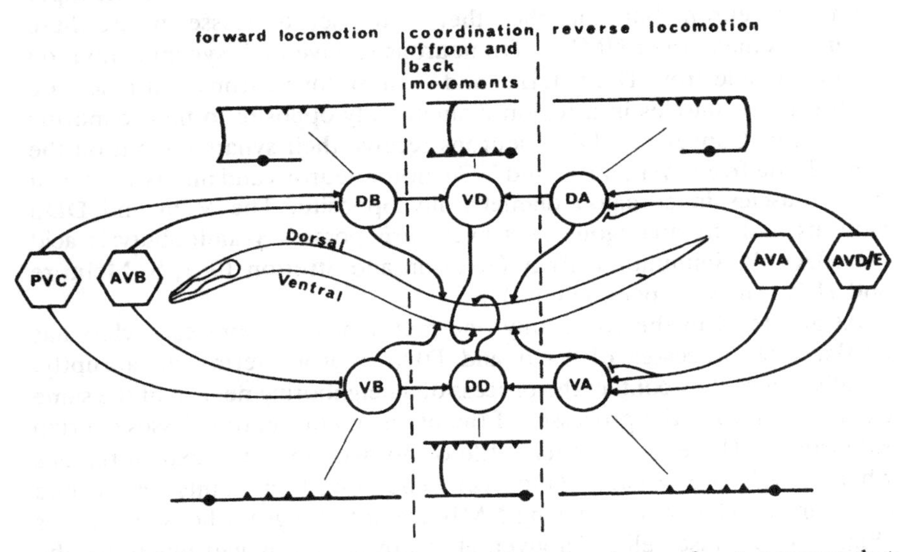

Figure 11 Ventral cord circuitry. There are six major classes of motor neuron that innervate the body musculature (circles), three innervate dorsal muscles (DB, DD, and DA), and three innervate ventral muscles (VB, VD, and DB). Class members are evenly distributed along the length of the ventral cord (Fig. 12). Forward movement is mediated by the activation of the DB and VB motor neurons via their associated interneurons AVB and PVC (hexagons). Both these motor neuron classes have posteriorly direct axons. That of VB runs in the ventral cord, whereas that of DB leaves its ventrally located cell body, runs around the animal as a commissure, and then runs in the dorsal cord. Backward movement is mediated by the activation of the DA and VA motor neurons by their associated interneurons AVA, AVD, and AVE. The disposition of the processes of DA and VA motor neurons is similar to those of DB and VB, except that they have anteriorly directed axons. The DA, VA, DB, and VB motor neurons are excitatory and probably contain the neurotransmitter acetylcholine. The DD and VD motor neurons are inhibitory and contain the neurotransmitter GABA. These latter classes do not receive their synaptic input from interneurons but rather from the other motor neuron classes. They synapse onto muscles situated on the opposite side of those that are innervated by their presynaptic partners and, thus, function as reciprocal inhibitors and are necessary for the coordination of front and back movements.

that of VAn in the ventral cord and that of DAn in the dorsal cord. ASn neurons are similar in morphology to DAn motor neurons and have a similar synaptic input, except for an additional class of connection. VBn and DBn neurons have similar morphologies to VAn and DAn, respectively, except that they have reversed polarity, with posteriorly directed axons. They also have their own characteristic pattern of innervation (Fig. 11). Studies of analogous neurons in *Ascaris* (see Sections II.D.3 and IV.A.1) suggest that all of these motor neurons make excitatory, cholinergic NMJs.

The VDn and DDn motor neurons do not receive their synaptic input from interneurons but from the other motor neuron classes where these latter cells make their NMJs. VDn neurons receive their synaptic input on the dorsal side from DAn, DBn, and ASn motor neurons and innervate ventral body muscles in a region diametrically opposite to their dendritic input field. Conversely, DDn neurons receive their synaptic input on the ventral side from VAn, VBn, and VCn motor neurons and innervate dorsal body muscles in a region diametrically opposite. The VDn and DDn neurons and the analogous cells in *Ascaris* possess γ-aminobutyric acid (GABA)-like immunoreactivity (Johnson and Stretton 1987; S. McIntire and H.R. Horvitz, pers. comm.).

At any level in the cord, only one motor neuron from each class has NMJs. The processes of VDn and DDn neurons terminate abruptly, usually where they contact the process of a neighboring neuron of the same class. In contrast, the processes of the other motor neuron classes overlap extensively. There is, however, little or no overlap in the axonal regions where NMJs are situated (Fig. 12). The transition points where one member of a class begins having NMJs are not in register between classes (Fig. 12). Because cells of a given class are joined by gap junctions, the territories where NMJs are made may be parceled out between the members of a class by intraclass competition.

The VCn neurons are specific to the hermaphrodite, their main function being to innervate the vulval muscles, although they also innervate ventral body muscles. Unusually, they have no obvious source of synaptic input (White et al. 1986). The pattern of VCn synapses suggests a role in egg laying (although animals missing the VCn cells can lay eggs). This hypothesized sex-specific function is further strengthened by the finding that these cells are replaced in males by the CAn and CPn motor neurons (Sulston et al. 1980).

Five pairs of interneurons (AVA, AVB, PVC, AVD, and AVE [Fig. 11]) synapse onto the ventral cord motor neurons in the adult (White et al. 1976, 1986). The AVE cells differ from the others in that they extend through only the first half of the ventral cord; their synaptic output is similar to the AVD cells. The AVA and AVB are large-diameter interneurons; the other three have smaller diameters.

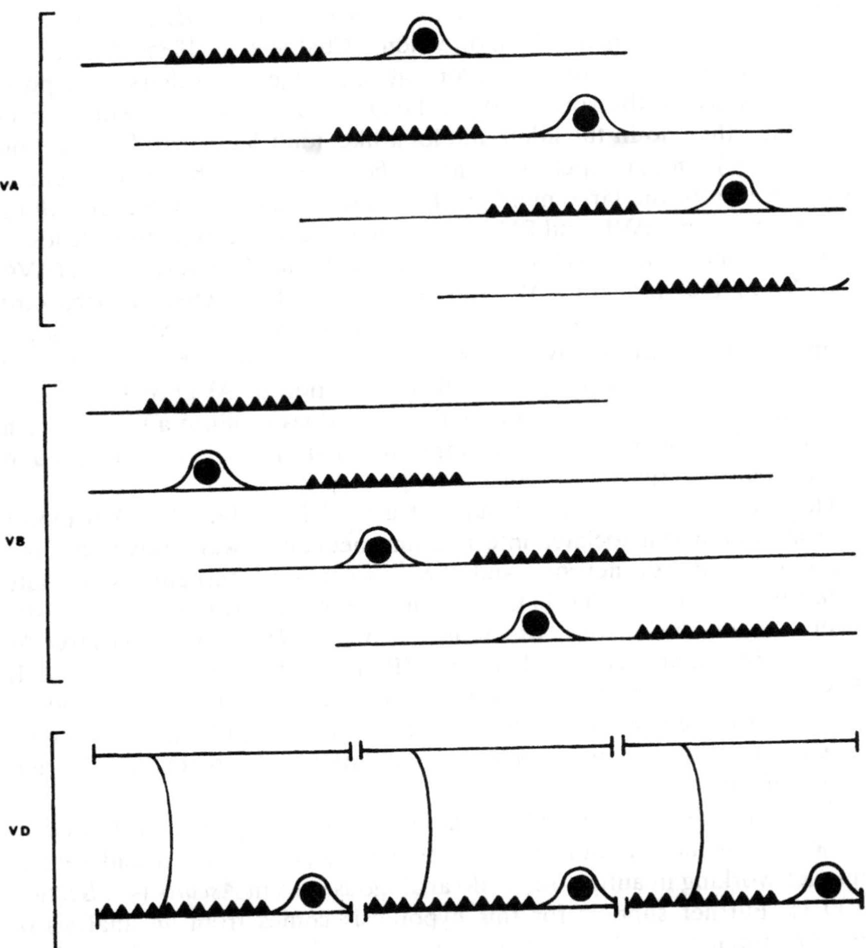

Figure 12 Schematic diagram showing the distribution of motor neuron classes along the length of the cord. Class members have been separated vertically for clarity. Processes of all motor neuron classes overlap, except DD and VD; however, there is little or no overlap in the regions of axon that are active in giving NMJs. Thus, at any given region along the cord there are representatives of one of each of the motor neuron classes actively giving NMJs. The transition points between adjacent class members of a particular class are not exactly in register with those of other classes.

2. *Function*

Laser ablation of motor neurons in the mid-cord regions of L1 larvae reveals that DBn neurons are required for normal movement forward,

DAn neurons for normal movement backward, and DDn neurons for coordinated movements in both directions (Chalfie et al. 1985). (These are the only ventral cord motor neurons at this stage; the others arise post-embryonically. In the young larvae, the DDn cells make different connections than they do in the adult and look like adult VDn cells [see Section III, below].) Similar experiments are difficult to do on the mature ventral cord because of the large number of ablations required. However, killing the interneurons (AVB and PVC) that innervate VB/DB neurons leads to defective forward movements in adults and ablating the interneurons (AVA and AVD) that innervate VA/DA neurons lead to defective backward movements (Chalfie et al. 1985). (Removal of AVA or AVB, the large-diameter interneurons, gives an obvious motor defect that is enhanced when the other interneurons are killed. Ablation of AVD or PVC alone does not detectably affect motor activity but does produce a loss of touch sensitivity [see Section V.E]. The latter interneurons may, thus, be used to modify motor activity.)

These observations strongly support a model in which DB/VB motor neurons and their associated interneurons mediate forward movement and the DA/VA motor neurons and their associated interneurons mediate backward movement (Fig. 11). All of these motor neurons have long distal extensions on their axons, which are devoid of NMJs; those on DA/VA axons point anteriorly and those on DB/VB axons point posteriorly. If these regions act as stretch receptors (see Section III.E), the opposite polarity of these classes may be important for wave propagation because activity of a motor neuron could be modulated by stretching an adjacent region of the body.

The morphology of the VD/DD motor neurons suggests that they could act as reciprocal inhibitors that function to keep the dorsal and ventral muscles working in antiphase, as do analogous cells in *Ascaris* (see Section II.D.3). Further support for this hypothesis comes from an analysis of *unc-30* mutants (J. White et al., unpubl.). The ventral cords in these animals are normal, except that the VDn and DDn neurons do not contain GABA and have very disorganized processes that make few, if any, NMJs. The mutants propagate forward and reverse waves of movement, but do so poorly. The most telling phenotype, however, is that the mutants shorten transiently before moving backward when tapped on the head, suggesting that both dorsal and ventral muscles are activated simultaneously.

3. *Electrophysiological Analysis of Behavior in* Ascaris

The ventral and dorsal nerve cords of *Ascaris* have been reconstructed and exhibit considerable similarities with the nerve cords of *C. elegans* (Stretton et al. 1978). There are seven classes of motor neurons that, by their structure, are close equivalents of the VAn, DAn, VBn, ASn, VDn, and DDn neurons. By directly stimulating commissures, Walrond et al. (1985)

found that the equivalents to DAn, DBn, and ASn motor neurons are excitatory and that the equivalents to the DDn and VDn motor neurons are inhibitory. The electrophysiological activity of the VA- or VB-like cells cannot be assessed directly. However, because of their many structural similarities to the DA-, DB-, and AS-like cells, it is believed that both cell types are excitatory.

Intracellular recording from the commissures has permitted an analysis of the electrical properties of individual neurons and of the synapses between neurons. Nerve processes have membrane resistances (60–300 $k\Omega cm^2$) that could permit signal transmission without classical all-or-none action potentials. Interestingly, although muscle cells exhibit action potentials, none have been observed in *Ascaris* motor neurons, nor has it been possible to evoke them. Moreover, although a strong depolarizing current produces a small, graded onset transient that is blocked by Co^{2+} and is thus likely to be Ca^{2+} current, no voltage-sensitive K^+ channels have been found. These findings suggest that signaling in the motor neurons is determined almost entirely by the passive membrane properties of the fibers (Davis and Stretton 1982; Stretton et al. 1985). Thus, signals in the motor neurons are graded, as is neuromuscular transmission for both excitatory and inhibitory motor neurons.

The VDn and DDn equivalents exhibit endogenous rhythmic electrical activity (Davis and Stretton 1981) that can be turned on or off by current injection or by synaptic input from excitatory motor neurons (Angstadt and Stretton 1985). The dorsal and ventral inhibitors oscillate in antiphase, as would be expected from their synaptic connections. Presumably, these oscillations are involved in the motor program that controls the periodic contractions of body muscle during locomotion.

III. DEVELOPMENT OF THE NERVOUS SYSTEM

A. Cell Lineages

Most of the neurons and their supporting cells arise from the AB blastomere, with additional cells arising from C and MS (see Chapter 5). There are three general periods of cell production. During the first half of embryogenesis, 222 cells (224 in the male) arise and provide the major components of the nervous system (Sulston et al. 1983). The second period of cell production occurs at the end of the L1 stage when most of the motor neurons of the adult ventral cord are generated (Sulston 1976; Sulston and Horvitz 1977). The accommodation of these new cells into an already functioning motor circuit is accompanied by the modification of some of the juvenile synapses (see below). A few other neurons arise from lineages immediately before (Q, G1, G2, H2), during (T), or after (V5.paa) the production of the ventral cord cells. The final period of cell production, during the L3 stage, is found only in the male and adds neurons and other cells that are required for male mating (Sulston et al. 1980).

Almost all bilaterally symmetric pairs of neurons arise from bilaterally symmetric cell lineages, but when a cell class has more than two members, for example, the motor neurons of the ventral cord, a variety of strategies can be used to generate the cells (Fig. 13; Sulston and Horvitz 1977; Sulston et al. 1980). Of the eight classes of motor neurons in the ventral cord, three (DAn, DBn, and DDn) arise during embryogenesis; the remaining five classes (VAn, VBn, VCn, ASn, VDn) arise postembryonically from 13 precursors, W and the Pna cells. W and Pna cells give rise to the

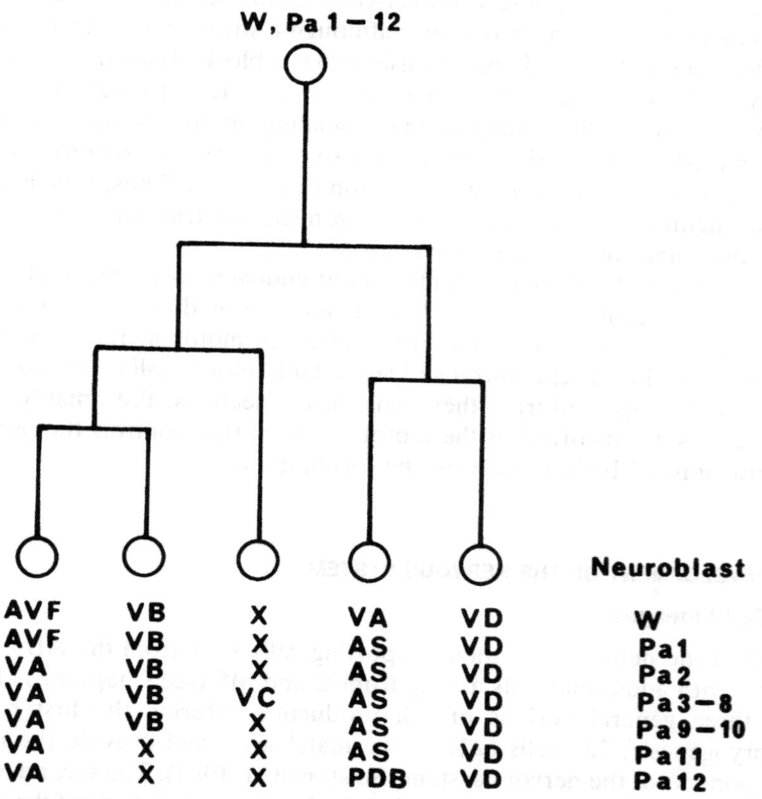

Figure 13 Cell lineages and fates of the postembryonically derived motor neurons. The late developing motor neurons are derived from a set of 13 neuroblasts (W and Pa1-12) that each produce five cells via an identical lineage. In general, the neuron classes are produced in stereotyped positions on these lineages, but there are exceptions such as the substitution of an AVF interneuron in the normal location of a VA in two cases and the substitution of a cell death (×) in the normal locations of VB (two cases) and VC (seven cases).

same pattern of cell divisions, and, for the most part, cells arising at particular positions in the pattern differentiate into the same type of neuron. Thus, the most posterior cell (pp) always becomes a VD motor neuron. However, the pattern is not perfect (Fig. 13). Exceptions include (1) the programmed cell death in some lineages (2) the differentiation of the aaa cell into an AVF interneuron in the W and Pla lineages and into a VA cell in the remaining lineages, and (3) the differentiation of the pa cell into a VA cell in the W lineage and into an AS cell in all of the other lineages.

The embryonic lineages giving rise to the DAn, DBn, and DDn cells show different patterns. The DDn cells arise in an almost clonal fashion from a pair of bilaterally symmetric precursors, but the DAn and the DBn motor neurons are not generated from characteristic patterns of cell division (although the DAn cells are either sister cells or the more posterior progeny of a division). Thus, a strict pattern of cell division need not be required to generate all types of cells. Some of the differences could, but need not, be a consequence of cell–cell interactions in the differentiation of the cells. Although cell interactions have been shown to play a relatively restricted role in the determination of cell division patterns (see Chapter 5), the extent of such interactions in cell differentiation is not yet known.

A number of mutations affect the lineages that give rise to the nervous system. Most of these affect the production of nonneuronal, as well as neuronal, cells and are discussed in Chapter 6. Mutations in a few genes appear to affect selectively neuronal lineages. Mutations in the gene *lin-32* affect the production and placement of a number of neurons (C. Kenyon and E. Hedgecock, pers. comm.). An examination of postembryonic lineages in these mutants indicates that the lateral neuroblasts are not generated; the cells remain hypodermal cells instead. Only two mutant alleles of this gene have been identified, and the null phenotype of the gene is not known. Nonetheless, these mutants appear to have the reverse phenotype of the neurogenic mutants of *Drosophila* (see, e.g., Campos-Ortega 1985). These *Drosophila* mutants, for example, *Notch*, appear to make additional nerve cells at the expense of hypoderm. A similar defect in the lineage of the T neuroblast, that is, no neuronal cells are made, is seen in *lin-17* mutants (E. Hedgecock, pers. comm.). In contrast, *lin-26* mutants have a phenotype resembling that of *Notch*. Lineages that normally produce ventral hypodermal cells give rise to neuronal cells instead (Sternberg and Horvitz 1984).

A single mutation in *lin-22* causes the repeated production of certain postembryonic sublineages in equivalent positions along the length of the body (Horvitz et al. 1982b; Sternberg and Horvitz 1984). These neuronal lineages are made at the expense of hypodermal cells, so this gene is superficially akin to the *Drosophila* neurogenic genes. However, it is more likely that the *lin-22* gene is important in directing where the sublineages are normally produced. As with the *lin-32* gene, more should be known when the null phenotype of the gene has been identified.

Mutations in *unc-86* cause reiterations within certain neuronal lineages and thus, cause some neurons to be present in multiple copies and others to be absent (Chalfie et al. 1981; Sulston et al. 1983).

It is intriguing that very few genes have been identified that can be mutated to cause lineage defects restricted to the nervous system. It is too early to know whether this sparsity reflects the relatively few mutant screens that have been done to identify animals with these defects or a real lack of this type of gene. It may be, and perhaps it is likely, that the nervous system is not generated, in large part, separately from the rest of the animal; such lineage mutants may be rare.

B. Programmed Cell Death

Two other features shape the final distribution of cells in the adult: Programmed cell death and cell migration (see Chapter 5). Of the two, cell death seems to play the more important role, as there is relatively little cell migration. One striking example of the involvement of cell death in the development of the nervous system is the production of sex-specific neurons in the embryo. The embryonic nervous systems of the two sexes differ superficially, in that the hermaphrodite has a pair of HSN cells and the male has four cephalic companion cells, CEMs (Sulston and Horvitz 1977). All six cells are made in embryos of both sexes, but the CEM cells die in hermaphrodites and the HSN cells die in males (Sulston et al. 1983). Thus, cell death appears to be a means of eliminating cells without extensive remodeling of cell lineages.

A similar example of this use of cell death is seen in the development of the postembryonic ventral cord (Fig. 13; Sulston 1976; Sulston and Horvitz 1977). Only in the P3a–P8a lineages does the Pnaap cell become a VC motor neuron. The aap cell in all the other lineages dies. Similarly, the Pnaaap cell differentiates as a VB neuron in all the lineages except those of P11a and P12a, where the cell dies. These deaths are, perhaps, appropriate, because VB cells, which normally project posteriorly, might be constrained by the posterior end of the ventral cord. It should be noted that this strategy of eliminating unnecessary copies of cells is not always used in *C. elegans*. Although the Pnaaa cells from these lineages usually differentiate into VA motor neurons, the first two cells of this type differentiate into AVF interneurons (Fig. 13) (White et al. 1986).

Mutations affecting cell death are described in Chapter 6. A somewhat surprising finding is that *ced-3* and *ced-4* mutants, animals in which programmed cell deaths do not occur, have apparently normal behavior (Ellis and Horvitz 1986). It is as if the extra nerve cells produced in these animals had no effect on the nervous system. However, electron micrographic studies of *ced-3* mutants have shown that these "undead" cells do, in fact, differentiate into neurons that form synapses onto other neurons and muscles (J. White et al., unpubl.).

C. Cell Migration

Although extensive cell migrations are rare in *C. elegans*, disruption of migration seems to affect the development of some cells, for example, the two postembryonic touch receptors AVM and PVM. These cells arise from identical positions in the QR and QL lineages, respectively (Sulston and Horvitz 1977). However, because of differences in the migration patterns of the precursor cells, AVM arises in a considerably more anterior position than PVM. The two cells share a number of features but differ in that only AVM branches and forms the synapses to ventral cord interneurons on the branch. As a result of these synaptic differences, only AVM can be stimulated to elicit touch-mediated movement (Chalfie and Sulston 1981). Laser killing of nearby cells can prevent the migration of the AVM precursor (Chalfie et al. 1983). Under these conditions, the cell no longer mediates a touch response.

Mutations can also alter the migration of these precursors. In *mab-5* mutants, the migration of the PVM precursors is similar to that of AVM, that is, they both move anteriorly (Chalfie et al. 1983). In these animals, the more anteriorly generated PVM grows a branch and can mediate touch-stimulated movement. Although these experiments appear to lend additional support for the hypothesis that the correct position and, therefore, correct migration of cells can be an important feature in determining the differentiated state of *C. elegans* neurons, the *mab-5* phenotype can be interpreted as a homeotic transformation of the PVM cell to an AVM cell (see Kenyon 1986). A number of other mutations affect the QR/QL migrations (Hedgecock 1987; see Chapter 6).

A second instance of extensive migration is the embryonic movement of the HSN cells (Sulston et al. 1983). In adult hermaphrodites, these cells are required for egg laying (Trent et al. 1983; J.G. Sulston and H.R. Horvitz, pers. comm.). Although no laser experiments have been done to interfere with these migrations, a subset of the egg-laying defective mutants have displaced HSN cells (see Chapter 6), suggesting that the proper position of the HSN cells is important for their correct function.

Another major migration is that of the nuclei of the precursors to the postembryonic ventral cord motor neurons (see Chapter 6). These nuclei, which are located laterally in the newly hatched animal, migrate into the cord at mid-L1. Mutations in *unc-83* and *unc-84* block this movement (Sulston and Horvitz 1981). The unmigrated precursors often fail to divide in these mutants but sometimes produce the normal number of cells laterally. Precursors may also be prevented from migrating by laser damage of adjacent regions. The ectopic cells produced by the division of these precursors have some but not all of the differentiated characteristics of neurons that are normally produced (J. White et al., unpubl.).

The phenotypes of mutants with alterations in cell migration help to emphasize that position is important in the differentiation of some *C. elegans* neurons. This importance is also seen when some cells are only slightly displaced. In particular, if the postembryonic precursor P1 is killed, the P2 neuronal precursor assumes a slightly more anterior position, and one of its progeny, P2aaa, differentiates into an AVF interneuron (J. Sulston and J. White, unpubl.). Usually it differentiates into a VA motor neuron (Fig. 13). Thus, cell–cell interactions are quite important in determining the differentiated states of these neurons.

D. Cell Outgrowth

Recently Durbin (1987) has examined the outgrowth of processes from motor neurons in the embryonic ventral cord. Of the three motor neurons, DAn, DBn, and DDn, the DDn cells send out the first processes. After the DDn ventral processes have grown out, all three motor neurons send commissures to the dorsal cord at the same time (the commissures of the DDn cells grow from the distal ends of their processes; those of the other cells arise from the cell bodies). Thus, there has been a delay in the outgrowth of processes from the DAn and DBn cells. The outgrowth of these commissural processes and their subsequent growth in the dorsal cord occur in a cell type-specific direction in the absence of other neuronal processes. In fact, the only physical correlate to the outgrowth of any of these processes is that the DAn commissures grow along the boundaries of the Pn cells. Durbin has also found that the tips of growing neurons are extended and flattened when the cells grow in the absence of other neuronal processes but are less specialized when the cells grow along other processes. Similar differences have been seen in the pattern of nerve growth in other organisms (see e.g., LoPresti et al. 1973).

Many behavioral mutants exhibit alterations in process placement (for review, Hedgecock et al. 1987). Such mutants have been characterized by vital fluorescein isothiocyanate (FITC) staining of sensory neurons (*unc-33*, *unc-44*, *unc-51*, *unc-76*, and *unc-6* [previously known as *unc-106*; Hedgecock et al. 1985]), by anti-horseradish peroxidase staining (*unc-13*, *unc-33*, *unc-44*, *unc-51*, *unc-61*, *unc-71*, *unc-73*, *unc-98*, and *unc-6* [S. Siddiqui and J. Culotti, pers. comm.]), by anti-GABA staining (*unc-5*, *unc-6*, *unc-51*, *unc-34*, *unc-40*, *unc-76*, *unc-62*, *unc-71*, *unc-73*, *unc-25*, *unc-30*, *unc-47*, and *unc-43* [S. McIntire, pers. comm.]), and by electron microscope reconstructions (*unc-3*, *unc-30* [J. White et al., unpubl.]; *unc-5* [S. Brenner et al., pers. comm.]; and *unc-6* [L. Nawrocki and N. Thomson, pers. comm.]). Although none of the neurons that are visualized with vital

FITC staining or anti-horseradish peroxidase staining are motor neurons, the fact that the misplaced processes were seen in *unc* mutants suggests that similar abnormalities to those seen in the stained neurons may also be present in the neurons of the motor circuitry.

Several mutants have commissures that fail to reach the dorsal cord (*unc-5*, *unc-6*, and *unc-51*). There are few, if any, processes in the dorsal cords of these mutants, the consequent lack of dorsal muscle innervation giving rise to a ventral coiler phenotype. It will be intresting to know whether the focus of action of these genes is in the motor neurons or in the hypodermal cells over which they pass.

All classes of motor neuron in the ventral cords of mutants of *unc-3* have highly disorganized processes. Motor neurons do make NMJs but often at ectopic sites, such as on the left-hand side of the hypodermal ridge. Processes of interneurons are relatively undisturbed from their wild-type locations. Motor neurons do not receive their normal class-specific synaptic inputs, however, because they rarely have their processes adjacent to the appropriate interneurons. The focus of action of this gene, as shown by mosaic analysis, is probably in the motor neurons and not in the hypodermis (Herman 1984 and pers. comm.).

In contrast to *unc-3*, mutants in *unc-30* exhibit process misplacements that are restricted to two motor neuron classes, VDn and DDn. Processes from these neurons wander aimlessly in the interior of the process bundle of the ventral cord, rather than at their normal location adjacent to the basement membrane in the motor end-plate region. The VDn and DDn neurons in these mutants also lack anti-GABA staining, whereas other GABA-ergic neurons stain normally. *unc-30* mutants have a characteristic shrinker phenotype (the body shortens when the animal is provoked to move backward or forward). This is consistent with the notion that the VDn and DDn neurons function as reciprocal inhibitors.

E. Patterns of Neural Differentiation

The development of the *C. elegans* nervous system is more complex than even an examination of the cell lineage would suggest. The time of cell production need not correspond to the time of cell maturation, as seen above in the delays in the development of the embryonic ventral cord. Even more dramatic delays are seen in the development of some sex-specific neurons. The HSN cells, for example, are needed for egg laying but arise embryonically. The sex muscles onto which these cells synapse are not even produced until the L3 stage (Sulston and Horvitz 1977). Moreover, the anterior HSN processes do not reach the nerve ring until the late L4 stage at the earliest (White et al. 1986). The VCn motor neurons arise in the late L1 stage and innervate the hermaphrodite sex muscles with a large number of branches (these are among the few examples of neurons with extensive arborizations in *C. elegans*). These neurons can be visualized with an antibody against the tetrapeptide Phe-Met-Arg-Phe-amide (FMRF-amide) but only in L4-stage, or older, animals. Interestingly, this branching onto the sex muscles is dependent on the presence of specialized vulval hypodermal cells that arise during the L3 stage, not the target muscles or the HSN neurons (C. Li and M. Chalfie, unpubl.).

In males, the CEM cells also appear to have a delayed maturation. These male sensory cells are embryonically derived and are thought to be important for chemotactic attraction to hermaphrodites in adults (see below). The nuclei of the cells enlarge and change in appearance in late L4 males (Sulston and Horvitz 1977). Interestingly, the sensory cilia of these cells develop embryonically (Sulston et al. 1983). Thus, there may be at least two distinct stages in the differentiation of these cells. Not all cells that are important for reproductive function are generated early and mature late. The neurons of the male tail are generated during the L3 stage (Sulston et al. 1980).

It is not clear why some cells are made early yet mature late. Perhaps, some neurons have (or had) some, as yet unknown, function in the early larvae. (It is easier to imagine a role of such modification in the differentiation of the CEM cells than the HSN cells.) Alternatively, the CEM and HSN cells may need to be produced at an early time to ensure their proper positioning within the nervous system. The importance of the timing of cell production and cell outgrowth is still unclear.

A great number of changes are seen in the development of the motor circuitry. The enlargement of the motor neuron pool at the end of the L1 stage has already been mentioned. In addition, some cells change their pattern of synapses. This was first demonstrated for the DD motor neurons (White et al. 1978). At hatching, these cells receive synapses from the DA and DB cells in the dorsal cord and synapse onto muscles in the ventral cord. In the adult, the DD cells have different synaptic partners: They receive input from the VA and VB cells in the ventral cord and have neuromuscular junctions in the dorsal cord. This rewiring is not contingent on the development of the adult nervous system, as mutants in which the adult cells are not made also rewire. The DD cells are not the only cells to rewire. Three neurons near the anterior of the ventral cord, the SAB cells, are motor neurons in newly hatched larvae but interneurons in adults (White et al. 1986).

Subtler changes may also occur in the development of the motor circuitry. A dominant mutation in the gene *deg-1* results in the degeneration of the PVC interneurons (among other cells; M. Chalfie, unpubl.). The death of the cells occurs midway through larval development. Before this time, the cells appear normal and are functional in that the animals have normal posterior touch sensitivity (which requires the PVC cells). The phenotype is temperature sensitive, and the temperature critical period is at the time of degeneration. These data suggest that this mutation and possibly the wild-type gene are expressed after functional PVC cells are made.

The motor circuitry is not the only part of the nervous system that undergoes enlargement. The anterior touch circuit is also modified by the addition of the AVM touch cell (Chalfie et al. 1985). This cell connects to the embryonic touch cells ALML/R with gap junctions to form a functional net of touch receptors. This joining of identical cells may be an important means of generating complex neuronal structures in *C. elegans*, an animal whose neurons show little branching. Additional changes have also been seen in the structure of some of the head sensory neurons when animals enter the dauer state (see Chapter 12).

F. Differentiation of Identified Cells

The identification of behavioral characteristics that are distinct features of particular classes of cells has permitted screens for mutations that act in a cell-specific manner. Mutations affecting the differentiation of specific classes of cells have been sought for a set of mechanosensory cells (ALML/R, PLML/R, AVM, and PVM; Chalfie and Sulston 1981) and for a pair of cells (HSN) required for egg laying (Trent et al. 1983; C. Desai and H.R. Horvitz, pers. comm.). A description of these mutants is given below, in Section V. In studies of both types of cells, a relatively small number of genes (~20) have been identified in genetic screens. Not all of these genes have effects solely on the cells of interest; for example, some genes affect the lineages that produce the cells. In fact, a very small number of genes act in a cell-specific fashion. In addition, certain aspects of the mature cells, such as the direction of process outgrowth (see Section III.D) are not affected by any putative cell-specific mutations. Because only two cell types have been examined in this way, it is, perhaps, premature to generalize on these results.

IV. NEUROTRANSMITTERS

A number of neurotransmitters have been implicated in the functioning of the *C. elegans* nervous system, but because electrophysiological measurements are not possible, evidence for the use of any specific neurotransmitter at a particular synapse is, for the most part, circumstantial. This evidence comprises four types: (1) pharmacological observations of effects of putative neurotransmitters, their agonists, and antagonists on intact or cut worms; (2) biochemical analysis of transmitters or the enzymes required for their biosynthesis or metabolism; (3) histological localization of putative transmitters in wild-type, mutant, and laser-ablated animals; and (4) analysis of mutants with defects in putative transmitter usage or metabolism. Additional data come from the analysis of neurotransmitters in *Ascaris*. Most information discussed below has been obtained on more traditional neurotransmitters (acetylcholine, GABA, and the biogenic amines). Investigators have only recently begun to study neuropeptides in

C. elegans. As indicated above, FMRF-amide-like immunoreactivity is exhibited by a number of neurons. In addition, antibodies against cholecystokinin and neuropeptide Y stain different subsets of neurons (S. McIntire and H.R. Horvitz, pers. comm.).

A. Acetylcholine

As is true for other nematodes, acetylcholine (ACh), ACh agonists, or inhibitors of the ACh degradation enzyme, acetylcholinesterase (AChE), cause contraction of the body-wall muscle cells when applied to cut preparations of *C. elegans* (Sulston et al. 1975; Lewis et al. 1980b). A few agonists (e.g., levamisole) and several AChE inhibitors are capable of penetrating the cuticular (or intestinal) lining and contract intact animals (Lewis et al. 1980b). Intact animals also respond by laying eggs (Trent et al. 1983) and decreasing pharyngeal pumping (Brenner 1974).

1. Physiological Effects of ACh in Ascaris

Our knowledge of the physiological basis of ACh-induced muscle contraction in nematodes stems, in large part, from work on *Ascaris*. Studies by del Castillo and his colleagues, following the identification of ACh in *Ascaris* by Mellanby (1955), demonstrated that (1) microiontophoretic application of ACh to neuromuscular synapses caused a muscle depolarization that was potentiated by AChE inhibitors and blocked by the ACh antagonist *d*-tubocurarine (del Castillo et al. 1963), and (2) stimulation of the nerve cord evoked depolarizing responses with the same pharmacological characteristics as the application of ACh (del Castillo et al. 1967). Because *d*-tubocurarine also causes hyperpolarization of resting muscles, del Castillo et al. (1963) also proposed that ACh was released tonically onto muscles from a presumably neuronal source.

Recent work by Johnson, Stretton, and their colleagues have refined our understanding of the action of ACh in the motor nervous system. Choline acetyltransferase (ChAT), the synthetic enzyme for ACh, has been shown by biochemical assay to be concentrated in three classes of excitatory motor neurons (equivalent to the DAn, DBn, and ASn cells of *C. elegans*) but not in two classes of inhibitory neurons (the DDn and VDn analogs; Johnson and Stretton 1985). Stimulation of these excitatory motor neurons leads to muscle cell depolarization that can be blocked by *d*-tubocurarine (Walrond et al. 1985). These excitatory cells also synapse onto inhibitory motor neurons (see Section II.D.3), and there is pharmacological evidence that these neuron–neuron synapses are cholinergic (Davis and Stretton 1983; Walrond and Stretton 1985). Additional cells have been identified as cholinergic by electrophysiological (Angstadt and Stretton 1983) or biochemical means (Johnson and Stretton 1980; C.D. Johnson, pers. comm.).

ACh may perform nonneural functions in *Ascaris* as well. For example,

ChAT has been identified in the hypodermal syncytium underlying the somatic musculature throughout the body (Johnson and Stretton 1985). Although the function of hypodermal ACh synthesis is unknown, if ACh were released by the hypodermis, it might cause tonic muscle contraction and thereby contribute to the generation of muscle tone and the accompanying "hydrostatic skeleton."

2. ChAT

Mutations resulting in resistance to AChE inhibitors can occur in *unc-1*, *unc-10*, *unc-11*, *unc-13*, *unc-17*, *unc-18*, *unc-29*, *unc-32*, *unc-36*, *unc-63–unc-65*, and *cha-1* (Brenner 1974; Rand and Russel 1984; S. Carr et al., pers. comm.). Examination thus far has shown that none of these mutations achieve resistance by altering AChE or by increasing its resistance to inhibitors in vivo. Rather, their resistance stems from the reduced effects of AChE inhibition on the behavior of the mutants. Alleles of one of these genes, *cha-1*, have been shown to harbor altered ChAT (Rand and Russell 1984), indicating that one mechanism for achieving resistance involves a decrease in ACh synthesis (the AChE inhibitors would have smaller effects in the presence of less ACh). It seems possible that other loci conveying resistance to AChE inhibitors may encode for components involved in other steps in ACh synthesis, packaging, or release.

There are two forms of ChAT in *C. elegans* (molecular weights 71,000 and 154,000; Rand and Russell 1985a). Both forms have similar pH dependence, thermal stability, and K_m for choline, and the large form can convert to the smaller by the addition of 500 mM NaI.

cha-1 appears to be the structural gene for ChAT or for a subunit of ChAT (Rand and Russell 1984). Drastic reductions of ChAT are needed before a visible phenotype is seen under standard culture conditions. Besides their resistance to AChE inhibitors, animals homozygous for severe *cha-1* alleles are small, slow-growing, and uncoordinated in a characteristic coiling, jerky manner. One allele was isolated in a screen for temperature-sensitive paralytic mutants (Hosono et al. 1985). The null phenotype of this gene is not known, but recently a lethal *cha-1* allele has been identified (T. Rogalski and J. Rand, pers. comm.).

All of the phenotypes of the *cha-1* mutants (except for the reduction of ChAT activity) are shared by mutants with defects in *unc-17*, a locus that lies just to the right of *cha-1*. Although most alleles of *cha-1* complement those of *unc-17*, three alleles fail to complement alleles of both genes (Rand and Russell 1984). One of the unusual alleles (*p1156*) has low levels of ChAT activity; the others (*e113, e876*) have at least half the wild-type level of enzyme activity. *p1156* maps between *cha-1* and *unc-17*, and *e113* and *e876* map to the right of *unc-17* (J. Rand, pers. comm.). From this unusual complementation pattern, Rand and Russell (1984) have proposed that *cha-1 unc-17* encodes a single polypeptide that forms multimers with two domains (one for enzymatic activity; the other for a second function, e.g., regulation or localization).

3. AChE

At least three classes of AChE (A, B, and C) have been distinguished on the basis of their K_m values for ACh, their selectivities for related substrates, and their sensitivities to cholinesterase inhibitors. More than one form of activity within a given class can be identified using sucrose density sedimentation and ion exchange chromatography (Johnson and Russell 1983; Kolson and Russell 1985a). Detergent is required for full solubilization of some of these forms. None appears to be collagen tailed, as with some vertebrate AChEs.

Each AChE class is selectively affected by mutations in one of the three *ace* genes (Culotti et al. 1981; Johnson et al. 1981; C.D. Johnson et al., pers. comm.). Dosage experiments for *ace-1*, *ace-2*, and *ace-3* suggest that they are, respectively, the structural genes for AChE classes A, B, and C. Additional evidence for *ace-1* being the structural gene for class-A activity comes from the finding of K_m and thermal sensitivity changes in the residual class-A activity in two leaky *ace-1* mutants (Kolson and Russell 1985b). Mutations affecting the distribution among the forms of class-A, -B, or -C activity have not been found, and the significance of the different forms within each class is not known.

Except for different drug sensitivities, none of the single *ace* mutants nor *ace-3;ace-1* or *ace-3;ace-2* double mutants have a significant behavioral phenotype. Clearly, no single activity is essential. (This conclusion is supported further by the finding that the histochemical localization of the AChE activity is essentially the same in wild-type, *ace-1*, and *ace-2* mutants [Culotti et al. 1981].) The *ace-2;ace-1* double mutant, however, is characterized by slow, uncoordinated forward movement and by hypercontraction and paralysis when the animals are stimulated to go backward (there is no histochemically detectable AChE activity in the double mutant). The phenotype of the triple *ace* mutant is more severe. When all three mutations are present, paralyzed but otherwise normal appearing larvae are produced that do not seem to be able to escape from the eggshell. When the shell (or remaining surrounding membrane) is broken, the animals fail to develop. Interestingly, strains that contain mutations in all three *ace* genes and in severe alleles of *cha-1* or *unc-17* are viable (although they grow extremely poorly; C.D. Johnson and J.B. Rand, pers. comm.). The lethality of the triple *ace* mutant thus appears to result from excessive buildup of ACh, rather than from the loss of some essential function of the *ace* genes themselves.

Histochemical staining for AChE activity in wild-type animals appears at the nerve ring, ventral ganglion, pharyngeal–intestinal valve, near the anal depressor muscles of the tail and, to a lesser extent, at the ventral and dorsal nerve cords (Culotti et al. 1981; some nonneuronal nuclei also stain, but the significance, if any, of this staining is unknown). *ace-1* mutants have a similar pattern of staining. The pattern is also similar in *ace-2* mutants, except that the staining is reduced in general and is missing altogether from the pharyngeal–intestinal valve. There is no detectable histochemical staining in *ace-2;ace-1* double mutants. These data suggest that the distribution of the *ace-1* and *ace-2* products are similar and, thus, give a histological basis for the lack of gross behavioral phenotypes in the single mutants. The distribution of the *ace-3* activity has been localized with an antiserum directed against the class-C activity (B. Stern and R.L. Russell, pers. comm.). This antibody binds to the nerve ring and nearby cell bodies as well as two lateral cells, the CAN cells.

Genetic mosaics in which *ace-2;ace-1* or *ace-2;ace-3;ace-1* cells are produced in a background that is *ace-2;ace-1(+)* or *ace-2;ace-3;ace-1(+)* have been used to analyze the tissue specificity of class-A AChE synthesis (Herman and Kari 1985; C.D. Johnson et al., pers. comm.). The data are consistent with autonomous expression of *ace-1* in muscle cells. Loss of *ace-1* function in the AB lineage (which produces 280 of the 282 nonpharyngeal neurons) is without apparent effect on coordinated locomotion, growth, or the AChE histochemical staining pattern. This result implies that any nerve–nerve cholinergic synapses in *C. elegans* do not require the expression of these three *ace* genes in either pre- or postsynaptic cells.

4. ACh Receptors

Brenner (1974) identified five loci that could be mutated to give resistance to the anthelmintic tetramisole. He suggested that because of its behavioral effects, tetramisole acted as a cholinergic agonist. Lewis et al. (1980a,b) have extended this work by identifying an additional seven loci that convey levamisole resistance (levamisole is the *l* isomer; tetramisole is a *d,l* mixture). Because these mutants were resistant to a number of cholinergic agonists and because the resistance could be mimicked by treating wild-type animals with mecamylamine, Lewis et al. suggested that defects in a nicotinic ACh receptor could convey levamisole resistance. Mutations at seven loci (*unc-29, unc-38, unc-50, unc-63, unc-74, lev-1*, and *lev-7*) produce Unc animals that can, nonetheless, grow in the presence of levamisole. Eight loci (*unc-29, unc-38, unc-63, unc-74, lev-1, lev-8–lev-10*) can be mutated to give resistant animals that are essentially wild type, and two others (*unc-22* and *lev-11*) can be mutated to produce resistant twitchers. Although the significance of this is unknown, it is intriguing that five of these genes (*unc-29, unc-38, unc-63, unc-74, lev-11*) are located near each other on chromosome I.

The binding activity of the levamisole receptor has been studied in these mutants using ^{3}H-labeled *meta*-aminolevamisole (Lewis et al. 1987). These experiments suggest that *unc-29, unc-50,* and *unc-74* are required for the levamisole-binding activity and that the other four genes affect the binding.

Although these mutants are resistant to levamisole and nicotine, they still respond to carbachol and AChE inhibitors, and they retain some motor activity (e.g., they are uncoordinated less severely than *cha-1*). These properties led Lewis et al. (1980b) to suggest that other ACh receptors existed that were not sensitive to nicotine. Culotti and Klein (1983) found saturable binding of [^{3}H]*N*-methylscopolamine and [^{3}H]quinudidinyl benzilate to *C. elegans* extracts. This binding activity can be blocked by a number of muscarinic antagonists, suggesting that the animal possesses muscarinic ACh receptors. No behavioral defects have been described for animals treated with these antagonists, even when applied to cut worms (Lewis et al. 1980b; Culotti and Klein 1983), and no mutants with altered binding properties have been identified.

B. GABA

It is likely that GABA is an inhibitory neuromuscular transmitter in nematodes. del Castillo et al. (1964) showed that GABA exerts a powerful hyperpolarizing action on *Ascaris* muscle cells by increasing the conductance of Cl^- ions. Electrophysiological analysis has identified two classes of inhibitory motor neurons in *Ascaris* (analogous to the VDn and DDn cells in *C. elegans*; see Section II.D.3). GABA and the biosynthetic enzyme that forms it, glutamic acid decarboxylase, have been detected in the nerve cords of *Ascaris* (S. Burden and A.O.W. Stretton; C.D. Johnson and A.O.W. Stretton; both pers. comm.). Antisera that react with GABA reveal GABA-like immunoreactivity in both classes of inhibitory motor neurons (Johnson and Stretton 1987). Additional cells, specifically the *Ascaris* analogs of the four RME neurons, the DVB neuron, and six unidentified neurons in the head ganglia, also stain.

Application of GABA to cut wild-type *C. elegans* (Lewis et al. 1980b) or of muscimol, a GABA agonist, to intact animals (M. Chalfie, unpubl.) causes a flaccid paralysis. When these animals are touched at either end, they contract, as if the normal antagonism of ventral and dorsal muscles is absent. Avermectin, an anthelmintic whose mode of action is thought to be related to GABA, also paralyzes *Ascaris* and *C. elegans* (Kass et al. 1980), although the drug does not generate a flaccid state. Electrophysiological effects of avermectin are seen at inhibitory neuromuscular synapses and at presumptive interneuron–motor neuron synapses in *Ascaris* (Kass et al. 1984). Hedgecock (1976b) found detectable levels of glutamic acid decarboxylase and GABA transaminase (the enzyme required for the metabolism of GABA) in wild-type animals.

On the basis of GABA-like immunoreactivity, GABA has been identified as a putative neurotransmitter in the VDn, DDn, RME, and DVB neurons and also in an additional, as yet unidentified neuron in the ventral ganglion (S. McIntire and H.R. Horvitz, pers. comm.). Several mutants have altered patterns of anti-GABA staining. Mutants in the gene *unc-25* have no staining, whereas mutants in *unc-47* have enhanced staining. The lack of staining could be due to a defect in the biosynthetic pathway for GABA, and the increased staining could perhaps be due to a failure of the synaptic release mechanisms for this neurotransmitter. These two mutants affect all anti-GABA staining cells, whereas mutants in *unc-30* have no anti-GABA staining in the VDn and DDn neurons and mutants in *unc-43* have reduced levels of staining in their VD neurons (S. McIntire and H.R. Horvitz, pers. comm.).

C. Biogenic Amines

There is evidence of dopamine, serotonin, and octopamine in *C. elegans*. Sulston et al. (1975) identified eight cells in the hermaphrodites (the four CEP cells and the two pairs of ADE and PDE cells) as dopaminergic by formaldehyde-induced fluorescence. Six additional dopaminergic cells were found in the male tail, which were later identified as being the paired ray neurons R5A, R7A, and R9A (Sulston and Horvitz 1977).

Lesions in four genes (*cat-1–cat-4*) affect the extent of dopamine staining (Sulston et al. 1975). These mutants show the reduction (*cat-3*) or loss (*cat-1*) of staining from cell processes or the reduction (*cat-4*) or absence (*cat-2*) of staining from cell bodies and processes. The cell-body staining of the *cat-1* mutants is phenocopied by reserpine. There is a reduction in dopamine content of the animals corresponding to the severity of the loss of formaldehyde-induced fluorescence; the *cat-2;cat-1* double mutant, for example, has no detectable dopamine. None of these mutants, however, is detectably altered in the activity of any of the enzymes needed for dopamine synthesis.

Somewhat disappointingly, neither reserpine treatment nor any of these mutations leads to gross behavioral defects. The only indication that dopamine may act as a transmitter is that the mating efficiency of *cat-1*, *cat-2*, and *cat-4* males is less than wild type (Hodgkin 1983a). These data correlate with the loss of dopamine staining in a subset of ray neurons and are consistent with the suggestion (Sulston et al. 1975) that dopamine has a functional role in these putative mechanosensory cells.

Formaldehyde-induced fluorescence has also been used to identify a pair of serotonergic neurons (the NSM cells) in the pharynx (Albertson and Thomson 1976; Horvitz et al. 1982b). More recently, antibodies against serotonin have revealed cross-reactive material in a number of cells (C. Desai et al., pers. comm.). In addition to the NSM cells, staining is seen in six cells in the head of both sexes, ten cells in the ventral cord and four to five cells in the tail of the male, and the two HSN cells of the hermaphrodite. Such antibodies also stain two NSM cells in the pharynx and five cells in the male ventral cord in *Ascaris* (C.D. Johnson, pers. comm.)

Unlike dopamine, exogenously added serotonin has profound behavioral effects on *C. elegans* (Croll 1975; Horvitz et al. 1982b; Trent et al. 1983). Treated animals move slowly, increase their rate of pharyngeal pumping, and increase egg laying. A number of egg-laying defective mutants have been shown to have reduced serotonin levels (C. Desai et al., pers. comm.). It is particularly striking that of these mutants, those of *egl-44* and *egl-46* show a cell-specific (HSN) loss of serotonin staining.

Octopamine has been identified biochemically in *C. elegans* but has not been localized to any particular cells (Horvitz et al. 1982b). In many respects, adding octopamine exogenously acts in opposition to the effects of serotonin; thus, it reduces pharyngeal pumping and inhibits egg laying. Mutations in three genes, *daf-10*, *che-3*, and *osm-3*, lack detectable levels of octopamine. Mutations in these genes have been shown to have ultrastructural defects in sensory cilia (Lewis and Hodgkin 1977; Albert et al. 1981; Perkins et al. 1986). In particular, the defects in *osm-3* are confined to the amphid and phasmid cilia (Perkins et al. 1986), suggesting that these chemosensory neurons may use octopamine or regulate neurons that use octopamine.

V. BEHAVIOR AND ITS GENETIC ANALYSIS

In this section we describe the analysis of those behaviors that are known to be disrupted by mutation. One striking feature of the *C. elegans* nervous system that eases the genetic analysis is that most of the neurons are nonessential in hermaphrodites under laboratory conditions. Only three neurons are required for normal development (CANL and CANR in the body [J. Hodgkin and J.E. Sulston, pers. comm.] and M4 in the pharynx [L. Avery and H.R. Horvitz, pers. comm.]). Thus, many mutations leading to defects in the nervous system need not have lethal phenotypes. *C. elegans* behaviors that have been studied by mutational analysis include coordinated movement, chemotaxis, thermotaxis, osmotic avoidance, male mating, egg laying, and mechanosensation. Other behaviors exist but have not been the focus of genetic analysis or are only just beginning to be studied. For example, mutations in *cha-1*, *unc-13*, and *unc-18* slow pharyngeal pumping, and mutations in *unc-31* cause the pharynx to pump constitutively (L. Avery and H.R. Horvitz, pers. comm.). Other behaviors that have been noted in the animal include a "precipice response," seen as a rapid backing when the animal comes to a sharply cut edge of a piece of agar (J. Culotti, pers. comm.), and a weak but significant response to light (Burr 1985).

A. Movement

Mutations in a large number of genes affect the normal movement of *C. elegans;* uncoordinated (Unc) mutants were among the first isolated in the animal (Brenner 1974). A list of the *unc* genes and a description of mutant phenotypes is given in Appendix 4. An Unc phenotype could result from defects in the muscles or their attachments to the body wall or from lesions affecting the nervous system. It is difficult to assess the focus of action (neuron, muscle, or hypodermis) for many of the *unc* genes because similar behavioral phenotypes may be produced by quite different underlying lesions. Although mutations in a significant number of *unc* genes produce well-characterized lesions in the muscle (see Chapter 10) or in the nervous system, it is not always clear where the focus of gene activity is situated. For example, a muscle degeneration phenotype might be a consequence of a failure of the innervating motor neurons to provide an essential trophic factor. Mosaic analysis (Herman 1984; Herman and Kari 1985; see Chapter 2) should facilitate the identification of the focus of action of some *unc* genes.

The alterations seen in the nervous systems of uncoordinated mutants can be categorized into four general classes, according to whether they are in the cell lineages that produce motor neurons, synaptic specificity, process placement, or neurotransmitter function (these last two are described in Sections III.D and IV). In many cases, the behavioral phenotype is consistent with the anatomical phenotype, given our current understanding of the operation of the ventral cord circuitry. This is comforting. Some of the anatomical lesions in the nervous system that have been observed are described here.

Because most of the ventral cord motor neurons arise after hatching, mutations affecting postembryonic lineages result in animals that are initially coordinated but become uncoordinated (see Chapter 6). One class of such mutations (in genes *unc-83* and *unc-84*; Sulston and Horvitz 1981) affects the migration of the ventral cord precursors, resulting in fewer ventral cord motor neurons. Other mutations produce defects in cell division. Such lineage mutants include *lin-6* (White et al. 1978; Sulston and Horvitz 1981), *lin-5* (Albertson et al. 1978), *unc-59*, and *unc-85* (White et al. 1982). Mutations in *lin-6* block postembryonic DNA replications; those in the other genes block all or some postembryonic cell divisions. The blocked cells in the late blocking mutants (*unc-59* and *unc-85*) differentiate into neurons that have the characteristics of only one of the daughters that are normally produced. Those of the early blockers, on the other hand (*lin-5*), differentiate into neurons that may have characteristics of more than one of the normally produced daughters. Interestingly, the blocked, polyploid cells in these animals have long branched processes, the total process lengths being similar to the sum of the lengths of the processes from the normal descendants of these cells.

Two mutants have been found to have changes in synaptic specificity. Mutants in the gene *unc-4* move forward reasonably well but coil with their dorsal sides innermost when provoked to move backward. Electron microscope reconstructions of these mutants have shown that the VAn neurons in mid-body do not receive their normal synaptic input from AVA and AVD interneurons but instead receive synaptic input from AVB interneurons (J. White et al., unpubl.). The probable explanation for the behavioral phenotype is that when the animals try to move backward and the AVA and AVD interneurons are activated, only the unaffected DAn neurons will operate and the animals will coil dorsally. Mutants in the gene *unc-55* often coil ventrally when trying to move in either direction. Electron microscope reconstructions of these animals have shown that the inhibitory VDn neurons innervate the dorsal muscles instead of the ventral muscles (L. Nawrocki and N. Thomson, pers. comm.). Thus, the ventral body muscles receive no inhibitory input and the dorsal muscles receive inhibition from both VDn and DDn neurons, which probably accounts for the ventral coiling behavior of these animals.

B. Mutations Affecting Behaviors Associated with the Head Sensory Neurons

The many sensory receptors in the head (see Section II.B.2) are thought to mediate a variety of behaviors including chemotaxis, thermotaxis, osmotic avoidance and dauer larva formation and recovery. Mutants defective in each of these behaviors have been isolated. Although mutations in a few genes affect only one type of behavior, many affect more than one. These latter mutations could be affecting cellular components that are used by different types of sensory cells (e.g., the sensory cilia; see Perkins et al. 1986) or downstream cells that are postsynaptic to a number of sensory receptors. Alternatively, such mutations may indicate that two behaviors require the same receptor. With the possible exception of the AFD cells, which are thought to mediate the thermotactic response (see Section V.B.2 below), and the ASH cells, which are required for osmotic avoidance (see Section V.B.3 below), the specific functions of the sensory cells in the head are unknown.

1. Chemotaxis

One of the first sensory behaviors examined in *C. elegans* was chemotaxis (Ward 1973). Ward found that the animals are attracted to a variety of compounds including cyclic nucleotides, some anions (e.g., Cl^-) and cations (e.g., Na^+), basic pH, some amino acids, and bacterial extracts. Similar observations were made and extended by Dusenbery (1974, 1975,

1976b, 1980a,c), who found additional attractants (pyridine, CO_2 at basic pH, O_2) and repellents (D-tryptophan, CO_2 at acid pH).

Two additional chemical responses have been noted in *C. elegans*. Golden and Riddle (1984a) identified a fatty acid pheromone that acts as a signal for dauer formation (see Chapter 12). H.R. Horvitz and J.E. Sulston (pers. comm.) found that males, but not hermaphrodites, are attracted to hermaphrodites. It is possible that this behavior is mediated via the four cephalic companion (CEM) sensory neurons that are only found in the male (Sulston and Horvitz 1977).

Mutants defective in chemotaxis have been identified by selecting animals that are unable to move toward an attractant (Na^+ or Cl^-), either on plates or in a countercurrent system, or by screening mutants defective in other behaviors (Dusenbery et al. 1975; Lewis and Hodgkin 1977). Dusenbery et al. (1975) identified six genes: *tax-1–tax-6* (see also Dusenbery et al. 1976b). Six genes (*che-1–che-3*, *che-5–che-7*) were defined by mutations identified by Lewis and Hodgkin (1977). It is difficult to know how many different genes are represented in these sets; the mutations defining *tax-1* and *che-1* fail to complement each other, but most of the others remain to be tested.

Certain dauer-defective mutations, a few of the thermotaxis mutations, and all of the osmotic avoidance mutations also result in chemotactic defects (see below). Except for the dauer-defective mutations and those mutations resulting in abnormal FITC-staining patterns, the genetic mapping of these mutations has proved to be difficult because the behavior is best seen in populations of animals. Mutants defective in 15 genes have abnormal FITC staining (Perkins et al. 1986): *che-2*, *che-3*, *daf-6*, *daf-10*, *daf-19*, *cat-6*, all the *osm* genes, and five genes, *che-10–che-14*, for which mutants were first isolated because of defects in FITC staining. These latter mutants are not attracted to NaCl, suggesting that the affected cells could be important for this chemotaxis.

All of the chemotactic mutants identified by abnormal FITC staining have sensory abnormalities in the head (Lewis and Hodgkin 1977; Albert et al. 1981; Perkins et al. 1986); however, it has been too difficult to determine which cells mediate the various chemotactic responses. Now that the embryonic lineage is known (Sulston et al. 1983), it should be possible to ablate subsets of sensory cells or their precursors in the embryo and assess the effect of the loss on various chemosensory behaviors.

2. *Thermotaxis*

When placed on a thermal gradient, *C. elegans* larvae and adults will migrate to the temperature at which they have been growing (Hedgecock and Russell 1975). This behavior can be modified by acclimation at a new temperature. This is one of the few known cases of a modifiable behavior in *C. elegans* (habituation or adaptation of chemotaxis [Ward 1973] and touch sensitivity [Chalfie and Sulston 1981] and sensitization of pharyngeal pumping by the addition of bacteria after starvation [L. Avery and H.R. Horvitz, pers. comm.] are others). Starved adults and dauer larvae, when grown at 20°C, however, migrate away from that temperature.

A number of mutants have been tested for thermotactic abnormalities, and additional mutants have been isolated as thermotactic defectives (*ttx*, Hedgecock and Russell 1975). Many of the mutants are athermotactic; others show reversal of the usual behavior. For example, some animals appear to seek colder temperatures regardless of their growth temperature, and others migrate to higher temperatures (except when raised at 25°C). Many, but not all, of the mutants have chemotaxis defects (see also Dusenbury and Barr 1980). Recently, Perkins et al. (1986) confirmed R. Ware's observation (pers. comm.) that the cryophilic mutant *ttx-1(p767)* lacks the many fingerlike projections on the AFD amphidial neurons.

3. *Osmotic Avoidance*

High concentrations of a variety of chemicals (e.g., NaCl, fructose, sorbitol, and Na acetate), which are chemoattractants at lower concentrations, act as repellents to wild-type animals (Ward 1973; Culotti and Russell 1978). Because of the lack of chemical specificity of this response, Culotti and Russell suggested that wild-type animals normally avoid solutions of high osmolarity. Mutants defective in this behavior were selected because they crossed a double barrier of NaCl or fructose. All the mutants (representing defects in six genes, now called *osm-1*, *che-3*, *osm-3*, *daf-10*, *osm-5*, and *osm-6*) failed to avoid both NaCl and fructose, irrespective of which compound was used for the initial isolation. The mutants are severely chemotactically defective (Dusenbery 1980b) and have ultrastructural defects in the ciliated endings of the amphidial neurons (Perkins et al. 1986). Laser ablation of the amphid sheath cells eliminates osmotic avoidance, and killing of the ASH amphid neurons (but not other types of amphidial neurons) greatly reduces the response (J. Thomas and H.R. Horvitz, pers. comm.).

4. *Dauer Formation*

Dauer formation is discussed in greater detail in Chapter 12. Here, we discuss those genes that appear to affect the structure and function of the nervous system. Golden and Riddle (1982) have partially purified a pheromone from *C. elegans* that induces dauer formation even under conditions of ample food and no crowding. It is not surprising, then, that some dauer-defective mutants would have defects in the chemosensory anatomy. In fact, mutations in *daf-6*, *daf-10*, *daf-19*, and *che-3* all lead to morphological abnormalities in the structure of sensory cells in the head (Albert et al. 1981; Perkins et al. 1986). The only consistent anatomical defect in all four mutants is in the amphids (they are also defective in FITC staining and chemotaxis response to NaCl; Perkins et al. 1986). Thus, it is likely that signals for dauer formation (pheromone, food signals, and temperature; see Golden and Riddle 1984a) are received by cells as yet unidentified in the amphids. Albert et al. (1981) have found that the wild-type activity and, thus, presumably normal sensory cells are needed for recovery from the dauer state. As noted above (see Chapter 12; Albert and Riddle 1983), the sensory anatomy is modified in dauer larvae.

C. Male Mating

Male mating is one of the most complex behaviors in *C. elegans* (see, e.g., Ward and Carrel 1979). As indicated above (Section V.B.1), males exhibit a chemotaxis toward hermaphrodites that is thought to be mediated by the four CEM cells. When a male's tail touches a hermaphrodite, the male places its copulatory bursa in contact with the hermaphrodite and begins moving backward along it. (The males do not seem to recognize the hermaphrodites as such; contact with other males or even a male's own head will produce this same behavior.) This backward movement, which may result in the male's encircling the length of the hermaphrodite many times, allows the male to find the vulva. Once the bursa has contacted the vulva, the male extends a pair of cuticle-covered projections called spicules. The attachment of the spicules secures the male to the hermaphrodite and opens the vulva. Sperm are then ejaculated into the hermaphrodite, the spicules are removed, and the male swims away from the hermaphrodite.

Mating behavior thus involves both chemosensory and mechanosensory input, as well as special motor control, and the male nervous system differs in a number of ways from that of the hermaphrodite (Sulston et al. 1980). In addition to the CEM cells, the males have a number of sensory neurons in the tail. These include the nine ray sensory sensilla on each side of the bursa, sensilla in the spicules, and sensilla in front of (the hook sensillum) and behind (two postcloaca sensilla) the cloaca. It is likely that the ray sensilla, and probably most of the other sensilla, serve a mechanosensory role. Each spicule has a putative proprioceptor (SPC), as well as two putative chemoreceptors (SPD and SPV). Laser ablation studies have shown that animals lacking the ray sensilla fail to respond to contact with hermaphrodites and do not make the characteristic backing movements of mating. Males lacking the hook are normal in this respect but endlessly search around the hermaphrodite and rarely locate the vulva (J.E. Sulston, pers. comm.).

There are additional differences in males, including the absence of the VCn and HSN cells that are seen in hermaphrodites. The VCn motor neurons are replaced by two classes of ventral cord motor neurons, CAn and CPn. The neuroanatomy of the male has not been reconstructed completely as it has for the hermaphrodite. Thus, it is not known, for example, how the male/hermaphrodite differences affect the organization and connectivity in the nerve ring.

Hodgkin (1983a) has tested the mating efficiency of males containing representative alleles of most nonlethal genes in *C. elegans* (see also Appendix 4B). As might be expected, mutations that affect the shape of the spicules or the copulatory bursa (some among the *dpy*, *lon*, *sma*, *bli*, and *vab* mutations) abolish or reduce the ability of males to mate. Although many *unc* males mate, it is also not surprising that a number of *unc* mutations lead to lower mating efficiency. Many of the *unc* mutants that never mate also have abnormal spicules or bursae. Some of the *che*, *daf*, and *osm* mutants have reduced or no ability to mate. As indicated above, many of these mutants are generally defective in sensory behavior. These mutants suggest that the sensory apparatus of the tail is important.

Hodgkin (1983a) screened for and characterized 17 strains with males that could not mate (see Chapter 9). Three strains carried mutations in *che-2* and *che-3*, genes that affect the development of many sensory systems. The other mutations defined ten *mab* genes (for *m*ale *ab*normal). All of the *mab* mutations result in demonstrable bursal or ray defects or in male-specific lineage defects. Only *mab-3*, *mab-6*, and *mab-9* may be truly male specific; the other mutations also produce defects in the hermaphrodite.

D. Egg Laying

The anatomy of the egg-laying system is described in Chapter 4. The vulval and uterine muscles control egg laying. Two types of neurons synapse onto these sex muscles: the two HSN cells and the six VC motor neurons. Laser ablation of both HSN neurons results in animals that fail to lay eggs normally and become severely bloated with eggs retained in the uterus (H.R. Horvitz and J.E. Sulston, pers. comm., cited in Trent et al. 1983). In contrast, ablation of all six VC neurons, or of the neurons that synapse onto the HSN neurons, does not have an obvious effect on egg laying. Thus, the HSN neurons are the only neurons of the egg-laying circuitry, as defined by synaptic connectivity, that have been shown to be required for normal egg laying.

Environmental stimuli can affect the rate of egg laying. One stimulus for egg laying is food: Hermaphrodites lay more eggs in the presence than in the absence of bacteria (Trent 1982). The effect on egg laying of removing animals from bacteria is very rapid: Hermaphrodites added to a buffered solution without *Escherichia coli* essentially stop laying eggs immediately (although some eggs are released later). The embryos within such hermaphrodites continue to develop and hatch within their parent. The resulting "bag of worms" is characteristic of animals severely blocked in egg laying.

Pharmacological agents that stimulate egg laying in the absence of bacteria have helped define the in vivo action of three neurotransmitter candidates (acetylcholine, octopamine, and serotonin) in the egg-laying system (Horvitz et al. 1982b; Trent et al. 1983). Drugs that potentiate cholinergic activity stimulate egg laying (see also Brenner 1974). Octopamine inhibits, and the octopaminergic antagonist phentolamine stimulates, egg laying (Horvitz et al. 1982b). Serotonin stimulates egg laying, as does imipramine, a drug that potentiates endogenous serotonin by preventing serotonin reuptake into presynaptic nerve terminals. The action of imipramine is dependent upon the presence of endogenous serotonin.

The HSN cells stain with antiserotonin antibody (see Section IV.C). Furthermore, when the HSN neurons are killed by laser ablation, animals still lay eggs in response to serotonin but do not lay eggs in response to imipramine (H.R. Horvitz and J.E. Sulston, pers. comm., cited in Trent et al. 1983). These experiments strongly suggest that the HSN neurons are the endogenous source of the serotonin that functions in egg laying. The sites of action of the other pharmacological agents that affect egg laying have not been identified. Both the cholinergic agonists and the octopaminergic antagonist fail to stimulate egg laying by HSN-deficient animals (H.R. Horvitz and C. Trent, pers. comm.).

Mutants isolated on the basis of a variety of defects (other than abnormalities in egg laying) have proved to be deficient in egg laying, including certain mutants of each of the following classes (Trent et al. 1983): (1) uncoordinated mutants abnormal in muscle structure (see Chapter 10), (2) other uncoordinated mutants, mechanosensory mutants, (3) cell lineage mutants, (4) dauer-constitutive mutants (temperature-sensitive, dauer-constitutive mutants when grown at the permissive temperature are egg-laying defective; see Chapter 12), and (5) sexual-transformation mutants (mutant hermaphrodites that are slightly sexually transformed fail to develop certain components of the egg-laying system; see Chapter 9).

In addition, based upon their bloated and, in many cases, bag of worms appearance, numerous mutants deficient in egg laying have been isolated directly. For example, a general screen for egg-laying defective mutants identified 40 new genes (Trent et al. 1983). These *egl* (*egg-laying* abnormal) mutants defined five classes, based upon their responses to serotonin and imipramine: (1) responsive to neither compound (14 genes)—many, and perhaps all, of these mutants are defective in the functioning of a component of the egg-laying system that acts after the HSN neurons, that is, either the sex muscles or the vulva; (2) responsive both to serotonin and to imipramine (seven genes)—some of these mutants are partially defective in essential components of the egg-laying system (vulva, sex muscles, or HSN neurons), whereas others appear to be defective in the system that senses and processes sensory stimuli (see below); (3) responsive to serotonin but not to imipramine (four genes)—these mutants are defective in the HSN neurons (see below); (4) responsive to imipramine but not to serotonin (one gene)— the cellular lesion of this mutant is unknown; and (5) variably responsive (14 genes). Thus, mutants defective in each of the three general components of the egg-laying system—vulva, sex muscles, and neurons—have been identified. Because most of the *egl* genes defined in this screen were represented by only a single mutant allele, it is likely that many other genes exist that can mutate to perturb egg laying.

Mutations of the second class described above, those that allow the stimulation of egg laying both by serotonin and by imipramine, could be defective in steps in egg laying that function prior to or in parallel with the HSN neurons. There are four dauer-constitutive mutants in this class, and because dauer formation appears to involve processing of chemosensory information, chemosensory neurons may be functionally important components of the egg-laying system.

Mutations of the third class described above, those that affect the functioning of the HSN neurons, are of particular interest, as they define genes that affect one neuron type. Subsequent experiments in which egg-laying-defective, serotonin-responsive, imipramine-nonresponsive mutants have been sought specifically have identified 16 genes (represented by 47 mutant alleles) that can mutate to give this phenotype (Trent et al. 1983; C. Desai and H.R. Horvitz, pers. comm.). Other mutants defective in HSN development have been identified by screening directly for abnormalities in HSN migration (using Nomarski microscopy) or in HSN cell morphology (using fluorescence microscopy to visualize serotoninergic cells) (G. Garriga et al., pers. comm.). To date, 26 genes have been identified that affect a variety of steps in the development and functioning of the HSN neurons.

Mutants defective in 4 of these 16 genes generate HSN neurons that then undergo programmed cell death. The apparent basis for this phenotype is a transformation in the sexual fate of the HSN cells, which undergo programmed cell death in wild-type males (Sulston et al. 1983; see Chapter 9). *egl-5* and *egl(n-1332)* mutants may be defective in the determination of the HSN fate, as these mutants generate cells that are mispositioned and lack serotonin. Mutants defective in any of six genes are blocked in the long-range embryonic migration of the HSN cells; interestingly, in these mutants, the displaced HSN neurons still synthesize serotonin but in many cases generate nerve processes that are abnormal in length and/or in morphology. In *egl-45* or *unc-86* mutants, the HSN neurons migrate properly but fail to undergo specific changes in size and shape that normally occur during the fourth larval stage. Mutants in four genes have HSN cell bodies that are normally positioned but HSN processes that are defective in outgrowth. *egl-44* or *egl-46* mutants have HSN neurons that contain low or undetectable levels of serotonin. Finally, mutants defective in six other genes have no abnormalities that have been detected with the light microscope; some of these mutants could be abnormal in synapse formation or function.

E. Mechanosensation

C. elegans exhibits a number of touch responses: (1) It recoils on bumping into an object (Croll 1976b); (2) it moves away from stroking with a fine hair (Sulston et al. 1975; Chalfie and Sulston 1981); and (3) if incapable of responding to gentle touch stimuli, it moves away from a stronger touch, for example, with a metal prod (Chalfie and Sulston 1981; Chalfie et al. 1985). Touch sensitivity may play other roles in *C. elegans*. Touching the animals often causes a transient cessation of pharyngeal pumping (Chalfie et al. 1985). In male mating behavior, the males touch the hermaphrodite in an attempt to locate the vulva. Because the receptors to gentle touch synapse onto the HSN cells, the touch may affect egg laying during mating.

Numerous ciliated neurons in the hermaphrodite send processes that are embedded within the cuticle but fail to pass through it (see Section II.B.2). Because of this anatomical feature, it is likely that these cells are mechanoreceptors. As noted in the discussion of chemosensation (Section V.B.1), a number of genes disrupt the sensory cilia of both the putative chemosensory and mechanosensory neurons. It is not known whether mutants with these defects respond when they bump into objects. However, many of the males with these mutations are incapable of mating (Hodgkin 1983a; Perkins et al. 1986), a behavior that probably requires the presumed mechanosensory ray neurons. The partial mating defects in the *cat* mutants might also be ascribed to a mechanosensory defect (Sulston et al. 1975).

The response to touching with a fine hair is much better understood. It is mediated by the six microtubule cells (ALML/R, PLML/R, AVM, and PVM; Chalfie and Sulston 1981). Laser ablation of these cells results in animals that are incapable of responding to the hair but do move if bumped on the tip of the head or prodded with a metal wire. Three of the cells are important for anterior touch sensitivity (ALML/R and AVM), and two are important for posterior touch sensitivity (PLML/R), as determined by laser ablation experiments. (The sixth cell, PVM, does not lead to touch-mediated movement in wild-type animals but does so when displaced more anteriorly in *mab-5* mutants; Chalfie et al. 1983.)

A simple three-neuron reflex (touch receptor–interneuron–motor neuron) has been proposed to mediate touch behavior (Chalfie et al. 1985). This model, derived from electron microscope reconstructions, has been tested by laser killing of the cells involved. Removal of the PVC interneurons abolishes touch sensitivity in the tail to both the hair and the wire. Similarly, killing the AVD interneurons causes the loss of anterior touch sensitivity in young larvae. These interneurons are coupled by gap junctions to the posterior touch cells or anterior touch cells, respectively. They also receive chemical synapses from the opposing set of touch receptors. (Interestingly, the posterior touch cells form gap junctions with a pair of small neurons, the LUA cells, which form chemical synapses onto the same interneurons as the touch cells. Thus, the LUA cells appear to act as connector cells in the circuit.) It appears that each set of touch receptors activates one set of interneurons and inactivates the other.

The microtubule cells form a number of other synapses that do not seem to be part of the circuit for touch-mediated movement. Some of these connections suggest that touch information is used in other areas of the nervous system that are not directly involved in the withdrawal response. In addition, as indicated above, the touch cells synapse onto the BDU cells, cells that are needed for the guidance of the AVM touch cell into the nerve ring.

The absence of the microtubule cells results in animals that no longer respond to the hair touch; without the AVD and PVC cells, the animals fail to respond to either the hair or the wire. Thus, mutants selected for an insensitivity to hair but not wire touch would be expected to have touch cell defects or be defective in interactions involving the touch cells. Over 350 mutations causing this selective touch insensitivity have been identified (Sulston et al. 1975; Chalfie and Sulston 1981; Chalfie et al. 1981; Chalfie and Thomson 1982; M. Chalfie and M. Au, unpubl.). Of these mutations 348 have been characterized and map to 18 complementation groups (*mec-1–mec-10*, *mec-12*, *mec-14–mec-17*, *egl-5*, *unc-86*, and *lin-32*). Because multiple alleles of almost all of these genes have been identified, it is likely that the map is near to being saturated for those genes that can be mutated to give a strong touch-insensitive (Mec) phenotype. (A number of mutations have more marginal effects on touch sensitivity. These weak *mec* mutations have not been mapped, and it is not known how many genes can convey this phenotype.)

Examination of representative *mec* mutants has allowed the characterization of these genes into three categories. The first category contains two genes, *lin-32* and *unc-86*, that are needed for the generation of the touch cells. Mutations in these genes result in altered cell lineages in which no touch cells are made (Chalfie et al. 1981; C. Kenyon and E. Hedgecock, pers. comm.). These genes, which are pleiotropic in their action, are discussed in Chapter 6. The second category contains a single gene, *mec-3*. *mec-3* mutants have wild-type lineages, but the cells produced lack the characteristic features of the touch cells. This gene, therefore, is required for the expression of the touch cell fate. The third category contains those genes that are needed for touch cell function; mutants defective in these genes have cells that are identifiable as touch cells. Mutations in four of these genes affect characteristic features of the differentiated cells, the 15-protofilament microtubules, and the specific extracellular matrix called the mantle. Examination of these mutants suggests that successful sensory transduction requires both the mantle (for appropriate attachment) and the special microtubules. Also in this last category are eight genes for which representative mutants have touch cells with no detectable abnormalities (although these animals have not been examined for any connectivity defects).

One gene, *mec-4*, in the last category deserves additional comment. Recessive mutations in *mec-4* result in touch-insensitive animals with normal appearing touch cells. Three dominant mutations in this gene, however, produce a striking phenotype: The touch cells degenerate soon after they arise. The degenerative deaths are not affected by any of the *ced* mutations, suggesting that they involve a process distinct from programmed cell death. Mosaic analysis on *mec-4* suggests that the mutant gene needs to be expressed in the microtubule cells for the degenerative deaths to occur (Herman 1987).

Although an extensive search has been done for mutations that selectively affect touch sensitivity, very few genes have been identified that appear to act in a cell-specific fashion on touch cell development. In fact, the *mec* genes represent an approximate upper limit to the number of genes that can be mutated to give this phenotype. Until mosaic analysis or appropriate in situ hybridization or antibody staining can be done to localize the activity of these genes, it is still not possible to say that even these genes are cell specific because the screen for touch mutants would exclude alleles of these having more general effects (although *mec-3, mec-4,* and *mec-7* [a gene that affects microtubule structure] are likely to be cell specific). Certainly the lineage mutants are not specific to the touch cells, and nonmicrotubule cell defects have been seen in *mec-1* (Lewis and Hodgkin 1977; Chalfie and Sulston 1981) and *mec-8* (Perkins et al. 1986) mutants. Because the function of most of the *C. elegans* neurons is not known, other cells could be affected in these mutants. Nonetheless, even if remaining *mec* genes are cell specific, they do not explain all of the features of these cells. For example, null mutations have not been found that selectively alter the course of the touch cell processes or that selectively eliminate only anterior or posterior touch sensitivity (the pleiotropic *lin-32* and *egl-5* mutations affect only posterior touch sensitivity). That such defects have not been found suggests that these properties of the cells are the consequence of more general controls in development.

Support for a more general control of development comes from the experiments using *mab-5* mutants to cause displacement of the PVM cells (see Section III.D). E. Hedgecock (pers. comm.) has identified several *unc* genes that cause the mispositioning of touch cell processes, when mutated. These experiments and the genetic data on the *mec* genes suggest that the six touch cells may be intrinsically identical cells that express different properties (possession of a branch or different sets of synapses), according to their placement and time of birth within the animal.

The touch insensitivity mutations identify a subset of the genes needed for the production and maturation of the touch cells. However, by affecting the generation, specification, and function of the cells, they help to define three stages in touch cell development. Molecular analysis of the genes in this developmental pathway is just beginning, but the initial results provide a model of how touch cell differentiation is controlled. Sequence analysis of *mec-3* DNA (J. Way and M. Chalfie, unpubl.) shows that the gene contains a sequence that is similar to the *Drosophila* homeobox sequences (see, e.g., Gehring 1987). This similarity suggests that the *mec-3* product may act as a DNA-binding protein. As such, the *mec-3* product could specify the differentiation of the touch cells by regulating (directly or indirectly) the expression of other touch cell genes.

VI. FUTURE DIRECTIONS

The structure and development of the *C. elegans* nervous system have been analyzed to a resolution and completeness that has not been possible for any other animal. The knowledge of the wiring diagram and the cell lineage will serve as the basis of future work. One major area of research will be the reconstruction of the detailed structure of the nervous system in younger animals, both early larval stages and the embryo. These studies, as well, perhaps, as those of the male nerve ring, should supply insights into how the *C. elegans* nervous system is assembled. Such reconstructions, some already begun (see above), should also allow us to identify cells that may have pivotal roles in the development of the nervous system. The importance of these cells can then be tested by laser ablation.

Laser ablation studies, particularly of cells in the embryo, should also help characterize activities associated with many more of the neurons. The identification of characteristic behaviors associated with identified cells should give a greater understanding of functional aspects of the circuitry. Studies investigating the sensory cells of the head should prove particularly informative. Additional cell characterization will come from the increased use of various labeling methods (e.g., FITC, antisera) and from comparative studies using *Ascaris*. Such studies serve as the basis both for the analysis of existing mutants and for mutational screens designed to identify additional genes acting in neuronal differentiation.

A large number of mutations have already been identified that affect nervous system function, but their analysis is incomplete. For example, reconstruction of the ventral cord of more of the *unc* mutants should provide insight into both the development and function of the motor system. In addition, a number of mutations result in a change of number, type, position, or the relative timing of appearance of various nerve cells. Reconstruction of the nervous systems of these mutants should provide insight into how the structure and synapses within the nervous system are affected by such alterations.

Parallel with these studies should be the continued genetic and molecular analysis of mutations affecting the nervous system. In particular, mosaic analysis will be important in identifying the focus of action of many of the genes. The molecular analysis of the genes should also provide a clearer picture of how genes affect behaviors.

Finally, additional methods will undoubtedly be used to study the *C. elegans* nervous system. Recently, for example, H. Bhatt and E. Hedgecock (pers. comm.) have developed procedures to study *C. elegans* neurons in tissue culture. This methodology, coupled with our knowledge of the cells in vivo, should help investigate the problem of the relative importance of intrinsic and extrinsic factors in nerve cell differentiation.

The small nervous system of *C. elegans* has proved to be quite complex. What is perhaps encouraging is that so much of the development and function of this nervous system has become amenable to study. Much should be learned in the future.

ACKNOWLEDGMENTS

We are grateful to Jay Burr, Dave Dusenberg, Ed Hedgecock, Bob Horvitz, Carl Johnson, Jim Lewis, Jim Rand, and Tony Stretton for their comments on the manuscript.

12
The Dauer Larva

Donald L. Riddle
Division of Biological Sciences
University of Missouri–Columbia
Columbia, Missouri 65211

I. INTRODUCTION

The dauer, or "enduring," larval stage of *Caenorhabditis elegans* is an example of facultative diapause. Environmental factors act as signals to a receptive developmental stage (the L1 larva), resulting in altered physiology and developmental potential that leads to formation of a third-stage larva specialized for dispersal and long-term survival. Dauer larvae are capable of active movement, but they do not feed. They have a unique morphology and resistance to stress, are altered in energy metabolism, and are arrested in development, as well as aging. Dauer larvae survive four to eight times the 3-week life span of animals that have bypassed the dauer stage. The dauer state itself has been considered to be nonaging, because the duration of the dauer stage does not affect postdauer life span (Klass and Hirsh 1976). As far as is known, the consumption of stored energy may be the major factor limiting dauer larva life expectancy.

Various forms of diapause are common among nematodes, and there may be interesting parallels between the *C. elegans* dauer larva and arrested developmental forms that have been documented as obligatory stages in the life cycles of a number of parasitic species (for review, see Michel 1974; Evans and Perry 1976). In many cases, these arrested stages are dispersal forms specialized to survive periods between host infections (Riddle and Bird 1985).

Dauer larvae exhibit behavior not observed in other stages. Pharyngeal pumping is completely suppressed, and the larvae often lie motionless.

  0-87969-307-X/88 $1 + .00.

Although lethargic, they do show a negative response to touch by rapid movement away from the stimulus. Dauer larvae also tend to crawl up objects that project from the substrate, stand on their tails, and wave their heads back and forth. In the natural soil environment, this behavior, called nictation (Croll and Matthews 1977), may permit attachment to passing insects so that dauer larvae may be carried to new locations.

Diapause may be ended in response to conditions improved for growth and reproduction. Developmental commitment to recovery (exit) from the dauer stage occurs within 1 hour after the animal is placed in a fresh environment with food. After 2–3 hours, it begins to feed, it resumes development, and molts to the L4 stage after an additional 8–10 hours (Fig. 1).

II. MORPHOLOGY AND FUNCTION

Dauer larvae are relatively thin and dense as a consequence of radial shrinkage of the body at the dauer-specific molt (Cassada and Russell 1975). Electron micrographs reveal that the volume of the hypodermis is

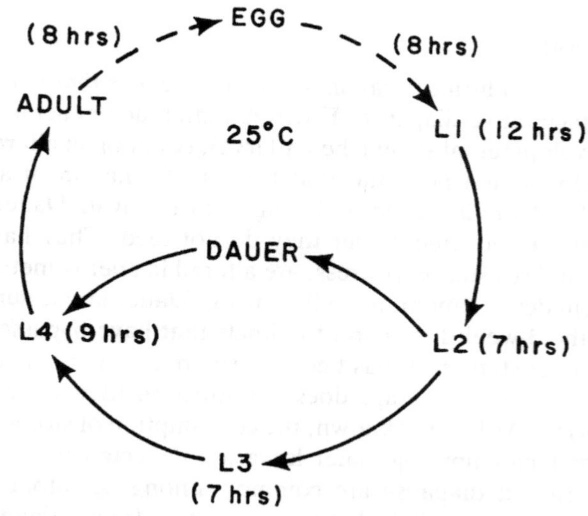

Figure 1 The life cycle of *C. elegans*. The duration of each developmental stage during growth at 25°C on petri plates seeded with *Escherichia coli* strain OP50 is given in parentheses. The solid arrows represent molts. Molting times were determined by monitoring pharyngeal pumping in synchronized populations. Wild-type dauer larvae obtained from starved populations initiate pharyngeal pumping 3–4 hr after being placed in food. The molt to the L4 stage occurs after an additional 8 hr. The 8 hr from egg to L1 is the interval between egg laying and hatching. The entire life cycle requires 3 days at 25°C.

preferentially reduced. About 1 hour after radial shrinkage, dauer larvae acquire resistance to detergent treatment (Swanson and Riddle 1981), presumably as a result of cuticle modification and the sealing of the buccal cavity by a cuticular block (Popham and Webster 1979a; Albert and Riddle 1983). The body-wall cuticle is relatively thick and, when viewed in transverse section, contains a radially striated inner layer not found in other stages (Popham and Webster 1978; Singh and Sulston 1978). The dauer cuticle also contains a dauer-specific form of collagen (Cox et al. 1981b; Cox and Hirsh 1985). The intestinal lumen of the nonfeeding dauer larva is shrunken, and microvilli are small and indistinct (Popham and Webster 1979a). The excretory gland, which exhibits an active secretory morphology in all growing stages, is inactive in the dauer stage, as evidenced by the absence of secretory granules (Nelson et al. 1983).

Electron microscopic examination of the anterior sensory anatomy of the dauer larva revealed an internal and external morphology unlike that of other larvae or adults (Albert and Riddle 1983). Externally, the dauer larva's lip region is relatively flat, and the sensilla surrounding the buccal cavity are not distinguishable (Fig. 2). Internal structural variation has been observed in seven classes of neuron affecting both chemosensory and mechanosensory organs. These organs include the amphids, inner labial sensilla, and deirids (see Chapter 11). For example, contact of the dauer inner labial neurons with the surface environment is relatively remote, via narrow angular channels through the thickened cuticle. This raises the possibility that these organs may have decreased sensitivity in comparison with other developmental stages. Such a structural difference could conceivably account for an altered behavior, such as the observed unresponsiveness to attractants in chemotaxis assays.

Dauer larvae do not reveal the prompt chemotactic response to bacteria seen with well-fed L2 larvae or adults (Albert and Riddle 1983). Although many dauer larvae do not direct their movement toward the bacteria, they do respond to the presence of food by initiating the recovery process. It appears that directed movement to food is delayed until the onset of pharyngeal pumping, which is correlated with the opening of the buccal capsule. It is possible that accompanying changes in the shape of the nonmovable lips that surround the mouth may facilitate chemotaxis. For example, the inner labial neurons may gain more immediate access to environmental stimuli. Scanning electron microscopy of recovering dauer larvae suggests that "swelling" of the lips is correlated with the resumption of feeding and the opening of the buccal capsule (S.J. Brown and D.L. Riddle, unpubl.).

In contrast to the inner labial sensilla, the amphid channels are not modified extensively in dauer larvae, and the amphid openings on the lateral sides of the dauer's nose are easily discernible in scanning electron micrographs. However, the positions of certain sensory dendrites (neurons

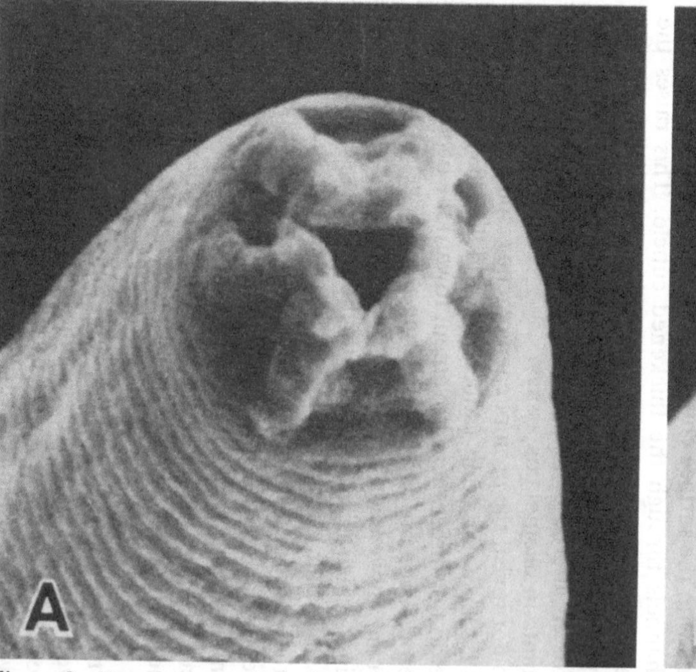

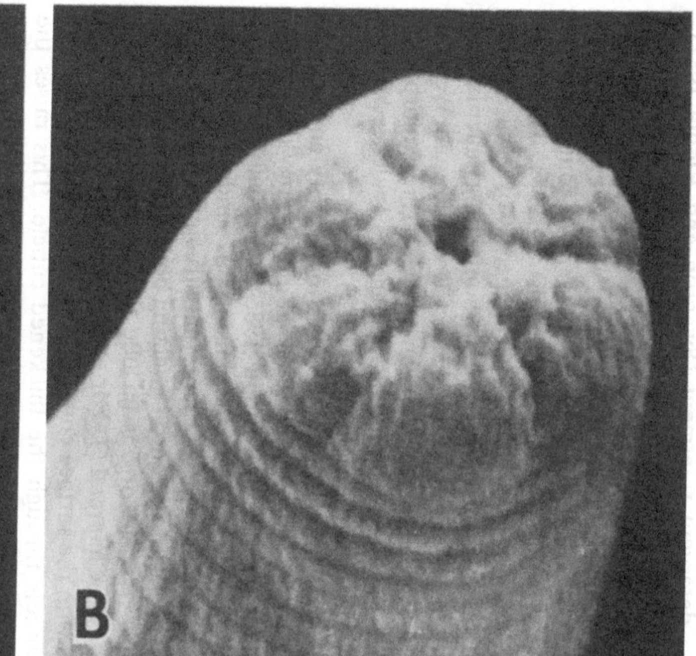

Figure 2 External morphological differences between an L2 larva (*A*) and a dauer larva (*B*) are shown by scanning electron micrographs of the nose. The field width in each micrograph is 7.2 μm. An internal cuticular block, which closes the mouth of the dauer completely (*B*), is revealed in transmission electron micrographs of transverse sections (not shown). In preparation for scanning electron microscopy, larvae were fixed overnight in cold 3% glutaraldehyde in 0.1 M phosphate buffer (pH 7.1) and postfixed for 6 hr in cold 1% buffered OsO_4. Specimens were dehydrated in a graded series of ethanol, critical point dried in CO_2, mounted on copper adhesive tape, and gold coated. Photographs were taken at 10 kV on a Jeol JSM-1 scanning electron microscope.

ASG and ASI) are displaced posteriorly within the channel, and the interface between neuron AFD (the "finger cell") and the sheath cell that forms the channel includes about twice as many AFD microvillar projections in dauer larvae as are present in L2 or postdauer L4 larvae (Albert and Riddle 1983).

Cassada and Russell (1975) observed that resistance of dauer larvae to 1% SDS required both the unique dauer cuticle and the suppression of pharyngeal pumping, but the physiological basis for the ability of chemosensory organs to retain function after exposure of the dauer larvae to SDS is not understood. The most likely mechanism for protection of the amphidial dendrites from detergent treatment is the secretion of material into the channels by sheath cells. Amphid secretions similar to those identified in a variety of other nematode species (McLaren 1976) were also observed in scanning electron micrographs of *C. elegans* dauer larvae (Albert and Riddle 1983).

The two deirids are mechanosensory organs that exhibit dauer-specific developmental alterations. In dauer larvae the sensory endings are oriented parallel to the body wall, and the electron-dense tips are not located within the cuticle but are held against it by a dauer-specific cuticular substructure shaped like a funnel. This structure might aid or amplify transduction of mechanical stimuli through the dauer larva's specialized cuticle and lateral alae. Postdeirids may be altered similarly in dauer larvae, but these sensilla have not been studied. Examination of postdauer L4 larvae showed that the morphology of the deirids, the inner labial neurons and amphidial cells, is similar to that of L2 larvae and adults. Hence, the alterations of these neurons in the dauer larva are reversible. This demonstrates a plasticity in sensory morphology not observed previously in nematodes.

The morphology of the excretory gland cell in feeding animals suggests that a large amount of material is synthesized within this cell, packaged into granules, secreted into the excretory duct, and transported to the exterior of the worm along with fluid from the excretory sinus (Nelson et al. 1983). The presence of secretory granules in all growing stages examined suggests that secretion is either constant or occurs at repeated intervals. The striking differences in gland cell morphology between dauer larvae and other stages suggest that release or degradation of secretory granules accompanies dauer formation and that secretory activity is not present in the dauer stage. However, inactivity of the gland in the dauer larva is apparently a consequence of the dauer state, not a regulator of dauer larva formation. Laser ablation of the gland cell nuclei does not result in commitment of the worm to form a dauer larva, nor does it prevent entry into, or exit from, the dauer stage (Nelson and Riddle 1984).

It is also unlikely that the excretory gland cell secretes the dauer-inducing pheromone described below (Golden and Riddle 1982). Dauer larvae produce the pheromone even though the gland is inactive. In addition, worms that have had their gland nuclei bilaterally ablated during the L2 stage still produce the pheromone, as determined by bioassay of individual adults (J.W. Golden and D.L. Riddle, unpubl.).

III. ENVIRONMENTAL CUES

The environmental factors that influence development of the dauer larva include a pheromone, the food supply, and temperature (Golden and Riddle 1982; Ohba and Ishibashi 1982; Golden and Riddle 1984a). The concentration of the dauer-inducing pheromone apparently serves as a measure of the population density. Higher concentrations of pheromone (a family of structurally related, stable, nonvolatile, fatty acidlike compounds) enhance dauer larva formation and inhibit recovery from the dauer stage. At least two *Caenorhabditis* species (*C. elegans* and *C. briggsae*) produce a common pheromone activity (Fodor et al. 1983), whereas other soil rhabditidae produce a pheromone not active on *Caenorhabditis* (Golden and Riddle 1982). The *Caenorhabditis* pheromone is produced throughout the life cycle, and efficient induction of dauer larva formation requires continuous exposure to the pheromone after the middle of the first larval stage. Dauer-inducing conditions prolong the L2 intermolt period and result in a morphologically distinct L2 larva designated L2d (Golden and Riddle 1984a).

The food signal is a neutral, carbohydratelike substance produced by *Escherichia coli* and found in yeast extract. It has effects opposite that of the pheromone; it inhibits dauer larva formation and enhances recovery (Golden and Riddle 1984b). The ratio of pheromone to food signal, and not the absolute amount of either stimulus, influences dauer larva formation and recovery. The behavioral assays for dauer larva recovery suggest that dauer larvae integrate the two competitive chemosensory cues in reaching a "decision" either to resume development or to remain in the dauer stage. Dauer larva recovery is not only affected by the externally applied cues. The age of the dauer larvae also has a dramatic influence on their response to the combination of pheromone and food signal. Over a 2-week period, they become less sensitive to the pheromone, more sensitive to food, or both (Golden and Riddle 1984a).

The dauer-inducing pheromone is assayed most conveniently by its inhibition of dauer larva recovery in the presence of a standardized amount of food signal in the wells of a microtiter dish (Golden and Riddle 1982). Pheromone activity is measured by testing a dilution series of a pheromone sample for the greatest dilution that inhibits recovery (Fig. 3). The bioassay is semiquantitative; fourfold differences in pheromone concentration have been reliably detected. The sample dilution series typically produces assay wells scored as positive or negative, separated by a single assay well

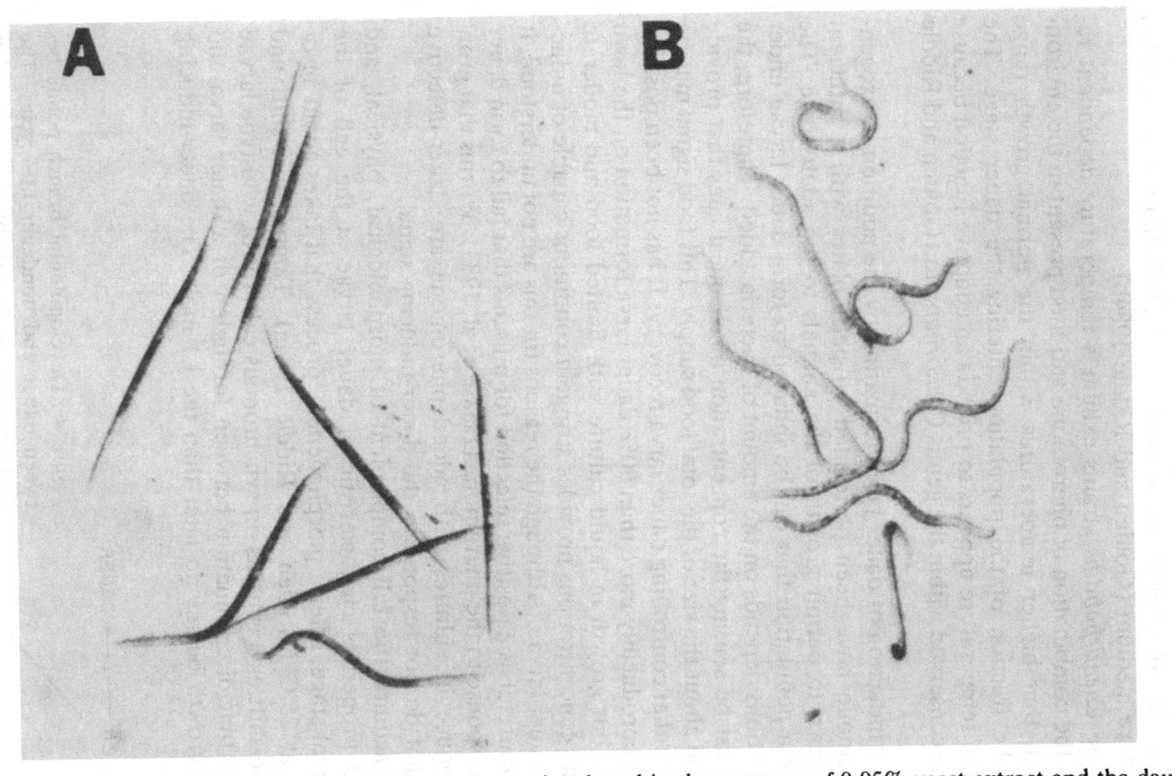

Figure 3 Dauer larva recovery bioassay. (*A*) Dauer larvae incubated in the presence of 0.05% yeast extract and the dauer-inducing pheromone are inhibited from recovering, retain their thin, dark appearance, and usually remain rigid and motionless. (*B*) Dauer larvae incubated in the absence of added pheromone recover to resume development within 4 hr, increase movement, and begin feeding (worms are ~400 μm in length.)

scored as a "borderline" response, in which some dauer larvae are inhibited from recovery and some are not. The number of units of pheromone activity in a sample can be calculated as the reciprocal of the final dilution that produces the borderline response. A recently improved bioassay allows detection of pheromone activity at a 1/2000 dilution of depleted liquid culture medium (Golden and Riddle 1984b).

Wild-type *Caenorhabditis* strains exhibit a temperature-dependent response to the combination of pheromone and food present in the environment, such that higher temperatures within the normal growth range increase the fraction of the population entering the dauer stage. The influence of temperature appears to provide an adaptive function because it has been conserved within the genus *Caenorhabditis* (Golden and Riddle 1984c).

Tests for induction of dauer larva formation by the addition of exogenous pheromone have been performed on petri plates with a sufficient bacterial lawn to permit growth of the animals to the adult stage. The fraction of a population that can be induced to form dauer larvae under these conditions depends on the amount of bacteria added. Therefore, the animals must be sensing the concentration of chemical cues that provide information about the size of the total food supply. This food signal may be the same as that controlling dauer larva recovery. It has not been possible to induce reproducibly more than 90% dauer larva formation on the lawn of *E. coli*, but growth in liquid culture with limited food and moderate pheromone concentrations provides stringent conditions capable of inducing 100% dauer larvae. Although the reasons for this are not understood, it has proved useful as a positive selection for mutants that fail to form dauer larvae in response to the pheromone (Haase et al. 1983). Worms that grow and reproduce in cultures with added pheromone are dauer-defective mutants that do not respond to the dauer-inducing signal.

Worms acquire the full complement of morphological, physical, and behavioral properties characteristic of dauer larvae at the end of the second larval stage, after a morphogenetic process that takes about 11 or 12 hours at 25°C (Golden and Riddle 1984a). Temperature-shift and pheromone-shift experiments on synchronous cultures of wild-type larvae established that discrimination between formation of a dauer larva and continued growth begins no later than the L1 molt. The dauer-inducing

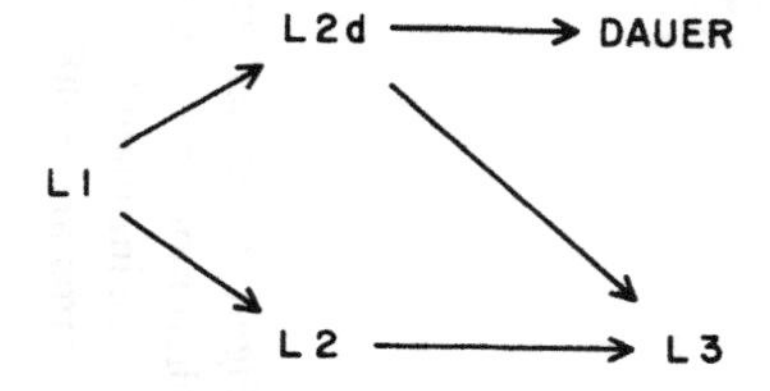

Figure 4 Developmental branch points between dauer and nondauer life cycles. The L2d larvae retain the potential to form either dauer or L3 larvae, whereas L2 larvae are committed to nondauer development.

influences of pheromone and elevated temperatures are effective at this time. Commitment to form an L3 larva at the following molt is essentially irreversible, whereas preparation for dauer larva formation can be reversed by changing one or more of the three environmental parameters prior to the second molt (Fig. 4).

IV. PHYSIOLOGY

Because dauer larvae are formed in response to overcrowding and limited nutrition, investigation of energy metabolism in L1, L2, predauer (L2d), and dauer larvae is a natural approach to the physiology of dauer larva development. The major phosphorus-containing metabolites in extracts of *C. elegans* have been studied using nuclear magnetic resonance (NMR). The relative concentrations of ATP, ADP, AMP, sugar phosphates, and other metabolites change during the life cycle, producing stage-specific phosphorus NMR spectra (W.G. Wadsworth and D.L. Riddle, in prep.). These spectra are consistent with assays of isocitrate dehydrogenase and isocitrate lyase (Fig. 5), indicating high activity of the glyoxalate pathway only during the L1 stage, whereas respiration during the L2, L3, and L4 stages occurs preferentially through the tricarboxylic acid (TCA) cycle.

Around the L1 molt, energy metabolism in animals destined to become dauer larvae diverges from that of animals committed to growth. Relative to the L1, the pre-L3 L2 larvae exhibit increased isocitrate dehydrogenase activity, as well as increased ATP and other high energy phosphates, but predauer (L2d) larvae exhibit declining enzyme activities and declining levels of high energy phosphates. The predominant phosphorus NMR signal in dauer larva extracts corresponds to inorganic phosphate. The higher energy state of growing L2, L3, and L4 larvae can be restored within 4 hours after wild-type dauer larvae resume feeding in bacteria. Thus, *C. elegans* metabolism responds to environmental change but is also developmentally regulated.

V. MUTANTS

The genetic study of dauer larva formation has been aimed at understanding how a subset of an animal's genes specifies a simple developmental sequence. Dauer larva formation is not essential for survival of the organism in the laboratory, and it lends itself well to methods of genetic selection. Thus, it is possible to use genetic analysis to identify genes affecting the switch into the dauer stage and genes influencing dauer larva morphogenesis. Mutants usually can be propagated in the homozygous state for developmental, physiological, and biochemical study. Two classes of recessive mutant have been analyzed. Dauer-constitutive mutants form dauer larvae independently of the environmental cues. Temperature-sensi-

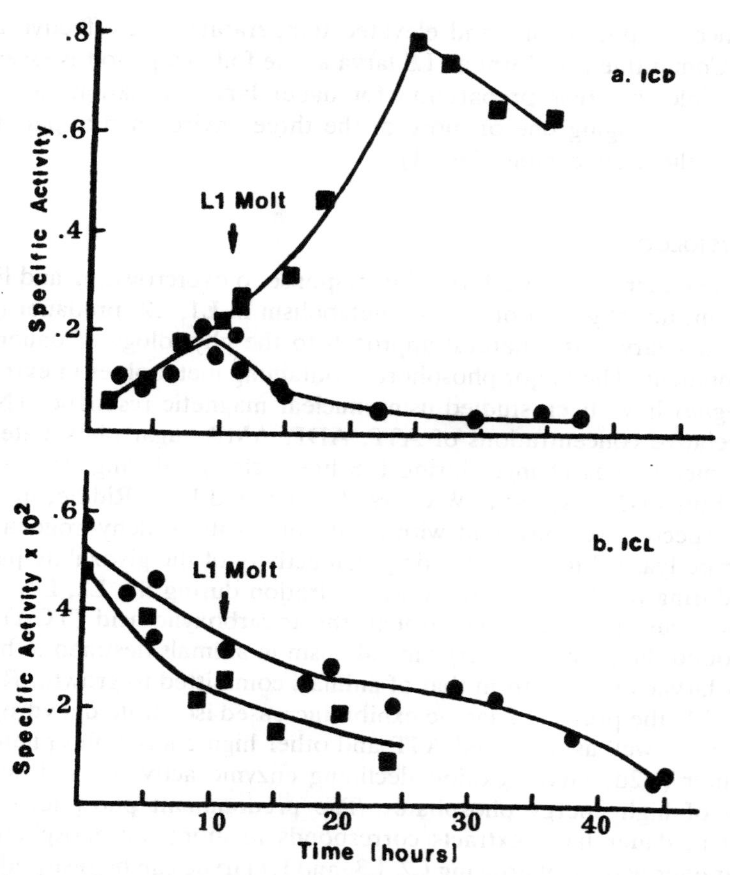

Figure 5 Specific activities of isocitrate dehydrogenase and isocitrate lyase during larval development. Synchronized L1 larvae were placed on lawns of *E. coli* at 25°C to initiate development, and samples were assayed for isocitrate dehydrogenase (ICD) activity (*a*) and isocitrate lyase (ICL) activity (*b*). Enzyme activity of N2 (■) represents the L1 (0–12 hr), L2 (13–20 hr), L3 (21–26 hr), and L4 (27–35 hr) stages. The enzyme activity of CB1372 *(daf-7)* (●) represents the L1 (0–12 hr), L2d (13–30 hr), and dauer stages.

tive dauer-constitutive mutants form dauer larvae at high frequency only at restrictive temperatures, and if such larvae are shifted to permissive temperatures, they exit from the dauer stage and resume growth (Swanson and Riddle 1981). Such mutants are easily selected as detergent-resistant larvae formed in abundant food (Cassada 1975; Riddle 1977).

A second type, called dauer-defective mutants, is unable to form dauer larvae. About half of such mutants exhibit sensory defects involving chemotaxis, male mating, or osmotaxis (Riddle 1977; Albert et al. 1981).

Several mutants defective in both dauer larva formation and chemotaxis have been examined ultrastructurally, and a variety of morphological abnormalities in the afferent endings of anterior and posterior sensory neurons have been observed (Lewis and Hodgkin 1977; Albert et al. 1981; Perkins et al. 1986). Of all the anterior sense organs examined in mutants that are defective in both dauer larva formation and chemotaxis, only the two amphids are affected consistently in their morphology (Figs. 6 and 7).

VI. GENETIC PATHWAY

A relationship between dauer-constitutive and dauer-defective mutations first was detected by selecting revertants of temperature-sensitive dauer constitutives, *daf-2, daf-4,* and *daf-7* (Riddle 1977). Homozygous recessive revertants, arising in the F2 generation after mutagenesis, grew and reproduced at 25°C, whereas parental-type larvae became dauer larvae and stopped growing. Not only did most revertants escape dauer larva formation, but they failed to form dauer larvae even when starved. Crossing the revertants with the parental strain revealed that they were double mutants that not only carried the parental dauer-constitutive mutation but also were homozygous for an epistatic, recessive dauer-defective mutation.

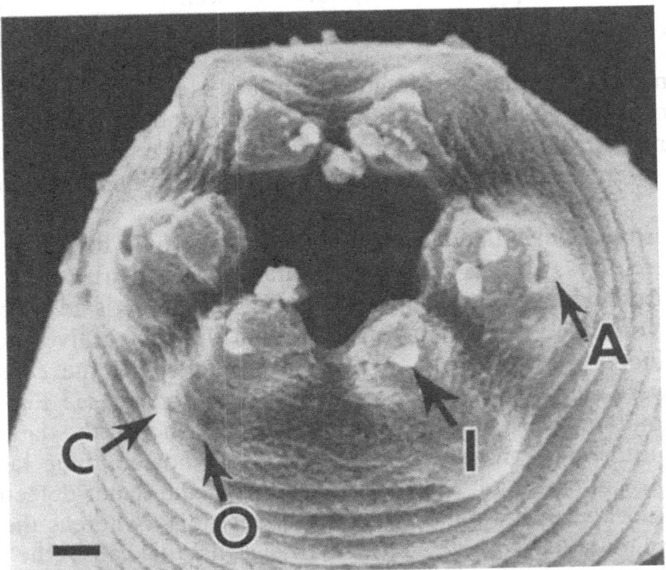

Figure 6 Scanning electron micrograph of a wild-type (N2) adult, showing the locations of major anterior sensory organs. A few bacterial cells are visible around the buccal cavity. (A) Amphid; (C) cephalic sensillum; (I) inner labial; (O) outer labial. Scale bar represents 1.0 μm.

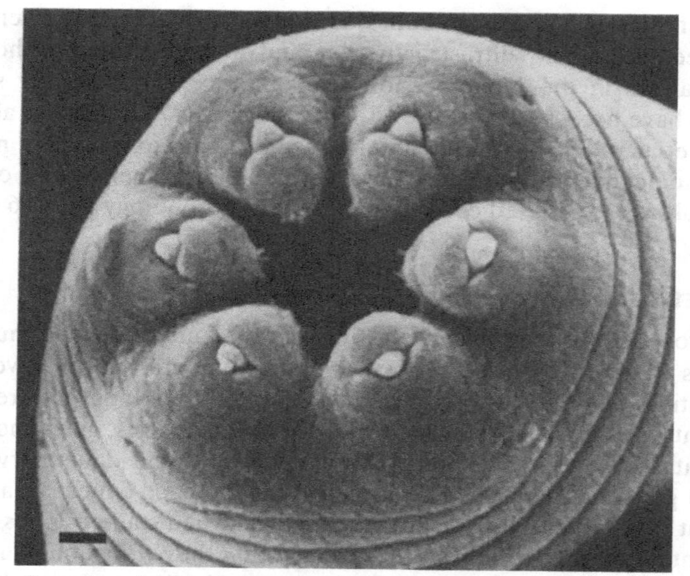

Figure 7 Scanning electron micrograph of a CB1377 (*daf-6*) adult. The amphids are not visible. The amphid channels are blocked by the sheath cells, preventing neuronal access to the external environment. Specimen orientation is similar to that in Fig. 3, dorsal side up. Scale bar represents 1.0 μm.

Thus, some dauer-defective mutations can act as epistatic suppressors of dauer-constitutive mutations. Initially, 25 independently isolated revertants were distributed among six complementation groups, and some revertants were found to be alleles of previously identified dauer-defective genes.

To determine which dauer-constitutive mutants could be suppressed by various defectives, a series of multiple mutants was constructed by genetic crosses (Riddle et al. 1981). The data suggest that there is an order to the functions controlled by the *daf* genes and that the constitutive and defective mutations affect a common sequence (Fig. 8). If the constitutive mutants were simply blocked in entry to the L3 stage (see Fig. 1) and defective mutants were blocked in a separate sequence to the dauer larva, the constitutive-defective double mutants could not continue larval development at all. Instead, the data indicate that the dauer-defective mutants are blocked in the pathway to the dauer stage, whereas the dauer-constitutive mutants generate a false internal signal that causes the mutants to form dauer larvae even in the absence of appropriate environmental cues. If the pathway is blocked after the false signal, a double mutant will be dauer defective. However, if the false signal is generated after the block, a double mutant will be dauer constitutive.

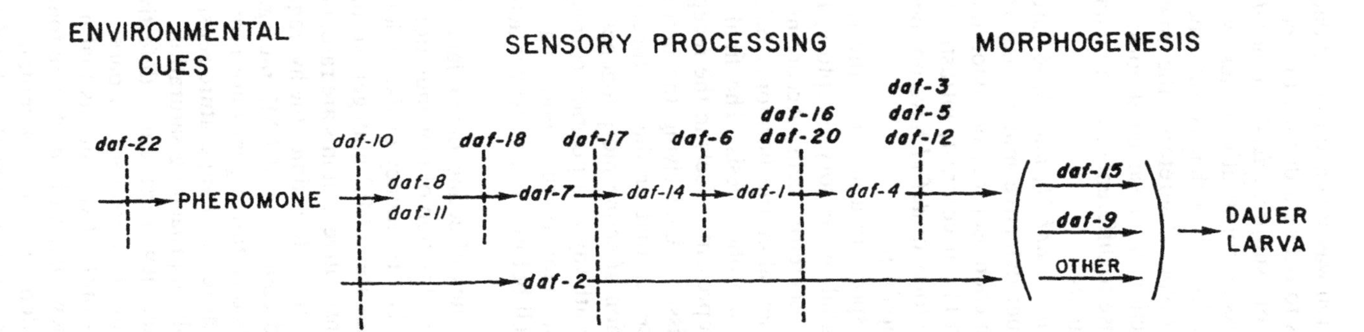

Figure 8 A genetic pathway for dauer larva formation (one of several possible representations) based on epistatic relationships between dauer-constitutive mutations and dauer-defective mutations. Gene names of dauer-constitutive mutations are given in the pathway itself to represent points where false signals may be initiated. Mutations in dauer-defective genes block the pathway at the positions shown by dashed lines. The *daf-10* mutation is drawn to block both branches of the pathway on the basis of indirect evidence. First, both *daf-8* and *daf-2* are epistatic to *daf-10* and, second, the *daf-10* mutation blocks recovery of both *daf-2* and *daf-8* dauer larvae. All mutations are epistatic to *daf-22*, which fails to produce the dauer-inducing pheromone. Genes shown in boldface type have been tested for epistatic relationships with *daf-9* and *daf-15*. For genes other than *daf-9* and *daf-15*, all possible constitutive-defective combinations have been constructed.

An unambiguous order of function was assigned to most dauer-constitutive genes. However, *daf-2* seems to define a distinct branch of the pathway. The *daf-16* and *daf-20* mutations suppress *daf-2* but not *daf-4,* whereas *daf-3, daf-5*, and *daf-12* suppress all constitutive genes except *daf-2.* A block in the upper branch of the pathway is both necessary and sufficient to produce a dauer-defective phenotype. Therefore, the lower pathway specified by *daf-2* must not function in these conditions. This second pathway may represent the response to different environmental signals, or it may function only in *daf-2* mutants.

Three steps, represented by *daf-10, daf-11*, and *daf-6,* are correlated with chemotaxis defects. The dauer-defective mutants *daf-10* and *daf-6* exhibit pleiotropic defects, both in chemosensory behavior and in sensory neuronanatomy (Albert et al. 1981). In contrast, ultrastructural abnormalities in the anterior sensory anatomy of *daf-11* have not been detected (P.S. Albert and D.L. Riddle, unpubl.).

The genetic pathway may correspond to a neural pathway involved in detection of an environmental signal by sensory receptors, followed by transduction of the signal to neurosecretory cells. According to this hypothesis, neurosecretion would then alter the hormonal balance in the nematode to initiate dauer larva morphogenesis. The final step in the sensory processing pathway may represent defects in the reception of the neurosecretory signal by target tissues. Underlying the neural pathway would be the developmental processes that produce a functional set of nerve cells. For example, a mutation affecting morphogenesis of specific sensory neurons would affect the pathway at a position corresponding to the function of those cells. Thus, many of the genes may exert their primary effects on development well before the pathway is actually set in motion physiologically.

There are approximately 100 genes affecting dauer larva formation, based on mutation frequency and the fraction of new mutations that recur in previously characterized genes. The frequency of dauer-defective mutants obtained by ethylmethanesulfonate (EMS) mutagenesis of wild-type *C. elegans* is 5%, whereas dauer-constitutive mutants are rare, occurring at an aggregate frequency of 0.2%, or about four times the average mutation frequency per gene under standard conditions of EMS mutagenesis. Almost all of the 13 identified dauer-constitutive genes are represented by more than one mutant allele. Therefore, the temperature-sensitive dauer-constitutive mutant class is probably approaching saturation.

Two other experiments support this view. First, dauer-defective mutants *(daf-16* and *daf-20)* late in the pathway were used as parent strains for selecting a restricted class of dauer-constitutive mutants that might define even later steps (see Fig. 8). Five new alleles of *daf-4* were found, but no new dauer-constitutive genes were identified. Second, mutants with maternal effects have not been overlooked. All eight alleles of *daf-1* share this property. Homozygous *daf-1* animals do not form dauer larvae if their maternal parent was heterozygous *(+/daf-1)*. Recently, maternal-effect mutants of two other genes, *daf-23 II* and *daf-25 I*, have been identified (P.S. Albert and D.L. Riddle, unpubl.).

VII. MUTANT RESPONSES TO PHEROMONE

Except for *daf-2*, the temperature-sensitive dauer-constitutive mutants hyperrespond to the dauer-inducing pheromone at their permissive temperature (Golden and Riddle 1984c), and they form dauer larvae independently of pheromone at their restrictive temperature (Golden and Riddle 1985). The temperature-sensitive periods (TSPs) for most temperature-sensitive dauer-constitutive mutants correspond to the timing of the temperature-sensitive response of wild-type worms grown in the presence of pheromone (Golden and Riddle 1984a). These experiments suggest that the initial discrimination between alternate developmental pathways is made around the L1 molt. An exceptional temperature-sensitive dauer-constitutive mutant is *daf-14* (Swanson and Riddle 1981), which has its major TSP within the first larval stage, although it has a minor TSP at about the time of the L1 molt. The *daf-14* mutant may be precocious in its discrimination between the alternative developmental pathways. *daf-2* is the only dauer-constitutive mutant that does not hyperrespond to the dauer-inducing pheromone, and it is the only such gene with both tmperature-sensitive and non-temperature-sensitive alleles. In contrast, all known mutations in the dauer-constitutive genes *daf-4* and *daf-7* produce a temperature-sensitive phenotype, even though some mutations appear to be amber nonsense alleles (Golden and Riddle 1984c). Thus, constitutive dauer larva formation may result from non-temperature-sensitive genetic defects that reveal the wild-type temperature-sensitive process.

Seven dauer-constitutive mutants are deficient in egg laying, and at least two *egl* mutants, *egl-4* and *egl-40*, overrespond to the dauer-inducing pheromone (Golden and Riddle 1984c). The egg-laying rate appears to be modulated by environmental factors such as the food supply, and some *egl* mutants may be defective in processing these environmental signals. The *egl* mutants have been classified into categories by their response to drugs that affect egg laying (Trent et al. 1983). One category includes seven *egl* genes and four dauer-constitutive genes. Interestingly, all alleles of the seven *egl* genes in this category are temperature sensitive, as are the dauer-constitutive alleles.

The analysis of dauer-defective mutants indicates that the dauer-inducing pheromone is necessary for the formation of dauer larvae. Dauer-defective mutants representing ten genes failed to respond to the pheromone. If another cue, such as starvation, were sufficient in itself for the induction of dauer larvae, some dauer-defective mutants might have been blocked in response to this cue and might have been expected to respond to pheromone.

VIII. PHEROMONE PRODUCTION

A mutant deficient in pheromone production has been isolated recently, and the mutant defect is sufficient to produce a dauer-defective phenotype. In contrast to the wild-type or to other dauer-defective strains (Golden and Riddle 1984c), pheromone activity was not found in the medium of growing *daf-22* cultures (Golden and Riddle 1985), nor was it found in organic solvent extracts of the worms themselves, using bioassays capable of detecting 0.1% of the wild-type activity. The absence of pheromone activity in the organic extracts suggests that the mutation blocks pheromone production rather than secretion. Addition of exogenous pheromone completely restores the mutant's ability to form dauer larvae. The *daf-22* mutant provides genetic evidence that dauer larva formation is not a response to nonspecific effects of overcrowding such as a buildup of metabolic waste products. This evidence is consistent with the observed species specificity of pheromone activity.

To determine whether pheromone production is necessary for expression of the dauer-constitutive phenotype at restrictive temperature, doubly mutant strains carrying *daf-22* and a dauer-constitutive mutation were constructed (Golden and Riddle 1985). The *daf-22* mutation did not affect the expression of any of the dauer-constitutive mutations (Fig. 8). The double mutants were pheromone deficient, and they formed dauer larvae constitutively at 25°C. Therefore, all of the dauer-constitutive mutations are sufficient to induce dauer larva formation even in the absence of detectable pheromone, a result consistent with other genetic data suggesting that dauer-constitutive mutants produce a false internal signal that induces dauer larva formation (Riddle et al. 1981).

The observations mentioned above lead to the following model for the sensory control of dauer larva formation in wild-type and *daf* mutants. In the wild type, the dauer-inducing pheromone and elevated temperature may act independently to enhance dauer larva formation, whereas food signal inhibits dauer larva formation. Normally, all three cues are influencing the decision to form a dauer larva simultaneously. The pheromone is normally necessary for dauer larva formation, and temperature and food signal have modulating effects. Dauer-defective mutants may have defects at any point in the signaling pathway, but they always underrespond to the pheromone. Some leaky mutants may exhibit temperature dependence. Mutants that have lost all chemosensory function will be dauer defective, because responsiveness to pheromone is necessary for dauer larva formation in a wild-type genetic background.

Dauer-constitutive mutants, on the other hand, produce an internal signal favoring dauer larva formation. This may result from a specific defect in their response to the food signal. In such a case, their over-response to pheromone and temperature would be an indirect result of their genetic predisposition toward dauer larva formation. Temperature alone (25°C) becomes sufficient to induce dauer larvae constitutively, even in the complete absence of pheromone. At lower temperatures, their hyperresponse to pheromone becomes evident.

IX. MUTANTS DEFECTIVE IN MORPHOGENESIS

Two non-temperature-sensitive dauer-constitutive mutants, *daf-9(el406) X* and *daf-15(m81) IV,* form dauerlike larvae that are abnormal in morphology and sensitive to detergent (Albert and Riddle 1988). The mutations affect the switch into the dauer developmental sequence, but dauer larva morphogenesis is incomplete. These larvae fail to reach reproductive maturity. The *daf-9* dauerlike larvae exhibit many dauer features, such as flattened nose, occluded buccal cavity, striated cuticle, and lateral alae, but sensory dendrite morphologies, characteristic of both dauer and L2 larvae, are observed. They are shrunken radially but have a more transparent appearance than do normal dauer larvae. A few individuals resume sporadic feeding and undergo some gonadal cell proliferation, but they are sterile. The *daf-15* dauerlike larvae are dark bodied and never pass through the L2 lethargus successfully but continue sporadic feeding, increasing in size to about 0.7 mm. The *daf-15* larvae are nondauer in cuticle structure, nose shape, and most aspects of sensory morphology, and they have an open mouth. However, the intestinal cells contain the electron opaque vesicles seen in N2 dauer larvae and not found in *daf-9*. Both mutants have L2-like intestinal lumens and excretory glands. Although *daf-9* is more dauerlike than *daf-15* by several criteria, these mutants seem to initiate different portions of dauer larva morphogenesis (Albert and Riddle 1988).

The *daf-9* and *daf-15* mutations are epistatic to all tested dauer-defective mutations and, therefore, represent "late" functions in the genetic pathway (Fig. 8). Because neither *daf-9* nor *daf-15* are temperature sensitive, they may function after the wild-type temperature-sensitive process. Dauerlike *daf-15;daf-9* larvae are more resistant to SDS than either parental mutant, and their appearance is somewhat like a mixture of *daf-9* and *daf-15* phenotypes, but morphogenesis is still incomplete. Therefore, other dauer pathways in addition to those induced by mutations in *daf-9* and *daf-15* must also be necessary to complete morphogenesis. Similarly, dauerlike *daf-9* or *daf-15* larvae homozygous for the dauer-constitutive (temperature-sensitive [ts]) mutation *daf-2(e1370ts),* or *daf-7(e1372ts),* are more resistant to SDS at 25°C than *daf-9* or *daf-15* alone. In this regard, the double mutants are somewhat more dauerlike, but the effects of *daf-9* and *daf-15* are not masked by the constitutive signal to form a complete dauer larva. Doubly mutant larvae are not capable of exit from the dauerlike state. It is possible that the *daf-9* and *daf-15* defects result in abnormal timing or execution of morphogenetic sequences, so that normal dauer larvae cannot be formed, even when a dauer-constitutive signal is generated.

Other genes do not affect the signal but do affect the subsequent process of dauer larva morphogenesis. A *daf-13* mutant forms morphologically normal dauer larvae in response to appropriate conditions, but the dauer larvae are sensitive to SDS treatment, possibly because of a defect in cuticle maturation (P.S. Albert and D.L. Riddle, unpubl.). When the *daf-13* mutation is combined with the dauer-constitutive mutation, *daf-2,* the constitutive phenotype is unaltered except that the constitutively formed dauer larvae are sensitive to SDS. The *daf-13* mutant is representative of a class of purely "process" mutations that do not affect the signal pathway and do not exhibit epistatic interactions with dauer-constitutive mutations.

X. EXIT FROM THE DAUER STAGE

There are few, if any, genes involved specifically in controlling exit from the dauer stage. Selection (by resistance to SDS) for mutants that are abnormally slow in dauer larva recovery produced no new mutant types. Dauer-constitutive mutants (which recover slowly) and leaky dauer-defective mutants (affected in both entry and exit) were the only ones found. This suggests that the genetic requirements for recovery from the dauer stage are similar to the requirements for entry. To determine whether all dauer-defective genes specifying the entry pathway also affect exit from the dauer stage, constitutive-defective double mutants were constructed in which the constitutive gene was epistatic. In the case of the *daf-10* and *daf-6* double mutants, recovery of constitutively formed dauer larvae (on temperature shift to 15°C) was inhibited almost completely, showing that function of these two dauer-defective genes is required for both entry and exit (Albert et al. 1981). In the case of the *daf-17* double mutant, dauer larva recovery was inhibited, at most, only marginally. Thus, the genes required to trigger exit from the dauer stage may be simply a subset of the genes required to trigger entry. This is consistent with the observation that both entry into and exit from the dauer stage is controlled by the dauer-inducing pheromone.

Dauer larvae initiate recovery when the environmental ratio of food to dauer-inducing pheromone is appropriately high (Golden and Riddle 1982). They become developmentally committed to recovery 50–60 minutes after being shifted from dauer-inducing conditions to a fresh lawn of bacteria (Golden and Riddle 1984a). Increased activity of the excretory system, along with twitching of pharyngeal muscles and the opening of the intestinal lumen, are the first visible signs that recovery from the dauer stage is under way (Nelson and Riddle 1984). It is interesting to speculate that, prior to opening of the mouth and initiation of feeding, the dauer larva may excrete metabolites involved in maintenance of the developmentally arrested state.

Ultrastructural analysis of mutants has correlated neuronal abnormalities with defects in both chemotaxis and dauer larva formation (Lewis and Hodgkin 1977; Albert et al. 1981). A chemosensory mutant *daf-10(e1387) IV*, affected in its ability to enter into or exit from the dauer stage, possesses abnormal amphidial neurons (Albert et al. 1981). Because the amphid channels appear to be open in the dauer stage, these neurons remain possible candidates for involvement in dauer-specific chemosensory functions. Receptors for the dauer-inducing pheromone and/or recovery stimuli may be located in these organs.

The generalization that genes affecting the behavior of dauer larva recovery may be simply a subset of the genes affecting the behavior of dauer larva formation may not apply to the developmental sequences that these behaviors set in motion. Indeed, the genetic control of developmental events during the L3 and L4 stages of nondauer development may be substantially different from the control of corresponding events during postdauer development. Upon feeding, dauer larvae complete the third larval stage, a third larval molt, fourth-stage development, and a fourth larval molt. The hypodermal cell lineages during postdauer development appear essentially identical to those of the corresponding stages (L3 and L4) of nondauer development. However, certain mutations in heterochronic genes (Ambros and Horvitz 1984) affect the two developmental pathways differentially. For example, mutations of *lin-14* or *lin-28* cause profound effects on nondauer development, including changes in the total number of molts and changes in the timing of vulval development and adult cuticle formation (Ambros and Horvitz 1987). In contrast, the postdauer development of these same mutants appears identical to that of the wild type. Thus, it appears that the normal activities of these genes are required for the proper temporal pattern of nondauer development but are not required for the corresponding stages of postdauer development. These pathways are not completely distinct; for example, *lin-29* activity seems to be required for the proper timing of both developmental sequences (V. Ambros and Z. Liu, pers. comm.).

At least one mutant abnormal in cuticle formation differentially affects adults developed from nondauer and postdauer pathways. Cox et al. (1980) observed that adult *sqt-2* animals derived from L3 juveniles displayed a squat phenotype, whereas adult *sqt-2* animals arising from dauer larvae manifested a roller phenotype. Because the cuticle is shed twice between the L3, or dauer, and adult stages, it is unlikely that this difference in adults is due to residual cuticle differences between the L3 and the dauer larva. These investigators suggested that published observations on the adult form should carefully specify whether the adult was derived from the L3 or the dauer larva.

XI. SUMMARY

In summary, genetic and developmental studies have led to an understanding of some of the complex behavior and physiology of dauer larvae. A variety of approaches has provided information on genetic relationships, developmental timing, and cellular specificity. The major challenge at present is to determine the functions of some of the *daf* genes.

ACKNOWLEDGMENTS

I thank M. Chalfie and J. Sulston for helpful comments on the manuscript. This work has been supported by National Institutes of Health grant HD-11239 and Research Career Development award HD-00367.

Appendices

APPENDIX 1
Parts List

Compiled by John Sulston and John White

MRC Laboratory of Molecular Biology
Cambridge, England

Appendix 1 provides a list of all terminally differentiated somatic cells, together with key blast cells (viz. embryonic founder cells and blast cells present at hatching).

First column: Functional name. This nomenclature is fairly arbitrary; its purpose is to allow cells to be classified by type. Note the use of appended symmetry operators (A, anterior; P, posterior; D, dorsal; V, ventral; L, left; R, right) to distinguish between equivalent cells in symmetrical groups.

Second column: Lineage name. This nomenclature formally summarizes the lineage by which the cell is derived from a key blast cell; it is described fully in Chapter 5.

Third column: Brief description. See appropriate chapters for more detail.

The Table following the Parts list provides nuclear counts.

PARTS LIST FOR THE NEMATODE *C. ELEGANS*

AB	P_0 a	Embryonic founder cell
ADAL	AB plapaaaapp	Ring interneurons
ADAR	AB prapaaaapp	
ADEL	AB plapaaaapa	Anterior deirids, sensory receptors in lateral alae, contain dopamine
ADER	AB prapaaaapa	
ADEshL	AB arppaaaa	Anterior deirid sheath cells
ADEshR	AB arpppaaa	"
ADEso	H2 aa L&R	Anterior deirid socket
ADFL	AB alpppppaa	Amphid neurons, dual ciliated sensory endings, probably chemosensory, enter ring via commissure from ventral ganglion, take up FITC
ADFR	AB praaappaa	
ADLL	AB alppppaad	Amphid neurons, dual ciliated sensory endings, probably chemosensory, project directly to ring, take up FITC
ADLR	AB praaapaad	
AFDL	AB alpppapav	Amphid finger cells, associated with amphid sheath
AFDR	AB praaaapav	
AIAL	AB plppaappa	Amphid interneurons
AIAR	AB prppaappa	"
AIBL	AB plaapappa	Amphid interneurons
AIBR	AB praapappa	"
AIML	AB plpaapppa	Ring interneurons
AIMR	AB prpaapppa	"
AINL	AB alaaaalal	Ring interneurons
AINR	AB alaapaaar	"
AIYL	AB plpapaaap	Amphid interneurons
AIYR	AB prpapaaap	"
AIZL	AB plapaaapav	Amphid interneurons
AIZR	AB prapaaapav	"
ALA	AB alapppaaa	Neuron, sends processes laterally adjacent to excretory canal and also along dorsal cord
ALML	AB arppaappa	Anterior lateral microtubule cells, touch receptor neurons
ALMR	AB arpppappa	
ALNL	AB plapappppap	Neurons associated with ALM, send processes into tailspike
ALNR	AB prapappppap	
AMshL	AB plaapaapp	Amphid sheath cells
AMshR	AB praapaapp	"
AMsoL	AB plpaapapa	Amphid socket cells
AMsoR	AB prpaapapa	"
AQR	QR ap	Neuron, rudimentary cilium, projects into ring
AS1	P1 apa	Ventral cord motoneurons, innervate dorsal muscles, no ventral counterpart, probably cholinergic, similar to VAn but receive additional synaptic input from AVB
AS2	P2 apa	
AS3	P3 apa	
AS4	P4 apa	
AS5	P5 apa	"
AS6	P6 apa	"
AS7	P7 apa	"
AS8	P8 apa	"
AS9	P9 apa	"
AS10	P10apa	"
AS11	P11apa	"

ASEL	AB alppppppaa	Amphid neurons, single ciliated endings, probably chemosensory; project into ring via commissure from ventral ganglion, make diverse synaptic connections in ring neuropil
ASER	AB praaapppaa	
ASGL	AB plaapapap	
ASGR	AB praapapap	
ASHL	AB plpaappaa	
ASHR	AB prpaappaa	")
ASIL	AB plaapapppa	")
ASIR	AB praapapppa	") Also take up
ASJL	AB alpppppppa	") FITC
ASJR	AB praaappppa	")
ASKL	AB alpppapppa	")
ASKR	AB praaaapppa	")
AUAL	AB alpppppppp	Neurons, processes run with amphid neuron dendrites but lack ciliated ending
AUAR	AB praaapppp	
AVAL	AB alppaaapa	Ventral cord interneurons, synapse onto VA, DA & AS motoneurons; formerly called alpha
AVAR	AB alaappapa	
AVBL	AB plpaapaap	Ventral cord interneurons, synapse onto VB, DB & AS motoneurons; formerly called beta
AVBR	AB prpaapaap	
AVDL	AB alaaapalr	Ventral cord interneurons, synapse onto VA, DA & AS motoneurons; formerly called delta
AVDR	AB alaaapprl	
AVEL	AB alpppaaaa	Ventral cord interneurons, same post-synaptic targets as AVD but processes restricted to anterior cord
AVER	AB praaaaaaa	
AVFL/R	P1 aaaa	Interneurons, processes in ventral cord and ring, few synapses
AVFL/R	W aaa	
AVG	AB prpapppap	Ventral cord interneuron, few synapses, sends process into tailspike
AVHL	AB alapaaaaa	Neurons, mainly presynaptic in ring and postsynaptic in ventral cord
AVHR	AB alappapaa	
AVJL	AB alapapppa	Neurons, mainly postsynaptic in ventral cord and presynaptic in ring
AVJR	AB alappppa	
AVKL	AB plpapapap	Ring and ventral cord interneuron
AVKR	AB prpapapap	"
AVL	AB prpappaap	Ring and ventral cord interneuron/ motoneuron, few synapses
AVM	QR paa	Anterior ventral microtubule cell, touch receptor
AWAL	AB plaapapaa	Amphid wing cells, neurons having ciliated sheet-like sensory endings closely associated with amphid sheath
AWAR	AB praapapaa	
AWBL	AB alpppppap	
AWBR	AB praaappap	"
AWCL	AB plpaaaaap	"
AWCR	AB prpaaaaap	"
B	AB prppppapa	Rectal cell, postemb. blast cell in male
BAGL	AB alppappap	Neurons, ciliated endings in head, not part of a sensillum, associated with ILso
BAGR	AB arappppap	
BDUL	AB arppaappp	Neurons, processes run along excretory canal and also in nerve ring, unique darkly staining synaptic vesicles
BDUR	AB arpppappp	
C	P_0 ppa	Embryonic founder cell
CA1	P3 aapa	Male specific cells, not reconstructed
CA2	P4 aapa	"
CA3	P5 aapa	"
CA4	P6 aapa	Male specific neurons, innervate dorsal muscles
CA5	P7 aapa	
CA6	P8 aapa	"

CA7	P9 aapa	"
CA8	P10aapa	Male specific cells in ventral cord,
CA9	P11aapa	neuron-like but lack synapses
CANL	AB alapaaapa	Process runs along excretory canal, no
CANR	AB alappappa	synapses seen, essential for survival
CEMDL	AB plaaaaaap	Male specific cephalic neurons (die in
CEMDR	AB arpapaaap	hermaphrodite embryo), open to outside,
CEMVL	AB plpaapapp	possibly mediate male chemotaxis towards
CEMVR	AB prpaapapp	hermaphrodites
CEPDL	AB plaaaaappa	Neurons of cephalic sensilla, contain
CEPDR	AB arpapaappa	dopamine
CEPVL	AB plpaappppa	"
CEPVR	AB prpaappppa	"
CEPshDL	AB arpaaaapp	Cephalic sheath cells, sheet-like
CEPshDR	AB arpaaapap	processes envelop neuropil of ring and
CEPshVL	AB plpaaapap	part of ventral ganglion
CEPshVR	AB prpaaapap	"
CEPsoDL	AB alapapppp	Cephalic socket cells
CEPsoDR	AB alapppppp	"
CEPsoVL	AB alppaappp	"
CEPsoVR	AB alaapappp	"
CP0	P2 aap	Male specific cells in ventral cord, not
CP1	P3 aapp	reconstructed
CP2	P4 aapp	"
CP3	P5 aapp	"
CP4	P6 aapp	Male specific motoneurons in ventral
CP5	P7 aapp	cord
CP6	P8 aapp	"
CP7	P9 aapp	"
CP8	P10aapp	Male specific interneurons, project into
CP9	P11aapp	preanal ganglion
D	P_0 pppa	Embryonic founder cell
DA1	AB prppapaap	Ventral cord motoneurons, innervate
DA2	AB plppapapa	dorsal muscles, probably cholinergic
DA3	AB prppapapa	"
DA4	AB plppapapp	"
DA5	AB prppapapp	"
DA6	AB plpppaaap	"
DA7	AB prpppaaap	"
DA8	AB prpapappp	"
DA9	AB plpppaaaa	"
DB1	AB plpaaaapp	Ventral cord motoneurons, innervate
DB2	AB arappappa	dorsal muscles, probably cholinergic
DB3	AB prpaaaapp	"
DB4	AB prpappapp	"
DB5	AB plpapappp	"
DB6	AB plppaappp	"
DB7	AB prppaappp	"
DD1	AB plppappap	Ventral cord motoneurons, probably
DD2	AB prppappap	reciprocal inhibitors, change their
DD3	AB plppapppa	pattern of motor synapses during
DD4	AB prppapppa	postembryonic development, contain GABA
DD5	AB plppapppp	"
DD6	AB prppapppp	"
DVA	AB prppppapp	Ring interneurons, cell bodies in dorso-
DVB	K p	rectal ganglion, DVA has a large vesicle-
DVC	C aapaa	filled process in ring, DVB innervates
		rectal muscles
DVE	B ppap	Male specific neurons
DVF	B ppppa	"
DX1/2	F lvda	Male specific neurons, darkly staining
DX1/2	F rvda	cell bodies in preanal ganglion, processes
DX3/4	U laa	penetrate basement membrane and contact

DX3/4	U raa	muscles
E	P_0 pap	Embryonic founder cell
EF1/2	F lvdp	Male specific neurons, large, cell bodies
EF1/2	F rvdp	in preanal ganglion, synaptic inputs from
EF3/4	U lap	ray neurons
EF3/4	U rap	"
F	AB plppppapp	Rectal cell, postemb. blast cell in male
FLPL	AB plapaaapad	Neurons, ciliated endings in head, not
FLPR	AB prapaaapad	part of a sensillum, associated with ILso
G1	AB prpaaaapa	Postemb. blast cells; excretory socket is
G2	AB plapaapa	G1 in embryo, G2 in L1, G2.p later
GLRDL	MS aaaaaal	Set of six cells that form a thin
GLRDR	MS aaaaaar	cylindrical sheet between pharynx and
GLRL	MS apaaaad	ring neuropil; no chemical synapses, but
GLRR	MS ppaaaad	gap junctions with muscle arms and RME
GLRVL	MS apaaaav	motoneurons, send processes to tip of head
GLRVR	MS ppaaaav	"
H0L	AB plaaappa	Seam hypodermal cells
H0R	AB arpapppa	"
H1L	AB plaaappp	Seam hypodermal cells, postemb. blast cell
H1R	AB arpapppp	"
H2L	AB arppaaap	"
H2R	AB arpppaap	"
HOA	P10pppa	Neuron)
HOB	P10ppap	Neuron) components of hook sensillum
HOsh	P10ppppp	Sheath) situated anterior to cloaca in
HOso	P10ppaa	Socket) males
HSNL	AB plapppappa	Hermaphrodite specific motoneurons (die in
HSNR	AB prapppappa	male embryo), innervate vulval muscles, contain serotonin
I1L	AB alpapppaa	Anterior sensory,)
I1R	AB arapappaa	input from RIP)
I2L	AB alpappaapa	Anterior sensory)
I2R	AB arapapaapa	") Pharyngeal
I3	MS aaaaapaa	") interneurons
I4	MS aaaapaa	)
I5	AB arapapapp	Posterior sensory)
I6	MS paaapaa	")
IL1DL	AB alapappaaa	Neurons of inner labial sensilla,
IL1DR	AB alappppaaa	ciliated endings with striated rootlets,
IL1L	AB alapaappaa	synapse directly onto muscle cells
IL1R	AB alaappppaa	"
IL1VL	AB alppapppaa	"
IL1VR	AB arappppaa	"
IL2DL	AB alapappap	Neurons of inner labial sensilla, ciliated
IL2DR	AB alappppap	endings without rootlets, open to the
IL2L	AB alapaappp	outside
IL2R	AB alaappppp	"
IL2VL	AB alppapppp	"
IL2VR	AB arappppp	"
ILshDL	AB alaaaparr	Inner labial sheath cells
ILshDR	AB alaaappll	"
ILshL	AB alaaaalpp	"
ILshR	AB alaapaapp	"
ILshVL	AB alppapaap	"
ILshVR	AB arapppaap	"
ILsoDL	AB plaapaaap	Inner labial socket cells, form
ILsoDR	AB praapaaap	protuberances at the tip of each labium
ILsoL	AB alaaapall	"
ILsoR	AB alaaapprr	"
ILsoVL	AB alppapapp	"
ILsoVR	AB arapppapp	"

K	AB plpapppaa	Rectal cell, postemb. blast cell	
K′	AB plpapppap	Rectal cell	
LUAL	AB plpppaapap	Interneurons, short processes in posterior	
LUAR	AB prpppaapap	end of ventral cord	
M	MS apaapp	Postemb. mesoblast	
M1	MS paapaaa	Pharyngeal motoneurons	
M2L	AB araapappa	"	
M2R	AB araappppa	"	
M3L	AB araapappp	Pharyngeal sensory/motoneurons	
M3R	AB araappppp	"	
M4	MS paaaaaa	Pharyngeal motoneurons	
M5	MS paaapap	"	
MCL	AB alpaaappp	Pharyngeal neurons that synapse onto	
MCR	AB arapaappp	the marginal cells	
MI	AB araappaaa	Pharyngeal motoneuron/interneuron	
MS	P_0 paa	Embryonic founder cell	
NSML	AB araapapaav	Pharyngeal neurosecretory motoneurons,	
NSMR	AB araapppaav	contain serotonin	
OLLL	AB alppppapaa	Neurons of outer labial sensilla, ciliated	
OLLR	AB praaapapaa	endings	
OLLshL	AB alpppaapd	Lateral outer labial sheath cells	
OLLshR	AB praaaaapd	"	
OLLsoL	AB alapaaapp	Lateral outer labial socket cells	
OLLsoR	AB alappappp	"	
OLQDL	AB alapapapaa	Neurons of quadrant outer labial sensilla,	
OLQDR	AB alapppapaa	ciliated endings with striated rootlets	
OLQVL	AB plpaaappaa	"	
OLQVR	AB prpaaappaa	"	
OLQshDL	AB arpaaaapa	Quadrant outer labial sheath cells	
OLQshDR	AB arpaaapaa	"	
OLQshVL	AB alpppaaap	"	
OLQshVR	AB praaaaaap	"	
OLQsoDL	AB arpaaaaal	Quadrant outer labial socket cells	
OLQsoDR	AB arpaaaaar	"	
OLQsoVL	AB alppaaapp	"	
OLQsoVR	AB alaappapp	"	
P_0		The single-cell zygote	
P_4	P_0 pppp	Embryonic founder cell: germ line	
P1/2	AB plapaapp	Postemb. blast cells for ventral cord	
P1/2	AB prapaapp	neurons, ventral hypodermis, vulva, male	
P3/4	AB plappaaa	preanal ganglion; form ventral hypodermis	
P3/4	AB prappaaa	in L1	
P5/6	AB plappaap	"	
P5/6	AB prappaap	"	
P7/8	AB plappapp	"	
P7/8	AB prappapp	"	
P9/10	AB plapapap	"	
P9/10	AB prapapap	"	
P11/12	AB plapappa	"	
P11/12	AB prapappa	"	
PCAL	Y plppd	Sensory neuron	)
PCAR	Y prppd	"	)
PCBL	Y plpa	Neuron, ending	)
PCBR	Y prpa	in sheath	)
PCCL	B arpaaa	Sensory neuron	) postcloacal sensilla
PCCR	B alpaaa	"	) in male
PChL	Y plaa	Hypodermal cell	)
PChR	Y praa	"	)

PCshL	Y plppv		Sheath)
PCshR	Y prppv		")
PCsoL	Y plap		Socket)
PCsoR	Y prap		")
PDA herm	Y		Motoneuron, cell body in preanal
PDA male	Y a		ganglion, few NMJ's to dorsal muscles
PDB	P12apa		Interneuron/motoneuron, posterior v. cord
PDC	P11papa		Male specific interneuron, preanal ganglion
PDE	V5 paaa	L&R	Neuron, dopaminergic) postdeirid
PDEsh	V5 papp	L&R	Sheath) sensillum,
PDEso	V5 papa	L&R	Socket) contains dopamine
PGA	P11papp		Male specific interneuron, preanal ganglion
PHAL	AB plpppaapp		Phasmid neurons, ciliated endings,
PHAR	AB prpppaapp		probably chemosensory, similar to amphids
PHBL	AB plapppappp		but situated in tail, take up FITC
PHBR	AB prapppappp		"
PHC	T pppaa	L&R	Neuron, striated rootlet in male, sends process into tail spike
PHshL	AB plpppapaa		Phasmid sheath cells
PHshR	AB prpppapaa		"
PHso1	T paa	L&R	Phasmid socket cells
PHso2	T pap	L&R	"
PLML	AB plapappppaa		Posterior lateral microtubule cells,
PLMR	AB prapappppaa		touch receptor neurons
PLN	T pppap	L&R	Interneuron, associated with PLM, sends process into tailspike
PQR	QL ap		Neuron, rudimentary cilium, projects into preanal ganglion
PVCL	AB plpppaapaa		Ventral cord interneurons, cell bodies in
PVCR	AB prpppaapaa		lumbar ganglion, synapses onto VB & DB motoneurons; formerly called delta
PVD	V5 paapa	L&R	Neuron, lateral process adjacent to excretory canal
PVM	QL paa		Posterior ventral microtubule cell, touch receptor
PVN	T appp	L&R	Interneuron, projects into nerve ring, has several branches, sends processes into vulval region
PVPL	AB plppppaaa		Interneurons, project along ventral cord
PVPR	AB prppppaaa		into nerve ring
PVQL	AB plapppaaa		Interneurons, project along ventral
PVQR	AB prapppaaa		cord to ring, darkly staining vesicles
PVR	C aappa		Asymmetric interneuron, projects along ventral cord into ring, sends process into tailspike
PVT	AB plpappppa		Interneuron, projects along ventral cord
PVV	P11paaa		Male specific motoneuron, ventral cord
PVW	T ppa	L&R	Interneuron, posterior ventral cord, few synapses
PVX	P12aap		Male specific interneurons, postsynaptic
PVY	P11paap		in ring and ventral cord

PVZ	P10ppppa		Male specific motoneuron, ventral cord	
QL	AB plapapaaa		Postemb. neuroblast, migrates anteriorly	
QR	AB prapapaaa		Postemb. neuroblast, migrates posteriorly	
R1A	V5 pppppaaa	L&R	Neuron	
R1B	V5 pppppapa	L&R	Neuron	
R1st	V5 pppppapp	L&R	Structural cell	
R2A	V6 papapaaa	L&R	etc	
R2B	V6 papapapa	L&R	"	
R2st	V6 papapapp	L&R	"	
R3A	V6 papppaaa	L&R	"	
R3B	V6 papppapa	L&R	"	male sensory rays:
R3st	V6 papppapp	L&R	"	
R4A	V6 pppapaaa	L&R	"	neuron type A has
R4B	V6 pppapapa	L&R	"	striated rootlet
R4st	V6 pppapapp	L&R	"	
R5A	V6 pppppaaa	L&R	"	neuron type B lacks
R5B	V6 pppppapa	L&R	"	striated rootlet, has
R5st	V6 pppppapp	L&R	"	darkly staining tip
R6A	V6 ppppaaaa	L&R	"	open to the outside
R6B	V6 ppppaapa	L&R	"	(except R6B, which
R6st	V6 ppppaapp	L&R	"	lacks dark tip and
R7A	T apappaaa	L&R	"	does not open to the
R7B	T apappapa	L&R	"	outside)
R7st	T apappapp	L&R	"	
R8A	T appaaaaa	L&R	"	R5A, R7A and R9A contain
R8B	T appaaapa	L&R	"	dopamine
R8st	T appaaapp	L&R	"	
R9A	T appapaaa	L&R	"	
R9B	T appapapa	L&R	"	
R9st	T appapapp	L&R	"	
RIAL	AB alapaapaa		Ring interneurons, prominent reciprocal synapses onto RMD and SMD	
RIAR	AB alaapppaa			
RIBL	AB plpaappap		Prominent branched ring interneurons	
RIBR	AB prpaappap		"	
RICL	AB plppaaaapp		Ring interneurons	
RICR	AB prppaaaapp		"	
RID	AB alappaapa		Motoneuron, projects around ring and along dorsal cord, innervates dorsal body muscles	
RIFL	AB plppapaaap		Ring interneurons	
RIFR	AB prppapaaap		"	
RIGL	AB plppappaa		Ring interneurons	
RIGR	AB prppappaa		"	
RIH	AB prpappaaa		Ring interneuron	
RIML	AB plppaapap		Prominent ring motoneurons/interneurons	
RIMR	AB prppaapap		"	
RIPL	AB alpapaaaa		Ring/pharynx interneurons, only direct connection between pharynx and ring, makes no chemical synapses	
RIPR	AB arappaaaa			
RIR	AB prpapppaa		Ring interneuron	
RIS	AB prpappapa		Ring interneuron	
RIVL	AB plpaapaaa		Ring interneurons	
RIVR	AB prpaapaaa		"	
RMDDL	AB alpapapaa		Ring motoneurons/interneurons, many synapses, probably reciprocal inhibitor	
RMDDR	AB arappapaa			
RMDL	AB alpppapad		"	
RMDR	AB praaaapad		"	
RMDVL	AB alppapaaa		"	
RMDVR	AB arapppaaa		"	

RMED	AB alapppaap	Ring motoneurons, probably contain GABA
RMEL	AB alaaaarlp	"
RMER	AB alaaaarrp	"
RMEV	AB plpappaaa	"
RMFL	G2 al	Ring motoneurons/interneurons
RMFR	G2 ar	
RMGL	AB plapaaapp	Ring motoneurons/interneurons
RMGR	AB prapaaapp	"
RMHL	G1 l	Ring motoneurons/interneurons
RMHR	G1 r	"
SAADL	AB alppapapa	Ring interneurons, have anteriorly
SAADR	AB arapppapa	directed processes that run sublaterally
SAAVL	AB plpaaaaaa	"
SAAVR	AB prpaaaaaa	"
SABD	AB plppapaap	Ring interneurons, have anteriorly
SABVL	AB plppapaaaa	directed processes that run sublaterally,
SABVR	AB prppapaaaa	synapse to anterior ventral body muscles in L1
SDQL	QL pap	Posterior lateral)Interneurons, processes
SDQR	QR pap	Anterior lateral)project into ring
SIADL	AB plpapaapa	Receive a few synapses in the ring, have
SIADR	AB prpapaapa	posteriorly directed processes that
SIAVL	AB plpapappa	run sublaterally
SIAVR	AB prpapappa	"
SIBDL	AB plppaaaaa	Similar to SIA
SIBDR	AB prppaaaaa	"
SIBVL	AB plpapaapp	"
SIBVR	AB prpapaapp	"
SMBDL	AB alpapapapp	Ring motoneurons/interneurons, have
SMBDR	AB arappapapp	posteriorly directed processes that run
SMBVL	AB alpapappp	sublaterally
SMBVR	AB arappappp	"
SMDDL	AB plpapaaaa	"
SMDDR	AB prpapaaaa	"
SMDVL	AB alppappaa	"
SMDVR	AB arappppaa	"
SPCL	B alpaap	Male specific sensory/motoneurons,
SPCR	B arpaap	innervate spicule protractor muscles
SPDL	B alpapaa	Sensory neurons of copulatory spicule
SPDR	B arpapaa	in male, ciliated, open to outside at
SPVL	B a(l/r)aalda	tip of spicule
SPVR	B a(l/r)aarda	"
SPshDL	B alpapap	Sheath cells of male copulatory spicules
SPshDR	B arpapap	"
SPshVL	B a(l/r)aaldp	"
SPshVR	B a(l/r)aardp	"
SPso1L	B a(l/r)pppl	Socket cells of male copulatory spicules
SPso1R	B a(l/r)pppr	"
SPso2L	B a(l/r)aald	"
SPso2R	B a(l/r)aard	"
SPso3L	B a(l/r)aalv	"
SPso3R	B a(l/r)aarv	"
SPso4L	B alpapp	"
SPso4R	B arpapp	"
TL	AB plappppp	Tail seam hypodermal cells, postemb. blast
TR	AB prappppp	cells, function as phasmid sockets in L1
U	AB plppppapa	Rectal cell, postemb. blast cell in male
URADL	AB plaaaaaaa	Ring motoneurons, non-ciliated endings in
URADR	AB arpapaaaa	head, associated with OLQ in embryo
URAVL	AB plpaaapaa	"

URAVR	AB prpaaapaa	"
URBL	AB plaapaapa	Neurons, presynaptic in ring, non-ciliated endings in head, associated with OLL in embryo
URBR	AB praapaapa	
URXL	AB plaaaaappp	Ring interneurons, non-ciliated endings in head, associated with CEPD in embryo
URXR	AB arpapaappp	
URYDL	AB alapapapp	Neurons, presynaptic in ring, non-ciliated endings in head, associated with OLQ in embryo
URYDR	AB alapppapp	
URYVL	AB plpaaappp	
URYVR	AB prpaaappp	"
V1L	AB arppapaa	Seam hypodermal cells, postemb. blast cells
V1R	AB arppppaa	
V2L	AB arppapap	"
V2R	AB arppppap	"
V3L	AB plappapa	"
V3R	AB prappapa	"
V4L	AB arppappa	"
V4R	AB arppppa	"
V5L	AB plapapaap	"
V5R	AB prapapaap	"
V6L	AB arppappp	"
V6R	AB arpppppp	"
VA1	W pa	Ventral cord motoneurons, innervate ventral body muscles
VA2	P2 aaaa	"
VA3	P3 aaaa	"
VA4	P4 aaaa	"
VA5	P5 aaaa	"
VA6	P6 aaaa	"
VA7	P7 aaaa	"
VA8	P8 aaaa	"
VA9	P9 aaaa	"
VA10	P10aaaa	"
VA11	P11aaaa	"
VA12	P12aaaa	", but also interneuron in preanal ganglion
VB1	P1 aaap	Ventral cord motoneurons, innervate ventral body muscles (VB1 receives additional synaptic input in the ring)
VB2	W aap	
VB3	P2 aaap	
VB4	P3 aaap	"
VB5	P4 aaap	"
VB6	P5 aaap	"
VB7	P6 aaap	"
VB8	P7 aaap	"
VB9	P8 aaap	"
VB10	P9 aaap	"
VB11	P10aaap	"
VC1	P3 aap	Hermaphrodite specific ventral cord motoneurons, innervate vulval muscles and ventral body muscles
VC2	P4 aap	
VC3	P5 aap	
VC4	P6 aap	"
VC5	P7 aap	"
VC6	P8 aap	"
VD1	W pp	Ventral cord motoneurons, innervate ventral body muscles, probably reciprocal inhibitors, contain GABA
VD2	P1 app	
VD3	P2 app	
VD4	P3 app	"
VD5	P4 app	"
VD6	P5 app	"
VD7	P6 app	"
VD8	P7 app	"
VD9	P8 app	"
VD10	P9 app	"
VD11	P10app	"
VD12	P11app	"
VD13	P12app	"
W	AB prapaapa	Postemb. neuroblast, analogous to Pn.a

XXXL	AB plaaapaa	Embryonic head hypodermal cells, no	
XXXR	AB arpappaa	obvious function later	
Y	AB prpppaaaa	Rectal cell at hatching, becomes PDA in hermaphrodite, postemb. blast cell in male	
Z1	MS pppaap	Somatic gonad precursor cell	
Z2	P4 p	Germ line precursor cell	
Z3	P4 a	"	
Z4	MS appaap	Somatic gonad precursor cell	
arc ant	AB X3	Interface between pharynx and hypodermis,	
arc post	AB X3	form anterior part of buccal cavity	
cc herm DL	M dlpa	Pair of postemb. coelomocytes in	
cc herm DR	M drpa	hermaphrodite	
cc male D	M dlpappp	Single postemb. coelomocyte in male	
ccAL	MS apapaaa	Embryonic coelomocytes	
ccAR	MS ppapaaa	"	
ccPL	MS apapaap	"	
ccPR	MS ppapaap	"	
e1D	AB araaaapap	Pharyngeal epithelial cells	
e1VL	AB araaaaaaa	"	
e1VR	AB araaaaapa	"	
e2DL	AB alpaapaap	"	
e2DR	AB araaapaap	"	
e2V	AB alpappapa	"	
e3D	AB araapaaaa	"	
e3VL	AB alpaaaaaa	"	
e3VR	AB arapaaaaa	"	
exc cell	AB plpappaap	Large H-shaped excretory cell	
exc duct	AB plpaaaapa	Excretory duct	
exc gl L	AB plpapapaa	Excretory glands, fused, send processes to	
exc gl R	AB prpapapaa	ring, open into excretory duct	
exc socket	G2 p	Excretory socket, links duct to hypodermis	
g1AL	MS aapaapaa	Pharyngeal gland cells	
g1AR	MS papaapaa	"	
g1P	MS aaaaapap	"	
g2L	MS aapapaa	"	
g2R	MS papapaa	"	
gon herm anch	Z1 ppp (5L) or Z4 aaa (5R)	Anchor cell, induces vulva	Hermaphrodite Gonad
gon herm dtc A	Z1 aa	Anterior) Distal tip cells, inhibit	
gon herm dtc P	Z4 pp	Posterior) meiosis in neighbouring germ cells, lead gonad during morphogenesis	
gon herm dish A	Z1 apa	Anterior) Epithelial sheaths of	
	Z1 paaa	") distal arms, no muscle	
gon herm dish P	Z4 pap	Posterior) fibres	
	Z4 appp	")	
gon herm prsh A	Z1 appa X4	Anterior) Epithelial sheaths of	
	Z1 paapa X4	") proximal arms, have	
gon herm prsh P	Z4 paap X4	Posterior) muscle fibres	
	Z4 appap X4	")	
gon herm spth A	Z1 appp X9	Anterior)	
	Z1 paapp X9	")	
	Z1 papaa X3	")	
	Z4 apaaa X3	") Spermathecae	
gon herm spth P	Z4 paaa X9	Posterior)	
	Z4 appaa X9	")	
	Z4 apapp X3	")	

Name	Lineage	Description	Group
	Z1 pappp X3	"	
gon herm sujn A	Z1 papa X2	Anterior	Spermathecal-uterine junctions
	Z1 ppaaa X2	"	Spermathecal-uterine junctions
	Z4 apaa X2	"	Spermathecal-uterine junctions
gon herm sujn P	Z1 papp X2	Posterior	Spermathecal-uterine junctions
	Z4 apap X2	"	Spermathecal-uterine junctions
	Z4 aappp X2	"	Spermathecal-uterine junctions
gon herm dut	Z1 pap X14	Dorsal uterus	
	Z4 apa X14	"	
gon herm vut	Z1 ppa X10)	Ventral uterus	
	Z4 aaa X10) (5L)	"	
	Z4 aap X12)	"	
	or		
	Z1 ppa X12)	"	
	Z1 ppp X10) (5R)	"	
	Z4 aap X10)	"	
gon male dtc	Z1 a	Distal tip cells, inhibit meiosis in	Male Gonad
	Z4 p	neighbouring germ cells	Male Gonad
gon male link	Z1 paa or	Linker cell, leads gonad during	Male Gonad
	Z4 aaa	morphogenesis and initiates union	Male Gonad
		with cloaca	Male Gonad
gon male sves	Z1 pap X1	Seminal vesicle	Male Gonad
	Z1 pp X10	"	Male Gonad
	Z4 (p/a)aa X1	"	Male Gonad
	Z4 aap X1	"	Male Gonad
	Z4 ap X10	"	Male Gonad
gon male vdef	Z1 pap X10	Vas deferens	Male Gonad
	Z4 aa X20	"	Male Gonad
	or		Male Gonad
	Z1 pa X20	"	Male Gonad
	Z4 aap X10	"	Male Gonad
hmc	MS appaaa	Head mesodermal cell, function unknown	
hyp1	AB X3	Cylindrical hypodermal syncytium in head	
hyp2	AB X2	"	
hyp3	AB X2	"	
hyp4	AB X3	"	
hyp5	AB X2	"	
hyp6	AB X6	"	
hyp7	AB X11	Embryonic	Large hypodermal syncytium
hyp7	C X12	"	Large hypodermal syncytium
hyp7	H1 L&R X6	Postemb.	Large hypodermal syncytium
hyp7	H2 L&R X8	"	Large hypodermal syncytium
hyp7	P1 p	"	Large hypodermal syncytium
hyp7	P2 p	"	Large hypodermal syncytium
hyp7	P9 p	"	Large hypodermal syncytium
hyp7	T L&R X4	"	Large hypodermal syncytium
hyp7	V1 L&R X14	"	Large hypodermal syncytium
hyp7	V2 L&R X14	"	Large hypodermal syncytium
hyp7	V3 L&R X14	"	Large hypodermal syncytium
hyp7	V4 L&R X14	"	Large hypodermal syncytium
hyp7	V5 L&R X4	"	Large hypodermal syncytium
hyp7	V6 L&R X10	"	Large hypodermal syncytium
hyp7 herm	Pn X8	Postemb., hermaphrodite	Large hypodermal syncytium
hyp7 herm	T L&R X2	"	Large hypodermal syncytium
hyp7 herm	V5 L&R X4	"	Large hypodermal syncytium
hyp7 herm	V6 L&R X4	"	Large hypodermal syncytium
hyp7 male	Pn X9	Postemb., male	Large hypodermal syncytium
hyp7 male	T L&R X8	"	Large hypodermal syncytium
hyp7 male	V5 L&R X6	"	Large hypodermal syncytium
hyp7 male	V6 L&R X8	"	Large hypodermal syncytium
hyp8/9	AB plpppapap	Tail ventral hypodermis	
hyp8/9	AB prpppapap	"	
hyp10	AB X2	"	
hyp11	C pappa	Tail dorsal hypodermis	

hyp P12	P12pa	Preanal hypodermis
hyp hook	P10papp	Hypodermis associated with hook sensillum
hyp hook	P11 X3	of male
int	E X20	Embryonic intestinal cells
int	In a & In p	Postemb. division of fourteen intestinal nuclei, without cell division
linker killer	U lp	One of these cells, sometimes fused with
linker killer	U rp	U l/ra, phagocytoses the male linker cell
m1DL	AB araapaaap	Pharyngeal muscle cells
m1DR	AB araappaap	"
m1L	AB araaaaaap	"
m1R	AB araaaaapp	"
m1VL	AB alpaaaapa	"
m1VR	AB arapaaapa	"
m2DL	AB araapaapa	"
m2DR	AB araappapa	"
m2L	AB alpaaapaa	"
m2R	AB arapaapaa	"
m2VL	AB alpaaaaap	"
m2VR	AB arapaaaap	"
m3DL	MS aaapaaa	"
m3DR	MS paaaapa	"
m3L	AB alpaapapp	"
m3R	AB arapaappa	"
m3VL	AB alpappppp	"
m3VR	AB arapapppp	"
m4DL	MS aaaaapp	"
m4DR	MS paaaapp	"
m4L	MS aaapaap	"
m4R	AB araaapapp	"
m4VL	MS aapaaaa	"
m4VR	MS papaaaa	"
m5DL	MS aaaapap	"
m5DR	MS paaappa	"
m5L	AB araapapap	"
m5R	AB araapppap	"
m5VL	MS aapaaap	"
m5VR	MS papaaap	"
m6D	MS paaappp	"
m6VL	MS aapappa	"
m6VR	MS papappa	"
m7D	MS aaaappp	"
m7VL	MS aapaapp	"
m7VR	MS papaapp	"
m8	MS aaapapp	"
mc1DL	AB alpaapapa	Pharyngeal marginal cells
mc1DR	AB araaapapa	"
mc1V	AB alpappppa	"
mc2DL	AB araapaapp	"
mc2DR	AB araappapp	"
mc2V	AB arapapppa	"
mc3DL	MS aaapapa	"
mc3DR	MS paapapa	"
mc3V	AB alpappapp	"
mu anal	AB plpppppap	Anal depressor muscle
mu bod	AB prpppppaa	Embryonic body wall muscles
mu bod	C X32	"
mu bod	D X20	"
mu bod	MS X28	"
mu bod	M X14	Postemb. body wall muscle
mu male diag	M X15	Diagonal muscles in male body wall
mu male gub	M X4	Male gubernacular muscles
mu int L	AB plppppaa	Intestinal muscles, attach to intestine
mu int R	MS ppaapp	and body wall anterior to anus

mu male long	M X10	Longitudinal muscles in male body wall
mu male obl	M X4	Oblique muscles in male body wall
mu sph	AB prpppppap	Sphincter muscle of intestino-rectal valve
mu male spic	M X8	Muscles of male copulatory spicules
mu herm ut	M X8	Hermaphrodite uterine muscles
mu herm vul	M X8	Hermaphrodite vulval muscles
proct	B X19	Male proctodeum, union of intestine and
proct	F X4	vas deferens, contains copulatory spicules
rect D	AB plpappppp	Rectal epithelial cells, adjacent to
rect VL	AB plppppaap	intestino-rectal valve, have microvilli
rect VR	AB prppppaap	"
rect hyp	K a	Rectal hypodermis, underlie cuticle
rect hyp	K'	"
se	H1 L&R X4	Seam hypodermal cells, remain separate
se	H2 L&R X2	from hypodermal syncytia, make alae;
se	V1 L&R X4	listed here are adult seam cells, which
se	V2 L&R X4	fuse together in L4; juvenile seam cells
se	V3 L&R X4	are listed under blast cell names
se	V4 L&R X4	"
se herm	T L&R X2	"
se herm	V5 L&R X2	"
se herm	V6 L&R X4	"
se male	V5 L&R X2	"
set	V5 L&R X2	Male tail seam, does not fuse with main
set	V6 L&R X8	seam, does not make alae
spike/death	AB plpppppppa	Used during embryogenesis to make tail
spike/death	AB prpppppppa	spike, then die
virL	AB prpappppp	Intestino-rectal valve
virR	AB prpappppa	"
vpi1	MS paapapp	Pharyngo-intestinal valve
vpi2DL	MS aapappp	"
vpi2DR	MS papappp	"
vpi2V	MS aappaa	"
vpi3D	MS aaappp	"
vpi3V	MS aappap	"
vulva	P5 X7	Hermaphrodite vulva
vulva	P6 X8	"
vulva	P7 X7	"

Table 1 Nuclear counts

	L1								
	AB	MS	E	C	D	P_4	total	blast	Adult total
Hypodermis									
(excluding rectum)	72			13			85	35	213^{211}
Body nervous system									
neurons	200^{202}			2			202^{204}		282^{361}
supporting cells	40						40		50^{86}
ring glia		6					6		6
Body mesoderm									
muscle	1	28		32	20		81		111^{136}
coelomocytes		4					4		6^{5}
excretory system	4						4		4
hmc		1					1		1
mesoblast (M)		1					1	1	
Alimentary tract									
arcade	9						9		9
pharynx									

structural	16	2					18		18
neurons	14	6					20		20
muscles	19	18					37		37
glands		5					5		5
valve p/i		6					6		6
intestine			20				20	14^{+}	34
valve i/r	2						2		2
muscle	3	1					4		4
rectum	9						9	1^{5}	8^{31}
Gonad									
somatic		2					2	2	143^{55}
germ line						2	2	2	indefinite
Total survivors	389^{391}	80	20	47	20	2	558^{560}	55^{59}	959^{1031} (+ germ)
Deaths	98^{96}	14		1			113^{111}		131^{148}
Total produced	487	94	20	48	20	2	671		1090^{1179} (+ germ)

Cell count is less than nuclear count because of cell fusions and postembryonic divisions in the intestine ($^{+}$) which are of nuclei only. Main entries are for hermaphrodite; male counts, where different, are shown as superscripts. Postembryonic blast cells (column 8) are included in L1 totals (column 7). Hypodermis includes XXXs, Qs and hermaphrodite vulva. Rectum becomes cloaca in adult male. Sex muscles and body muscles are placed in a single category. Rectal cell Y becomes a neuron during postembryonic development of hermaphrodite. (Reprinted, with permission, from Sulston et al. 1983.)

APPENDIX 2
Neuroanatomy

Compiled by John White, Eileen Southgate, and Richard Durbin

MRC Laboratory of Molecular Biology
Cambridge, England

Part A: Positions of Neuronal Processes and Cell Bodies

Part B: Neuronal Connectivity Diagrams

  0-87969-307-X/88 $1 + .00.

Part A: Positions of Neuronal Processes and Cell Bodies

The processes and cell body positions of all the neurons in *C. elegans* are listed. The pharyngeal neurons are grouped together at the end. See Appendix 1 for an explanation of cell nomenclature.

ADEL
ADLL
AIAL
AIML
AIYL
ALA
ADAL
ADFL
AFDL
AIBL
AINL
AIZL

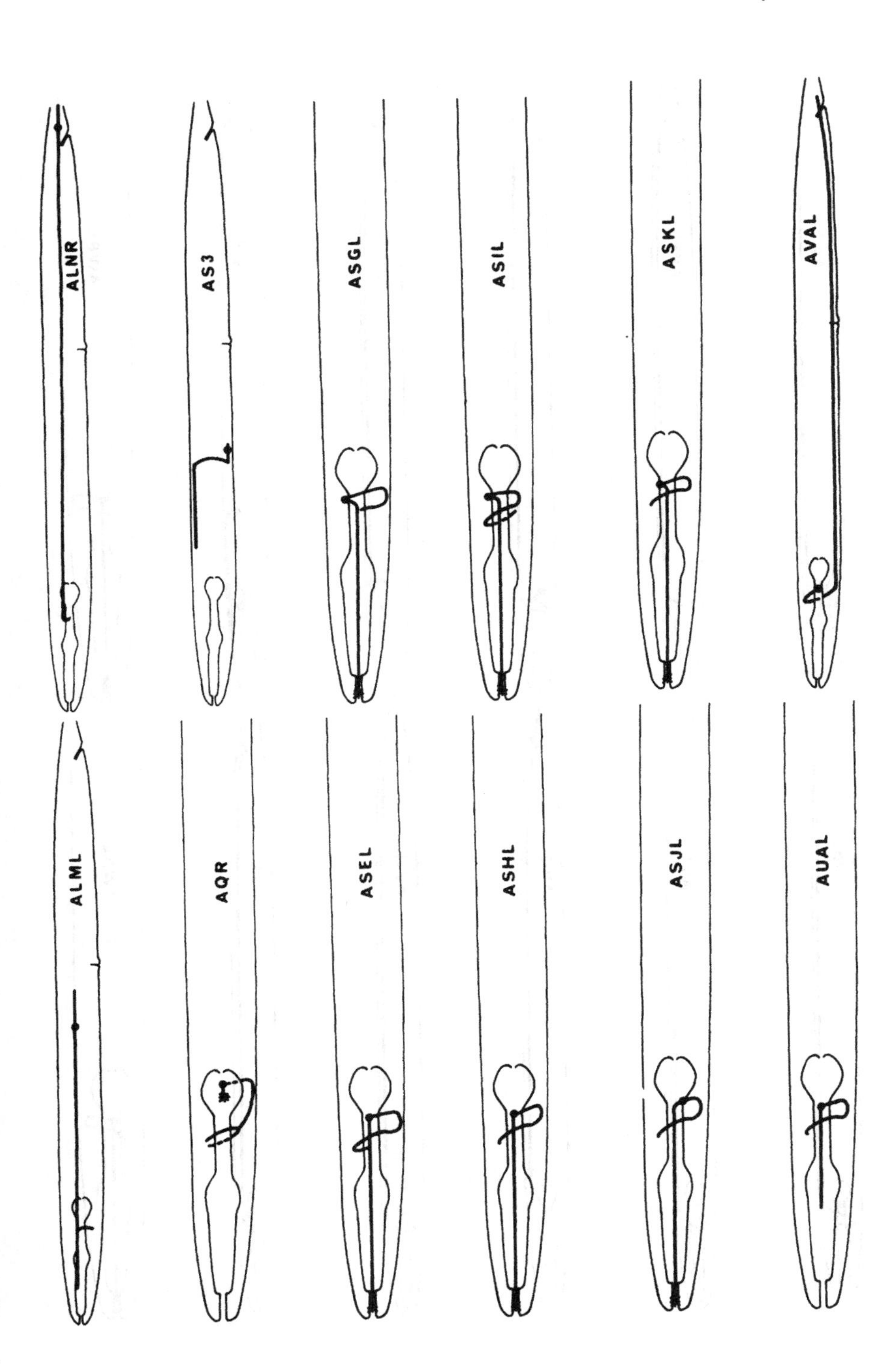
ALNR
AS3
ASGL
ASIL
ASKL
AVAL
ALML
AQR
ASEL
ASHL
ASJL
AUAL

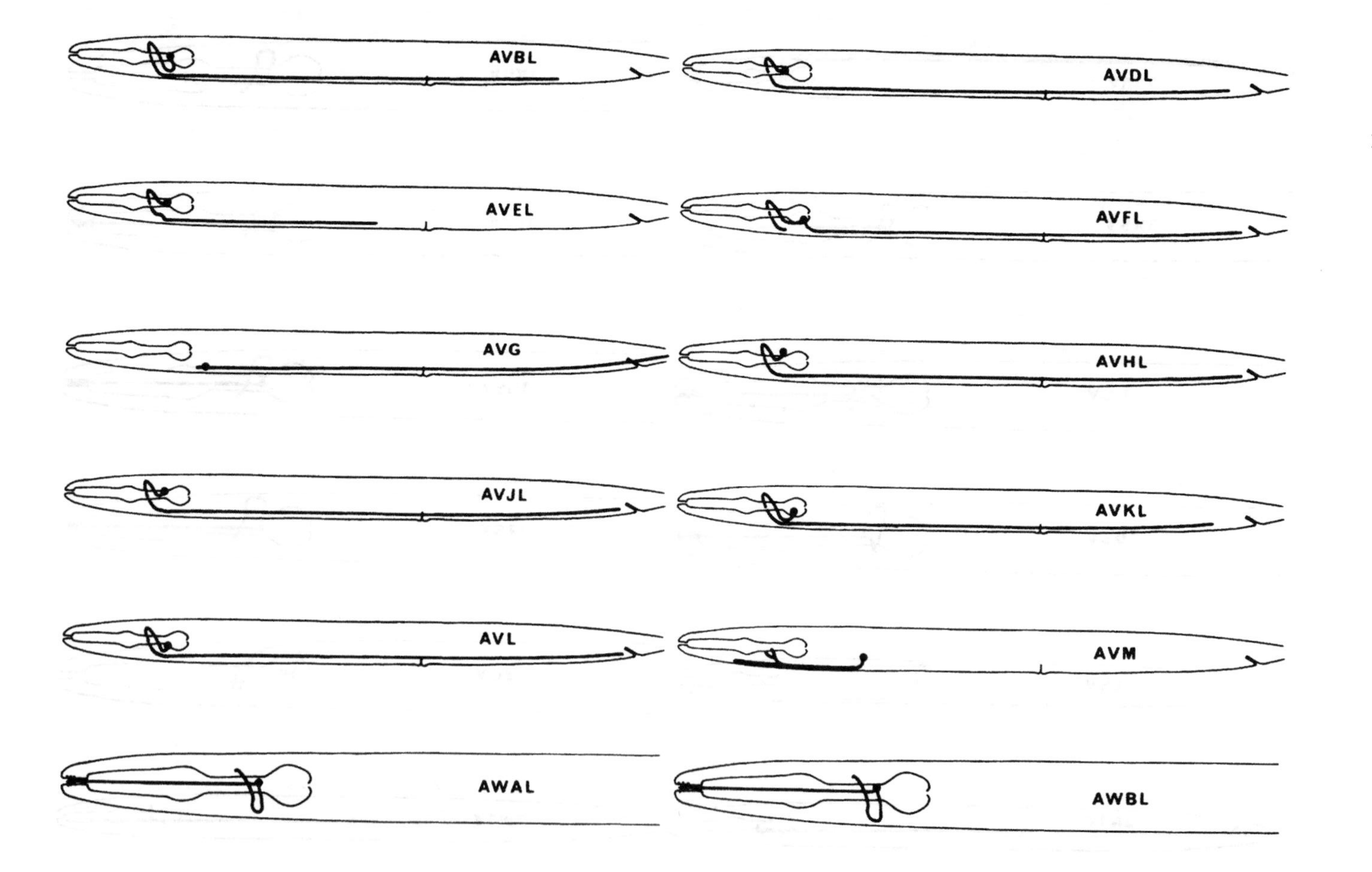
AVBL
AVDL
AVEL
AVFL
AVG
AVHL
AVJL
AVKL
AVL
AVM
AWAL
AWBL

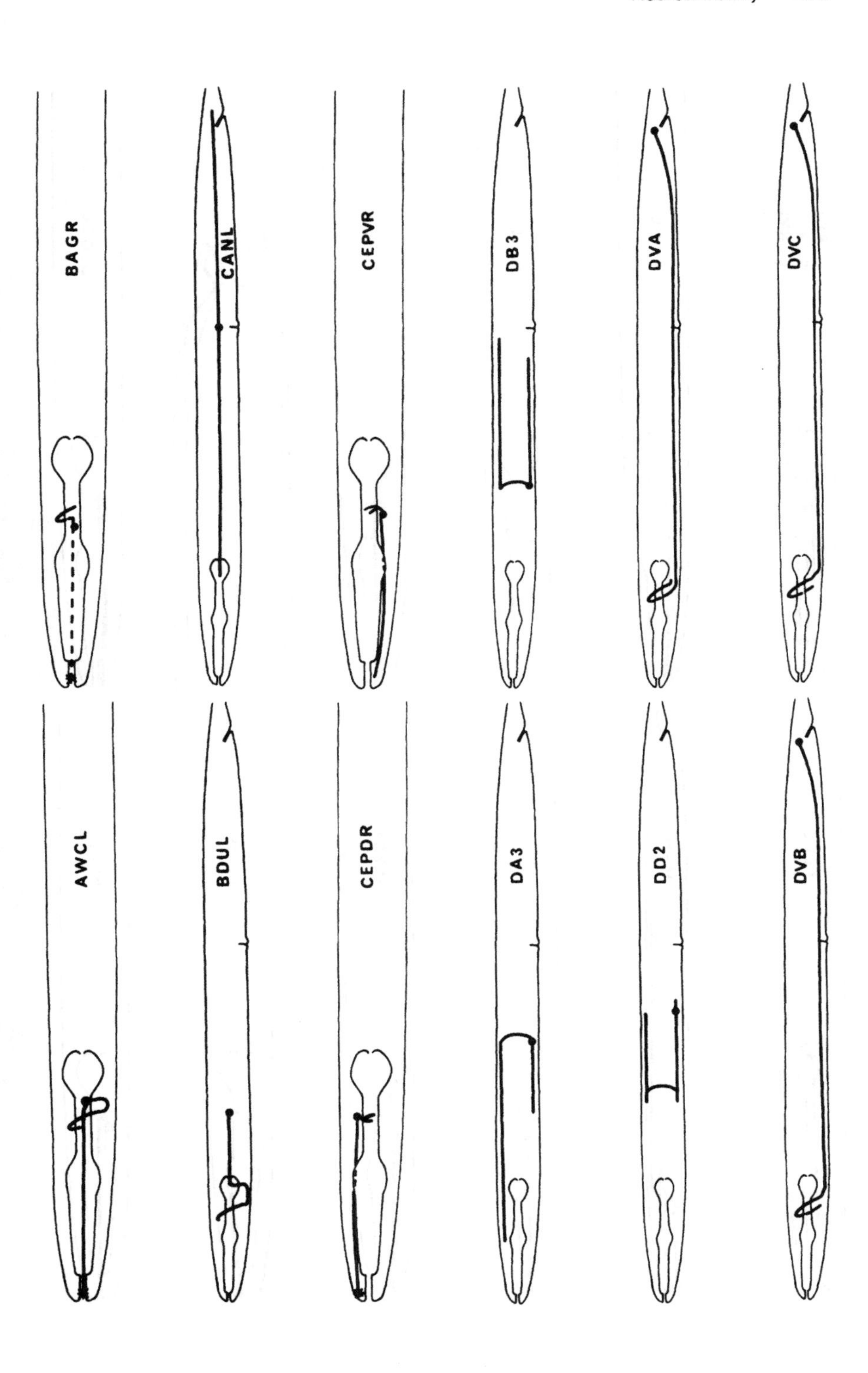
BAGR
CANL
CEPVR
DB3
DVA
DVC
AWCL
BDUL
CEPDR
DA3
DD2
DVB

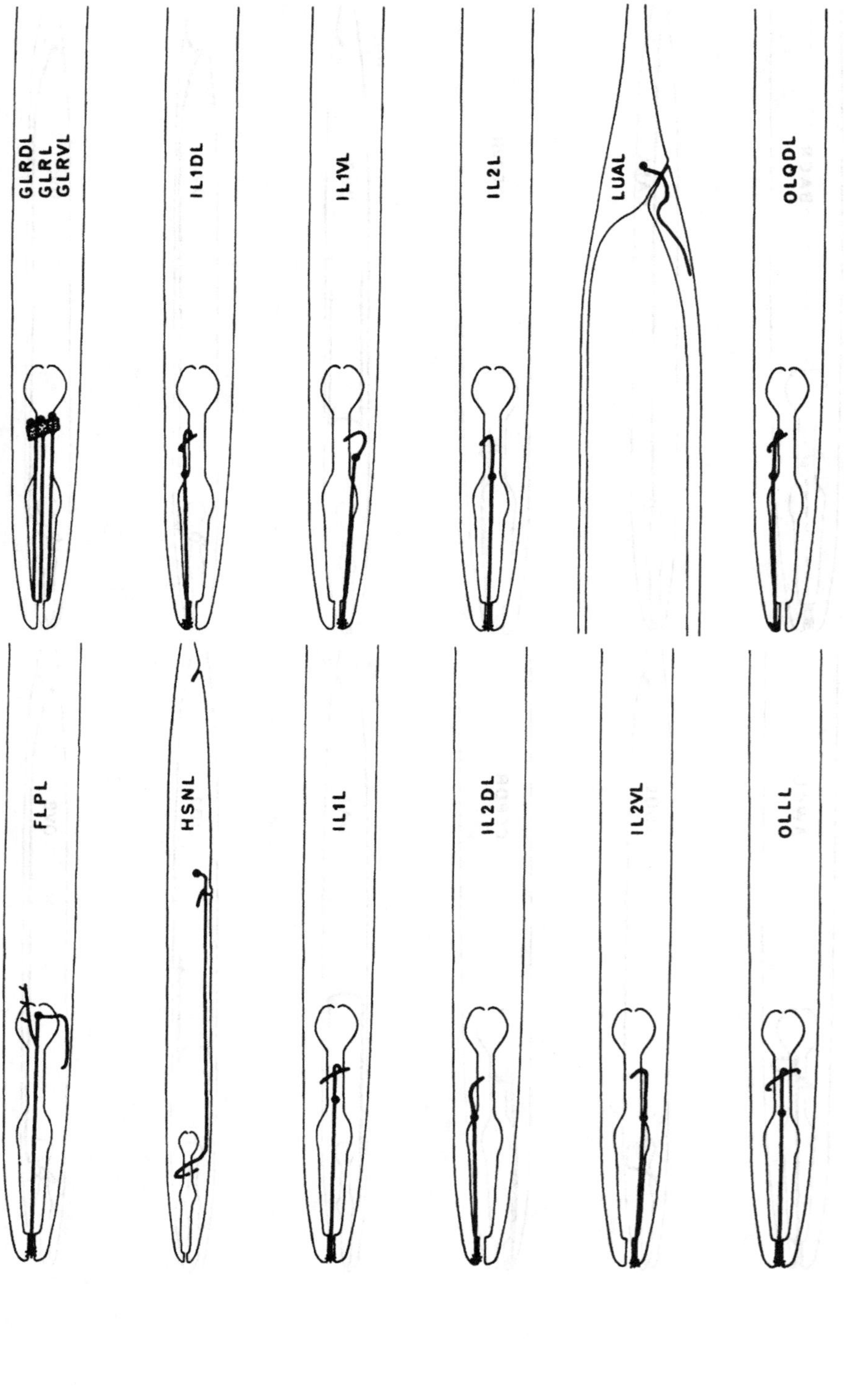

GLRDL
GLRL
GLRVL
IL1DL
IL1VL
IL2L
LUAL
OLQDL
FLPL
HSNL
IL1L
IL2DL
IL2VL
OLLL

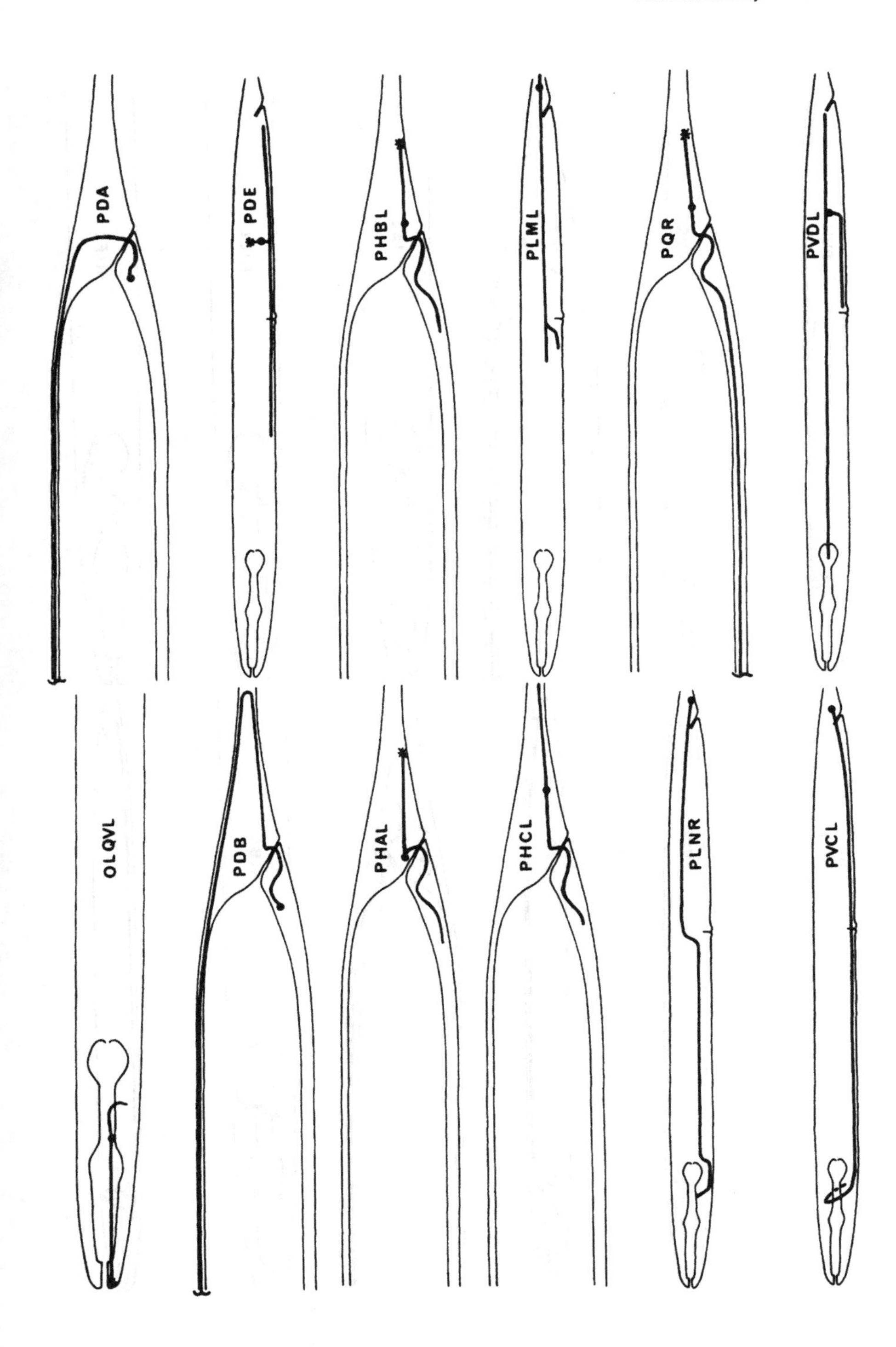
PDA
PDE
PHBL
PLML
PQR
PVDL
OLQVL
PDB
PHAL
PHCL
PLNR
PVCL

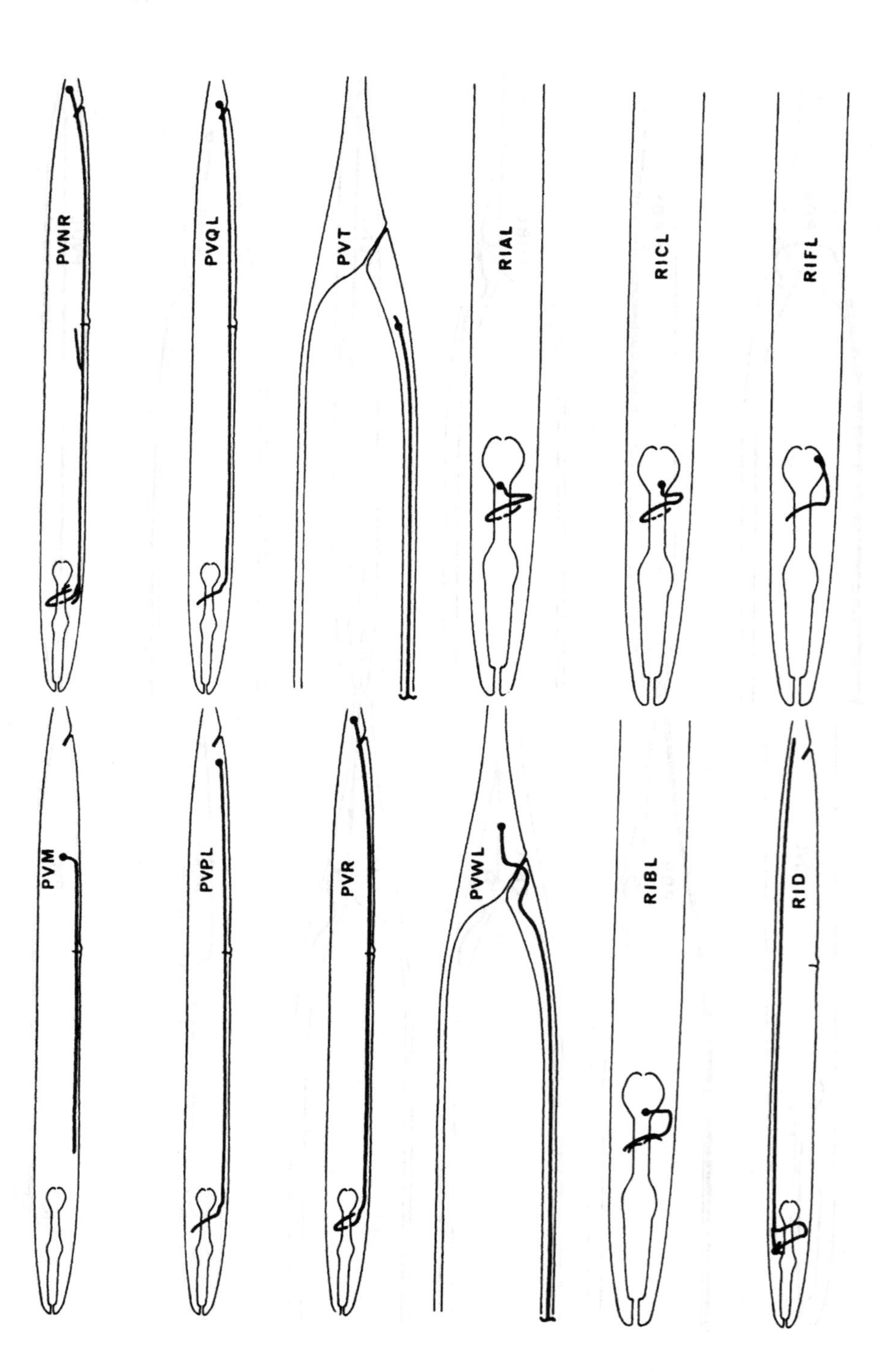
PVNR
PVQL
PVT
RIAL
RICL
RIFL
PVM
PVPL
PVR
PVWL
RIBL
RID

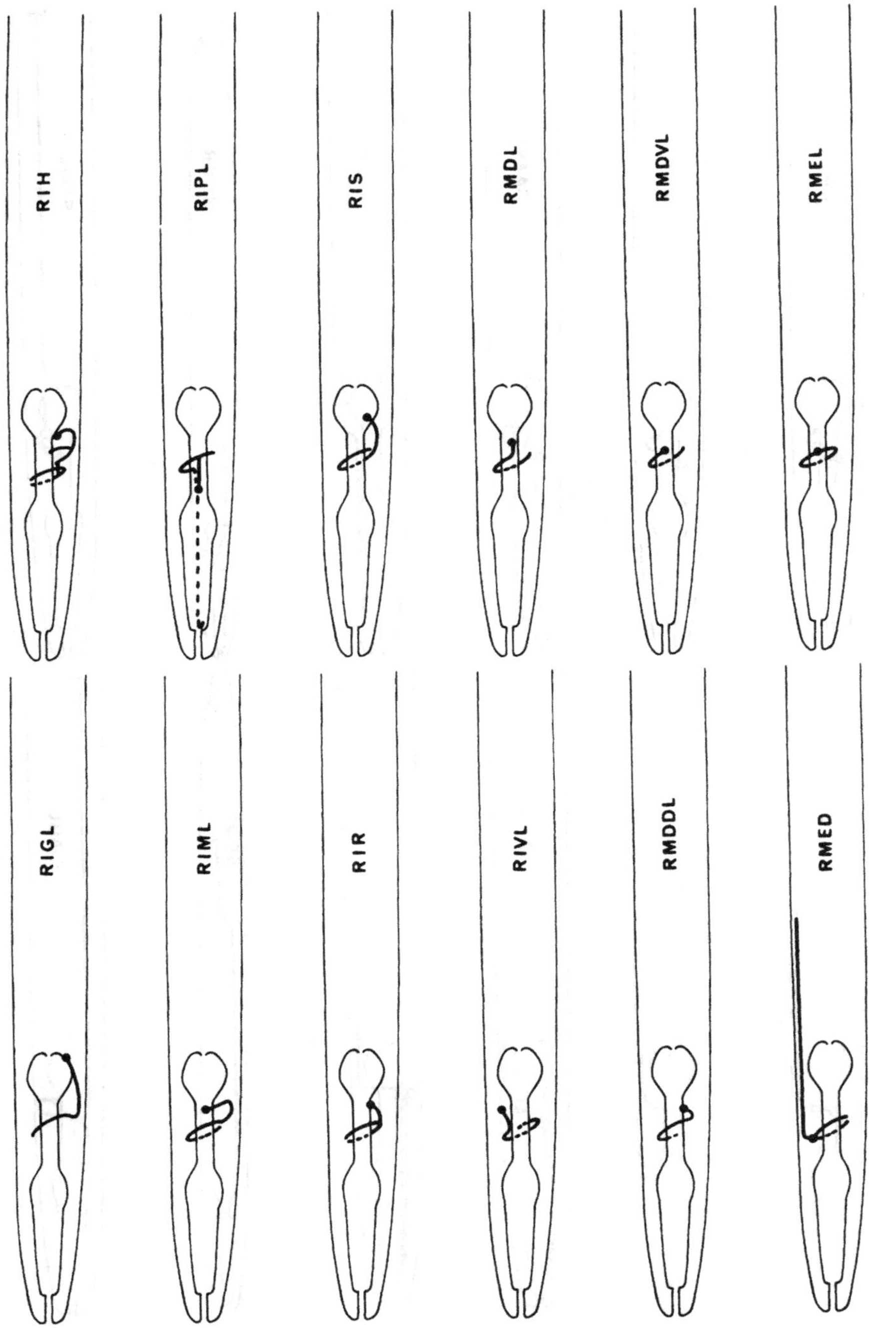

RIH
RIPL
RIS
RMDL
RMDVL
RMEL
RIGL
RIML
RIR
RIVL
RMDDL
RMED

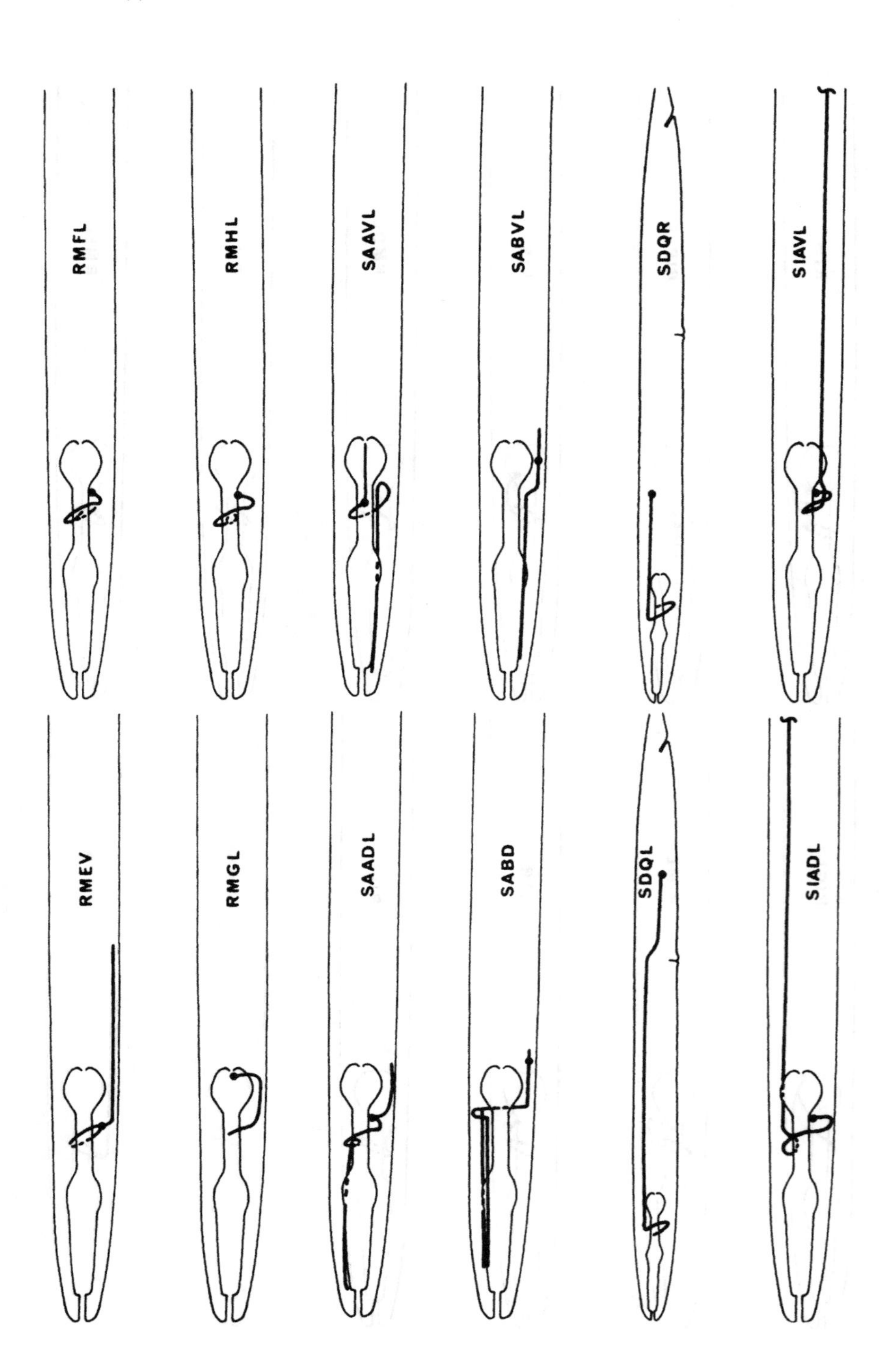
RMFL
RMHL
SAAVL
SABVL
SDQR
SIAVL
RMEV
RMGL
SAADL
SABD
SDQL
SIADL

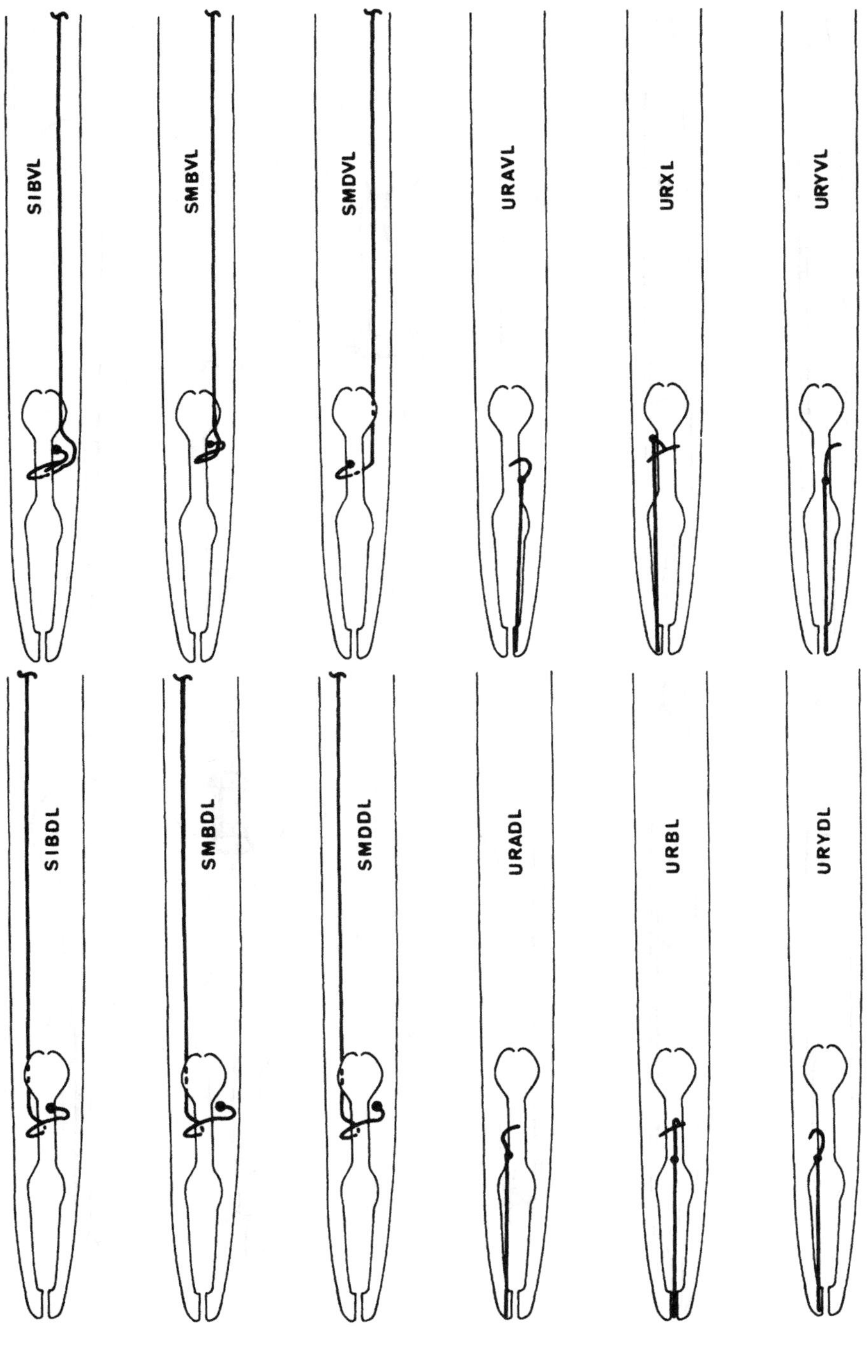

SIBVL
SMBVL
SMDVL
URAVL
URXL
URYVL
SIBDL
SMBDL
SMDDL
URADL
URBL
URYDL

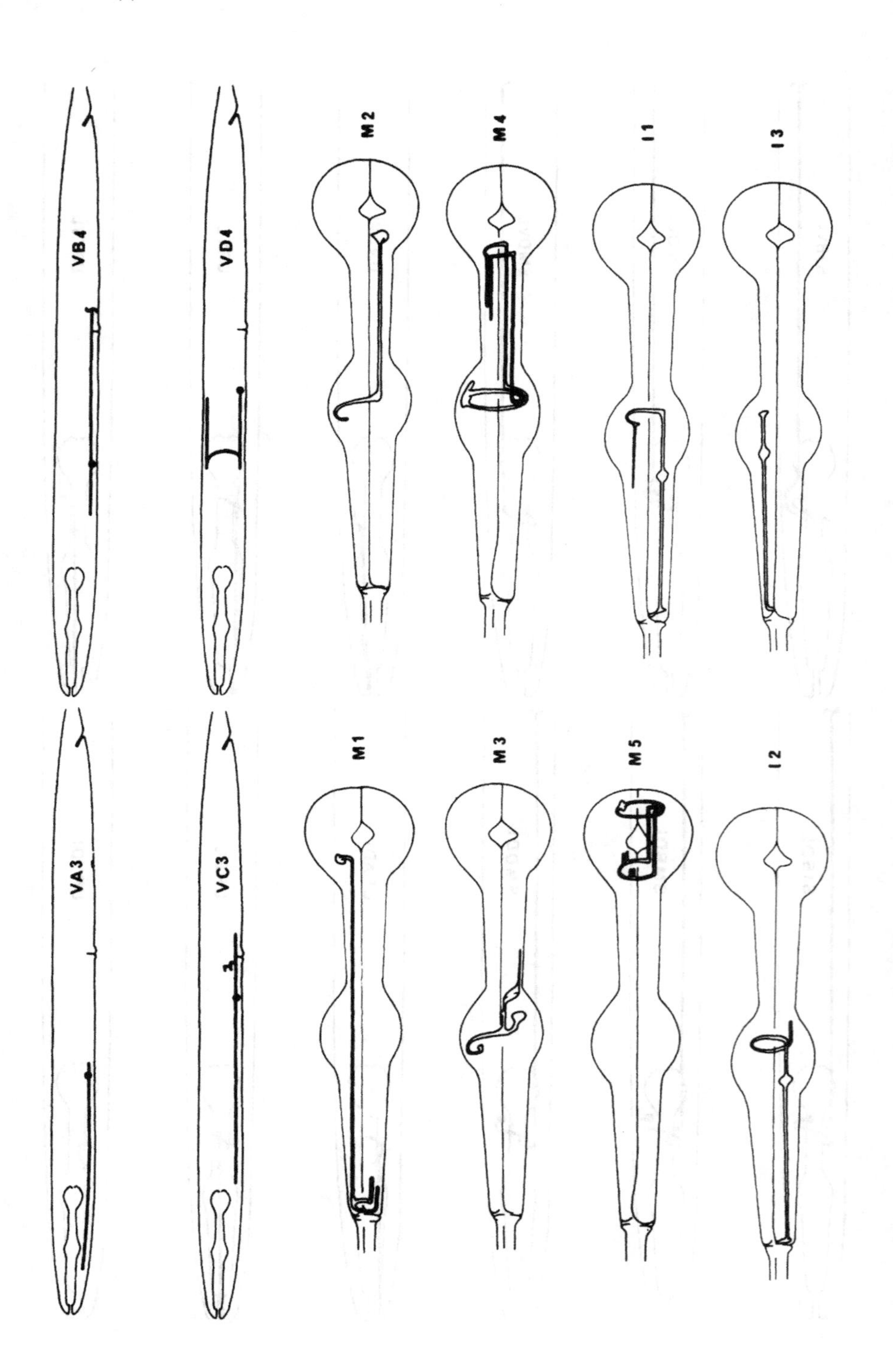
VB4
VD4
M2
M4
I1
I3
VA3
VC3
M1
M3
M5
I2

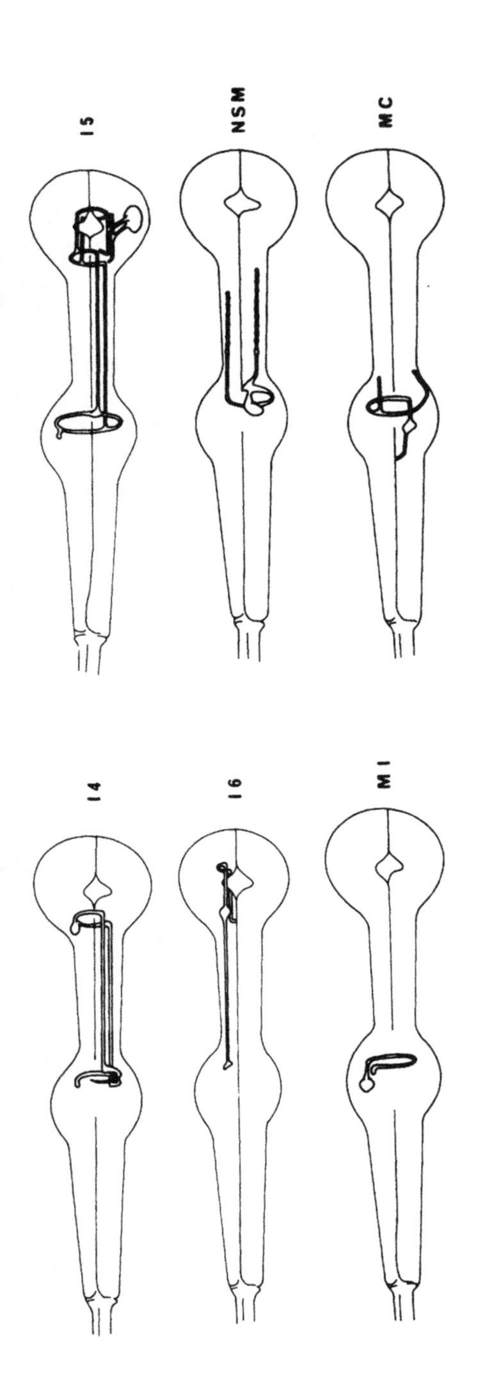
I5
NSM
MC
I4
I6
MI

Part B: Neuronal Connectivity Diagrams

The following diagrams show the pattern of synaptic connectivity within the nervous system excluding the pharynx. Classes of sensory neurons are represented by triangles, interneurons by hexagons, and motor neurons by circles. Connections between neurons are mediated by chemical synapses (arrows) and gap junctions (Ts). The relative anatomical prominence of a chemical synapse is indicated by the thickness of the line showing the connection. To simplify the presentation, the circuitry has been split into six units. Certain classes have been included in more than one unit. Connections made by classes within a unit to classes that are not included are shown in the lists associated with each unit. See Appendix 1 for an explanation of cell nomenclature.

	PRE	POST	EJ
CEP	AVE OLL OLQ	OLL ALM RIH ADE	OLQ RIH
ILI			ILI OLL RMC
IL2	RIH OLQ ADE		
RIA		ADF AIZ RIH AIY AUA BAG AQR ASH AWB	
RIB	AVE AIB AIZ AVA AIN	OLL AUA AIY BAG OLQ AIB RIH	RIB SMB DVB AVB
RIC	AVA	OLQ	ASH AVK
RIG	RMH AVE AIZ AVK BAG	ADE DVC PVP BAG DVB	RIG OLQ OLL BAG AQR AVK
RIM	SMD AVB	AIB RIF AIZ	AVE AVA AIY
RIP	OLQ?	RIH PVR	PHARYNX
RIS	AVE AVK	SDQ PVC	SMD AVJ AIB
RIV	RMD SAA SMD SIA		SDQ AIB RIV SMD
RMD	URY	OLQ ADE OLL OLQ RIV RMD	RMD AVE

	PRE	POST	EJ
RME	SMD	OLL URA	RME AVE GLR
RMF	AVK	DVB DVC	
RMG	AVE	ADE	URX IL2 AWB ASK ASH ADL
RMH		RIG	
SAA	AVA	RIV ALN VB1	SMB
SMB		AIZ DVA ALN ADF AVH	RIB AVK SAA
SMD	SMD	RIM RIV RME SAA OLL SMD	RIS RIV SMD
URA	RME		
URX	AUA AVE AVB		RMC AUA
URY	AVE		AVA

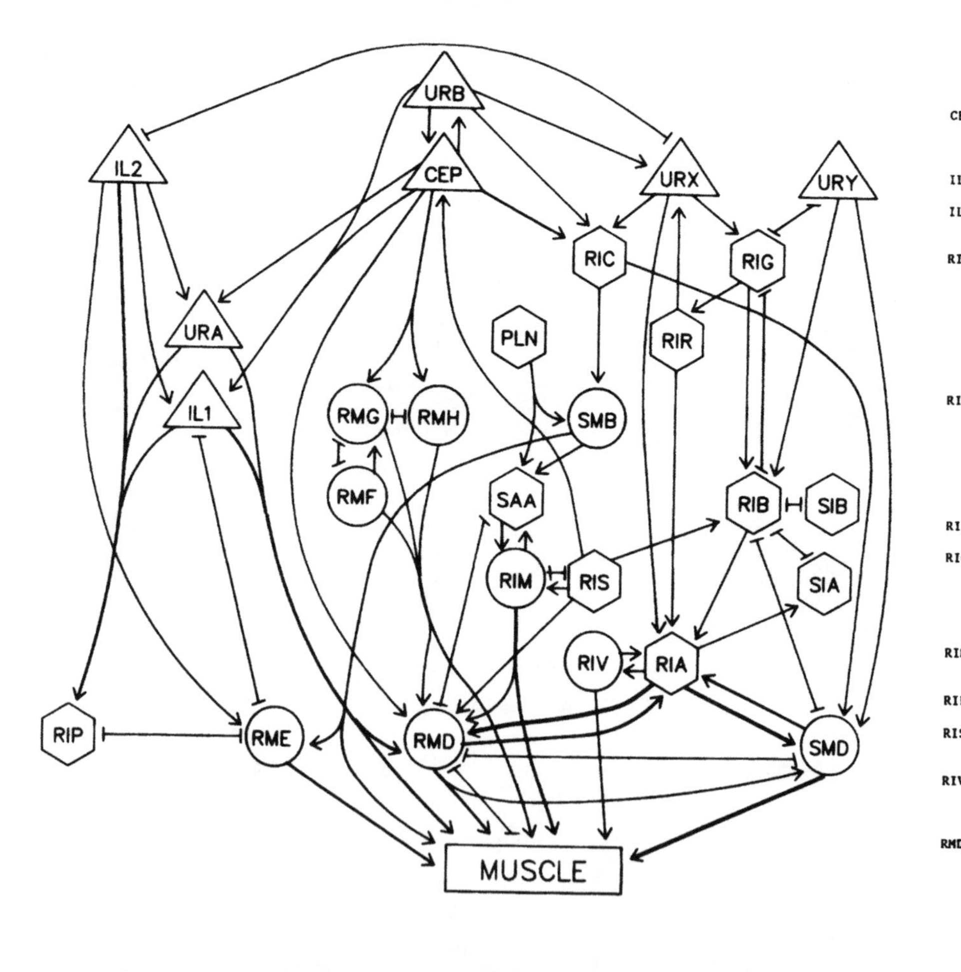

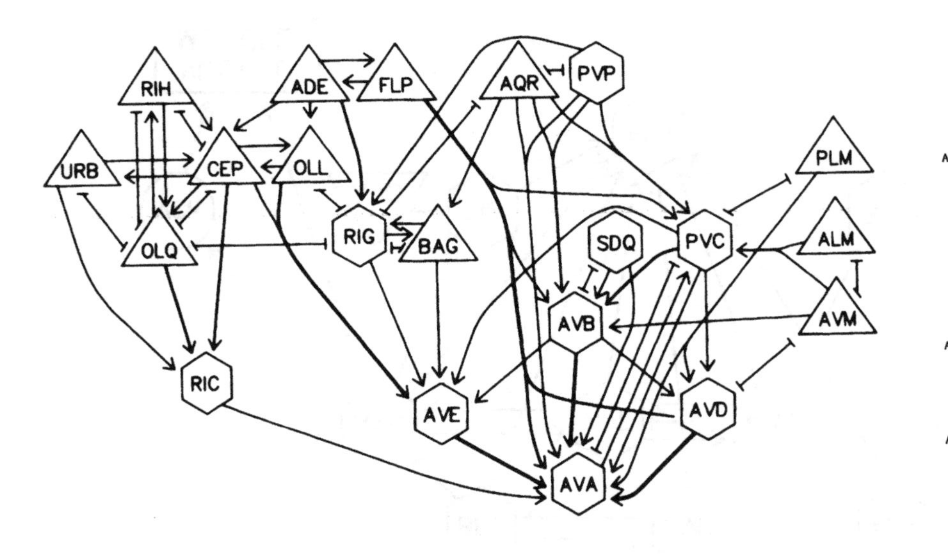

	PRE	POST	EJ .
ADE	RMD	AVM	
		BDU	
	AVA	IL2L/R	AVK
ALM	BDU		PVR
	CEP		
AQR	RIA	DVA	AVK
	AVD		RIG
AVA	VAn	ADE	AVA
	DAn	SAA	RIM
	ASn	DVC	URY
	SAB	AUA	VAn
		ASH	DAn
		ADL	ASn
		RIB	SAB
		PHB	
		PQR	
		DVA	
		LUA	
AVB	ASn	RIF	AVB
	HDC	RIM	RIB
		ASH	SIBV
		PVR	DVA
		ADA	RID
		AIB	VBn
		AVF	DBn
		ADL	PVN
		URX	
		PVN	
		AVG	
		AVF	
AVD	SAB	AQR	FLP
	VAn	ASH	AVJ
	DAn	ADL	ADA
	ASn	PQR	
		LUA	
		PVN	
AVE	SAB	FLP	RIM
	VAn	URY	RME
	DAn	URX	RMD
	ASn	DVA	AVE
		RIB	
		RIS	
		AUA	
		ALA	
		AVK	
		AIZ	
		RMG	
		AVJ	
AVM	ADE		
	BDU		
	PVR		
BAG	RIA	AIN	RIR
	RIB		BAG
CEP	ILI	ALM	
	RMH	RIS	

	PRE	POST	EJ
FLP	AVE		RIH
	AIB		FLP
			AVD
OLL	SMD		OLL
	RIB		IL1
	RMD		
OLQ	RMD	IL2	RIB
	SIB		RMD
PLM	HSN		LUA
	DVA		PVR
	PDE		
PVC	RID	AVJ	PVC
	DBn	PHB	AVJ
	VBn	PHC	PVW
		VA12	PDE
		LUA	
		PVM	
		DVA	
		PVN	
		PVD	
PVP	AVH	RIF	PVP
	HDC	AVH	DVC
	AVL		PQR
			PHA
			AVK
			VDn
RIC	SMD	URX	ASH
	SMD		AVK
RIG	RIR	DVB	URY
	RMH	DVC	RIB
	AIZ	URX	AVK
	AVK		RIG
	RIB		
RIH	RIA	IL2	ADF
	AIZ		FLP
	RIB		
	RIP		
SDQ	DVA		RIV
	AIB		SDQ
	RIS		

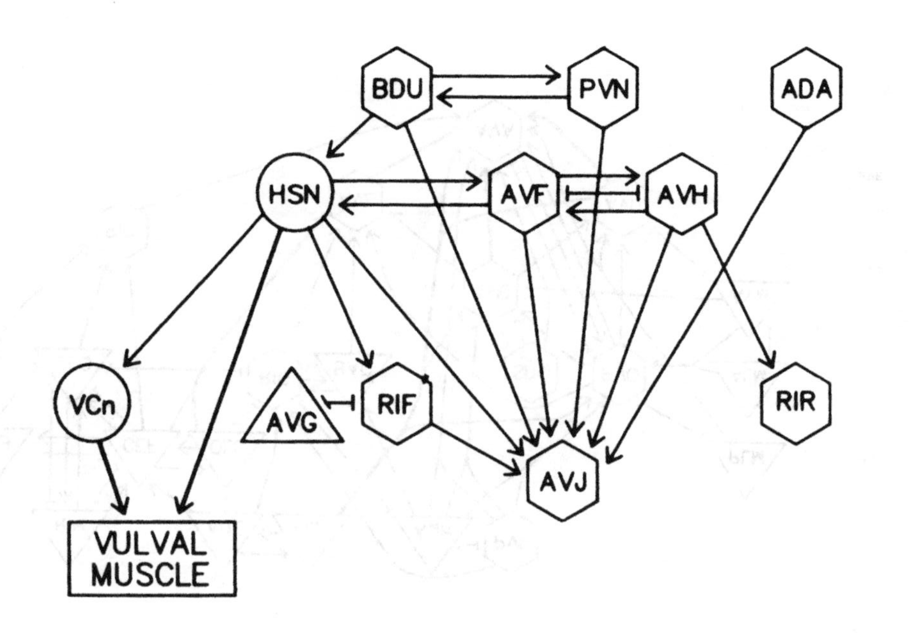

	PRE	POST	EJ
ADA	AVB		ADA
	RIM		AVD
	SMD		PVQ
	RIP		ASH
			ADF
AVF	AVB	AIM	AVF
	NMJ(VC)		
AVG	AVB	PHA	
	PVN		
AVH	PVP	PHA	
	SMB		
	ADF		
AVJ	PVC	ADL	AVJ
	AVB	PVR	RIS
	RIS	PVC	PVC
	AVE		AVD
	AVD		
BDU	ADE	ALM	
		HSN	
		AVM	
HSN	AWB	PLM	HSN
	BDU		
	AIZ		
PVN	AVD	AVG	AVB
	PQR	PDA	HSN
	NMJ(VC)		PVQ
	AVA		
	AVL		
	PVC		
	VDn		
	PVT		
	DDn		
	PVW		
RIF	AVB	AIA	
	RIM		
	PVP		
	ALM		
RIR	RIA	RIG	BAG
	AIZ	ADF	
	URX	DVA	
VCn	VCn	VCn	VCn
	DDn		
	VDn		
	NMJ(VC)		

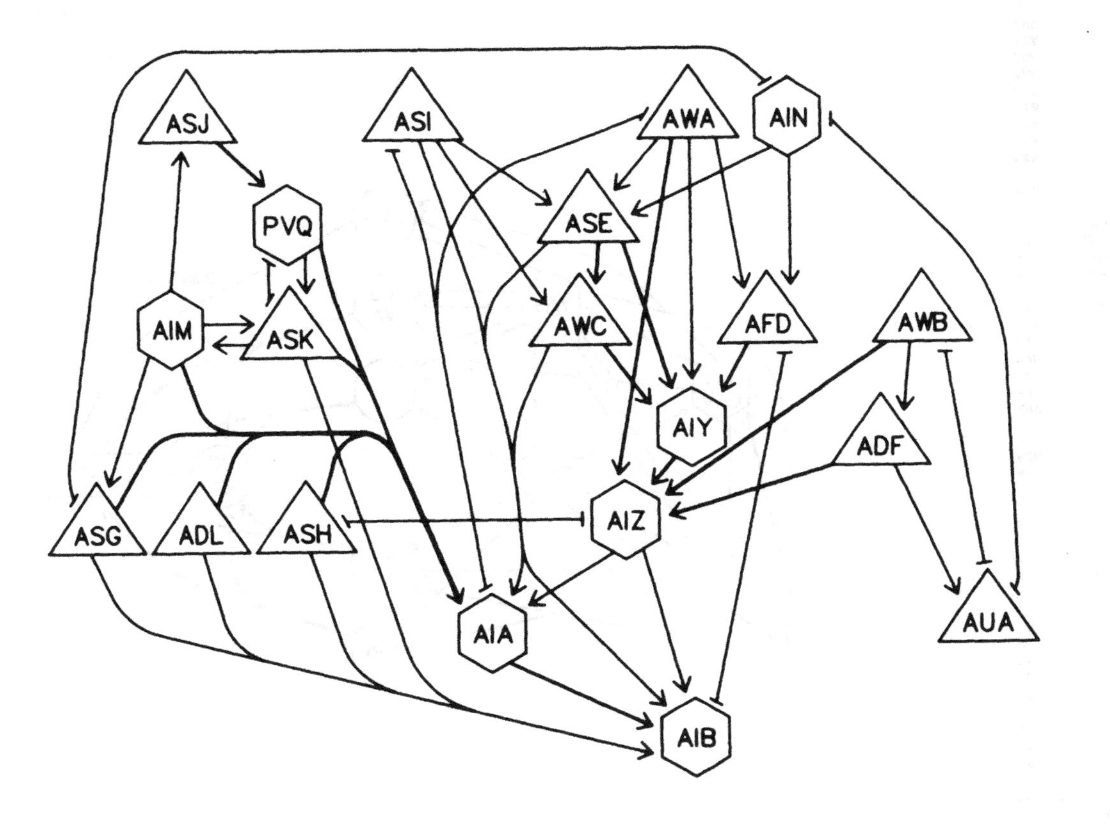

	PRE	POST	EJ
ADF	RIA		ADF
	SMB		ADA
	RIR		RIH
			AIA
ADL	AVD		ADL
	AVB		OLQ
	AVA		RMG
AFD			AFD
AIA	AWC		AIA
	RIF		ADF
AIB	RIM	DVB	DVC
	AVB	DVC	DVB
	RIB	RIM	RIG
	SAAD	RIB	RIS
		FLP	RIV
AIM	AVF		AIM
	CEPsh		SIBD
AIN	BAG		AIN
	RIB		
	CEPsh		
AIY	RIA		AIY
	RIB		RIM
AIZ	RIA	AWA	AIZ
	SMB	RIR	ASG
	RIM	RIH	
	AVE	AWB	
		HSN	
ASE	RIA		
ASG			AIZ
ASH	RIA		ASH
	AVB		RMG
	AVA		ADA
	AVD		RIC
ASI			ASI
ASK		ASJ	ASK
			RMG
AUA	RIB	URX	AUA
	RIA		URX
	AVE		
	AVA		
AWA			AWA
AWB	RIA		AWB
	AVB		RMG
AWC		AIA	
PVQ	AVF	PHA	PVQ
	HSN		ADA

	PRE	POST	EJ
AVA		SAA	RIM
		FLP	URY
		RIC	AVA
		DVC	SAB
		PVP	AVJ
		AUA	
		ASH	
		AQR	
		ADL	
		RIB	
		SDQ	
		ADE	
		PLM	
		PHB	
		PQR	
		LUA	
AVB	HDC	RIF	AVB
		RIM	RIB
		AVM	SIBV
		ASH	SDQ
		PVR	PVN
		PVP	
		SDQ	
		ADA	
		AQR	
		FLP	
		AIB	
		AVF	
		ADL	
		URX	
		PVN	
		AVG	
AVD/E	SAB	PVN	ADA
		ASH	FLP
		ADL	AVJ
		FLP	AVM
		AQR	RIM
		PQR	RME
		LUA	RMD
		PLM	
		PHB	
		OLL	
		URY	
		CEP	
		URX	
		RIB	
		RIS	
		BAG	
		AUA	
		ALA	
		AVK	
		AIZ	
		RIG	
		RMG	
		AVJ	

	PRE	POST	EJ
DVA	RIR	AIZ	PVR
	AQR	SDQ	PVC
	SMB	PDE	
		AIZ	PHC
	AUA	PLM	
	PVR	PHA	
		PVD	
PVC		PVP	PVC
		ALM	AVJ
		AVM	PVW
		AVJ	PLM
		AQR	PDE
		PHB	DVA
		PHC	PHC
		VA12	
		LUA	
		PVM	
		PVN	
		PVD	
VA/DA			VA/DA
VB/DB			VB/DB
VD/DD			PVP
			VD/DD

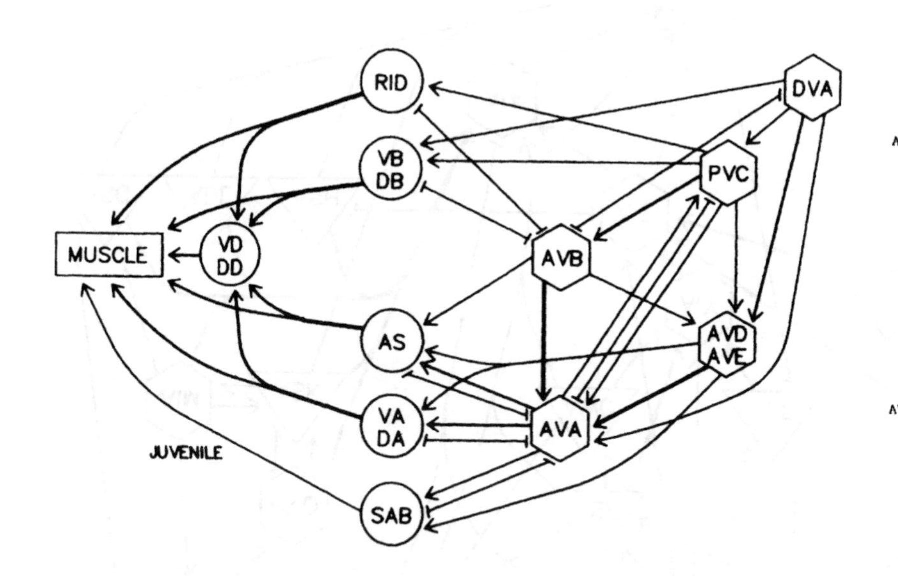

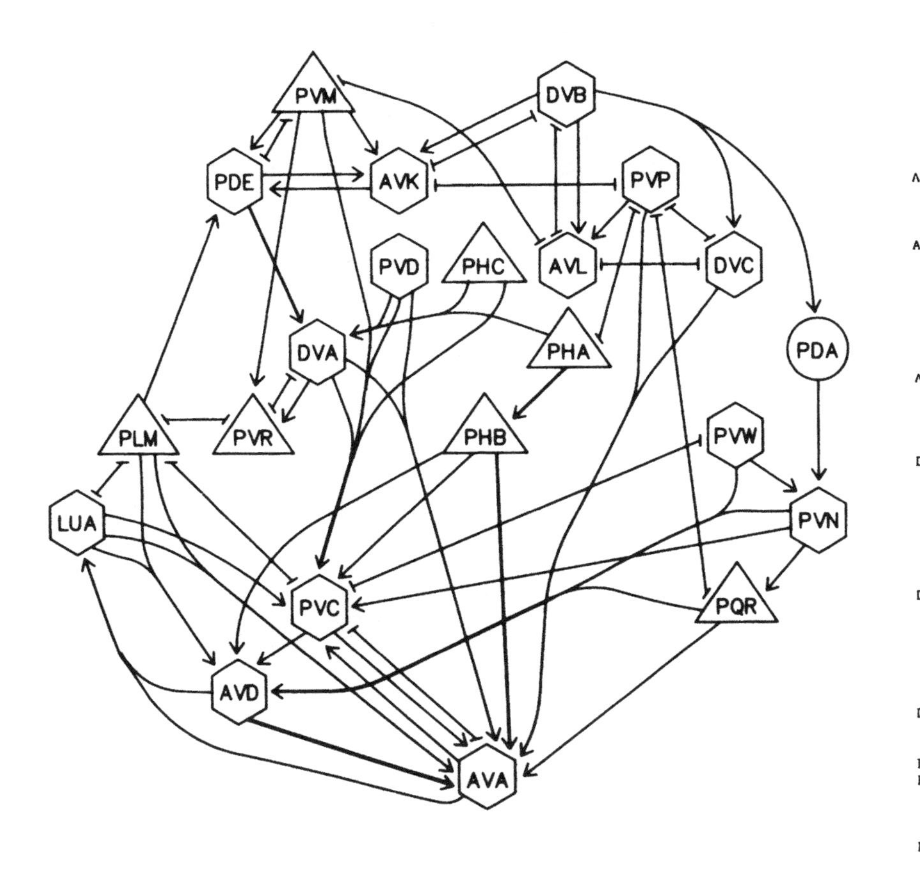

	PRE	POST	EJ
AVA	SAB	SAA	AVA
	VAn	FLP	LUA
	DAn	RIC	
	ASn	AUA	RIM
		ASH	URY
		AQR	VAn
		SDQ	DAn
		ADL	ASn
		RIB	SAB
		AVE	
		AVB	
		ADE	
AVD	SAB	ASH	ADA
	VAn	ADL	FLP
	DAn	FLP	AVJ
	ASn	AQR	AVM
AVK	RIM	RMF	AVK
	AVE	RIG	SMB
	SMD		AQR
	HDC		RIC
			ADE
			RIG
AVL	NMJ (VC)		
	SAB		
	VD12		
DVA	AVE	PLM	AVB
	RIR	AIZ	PVC
	AQR	SDQ	
	SMB		
	AIZ		
	AUA		
	DBn		
	VBn		
DVB	AIB		RIB
	RMF		AIB
	RIG		
	DA8		
	DD6		
	NMJ		
	(Anal)		
DVC	RIG		AIB
	AIB		VDI
	RMF		
LUA	AVJ		AVA
PDA	DA9		
	HDC		
	NMJ (DC)		
	DD6		
PDE	HDC		PVC
			PDE

	PRE	POST	EJ
PHA	PHA	PHA	PHA
	AVG		
	PVQ		
	AVF		
	AVH		
PHB	VA12		PHB
			AVH
PHC	DA9		PHC
			VA12
			DA9
PLM	HSN		
	DVA		
PVC	AVB	PVP	PDE
	RID	ALM	PVC
	AVE	AVM	AVJ
	VBn	AVJ	DVA
	DBn	AQR	
		VA12	
PVN	NMJ (VC)	BDU	AVB
	AVA	AVG	HSN
	AVJ		PVQ
	AVL		
	DDn		
	VDn		
	AVJ		
	PVT		
	BDU		
PVP	PVC	RIF	PVP
	AVB	AVH	AQR
	RIG		VDn
	AVH		
	HDC		
PVR	AVB	AVM	ALM
	RIP		
	AVJ		

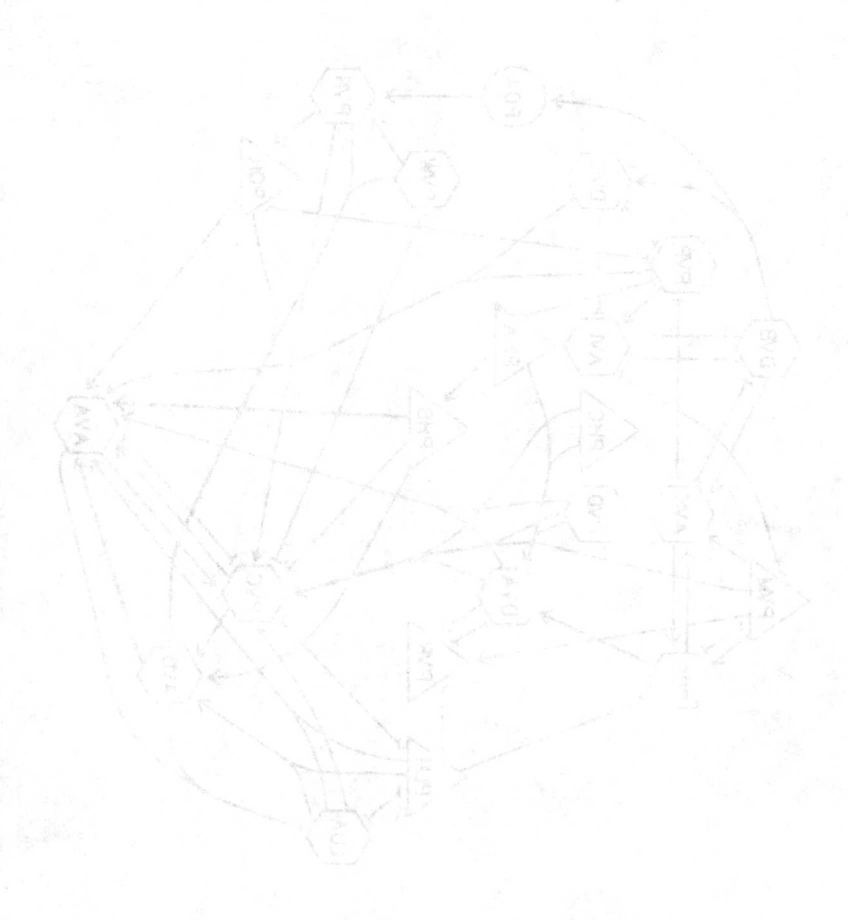

APPENDIX 3
Cell Lineage

Compiled by John Sulston,* H. Robert Horvitz,† and Judith Kimble‡

*MRC Laboratory of Molecular Biology,
Cambridge, England
†Massachusetts Institute of Technology,
Cambridge, Massachusetts
‡University of Wisconsin, Madison, Wisconsin

Part A: Lineage Charts

Part B: Positions of Nuclei

Part A: Lineage Charts

Figure 1 Embryonic cell lineage of *C. elegans*. All interconnecting lines between the separate panels have been drawn, so that the pages can be copied, trimmed, and pasted together to give a complete chart. Vertical axis represents time at 20°C, from 0 min at first cleavage to 800 min at hatching. Many of the observations were made on eggs that were developing at slightly different rates (due to temperature variation and the effect of prolonged illumination); these primary results were normalized, by means of certain prominent cell divisions, to the course of events in eggs kept at 20°C and viewed infrequently. Precise times of individual events should not be taken too seriously; likely error varies from ±10% at the beginning of the lineage to ±2% at 400 min. Horizontal axis represents the direction of cell division. The majority of divisions have a marked anterior-posterior bias and are shown with anterior to the left and posterior to the right, without any label. Only when this would lead to ambiguity in naming the daughters is an alternative direction indicated (l = left; r = right; d = dorsal; v = ventral); thus, the system is taxonomic, rather than fully descriptive. The natural variation seen suggests that the precise direction of cell divisions is unimportant, at least in later development. Note that the daughters of a left-right division are not necessarily bilaterally symmetrical: For example, all of the cells derived from ABalaaapa lie on the left of the animal and the right-hand daughters lie nearer the midline. Each terminal branch of the embryonic lineage is labeled either with X (indicating cell death; the position of the X on the time axis indicates the time of maximum refractility) or with a lineage name followed by a functional name. Large arrowheads denote cells that divide postembryonically, and small arrowheads denote nuclei that divide postembryonically. O and --- link precursors that give rise to bilaterally symmetrical groups of cells, the symbol ~ being included for cases of imperfect symmetry. (cord) Ventral cord; (gang) ganglion; (lumb) lumbar; (d-r) dorsorectal; (p-a) preanal; (r-v) retrovesicular; (lat neur) isolated neuron lying laterally. (Reprinted, with permission, from Sulston et al. 1983.)

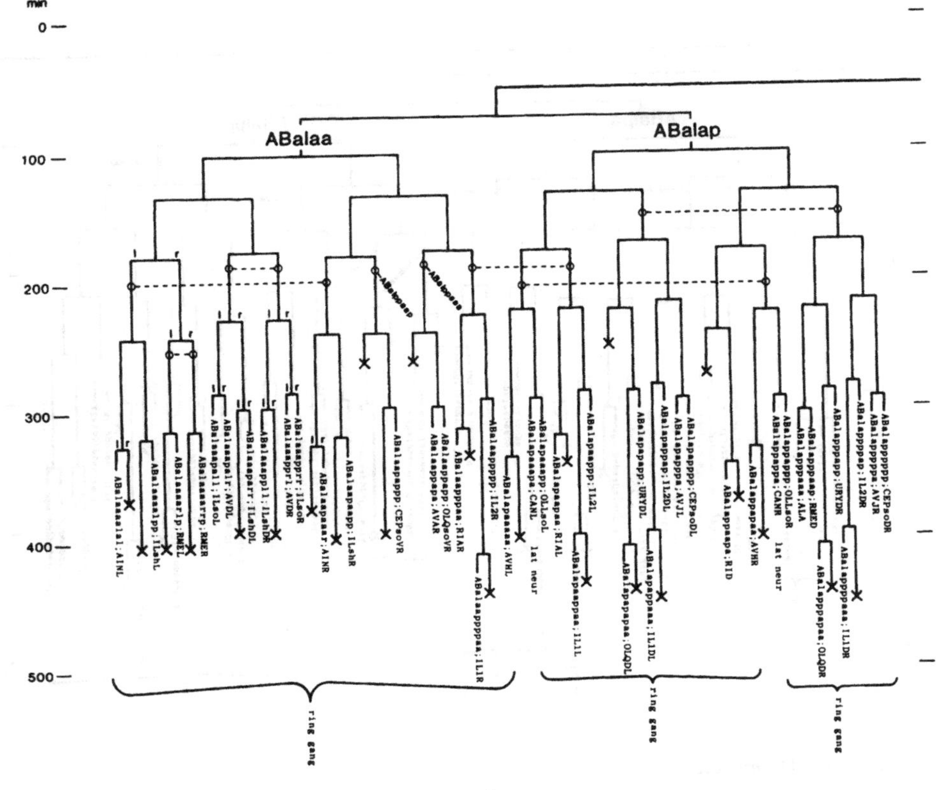

Figure 1 (*Continued on following pages.*)

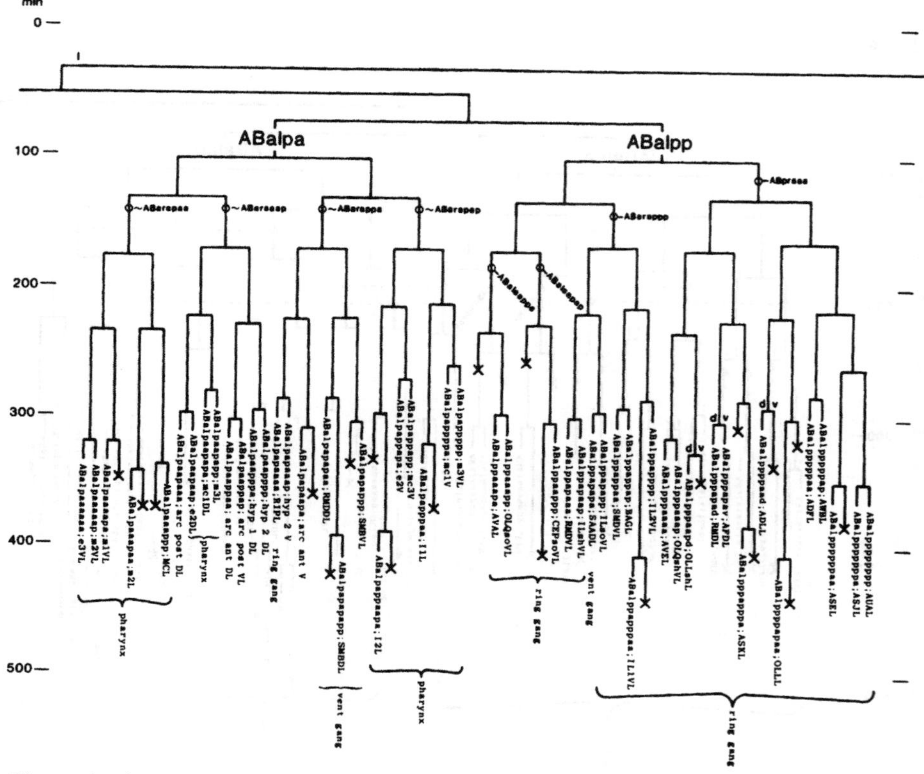

Figure 1 (*Continued.*)

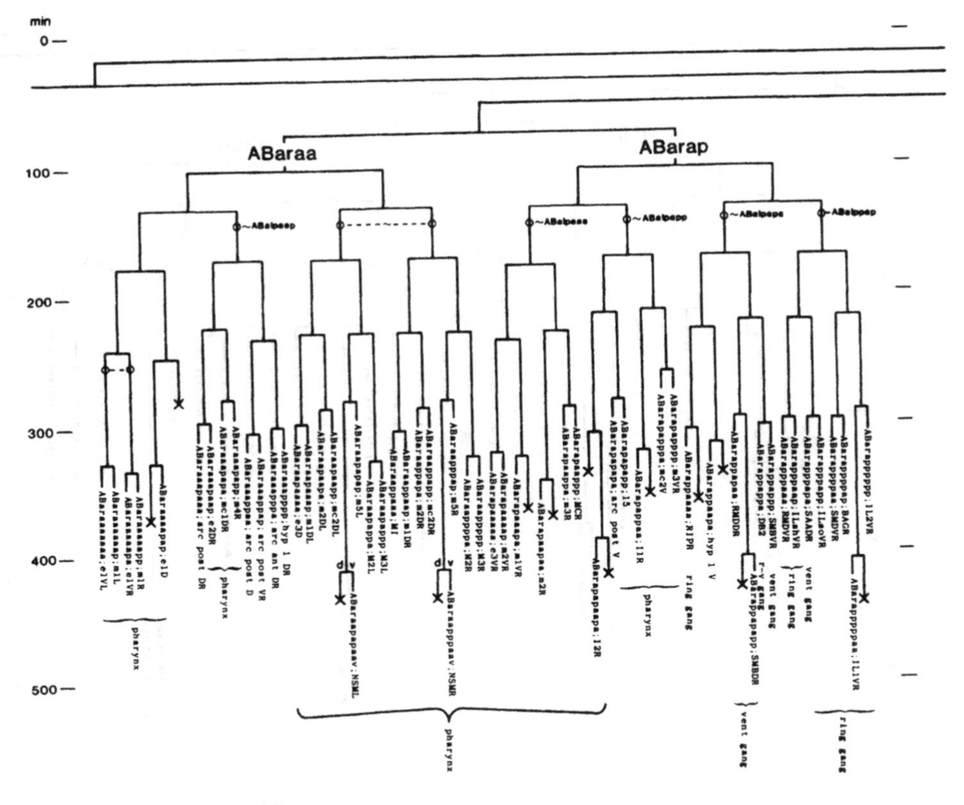

Figure 1 (*Continued.*)

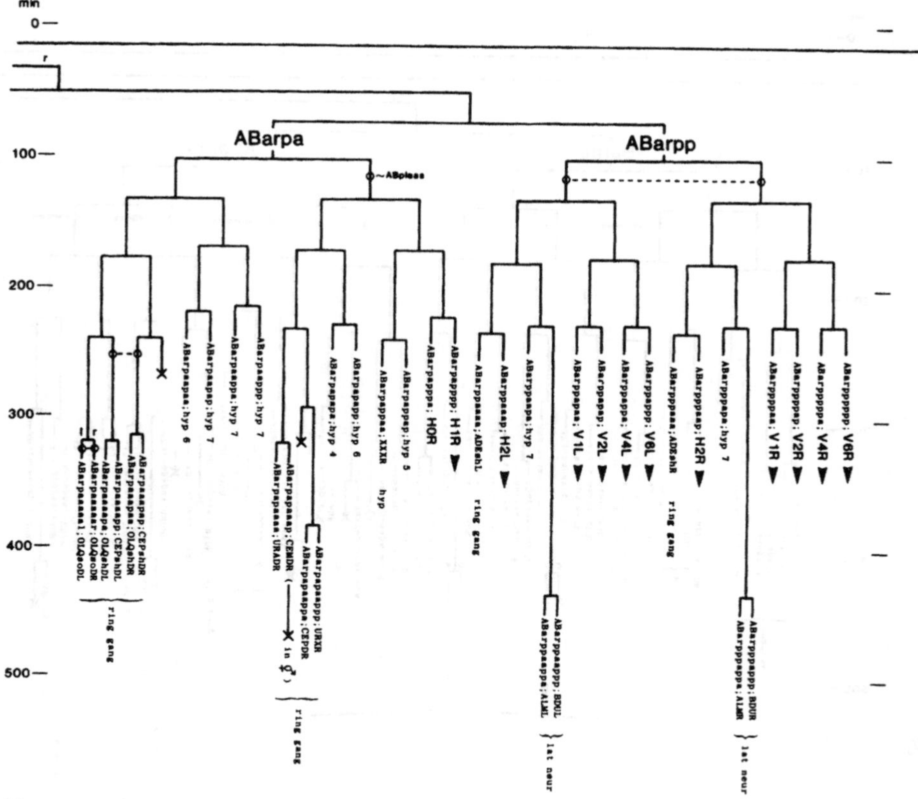

Figure 1 (*Continued.*)

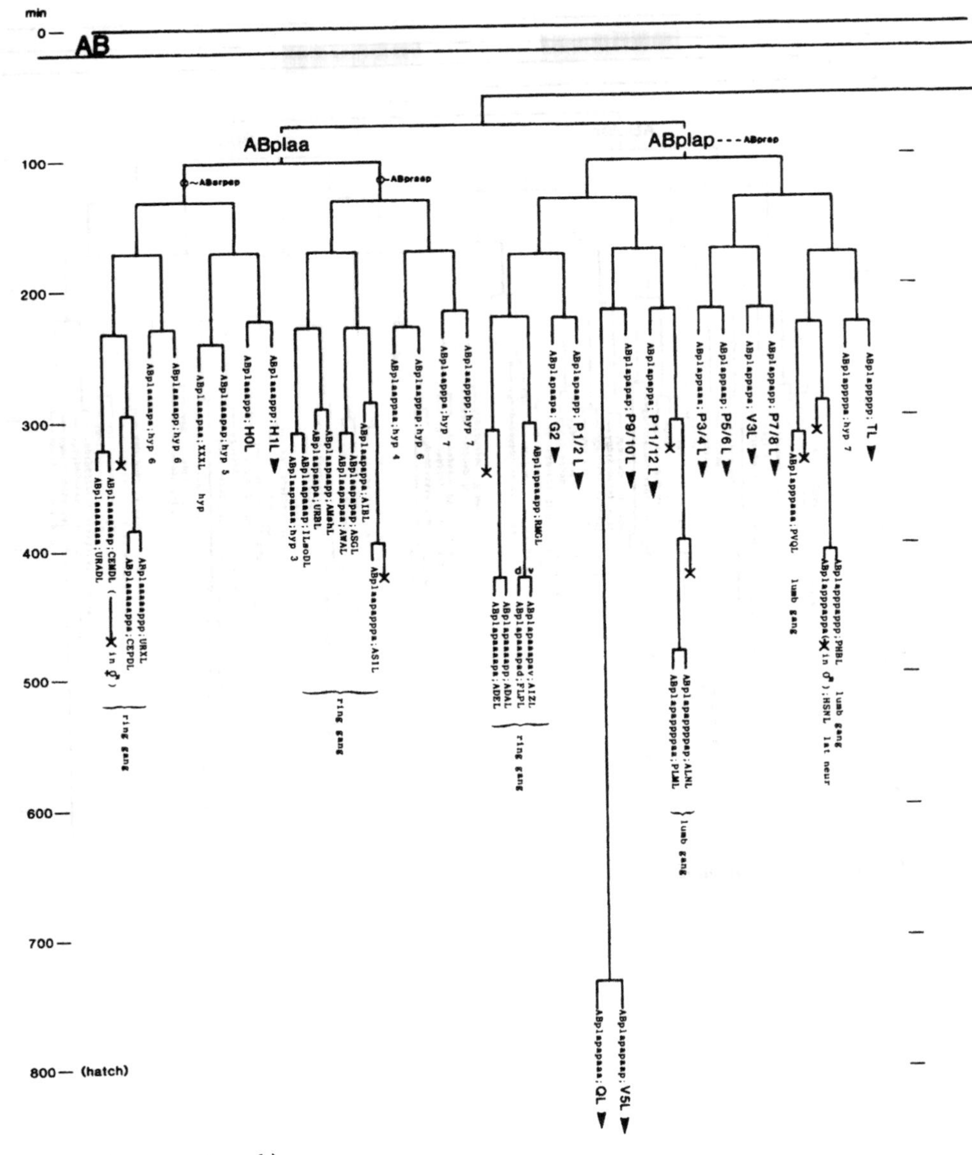

Figure 1 (*Continued.*)

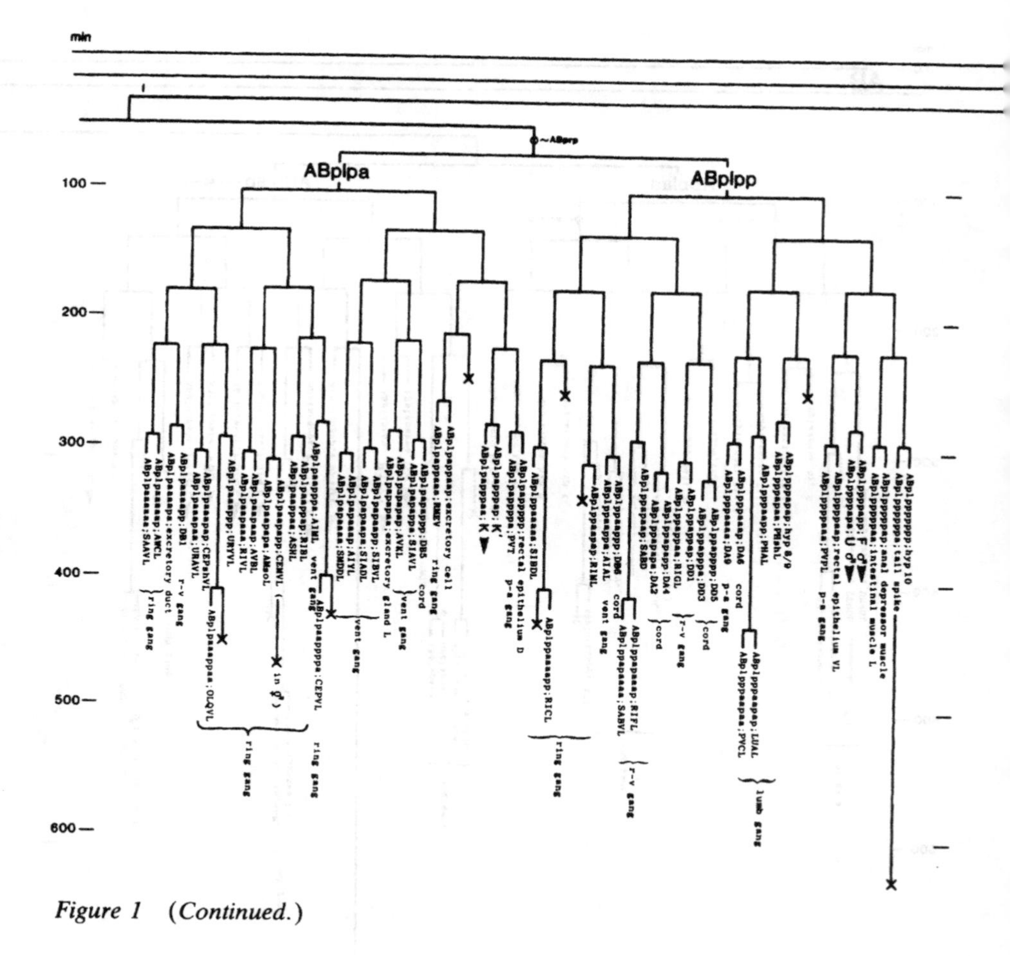

Figure 1 (*Continued.*)

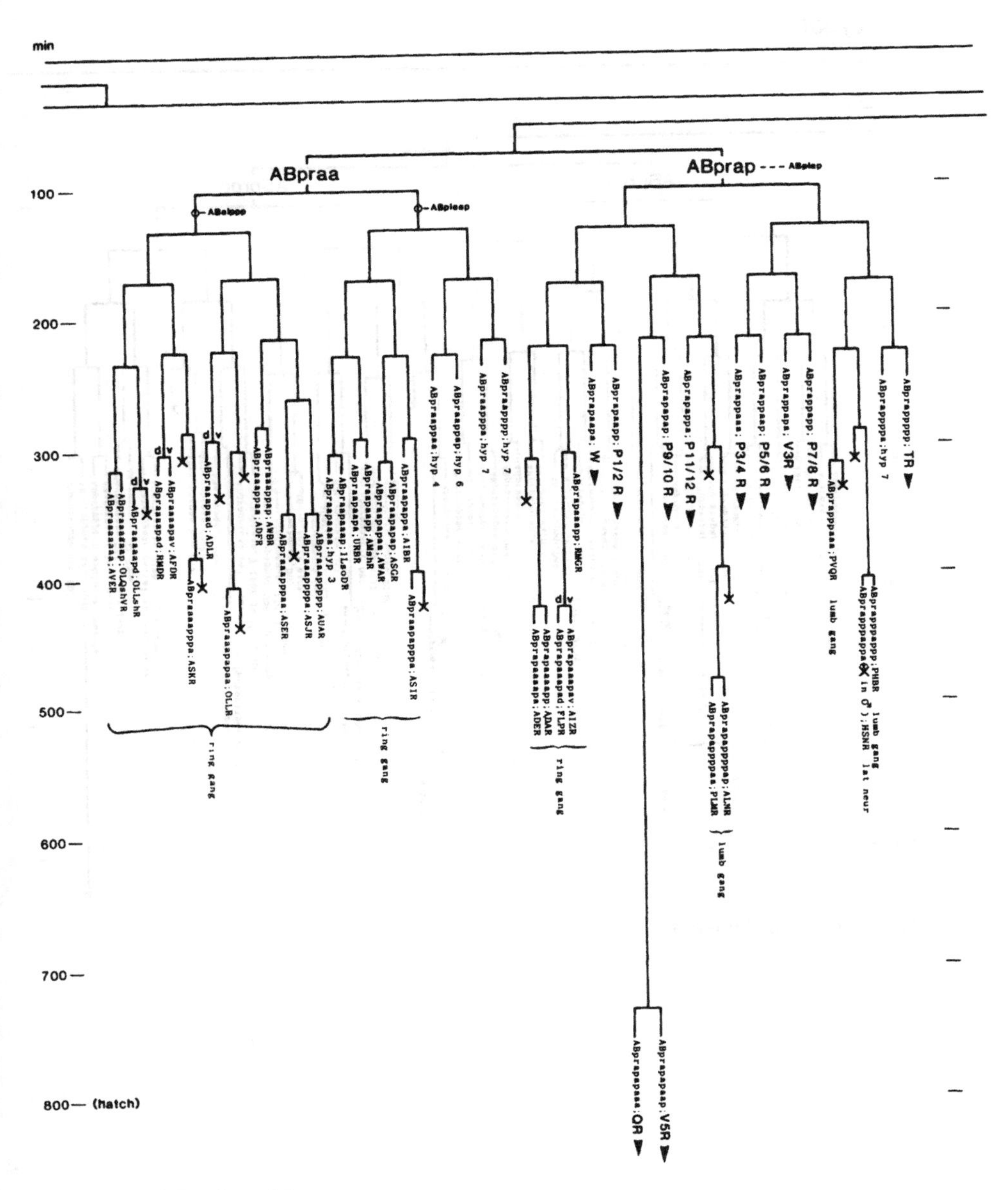

Figure 1 (*Continued.*)

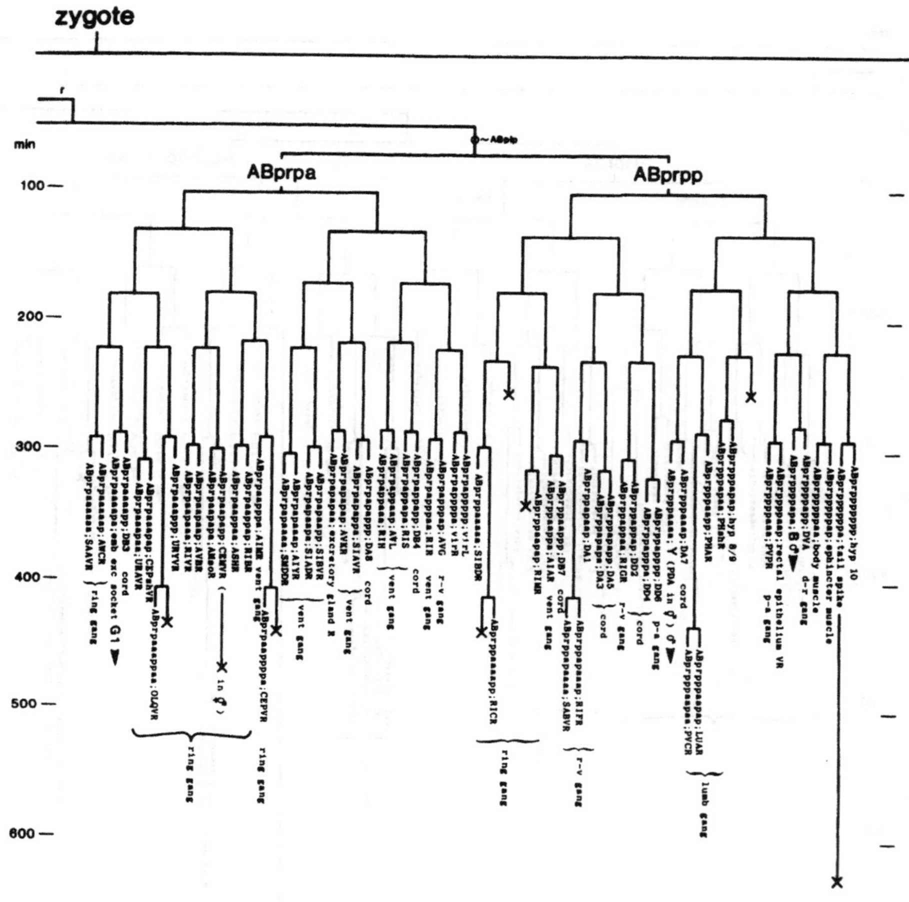

Figure 1 (*Continued.*)

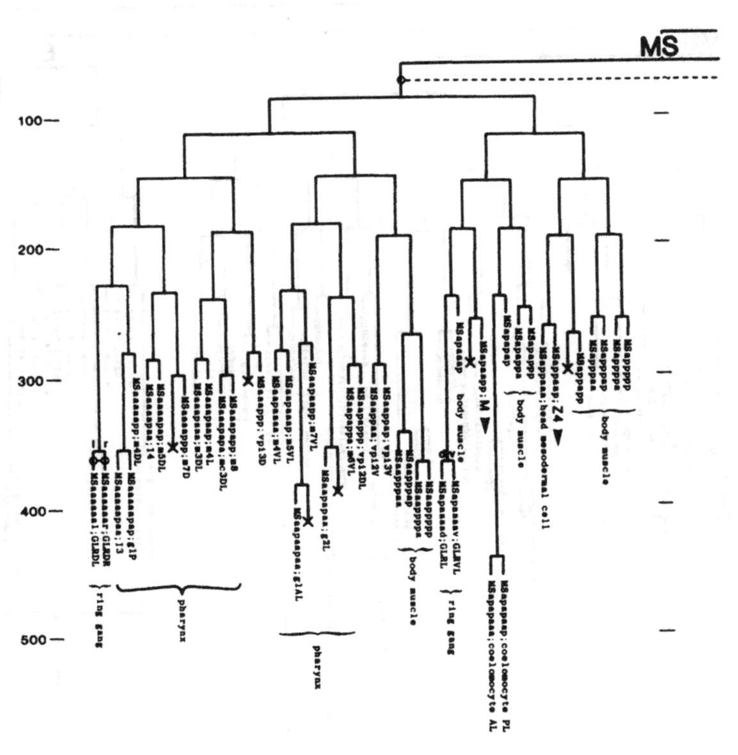

Figure 1 (*Continued.*)

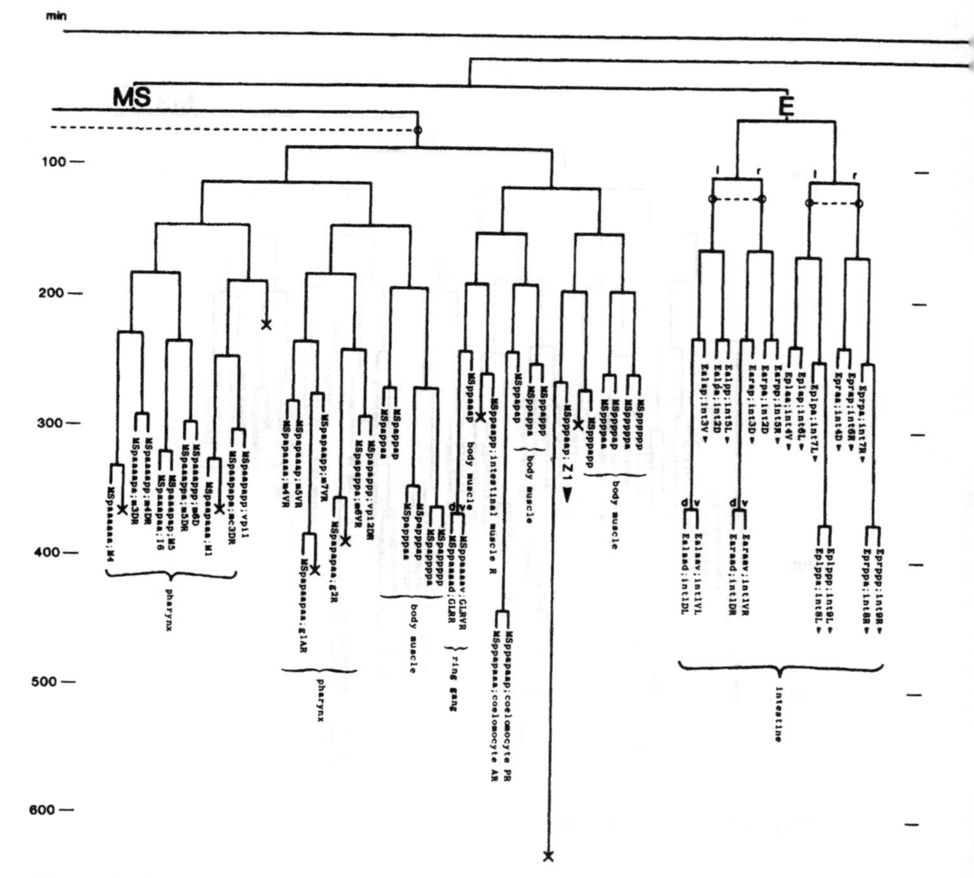

Figure 1 (*Continued.*)

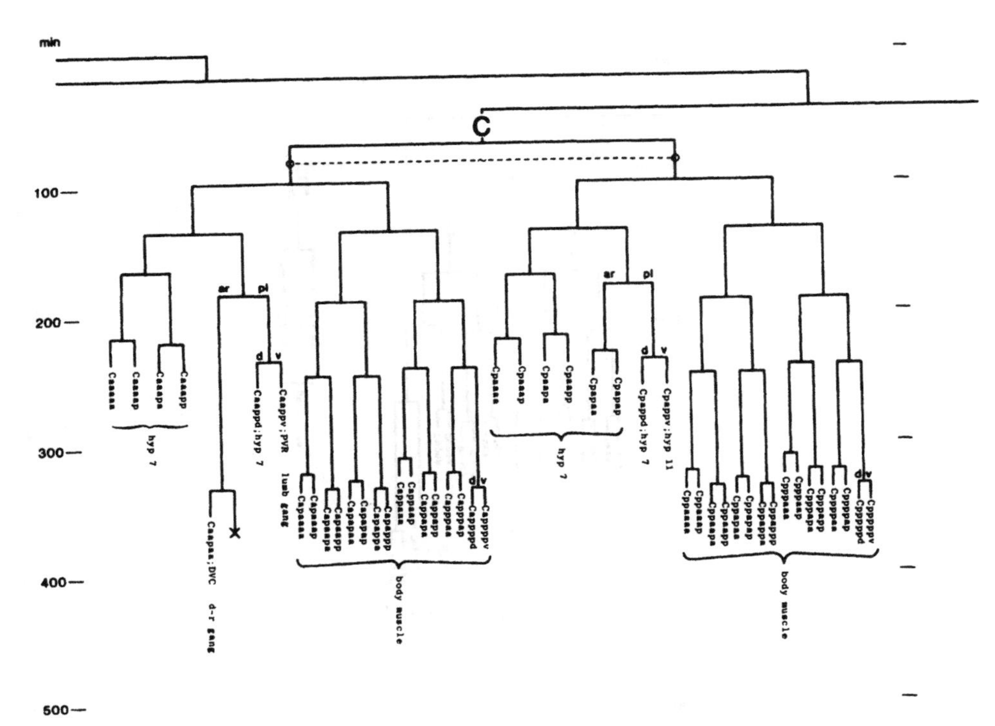

Figure 1 (*Continued.*)

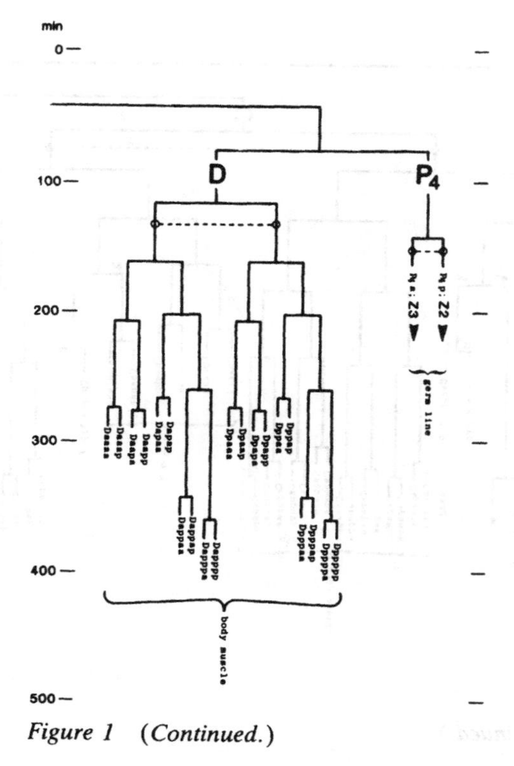

Figure 1 (*Continued.*)

Figures 2–8 Postembryonic cell lineages of *C. elegans*. All divisions are anterior-posterior unless otherwise indicated. Molts are indicated by solid lines on the time axis, the adjacent stippling indicating the period of lethargus. (Reprinted, with permission: Figs. 2–6 are from Sulston and Horvitz [1977]; Figs. 7–8 are from Kimble and Hirsh [1979].)

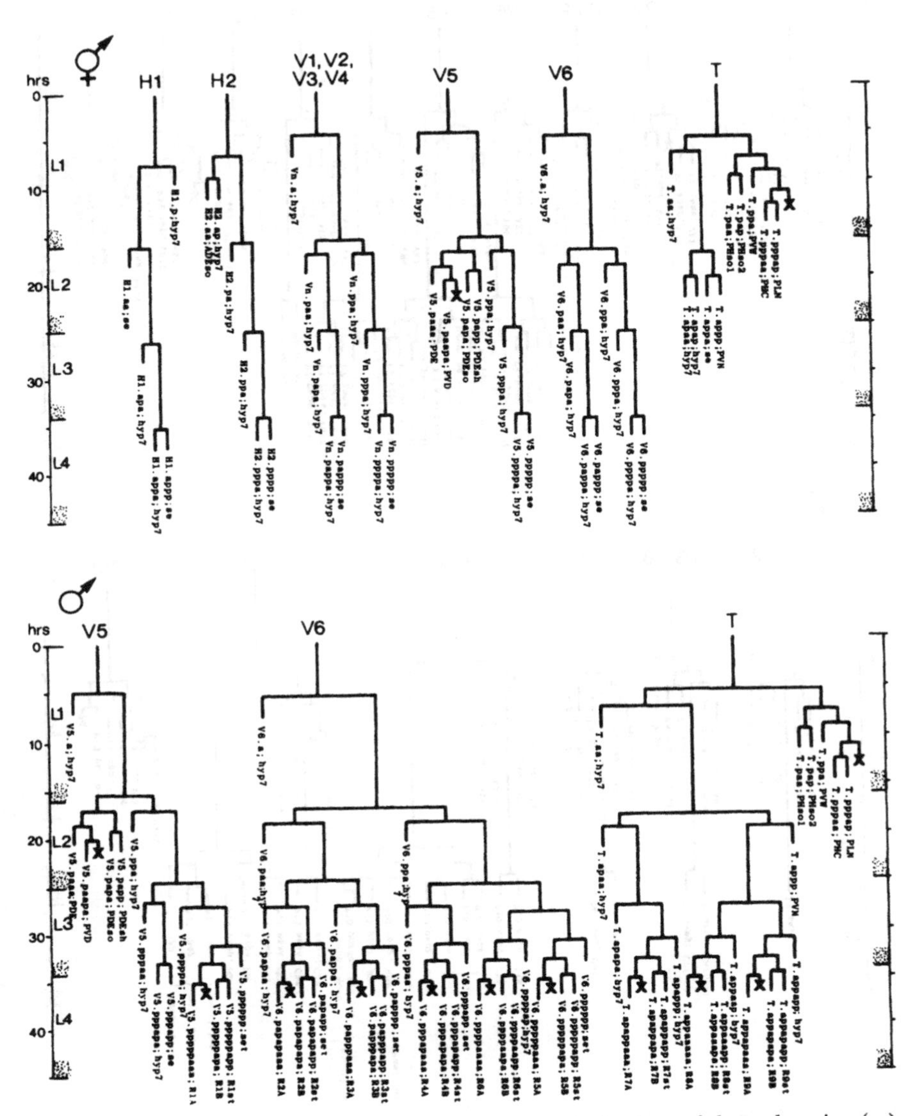

Figure 2 H, V, and T lineages: development of the lateral hypodermis. (se) Seam; (set) tail seam.

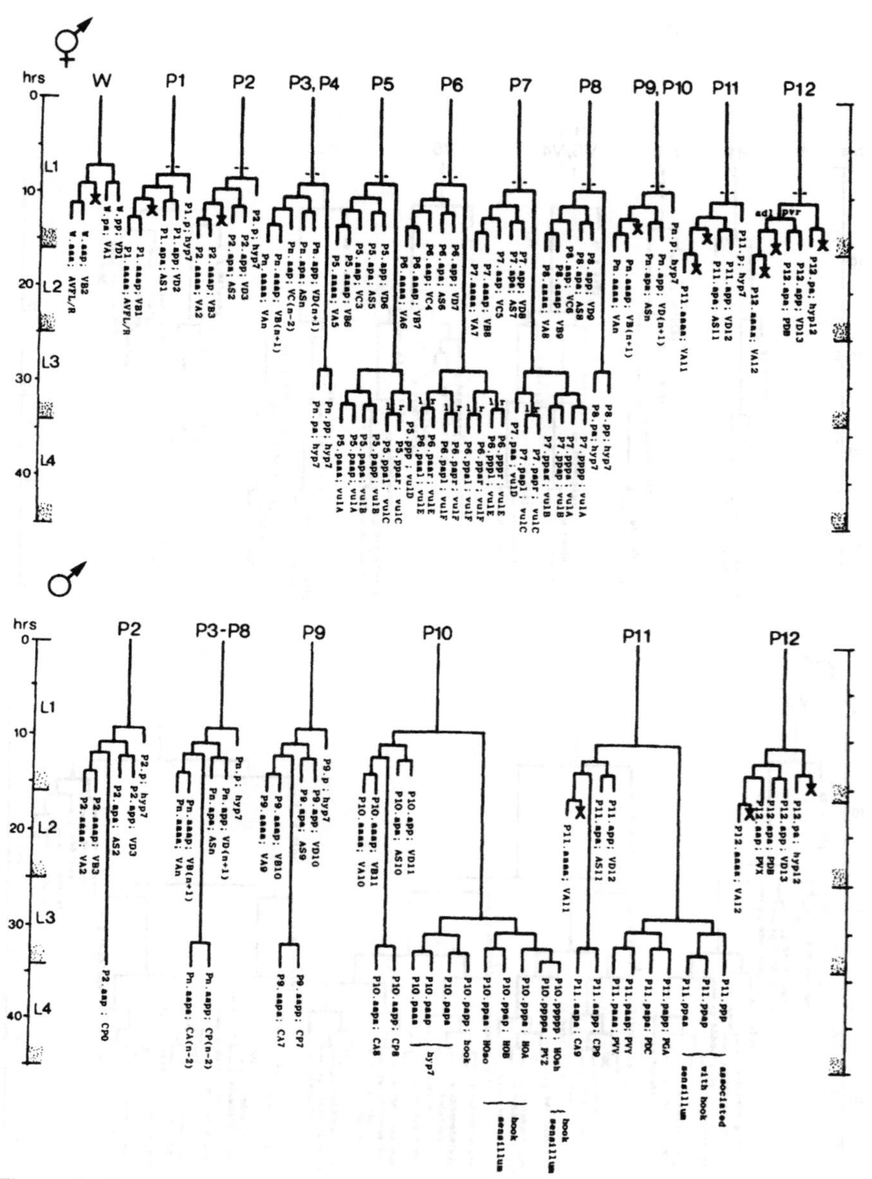

Figure 3 P lineages: development of the ventral nervous system. Dotted lines indicate the times at which nuclei migrate into the ventral cord.

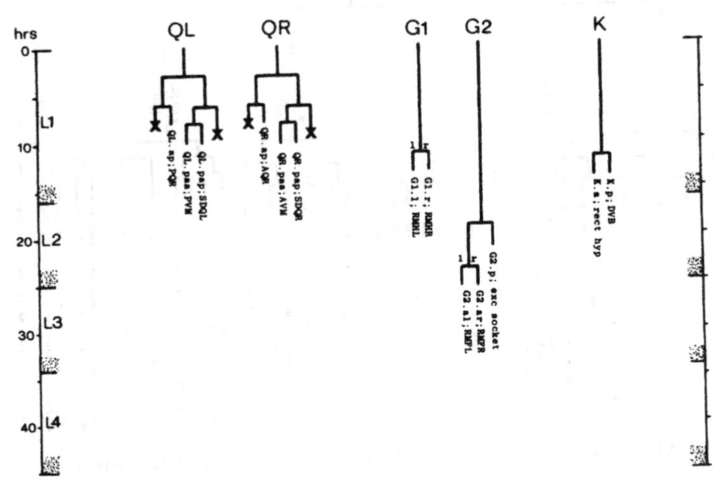

Figure 4 Q, G, and K lineages: other neuroblasts.

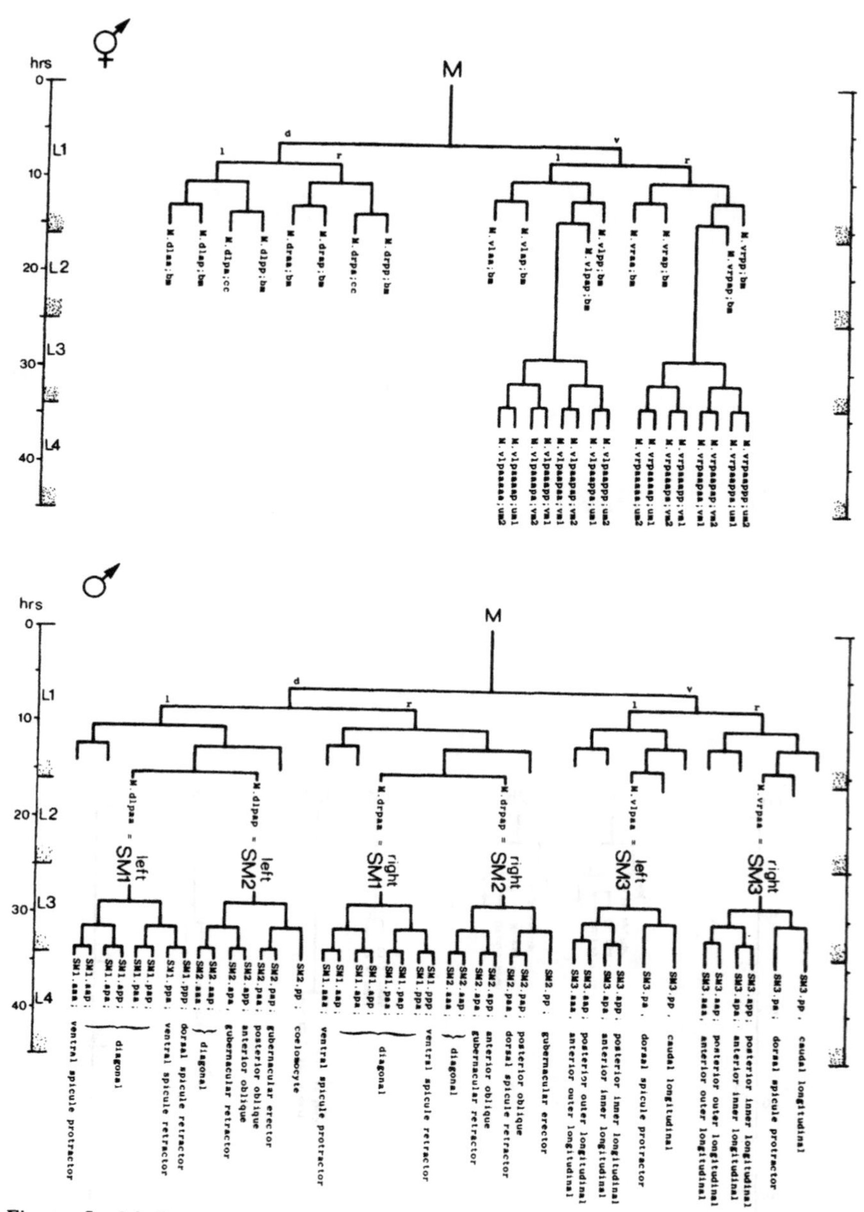

Figure 5 M lineage: mesodermal development. (bm) Body muscle; (cc) coelomocyte.

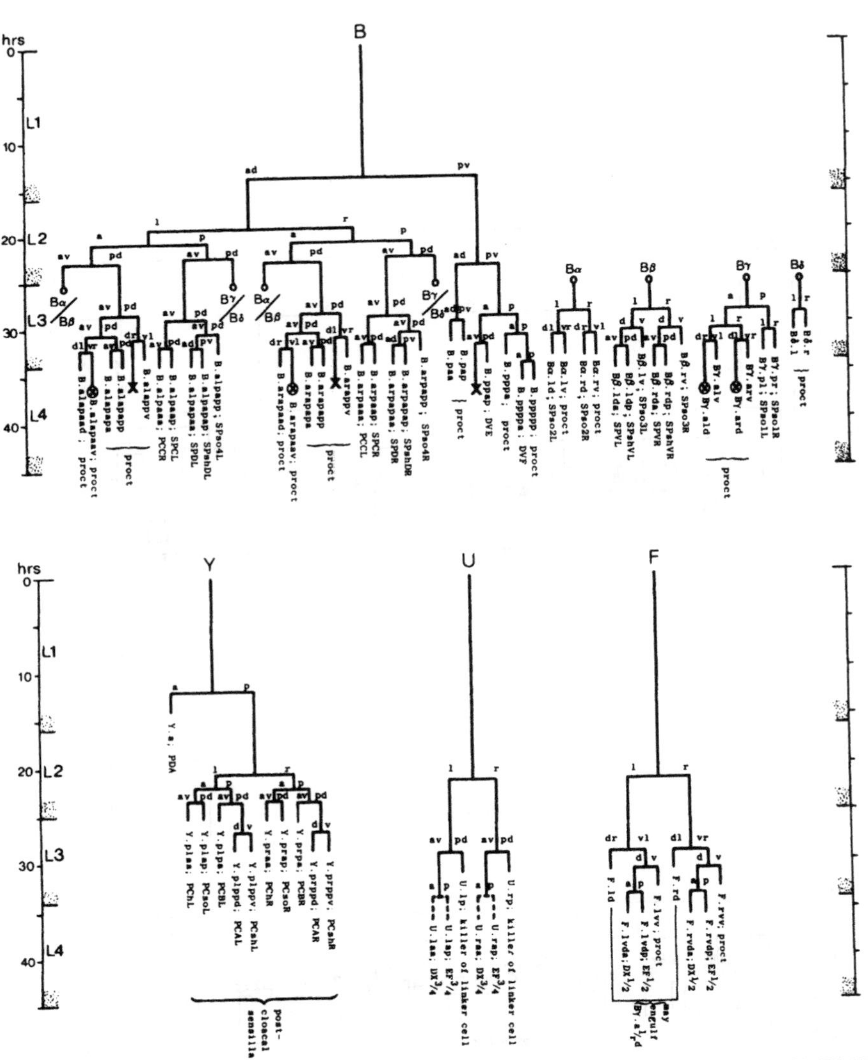

Figure 6 B, Y, U, and F lineages: ectodermal development in the male tail. Y, U were formerly C, E, respectively.

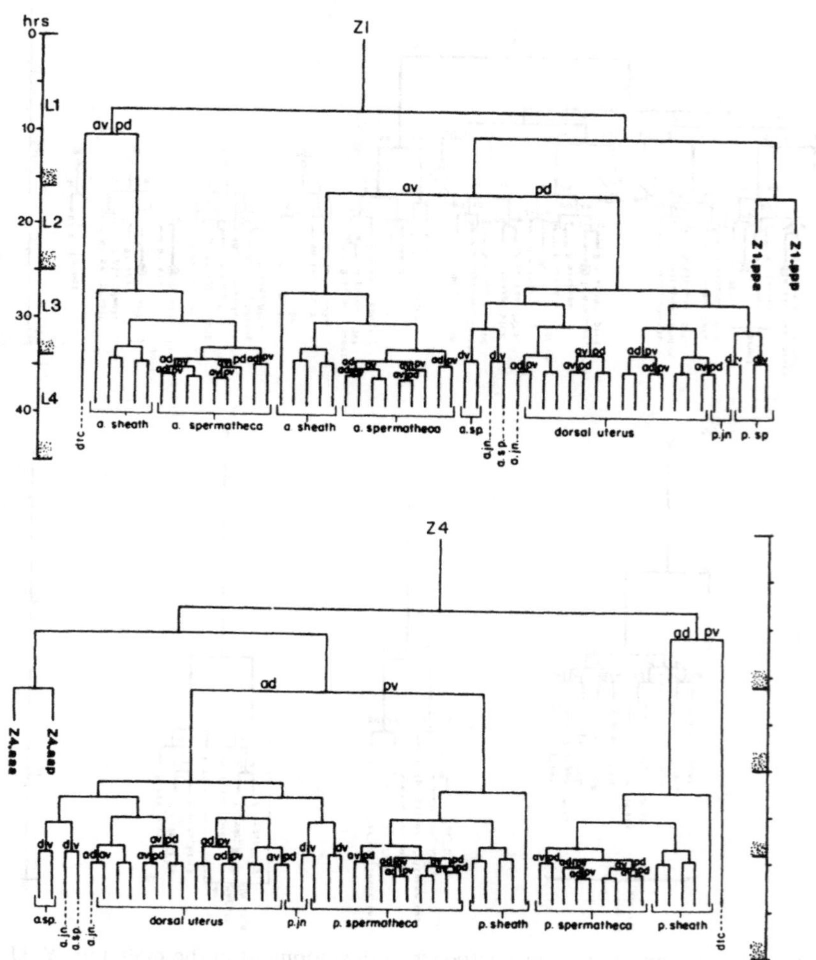

Figure 7 (*Continued on facing page.*)

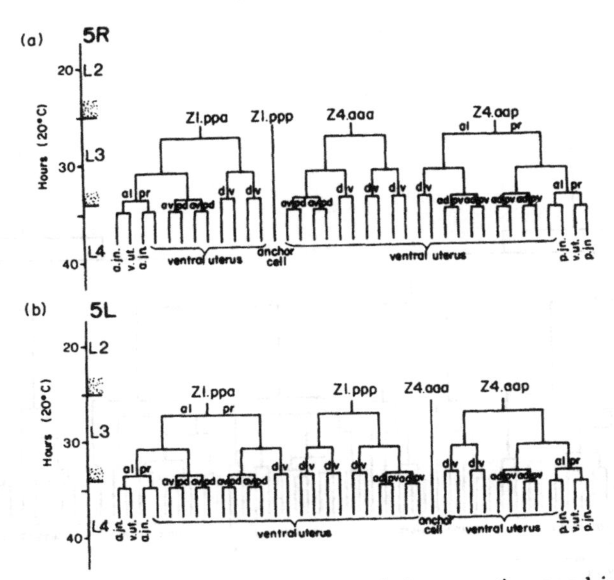

Figure 7 Z1 and Z4 lineages: development of the somatic gonad in the hermaphrodite. (*Left-hand page*) Invariant lineages; (*above*) the two alternative lineages followed by Z1.ppa, Z1.ppp, Z4.aaa, and Z4.aap. (a) Anterior; (p) posterior; (dtc) distal tip cell; (sp) spermatheca; (jn) spermathecal-uterine junction; (v.ut.) ventral uterus.

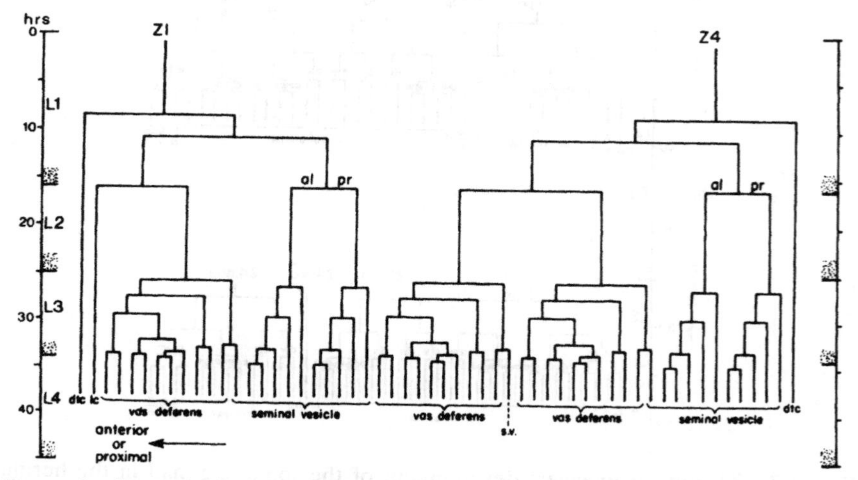

Figure 8 Z1 and Z4 lineages: development of the somatic gonad in the male. One of the two alternative lineages is shown here; in the other, the fates of Z1.paa and Z4.aaa are exchanged. (dtc) Distal tip cell; (lc) linker cell; (s.v.) seminal vesicle.

Part B: Positions of Nuclei

Figures 9–11 Drawings of embryos. Circles and ovals represent nuclei, traced by means of a camera lucida, the thickness of the lines being inversely related to depth; outlines of the egg, embryo, and internal structures are traced with thin lines (regardless of depth). Anterior is toward the top of the page. Dying nuclei are stippled. (Reprinted, with permission, from Sulston et al. 1983.)

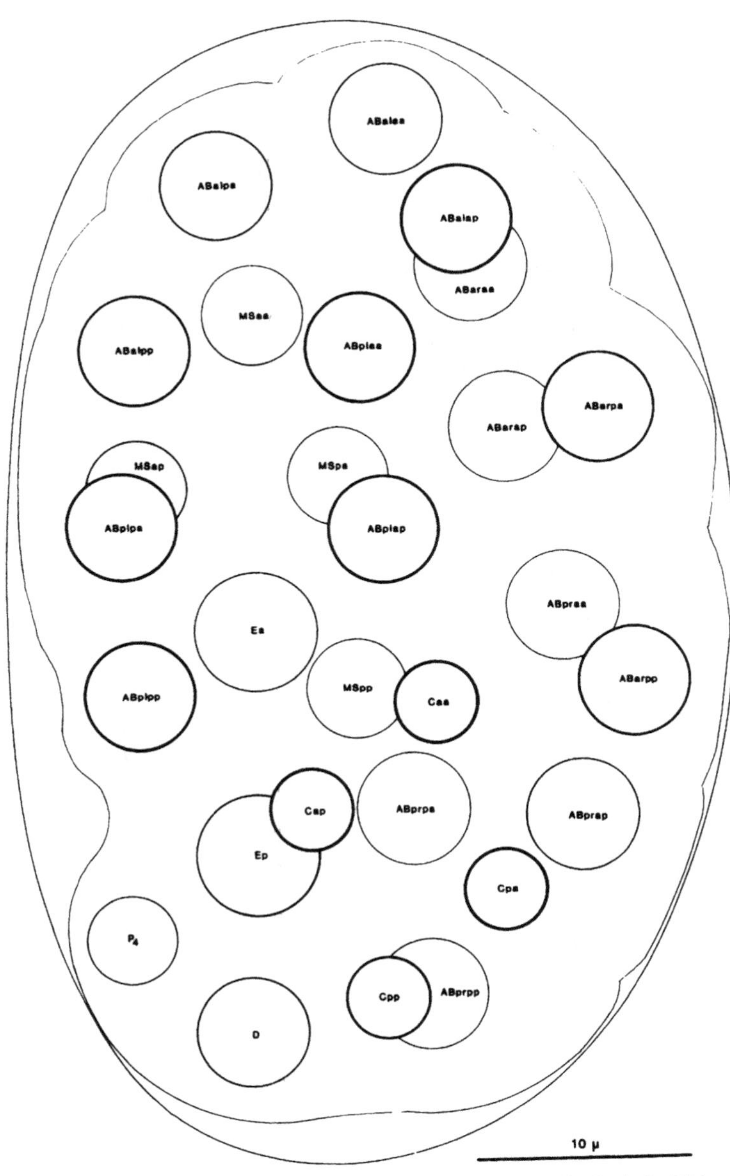

Figure 9 Embryo, 100 min, left dorsal aspect; all nuclei included. This stage has been well characterized, and the observer quickly learns to recognize all the nuclei; it is a useful starting point both for lineages and for ablation experiments. An embryo in the orientation shown will present a dorsal aspect until it turns at 350 min; an embryo with the MS cells uppermost will present a ventral aspect. The intestinal precursors are entering the interior, leaving a characteristic depression on the ventral side.

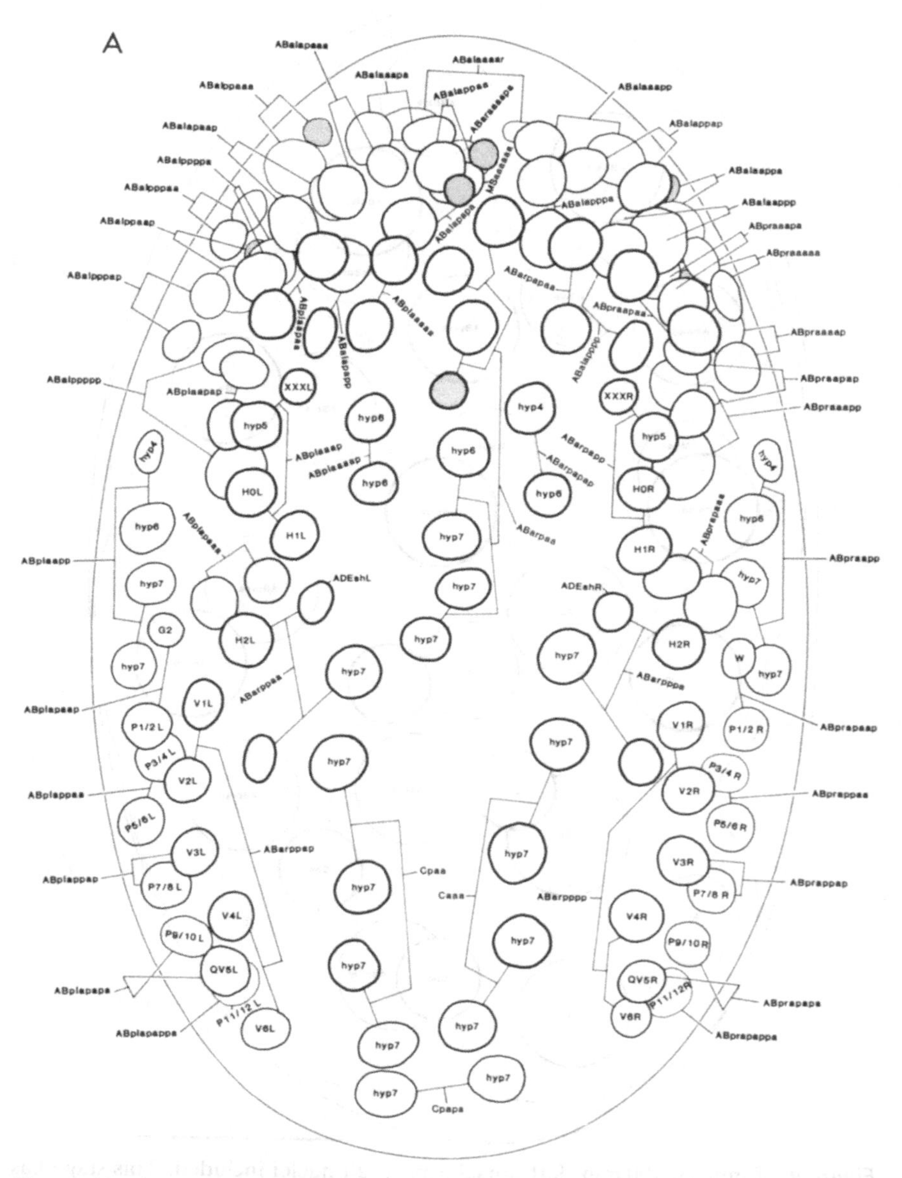

Figure 10 (*A*) Embryo, 260 min, dorsal aspect, superficial nuclei. Landmarks: nuclei of hyp4-hyp7, cell deaths; time points: division of various neuroblasts. Dorsal hypodermal cells have very granular cytoplasm and form prominent transverse ridges. Here and in Figure 10B some licence has been allowed in depicting cell deaths, because of their importance in pattern recognition; in fact, they do not all become refractile simultaneously.

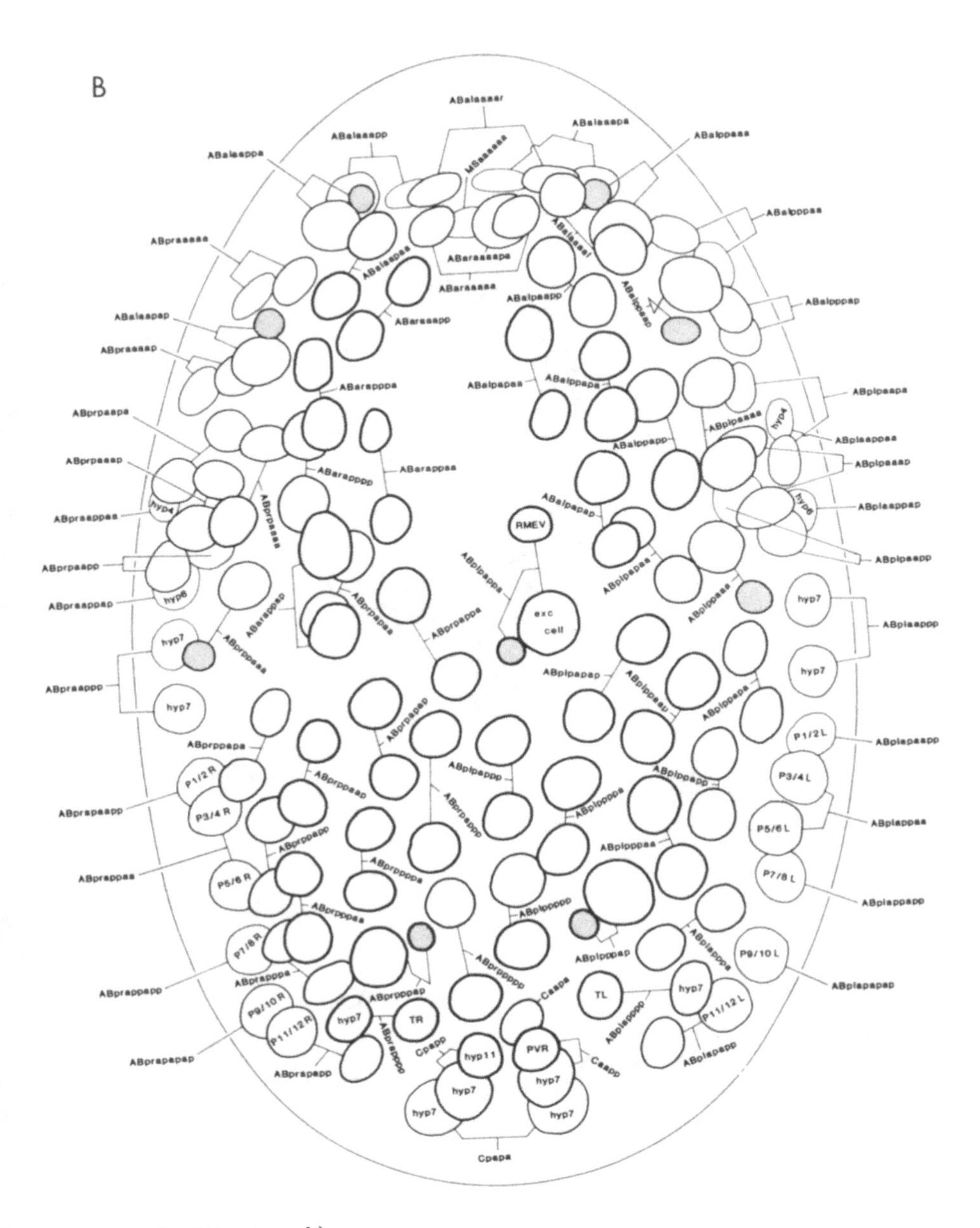

Figure 10 (*Continued.*)
(*B*) Embryo, 270 min, ventral aspect, superficial nuclei. Landmarks: excretory cell, cell deaths; time point: division of mother of excretory cell. The gap anterior to the excretory cell contains pharyngeal and buccal precursors which are entering the interior.

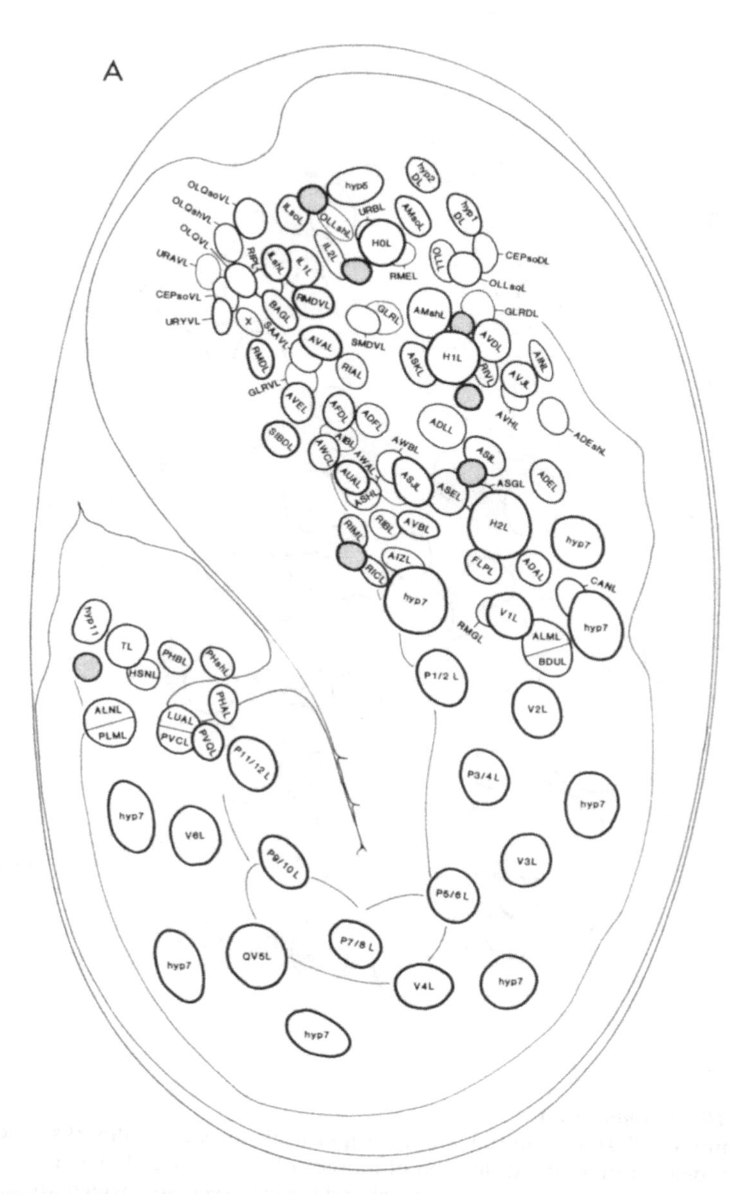

Figure 11A (*See facing page for part B.*)

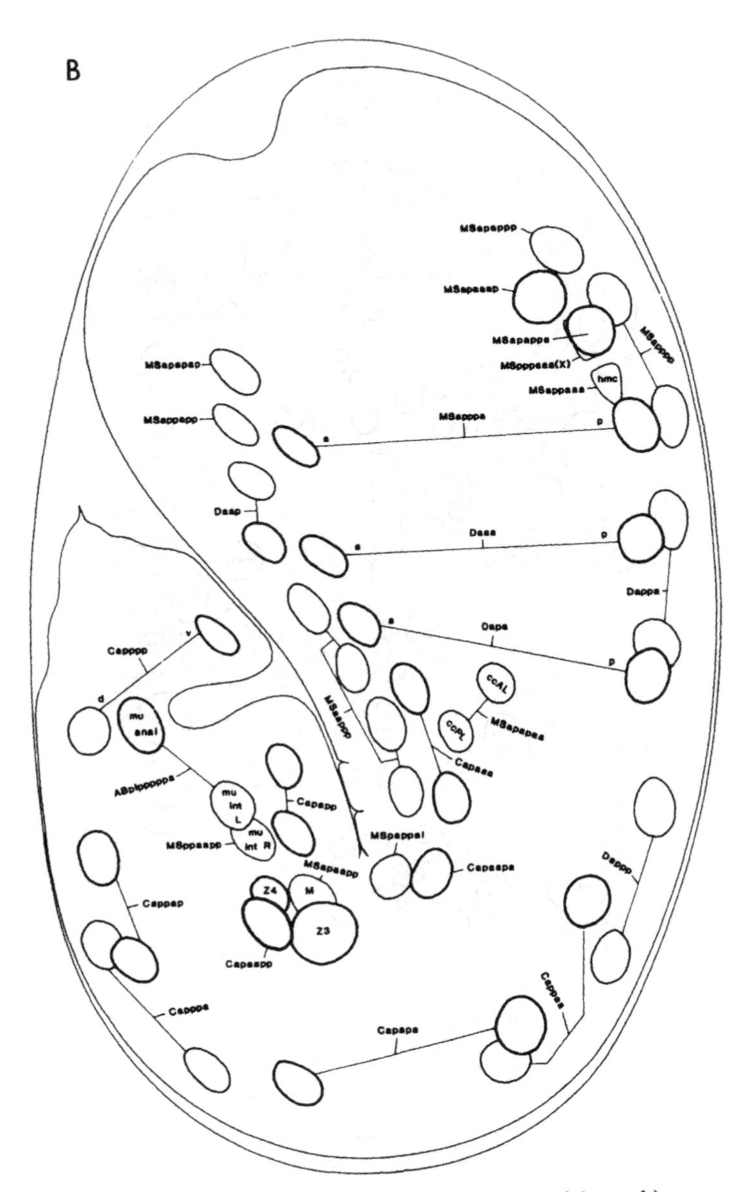

Figure 11B (*See following pages for part C and legend.*)

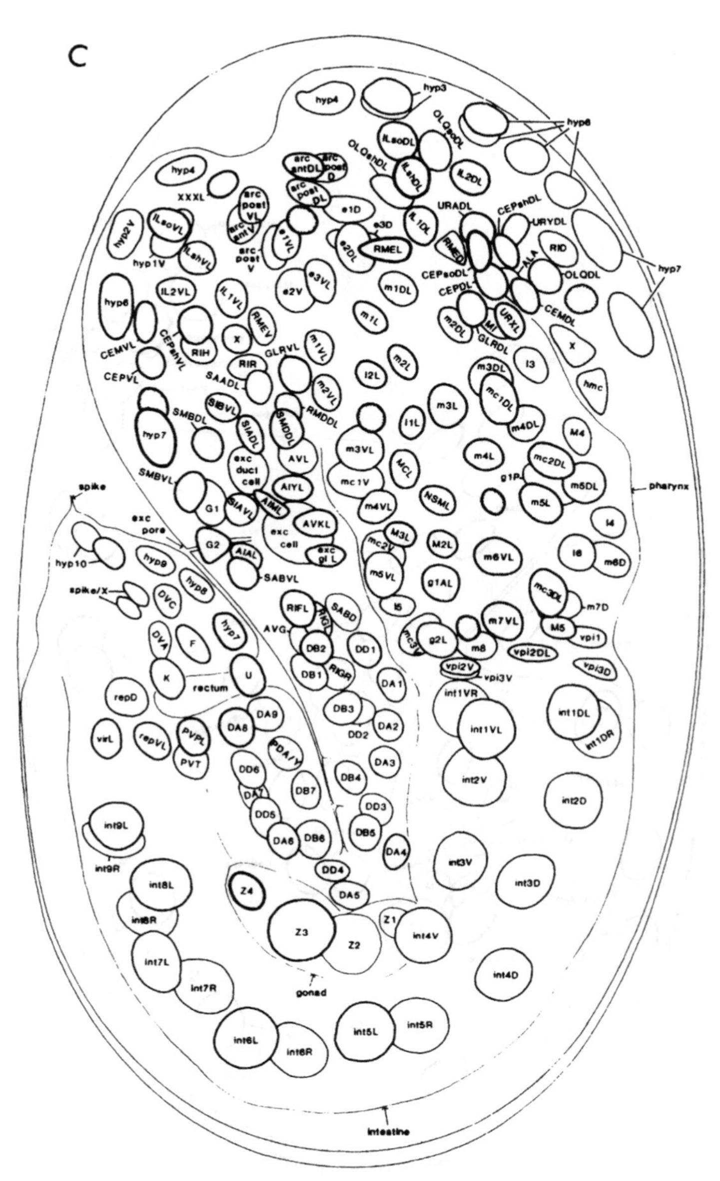

Figure 11C (*See facing page for legend.*)

Figure 11 This figure illustrates the arrangement of all left and central nuclei at 430 min after first cleavage; the right-hand side is a mirror image of the left, except where otherwise indicated. The three parts roughly represent three planes of focus (from superficial to central), but there is considerable overlap between them. The anterior sensory depression (not a mouth opening) is at the top and the lengthening tail, terminating in the spike, curves round to the left. On the ventral side of the tail, the rectal opening has appeared, and on the ventral side of the head the excretory duct leads to the excretory pore. (*A*) Left lateral ectoderm. No mid-plane nuclei are included, but the outlines of the pharynx, intestine, and gonad are shown for reference. Nuclei that will soon divide are labeled with the names of both presumptive daughters. The parent of QL and V5L is named QV5L. The pattern on the right is identical, except that: the homolog of hyp11 is PVR; an extra hyp7 nucleus (ABarpaappp) lies dorsal to H2R; there is no hyp2 DR. (*B*) Left and central mesoderm (excluding pharynx). M, mu int R, hmc, and MSpppaaa lie in the mid-plane. Unlabeled nuclei are in body muscles. The pattern on the right is identical, except that ABprpppppaa and ABprpppppap become a body muscle and the sphincter muscle, respectively, and lie slightly more anteriorly than their left-hand homologs. (*C*) Intestine, gonad, left central pharynx, and ectoderm. The pattern on the right is identical, except that: the homolog of G2 is W, of I6 is M1, of U is B, and of K is K′; an asymmetric neuron RIS (ABprpappapa) lies anterior to AVKR.

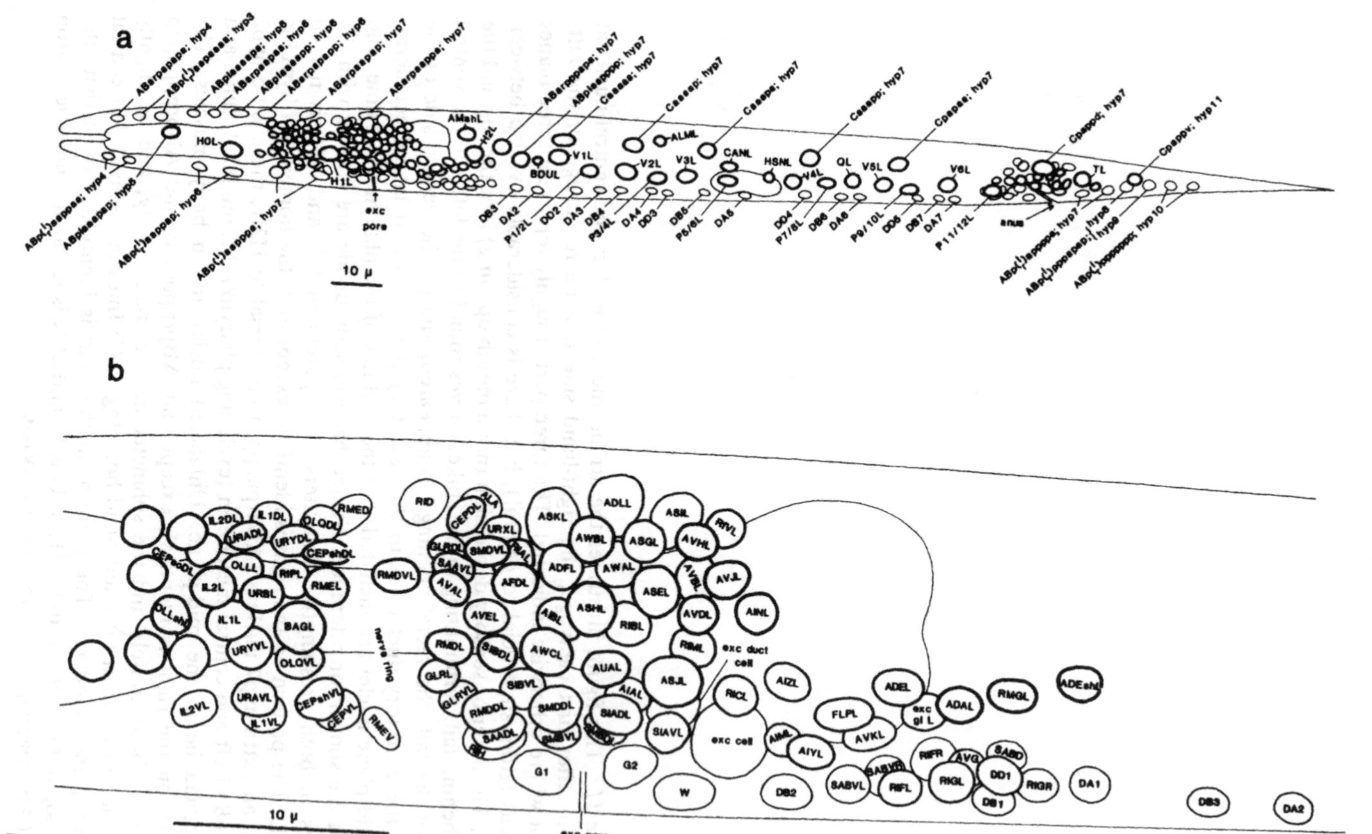

Figure 12 (*See facing page for legend and parts c and d*)

c

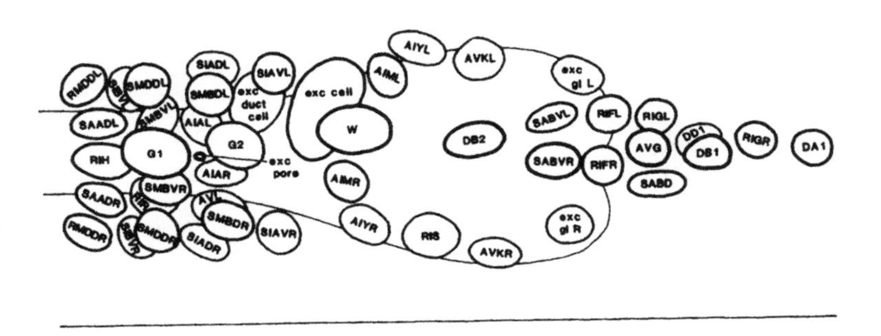

10 µ

d

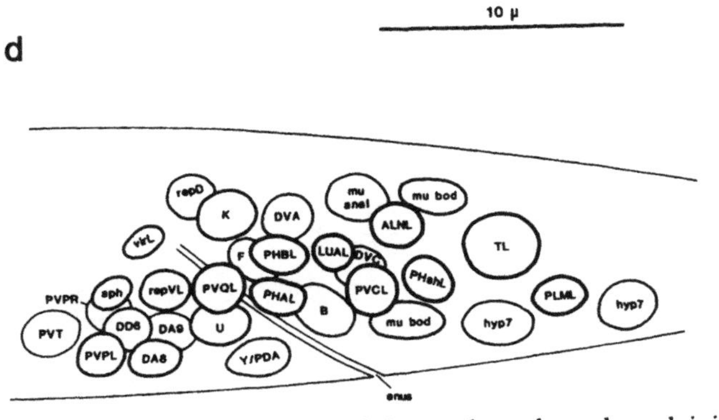

Figure 12 Arrangement of neuronal and larger hypodermal nuclei in newly hatched L1; based on camera lucida drawings. (*a*) Entire animal; left lateral aspect. Pattern on the right is identical, except that: an additional hyp7 nucleus (ABarpaappp) lies dorsal to H2R; homolog of QL is QR, of hyp11 is PVR. (*b*) Ring, ventral, and retrovesicular ganglia; left lateral aspect. Note that arrangement of ring ganglion cells around posterior bulb of pharynx is very variable at this stage. Anatomy anterior to the ring is not wholly known, since cells in this region were mostly identified by their processes in the embryo. (*c*) Ventral and retrovesicular ganglia; ventral aspect. (*d*) Preanal and left lumbar ganglia, rectal cells; left lateral aspect. (Corrected, with permission, from Sulston et al. 1983.)

APPENDIX 4
Genetics

Compiled by Jonathan Hodgkin,* Mark Edgley,†
Donald L. Riddle,† and Donna G. Albertson*
with the assistance of numerous other *C. elegans* investigators

*MRC Laboratory of Molecular Biology,
Cambridge, England
†University of Missouri, Columbia, Missouri

Part A: Genetic Nomenclature

GUIDELINES

Guidelines for *C. elegans* nomenclature have been published elsewhere (Horvitz et al. 1979); a brief account of the main conventions follows.

Gene Names

Genes are given names consisting of three italicized letters, a hyphen, and an arabic number, e.g., *dpy-5* or *let-37*. These gene names refer to the mutant phenotype originally detected and/or most easily scored: dumpy in the case of *dpy-5*, and lethal in the case of *let-37*. The gene name may be followed by an italicized Roman numeral, to indicate the linkage group on which the gene maps, e.g., *dpy-5 I* or *let-37 X*.

Mutation Names

Every mutation has a unique designation. Mutations are given names consisting of one or two italicized letters followed by an italicized Arabic number, e.g., *e61* or *mn138*. The letter prefix refers to the laboratory of isolation (listed below), as registered with the *Caenorhabditis* Genetics Center (CGC). Mutations known to be chromosomal rearrangements, rather than intragenic lesions, are named somewhat differently.

When gene and mutation names are used together, the mutation name is included in parentheses after the gene name, e.g., *dpy-5(e61)*, *let-37(mn138)*. When unambiguous (e.g., if only one mutation is known for a given gene), gene names are used in preference to mutation names (*let-37* rather than *mn138*).

Suffixes indicating characteristics of a mutation can follow a mutation name: These are usually two-letter nonitalicized letters, e.g., *e61*sd, where sd stands for semidominant.

The wild-type allele of a gene is defined as that present in the Bristol N2 strain, stored frozen at the CGC and other locations. Wild-type alleles can be designated by a plus sign immediately after the gene name, *dpy-5+*, or by including the plus sign in parentheses, *dpy-5(+)*. The widely understood convention of a superscript plus sign, $dpy\text{-}5^{+}$, has also been used.

Multiple Mutations

Mutants carrying more than one mutation are designated by sequentially listing mutant genes or mutations according to the left-right (=up-down) order on the genetic map. Different linkage groups are separated by a semicolon and given in the order *I, II, III, IV, V, X*. Heterozygotes, with allelic differences between chromosomes, are designated by separating mutations on the two homologous chromosomes with a slash. Where unambiguous, wild-type alleles can be designated by a plus sign alone, or even omitted. For example, *dpy-5(e61) unc-13(+)/dpy-5(+) unc-13(e51)* can also be written *dpy-5 + /+ unc-13* or *dpy-5/unc-13*.

Chromosomal Aberrations

Duplications (*Dp*), deficiencies (*Df*), and translocations (*T*) are known in *C. elegans* cytogenetics; these are given italicized names consisting of the laboratory mutation prefix, the relevant abbreviation, and a number, optionally followed by the affected linkage groups in parentheses (e.g., *eT1(III;V)*, *mnDp5(X;f)*, where *f* indicates a free duplication). Chromosomal balancers of unknown structure can be designated using the abbreviation *C*.

Strain Names

A strain is a set of individuals of a particular genotype with the capacity to produce more individuals of the same genotype. Strains are given nonitalicized names consisting of two uppercase letters followed by a number. The letter prefixes refer to the laboratory of origin and are different from mutation letter prefixes (both listed below).

Phenotypic Abbreviations

Phenotypic characteristics can be described in words, e.g., dumpy nonuncoordinated animals. If more convenient, a nonitalicized three-letter abbreviation, which usually corresponds to a gene name, may be used. The first letter of a phenotypic abbreviation is capitalized, e.g., Dpy non-Unc; in this example, the wild-type phenotype (with regard to coordination) is indicated by non. Abbreviations that do not correspond to gene names can also be used, e.g., Muv for multiple vulval development.

CAENORHABDITIS GENETICS CENTER LABORATORY DESIGNATIONS

AB	aa	Bird, Alan	CSIRO Adelaide, Australia
AF	sz	Fodor, Andras	Hungarian Academy of Science, Szeged
BA	hc	Ward, Sam	Carnegie Institution, Baltimore, MD
BC	s	Baillie, David	Simon Fraser University, Vancouver, B.C.
BE	sc	Edgar, Robert	University of California, Santa Cruz, CA
BH	hb	Honda, Barry	Simon Fraser University, Vancouver, B.C.
BL	in	Blumenthal, Tom	Indiana University, Bloomington
BM	bb	Mitchell, David	Boston Biomedical Research Institute, Boston, MA
BW	ct	Wood, William B.	University of Colorado, Boulder
CB	e	Hodgkin, Jonathan	MRC-LMB, Cambridge, England
CD	dc	Johnson, Carl	University of Wisconsin, Madison
DD	d	Otsuka, Anthony	University of California, Berkeley
DH	b	Hirsh, David	University of Colorado, Boulder
DR	m	Riddle, Don	University of Missouri, Columbia
EM	bx	Emmons, Scott	Albert Einstein University, Bronx, NY
FF	f	*(Brun, J)	Universite Claude Bernard, Lyons, France
FH	ec	Meneely, Phil	Fred Hutchinson Cancer Res. Center, Seattle, WA
FS	tf	Roberts, Tom	Florida State University, Tallahassee
GG	g	*(vonEhrenstein)	Max-Planck Institute, Gottingen
GT	a	Dusenbery, David	Georgia Institute of Technology, Atlanta
HE	su	Epstein, Henry	Baylor College of Medicine, Houston, TX
HH	hs	Hecht, Ralph	University of Houston, TX
JK	q	Kimble, Judith	University of Wisconsin, Madison
JM	ca	McGhee, Jim	University of Calgary, Alberta
JP	gn	Nelson, Greg	Jet Propulsion Laboratory, Pasadena, CA
KK	it	Kemphues, Ken	Cornell University, Ithaca, NY
KR	h	Rose, Ann	University of British Columbia, Vancouver
MJ	k	Miwa, Johji	Institute for Biomed. Research, Osaka, Japan
MK	hx	Klass, Michael	University of Houston, TX
MS	bz	Swanson, Margaret	Montana State University, Bozeman
MT	n	Horvitz, Bob	Massachusetts Institute of Technology, Cambridge
NA	gb	Bazzicalupo, Paolo	IIGB, Napoli, Italy
NE	rp	Pertel, Ruth	FDA, Washington, D.C.
NJ	rh	Hedgecock, Ed	Roche Institute, Nutley, NJ
NW	ev	Culotti, Joe	Northwestern University, Evanston, IL
PH	hf	Hartman, Phil	Texas Christian University, Fort Worth
PJ	j	Jacobson, Lew	University of Pittsburgh, PA
PR	p	Russell, Dick	University of Pittsburgh, PA
RC	g	Cassada, Randy	Universitat Freiburg, WG
RW	st	Waterston, Bob	Washington University, St. Louis, MO
SP	mn	Herman, Bob	University of Minnesota, St. Paul
TD	tc	Lew, Ken	Forsyth Dental Center, Boston, MA
TN	cn	Hosono, Ryuji	Kanazawa University, Kanazawa, Ishikawa, Japan
TJ	z	Johnson, Tom	University of California, Irvine, CA
TR	r	Anderson, Phil	University of Wisconsin, Madison
TT	t	Babu, P	Tata Institute, Bombay, India
TU	u	Chalfie, Marty	Columbia University, New York, NY
TY	y	Meyer, Barbara	Massachusetts Institute of Technology, Cambridge
VT	ma	Ambros, Victor	Harvard University, Cambridge, MA
WW	w	Samoiloff, Martin	University of Manitoba, Winnipeg
ZZ	x	Lewis, Jim	University of Missouri, Columbia

*Deceased.

Part B: List of Mapped Genes

EXPLANATION OF LIST OF MAPPED GENES AND MUTANT PHENOTYPES

Genes are listed in alphabetical order of gene names. Most genes have been defined by the isolation of one or more mutant alleles, but an increasing number have been defined only by the cloning and characterization of the wild-type gene. Several genes in this category have been included in this list.

Each *gene name* category (e.g., *dpy*) is preceded by an explanation of the gene name (DumPY: shorter than wild type) and by the strain designation of the laboratory assigning gene names within that category (CB = Cambridge, England).

The approximate *location* of each gene on the genetic map is given by dividing each chromosome into five arbitrary and not particularly equal regions L, LC, C, RC, R, where C stands for cluster or center. This is a compromise between merely giving the chromosome and giving a more precise (but possibly misleading) map location. Better information can be gleaned from the map (Appendix 4C), which has been arranged with the left arm of each chromosome "up" and the right arm "down."

The *reference allele* for each gene is given: This is usually the mutation that has been studied in most detail, and a brief phenotypic description of that allele follows. Descriptive abbreviations (e.g., ts, pdi) may be appended to the allele and are explained below. Other abbreviations are:

ES (*Ease of Scoring*): This indicates (very approximately) how easy it is to recognize a particular mutant phenotype by visual inspection under the dissecting microscope. ES3 means easy to score, ES2 means hard to score (may become easier with practice), ES1 means very hard to score except by special means (such as enzyme assay), ES0 means impossible to score, which may be the case for particular stages, sexes, backgrounds, etc. In general, the ES score refers to the ease of scoring at the stage when the mutation is maximally expressed. ES scores have been omitted for genes such as *let*.

ME (*Male Mating Efficiency*): This is recorded, where known, by an ME score, where ME0 = never mating, ME1 = almost never mating, ME2 = poor mating, ME3 = fair-to-excellent mating. Hermaphrodite mating efficiency is recorded where relevant (e.g., for vulvaless mutants) by an HME score (same conventions).

Penetrance is 100% unless otherwise noted.

Mutations have been induced by EMS (ethylmethanesulfonate) unless otherwise noted.

NA (*Number of Alleles*): Total number of known mutations for a given gene is indicated by an NA score. These numbers are minima. Alleles that have been lost are not included, except for a few cases in which the only known allele for a gene has been lost. Notes on additional alleles are added where relevant.

References

References have been kept to a minimum, favoring recent publications; earlier papers can be tracked down by literature search or via the CGC (*Caenorhabiditis* Genetics Center). Published references are by CGC number for brevity; a partial key is appended. The large amount of information that is based on unpublished observations is acknowledged by listing the laboratory designation for the source of the information.

Complicated Entries

The vast majority of mutations in *C. elegans* probably result in recessive loss of function, and this is an implicit assumption (unless otherwise stated) for all the named alleles. However, genes can also be mutated to various gain-of-function states. To accommodate such mutations, alleles have usually been assigned to one of three classes. The vast majority are:

Class 1 (usually implicit): reduction or loss of wild-type function. Where there is reason to believe that a particular mutation is null (i.e., gene activity completely absent), this has often been indicated.

There are also:
Class 2: putative regulatory mutations, resulting in overproduction, inappropriate production, or constitutive production of an otherwise unaltered gene product.
Class 3: gene product with novel activity.

Class 1 alleles are usually but not always recessive, Class 2 and Class 3 alleles are usually but not always dominant. It is probably impossible to produce a universally satisfactory classification; this one is an attempt to be consistent and helpful. The classical distinctions into amorph, hypermorph, and so on have been avoided here, although there is a rough correspondence (Class 1 = amorph and hypomorph, Class 2 = hypermorph, Class 3 = neomorph and antimorph).

ABBREVIATIONS

For a description of ES, ME, HME, and NA, see pp. 497 and 498. Abbreviations for identified cells and neurons are listed in Appendix 1 and Appendix 2.

aci	acetaldehyde-induced
cs	cold-sensitive (phenotype stronger at 15°C than at 25°C)
des	diethylsulfate-induced
dm	dominant
GABA	γ aminobutyric acid
gri	γ-ray-induced
icr	induced by ICR
ird	intragenic revertant of dominant
mat	maternal effect
mm	maternal expression of wild-type allele required for viability
mn	either maternal or embryonic expression of wild-type allele sufficient for viability
mnp	paternal expression of wild-type allele sufficient for viability
mnz	paternal expression of wild-type allele insufficient for viability
nm	both maternal and embryonic expression of wild-type allele required for viability
nn	embryonic expression of wild-type allele required for viability
pat	paternal effect
pdi	induced by ^{32}P decay
pka	previously known as
sd	semidominant (incompletely dominant, *a*/+ distinguishable from *a*/*a* and +/+)
spo	spontaneous
ts	temperature (i.e., heat)-sensitive (phenotype stronger at 25°C than at 15°C)
xri	X-ray-induced
FITC	fluorescein isothiocyanate
MMS	methyl methane sulfonate
Muv	multivulva
TSP	temperature-sensitive period
Vul	vulvaless
WT	wild type

REFERENCES (*CAENORHABDITIS* GENETICS CENTER KEY NUMBERS)

16 Babu. 1974 Mol. Gen. Genet. 135: 39.
31 Brenner. 1974 Genetics 77: 71.
82 Culotti, Russell. 1978 Genetics 90: 243.
128 Epstein, Thomson. 1974 Nature 250: 579.
163 Herman. 1978 Genetics 88: 49.
164 Herman et al. 1979 Genetics 92: 419.
168 Higgins, Hirsh. 1977 Mol. Gen. Genet. 150: 63.
178 Hodgkin, Brenner. 1977 Genetics 86: 275.
179 Hodgkin et al. 1979 Genetics 91: 67.
205 Klass et al. 1976 Dev. Biol. 52: 1.
214 Lewis, Hodgkin. 1977 J. Comp. Neurol. 172: 489.
241 Meneely, Herman. 1979 Genetics 92: 99.
245 Moerman, Baillie. 1979 Genetics 91: 95.
252 Nelson et al. 1978 Dev. Biol. 66: 386.
318 Riddle, Brenner. 1978 Genetics 89: 299.
321 Rose, Baillie. 1979 Nature 281: 599.
363 Sulston. 1976 Phil. Trans. Roy. Soc. 257B: 287.
365 Sulston et al. 1975 J. Comp. Neurol. 163: 215.
387 Ward. 1973 PNAS 70: 217.
393 Ward, Miwa. 1978 Genetics 88: 285.
397 Waterston, Brenner. 1978 Nature 275: 715.
399 Waterston et al. 1977 J. Mol. Biol. 117: 679.
426 Wood et al. 1980 Dev. Biol. 74: 446.
427 Rose, Baillie. 1980 Genetics 96: 639.
443 Herman et al. 1980 Genetic Maps 1:183.
448 Schierenberg et al. 1980 Dev. Biol. 76: 141.
449 Miwa et al. 1980 Dev. Biol. 76: 160.
461 Waterston et al. 1980 Dev. Biol. 77: 271.
464 Lewis et al. 1980 Neuroscience 5: 697.
465 Cox et al. 1980 Genetics 95: 317.
470 Siddiqui, Babu. 1980 Science 210: 330.
484 Lewis et al. 1980 Genetics 95: 905.
486 Greenwald, Horvitz. 1980 Genetics 96: 147.
488 Zengel, Epstein. 1980 Cell Motility 1: 73.
493 Bhat, Babu. 1980 Mol. Gen. Genet. 180: 635.
495 Argon, Ward. 1980 Genetics 96: 413.
496 Horvitz, Sulston. 1980 Genetics 96: 435.
497 Meneely, Herman. 1981 Genetics 97: 65.
498 Hodgkin. 1980 Genetics 96: 649.
499 Moerman, Baillie. 1981 Mut. Res. 80: 273.
500 Sulston, Horvitz. 1981 Dev. Biol. 82: 41.
501 Cassada et al. 1981 Dev. Biol. 84: 193.
502 Chalfie, Sulston. 1981 Dev. Biol. 82: 358.
503 Albert et al. 1981 J. Comp. Neurol. 198; 435.
504 Riddle et al. 1981 Nature 290: 668.
505 Swanson, Riddle. 1981 Dev. Biol. 84: 27-40.
507 Chalfie et al. 1981 Cell 24: 59.
515 Waterston. 1981 Genetics 97: 307.
518 Babu, Brenner. 1981 Mut. Res. 82: 269.
529 Ward et al. 1981 J. Cell Biol. 91: 26.
533 Siddiqui, Babu. 1980 Mol. Gen. Genet. 179: 21.
550 Chalfie, Thomson. 1981 J. Cell Biol. 93: 15.
552 Horvitz et al. 1982 Science 216: 1012.
554 Rosenbluth, Baillie. 1981 Genetics 99: 415.
558 Roberts, Ward 1982 J. Cell Biol. 92: 113.
562 Hedgecock, Thomson. 1982 Cell 30: 321.
564 Greenwald, Horvitz. 1982 Genetics 101: 211.
565 Hartman, Herman. 1982. Genetics 102: 159.
571 Moerman et al. 1982. Cell 29: 773.
574 Hecht et al. 1982 Dev. Biol. 94: 183.
577 Herman et al. 1982 Genetics 102: 379.
579 Waterston et al. 1982 J. Mol. Biol. 158: 1.
592 Rogalski et al. 1982 Genetics 102: 725.
608 Hodgkin. 1983 Genetics 103: 43.
609 Hosono et al. 1982 J. Exp. Zool. 224: 135.
620 Ambros, Horvitz. 1984 Science 226: 409.

627 Hedgecock et al. 1983 Science 220: 1277.
630 Hodgkin. 1983 Nature 304: 267.
632 Tabuse, Miwa 1983 Carcinogenesis 4: 783.
635 Trent et al. 1983 Genetics 104: 619.
642 Johnson, Russell. 1983 J. Neurochem. 41: 30.
646 Greenwald et al. 1983 Cell 34: 435.
654 Sanford et al. 1983 J. Biol. Chem. 258: 12804.
663 Hartman. 1984 Photochem. & Photobiol. 39: 169.
665 Meneely, Wood. 1984 Genetics 106: 29.
666 Hodgkin. 1983 Mol. Gen. Genetic. 192: 452.
672 Horvitz et al. 1983 CSH Symp. Quant. Biol. 48: 453.
675 Rand, Russell. 1984 Genetics 106: 227.
676 Goldstein. 1984 Can. J. Genet. Cytol. 26: 13.
680 Golden, Riddle. 1984 PNAS 81: 819.
689 Denich et al. 1984 Roux's Arch. Dev. Biol. 193: 164
696 Albertson. 1984 EMBO J. 3: 1227.
710 Anderson, Brenner. 1984 PNAS 81: 4470.
712 Herman. 1984 Genetics 108: 165.
714 Kimble et al. 1984 Dev. Biol. 105: 234.
715 Sigurdson et al. 1984 Genetics 108: 331.
724 Doniach, Hodgkin. 1984 Dev. Biol. 106: 223.
729 Bolten et al. 1984 PNAS 81: 6784.
731 Sternberg, Horvitz. 1984 Ann. Rev. Genet. 18: 489.
733 Hartman. 1985 Genetics 109: 81.
739 Goldstein. 1984 Mutat. Res. 129: 337.
741 Waterston et al. 1984 J. Mol. Biol. 180: 473.
742 Landel et al. 1984 J. Mol. Biol. 180: 497.
748 Cox, Hirsh 1985 Mol. Cell. Biol. 5: 363.
750 Rosenbluth et al. 1985 Genetics 109: 495.
751 Cox et al. 1985 Genetics 109: 513.
762 Ferguson, Horvitz. 1985 Genetics 110: 17.
766 Golden, Riddle. 1985 Mol. Gen. Genet. 198: 534.
768 Wills et al. 1983 Cell 33: 575.
769 Brown, Riddle. 1985 Genetics 110: 421.
779 Dibb et al. 1985 J. Mol. Biol. 183: 543.
793 Hedgecock et al. 1985 Dev. Biol. 111: 158.
797 Hodgkin. 1985 Genetics 111: 287.
801 Albertson. 1985. EMBO Journal 4: 2493.
803 Spieth et al. 1985 Nucleic Acids Res. 13: 7129.
808 Spieth, Blumenthal 1985 Mol. Cell. Biol. 5: 2495.
812 Greenwald. 1985 Cell 43: 583.
813 Rogalski, Baillie. 1985 Mol. Gen. Genet. 210: 409.
814 Nelson, Honda. 1985 Gene 38: 245.
859 McGhee, Cottrell. 1985 Mol. Gen. Genet. 202: 30.
861 Goldstein. 1985a Chromosoma 93: 256.
862 Sebastiano et al. 1986. Genetics 112: 459.
864 Sigurdson et al. 1986 Mol. Gen. Genet. 202: 212.
865 Ellis et al. 1986 Nucleic Acids Res. 14: 2345.
870 Ellis, Horvitz. 1986 Cell 44: 817.
873 Heine, Blumenthal. 1986 J. Mol. Biol. 188: 301.
876 Miller et al. 1986 PNAS 83:2305.
877 Moerman et al. 1986 PNAS 83: 2579.
883 Fodor, Deak. 1985 J. Genet. 64: 143.
898 Greenwald, Horvitz. 1986 Genetics 113: 63.
906 Kusch, Edgar. 1986 Genetics 113: 621.
912 Salvato et al. 1986 J. Mol. Biol. 190: 281.
913 Kenyon. 1986 Cell 46: 477.
914 Park, Horvitz. 1986 Genetics 113: 821.
915 Park, Horvitz. 1986 Genetics 113: 853.
922 Hodgkin. 1986 Genetics 114: 15.
923 Doniach. 1986 Genetics 114: 53.
941 Meyer, Casson. 1986 Cell 47: 871.
943 Barton et al. 1987 Genetics 115: 107.
944 Villeneuve, Meyer. 1987 Cell 48: 25.

LIST OF MAPPED GENES AND MUTANT PHENOTYPES

Gene	Location	Reference allele, phenotype, comments, other alleles.
ace:		abnormal ACEtylcholinesterase. PR
ace-1	XR	p1000: class A acetylcholinesterase reduced 100%; no behavioral phenotype alone (ES1, ME3), but ace-2;ace-1 is uncoordinated (hypercontracted), and ace-2;ace-3;ace-1 is L1 lethal. NA7. Ref. 464, 642, CD, PR.
ace-2	IL	g72: class B acetylcholinesterase reduced 98%; no behavioral phenotype alone (see ace-1). ES1, ME3. NA12 (g73 etc.). Ref. 464, 642, PR.
ace-3	IIR	dc2: class C acetylcholinesterase reduced >95%; no behavioral phenotype (see ace-1). ES1.NA2 (dc3). Ref. CD.
act:		ACTin. DH.
act-1 act-2 act-3	VRC	st15sd: (pka unc-92): both st15 and st15/+ small, slow-growing, uncoordinated, almost paralysed, abnormal thin filament ultrastructure in muscle. ES3, ME0. NA(sd)5 (st22ts (25°phenotypes similar to st15), st119 (dominant paralysed, recessive larval lethal), etc.). All are probably dominant neomorphic mutations of one of the three clustered actin genes. Double mutant intragenic revertants are wildtype in phenotype even if act-1 or act-3 is disrupted (e.g. st22st283, internal deficiency in act-3). Ref. 741, 742, RW.
act-4	X	No mutations known. Probable main gene for cytoplasmic actin. Ref. 801, DH.
ali:		abnormal lateral ALAE. CB
ali-1	VRC	e1934: faint or invisible alae in L1. ES1,ME3. NA1. Ref. CB.
ama:		AMAnitin resistant. DR.
ama-1	IVLC	m118sd: heterozygote m118/+ resistant to 400μg/ml α-amanitin, altered RNA polymerase II, no other phenotype. ES1. NA>10 (m235 (late larval lethal), m237 (embryonic lethal), m238ts (sterile adult), etc.) Ref. 654, DR.
ama-2	VLC	m323dm: heterozygote m323/+ resistant to 100 μg/ml α-amanitin; recessive lethal. ES1. NA1. Ref. DR.
anc:		abnormal nuclear ANChorage. CB.

anc-1 IC e1753amber: nuclei of hypodermal cells not elastically anchored, other cytoskeletal abnormalities; no gross phenotype. ES1. NA5 (e1802, e1873 etc.(all resemble e1753)). All alleles tend to revert spontaneously by intragenic reversion. Ref. 562, NJ.

ben: BENzimidazole resistant. TU.

ben-1 IIILC e1880ts,sd: resistant to 14μM benomyl, no other phenotype. ES1. NA>20 (e1910, u102 etc.: all somewhat ts and sd). Ref. TU.

bli: BLIstered cuticle. CB.

bli-1 IIRC e769: adult blistered, especially head. ES3 (old adult), ME2. NA7 (e770, e935 (weaker allele), e993spo, etc.). Spontaneous mutations frequent. Ref. 31, 465.

bli-2 IILC e768: adult blistered, especially head; slightly small. ES3(old adult), ME2. NA4 (e107, e527, etc.). Ref. 31.

bli-3 IL e767: small, irregular shape, variable slight blistering in adult. ES2(adult,late larvae) ME1. NA1. Ref. 31.

bli-4 IC e937: adult blistered, especially head. ES3(old adult), ME1. NA1. Ref. 31.

bli-5 IIIR e518: small, dumpyish, adult blistered especially around pharynx; abnormal bursa in adult male. ES3(adult,late larvae), ME0. NA2. Ref. 31, BC,BE.

bli-6 IVC sc16: adult blistered, both head and body, often small blisters. ES3(old adult). NA2 (n776sd). Ref. 914,BE.

cad: abnormal CAthepsin D. PJ.

cad-1 IIR j1: 90% reduced cathepsin D. Ref PJ.

caf: abnormal CAFfeine resistance. PH

caf-1 IVL hf3: resistant to 30mM caffeine; no gross phenotype. ES1. NA8. Ref. PH.

caf-2 IRC hf5: resistant to 30mM caffeine; no gross phenotype. ES1. NA3. Ref. PH.

cal: CALmodulin related genes.

cal-1 IV No mutants known. Sequence related to calmodulin and troponin C. Ref. 912.

cat: abnormal CATecholamine distribution. CB.

cat-1 XRC e1111amber: dopamine and serotonin absent from neuron processes, present in cell bodies. ES1,ME1. NA1. Ref. 365, 397, MT.

cat-2 IIL e1112: dopamine reduced > 95%; serotonin normal. ES1, ME3, NA1. Ref. 365, MT.

cat-3 IIILC e1333(lost): dopamine reduced in neuron processes. Ref. 365.

cat-4 VC e1141: dopamine reduced > 90%. ES1, ME2. NA1. Ref. 365.

cat-5 VL e1334(lost): variable displacement of CEPD cell bodies and other cells. Ref. 365.

cat-6 V e1861: sensory neurons CEP, ADE, PDE take up FITC; tubular body of CEP cilia disrupted; slightly defective dauer formation. ES1, ME3. NA1. Ref. 793, NJ.

ced: CEll Death abnormality. CB.

ced-1 IR e1735: programmed cell deaths abnormal, dying cells arrest at highly refractile stage, killer cells fail to engulf target cells; no gross phenotype. ES1. NA9 (e1754, n691 etc: all alleles resemble e1735). Ref. 627.

ced-2 IVL e1752: programmed cell deaths abnormal, phenotype identical to ced-1(e1735). ES1. NA1. Ref. 627.

ced-3 IVR n717: programmed cell deaths fail to occur; epistatic to ced-1, ced-2, and nuc-1; recessive suppressor of egl-1 and of egg-laying defect of egl-41 homozygotes; semidominant suppressor of egl-1(n487)/+ and egl-41(n1069)/+ heterozygotes; no gross phenotype. ES1 (ES2 in egl-1 background), ME3. NA7 (n718 etc.). Ref. 672, 870.

ced-4 IIILC n1162: programmed cell deaths fail to occur; recessive suppressor of egl-1 and egl-41 homozygotes; no gross phenotype. ES1 (ES2 in egl-1 background), ME3. NA1. Ref. 870.

cha: abnormal CHoline Acetyltransferase.

cha-1 IVLC p1152: 99% reduced choline acetyltransferase, slow-growing, uncoordinated (coiler), resistant to aldicarb, trichlorfon. ES3. NA6 (b401ts (ME2), p1154 etc.) One cha-1 allele p1156 fails to complement unc-17, and unc-17 alleles e113 and e876 fail to complement cha-1. See also unc-17. Ref. 675, PR.

che:		abnormal CHEmotaxis. CB.
che-1	IC	e1034 non-chemotactic to sodium ion; abnormal sensory neuroanatomy, especially AFD, IL2 cells. ES1, ME2. NA2 (pka tax-1). Ref. 82, 214, 393.
che-2	X	e1033: non-chemotactic to sodium ion, slightly small, defective osmotic avoidance, defective dauer formation, males impotent; no FITC uptake; ciliated neurons have abnormal stunted ultrastructure. ES2, ME0. NA1. Ref. 214, NJ.
che-3	IC	e1124: non-chemotactic to sodium ion, slightly small, defective osmotic avoidance, defective dauer formation, males impotent; ciliated neurons have abnormal stunted ultrastructure, no FITC uptake; octopamine deficient. ES2, ME0. NA5 (e1379, e1253 (pka che-8), p801 (pka osm-2),etc.). Ref. 214, 82, 503, 552, 793, NJ.
che-4		See mec-1.
che-5	IV	e1073: poor chemotaxis to sodium ion, erratic movement. ES1, ME3. NA1. Ref. 214.
che-6	IV	e1126: non-chemotactic to chloride ion; abnormal IL2 basal bodies. ES1, ME3. NA1. Ref 214.
che-7	V	e1128: non-chemotactic to chloride ion, small. ES1, ME3. NA1. Ref. 214.
che-8		See che-3.
che-9		See mec-2.
che-10	IILC	e1809: no FITC uptake by amphids or phasmids; striated ciliary rootlets missing from OLQ,IL1, BAG sensory neurons; defective osmotic avoidance; males impotent. ES1, ME0. NA1. Ref. 793, NJ.
che-11	VRC	e1810: no FITC uptake by amphids or phasmids; defective in osmotic avoidance and dauer formation. ES1, ME0. NA2(e1815). Ref. 793, NJ.
che-12	VRC	e1812: weak FITC uptake by amphids and phasmids; amphid sheath cells fail to secrete matrix material; defective osmotic avoidance. ES1, ME3. NA1. Ref. 793, NJ.
che-13	IC	e1805: no FITC uptake by amphids or phasmids; severely shortened axonemes and ectopic assembly of ciliary structures and microtubules in many sensory neurons; defective in osmotic avoidance and dauer formation. ES1, ME0. NA1. Ref. 793, NJ.
che-14	I	e1960: some FITC uptake by amphids but not by phasmids; abnormal uptake by CEP, ADE, PDE; abnormal amphid channel due to misjoining of sheath and socket cell. ES1. NA1. Ref. NJ.

clr: CLeaR. CB.

clr-1 IILC e1745ts: starved, translucent appearance at 20°, facilitating Nomarski visualization of neuron processes; phenocopied by growth on 1mM orthovanadate. Inviable at 25°. ES3 (25°), ES2 (20°). NA1. Ref. CB,NJ.

col: COLlagen structural genes. DH. At least forty genes exhibiting homology to vertebrate collagen genes; many characterized but unmapped. No mutants known as yet.

col-2 IVC Transcript present in dauer larvae. Ref. 569, 748, 751.

col-3 IVC Transcript present in all stages. Ref. 748, 751.

col-4 IVC Ref. 751.

col-5 IVC Ref. 751.

col-6 II Transcript present in dauer larvae. Ref. 748, 751.

col-8 III Transcript present in dauer larvae and adults. Ref. 748, 751.

col-9 X Ref. 751.

daf: abnormal DAuer Formation (dauer larvae are resistant L3 larvae). DR.

daf-1 IVL m40ts, mat: constitutive dauer formation at 25°, reversible by shift to 15°. ES3 (L3), N11 (n690 (Egl, Type C) etc.). Ref. 504, 505, 680, 635, 766.

daf-2 IIIL e1370ts: constitutive dauer formation at 25°; reversible by shift to 15°. ES3 (L3). NA19. Ref. 505, 680, 766.

daf-3 XL e1376: defective dauer formation (daf-2 suppressible). ES1, ME3, NA8. Ref. 504.

daf-4 IIIC m63ts: constitutive dauer formation at 25°; reversible by shift to 15°; at all temperatures small and defective in intestinal endocytosis; Type E Egl. ES3(L3,25°); ES2(other stages). NA8(e1364, m72amber etc; all alleles resemble m63). Ref 505, 635, 680, 766, PJ.

daf-5 IIR e1386: defective dauer formation (daf-2 suppressible). ES1, ME3. NA11. Ref. 504.

daf-6 XR e1377: defective dauer formation (daf-1, daf-2 suppressible), abnormal chemotaxis, defective osmotic avoidance, poor male mating; abnormal sensory anatomy especially amphidial sheath cells; no FITC uptake by amphids or phasmids. ES1, ME2. NA1. Ref. 503, 504,712.

daf-7 IIIL e1372ts: constitutive dauer formation at 25°, reversible at 15°; Type C Egl at all temperatures. ES3(L3). NA6 (m70, m62amber, n696 etc.; all alleles Egl, ts Daf). Ref. 505, 680, 635, 766.

daf-8 IRC e1393ts: constitutive dauer formation at 25°, reversible by shift to 15°; Type C Egl at all temperatures. ES3(L3). NA1. Ref. 504, 505, 635, 680.

daf-9 XL e1406: constitutive formation of abnormal dauer-like larvae, irreversible (genetically lethal). ES3 (L3), NA1. Ref. DR.

daf-10 IVC e1387: defective dauer formation (daf-8, daf-2 suppressible), abnormal chemotaxis, abnormal osmotic avoidance, males impotent; abnormal sensory anatomy, especially amphidial neurons and sheath cells, cephalic neurons; octopamine deficient. ES1. ME0. NA2 (m79, p821 pka osm-4). Ref. 503, 504, 82, 552.

daf-11 VRC m47ts: constitutive dauer formation at 25°, reversible by shift to 15°. ES3(L3). NA7. Ref. 504, 766.

daf-12 XC m20: defective dauer formation (daf-2 suppressible). ES1, ME3. NA1. Ref. 504.

daf-13 X m66: formation of SDS-sensitive dauer larvae. ES1, ME3. NA1. Ref. 504, DR.

daf-14 IVC m77ts: constitutive dauer formation at 25°, incomplete penetrance, reversible by shift to 15°; Type C Egl. ES3(L3). NA1. Ref. 505, 635, 766.

daf-15 IVC m81: constitutive formation of abnormal dauer-like larvae, irreversible (genetically lethal). ES3(L3). NA1. Ref. DR.

daf-16 IRC m26: defective dauer formation (daf-4 suppressible). ES1. ME3. NA1. Ref. 504.

daf-17 IR m27: defective dauer formation (daf-14 suppressible). ES1, ME3. NA5. Ref. 504, DR.

daf-18 IVL e1375: defective dauer formation (daf-7, daf-2 suppressible). ES1, ME3. NA1. Ref. 504.

daf-19 IIC m86ts: constitutive dauer formation at 25°, reversible by shift to 15°; defective chemotaxis and osmotic avoidance; fails to take up FITC at 15° or 25°; cilia but not ciliary rootlets missing from sensory neurons. ES3(L3), ME0. NA1. Ref. DR, NJ.

daf-20 X m25: defective dauer formation (daf-4 suppressible). ES1, ME3. NA2. Ref. 504.

daf-22 IIR m130: defective dauer formation (suppressible by exogenous dauer pheromone), defective dauer pheromone production. ES1. NA1. Ref. 766, DR.

deg: DEGeneration of certain neurons. TU.

deg-1 XC u38ts, dm: touch-insensitive, prod-insensitive, only in tail. PVC interneurons degenerate at L1/L2, certain other neurons die at hatching (some IL1,also probably AVG) or L4 moult (probably AVD). ES2. NA1. Intragenic revertants (e.g u38u175) are wildtype (probable null phenotype). Ref. TU.

dpy: DumPY: shorter than wild type. CB.

dpy-1 IIIL e1: strong dumpy. ES3, ME1. NA >10 (e6, e830xri, e874pdi, e1177icr etc.). High forward mutation frequency. Ref. 31.

dpy-2 IIC e8: dumpy, left roller; early larvae non-dumpy. ES3, ME1. NA >10 (e115, sc38ts, sc78sd, e489 (pka rol-2), etc.). Ref. 31, 465, BE.

dpy-3 XL e27: medium dumpy (non-roller), L1 non-dumpy. ES3 (all stages), ME1. NA >10 (e2079, sc26: left roller dumpy, m39ts; e182 (pka dpy-12) etc.). Ref. 31, 465, BE, SP.

dpy-4 IVR e1166sd,icr: large dumpy; e1166/+ is slightly dumpy. ES3 (adult), ES1 (larvae). ME3. NA2 (e1158sd, icr). Ref. CB.

dpy-5 IC e61: strong dumpy, early larvae non-dumpy; e61/+ is very slightly dumpy. ES3, ME1. NA3 (e565, e907 pdi). Ref. 31.

dpy-6 XC e14: small dumpy. ES3, MEO. NA2 (e1502aci). Ref. 31.

dpy-7 XC e88: dumpy (non-roller) ES3. ME1. NA8 (e1324ts, sc27ts: left roller dumpy, etc.). Ref. 31, 465.

dpy-8 XLC e130: medium dumpy (non-roller) ES3, ME1. NA >10 (e1281ts, sc24ts (left roller dumpy), etc.). Ref. 31, 465.

dpy-9 IVL e12: large dumpy. ES3 (adult), ES1 (larvae). ME2. NA4 (e424, e858, e1164icr). Ref. 31.

dpy-10 IIC e128: small dumpy (non-roller); sometimes poor freezing. ES3 (all stages), ME1. NA>10 (e223, sc48 (left roller dumpy), sc30ts (left roller dumpy at 25°), sc? (pka rol-7), etc.). Ref. 31, 465.

dpy-11 VC e224: medium dumpy. ES3 (adult), ES2 (larvae). ME2. NA >10 (e33, e1180icr (these and most other alleles have much stronger dumpy ('piggy') phenotype), etc.). Ref. 31, CB.

dpy-12 | | See dpy-3.

dpy-13 IVLC e184sd: strong dumpy; e184/+ is medium dumpy. ES3 (all stages). ME2. NA7. (e458, e1165icr, e225 (pka dpy-16: recessive medium dumpy)). Ref. 31, CB.

dpy-14 IC e188ts: medium dumpy adult, strong dumpy L1(20°); ts lethal (25°). ES3 (all stages, 20°). Synthetic lethal with some mutations at 20°. ME2. NA1. Ref. 31, 427.

dpy-15 See sqt-3.

dpy-16 See dpy-13.

dpy-17 IIILC e164: medium dumpy, spindle-shaped adult; old adults sometimes less dumpy; strong dumpy L1. ES3 (all stages). ME2. NA7 (e1345ts, e905pdi, e1295icr, etc.). Ref. 31.

dpy-18 IIIR e364amber: medium dumpy. ES3 (adult), ES2 (larvae). ME2. NA3 (e499, e1270). Ref. 31, 515.

dpy-19 IIIC e1259ts, des, mat: dumpy (20°, 25°), wildtype (15°). ES3 (adult, 25°). Weaker phenotype if mother dpy-19/+. MEO (25°), ME3 (15°). NA2 (e1314ts). Ref. CB, MT.

dpy-20 IVC e1282ts: medium dumpy (20°), weak dumpy (15°), strong dumpy, round-nosed (25°). ES3 (25°, all stages). ME2 (20°). NA4 (other alleles all strong dumpy (20°), almost inviable (15°): e1362, e1415, e2017amber). Ref. 609, 797.

dpy-21 VR e428: weak dumpy (XX), non-dumpy (XO); lethal to 2A;3X; some X chromosome transcript levels elevated. ES3 (adult), ES1 (larvae). ME3. NA3 (e459 (resembles e428), ct16 (very weak dumpy)). Ref. 665, 666, 941.

dpy-22 XC e652: variable scrawny dumpy, slow-growing (XX); very abnormal small or inviable male (XO). ES2 (all stages, XX), ES3 (all stages, XO). MEO. NA1. Ref. 666.

dpy-23 XLC e840: variable dumpy, head swollen around pharynx, inviable at 15° (XX); very abnormal or inviable males (XO). ES2 (XX), ES3 (XO). MEO. NA1. Ref. 666.

dpy-24 IRC s71: weak dumpy. ES3 (adult), ES1 (larvae). NA1. Ref. 427.

dpy-25 IIL e817sd: strong dumpy, inviable 15°; e817/+ is medium dumpy. ES3 (all stages). ME1. NA1 Ref. CB.

dpy-26 IVRC n199mat: XX daughters of n199/+ mothers are maternally rescued, weak dumpy phenotype with protruding vulva (ES2), 4% Him; XX daughters of n199/n199 mothers are severely dumpy and Him (2% of brood) or die as embryos or young larvae (98% of XX brood). XO n199 animals are non-dumpy, almost wildtype males (ME3). Similar or slightly stronger phenotypes in n199/Df. NA3 (n198 (resembles n199), y6 (weaker phenotypes). Ref. 666, TY.

dpy-27 IIIL rh18mat: XX phenotype is zygotic weak dumpy, maternal effect dumpy/near lethal, like dpy-26 but non-Him; some X chromosome transcript levels elevated in XX. ES2. XO phenotype nearly wildtype male (ME3). NA1. Ref. 941, CB.

dpy-28 IIIRC y1mat, ts: at 20°XX phenotype is zygotic weak dumpy, maternal effect dumpy/near lethal, like dpy-26 but non-Him; some X chromosome transcript levels elevated in XX. At 15° viable. ES2. XO phenotype nearly wildtype male (ME3). NA1. Ref. 941, TY.

egl: EGg Laying defective. MT. Drug response categories:
Type A, resistant (i.e., fails to lay eggs in response) to serotonin and imipramine (these mutants presumably defective in vulva or in sex muscles, probably latter if hermaphrodites can mate well) ;
Type B, serotonin-sensitive and imipramine-resistant;
Type C, serotonin-sensitive and imipramine-sensitive;
Type D, serotonin-resistant and imipramine-sensitive;
Type E, variable response.
Egl-c: alleles of several unc genes, etc., are egg-laying constitutive.

egl-1 VRC n487sd,ts: Type B Egl, transient bloating; HSN cells undergo programmed cell death, hence ced-3 suppressible. No phenotype in male. n487/+ is ts Egl (penetrance 85% at 25°, 60% at 20°). ES2 (adult). ME3. NA4 (n986, n987 (stronger alleles, non-ts: n987/+ penetrance >95%), n1084 (weaker than n487), etc.). Ref. 635, MT.

egl-2 VL n693dm: Type D Egl, severe bloating; weak uncoordinated kinker phenotype. ES3 (adult), ME1. NA1. Ref. 635.

egl-3 VC n150ts: Type C Egl, moderate bloating, coiler phenotype. ES3 (adult, 25°), ES2 (larvae, males). NA4 (n588: weakly semidominant, n729, etc.). Ref. 635.

egl-4 IVL n478: Type E Egl, transient bloating; suppressible by daf-3. ES2 (adult). NA5 (n477, n479ts). Ref. 635, 680.

egl-5 IIIRC n486: Type B Egl, moderate bloating, uncoordinated coiler phenotype; HSN cell bodies absent or displaced and serotonin-negative; not suppressed by ced-3. ES3 (adult), ME0. NA6 (n945, n988 etc.). Ref. 635, MT.

egl-6 XC n592sd: transient bloating, Type E, uncoordinated weak kinker phenotype. ES2, ME2. NA1. Ref. 635.

egl-7	IIILC	n575ts, sd: moderate bloating at 25°, some bloating at 15°. Type C ES2 (adult 25°). NA1. Ref. 635.
egl-8	VL	n488: transient, variable bloating, Type E. ES2 (adult). NA1. Ref. 635.
egl-9	VC	n586ts: transient bloating at 25°, Type E. ES2 (adult, 25°). NA2 (n571 (non-ts)). Ref. 635.
egl-10	VRC	n692sd, ts: transient bloating, sluggish, weak kinker. Type B/C. Weakly semidominant, partially ts. ES3 (adult). ES2 (other stages). ME2. NA10 (n480ts (slight bloating, Type C), n1068 etc.). Ref. 635, MT.
egl-11	VC	n587ts: transient bloating at 25°, Type E. ES2 (adult, 25°). NA1. Ref. 635.
egl-12	VR	n602sd: transient bloating, Type E, weakly semidoninant. ES2 (adult). NA2 (n599 (recessive)). Ref. 635.
egl-13	XC	n483: penetrance 70%, severe bloating or bag-of-worms phenotype (Type A); minority WT. ES3/0 (adult), ME3. NA2 (e1447). Ref. 635.
egl-14	XC	n549: transient bloating. Type E. Males abnormal. ES2 (adult), ME2. NA1. Ref. 635.
egl-15	XRC	n484: moderate to severe bloating, 60% form bag-of-worms, Type A. Vulval and uterine muscles defective because of defects in sex myoblast migrations. ES3 (adult), ME3. NA1. Ref. 635, MT.
egl-16	XR	n485mat : very variable bloating; some animals form bags-of-worms (Type A) or explode at vulva; abnormal vulval morphology; some masculinization of XX animals (especially at 15°); elevated X chromosome expression in XX. Weaker phenotype if mother n485/+. ES2/3 (adult). ME3. XX progeny of n485/Df mothers have incompletely penetrant partially male (Tra) phenotype. NA2 (y4). Ref. 635, 944.
egl-17	XL	e1313: moderate to severe bloating, 30% form bag-of-worms. Type A. ES3(adult). NA1. Ref. 635.
egl-18	IVL	n162: variable bloating, a few form bag-of-worms, Type E; vulval abnormalities. ES3/1 (adult). NA3 (n474, n475). Ref. 635.
egl-19	IVC	n582: moderate bloating, Type D?; slow and floppy; long. ES3 (adult), ES2 (other stages). NA1. Ref. 635.
egl-20	IVC	n585ts: moderate bloating at 25°, Type E. ES3 (adult). NA1. Ref. 635.
egl-21	IVC	n611ts: transient bloating at 25°, Type E, partially temperature sensitive, uncoordinated weak coiler. ES2 (adult). NA3 (n476, n576). Ref. 635.

egl-22 See unc-31.

egl-23 IVR n601dm: severe bloating (Type A), some animals form bag-of-worms; uncoordinated sluggish phenotype (recessive). ES3 (adult). ES1 (larvae). ME3. NA1. Ref. 635.

egl-24 IIIC n572: variable bloating (Type A), some animals WT. ES2 (adult). NA1. Ref. 635.

egl-25 IIIC n573: variable bloating, Type A, variable abnormal tail morphology. ES2 (adult), NA1. Ref. 635.

egl-26 IIL n481: variable bloating, Type A, many animals form bag-of-worms (60%); abnormal vulva. ES2 adult. NA1. Ref. 635.

egl-27 IIC n170: variable bloating, moderate to severe, Type E; possibly abnormal vulva; males very abnormal: reduced rays, short spicules, swollen bursa, hermaphroditic tail spike. ES2 (adult). ES3 (adult male). ME0. NA1. Ref. 635, CB.

egl-28 IIC n570ts: transient bloating at 25°, Type E, partially temperature sensitive. ES2 (adult). NA1. Ref. 635.

egl-29 IIC n482: variable, some animals WT, some animals form bag-of-worms (80%); Type A; variably abnormal vulva (protrusive). ES2 (adult). ME3. NA1. Ref. 635.

egl-30 IL n686sd: moderate bloating, Type E; uncoordinated slow phenotype. ES3 (adult), ES2 (other stages). NA2 (n715sd: uncoordinated paralysed phenotype, n715/+ has Type C bloating). Ref. 635.

egl-31 IRC n472: moderate to severe bloating, Type A, some animals form bag-of-worms (20%) or explode at vulva; sex muscles variably defective because of abnormalities in early M lineage; uncoordinated poor backing phenotype. ES2 (adult), ES2 (other stages). NA1. Ref. 635.

egl-32 IC n155ts: moderate bloating, Type E, partially temperature-sensitive; suppressible by daf-3. ES2 (adult). ME3. NA1. Ref. 635, 680.

egl-33 IC n151ts: moderate to severe bloating at 25°, Type A, some animals form bag-of-worms (40%) or explode at vulva; uncoordinated kinker phenotype at 25°. ES3 (adult, 25°), ES2 (larvae, 25°). NA1. Ref. 635.

egl-34 ILC n171: variable bloating, Type E, some animals WT, some form bag-of-worms (10%). ES2 (adult). NA2 (e1452). Ref. 635.

egl-35 IIIR n694ts: transient bloating at 25°, Type C. ES2 (adult, 25°). ME3. NA1. Ref. 635.

egl-36 XRC n728dm: severe bloating, Type A, some animals form bag-of-worms (50%). ES2 (adult). ME3. n728/+ has similar phenotype. NA1. Ref. 635.

egl-37 IIC n742ts: transient bloating at 25°, Type E. ES2 (adult, 25°). ME3. NA1. Ref. 635.

egl-38 IVC n578: severe bloating, Type A, almost no egg-laying, almost all animals form bag-of-worms (>90%); uncoordinated very sluggish phenotype; abnormal vulva. ES3 (all stages). NA1. Ref. 635.

egl-39 IIIRC n730ts: transient bloating at 25°, Type C; uncoordinated sluggish, weak coiler phenotype. ES3 (adult, 25°), ES2 (other stages, 25°). NA1. Ref. 635.

egl-40 IVC n606ts, sd: transient bloating. Type C, partially temperature sensitive. ES2 (adult, 25°). ME3. NA1. Ref. 635, 680.

egl-41 VRC n1077cs, sd: moderate bloating. Type B, HSN cells absent in adult (ced-3 suppressible), other indications of weak masculinization of XX animals. Also XO animals sometimes weakly feminized. Penetrance of n1077/+ XX: 60% Egl at 20°. ES2 (adult). ES1 (other stages). NA4 (all similar phenotype: n1074, n1069, e2055). Ref. 923, MT.

egl-42 IIC n995sd: moderate bloating, Type B; slight bloating in n995/+. ES2 (adult). NA2 (n996sd). Ref. MT.

egl-43 IIRC n997: moderate bloating, Type B. HSN cell bodies serotonin-positive but misplaced posteriorly, along path of HSN migration; aberrant variable HSN processes. ES2 (adult), ES1 (other stages). NA2 (n1079). Ref. MT.

egl-44 IILC n1080: moderate bloating, Type B; HSN cell bodies do not contain serotonin. ES2(adult). NA3 (n998, n1087). Ref.MT.

egl-45 IIIC n999: moderate bloating, Type B, HSN cell bodies sometimes degenerate in L4, absent in adult and never contain serotonin. ES2 (adult), ES1 (other stages). NA1. Ref. MT.

egl-46 VC n1127: moderate bloating, Type B; HSN cell bodies do not contain serotonin. ES2 (adult), ES1 (other stages). NA4 (n1075, n1076 etc.). Ref. MT.

egl-47 VC n1081dm: moderate bloating, Type B. ES2 (adult), NA2 (n1082dm). Ref. MT.

egl-48 IIL e1952: moderate bloating; abnormal vulva; HSN cells present. Weak suppressor of some tra mutations. ES2(adult), ME3. NA1. Ref. 922, CB.

emb: abnormal EMBryogenesis. GG, RC. See also let, ooc, par and zyg entries.

emb-1 IIILC hc57ts,mm: at 25° one to 24 cell arrest, no pseudocleavage, defective cytoplasmic streaming. Normal and defective execution before one cell stage. Viable at 16°. NA2 (hc62ts: similar phenotype). Ref. 448,449, HH.

emb-2 IIILC hc58ts,mnz: at 25°lima bean arrest; abnormal blastocoel; division rates slowed about 20%. Normal execution in oogenesis or early cleavage; defective execution later. Viable at 16°, slightly slow divisions. NA1. Ref. 448,449.

emb-3 IVC hc59ts,mm: at 25°lima bean arrest; early nuclear and cytoplasmic abnormalities, abnormal blastocoel; division rates slightly faster. Normal and defective execution before fertilization. NA1. Ref. 448,449.

emb-4 VR hc60ts,mm: at 25° lima bean arrest; slow division rates, especially E lineage. Shift to 25° in L1 results in defective gonadogenesis. At 16° subviable: 60% of eggs fail to hatch, many larvae die or have gross morphological abnormalities. NA1. Ref. 448,449.

emb-5 IIIC hc61ts,mm: at 25°lima bean arrest, misplaced gut granule birefringence; abnormal gastrulation, division rates slowed (except E lineage), Ea divides a/p. Normal and defective execution at 24 cell stage. Shift to 25° in L1 results in defective gonadogenesis. Viable at 16°. NA4 (g16ts, g65ts: similar to hc61; hc67ts: division rates faster than WT, otherwise like hc61). Ref. 448,449,689.

emb-6 ILC hc65ts,mm: at 25°14 cell arrest, no gut granule birefringence, division rates very slow. Normal and defective execution before first cleavage. Viable at 16°. NA2 (g36ts: similar to hc65). Ref. 448,449,689.

emb-7 IIIC hc66ts,mm: at 25°lima bean arrest, division rates slightly slowed. Normal and defective execution before first division. At 16° viable, but slightly abnormal division rates. NA2 (b84ts (pka zyg-4: 25°embryonic arrest <100 nuclei; L1 temperature shift-up affects gonadogenesis)). Ref. 448,449,426.

emb-8 IIILC hc69ts,mm: at 25° eggs osmotically sensitive, variable arrest (early to mid-cleavage), division rates slowed. At 16° viable but eggs are fragile. NA1. Ref. 448,449.

emb-9 IIIRC hc70sd,ts,nn: at 25°early pretzel arrest, a few animals hatch and die in L1; ts for larval growth (arrest shortly after any shift to 25°). Normal and defective execution in late embryogenesis. NA5 (all similar phenotypes: g23ts, g34ts, b117ts, b189ts (pka zyg-6)). Ref. 426,449,689.

emb-10 IC hc63ts. Ref.443, JM

emb-11 IVC g4ts,mm: osmotically sensitive eggs at 25.6°(leaky at 25°), 94% arrest during early proliferation; escapers mostly arrest in L1. Temperature shift-up in L1 or L4 results in abnormal gonadogenesis. Adults dumpy at 25°, uncoordinated at 16°. NA2 (g1ts (similar phenotypes but lower penetrance)). Ref. 501.

emb-12 IC g5ts,mm: osmotically sensitive eggs at 25.6°(leaky at 25°), 72% arrest during early proliferation; escapers have abnormal gonads. Temperature shift-up in L1 results in incompletely penetrant gonad defects. Viable at 16°. NA1. Ref. 501.

emb-13 IIIRC g6ts,nm: at 25° 99% eggs arrest at lima bean, no gut granule birefringence. Reduced cytoplasmic streaming before first cleavage, slow divisions; E,D and P lineages early. Escapers arrest as larvae, abnormal gonadogenesis; L1 temperature shift-up results in larval arrest, abnormal gonad. At 16° viable, variably long. NA1. Ref. 501,689.

emb-14 IC g43ts,nm: osmotically sensitive eggs at 25°, 100% eggs arrest during proliferation. Escapers arrest as larvae; L1 temperature shift-up results in some gonad abnormality. At 16° viable, poor male mating. NA2 (g14ts,mm? (similar phenotypes, lower penetrance)). Ref. 501.

emb-15 XRC g15ts,mnz: at 25° 97% eggs arrest at pretzel; prolonged mitoses during cleavage. Escapers arrest L1-L4. Viable 16°. NA1. Ref. 501,689.

emb-16 IIIRC g19ts,mm: at 25° 91% eggs arrest at lima bean, misplaced gut granule birefringence. E lineage early, MS delayed. Escapers arrest in L1, abnormal gonadogenesis. L1 temperature shift-up affects gonadogenesis. At 16° viable, somewhat uncoordinated. NA1. Ref. 501,689.

emb-17 IC g20ts,mnz: at 25° 100% eggs arrest at lima bean, normal gut granule birefringence; eggs variably irregular in shape. Some lineages early, some delayed. L1 or L4 temperature shift-up affects gonadogenesis. At 16° uncoordinated, synthetic lethal with many dpy mutations. NA1. Ref. 501,689.

emb-18 VC g21ts,mm: at 25° 100% eggs arrest at lima bean, normal gut granule birefringence; eggs variably round in shape. C and E lineages late, P4 early. Escapers arrest L2-L4. L1 temperature shift-up affects gonadogenesis. At 16° slightly long. NA1. Ref. 501,689.

emb-19 IC g22ts,mm: osmotically sensitive eggs at 25°, 100% arrest during early proliferation. Escapers arrest L1-L4. L1 temperature shift-up affects gonadogenesis. At 16° viable. NA1. Ref. 501.

emb-20 IR g27ts,mm: osmotically sensitive eggs at 25.6°(leaky at 25°), 96% arrest during early proliferation. Escapers viable, F1 Emb. At 16° viable. NA1. Ref. 501.

<u>emb-21</u> IIRC <u>g31</u>ts,mm: at 25° 90% eggs arrest 26-30 cell stage, no gut granule birefringence; variable round eggs; large ooplasmic granules, prolonged mitoses, especially AB. Escapers arrest L2-L3, abnormal gonadogenesis. L1 temperature shift-up affects gonadogenesis. NA1. Ref. 501,689.

<u>emb-22</u> VC <u>g32</u>ts,nm: at 25° 100% eggs arrest at lima bean, misplaced gut granule birefringence. Strong ooplasmic streaming, erratic pronuclear migration, no polar body formation, all mitoses prolonged. Temperature shift-up at all stages after lima bean results in immediate arrest. At 16° viable. NA1. Tightly linked to, and possibly identical with, <u>act-1/2/3</u> (q.v.). Ref. 501,689, RC.

<u>emb-23</u> IIRC <u>g39</u>ts,mm: at 25° 100% eggs arrest at lima bean, misplaced gut granule birefringence. Very large ooplasmic granules; MS, C and P_2 divisions late. Escapers arrest L1-L2, abnormal gonads. L1 temperature shift-up results in L1-L2 arrest, abnormal gonads. At 16° viable. NA1. Ref. 501,689.

<u>emb-24</u> IIIC <u>g40</u>ts,mnz: at 25° 100% eggs arrest at pretzel, normal gut granule birefringence. E lineage late. L1 temperature shift-up results in L1-L3 arrest. At 16° viable but many defective embryos. NA1. Ref. 501,689.

<u>emb-25</u> IIIRC <u>g45</u>ts,mm: at 25° 100% eggs arrest at 100-250 cells, normal gut granule birefringence; eggs variably long in shape. Later divisions slow or incomplete. Escapers arrest L1-L2. L1 temperature shift-up results in L1-L2 arrest. At 16° viable. NA1. Ref. 501,689.

<u>emb-26</u> IVC <u>g47</u>ts,mm: at 25.6° (leaky at 25°) 80% eggs arrest at lima bean, abnormal gut granule birefringence. Slow ooplasmic streaming, cleavage timing alterations. Escapers arrest L1-L4, abnormal gonads. L1 temperature shift-up affects gonadogenesis. At 16° viable, slightly long, poor male mating. Meiotic phenotype at 20°(= <u>him-12</u>): self-progeny 8% XO male, low brood size (65% unhatched eggs), males sire many inviable zygotes. ES3(progeny), ME2. NA1. Ref. 501,666,689, CB.

<u>emb-27</u> IIC <u>g48</u>ts,mm,pat: at 25° 100% eggs arrest at one cell stage. No polar body formation, no pseudocleavage, 30% fail to reform procucleus; pronuclear migration and fusion are slow or absent, pronuclei disintegrate. Escapers viable adults, F1 Emb. Males crossed with wildtype hermaphrodites sire progeny that arrest at 100 cell stage. L1 temperature shift-up results in viable adult, F1 Emb. At 16° viable. NA1. Ref. 501,689, RC.

<u>emb-28</u> VR <u>g49</u>ts,mm: at 25.6° (leaky at 25°) 91% arrest at lima bean, misplaced gut granule birefringence. Very large ooplasmic granules, slow and abnormal first cleavage, later divisions slow. Escapers viable with abnormal

gonad. L1 temperature shift-up results in viable adult, F1 Emb. At 16° viable. NA1. Ref. 501,689.

emb-29 VL g52ts,nn: at 25° 100% eggs arrest before mitosis during 150 - 200 cell stage; misplaced gut granule birefringence. Some divisions delayed. Escapers arrest L2-L4. L1 or L4 temperature shift-up affects gonadogenesis. Probably general G2/M cell cycle arrest. At 16° viable. NA2 (b262: similar phenotypes). Ref. 501,689, HH.

emb-30 IIIC g53ts,mm,pat: round osmotically sensitive eggs at 25°; 96% eggs arrest at one cell stage. Ooplasmic streaming somewhat abnormal, only one polar body, no pseudocleavage, no pronuclear migration or fusion, some endomitosis. Escapers are viable adults, F1 Emb. Males crossed with wildtype hermaphrodites sire progeny that arrest at 100-cell stage. L1 temperature shift-up results in viable adult, F1 Emb. At 16° viable, slightly uncoordinated. NA1. Ref. 501,689, RC.

emb-31 IVC g55ts,mm: at 25° 100% eggs arrest at lima bean, normal gut granule birefringence. Abnormal blastocoel and gastrulation. Escapers sometimes arrest in L1-L2, abnormal gonads. L1 temperature shift-up affects gonadogenesis. At 16° viable. NA1. Ref. 501,689.

emb-32 IIIL g58ts,mnz: at 25.6° (leaky at 25°) 97% eggs arrest at lima bean, normal gut granule birefringence. Ooplasmic streaming slow, cleavage timing altered, late divisions slow. Escapers arrest in L2. L1 temperature shift-up results in L2 arrest, abnormal gonadogenesis. At 16° viable, dumpy. NA1. Ref. 501,689.

emb-33 IIIC g60ts,mm: at 25° 100% eggs arrest at lima bean or earlier, misplaced gut granule birefringence. Escapers have abnormal gonadogenesis. L1 temperature shift-up results in viable adult, F1 Emb. At 16° viable. NA1. Ref. 501,689, HH.

emb-34 IIIRC g62ts,nm: at 25° 100% eggs arrest at lima bean, abnormal gut granule birefringence; egg shape variable. Some pseudocleavage failure, erratic pronuclear migration, first mitosis prolonged, later divisions prolonged and skewed. L1 temperature shift-up results in L1 arrest. At 16° viable. NA1. Ref. 501,689.

emb-35 IVL g64ts,mnz: at 25.6° (leaky at 25°) 97% eggs arrest at lima bean, normal gut granule birefringence. Later divisions delayed and prolonged, P_4 early. Escapers small, arrest L1-L4. L1 temperature shift-up results in some larval arrest, dumpy adults, F1 Emb. At 16° viable. NA1. Ref. 501,689.

enu: ENhancer of Uncoordination. CB. See also sus.

enu-1 II ev419sd: no phenotype alone, but enhances phenotype of unc-107(ev411)V (q.v.). ES3 in unc-107 background, ES0 without unc-107. NA1. Ref. NW.

fem: FEMinization of XX and XO animals. CB.

fem-1 IVLC e1965mat: XO animals transformed into fertile females if mother homozygous e1965, into intersexes if mother heterozygous e1965/+; XX animals fertile females if mother homozygous, fertile females or hermaphrodites if mother heterozygous. ES3 (XO adult), ES2 (XX adult), MEO. NA >20 (e1991amber (resembles e1965) etc; also many ts alleles eg hc17ts (pka isx-1: causes only partial XO feminization at 25°, XX self fertile at 20°). Ref. 252, 714, 724.

fem-2 IIIL e2105ts,mat: XO animals transformed into fertile females (25°) or intersexes (20°)if mother homozygous e2105, into abnormal males if mother e2105/+; XX animals fertile females if mother homozygous, hermaphrodite if mother heterozygous; XX phenotype non-ts. ES3 (XO adult), ES2 (XX progeny). MEO. NA>10 (e2102, etc.: all ts for XO phenotype; b245ts (pka isx-2: weaker allele, causes only partial XO feminization at 25°, XX self-fertile at 20°). Ref. 714, 922.

fem-3 IVLC Class 1 (loss-of-function): e1996sd,mat: XO animals transformed into fertile females if mother homozygous e1996, into intersexes if mother heterozygous e1996/+; XX animals fertile females. Some e1996/+ XX animals female. Some e1996/+ sons of e1996 mothers are feminized. ES3 (XO adult), ES2 (XX adult). MEO. NA > 20 (e1950, e2006ts: XX self-fertile at 20°). Ref. 922.
Class 2(gain-of-function): q20sd,ts: at 25° XX germline makes only sperm, at 15° XX germline makes oocytes and excess sperm. ES3(progeny), ME3. NA9 (q24 etc.: all ts). Ref. 943.

fer: FERtilization defective (abnormal sperm). See also spe. The term fer will be retained for previously described fertilization-defective mutants, but new genes in this category will be termed spe. BA.

fer-1 IRC hc1ts: hermaphrodites and males grown at 25° produce nonfunctional nonmotile sperm with short pseudopods, membranous organelles fail to fuse with sperm plasma membrane; TSP in L4. Similar phenotype in hc1/Df. ES2 (adult). Self-fertility <1%(25°), 100% (16°). Sperm normal if grown between 16° and 24.5°. NA8 (hc80 (nonconditional), hc24 (leaky ts), b232, etc.). Ref. 393,495,529, BA.

fer-2 IIIR hc2ts: hermaphrodites and males grown at 25° produce nonfunctional nonmotile sperm with aberrant pseudopods, perinuclear tubules. TSP L2-L4. Self fertility 2%

(25°), 70% (16°). ES2 (adult). NA1. Ref. 495,529.

fer-3 IIL hc3ts: hermaphrodites and males grown at 25° produce nonfunctional nonmotile sperm with some aberrant pseudopods, perinuclear tubules. TSP L3-L4. Self-fertility 1% (25°), 60% (16°). ES2 (adult). NA1. Ref. 495,529.

fer-4 VC hc4ts: hermaphrodites and males grown at 25° produce nonfunctional nonmotile sperm with aberrant pseudopods, perinuclear tubules. Self-fertility 3% (25°), 50% (16°). ES2 (adult). NA1. hc4/Df hermaphrodites have abnormal gonads, with no oocyte production, no rescue by wild-type sperm. Ref. 495,529, BA.

fer-5 See fer-6.

fer-6 IRC hc6ts: hermaphrodites and males grown at 25° produce few sperm with defective pseudopods and retention of fibrous bodies. Self-fertility 1% (25°), 50% (16°). TSP L4. ES2 (adult). NA1 (hc23ts (pka fer-5) is probably reisolate of hc6). Ref. 495,529, BA.

fer-7 IR hc34ts: hermaphrodites and males grown at 25° produce nonfunctional nonmotile sperm. Self-fertility 6% (25°), 40% (16°). TSP L4. ES2 (adult). NA1. Ref. 495.

fer-14 XC hc14ts: hermaphrodites and males grown at 25° produce nonfunctional spermatozoa. Self-fertility <0.5% (25°), 8% (16°). ES2 (adult). NA1. Ref. 558, BA.

fer-15 IIC hc15ts: hermaphrodites and males grown at 25° produce spermatids that fail to activate into spermatozoa. Self-fertility <1% (25°), 100% (16°). ES2 (adult). Similar phenotype in hc15/Df. NA3 (hc89, b26). Ref. 558, BA.

flu: abnormal FLUorescence under ultraviolet illumination. TT.

flu-1 VL e1002sd: increased gut fluorescence, bluish purple; low orthoaminophenol content, low kynurenine hydroxylase. ES1. ME2. NA6. Ref. 16,533.

flu-2 XLC e1003 : reduced gut fluorescence dull green; high orthoaminophenol content, low kynureninase levels, enhanced mutagen sensitivity. ES1. ME3. NA5. Ref. 16,493.

flu-3 IIC e1001 : increased gut fluorescence, purple. ES1. ME2. NA3. Ref. 16,470.

flu-4 XRC e1004 : increased gut fluorescence, blue. ES1. ME3. NA1. Ref. 16.

fog: Feminization Of Germline. CB.

fog-1 IL e2121sd: homozygous XX animals transformed into fertile females; homozygous and heterozygous XO animals are somatically male with both sperm and oocytes in germline. ES2 (adults). NA3. Ref. CB.

fog-2 VR q71: XX animals transformed into females, XO animals wildtype males. ES3(adult XX), ME3. NA6. Ref. JK.

ges: abnormal Gut ESterase. JM.

ges-1 VL ca1: electrophoretic variant of major gut esterase. No other phenotype. ES1, ME3. NA6 (ca4 etc.: all electrophoretic variants). Ref.859.

glp: abnormal Germ Line Proliferation. JK.

glp-1 IIIC q46: sterile; germ cells divide only a few times in both XX and XO, forming 10-20 sperm; somatic gonad superficially wildtype. ES3(adult). NA6 (q35, q50 (both have weaker phenotype than q46: some oocyte production, fertilized eggs arrest as late embryos), e2072mat (weak allele, maternal effect late embryonic lethal) etc). Ref JK, CB.

gus: abnormal GLucuronidaSe. NA.

gus-1 IR b405: β-glucuronidase activity reduced >99%. No gross phenotype. ES1. NA11(b410 (slight activity), gb25 etc.). Also intragenic revertants, e.g. gb94 b410. Probable structural gene. Ref. 862.

hch: defective HatCHing. CB.

hch-1 XR e1734: delayed hatching from eggshell, rescued by protease or wildtype hatching fluid; QL and descendant cells migrate forward instead of backward. ES1. NA2 (e1907: same phenotypes). Ref. NJ.

her: HERmaphroditization of XO animals. CB.

her-1 VC Class 1 (loss-of-function): e1518: XX animals wildtype, XO animals transformed into fertile hermaphrodites. ES3 (XO). Similar phenotype in e1518/Df. NA > 12 (e1520, e1561ts, etc.). Ref. 498,666.
Class 2 (gain-of-function): n695sd,ts: XO animals wildtype, XX animals partly masculinized egglaying defective hermaphrodites; n695/+ XX animals slightly masculinized. ES2 (XX). NA1. Intragenic revertants have Class 1 phenotype. NA(ird)>4. Ref. 635.

her-2 See tra-1.

him: High Incidence of Males (increased X chromosome loss) CB.

him-1 IC e879sd: self-progeny 21% XO male, 5% XXX hermaphrodite; 8% ova are nullo-X as a result of reductional meiotic nondisjunction; X chromosome recombination specifically reduced 40%. e879/+ self-progeny 1% XO. ES3 (progeny). ME2. NA1. Ref. 179.

him-2 IRC e1065: self progeny 2% XO male; 1.5% ova are nullo-X. ES2 (progeny). ME3. NA1. Ref. 179.

him-3 IVC e1147 : self-progeny 3.5% XO male, 2% ova are nullo-X. ES2 (progeny). ME3. NA2 (e1256, stronger allele: 11% Him, many unhatched eggs (71%), males sire inviable zygotes, probably generalized meiotic nondisjunction). Ref. 179.

him-4 XRC e1267icr: self-progeny 6% XO male, very low brood size, frequent gonad eversion with rupture at vulva; 2% nullo-X ova. Male gonad abnormal, testis fails to connect with proctodeum, may be twice reflexed. ES2 (adult). ME0. NA2 (e1266icr (same phenotypes)). Ref. 179.

him-5 VRC e1467ts: at 20°self-progeny 16% XO male, 3% XXX hermaphrodite; 11% ova are nullo-X, as a result of reductional meiotic nondisjunction; X chromosome recombination specifically reduced 50%. ES3 (progeny). ME3. NA2 (e1490 (stronger allele: 33% Him at 20°)). Ref. 179.

him-6 IVC e1423: self-progeny 15% XO male, 6% XXX hermaphrodite, low brood size (78% unhatched eggs), 8% nullo-X ova as a result of reductional meiotic nondisjunction; autosomal nondisjunction; males sire inviable zygotes. ES3 (progeny). ME3. NA2 (e1104 (weaker allele: 5% Him)). Ref. 179.

him-7 VL e1480: self progeny 3% XO male, 0.6% nullo-X ova; males make slight excess of nullo-X sperm, sire some inviable zygotes. ES2 (progeny). ME3. NA1. Ref. 179.

him-8 IVC e1489: self-progeny 37% XO male, 6% XXX hermaphrodite; 38% nullo-X ova as a result of reductional meiotic nondisjunction, X chromosome recombination specifically reduced 90%. ES3 (progeny). ME2. NA3 (g203, mn253 (both >30%Him)). Ref. 179, JC, SP.

him-9 IIRC e1487aci: self-progeny 5% males, 2% nullo-X ova. ES3 (progeny). ME3. NA1. Ref. 179.

him-10 IIILC e1511ts: self progeny 2% XO male (15°), 12% male (20°), 27% male (25°), low brood size, many unhatched eggs at 25°. ES3 (progeny 25°). ME3 (20°). NA1. Ref. 565.

him-11 IIIC n318: self progeny 5% XO male. ES3 (progeny). NA1. Ref. CB.

him-12 See emb-26.

him-13 IC e1742: self progeny 5% XO male. ES3 (progeny). NA1. Ref. CB.

isx: See fem.

let: LEThal. MN.

let-1 XR mn119: early larval lethal. NA6 (mn124, mn102 (late larval lethal), mn115 (hermaphrodite sterile, lays fertilized eggs, male sterile), etc.). Ref. 241,497.

let-2 XR mn153amber: embryonic lethal. NA>15 (almost all ts, complex interallelic complementation pattern; mn114ts (pka let-8), b246ts, e1470ts etc). High forward mutation frequency. Ref. 241,497.

let-3 XR mn104amber: early larval lethal, incompletely suppressible by sup-5. NA1. Ref. 241,497.

let-4 XR mn105amber: early larval lethal. NA1. Ref. 241,497.

let-5 XR mn106amber, mat,mnz: progeny grow slowly, fail to reach adulthood. NA2 (mn132 (same phenotype)). Ref. 241,497.

let-6 XR mn130: early larval lethal. NA4 (mn108, etc.). Ref. 241,497.

let-7 XR mn112: late larval lethal, XX animals die at early L3, XO die at late L4. NA1. Ref. 241,497.

let-8 See let-2.

let-9 XR mn107: defective sperm. NA1. Ref. 241,497.

let-10 XR mn113: early larval lethal. NA2 (mn118). Ref. 241,497.

let-11 XR mn116: early larval lethal. NA1. Ref. 241,497.

let-12 XR mn121amber: early larval lethal. NA2 (mn125). Ref. 241,497.

let-14 XR mn120: early larval lethal. NA1. Ref. 241,497.

let-15 XR mn127amber: early larval lethal. NA3 (mn123, e1471). Ref. 497.

let-16 XR mn117: early larval lethal. NA1. Ref. 241,497.

let-18 XR mn122amber: embryonic lethal, incompletely suppressible by sup-5. NA3 (mn142, mn136 (late larval lethal)). Ref. 497.

let-19 IIC mn19: larval lethal. NA1. Ref. 163,715.

let-20 II mn20 (= mnDf71): larval lethal. NA1. Ref. 163.

let-21 IIC e1778: sterile, uncoordinated, polynucleate oocytes. NA1. Ref. NJ.

let-22 IIC mn22: embryonic lethal. NA1. Ref. 163,715.

let-23 IIC mn23: early larval lethal. NA4 (mn215, mn224 (early larval lethal), n1045 (homozygous viable, cold sensitive vulvaless phenotype)). Ref. 163,715,762,MT.

let-24 IIC mn24: larval lethal. NA1. Ref. 163,715.

let-25 IIC mn25: larval lethal. NA1. Ref. 163,715.

let-26 IIC mn26: larval lethal. NA1. Ref. 163,715.

let-27 IILC mn27: larval lethal. NA1. Ref. 163,715.

let-28 IIR mn28: larval lethal. NA2. (mn212). Ref. 163,715.

let-29 IIC mn29: larval lethal. NA2. (mn182). Ref. 163,715.

let-30 IILC mn30: larval lethal. NA2. (mn239). Ref. 163,715.

let-31 IIC mn31: larval lethal. NA2. (mn223). Ref. 163,715.

let-32 IIC mn32: lethal. NA1. Ref. 163.

let-33 XR mn128mat,mm: homozygotes lay fertilized eggs that do not hatch, males fertile; mn128/Df is sterile. NA1. Ref. 497.

let-34 XR mn134: early larval lethal. NA1. Ref. 497.

let-35 XR mn135: early larval lethal. NA1. Ref. 497.

let-36 XR mn140: late larval lethal or adult producing inviable progeny; mn140/Df is early larval lethal. NA1. Ref.497.

let-37 XR mn138: early larval lethal. NA1. Ref. 497.

let-38 XR mn141mat, mnz: homozygotes lay fertilized eggs that do not hatch, males fertile; mn141/Df is sterile. NA1. Ref. 497.

let-39 XR mn144amber: early larval lethal, incompletely suppressible by sup-5. NA1. Ref. 497.

let-40 XR mn150ts: early larval lethal at 25°, sterile with abnormal gonads at 20°; males fertile; mn150/Df is early larval lethal at 20°. NA1. Ref. 497.

let-41 XR mn146amber: early larval lethal, incompletely suppressible by sup-5. NA1. Ref. 497.

let-49 IR st44: mid-larval lethal. NA1. Ref. 579,RW.

let-50 IR st33: early larval lethal. NA1. Ref. 579,RW.

let-51 IVC s41: embryonic lethal. NA1. Ref. 499,592.

let-52 IVC s42: early larval lethal. NA1. Ref. 592.

let-53 IVC s43: late larval lethal. NA1. Ref. 499,592.

let-54 IVLC s44: early larval lethal (L1-L2 molt). NA2 (s53). Ref. 592.

let-55 IVC s45: early larval lethal. NA1. Ref. 592.

let-56 IVC s46: late larval lethal, s46/Df is mid-larval lethal. NA4 (s168, s173 (both mid-larval lethal, s173/Df is early larval lethal)). Ref. 499,592, BC.

let-58 IVC s48(lost): embryonic or early larval lethal. Ref. 499,592.

let-59 IVC s49: early larval lethal. NA2. (s172). Ref. 499,592.

let-60 IVC s59: mid larval lethal (leaky). NA1. Ref. 499,592.

let-61 IVC s65: late larval lethal. NA1. Ref. 592.

let-62 IVC s175: lethal. NA1. Ref. 592.

let-63 IVC s170: mid-larval lethal. NA2 (s679:late larval lethal). Ref. 592, BC.

let-64 IVC s216mat: sterile adult, progeny not rescued by wildtype sperm. NA2(s171: leaky sterile adult). Ref. 592, BC.

let-65 IVC s254: mid-larval lethal. NA3 (s174; s694: sterile adult). Ref. 592, BC.

let-66 IVC s176: early larval lethal. NA1. Ref. 592.

let-67 IVC s214mat,mnz: sterile adult. NA1. Ref. 592.

let-68 IVC s680: sterile adult. NA3 (s693, s696). Ref. 813.

let-69 IVC s684: late larval lethal. NA1. Ref. 813.

let-70 IVC s689: mid-larval lethal. NA1. Ref. 813.

let-71 IVC s692: sterile adult. NA1. Ref. 813.

let-72 IVC s52: late larval lethal. NA2 (s695: mid-larval lethal). Ref. 813.

let-73 IVC s685: sterile adult. NA1. Ref. 813.

let-74 IVC s697: late larval lethal (leaky). NA1. Ref. 813.

let-75 IC s101: early larval lethal. NA1. Ref. 427, KR.

let-76 IC s80: L1 lethal. NA1. Ref. 427.

let-77 IC s90: late larval lethal. NA1. Ref. 427.

let-78 IC s82: late larval lethal. NA1. Ref. 427.

let-79 IC s81: L1 lethal. NA1. Ref. 427.

let-80 IC s96: L1 lethal. NA1. Ref. 427.

let-81 IC s88: L1 lethal. NA1. Ref. 427.

let-82 IC s85: L1 lethal. NA1. Ref. 427.

let-83 IC s97: L1 lethal. NA1. Ref. 427.

let-84 IC s91: late larval lethal. NA1. REf. 427.

let-85 IC s142: L1 lethal. NA1. Ref. 427.

let-86 IC s141: L1 lethal. NA1. REf. 427.

let-87 IC s106: semi-viable, L1 lethal with dpy-14(e188). NA2 (s87 (L1 lethal)). Ref. 427.

let-88 IC s132: L1 lethal. NA1. Ref. 427.

let-89 IC s133: L1 lethal. NA1. Ref. 427.

let-90 IC s140: L1 lethal. NA1. Ref. 427.

let-91 IVC s678: mid-larval lethal. NA1. Ref. 813.

let-92 IVC s504: early larval lethal. NA2 (s677). Ref. 813.

let-201 IR e1716: lethal. NA1. Ref. 710.

let-202 IR e1720: lethal. NA1. Ref. 710.

let-203 IR e1717: lethal. NA1. Ref. 710.

let-204 IR e1719: lethal. NA1. Ref. 710.

let-205 IR e1722: lethal. NA1. Ref. 710.

let-206 IR e1721: lethal. NA1. Ref. 710.

let-207 IR e1723: lethal. NA1. Ref. 710.

let-208 IR e1718: lethal. NA1. Ref. 710.

let-209 See rrn-1.

let-236 IIC mn88: larval lethal. NA1. Ref. 715.

let-237 IIC mn208: larval lethal. NA1. Ref. 715.

let-238 IIC mn229: larval lethal. NA1. Ref. 715.

let-239 IIC mn217: embryonic lethal. NA2 (mn93 (embryonic lethal)). Ref. 715.

let-240 IIC mn209: larval lethal. NA1. Ref. 715.

let-241 IIC mn228: larval lethal. NA1. Ref. 715.

let-242 IIC mn90: larval lethal. NA1. Ref. 715.

let-243 IIC mn226: larval lethal. NA1. Ref. 715.

let-244 IIC mn97: larval lethal. NA1. Ref. 715.

let-245 IIC mn185: larval lethal. NA2 (mn221). Ref. 715.

let-246 IIC mn99: larval lethal. NA1. Ref. 715.

let-247 IIC mn211: larval lethal. NA1. Ref. 715.

let-248 IIC mn237: larval lethal. NA1. Ref. 715.

let-249 IIC mn238: larval lethal. NA1. Ref. 715.

let-250 IIC mn207: larval lethal. NA1. Ref. 715.

let-251 IIC mn95: larval lethal. NA1. Ref. 715.

let-252 IIC mn100: larval lethal. NA1. Ref. 715.

let-253 IIC mn181: larval lethal. NA2 (mn184). Ref. 715.

let-254 IIR mn214: larval lethal. NA1. Ref. 715.

let-255 IIR mn186: larval lethal. NA2 (mn236). Ref. 715.

let-256 IIR mn231: larval lethal. NA1. Ref. 715.

let-257 IIR mn235: larval lethal. NA1. Ref. 715.

let-258 IIR mn206: larval lethal. NA1. Ref. 715.

let-259 IIR mn210: larval lethal. NA1. Ref. 715.

let-260 IILC mn232: larval lethal. NA1. Ref. 715.

let-261 IILC mn233: larval lethal. NA1. Ref. 715.

let-262 IILC mn87: embryonic lethal. NA1. Ref. 715.

let-263 IIC mn240: larval lethal. NA1. Ref. 715.

let-264 IIC mn227: larval lethal. NA1. Ref. 715.

let-265 IIC mn188: larval lethal. NA1. Ref. 715.

let-266 IIC mn194: larval lethal. NA1. Ref. 715.

let-267 IILC mn213: larval lethal. NA1. Ref. 715.

let-268 IIC mn189: larval lethal. NA2 (mn198). Ref. 715.

let-269 IIR mn201: larval lethal. NA1. Ref. 715.

let-270 IILC mn191: larval lethal. NA1. Ref. 715.

let-271 IILC mn193: larval lethal. NA1. Ref. 715.

let-272 IVL m243: mid-larval lethal. NA2 (m266). Ref. DR.

let-273 IVLC m263: embryonic lethal. NA1. Ref. DR.

let-274 IVLC m256: embryonic lethal. NA1. Ref. DR.

let-275 IVLC m245: mid-larval lethal. NA2 (m257). Ref. DR.

let-276 IVLC m240: embryonic lethal. NA6 (m239 etc.). Ref. DR.

let-277 IVLC m262: mid-larval lethal. NA1. Ref. DR.

let-278 IVLC m265: sterile adult. NA1. Ref. DR.

let-279 IVLC m261: sterile adult. NA1. Ref. DR.

let-280 IVLC m259: sterile adult. NA1. Ref. DR.

let-281 IVLC m247: sterile adult. NA1. Ref. DR.

let-282 IVLC m270: sterile adult. NA1. Ref. DR.

let-283 IVLC m306: sterile adult. NA1. Ref. DR.

let-284 IVLC m244: sterile adult. NA1. Ref. DR.

let-285 IVLC m248: sterile adult. NA1. Ref. DR.

let-286 IVLC m269: sterile adult. NA1. Ref. DR.

let-326 to let-350 VL Essential genes on left arm of LG V. Ref. BC.

let-351 to let-372 I Essential genes on LGI. Ref. KR.

let-376 to let-392 I Essential genes on LGI. Ref. KR.

let-401 to let-410 V Essential genes on LGV. Ref. BC.

lev: LEVamisole resistant. ZZ. See also unc-29, unc-38, unc-50, unc-63, unc-74.

lev-1 IVR — e211 (pka tmr-1): almost normal movement in absence of drug, uncoordinated but not hypercontracted in 1 mM levamisole. ES2. ME3. NA>10 (x22 etc.). Also two rare semidominant alleles with uncoordinated phenotype, x21: (uncoordinated, Egl-c, ME3), x63. Ref. 484, MT.

lev-7 IR — x13: slightly uncoordinated, poor backing in absence of drug, sensitive to hypo-osmotic shock, resistant to 1 mM levamisole. ES3, ME3. NA1. Ref. 484.

lev-8 XC — x15: almost normal movement in absence of drug, weakly resistant to 1 mM levamisole: body contracts but head does not; more resistant at 25°. ES2. NA1. Ref. 484.

lev-9 XC — x16: almost normal movement in absence of drug, weakly resistant to 1 mM levamisole. ES2. NA3. Ref. 484.

lev-10 IR — x17: almost normal movement in absence of drug, weakly resistant to 1 mM levamisole. ES2. NA1. Ref. 484.

lev-11 IR — x12: slightly long, uncoordinated mild twitcher phenotype, grows well in 1 mM levamisole; strong semidominant suppressor of unc-90(e1463). ES3, ME2. NA3 (x1: twitcher, e1724: lethal). Ref. 484, RW.

lin: abnormal cell LINeage.

lin-1 IVL — e1777amber: adult hermaphrodite has multiple (one to four) vulval protrusions, often bursts at abnormal vulva during L4 molt; ES3 (adult)/ES1 (larvae); HME2; adult male has rudimentary ectopic hooks, ES1, ME0. NA16 (e1026, n431 amber (phenotype resembles e1777), e1275 (weaker, slightly ts phenotype, ME1), etc.). Ref. 496, 500, 762.

lin-2 XRC — e1309: adult hermaphrodite vulvaless (penetrance 93%); ES2 (adult)/ES1 (larvae); HME2; adult male wildtype, ME3. NA13 (e1453amber (non-null, lower penetrance than e1309, 50% vulvaless), n105ts (penetrance 24% at 15°, 90% at 25°), n768 (weak allele, multiple vulval protrusions, penetrance 21%), etc.). Ref. 496, 500, 762.

lin-3 IVC — e1417: adult hermaphrodite vulvaless (penetrance 89%); HME2; adult male wildtype, ME2. NA4 (n378 (resembles e1417); n1059 (early larval lethal); n1058 (early larval lethal/sterile adult, ME3)). Ref. 496, 500, 762, MT.

lin-4 IIC — e912pdi: heterochronic, retarded, adult hermaphrodites long, abnormal movement, vulvaless, lack adult cuticle, ES3 (adult), ES1 (larvae); HME0 (20°)/HME1 (15°), adult males lack copulatory structures, ME0. NA1 (rare mutation). Ref. 507, 608, 620, 762.

lin-5 IIC e1348: thin, sterile and uncoordinated after L1, all postembryonic divisions fail from defective cytokinesis although DNA replication continues. No sexual maturation in either sex. ES3, ME0. NA2 (e1457 (similar phenotypes)). Ref. 8, 496, 500.

lin-6 IL e1466: thin, sterile and uncoordinated after L1, most postembryonic divisions absent or defective from absence of DNA synthesis; Q and I divisions normal. No sexual maturation in either sex. ES3, ME0. NA1. Ref. 496, 500.

lin-7 IIR e1413amber: adult hermaphrodite vulvaless (penetrance 98%), HME2; adult male wildtype, ME3. NA13 (e974 amber, n308 (both resemble e1413), n308cs (penetrance 95% at 15°, 28% at 25°),etc.). Ref. 496, 500, 762.

lin-8 IIL n111: adult hermaphrodite wildtype, Muv in homozygotes with lin-9, lin-35, lin-36, or lin-37; male wildtype, ME3. NA1. Ref. 496, 762, MT.

lin-9 IIIC n112: adult hermaphrodite wildtype, Muv in homozygotes with lin-8 or lin-38. NA3 (n942, n943 (sterile non-Muv alone, sterile Muv with lin-8 etc.). Ref. 762, MT.

lin-10 IC e1439: adult hermaphrodite vulvaless (penetrance 95%), HME2; adult male wildtype, ME3. Similar phenotype in e1439/Df. NA3 (e1438, n299). Ref. 762.

lin-11 IRC n389: adult hermaphrodite vulvaless (penetrance 100%), HME0, slightly uncoordinated; adult male slightly uncoordinated, wildtype morphology, ME1. NA4 (n382 (resembles n389), n672sd (n672/+ is 5% Vul), n566 (weaker allele, HME2)). Ref. 762.

lin-12 IIIC Class 1 alleles (loss-of-function) n941: abnormal vulva, sterile, small, many lineage transformations; putative null. ES3, ME0. NA1.
Class 2 alleles (gain-of-function) n137sd: adult hermaphrodite Muv, many lineage transformations. ES3, ME1 (n137/+ ME3), HME1. NA sd>15 (n177 etc. (resembles n137); n676, n302, n379 (these alleles have semidominant Vul phenotype, probably less hypermorphic)).
Also intragenic revertants of dominant alleles, NA(ird)>50: most have Class 1 phenotype e.g. n137n720 , n676n909amber (resemble n941); n137n460ts (wildtype at 25°, Muv at 15°); many other non-null alleles with tissue-specific effects, also Tc1-insertion null alleles. Ref. 646, 762, 812.

lin-13 IIIC n387ts,mat: adult hermaphrodite Muv and sterile at 25°, maternally rescued at 15° (n387 progeny of n387/+ are wildtype but produce sterile F2 only). Males have occasional ventral protrusions, ME0. n387/Df at 25° is early larval lethal. NA2 (n388 (resembles n387)). Ref. 762.

lin-14 XRC Class 1 alleles (loss of function), n536n540ird: precocious heterochronic lineage alterations in ectoderm; abnormal development of vulva, endoderm and mesoderm; abnormal cuticle formation; ES2; male more severely affected, only three molts, gonadal development abnormal. Similar phenotype in n536n540/Df. NA>10 (n179ts etc; also Class 1a, n355n679ird,ts (only early events precocious); Class 1b, n360ts (only late events precocious).
Class 2 alleles (gain of function), n536sd: retarded heterochronic alterations in many lineages, abnormal vulval development, cuticle formation, supernumerary molts, extra divisions in sex mesoblasts, intestine etc; gonadal lineages normal. ES3. MEO? NA2 (n355sd).
For both Class 1 and Class 2 mutations, late lineages are less affected if animal develops via dauer stage.
Ref. 672, 620, VT.

lin-15 XR n309: adult hermaphrodite Muv, vulva either normal or nonfunctional, two to six ventral protrusions. Some animals rupture during L4 molt. Adult males have one to three ventral protrusions, rudimentary ectopic hook. MEO. Possibly enhanced phenotype in n309/Df. NA12 (n765ts (wildtype at 15°, Muv at 20°but displays maternal effect, Muv at 25°); n767 (weaker allele, wildtype alone, Muv in homozygotes with lin-9, lin-35, lin-36, or lin-37); n744 (weaker allele, wildtype alone, Muv with lin-8 or lin-38); etc.). Ref. 672, 762.

lin-16 III e1743: thin, sterile, uncoordinated after L1, extensive failure of postembryonic divisions, no polyploid cells, most Vn.p cells join syncytium. Ref. CB, TU.

lin-17 IL n671: adult hermaphrodite slightly uncoordinated, long irregularly shaped tail, vulva may be nonfunctional, many hermaphrodites (50%) have single small protrusion posterior to vulva, some gonadal abnormality and sterility. ES2. Male tail grossly abnormal, may rupture during L2 molt, MEO. NA5 (n677 (resembles n671); n669 (less penetrant for Muv phenotype), etc.). Ref. 762.

lin-18 XL e620: some (<50%) hermaphrodites have single small protrusion posterior to vulva, occasional vulval rupture; slight maternal effect, slightly ts. ES2. Males phenotypically wildtype, ME3. NA2 (n1051amber,ts (resembles e620)). Ref. 762.

lin-19 III e1756: extra cell divisions in most post-embryonic lineages; sterile. ES3(adult), MEO. NA1. Ref. NJ.

lin-20 X e1796: abnormal lineages for H0, H1, G1 and G2 . NA1. Ref. NJ.

lin-21 III e1751sd: uncoordinated, abnormal migration of Q neuroblasts, both QL and QR migrate posteriorly. NA1. Ref. NJ.

lin-22 IV n372: V1-V4 divide like V5, resulting in multiple ectopic postdeirids in hermaphrodites and males, and multiple ectopic rays in males. Males also have hermaphroditic Pn.p's and are mostly missing lateral alae. ES1(hermaphrodite), ES2(male). ME2. NA1. Ref. 672, MT.

lin-23 II e1883: extra cell divisions in all postembryonic lineages; sterile. ES3(adult), ME0. NA3 (e1924, e1925) Ref. NJ.

lin-24 IVC n432sd,probably neomorphic: adult hermaphrodite vulvaless (n432 95% Egl, n432/+ 55% Egl). ES2. Male wildtype, ME3. n432/nDf27 is 54% Egl, +/nDf27 is wildtype. NA2 (n1057amber (non-null: n1057 and n1057/Df are wildtype, n1057/+ is 33% Egl)). Ref. 762.

lin-25 VC e1446: adult hermaphrodite vulvaless, 11% sterile. ES2. HME1. Males morphologically wildtype, ME0. NA2 (n545ts (at 25° resembles e1446, at 15° 8% Vul)). Ref. 762.

lin-26 IIC n156: adult hermaphrodites vulvaless (>99%), slightly small and fat, HME1. ES2. Males very small, scrawny, rounded tail, ME0. n156/Df is larval lethal. NA1. Ref. 762.

lin-27 I b151ts,mat,mn: at 25° hermaphrodite sterile with ball shaped gonad, other lineage abnormalities (e.g. M descendants). Male sterile with normal gonad morphology. NA1. Ref. JK.

lin-28 IC n719: precocious heterochronic alterations in many ectodermal lineages, more severe than lin-14(null), precocious abnormal vulva development, Egl, abnormal cuticle formation, only three molts, hermaphrodite and male V5.pa make ray cells in L2; gonadal development normal. Late lineages less affected if animal develops via dauer stage. ES3, ME0. NA4(n947, n1119, n1120: all resemble n719). Ref. 620.

lin-29 IIRC n333: retarded heterochronic alterations in L4 seam cells, no adult alae formed, supernumerary divisions; also Egl, protruding vulva; phenotype not affected by development via dauer stage. ES3. NA3 (n546, n836). Ref. 620.

lin-30 III e1908: many late embryonic and postembryonic lineage failures due to abnormal cytokinesis; variable; homozygous viable. NA1. Ref. NJ, CB.

lin-31 IIL n301: adult hermaphrodite Muv, vulva either functional or nonfunctional protrusion, 0-4 small ventral protrusions. ES3. Male gross morphology wildtype, ME0. Similar phenotype in n301/Df. NA11 (e1750, n376 etc.). Ref. 635, 762.

lin-32 XL e1926: many neuroblasts adopt hypodermal fates (Q, postdeirid, ray neuroblasts make only seam or syncytial nuclei); Q/V5 embryonic division delayed, various embryonic sensory neurons absent or displaced. Mutant phenotypes more severe in e1926/Df. ES1 (hermaphrodite), ES3 (adult male). NA2 (u282: stronger allele, lacks PLM, AVM, PVM neurons, hence touch-insensitive in tail). Ref. CB, NJ, TU.

lin-33 IVLC n1043sd: adult hermaphrodite vulvaless (n1043 95% Egl, n1043/+ 77% Egl). ES2. Male gross morphology wildtype, ME3. NA2 (n1044sd (resembles n1043)). Ref. 762.

lin-34 IVC n1046sd,amber,non-null: adult hermaphrodite Muv (penetrance 57%), normal vulva with one to three large additional protrusions, n1046/+ 17% Muv. ES2. Male gross morphology wildtype, ME1. NA1. Ref. 762.

lin-35 IC n745: wildtype alone, Muv in homozygotes with lin-8, lin-38, or lin-15(n767). ES2. NA2 (n373 (similar phenotype)). Ref. MT.

lin-36 IIIC n766: wildtype alone, Muv in homozygotes with lin-8, lin-38, or lin-15(n767). ES2, ME3. NA4 (n772 (similar phenotype) etc.). Ref. MT.

lin-37 IIIC n758: wildtype alone, Muv in homozygotes with lin-8, lin-38, or lin-15(n767). ES2. NA1. Ref. MT.

lin-38 IIR n751: wildtype alone, Muv in homozygotes with lin-9, lin-35, lin-36, lin-37, or lin-15(n744). ES2. NA2 (n761 (similar phenotype)). Ref. MT.

lin-39 IIIC n709ts: some or all P3.aap - P8.aap (presumptive VC neurons) die in hermaphrodites (ES1),; variably Egl (ES2); variably abnormal vulval divisions. ME3. NA1. Ref. MT.

lin-40 VL e2173: small, uncoordinated, abnormal vulva and gonad, sterile. ES3. NA1. Ref. CB.

lon: LONg. CB.

lon-1 IIIC e185: about 50% longer than wildtype at all stages, markedly tapering tail and head; low penetrance tendency to form constriction behind head, may result in auto-decapitation. ES3. Male bursa elongated, ME0. NA6 (e43, e1820 etc (some have stronger phenotype than e185, all show some decapitation)). Ref. 31, CB, MT.

lon-2 XLC e678: about 50% longer than wildtype at all stages. ES3. Male bursa slightly elongated, ME3 (ME0 at 25°?). NA5 (e405, e434 etc.: similar phenotype to e678). Ref. 31, CB, TU.

lon-3 VRC e2175spo: adult about 50% longer than wildtype. ES3(adult), ME3. NA1. Ref. CB.

mab: Male ABnormal. CB.

mab-1 IC e1228: adult male has swollen bursa, ES3(adult male), ME1. Adult hermaphrodite has protruding vulva. Weak suppressor of some tra mutations. ES2. NA4 (e1233, e2134spo: similar phenotype). Ref. 608, 922, CB.

mab-2 IC e1241: adult male has 6-18 missing copulatory rays as a result of variable failures of V and T lineages, bursa often grossly distorted. ME0/ME1. Hermaphrodite gross phenotype normal, late hypodermal V and T divisions variably defective. ES1 (hermaphrodite), ES3 (adult male). NA1. Ref. 608.

mab-3 IIRC e1240: adult male tail morphology grossly abnormal, often with hermaphrodite tail spike; adult male synthesizes yolk proteins; V and possibly other male lineages abnormal. ES3(adult male),ME0. Hermaphrodite phenotype is wildtype. NA5 (e1921ts, e2093 (similar but slightly weaker phenotypes than e1240)). Ref. 608, CB.

mab-4 IIIR e1252: adult male has swollen bursa, ES3, ME1. Adult hermaphrodite has protruding vulva. ES2. NA1. Ref. 608.

mab-5 IIIC e1239: postembryonic lineages and migrations in pre-anal region generally abnormal; adult male tail morphology thin, grossly abnormal, V rays missing, T rays present, sex mesoblast lineages etc., abnormal. ES3(male), ME0. Hermaphrodite gross phenotype is wildtype but V divisions, Q migrations, coelomocytes all abnormal. ES1 (hermaphrodite). Similar mutant phenotypes in e1239/Df. NA4 (e1936, e2011, e2088 (all resemble e1239)). Ref. 608, 629, 913.

mab-6 IIC e1249: adult male has very swollen bursa, ES2, ME1. Hermaphrodite phenotype wildtype NA1. Ref. 608.

mab-7 XL e1599: adult male slightly small, all bursal rays swollen, bursal fan reduced. ES2, ME1. Hermaphrodite slightly dumpy. ES2. NA1. Ref. 608.

mab-8 IIC e1250: adult male has swollen bursa. Adult hermaphrodite has protruding vulva. ES2, ME1. NA1. Ref. 608.

mab-9 IIL e1245: male tail morphology and development grossly abnormal, B lineage abnormal, many males die or rupture during L4 molt. ES3, ME0. Hermaphrodite phenotype wildtype. NA1. Ref. 608, CB.

mab-10 II e1248: adult male has slightly swollen bursa, reduced fan, thin rays; some adult males have supernumerary molt. ES2, ME0. Hermaphrodite has slightly protruding vulva, ES2. NA1. Ref. 608, CB.

mab-11 IL e2008: adult male has swollen bursa, ES3, ME1. Adult hermaphrodite has protruding vulva, ES2. Weak suppressor of some tra mutations. NA4. Ref. 922, CB.

mec:		MEChanosensory abnormality: animals insensitive to light tactile stimulation. Most, possibly all, have abnormalities in the microtubule cells (ALML/R, AVM, PVM, PLML/R). TU.
mec-1	VC	e1066(pka che-4): touch insensitive, lethargic; microtubule cells lack extracellular mantle, often displaced; some amphidial neurons also displaced. ES2, ME2. NA>50 (e1292, u39amber etc.). Most alleles resemble e1066 but may lack amphidial neuron displacement. Ref. 214, 502, TU.
mec-2	XLC	u8amber: touch insensitive, lethargic. ES2. ME? NA>40 (e75 (pka che-9: recessive, ME3), u7cs, e1084, e1514 etc.). Alleles vary in phenotype from weak recessive to strong dominant Mec. Ref. 214, 502, TU.
mec-3	IVC	e1338: touch insensitive, lethargic, microtubule cells small and lacking processes, ALM and PLM cells displaced. ES2, ME2. NA8 (e1498, e1612 etc.). Ref. 502, TU. Also Class 2 revertant allele, e1498u124: ALM and PLM processes abnormally long. Ref. TU.
mec-4	XR	u52amber: touch insensitive, lethargic. ES2, ME2. NA>50 (e1339 (weak allele, sometimes touch sensitive in tail); e1497, u45ts etc.). Also Class 3 alleles: e1611dm (pka mec-13: touch insensitive, microtubule cells become vacuolated and die in e1611 and e1611/+; probably neomorphic). ES2, ME3. NA(dominant)3. Intragenic revertants (e.g. e1611e1879) resemble u52. Ref. 502, TU, NJ.
mec-5	XR	e1340: touch insensitive, lethargic; mantle of microtubule cells not stained by peanut lectin. ES2, ME3. NA>30 (e1503ts, e1504 etc.: many are ts). Ref. 502, TU.
mec-6	IC	e1342: touch insensitive, lethargic. ES2, ME3. NA8 (e1609spo, u247ts (may be Mec only in head), etc.). Ref. 501, TU.
mec-7	XC	e1343sd,ts: touch insensitive, lethargic at 25°; microtubule cells lack microtubules; e1343/+ variably touch insensitive at 25°; wildtype at 15°. ES2, ME3. NA>30 (n434dm (dominant at all temperatures), e1506 (recessive at all temperatures),etc.). Most alleles incompletely dominant and ts. Ref. 502, 550, TU.
mec-8	IC	e398amber: touch insensitive, lethargic; disrupted fasciculation of amphid channel cilia. ES2, ME2. NA6. Ref. 502, 793.
mec-9	VC	u27amber: touch insensitive, lethargic. ES2, ME2. NA>20 (e1494, u164amber etc.). Ref. 502, TU.
mec-10	XC	e1515: touch insensitive, lethargic. ES2, ME3. NA5 (e1715 etc.). Ref. 502, TU.

mec-12 IIIL e1605: touch insensitive, lethargic. ES2, ME3. NA10 (e1607 (weaker allele), u94sd (u94/+ touch insensitive in head only) etc.). Weak alleles often lack synaptic branch of AVM; strong alleles have few microtubules in microbule cell processes. Ref. 502, TU.

mec-13 See mec-4.

mec-14 III u55: touch insensitive, lethargic. ES2. NA8. Ref. TU.

mec-15 II u215ts: touch insensitive, lethargic. ES2 (25°). NA5 (all alleles partially ts). Ref. TU.

mig: abnormal cell MIGration. CB.

mig-1 I e1787: abnormal migration of Q neuroblasts (both QL and QR migrate anteriorly); HSN migration also abnormal. ES1. NA1. Ref. NJ.

mig-2 X rh17: variable abnormal migration of Q neuroblasts, HSN cells, CAN cells. Ref. NJ, BW.

mig-3 V ct73: abnormal migration of CAN cells. Ref. BW.

mor: MORphological: rounded nose. CB.

mor-1 IIIC e1071: head less pointed than wildtype, slightly rounded nose, especially at 25°. ES2 (adult) ME3. NA1. Ref. CB.

mor-2 IVC e1125: head less pointed than wildtype, rounded nose. ES2 (adult). ME3. NA2 (e2015: similar phenotype). Ref. CB.

msp: Major Sperm Proteins. BA. A family of at least fifty genes, at least some of which are structural genes for major (16kD) sperm protein. Many members are pseudogenes. The genes are mostly in clusters.

msp-3 IIL Cluster of three genes including msp-3 (transcribed). Ref. BA.

msp-24 IVC Cluster of three genes, three pseudogenes including msp-24 (transcribed pseudogene). Ref. BA.

msp-45 IIL Cluster of four genes, one pseudogene, including msp-45. Ref. BA.

msp-56 IV Cluster of four genes including msp-56, msp-10 (transcribed). Ref. BA.

msp-113 IVL Cluster of six genes including msp-19 and msp-113 (both transcribed). Ref. BA.

msp-142 IIL Cluster of two genes, four pseudogenes, including msp-142 (transcribed). Ref. BA.

msp-152 IIL — Cluster of five genes including msp-152 (transcribed). Ref. BA.

myo: — MYOsin heavy chain structural genes. CB.

myo-1 IC — No mutations known. Pharyngeal myosin. Ref. 801, 876.

myo-2 XRC — No mutations known. Pharyngeal myosin. Ref. 801, 876.

myo-3 VRC — Structural gene for minor body-wall myosin heavy chain. Mutation st378 is probable null for myo-3: late embryonic lethal, very abnormal muscle ultrastructure. See also sup-3. Ref. 801, 876, RW.

myo-4 — See unc-54.

ncl: — abnormal NuCLeoli. CB.

ncl-1 IIILC — e1865: abnormal large nucleoli in all cells. ES1. NA2 (e1942: similar phenotype). Ref. CB, NJ.

ncl-2 IV — e1896: abnormal refractile nucleoli, especially in germ cells; sterile. ES2. NA1. Ref. CB, NJ.

nuc: — abnormal NUClease.

nuc-1 XRC — e1392amber: major endodeoxyribonuclease reduced >95%; condensed chromatin persists after programmed cell death, DNA in intestinal lumen not degraded. ES1, ME3. NA3 (n887 (resembles e1392), n334(weaker allele)). Ref. 363, 397, MT.

ooc: — abnormal OOCyte formation. SP.

ooc-1 IIRC — mn250mat, mm: hermaphrodites lay fertilized eggs that do not hatch. ES3(progeny). NA1. Ref. 715.

ooc-2 IIC — mn249mat, mm: hermaphrodites lay fertilized eggs that do not hatch. ES3(progeny). NA1. Ref. 715.

ooc-3 IIRC — mn241mat, mm: hermaphrodites lay fertilized eggs that do not hatch. ES3(progeny). NA1. Ref. 715.

ooc-4 IIIRC — e2078: hermaphrodite sterile, defective gametogenesis, especially oogenesis. ES3(adult), ME1. NA1. Ref. CB.

osm: — defective avoidance of high concentrations of sugars and salts. PR.

osm-1 XR p808: fails to avoid 4M fructose or 4 M NaCl; poor chemotaxis to NaCl; normal thermotaxis; fails to take up FITC; severely shortened axonemes,ectopic assembly of ciliary structures and microtubules in many sensory neurons. ES1, ME2. NA2 (p816 (similar phenotype, lower fertility)). Ref. 82, 793, NJ.

osm-2 See che-3.

osm-3 IVLC p802: fails to avoid 4M fructose or 4M NaCl; non-chemotactic; defective dauer formation; normal thermotaxis; fails to take up FITC; eliminates distal segment of amphid channel cilia, other ciliated neurons normal; octopamine-deficient.. ES1, ME3. NA1. Ref. 82, 552, 793.

osm-4 See daf-10.

osm-5 XL p813: same phenotypes as osm-1(p808), also males suicidal, poor mating. ES1, ME1. NA1. Ref. 82, 793.

osm-6 VC p811: same phenotypes as osm-1(p801). ES1, ME1. NA1. Ref. 82, 793, NJ.

par: abnormal cytoplasmic PARtitioning. KK.

par-1 V b274mat: homozygous hermaphrodites produce embryos that arrest with many differentiated cells, no detectable gut granules; first cleavage symmetrical, defective P granule localization. ES3(progeny). NA3 (e2012 , it32). Ref. KK.

par-2 IIIC it46ts,mat: homozygous hermaphrodites produce 90%(16°)to 99%(25°) arrested embryos, escapers sterile; abnormal early cleavages, defective P granule localization. ES3(progeny). NA4 (it5ts:weaker phenotype, e2030:grandchildless). Ref. KK.

plg: copulatory PLuG formation. CB.

plg-1 IIIRC e2001dm,spo: males lay down gelatinous blob over vulva of mated hermaphrodites; allele isolated from natural C. elegans strain Sta-5. ES1, ME3. NA1. Ref. CB.

rad: abnormal sensitivity to ultraviolet and/or ionizing radiation. PH.

rad-1 ICR mn155: extremely hypersensitive to UV, X, and gamma irradiation. ES1, ME3. NA1. Ref. 565, 663, 733.

rad-2 VC mn156: extremely hypersensitive to UV, X, and gamma irradiation, some sensitivity to chronic MMS treatment. ES1, ME3. NA1. Ref. 565, 663, 733.

rad-3 IC mn157: extremely hypersensitive to UV, not to X irradiation, hypersensitive to chronic MMS treatment (L1 arrest in 0.1 mM); reduced brood size (58% of wildtype). ES1, ME3. Ref. 565, 663.

rad-4 VL mn158cs: hypersensitive to UV, and to acute MMS treatment, not to X irradiation; reduced X chromosome nondisjunction (partly suppresses some him mutations); reduced brood size (34% of wildtype at 20°); <25% eggs hatch at 15°. ES1, ME2. NA1. Ref. 565.

rad-5 IIILC mn159ts: hypersensitive to UV and X irradiation, not to MMS; reduced brood size at 20° (7% of wildtype), sterile at 25°; increased spontaneous mutability. ES2, ME2. NA1. Ref. 565.

rad-6 IIIC mn160: hypersensitive to UV and X irradiation, not to MMS; hermaphrodites less viable (<50%) than males at 25°, equal viability at 15°. ES1, ME3. NA1. Ref. 565.

rad-7 IVC mn161: hypersensitive to UV, not to X irradiation or MMS. Reduced brood size (25% of wildtype). ES1, ME3, NA1. Ref. 565, 663, 733.

rad-8 IC mn163: hypersensitive to UV, not to X irradiation or MMS. Reduced brood size (7% of wildtype), young hermaphrodites lay many inviable eggs. ES1, ME0. NA1. Ref. 565.

rad-9 IIILC mn162: hypersensitive to UV, not to X irradiation or MMS. Reduced brood size (9% of wildtype). ES1, ME1, NA1. Ref. 565.

rec: abnormal RECombination. KR.

rec-1 IR s180: general enhancer of meiotic recombination. ES1. Ref. 321, KR.

rol: ROLler: helically twisted body, animals roll when moving. CB.

rol-1 IIR e91: adults left-handed rollers. ES3 (adult), ME1. NA2 (sc22ts: left roller, slightly dumpy adult at 25°). Ref. 168, 465.

rol-2 See dpy-2.

rol-3 VC e754: adults left-handed rollers. ES3 (adult), ME1. NA1. Ref. 168, 465.

rol-4 VRC sc8: adults and L4 larvae left-handed rollers. ES3 (adult), ME2. NA3 (b238ts: at 25° rolls only in adult, at 16° wildtype). Ref. 465.

rol-5 See sqt-1.

rol-6	IIC	e187: adults, L3 and L4 larvae right-handed rollers. ES3 (adults, late larvae). MEO. Similar phenotype in e187/Df. NA(recessive)7 (sc90: no phenotype alone but fails to complement e187). Null phenotype may be wildtype. Also Class 3 (neomorphic) allele, su1006dm: both su1006 and su1006/+ severe right-handed rollers in L3, L4, adult; phenotype more extreme than e187. Ref. 465, 713.
rol-7		See dpy-10.
rol-8	IIC	sc15: adults left-handed rollers. NA2. Ref. SC.
rrn:		Ribosomal RNA. CB.
rrn-1	IR	Structural locus for rRNA (pka NOR). Approximately 55 tandem copies of 18S, 5.8S and 28S rRNAs. Mutation let(e2000) (pka let-209) is associated with partial deletion; also eDp20(I;II) duplicates this locus. Ref. 865(sequence), 696(mapping).
rrs:		Ribosomal RNA, Small (5S). CB.
rrs-1	VR	Structural locus for 5S RNA. Approximately 110 tandem copies of 1 kb sequence coding for 5S RNA. No mutations known. Ref. 814.
sdc:		Sex and Dosage Compensation. See egl-16.
sma:		SMAll. CB.
sma-1	VRC	e30: short, round-headed, especially in early larvae; adults have slight rolling tendency. ES3. NA8 (e656, e2075 (pka sma-7, ME2), etc.). Ref. 31, CB.
sma-2	IIIC	e502: short, somewhat dumpy; males have crumpled spicules; synthetic lethal with some morphological mutants. ES3 (adult), MEO. NA4 (e172, e297 etc.). Ref. 31, CB.
sma-3	IIIC	e491: short, somewhat dumpy; males have crumpled spicules. ES3 (adult), MEO. NA3 (e637, e958). Ref. 31.
sma-4	IIILC	e729: short, non-dumpy; males have crumpled spicules. ES3 (adult), MEO. NA2 (e805). Ref. 31.
sma-5	XRC	n678: very small adult, slow-growing; small in L1 and L2. ES3. NA1. Ref. MT.
sma-6	IIC	e1482: short adult and late larvae. ES2, ME3. NA1. Ref. CB.
sma-7		See sma-1.

sma-8 VRC sma-8(e2111dm): short, blunt head; e2111/+ similar phenotype. ES3. NA2 (n716sd: recessive lethal). Ref. 914, CB.

spe: defective SPErmatogenesis. SP. See also fer.

spe-1 IIC mn147: hermaphrodites and males produce aberrant spermatids which fail to activate into spermatozoa. Spermatids exhibit unusual and rapid cytoplasmic movement. ES2 (adult). NA1. Ref. 715, BA.

spe-2 IIC mn63pat: hermaphrodites lay self-fertilized eggs that do not hatch, but eggs are viable if cross-fertilized by mn63 sperm from mn63/+ males. ES3(progeny). NA1. Ref.715.

spe-3 IILC mn230pat: same phenotype as spe-2(mn63). ES3(progeny). NA1. Ref. 715.

spe-4 IC hc78: hermaphrodites and males produce primary spermatocytes that contain 4 haploid nuclei; males retain these spermatocytes while hermaphrodites resorb them when oogenesis begins. Self-fertility 0%. Similar phenotype in hc78/Df. ES2(adult). NA2 (hc81). Ref. BA.

spe-5 IL hc93: hermaphrodites and males produce primary spermatocytes that occasionally begin first meiotic division but do not differentiate further. Self-fertility 0%. ES2 (adult). NA2 (hc110). Ref. BA.

spe-6 IIIR hc49: hermaphrodites and males produce primary spermatocytes that do not differentiate further. ES2 (adult). NA2 (hc92). Ref. BA.

spe-7 IIR mn252: hermaphrodites and males produce primary spermatocytes that do not differentiate further. ES2 (adult). NA1. Ref. BA.

spe-8 IL hc40: hermaphrodites and males produce nonfunctional nonmotile sperm with uniformly aberrant pseudopods. Self-fertility <1% (16° or 25°). ES2 (adult). NA6 (hc50, hc53 etc.). Ref. BA.

spe-9 IR hc52ts: hermaphrodites and males grown at 25° produce motile sperm that are fertilization defective. Self-fertility <1% (25°), 100% (16°). ES2 (adult). NA2 (hc88ts). Ref. BA.

spe-10 VC hc104ts: hermaphrodites and males produce nonfunctional, but motile, sperm with short pseudopods. ES2 (adult). NA1. Ref. BA.

spe-11 ILC hc77ts: hermaphrodites and males grown at 25° produce motile sperm that are fertilization defective. Self-fertility 3% (25°), 100% (16°). Similar phenotype in hc77/Df. ES2 (adult). NA2 (hc90 (nonconditional sterile)). Ref. BA.

spe-12 IRC hc76: hermaphrodites produce nonfunctional sperm, fertilization defective; males are fertile. Similar phenotype in hc76/Df. Self-fertility <1% (16° or 25°). ES2 (adult). NA1. Ref. BA.

sqt: SQuaT: short dumpyish animals, roll as heterozygotes. SC.

sqt-1 IIRC sc1: homozygotes short, dumpyish in adult and L4; roll in L3; sc1/+ non-dumpy right-handed rollers in adult, L4, L3. ES3, ME1. NA(dominant)>10 (su1005, e1584 (resemble sc1) e1350 (recessive dumpy) etc.).
Also recessive alleles, e.g. sc13 (pka rol-5): adults, L3 and L4 larvae left-handed rollers. ES3(adult, late larvae). NA>15 (sc101 (long), sc33ts (at 25° rolls only in adult, at 16° wildtype), etc. Complex complementation patterns. Null phenotype probably wildtype. Ref.465, SC.

sqt-2 IIL sc3: homozygotes slightly dumpyish in adult, late larvae; roll in L3; sc3/+ non-dumpy right-handed roller in adult, L4, L3. ES3(adult). NA7 (sc14 (stronger phenotype than sc3), etc.). Ref. 465.

sqt-3 VRC sc63ts: homozygotes at 25° are non-roller dumpy, heterozygotes sc63/+ are non-dumpy left-handed roller in adult, high penetrance deformed tail at 25°. ES3, ME0 (25°). At 15° slightly dumpy. NA>8 (sc80ts (resembles sc63), e24sd,ts (pka dpy-15: dumpy, inviable at 25°, e24/+ dumpy at 25°), etc.). Ref.465, CB, MT.

sup: SUPpressor. CB.

sup-1 IIIR e995dm: dominant suppressor of unc-17(e245); does not suppress unc-17(e876); no phenotype alone. ES3 in presence of e245, ME3. NA>5 (e995xr1, etc.: all resemble e995). Ref. CB.

sup-2 XC e997dm: dominant suppressor of unc-17(e245), does not suppress unc-17(e876), no phenotype alone. ES3 in presence of e245, ME3. Slightly weaker suppressor than sup-1. NA1. Ref. CB.

sup-3 VRC e1407sd: partial suppressor of certain mutations of muscle genes unc-15, unc-54, unc-87, e.g. unc-15(e73), unc-54(e1315); no phenotype alone; associated with alteration of minor body wall myosin gene myo-3 V, and causes increased levels of myo-3 product. ES3 in presence of e73, ME3. NA>10 (e1390 etc.).
Also super-suppressor alleles (unstable) and antisuppressor alleles derived from e1407 etc. Ref. 318, 769.

sup-4 IIIR e1877dm: dominant suppressor of unc-51(e369), does not suppress other alleles of unc-51. ES3 in presence of e369. NA2 (e1864). Ref. CB, NJ.

sup-5 IIIC e1464sd: amber suppressor, partially or completely suppresses amber alleles of more than forty different genes; slow growing, cs (sterile, at 15°) otherwise no phenotype alone; alteration in $tRNA^{Trp}$ anticodon. ES3 in presence of unc-13(e450) etc., ME1. NA>20, all identical to e1464. Ref. 397, 515, 729, 768.

sup-6 IIR st19dm: dominant suppressor of unc-13(e309), recessive lethal; does not suppress amber or ochre alleles of unc-54. ES3. Ref. CB, RW.

sup-7 XRC st5sd: strong amber suppressor, partially or completely suppresses amber alleles of more than forty different genes; slow growing or sterile at <22°, grows best at 23°, suppression strongest at low temperature; otherwise no gross phenotype alone; alteration in $tRNA^{Trp}$ anticodon. ES3 in presence of unc-13(e450) etc., ME2. NA6, all identical to st5. Ref. 515, 729, 768.

sup-8 VC e1563dm,uvi: dominant suppressor of unc-17(e245); no phenotype alone. ES3 in presence of e245. NA1. Ref. CB.

sup-9 IIL n180: recessive suppressor of unc-93(e1500), no phenotype alone; probably null allele. ES3 in presence of e1500, ME3. NA>20 (n180spo, n192uvi, etc.; all resemble n180). Ref. 486, 564.

sup-10 XR n183: recessive suppressor of unc-93(e1500), no phenotype alone; probably null allele. ES3 in presence of e1500, ME3. NA>20 (n181spo, n245gri, etc.). Also Class 3 (neomorphic) allele, n983: uncoordinated, rubber band paralysis like unc-93(e1500); suppressed by intragenic revertants, unc-93 null alleles etc. Ref. 486, 564, 898.

sup-11 IL n403sd: dominant suppressor of unc-93(e1500), recessive small scrawny slow growing phenotype. ES3.
NA(dominant)10: all resemble n403.
Also alleles derived from n403, e.g. n403n406: recessive suppressor of unc-93(e1500), no phenotype alone.
NA(recessive)8.
Also putative null alleles derived from n403 e.g. n403n682: no suppressive effect, recessive lethal arrested in late embryogenesis. NA4 (n403n681amber: partly suppressed by sup-5, L1 lethal). Ref. 564.

sup-12 XL st89: recessive suppressor of unc-60(e677) and at least two other unc-60 alleles (e723, e890);abnormal gonad and minor changes in body wall muscle ultrastructure alone; ES3 in presence of e677. NA>10. Ref. RW.

sup-13 III st210: strong recessive suppressor of unc-78(e1217) and at least two other unc-78 alleles (e1221, st43); no phenotype alone. ES3 in presence of e1217. Ref. RW.

sup-16 IVR st500sd: semidominant suppressor of unc-52(e669) and at least one other unc-52 allele (e1421); does not suppress unc-13(e1091); no phenotype alone. ES3 in presence of e669. Ref. RW.

sup-17 IR n316 weakly sd: allele and lineage non-specific suppressor of lin-12 dominant mutatations; phenotype alone is recessive Dpy, recessive Egl; abnormal vulval cell lineages; male tail abnormal; Egl and Dpy phenotypes stronger at 25°. ES2, ME0. Similar degree of suppression of lin-12(dom) in nDf24/+ and n316/+; also n316/nDf24 is maternal effect lethal. NA2 (n1260 weakly sd, ts (15° phenotype non-Dpy, non-Egl, very weak suppressor, ME3; 25° phenotype similar to n316: Dpy, Egl, abnormal vulva, ME0)). Ref. MT.

sup-18 IIIC n463: strong suppressor of sup-10(n983), weak suppressor of unc-93(e1500). Ref. MT.

sup-19 V m210. Ref. DR.

sup-20 XR n821: suppressor of unc-105(n490); null phenotype is recessive lethal. Ref. MT.

sup-21 XRC e1957sd: amber suppressor, partially or completely suppresses amber mutations in many genes but not in unc-13; no phenotype alone. ES3 in presence of tra-3(e1107) etc., ME3. NA4 (e1958, e2061 etc.(all weaker suppressors)). See also sup-28. Ref. 797, RW.

sup-22 IVLC e2057: weak amber suppressor, suppresses amber mutations in a few genes (tra-3, dpy-20); no phenotype alone. ES3 in presence of tra-3(e1107), ME3. NA1. See also sup-29. Ref. 797, RW.

sup-23 IVC e2059: weak amber suppressor, suppresses amber mutations in a few genes (tra-3, dpy-20, lin-1); no phenotype alone. ES3 in presence of tra-3(e1107), ME3. NA1. Ref. 797.

sup-24 IVR st354: weak amber suppressor, suppresses unc-52(e669), weakly suppresses unc-13(e450); no gross phenotype alone; alteration in $tRNA^{Trp}$ gene. ES3 in presence of unc-52(e669). NA1. Ref. RW.

sup-25 V e1956: suppressor of lin-12(n137). Ref. CB.

sup-26 III n1091sd: suppressor of her-1(n695sd); no phenotype alone. NA2 (ct49sd: similar phenotype). Ref. BW.

sup-27 V n1092sd: suppressor of her-1(n695sd); no phenotype alone. NA2 (n1102sd: similar phenotype). Ref. BW.

sup-28 XRC e2058 (pka sup-21): amber suppressor, suppresses tra-3(e1107), dpy-20(e2017), but not unc-13(e450); no gross phenotype alone; alteration in $tRNA^{Trp}$ gene. ES3 in presence of tra-3(e1107), ME3. NA1. Ref. 797, RW.

sup-29 IVLC e1986 (pka sup-22): very weak amber suppressor, suppresses tra-3(e1107); no phenotype alone; alteration in $tRNA^{Trp}$ gene. ES3 in presence of tra-3(e1107), ME3. NA1. Ref. 797, RW.

sus: SUppressor of Suppressor. DR. See also enu.

sus-1 IIIR m155: reduces suppression by sup-3 alleles; enhances paralysed phenotype of unc-15(e73), unc-54(e190); no phenotype alone. ES1. NA1. Ref. 769.

tmr: See lev.

tpa: TPA (tetradecanoyl phorbol acetate) resistant. MJ.

tpa-1 IVL k501sd: resistant to 0.1 micromolar TPA. Ref. 632.

tra: TRAnsformer: XX animals transformed into males. CB.

tra-1 IIIR e1099: XX animals transformed into low fertility males; gonad morphology variable, testes reduced in size, few sperm made; XO phenotype male. ES3, ME1 (XX). NA(recessive)>15 (e1781amber (soma male, gonad contains apparent oocytes), e1488 (gonad and intestine hermaphrodite (self-fertile), rest of body male), e1732 (gonad more masculinized than body), etc.).
Also Class 2 (gain-of-function) alleles, e1575sd (pka her-2): both XX and XO animals transformed into fertile females by e1575 or e1575/+. ES3, ME0. NA(dominant)>15 (some weaker than e1575, cause incomplete feminization).
Also intragenic revertants of dominants, resemble Class 1 alleles. NA>30. Ref. 178, 498, 630, CB.

tra-2 IIC e1095sd: XX animals transformed into infertile males with abnormal tail anatomy; gonad morphology male, normal spermatogenesis; XO phenotype wildtype male; e1095/+ is Egl hermaphrodite. ES3, ME3 (XO), ME0 (XX). NA(recessive)>20 (e1425amber (resembles e1095); f70 (masculinized self-fertile hermaphrodite), e1875 (slightly Egl hermaphrodite, very weak allele), b202ts: self-fertile hermaphrodite at 16°, Tra-2 at 25°), etc.).
Also Class 2 (gain-of-function) alleles, e2020dm: XX animals completely feminized, XO animals fertile males, slightly feminized. NA(dominant)6: most are weaker than e2020.
Also intragenic revertants of dominants, resemble Class 1 alleles. NA>10. Ref. 178, 205, 923.

tra-3 IVR e1107amber,mat: e1107 daughters of e1107/+ parents are wildtype hermaphrodite, e1107 XX progeny of e1107 parents are abnormal sterile males or intersexes, occasionally self-fertile, especially at 15° (15° brood 1% of wildtype); XO phenotype wildtype. ES3 (progeny testing),

MEO (XX), ME3 (XO). NA4 (e1525amber, e1767 (non-amber),etc: all alleles resemble e1107). Amber alleles very efficiently suppressed. Ref. 178, 498, 797.

unc: UNCoordinated. CB.

unc-1 XL Class 1 (loss-of-function), e719: putative null, recessive kinker. ES3. NA(recessive)>10 (e538 etc). Class 3 (neomorph or antimorph), e94sd (former reference allele): recessive kinker, e94/+ is weak coiler, ES2, ME2; e1598dm (pka unc-102): both homozygote and e1598/+ are strong coilers, resistant to aldicarb (cholinesterase inhibitor); intragenic revertants of e1598 have Class 1 phenotype. ES3, MEO. NA(dominant)>8. Complex complementation among unc-1 alleles. Ref. 31, 914, NW.

unc-2 XL e55: weak kinker, sluggish, thin. ES3, ME2. NA3 (e97, e129). Ref. 31, CB.

unc-3 XR e151: weak coiler, tends to coil tail, active; very disorganized ventral nerve cord; mutant focus in ABp descendants; ES3, ME1. Similar phenotype in e151/Df. NA4 (e54 (weaker allele), e95 etc.). Ref. 497, 712, CB.

unc-4 IIC e120: large, healthy, active, moves forward well, cannot back; ventral cord VA motoneurons have normal anatomy but most have synaptic inputs appropriate to VB motoneurons. ES3, MEO. NA5 (e26 etc.). Ref. 31, 715, CB.

unc-5 IVLC e53: severe coiler, grows well; L1 also severe coiler; dorsal hypodermal cells abnormal, dorsal nerve cord absent or almost absent, cord commissures fail to reach targets, muscle arms misdirected; distribution of cell bodies in ventral cord is disorganized; gonad arms may fail to reflex; other anatomical abnormalities; Egl-c. ES3, MEO. NA7 (e553 (resembles e53), e152 (weaker phenotype behaviorally and anatomically, dorsal cord partially formed), etc.). Ref. 31, CB, MT, NJ.

unc-6 XC e78: slight kinker, poor backing, large, healthy, slightly fat; dorsal extensions of DD and VD neurons grow in aberrant directions, fail to reach dorsal cord; ventral cord disorganized, etc. ES3, ME1. NA8 (e7, e181, ev400 (pka unc-106: uncoordinated, high frequency of phasmid axon displacement; gonad arms may fail to reflex; other anatomical abnormalities), rh45 etc.). Ref. 31, MT, NW, NJ.

unc-7 XR e5: moves backward better than forward, kinker in forward movement, active, healthy, slightly thin. ES3, MEO. NA5 (e42, e65 (pka unc-12), etc.). Ref. 31.

unc-8 IVC e49: moves well but slowly and irregularly, often kinking, both forward and backward; e49/+ very slightly uncoordinated. ES2, ME1. NA(recessive)1. Also dominant coiler alleles, e15sd (pka unc-28): ES3,

MEO; n491sd (homozygotes strong coilers, n491/+ coiler, slightly weaker phenotype). Intragenic revertants of n491 (e.g. n491n1192), are wildtype, thus unc-8 null phenotype may be wildtype. NA(dominant)4. Ref. 31, 914, CB.

unc-9 XRC e101: moves backward better than forward, slight kinker in forward movement, active, healthy; larvae more severely uncoordinated; male fan slightly reduced, spicules tend to protrude. ES2, MEO. NA3 (e111). Ref. 31.

unc-10 XC e102: weak coiler, tends to back, loopy movement in reverse; fairly active, slightly small and thin. ES3, ME2. NA2 (e126). Ref. 31.

unc-11 ILC e47: kinker, jerky ratchet-like movement especially in reverse; slow pharyngeal pumping; slightly small and thin. ES3, ME2. NA3 (e511, e1054). Ref. 31.

unc-12 See unc-7.

unc-13 IC e51: paralysed, kinky, small, irregular pharyngeal pumping, able to lay eggs. ES3, MEO. NA>30 (e450amber, e312amber (non-null), e309 (see sup-6) etc.; all similar to e51 or slightly weaker). Ref. 31, 797, CB.

unc-14 IC e57: very sluggish, almost paralysed, small and dumpyish, tends to coil; some egg retention. ES3, MEO. NA6 (e157 etc.). Ref. 31, CB.

unc-15 IC Structural gene for paramyosin. Reference allele e73: limp paralysed phenotype, larvae move slightly better, Egl; disorganized muscle structure; suppressed by sup-3; e73/+ slightly slow; ES3, MEO. NA>5 (e1214amber (more severe phenotype than e73, not suppressed by sup-3), e1402ts (moves well at 15°, paralysed at 25°), etc.). Ref. 31, 318, 399, TR, RW.

unc-16 IIIRC e109: very sluggish, small; males more active. ES3, ME2. NA1. Ref. 31.

unc-17 IVL e245: severe coiler at all stages, rather small and thin; resistant to 0.1 mM lannate; suppressed by sup-1, sup-2, and sup-8; normal ChAT (choline acetyltransferase) levels. ES3, MEO. NA>10 (most alleles resemble e245 or have slightly weaker phenotype; also anomalous alleles e113 (less uncoordinated phenotype, drug sensitive, loopy movement, reduced ChAT levels, Egl-c) and e876 (similar)).
See also cha-1: cha-1/unc-17 is probably a complex locus. Ref. 31, 675, PR, MT.

unc-18 XC e81: paralysed, kinky, thin at all stages; able to lay eggs. ES3, MEO, NA2 (e174 (pka unc-19: similar phenotype)). Ref. 31.

unc-19 See unc-18.

unc-20 XL e112ts: at 25° severe kinker, some coiling; active, healthy. ES3 (25°), ME0 (25°); at 15° L1 coiler, adult wildtype. NA1. Ref. 31, CB.

unc-21 See unc-29.

unc-22 IVC e66: twitcher at all stages; (moves slowly with constant trembling); thin; unable to hypercontract and therefore levamisole resistant; Egl; e66/+ twitches in 1% nicotine; ES3, ME0. NA>200 (s32amber (resembles e66); m52dm; many Tc1 insertions etc.). Most unc-22/+ heterozygotes twitch in nicotine; abnormal muscle structure in most homozygotes; near normal muscle for alleles e105, s12, Tc1 induced alleles. See also unc-54. Ref. 245, 571, 877, BC, RW, TR.

unc-23 VLC e25: 'benthead', progressive dystrophy of head musculature so that adult head is bent dorsally or ventrally, hence very poor forward movement, good reverse movement. Egl. ES3 (adult), ME1. NA8 (e324: resembles e25; e611 (pka vab-4: incompletely penetrant allele), etc.). Ref. 31, 387, 461, CB.

unc-24 IVC e138amber: weak kinker, tends to back, often forms omega shape, fairly active, healthy. ES3, ME1. NA4 (e448, e927 etc.). Ref. 31, 318, 797, CB.

unc-25 IIIR e156: shrinker, contracts both dorsally and ventrally when prodded; slow; poor backing; slightly small; deficient in GABA. ES3, ME2. NA4 (e265, e591). Ref. 31, MT, CB.

unc-26 IVR e205: severe kinker, small, scrawny, flaccid, little movement; slow pharyngeal pumping. ES3, ME0, NA 9(e176, e345 (pka unc-48) etc. Ref. 31.

unc-27 XC e155: sluggish, poor backing, slightly dumpy; abnormal body muscle; ES3, ME2. NA5 (su142sd (pka unc-99: very slow, almost paralysed, faint muscle birefringence, abnormal muscle ultrastructurewith collections of thick or thin filaments), su195 (forms thin filament paracrystals) etc.). Ref. 31, 488, CB, RW.

unc-28 See unc-8.

unc-29 IRC e193: very sluggish L1, moves better as adult, weak kinker, head region stiff, moves better in reverse, fairly active; resistant to 1 mM levamisole; sensitive to hypoosmotic shock, ES3, ME2. NA>50 (e1072amber, e330 (pka unc-21), e403 (pka unc-56)). Ref. 484, ZZ, DR, MT.

unc-30 IVR e191: shrinker, contracts both dorsally and ventrally when touched; slow, good forward movement but poor backing, rather small; VD and DD motoneuron axons are displaced and fail to stain with anti-GABA antibodies. ES3, ME2, NA8 (e165, e318 etc.). Ref. 31, CB, MT.

unc-31 IVRC e169: very slow and sluggish, moves better in absence of bacteria; insensitive to prodding; Egl; constitutive pharyngeal pumping. ES2, ME1, NA>10 (n1304 (resembles e169), n422ts (pka egl-22, Type C Egl), e69, e86, u280, etc.). Ref. 31, 635, CB, MT, TU.

unc-32 IIIC e189: severe coiler, little movement in adult; moves well in L1 but coils in response to touch in L2 and later stages; rather small and thin; weakly Egl-c. ES3, MEO, NA1 (rare mutation). Ref. 31, CB, MT.

unc-33 IVL e204: very slow, almost paralysed, tends to curl, dumpyish; weak FITC uptake; amphid, phasmid, PDE axons abnormal; abnormal neuronal microtubules. ES3, MEO. NA5 (e572, e735). Ref. 31, NJ, NW.

unc-34 VL e315ts: at 20° fairly severe coiler, somewhat active, grows well; male has crumpled copulatory spicules; variable defects in VD and DD commissures; at 15° moves much better. ES3 (20°), MEO. NA3 (e566, e951pdi: non-ts alleles, coilers at all temperatures; e566 strongest allele). Ref. 31, CB, MT.

unc-35 IL e259: loopy irregular forward movement, poor backing; active; slightly thin. ES2. ME2. NA1. Ref. CB.

unc-36 IIIC e251: very slow, almost paralysed, loopy at rest, thin; normal ventral nerve cord ultrastructure. ES3, MEO. NA5 (e418 (resembles e251); eT1(III;V) (pka unc-72) fails to complement e251, (ME1)). Ref. 31, 554, CB.

unc-37 IC e262: weak coiler, fairly active. ES2, ME1. NA1. Ref. 31.

unc-38 ILC e264: weak kinker, sluggish, slightly dumpyish, sometimes Egl, resistant to 1 mM levamisole in body but not in head; sensitive to hypoosmotic shock. ES3, ME2. NA>40 (e213, x20 etc.). Ref. 31, 484, ZZ.

unc-39 VC e257: fairly severe kinker, can move forward and backward, fairly active, slightly dumpy. ES2, ME1. NA1. Ref. 31.

unc-40 IC e271: weak kinker, dumpyish, slow but fairly active, Egl; variable defects in VD and DD commissures. ES2, MEO. NA4 (e1430(pka unc-91), n324, n473). Ref. 31, 635, MT.

unc-41 VC e268: weak kinker, irregular jerky movement, slightly small; VD1, VD2, DD1 cell bodies mispositioned,usually anteriorly; RVG organization disrupted. ES3, MEO. NA9 (e252, e399 etc.). Similar phenotype in e268/Df. Ref. 31, 318, MT.

unc-42 VC e270: medium kinker, tends to back when touched on tail, slightly small, fairly active, grows well, Egl-c. ES3, ME1. Similar phenotypes in e270/Df. NA3 (e419, e623). Ref. 31, 318.

unc-43 IVC e408: slow, slightly rippling movement, poor backing, thin; old adults Egl, almost paralysed, tend to shrink and relax when prodded; larvae move better. ES3 (adult), MEO. NA3 (e266, e755). Ref. 31, 592.
Also gain-of-function allele, n498sd: paralysed, n498/+ less severe; revertants e.g. n498n1179 have recessive uncoordinated phenotype. Ref. 914.

unc-44 IVC e362: paralysed coiler, dumpy, tends to curl; weak FITC uptake; amphid, phasmid and PDE axons abnormal. ES3, MEO. NA11 (e427, e638 etc.). Ref. 31, NJ, NW.

unc-45 IIIL e286ts: at 25° limp, paralysed, Egl: muscle ultrastructure defective, few thick filaments; males move better; at 20° slow, at 15° wildtype. ES3 (adult, 25°) MEO. NA>5 (m94ts, r450ts: all alleles are temperature sensitive but some have stronger 20° phenotype than e286). Null phenotype probably lethal. Ref. 31, 128, 461, TR, RW.

unc-46 VLC e177: shrinker, contracts both dorsally and ventrally when prodded, slow, good forward movement, poor backing; slightly small. ES3, ME2. NA3 (e300, e642). Ref. 31.

unc-47 IIIRC e307: very poor backing, good slow forward movement, slight shrinker (contracts both dorsally and ventrally when prodded), small; abnormal staining with anti-GABA antibodies. ES3,ME2. NA3 (e542, e707). Ref. 31, CB, MT.

unc-48 See unc-26.

unc-49 IIIRC e382: shrinker, contracts both dorsally and ventrally when prodded, slow, poor backing, slightly small. ES3, ME2. NA6 (e407, e468). Ref. 31.

unc-50 IIIRC e306: weak kinker, slow but fairly active, resistant to 1 mM levamisole, sensitive to hypoosmotic shock. ES3, ME1. NA7 (e425 etc.: all similar to e306). Ref. 31, 484.

unc-51 VR e369: paralysed, dumpy, tends to curl; Egl; dorsal extensions of DD and VD neurons grow in aberrant directions, fail to connect to dorsal nerve cord; amphid, phasmid, PDE axons abnormal; suppressed by sup-4. ES3, MEO. NA6 (e389, e432 etc.). Ref. 31, MT, NJ.

unc-52 IIR e444: adults limp, paralysed except for head region, thin, Egl; larvae move well; progressive dystrophy, posterior muscles fail to accumulate myofilaments. ES3 (adult), MEO. NA>10 (e669 amber (well suppressed), e998 (stronger phenotype), r290 etc.). Ref.31,461,488,515,TR.

unc-53 IIRC e404: sluggish, poor backing, dumpyish, somewhat Egl; males have abnormal bursal anatomy. ES2, ME0. NA4 (n152, n166, n569). Ref. 31, MT.

unc-54 IR Structural gene for major body wall myosin heavy chain. Reference null allele e190: limp paralysed phenotype at all stages, larvae can move slightly more than adults, Egl; muscle ultrastructure very disorganized, few thick filaments. ES3, ME0. NA(recessive)>50 including e1108 amber; e1301ts; e675sd and s291 (in frame internal deletion mutants with heavy chain fragment, behavioral phenotype almost paralysed, slight twitcher); also Tc1 insertion mutants; etc.
Also unusual suppressor alleles (NA>15) e.g. s74: dominant suppressor of unc-22(s12), no other dominant phenotype, recessive phenotype slow, stiff, normal muscle ultrastructure; alteration in S1 portion of protein.
Also Class 3 (dominant) alleles, reference allele e1152sd: severe rigid paralysis, small; e1152/+ paralysed, weaker phenotype. ES3, ME0. NA(dominant)>10: e1153, n489 etc.; resemble e1152 or more severe, recessive lethal. Intragenic revertants have Class 1 phenotypes. Ref. 779 (sequence), 461 (recessive mutant phenotypes), 579 (intragenic mapping), 571 (unc-22 suppressors), 710 (dominant alleles), CB, MT, RW, TR etc.

unc-55 IC e402: slow, very poor backing, tends to coil, healthy; abnormal VD neurons. ES3, ME0. NA2 (e523). Ref. 31, CB.

unc-56 See unc-29.

unc-56 ILC e406: strong kinker, active, small and thin; slow pharyngeal pumping; tends to hypercontract. ES3, ME2. NA4 (e590, e957 etc.). Ref. 31, 484.

unc-58 XRC e665dm: 'shaker': animals short, rigidly paralysed with constant shaking of body; e665/+ phenotype similar but weaker, animals slightly longer and less rigid. ES3, ME0. NA(dominant)5 (e415, n495 (similar phenotypes), e778 (weaker phenotype)). Dominant alleles revert intragenically, e.g. e665e2112: recessive weak Unc (probable null phenotype). Ref. 31, 565, 914.

unc-59 IR e261: poor backward movement, forward better; thin; vulva variably abnormal, often protrusive, sometimes ruptured; many postembryonic lineage abnormalities resulting from variable failures in cytokinesis; gonad lineages sometimes defective; males have very abnormal tail anatomy. ES2, ME0. NA2 (e1005 (pka unc-88)). Ref. 500.

unc-60 VL e723: limp paralysed or very slow; thin; Egl; abnormal muscle ultrastructure with large aggregates of thin filaments; recessively suppressed by sup-12. ES3, ME0. NA>10 (e677 (pka unc-66), e890 (both resemble e723), etc.). Ref. 31, 461, 488, RW.

unc-61 VRC e228: poor backing, irregular waveform in forward movement; thin; protrusive vulva; male tail very abnormal (rays absent, spicules reduced etc.). ES3, ME0. NA1. Ref. 31, CB.

unc-62 VLC e644: slightly slow, irregular, sometimes rippling movement, especially in reverse; slightly dumpy; variable abnormalities in VD and DD commissures; male tail abnormal, bursa small, fan reduced, rays variably absent. ES2, ME0. NA1. Ref. 31, CB, MT.

unc-63 ILC e384: weak kinker, thin, inactive, resistant to 1 mM levamisole, sensitive to hypoosmotic shock. ES3, ME2. NA>50 (x18 etc.: most alleles resemble e384) Also rare exceptional alleles x26 (almost normal movement, slight levamisole resistance); b404 (resistant to cholinesterase inhibitor trichlorfon, slightly uncoordinated, slight levamisole resistance); x33 (more resistant in body than in head). Ref. 31, 484, ZZ.

unc-64 IIIR e246: sluggish, will move either forward or backward if prodded but almost immediately jams up; partially lannate resistant; healthy. ES2, ME0. NA1. Ref. 31.

unc-65 VRC e351: slow, moves well, slightly poor backing; partially lannate resistant; healthy; sometimes Egl. ES2, ME2. NA2 (e355). Ref. 31.

unc-66 See unc-60.

unc-67 IL e713: sluggish, can move well both forward and backward but frequently pauses or jams up. ES2, ME2. NA1. Ref. 31, 554.

unc-68 VC e540: weak kinker, slow, thin; head region but not body resistant to 1 mM levamisole, ouabain. ES3. ME2. NA8 (x14, x24 etc.: resemble e540). Ref. 484, ZZ.

unc-69 IIIRC e587: medium coiler, inactive, small; abnormalities in VD and DD commissures. ES3, ME0. NA2 (e602). Ref. 31, CB, MT.

unc-70 VC e524ts, sd: irregular, loopy movement, sometimes coiling, active, grows well. ES2, ME2. NA2 (n493sd: curly uncoordinated, non-ts). Revertants e.g n493n1170ird have recessive lethal phenotype (probable null phenotype). NA(ird)8. Ref. 31, 914.

unc-71 IIIR e541: strong kinker, especially in reverse, fairly active; commissures sometimes on wrong side. ES3, ME2. NA1. Ref. 31, CB, MT.

unc-72 See unc-36.

unc-73 ILC e936: coiler, small, dumpyish, inactive; commissures often on wrong side; male copulatory spicules short and crumpled. ES3, ME0. NA1. Ref. 31, CB, MT.

unc-74 ILC e883: weak kinker, slow, slightly small, resistant to 1 mM levamisole, sensitive to hypoosmotic shock; L1 movement very poor, no backing. ES3, ME2. NA>25 (x19 (resembles e883); etc.). Ref. 484, ZZ, MT.

unc-75 IR e950: weak coiler, especially in reverse; moves forward well; sluggish; short. ES3, ME2. NA1. Ref. 31.

unc-76 VR e911pdi: inactive, tends to curl, dumpyish; Egl-c; amphid, phasmid, PDE axons abnormal; dorsal extensions of DD and VD neurons frequently grow in abnormal directions, often fail to connect to dorsal nerve cord. ES3, ME0. NA1. Ref. 31, 793, MT, NJ, NW.

unc-77 IVL e625: irregular loopy movement, both forward and reverse; active; thin; sometimes protrusive vulva. Slight movement abnormality in e625/+. ES2, ME2. NA1. Ref. 31.

unc-78 XL e1217: slow, abnormal body muscle birefringence, abnormal muscle ultrastructure with large aggregates of thin filaments; suppressed to wildtype by sup-13. ES2, ME0. NA>5 (e1221, st43 etc.). Ref. 461, 488.

unc-79 IIILC e1068: irregular movement, somewhat sluggish. ES2, ME3. NA4 (e1031 etc.). Ref. 608.

unc-80 V e1272: slightly sluggish, tends to pause, good movement. ES1, ME3. NA2 (e1069). Ref. 608.

unc-81 IIIR e1122: slightly irregular movement. ES1, ME3. NA1. Ref. CB.

unc-82 IVC e1220: slow, good movement, abnormal reduced muscle birefringence, abnormal body muscle ultrastructure with enlarged thick filaments etc. ES2, ME2. NA9 (e1323, r194 etc.: all resemble e1220). Ref. 461, 488, TR.

unc-83 VLC e1408ts: reverse kinker as adult, variably Egl; L1 moves well; variable failures in postembryonic migration of Pn nuclei into ventral cord and (some alleles) embryonic migrations of hyp-7 hypodermal nuclei; all alleles are ts for Pn defect, non-ts for hyp-7 defect; unmigrated Pn nuclei mostly fail to divide; adult male phenotype variable, some can mate at 25°, all can mate at 15°. ES3, ME3 (15°). NA11 (n331amber,ts; e1409amber,ts (suppressible only for hyp-7 defect)). Ref. 500, MT.

unc-84 XR e1410ts: reverse kinker as adult, L1 moves well, variable failures in nuclear migrations: same phenotypes as unc-83(e1408). ES3, ME3 (15°). NA16 (e1174 etc.: all alleles ts; complex complementation pattern; null phenotype probably incompletely penetrant). Ref. 500, MT.

unc-85 IILC e1414amber: kinker in adult, cannot back, L1 moves well; many postembryonic lineages abnormal as a result of defective cytokinesis in cell division; all phenotypes similar to or slightly weaker than those of unc-59(e261). ES3, ME0. NA3 (n319, n471). Ref. 500.

unc-86 IIIC e1416: lethargic; Mec (touch cells absent); Egl (HSN cells fail to differentiate); non-chemotactic to NaCl; lineage abnormalities involving reiterative divisions of neuroblasts, hence supernumerary neurons and missing neurons; 2% Him. ES3, ME2. NA>30 (e1507 (Mec, Egl, non-Him, ME3); n848ts (superficially wildtype at 20°, Mec and Egl at 25°), etc.). Alleles form graded series; e1416/Df is ME0; null phenotype uncertain. Ref. 179, 507, MT.

unc-87 IC e1216: limp paralysed or very sluggish; somewhat Egl; larvae more paralysed; abnormal body muscle birefringence and ultrastructure, thin filaments in small bundles. ES3, ME0. NA>10 (e1459amber (phenotype more severe than e1216), st39, r320 etc.). Ref. 461, 488, DD, TR.

unc-88 See unc-59.

unc-89 ILC e1460: good movement, slightly small, thin and transparent; abnormal reduced birefringence in body wall and pharyngeal muscle; thick filaments disarranged, no M-line. ES1, ME3. NA>10 (st85amber, r291 etc.). Ref. 461, 488, RW, TR.

unc-90 XRC Class 3 (neomorphic) allele, e1463sd: homozygotes small, rigidly paralysed (hypercontracted), mostly sterile (not viable); abnormal pharyngeal pumping, abnormal muscle birefringence and ultrastructure; heterozygote e1463/+ is slowish, somewhat dumpy, muscle disorganization similar to but weaker than homozygote. Suppressed semidominantly by lev-11(x12); also unc-54 null alleles are epistatic to e1463. ES3(homozygote). NA(sd)1. Class 1 (loss-of-function) alleles: intragenic revertants of e1463, wildtype phenotype (probable null phenotype). Ref. 461, RW.

unc-91 See unc-40.

unc-92 See act-1.

unc-93 IIIL Class 3 allele, e1500sd: adults paralysed, Egl; slightly long; abnormal muscle; contract and relax when prodded ('rubber-band' phenotype). ES3, ME0. Poor mating in e1500/+ males. NA(Class 3)2 (n200: resembles e1500 but less severe phenotype). Class 1 (loss-of-function) alleles, n392: no phenotype alone but fails to complement e1500. Intragenic revertants of e1500 all have Class I phenotype, probably null alleles, e.g. e1500n234amber. NA(Class 1)>30. See also sup-9, sup-10, sup-11, sup-18. Ref. 486, 564, 898.

unc-94 IC — su177: slow, especially in adult; abnormal patchy muscle birefringence and ultrastructure, collections of thin and 'intermediate' filaments (birefringent clumps do not stain with anti-intermediate filament antibodies). ES2, ME2. NA1. Ref. 488, RW.

unc-95 IR — su106sd: very slow or paralysed, Egl; abnormal variable muscle birefringence, disorganized sarcomeres with collections of thin filaments. ES3. NA4 (su33amber (similar phenotypes, MEO), etc.). Ref. 488, RW.

unc-96 XL — su151: very slightly slow, slightly Egl, moves well; abnormal muscle birefringence and ultrastructure, collections of thin filaments near ends of cells. ES1, ME3. NA1. Ref. 488.

unc-97 XC — su110: limp paralysed adults, larvae paralysed, slightly curled; Egl; variable abnormal muscle birefringence; small, easily disrupted sarcomeres. ES3, MEO. NA1. Ref. 488.

unc-98 XC — su130: slow, somewhat Egl; abnormal muscle birefringence and ultrastructure, collections of thin filaments, poorly organized A and I bands. ES2, ME1. NA1. Ref. 488.

unc-99 — See unc-27.

unc-100 I — su149sd: slow, small, dumpyish; male bursa swollen; su149/+ is slightly slow and small. ES3, MEO. NA3 (su115 etc.). Ref. 488, CB.

unc-101 IR — m1des: coiler, very sluggish, moves poorly, slightly Egl, slightly short; resembles unc-75. Ref. DR.

unc-102 — See unc-1.

unc-103 IIIL — Class 3 allele, e1597sd: severe kinker, little movement, slight Egl; e1597/+ somewhat less severe kinker. ES3, MEO. NA2 (n500).
Class 1 (loss-of-function) alleles: intragenic revertants of e1597 e.g. e1597n1212, wildtype phenotype.
Ref. 914, CB.

unc-104 IILC — e1265: severe coiler; e1265/Df probably lethal. ES3, MEO. NA1. Ref. 715,DR.

unc-105 IIC — Class 3 (neomorphic) alleles, n490sd: small, hypercontracted rigid paralytic; very poor growth; abnormal muscle structure; n490/+ similar, less severe phenotype. ES3. NA(dominant)3 (n506 (MEO), n1274).
Class 1 (loss-of-function) alleles: intragenic revertants of Class 3 alleles are probably null alleles, e.g. n490n804: wildtype phenotype, no behavioral or muscular abnormality. NA(Class 1)>10.
Class 3 alleles also suppressed by mutations in many muscle genes e.g. unc-15, unc-22, unc-45, unc-54 etc.
Ref. 715, 914, 915, MT, TR.

unc-106 — See unc-6.

unc-107 V — ev411: slightly uncoordinated, partially shortened phasmid axons. Phenotype enhanced by one or two doses of enu-1(ev419): ev411; ev419 is severely uncoordinated, very short phasmid axons, cannot back. Ref. NW.

unc-108 I — n501dm: sluggish, poor backing. Similar phenotype in n501/+. ES3. NA1. Possibly haplo-insufficient locus. Ref. 914.

unc-109 I — n499dm: recessive lethal; n499/+ is uncoordinated.

unc-110 XRC — e1913dm: recessive lethal; e1913/+ is uncoordinated, thin, very sluggish, tends to shrink and relax when prodded. ES3, NA1. Ref. CB.

unc-111 VC — r346: moves well; disorganized body wall muscle; partial suppressor of unc-105(n490). ES1. NA1. Ref. TR.

unc-112 VC — r367: adults paralysed and thin, except for head region; Egl; disorganized body wall muscle; partial suppressor of unc-105(n490); young larvae nearly wild type (phenotypes very similar to unc-52). ES3(adult), ME0. NA1. Ref. TR.

unc-113 VC — r449: slightly slow; disorganized body wall muscle; partial suppressor of unc-105(n490). ES3. NA1. Ref. TR.

unc-114 VC — r476: paralysed; Egl; disorganized body wall muscle; partial suppressor of unc-105(n490). ES3, ME0. NA1. Ref. TR.

uvt: — Unidentified Vitellogenin-linked Transcripts. BL. No mutants known.

uvt-1 X — Abundant transcript present in larvae, adults. Ref. 873.

uvt-2 X — Rare transcript present in larvae, adult hermaphrodites. Ref. 873.

uvt-3 X — Transcript present in embryos, larvae. Ref. 873.

uvt-4 X — Abundant transcript present at all stages, less abundant in adults. Ref. 873.

uvt-5 X — Rare transcript present in larvae, adults. Ref. 873.

uvt-6 X — Rare transcript present in adult hermaphrodites. Ref. 873.

uvt-7 X — Transcript present in embryos, adult hermaphrodites. Ref. 873.

vab:		Variable ABnormal morphology.
vab-1	IILC	e2: notched head, variable dystrophy of ventral cephalic region, especially in L1; penetrance <70%. ES3/O, ME3. NA8 (e1063: penetrance 35%; e2027spo: penetrance >90%). Ref. CB.
vab-2	IVL	e96: notched head, especially in L1; resembles vab-1 alleles; penetrance 65%. ES3/O, ME3. NA3 (e141, e1208: penetrance 50%). Ref. CB.
vab-3	XRC	e648: notched head, especially in L1; 100% penetrance, dystrophy of ventral head regions, disorganized hypodermal and anterior sensory anatomy, non-chemotactic to NaCl; adult male tail variably deformed. ES3, MEO. NA2 (e1062: resembles e648). Ref. 214.
vab-4		See unc-23.
vab-6	IIIL	e697: dumpyish, lumpy appearance, especially in L1; sometimes twisted body. ES3 (L1), ME2. NA2 (e1023: resembles e697). Ref. 31, 608, CB.
vab-7	IIIR	e1562: hermaphrodite tail abnormal, sometimes twisted, tail whip never normal, often bobbed; sometimes uncoordinated with bent tail; adult male tail deformed. ES3, MEO. NA1. REf. 608.
vab-8	VRC	e1017: posterior half thin, pale, uncoordinated, anterior half normal; adult male tail anatomy vestigial; failure of posterior migration of CAN cells; rare animals have normal posterior morphology and one correctly placed CAN cell. ES3, MEO. Similar phenotypes in e1017/Df. NA2 (ct33 (resembles e1017, sometimes abnormal HSN migration)). Ref. CB, BW.
vab-9	IIC	e1744: slightly dumpy, tail whip knobbed at all stages except adult male (adult male tail tip slightly swollen); variably Egl. ES3 (larvae), ME2. NA3 (e1775, e2016: resemble e1744, slightly dumpier). Ref. CB, MT.
vab-10	IR	e698: degenerate head, bent dorsally or ventrally, penetrance 50%; generally poor growth; adult male tail invariably thin, fan and rays reduced in size. ES3/1, MEO. NA1. Ref. 608, CB.
vit:		VITellogenin structural genes (yolk protein genes). BL. No mutants known.
vit-1	X	Pseudogene, 95% homologous to vit-2. Ref. 803, 873.
vit-2	X	Structural gene for vitellogenin yp170B. Ref. 803, 873.
vit-3	X	Structural gene for vitellogenin yp170A. Ref. 803, 873.

vit-4 X Structural gene for vitellogenin yp170A; adjacent to vit-3. Ref. 803, 873.

vit-5 X Structural gene for vitellogenin yp170A; completely sequenced. Ref. 803, 873.

vit-6 IV Structural gene for vitellogenins yp115 and yp88 (cleaved from precursor protein VIT180). Ref. 808.

zyg: ZYGote defective: embryonic lethal. See also emb, let, ooc. BW.

zyg-1 IILC b1ts,mm: L1 shift-up results in embryonic arrest at 2-20 nuclei; first cleavage plane distorted, endomitosis. L1 shift-up results in abnormal gonadogenesis, male sterility. Viable at 16°. NA1. Ref. 373, 426, 574.

zyg-2 IC b10ts,mm: L4 shift-up results in embryonic arrest at 25 - 100 nuclei, early cleavage slow, some endomitosis; TSP before fertilization. L1 shift-up results in abnormal gonadogenesis, male sterility. Viable at 16°, male fertility low at 16°. NA1. Ref. 373, 426, 574.

zyg-3 IILC b18ts,sd,mn: L4 shift-up results in embryonic arrest at >200 nuclei, eggs lyse in vitro. TSP short, during early cleavage. L1 shift-up results in abnormal gonadogenesis, male sterility. Viable at 16°. NA1. Ref. 373, 426.

zyg-4 See emb-7.

zyg-5 II b89ts,nm: L4 shift-up results in embryonic arrest at <25 nuclei, few early cleavages, eggs lyse in vitro; TSP during cleavage; L1 shift-up results in larval arrest (L1 to L3). Slight dominant reduction in fertility. Viable at 16°. NA1. Ref. 426.

zyg-6 See emb-9.

zyg-7 IIIC b187ts,mn: L4 shift-up results in embryonic arrest?; L1 shift-up results in some L4 arrest, males fertile. Viable at 16°. NA1. Ref. 426.

zyg-8 IIIR b235ts,mnp: L4 shift-up results in embryonic arrest at 250 - 500 nuclei, first cleavage abnormal giving 3 or 4 blastomeres; phenotype paternally rescued by b235/+ sperm cytoplasm. L1 shift-up results in viable adult, F1 embryonic arrest, male sterility. Viable at 16°. NA1. Ref. 426, HH.

zyg-9 IIC b244ts,mm: L4 shift-up results in embryonic arrest at 1 - 200 nuclei; first cleavage variably abnormal, skewed; some endomitosis. TSP short, approximately during first cleavage. L1 shift-up results in viable adult, F1 embryonic arrest, does not affect male fertility. Viable at 16°. NA1. Ref. 426, 574.

zyg-10 IIILC b261ts,mn?: L4 shift-up results in embryonic arrest, abnormal skewed first cleavage, small P1 blastomere; TSP during oogenesis; L1 shift-up results in viable adult, F1 embryonic arrest, does not affect male fertility except in cross with b261. Viable at 16°. NA1. Ref. 426.

zyg-11 IIC b2ts,mm: L4 shift-up results in embryonic arrest at 1 - 150 nuclei, symmetrical first cleavage; TSP before fertilization; L1 shift-up results in F1 embryonic arrest. Viable at 16°. NA5 (mn40amber (non-ts oocyte-defective, eggs fertilized but do not hatch), etc.). Ref. 426, 574.

zyg-12 IILC b?ts,mm,spo(pka ber): L4 shift-up results in embryonic arrest at >200 nuclei, normal early cleavages; TSP before fertilization; L1 shift-up results in abnormal gonadogenesis, male sterility. Viable at 16°, 20°. Isolated from Bergerac strain BO. NA1. Ref. 426.

LIST OF REARRANGEMENTS OTHER THAN DUPLICATIONS AND DEFICIENCIES*

eT1(III;V) Reciprocal; the half-translocation eT1(III) (which disjoins from normal chromosome III) consists of the left half of III, including sma-3, joined to the left half of V, including dpy-11; the reciprocal half-translocation eT1(V) (which disjoins from normal chromosome V) consists of the right half of III, including sma-2, joined to the right half of V, including unc-42; the III breakpoint is probably in unc-36; translocation homozygotes are Unc-36; crossing over is suppressed on right half of III and left half of V in translocation heterozygotes. Ref. 554, 646, 750, 762.

mnC1 Putative intrachromosomal rearrangement of II; suppresses crossing over in heterozygotes over the dpy-10-unc-52 interval. Ref. 163, 164, 715, 762.

mnT2(II;X) Reciprocal; homozygotes rare or inviable; mnT2-bearing males are fertile; the smaller half-translocation, which can be maintained as a free duplication called mnDp11, consists of left tip of X and right end of II; reciprocal half-translocation is called mnT11, can be maintained in absence of mnDp11 in heterozygotes, disjoins from II, is dominant Him (32% males) and suppresses crossing over on X. Ref. 577, 696.

mnT3(II;X) Homozygotes rare or inviable; mnT3-bearing males are fertile; dominant Him (~35% males); dominant crossover suppressor of X. Ref. 577.

mnT6(II;X) Homozygotes rare or inviable; mnT6-bearing males are fertile; dominant Him (~35% males); dominant crossover suppressor of X. Ref. 577, 676, 739, 861.

*For explanation of nomenclature, see Part A; reference numbers refer to the list of Part B. Compiled by R. Herman.

mnT7(IV;X)	Homozygotes inviable; mnT7-bearing males fertile; dominant Him (~35% males). Ref. 577.
mnT8(V;X)	Homozygotes inviable; mnT8-bearing males fertile; dominant Him (~35% males). Ref. 577.
mnT9(I;X)	Homozygotes inviable; mnT9-bearing males fertile; dominant Him (~35% males); fails to complement unc-54. Ref. 577.
mnT10(V;X)	Reciprocal; heterozygotes Him (~20% male); homozygotes viable and Him; either half-translocation can exist in heterozygous form in absence of the other; mnT10(V) consists of left tip of X and all but left end of V. Reciprocal half-translocation mnT10(X), is responsible for Him trait. Ref. 577.
mnT11(II;X)	See mnT2(II;X). Ref. 577, 696.
mnT12(IV;X)	Fusion chromosome, with breakpoints near left end of IV and right end of X. Homozygotes viable. Heterozygotes show nondisjunction of X (27% males) and IV. Ref. 696, 864.
nT1(IV;V)	Homozygotes viable; recessive vulvaless (and hence Egl); crossing over is suppressed on right arm of IV(unc-17 to dpy-4) and left arm of V(unc-60 to dpy-11) in translocation heterozygotes. Ref. 654, 762.
sT1(III;X)	Homozygotes inviable; sT1-bearing males are fertile; crossing over is suppressed on right of III in translocation heterozygotes; translocation is marked with dpy-18(e364). Ref. 750.
sT2(IV;V)	Homozygotes inviable; crossing over suppressed on left arm of V in translocation heterozygotes; translocation is marked with unc-46(e177). Ref. 750.
sT3(III)	A half-translocation involving III and V (maintained in association with eT1); marked with unc-46(e177). Ref. 750.
sT4(III)	A half-translocation involving III and V (maintained in association with eT1); marked with unc-46(e177). Ref. 750.
szT1(I;X)	Dominant Him (~8% males); marked with lon-2(e678); homozygous inviable; szT1-bearing males are viable and weakly fertile; dominant suppressor of crossing over for left half of I and dpy-8-let-2 interval of X. Ref. 576, 883.

Part C: Genetic Map

Each of the six linkage groups is divided into two or three sections (that may be duplicated, trimmed at the black diamond symbols, and joined to make a complete chromosome). The gene cluster area of each autosome (marked by a "black box") is drawn on an enlarged scale on a separate page. Many genes that lie within the cluster appear only on the cluster enlargement. In addition to the main map and cluster sections, an additional page shows a section of LG I containing a set of lethal genes located by 2- and 3-factor mapping in the neighborhood of *dpy-5*. Other features of the map are as follows:

1. The scale of the map is 1 map unit to 0.5 centimeter for the main map, and 0.1 map unit to 0.7 centimeter for the autosomal clusters, except the LG V cluster (0.2 map unit to 0.7 centimeter), and the supplemental section of LG I (0.5 map unit to 0.7 centimeter). One map unit = 1% recombination. Each page of the map is identified by the appropriate Roman numeral or X, and includes a map scale bar. LG II and LG IV are not drawn at full length, but the positions of loci at the top end are given in a note on the map.
2. Two restriction fragment length polymorphisms (RFLPs) and the 5S RNA genes appear on the map. RFLPs are named with the CGC allele designation of the originating laboratory, followed by a capital P and the isolation number (e.g., *sP1*). 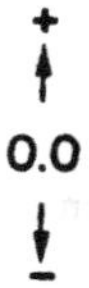
3. Linkage groups are displayed vertically, with the upper or + arms, and the lower, or − arms, extending from zero points defined by standard reference markers. Positions on the map (in map units) are referred to according to this convention. 
4. Duplications (Dp) and deficiencies (Df) are drawn to the left of the main map line; duplications are indicated by a double bar, and deficiencies are indicated by a single bar. Endpoints are placed within the position of flanking markers on the main map line as determined by complementation testing. Dotted lines within the Df or Dp intervals correspond to positions of genes on the line that have not been tested for complementation.
5. A few commonly used mapping markers positioned on the line are reproduced at the far right margin of the map at their correct plus-minus positions. These allow placement of a ruler across the map to facilitate reading endpoints of intervals off the line.

Conventions for gene placement and map symbols:

1. Genes are positioned on the line only when they have been

ordered with respect to their nearest neighbors on the line by 3-factor (3F), Df, or Dp data.

2. Distances between genes on the line are taken from the best available 2-factor (2F) data. If 3F mapping positions a gene between two markers and no 2F data exist, the gene is placed midway between the two markers, or the position is determined by interpolation from 3F data.
3. Genes are stacked at a single position on the line within a brace if they map to the same position by 3F or deficiency mapping, but have not been ordered with respect to each other (example: *unc-11* and *unc-73* on LG I).
4. Genes that have been ordered with respect to each other, but for which no relevant 2F data exist, are shown separated in their correct order on a branched line pointing to a single position on the line (example: *bli-3, unc-35,* and *egl-30* on LG I).

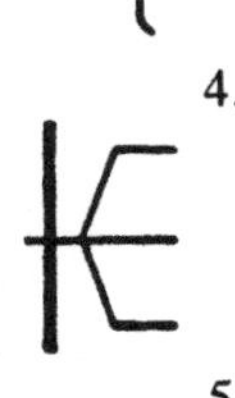

5. Genes placed off the line to the right are indicated on bars representing intervals with endpoints of four types:

   a. A solid single cross line is an endpoint determined by a 3F result with a gene on the line. The endpoint is placed *within* the position of the line marker (example: *anc-1 I* has been shown to lie between *unc-11* and *dpy-5* but has not been ordered relative to any intervening genes).

   b. A double cross line is an endpoint determined by complementation tests with deficiencies or duplications, placed *at* the position of the corresponding Df/Dp endpoint (example: *egl-31 I* has been shown to lie within *nDf24;* therefore, the double-line endpoints of the *egl-31* bar correspond to the *nDf24* endpoints).

   c. An arrow indicates that the gene in question was not separated from a marker on the line in a 3F cross (example: *fer-6 I* was shown to be below (right of) *dpy-5,* and the arrow at *unc-75* indicates that *fer-6* is either close above *unc-75* or below *unc-75*.
   d. A solid dot represents a limit of a 2F 95% confidence interval. 2F dots are used both to represent genes placed by 2F data in conjunction with very general 3F data, and to modify gene placement on 3F or Df/Dp bars. For example, *ace-2 I* has been shown to be above *dpy-5* but not ordered relative to any other marker, so it is placed on a 2F bar based on its distance from *dpy-5*. In a different case, *emb-19 I* has been shown to be below *dpy-5* and either very close above *unc-13* or below *unc-13*. 2F data indicate, with 95% confidence, that it

could be only as far below *unc-13* as the 2F dot shown. In a third case, *zyg-12 II* has been shown by 3F to lie between *unc-85* and *dpy-10,* and the 2F confidence interval is drawn overlapping this 3F interval. A 2F dot is superimposed on the lower 3F endpoint, even though the 2F confidence interval extends past it.

6. Genes with 95% confidence intervals as large as half a linkage group are noted in the CGC Gene Map Positions list but are left off the map figure.

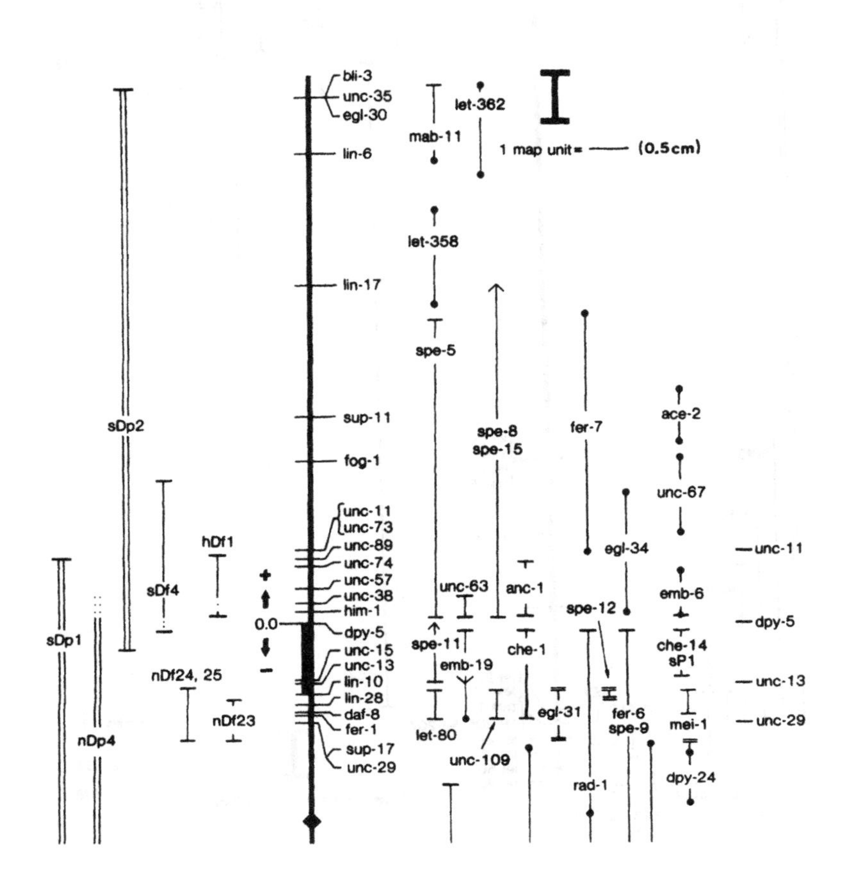
bli-3
unc-35
egl-30
lin-6
lin-17
sup-11
fog-1
unc-11
unc-73
unc-89
unc-74
unc-57
unc-38
him-1
0.0
dpy-5
unc-15
unc-13
lin-10
lin-28
daf-8
fer-1
sup-17
unc-29
sDp2
sDp1
nDp4
sDf4
hDf1
nDf24, 25
nDf23
mab-11
let-362
1 map unit = —— (0.5cm)
let-358
spe-5
spe-8
spe-15
fer-7
ace-2
unc-67
egl-34
emb-6
unc-63
anc-1
spe-12
spe-11
emb-19
che-1
che-14
sP1
let-80
unc-109
egl-31
fer-6
spe-9
mei-1
rad-1
dpy-24
unc-11
dpy-5
unc-13
unc-29

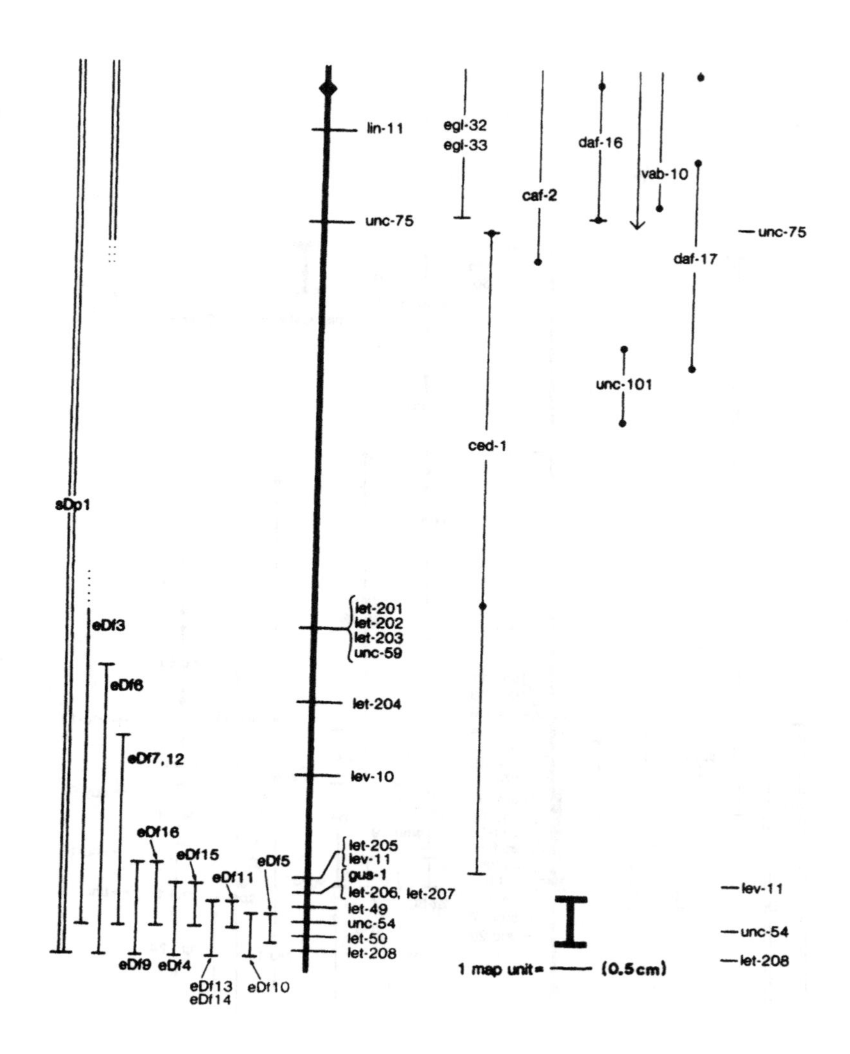
lin-11
unc-75
let-201
let-202
let-203
unc-59
let-204
lev-10
let-205
lev-11
gus-1
let-206, let-207
let-49
unc-54
let-50
let-208
egl-32
egl-33
caf-2
daf-16
vab-10
unc-75
daf-17
unc-101
ced-1
sDp1
eDf3
eDf6
eDf7,12
eDf16
eDf15
eDf11
eDf5
eDf9
eDf4
eDf13
eDf14
eDf10
lev-11
unc-54
let-208
1 map unit = (0.5 cm)

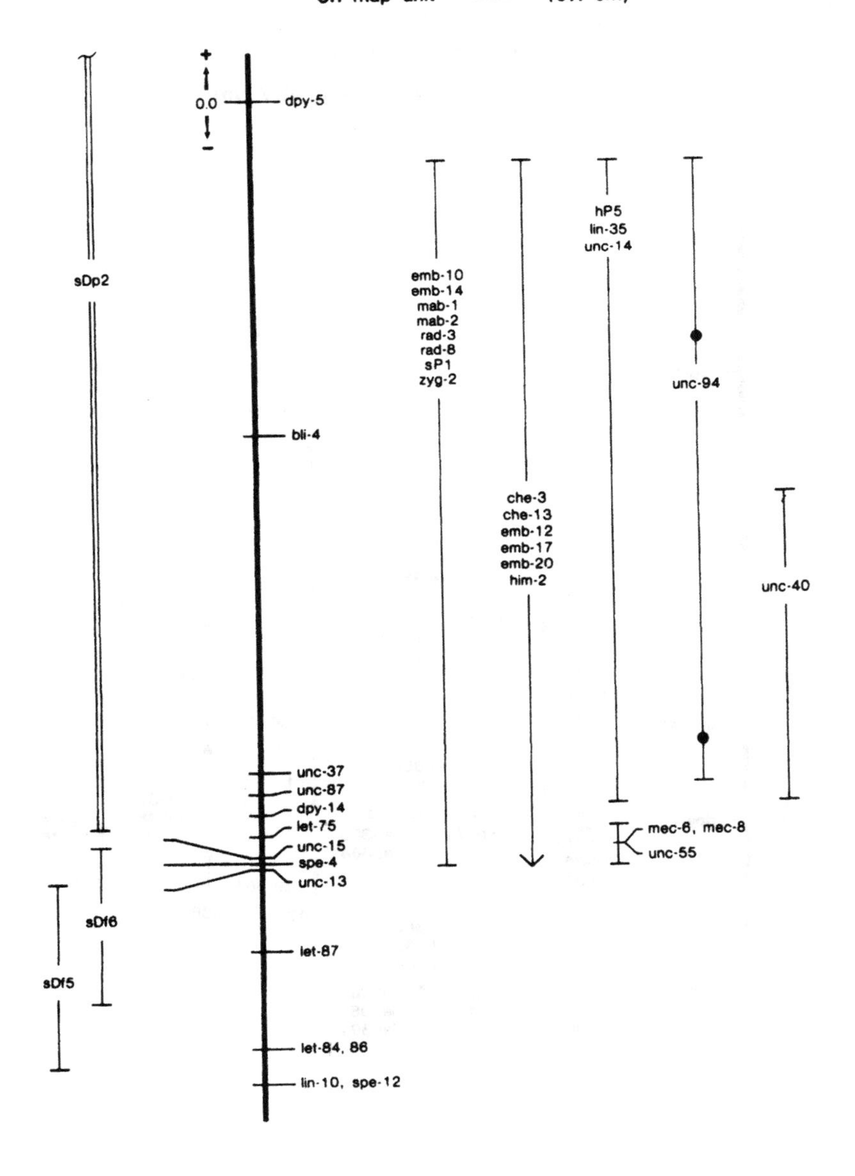
LG I CLUSTER
0.1 map unit = —— (0.7 cm)
+
0.0
−
dpy-5
sDp2
emb-10
emb-14
mab-1
mab-2
rad-3
rad-8
sP1
zyg-2
hP5
lin-35
unc-14
unc-94
bli-4
che-3
che-13
emb-12
emb-17
emb-20
him-2
unc-40
unc-37
unc-87
dpy-14
let-75
unc-15
spe-4
unc-13
mec-6, mec-8
unc-55
sDf6
sDf5
let-87
let-84, 86
lin-10, spe-12

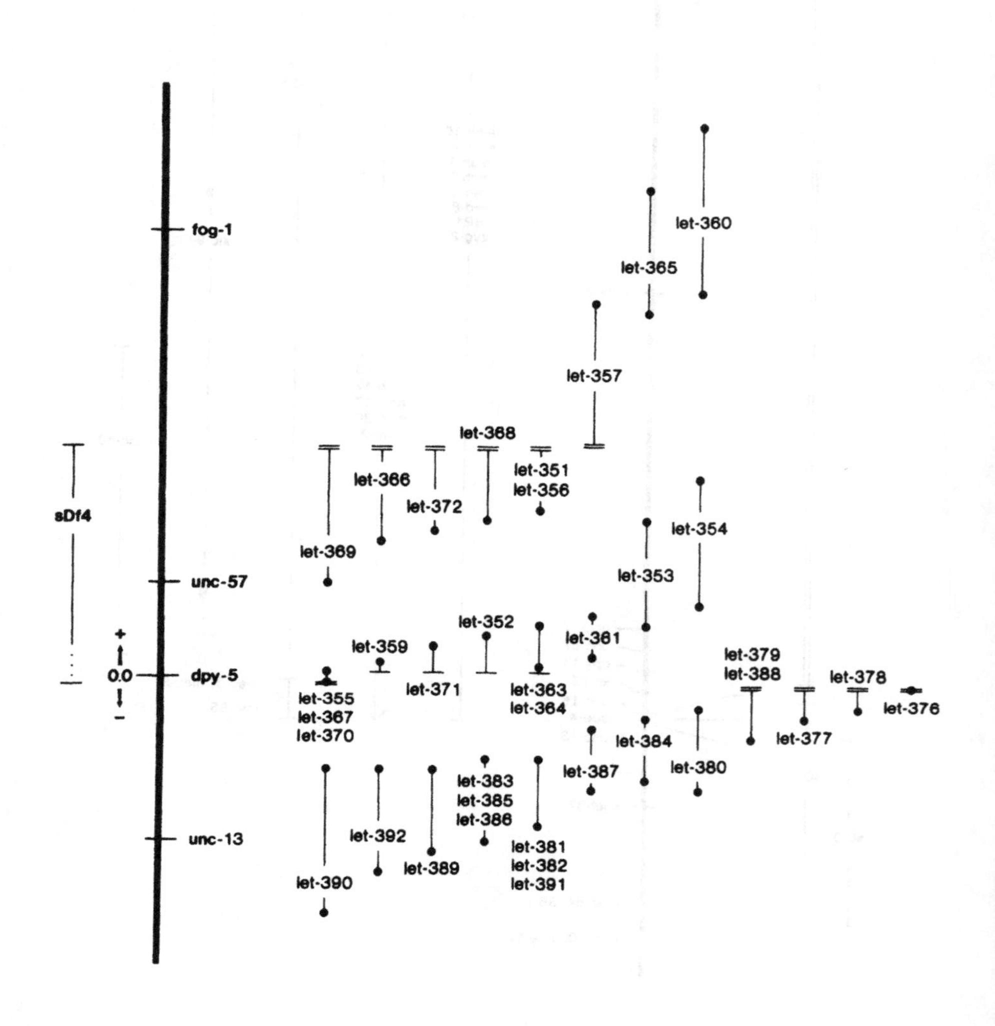
LG I
Lethals mapped to dpy-5
0.5 map unit = (0.7 cm)
fog-1
unc-57
dpy-5
unc-13
0.0
sDf4
let-360
let-365
let-357
let-368
let-366
let-372
let-351
let-356
let-369
let-354
let-353
let-352
let-359
let-361
let-371
let-363
let-364
let-379
let-388
let-378
let-376
let-355
let-367
let-370
let-377
let-384
let-380
let-387
let-383
let-385
let-386
let-392
let-389
let-381
let-382
let-391
let-390

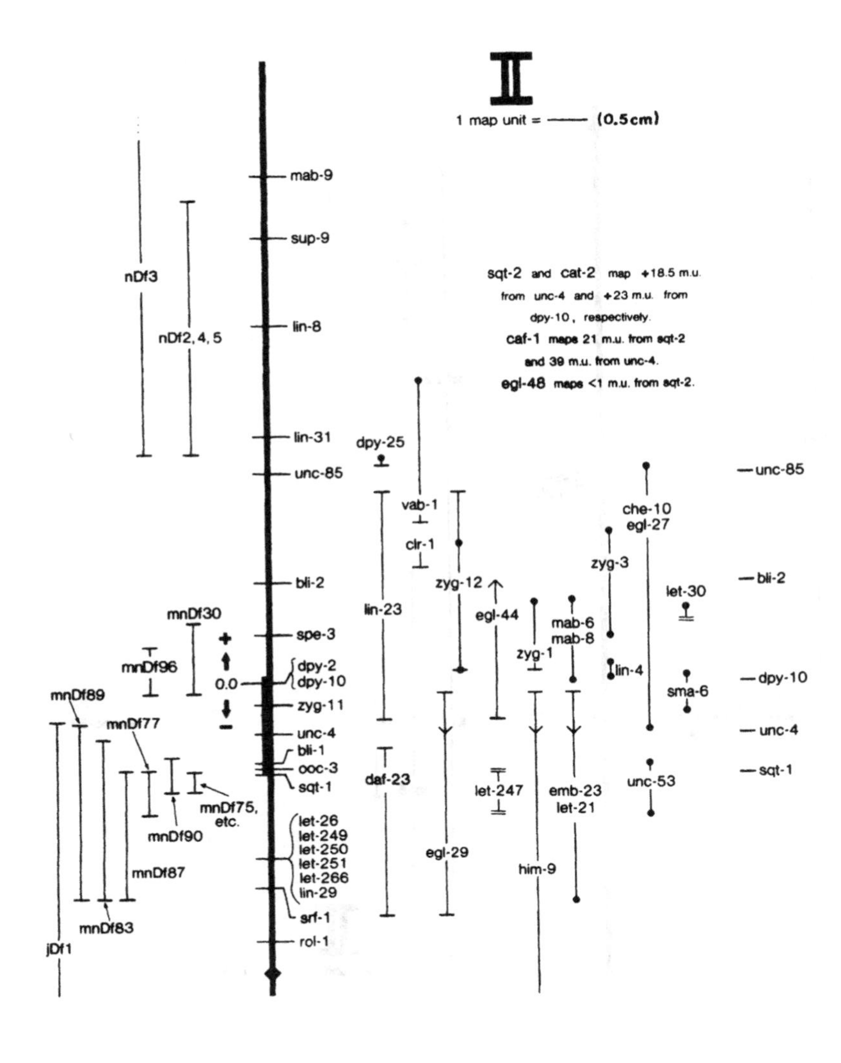
II
1 map unit = —— (0.5cm)
sqt-2 and cat-2 map +18.5 m.u. from unc-4 and +23 m.u. from dpy-10, respectively.
caf-1 maps 21 m.u. from sqt-2 and 39 m.u. from unc-4.
egl-48 maps <1 m.u. from sqt-2.
mab-9
sup-9
nDf3
lin-8
nDf2, 4, 5
lin-31
dpy-25
unc-85
vab-1
clr-1
che-10
egl-27
zyg-3
bli-2
zyg-12
egl-44
let-30
lin-23
mnDf30
spe-3
mab-6
mab-8
zyg-1
mnDf96
dpy-2
dpy-10
lin-4
sma-6
0.0
mnDf89
zyg-11
mnDf77
unc-4
bli-1
ooc-3
sqt-1
daf-23
let-247
emb-23
let-21
unc-53
mnDf75, etc.
mnDf90
let-26
let-249
let-250
let-251
let-266
lin-29
egl-29
him-9
mnDf87
srf-1
mnDf83
rol-1
jDf1

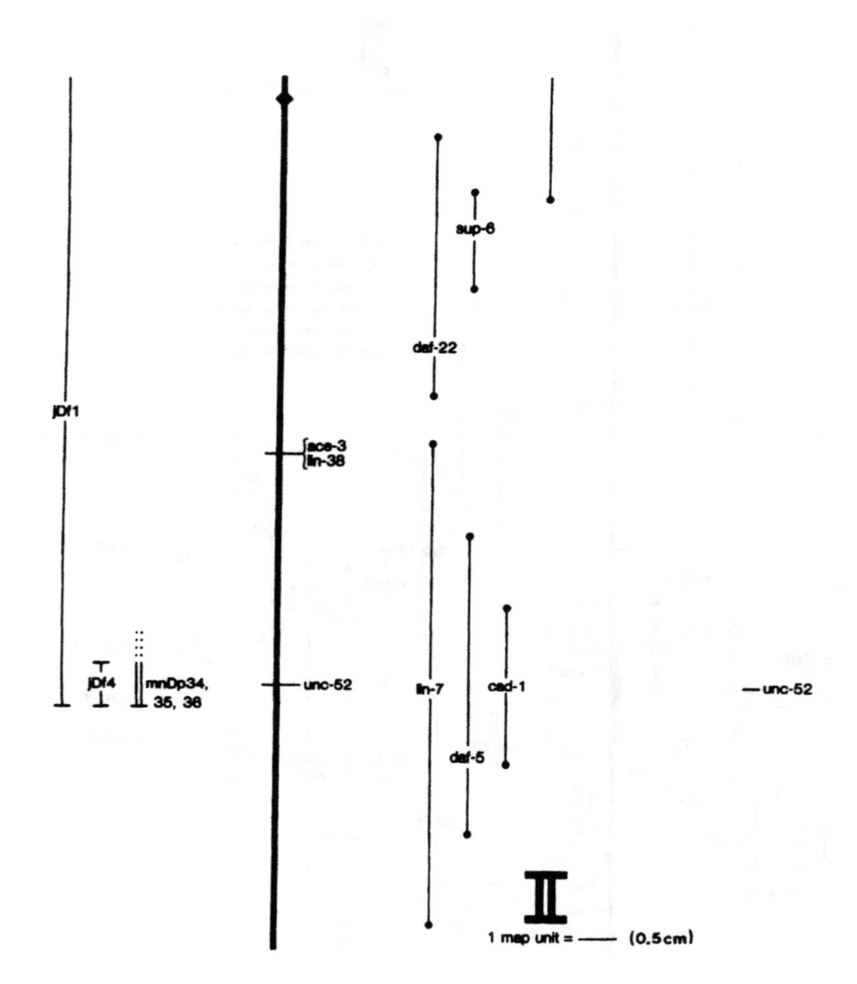
sup-6
daf-22
jDf1
ace-3
lin-38
jDf4
mnDp34,
35, 36
unc-52
lin-7
cad-1
unc-52
daf-5
II
1 map unit = —— (0.5cm)

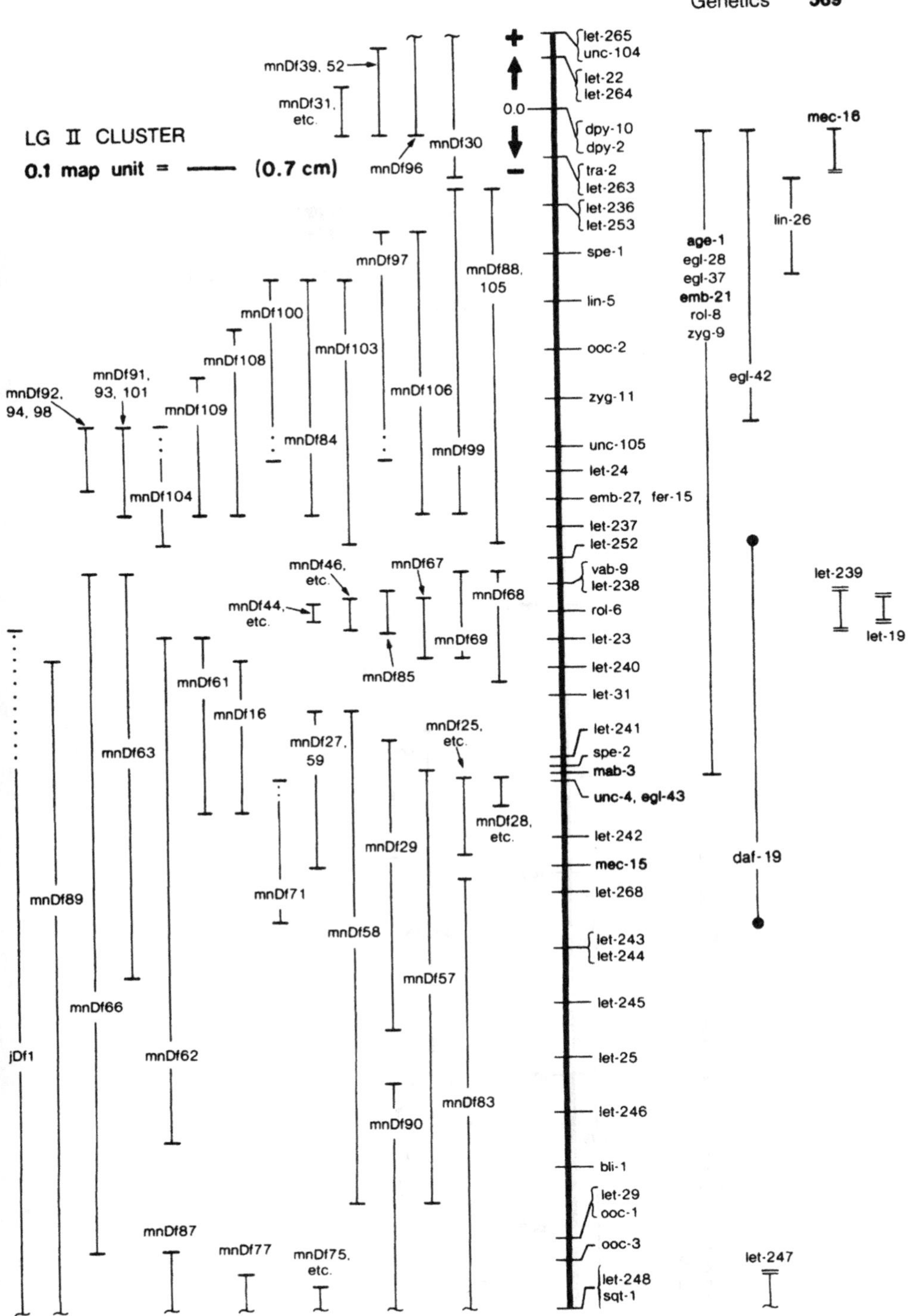
LG II CLUSTER
0.1 map unit = —— (0.7 cm)
+
0.0
−
let-265
unc-104
let-22
let-264
dpy-10
dpy-2
tra-2
let-263
let-236
let-253
spe-1
lin-5
ooc-2
zyg-11
unc-105
let-24
emb-27, fer-15
let-237
let-252
vab-9
let-238
rol-6
let-23
let-240
let-31
let-241
spe-2
mab-3
unc-4, egl-43
let-242
mec-15
let-268
let-243
let-244
let-245
let-25
let-246
bli-1
let-29
ooc-1
ooc-3
let-248
sqt-1
mec-16
lin-26
age-1
egl-28
egl-37
emb-21
rol-8
zyg-9
egl-42
daf-19
let-239
let-19
let-247
mnDf39, 52
mnDf31, etc.
mnDf96
mnDf30
mnDf88, 105
mnDf97
mnDf100
mnDf103
mnDf108
mnDf106
mnDf91, 93, 101
mnDf92, 94, 98
mnDf109
mnDf84
mnDf99
mnDf104
mnDf46, etc.
mnDf67
mnDf44, etc.
mnDf68
mnDf69
mnDf85
mnDf61
mnDf16
mnDf63
mnDf27, 59
mnDf25, etc.
mnDf28, etc.
mnDf29
mnDf89
mnDf71
mnDf58
mnDf57
mnDf66
jDf1
mnDf62
mnDf83
mnDf90
mnDf87
mnDf77
mnDf75, etc.

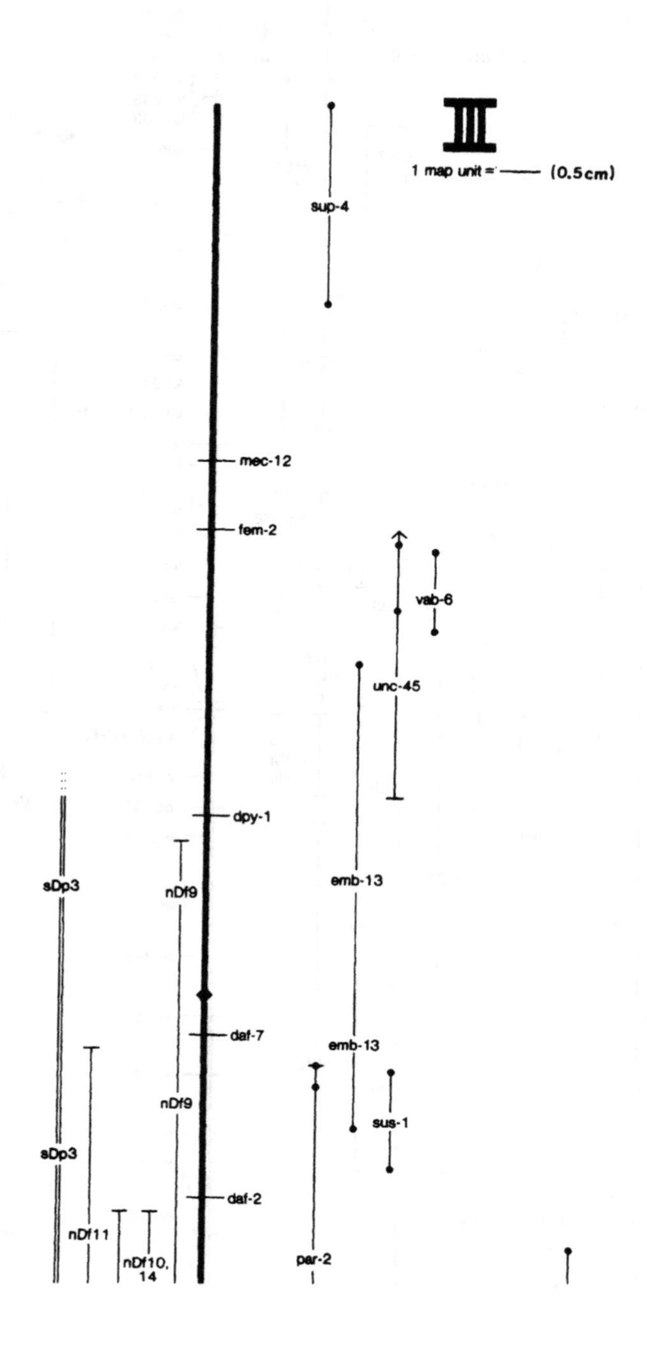
III
1 map unit = (0.5cm)
sup-4
mec-12
fem-2
vab-6
unc-45
dpy-1
sDp3
nDf9
emb-13
daf-7
emb-13
nDf9
sus-1
sDp3
daf-2
nDf11
nDf10, 14
par-2

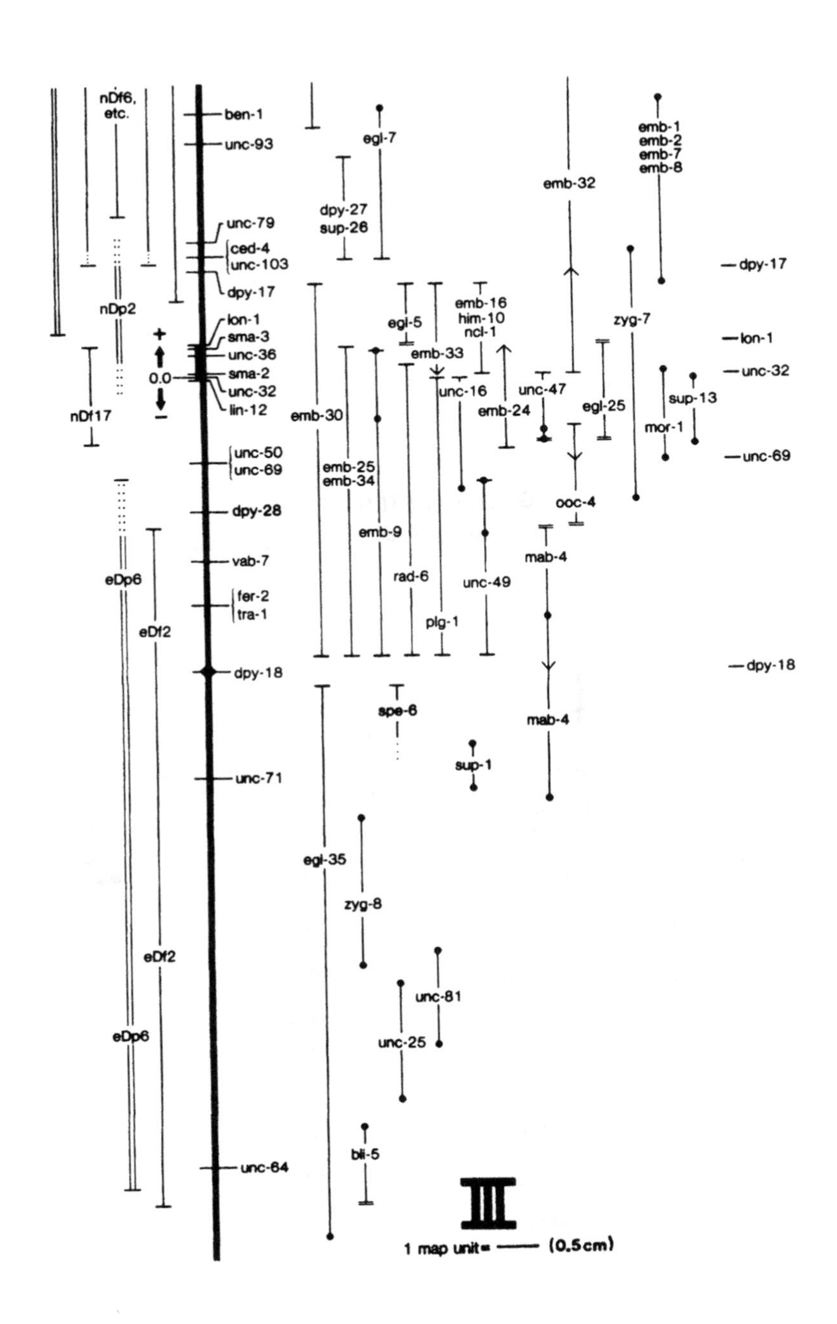
nDf6, etc.
ben-1
unc-93
egl-7
dpy-27
sup-26
emb-32
emb-1
emb-2
emb-7
emb-8
unc-79
ced-4
unc-103
dpy-17
nDp2
lon-1
sma-3
unc-36
sma-2
unc-32
lin-12
0.0
+
−
nDf17
emb-30
egl-5
emb-33
emb-16
him-10
ncl-1
unc-16
emb-24
unc-47
egl-25
zyg-7
sup-13
mor-1
dpy-17
lon-1
unc-32
unc-69
unc-50
unc-69
emb-25
emb-34
dpy-28
emb-9
ooc-4
unc-49
rad-6
mab-4
vab-7
eDp6
fer-2
tra-1
eDf2
plg-1
dpy-18
dpy-18
spe-6
mab-4
sup-1
unc-71
egl-35
zyg-8
eDf2
unc-81
unc-25
eDp6
bli-5
unc-64
III
1 map unit = —— (0.5cm)

LG III CLUSTER
0.1 map unit = —— (0.7 cm)

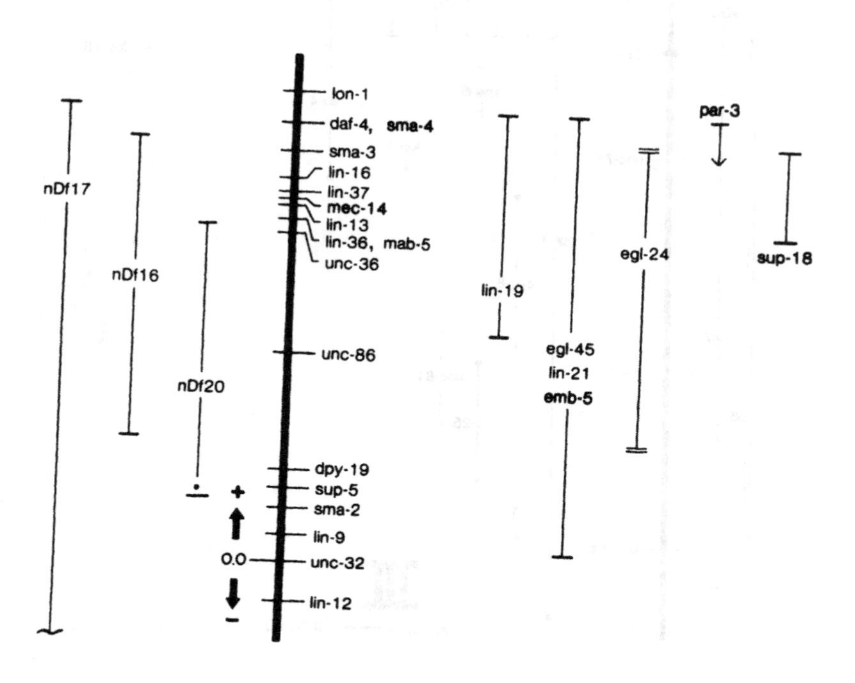

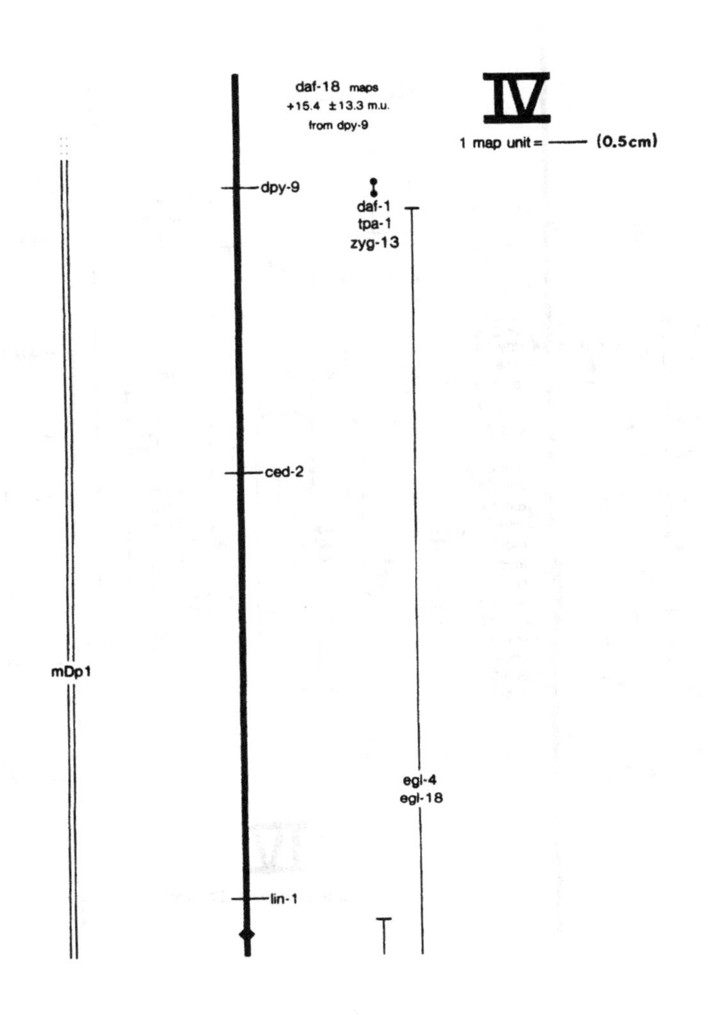
daf-18 maps
+15.4 ±13.3 m.u.
from dpy-9
IV
1 map unit = —— (0.5cm)
dpy-9
daf-1
tpa-1
zyg-13
ced-2
mDp1
egl-4
egl-18
lin-1

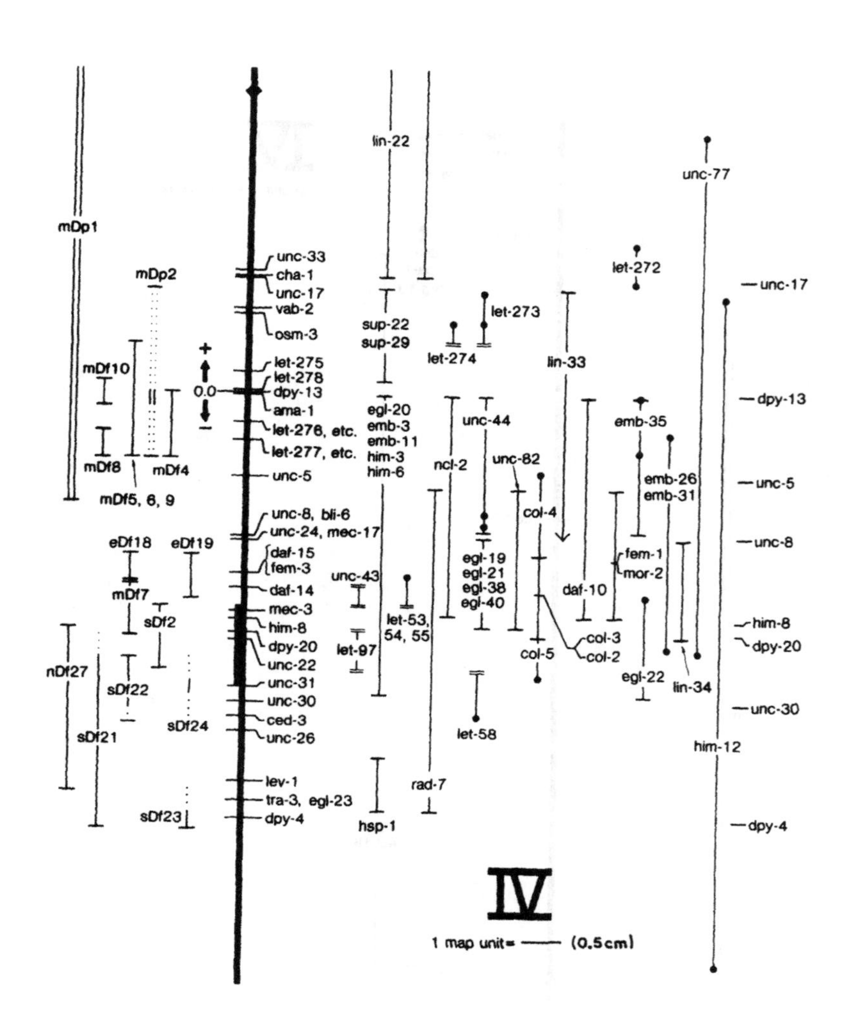

mDp1
mDp2
mDf10
mDf8
mDf4
mDf5, 6, 9
eDf18
eDf19
mDf7
sDf2
nDf27
sDf22
sDf21
sDf24
sDf23
+
0.0
−
unc-33
cha-1
unc-17
vab-2
osm-3
let-275
let-278
dpy-13
ama-1
let-276, etc.
let-277, etc.
unc-5
unc-8, bli-6
unc-24, mec-17
daf-15
fem-3
daf-14
mec-3
him-8
dpy-20
unc-22
unc-31
unc-30
ced-3
unc-26
lev-1
tra-3, egl-23
dpy-4
lin-22
sup-22
sup-29
egl-20
emb-3
emb-11
him-3
him-6
unc-43
let-97
let-53,
54, 55
hsp-1
rad-7
let-274
let-273
ncl-2
unc-44
egl-19
egl-21
egl-38
egl-40
let-58
unc-82
col-4
col-5
col-3
col-2
lin-33
daf-10
let-272
emb-35
emb-26
emb-31
fem-1
mor-2
egl-22
lin-34
unc-77
him-12
unc-17
dpy-13
unc-5
unc-8
him-8
dpy-20
unc-30
dpy-4
IV
1 map unit= —— (0.5cm)

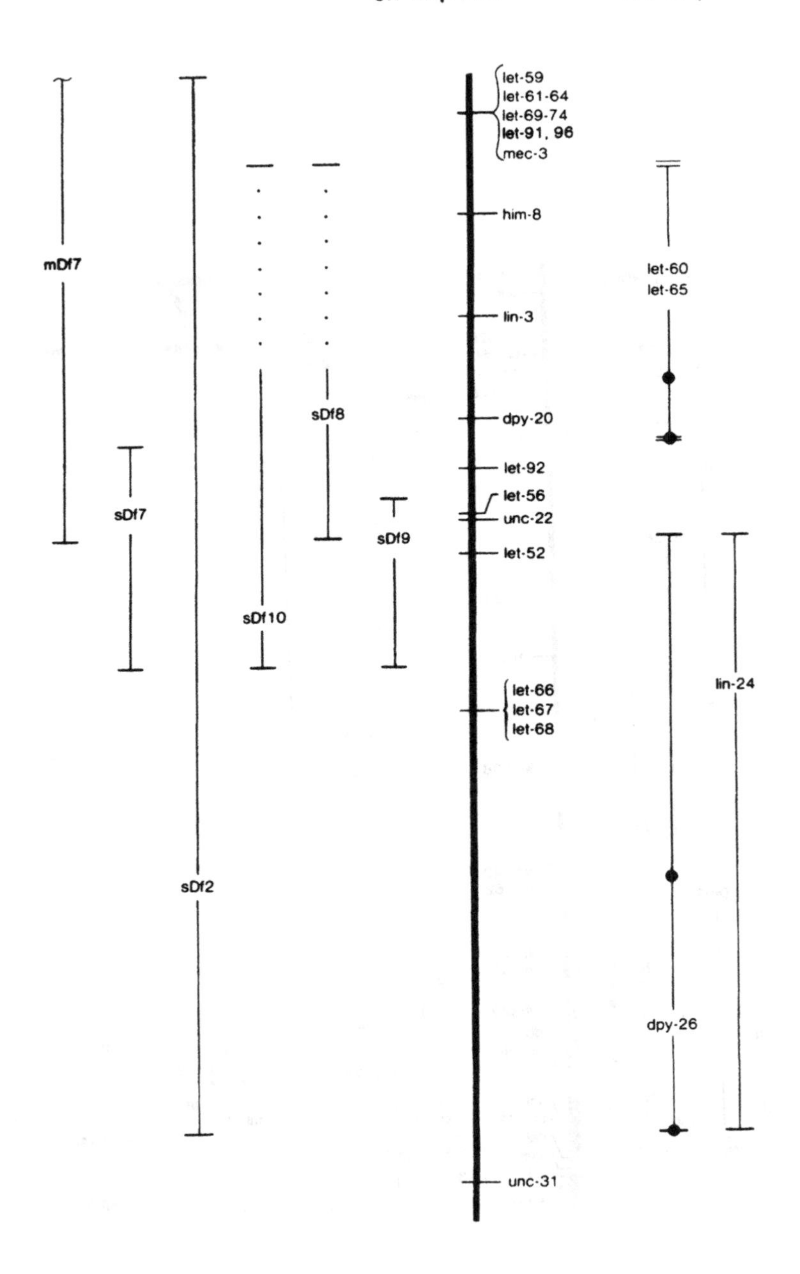
LG IV CLUSTER
0.1 map unit = —— (0.7 cm)
let-59
let-61-64
let-69-74
let-91, 96
mec-3
him-8
lin-3
dpy-20
let-92
let-56
unc-22
let-52
let-66
let-67
let-68
unc-31
mDf7
sDf2
sDf7
sDf8
sDf9
sDf10
let-60
let-65
dpy-26
lin-24

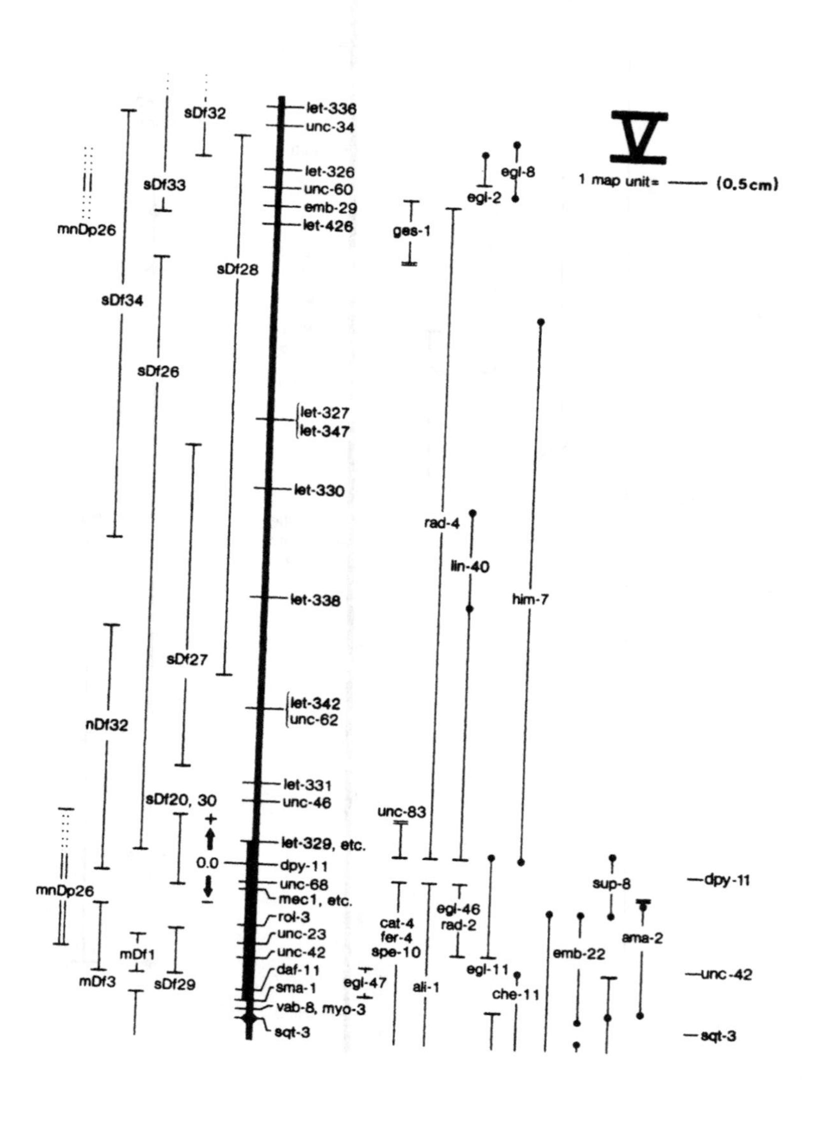
V
1 map unit= (0.5cm)
let-336
unc-34
let-326
unc-60
emb-29
let-426
let-327
let-347
let-330
let-338
let-342
unc-62
let-331
unc-46
let-329, etc.
dpy-11
unc-68
mec1, etc.
rol-3
unc-23
unc-42
daf-11
sma-1
vab-8, myo-3
sqt-3
0.0
sDf32
sDf33
mnDp26
sDf34
sDf28
sDf26
sDf27
nDf32
sDf20, 30
mDf1
mDf3
sDf29
ges-1
egl-2
egl-8
rad-4
lin-40
him-7
unc-83
cat-4
fer-4
spe-10
egl-47
ali-1
egl-46
rad-2
egl-11
che-11
emb-22
sup-8
ama-2
—dpy-11
—unc-42
—sqt-3

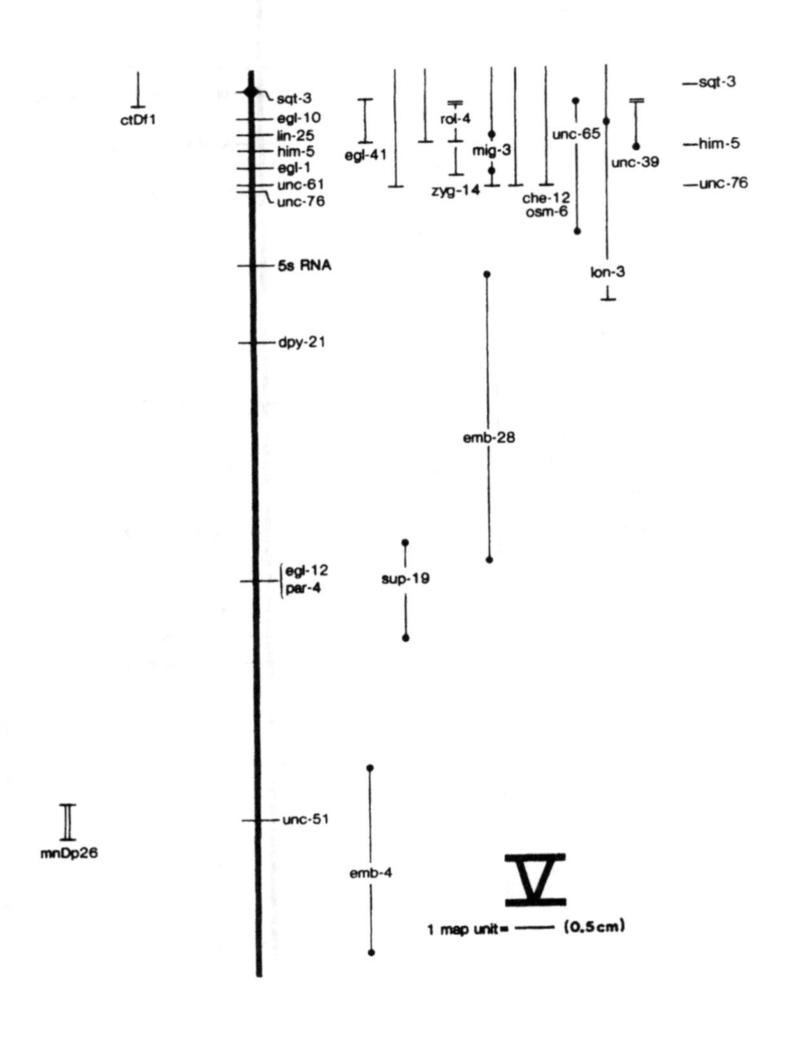
ctDf1
sqt-3
egl-10
lin-25
him-5
egl-1
unc-61
unc-76
5s RNA
dpy-21
egl-12
par-4
unc-51
mnDp26
egl-41
rol-4
mig-3
zyg-14
che-12
osm-6
unc-65
unc-39
lon-3
emb-28
sup-19
emb-4
—sqt-3
—him-5
—unc-76
V
1 map unit = —— (0.5cm)

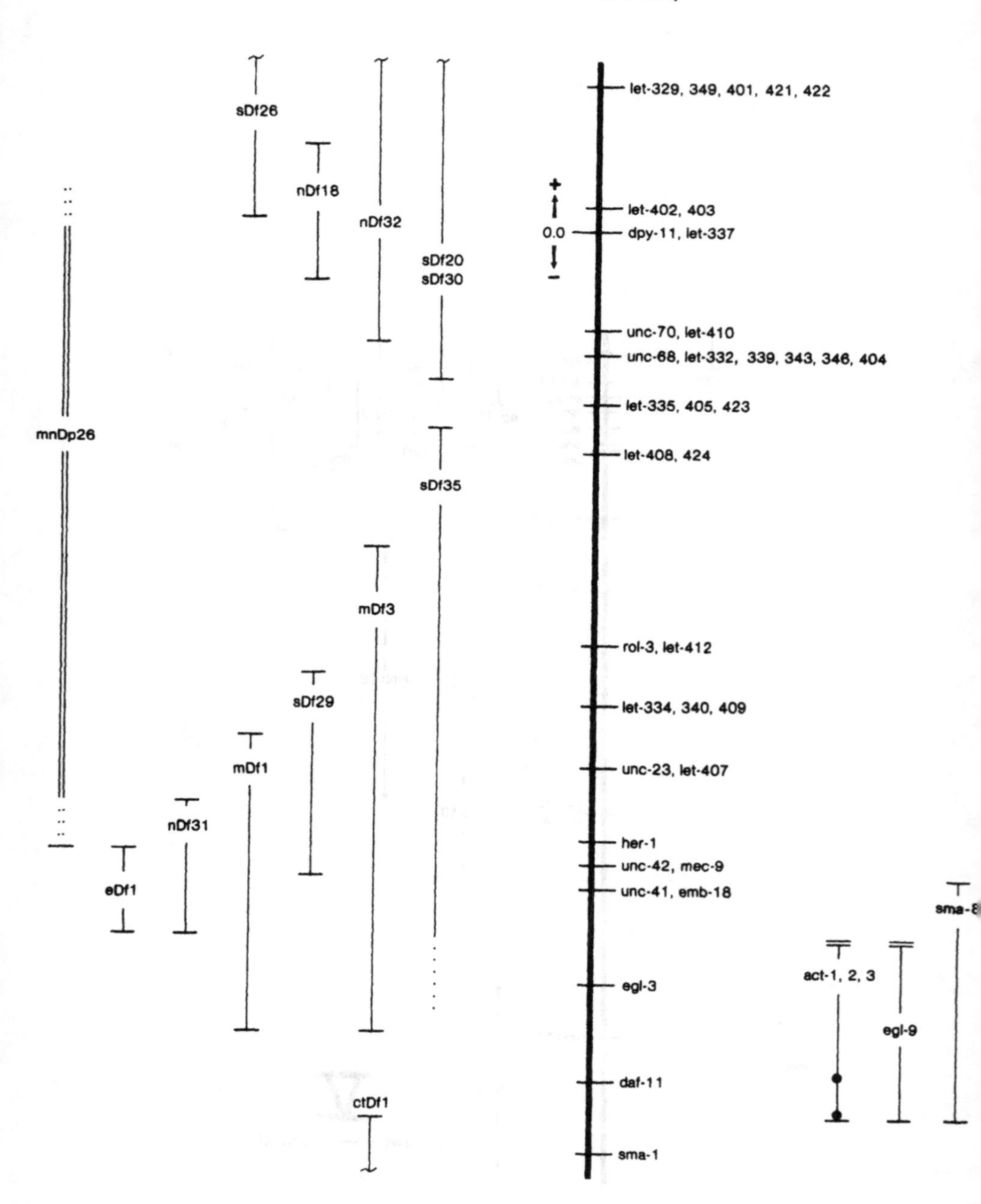
LG V CLUSTER
0.2 map unit = —— (0.7 cm)
let-329, 349, 401, 421, 422
let-402, 403
dpy-11, let-337
0.0
+
−
unc-70, let-410
unc-68, let-332, 339, 343, 346, 404
let-335, 405, 423
let-408, 424
rol-3, let-412
let-334, 340, 409
unc-23, let-407
her-1
unc-42, mec-9
unc-41, emb-18
egl-3
daf-11
sma-1
sDf26
nDf18
nDf32
sDf20
sDf30
sDf35
mDf3
sDf29
mDf1
nDf31
eDf1
mnDp26
ctDf1
act-1, 2, 3
egl-9

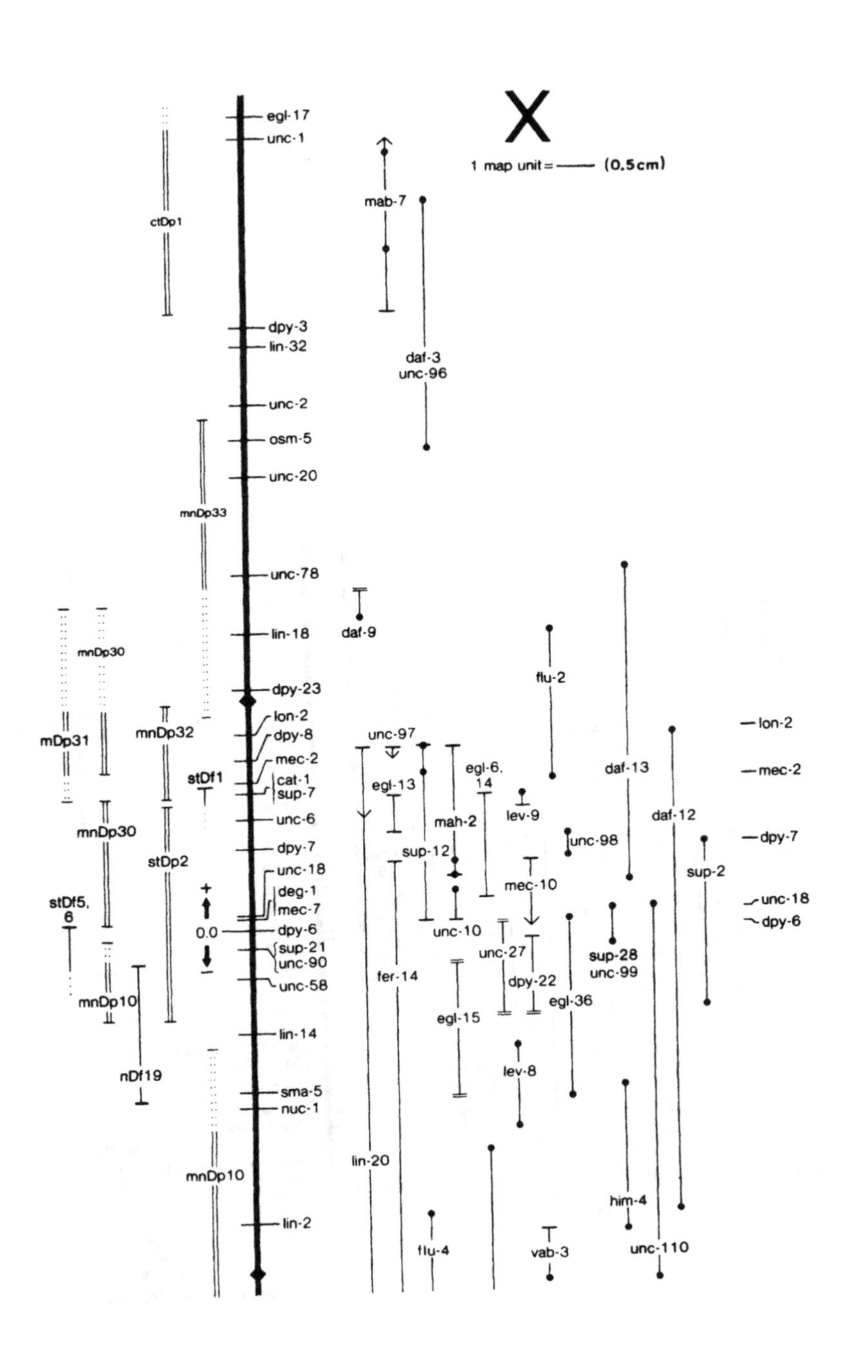
X
1 map unit = —— (0.5cm)
egl-17
unc-1
ctDp1
mab-7
dpy-3
lin-32
daf-3
unc-96
unc-2
osm-5
unc-20
mnDp33
unc-78
daf-9
lin-18
mnDp30
flu-2
dpy-23
lon-2
dpy-8
mec-2
cat-1
sup-7
unc-6
dpy-7
unc-18
deg-1
mec-7
dpy-6
sup-21
unc-90
unc-58
mDp31
mnDp32
stDf1
mnDp30
stDp2
stDf5, 6
0.0
mnDp10
nDf19
unc-97
egl-13
egl-6, 14
mah-2
lev-9
sup-12
unc-98
mec-10
unc-10
unc-27
dpy-22
egl-15
fer-14
sup-28
unc-99
egl-36
daf-13
daf-12
sup-2
lon-2
mec-2
dpy-7
unc-18
dpy-6
lin-14
lev-8
sma-5
nuc-1
mnDp10
lin-20
him-4
lin-2
flu-4
vab-3
unc-110

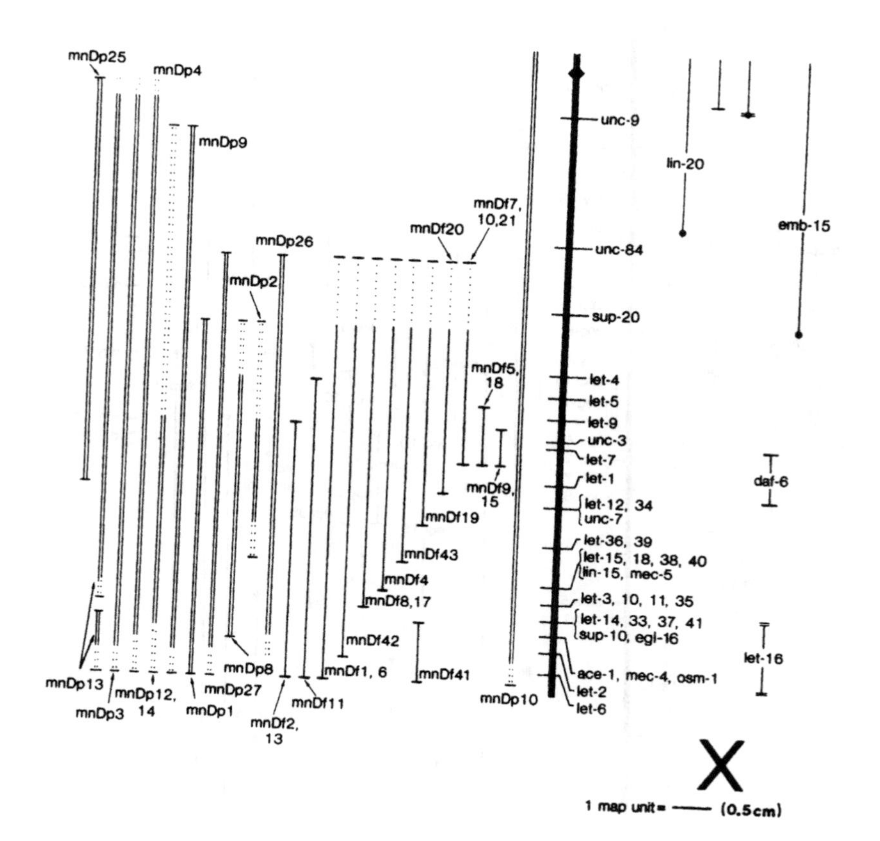
mnDp25
mnDp4
mnDp9
mnDp26
mnDp2
mnDf20
mnDf7, 10,21
mnDf5, 18
mnDf9, 15
mnDf19
mnDf43
mnDf4
mnDf8,17
mnDf42
mnDf1, 6
mnDf41
mnDp13
mnDp3
mnDp12, 14
mnDp1
mnDp27
mnDp8
mnDf2, 13
mnDf11
mnDp10
unc-9
unc-84
sup-20
let-4
let-5
let-9
unc-3
let-7
let-1
let-12, 34
unc-7
let-36, 39
let-15, 18, 38, 40
lin-15, mec-5
let-3, 10, 11, 35
let-14, 33, 37, 41
sup-10, egl-16
ace-1, mec-4, osm-1
let-2
let-6
lin-20
emb-15
daf-6
let-16
X
1 map unit = —— (0.5cm)

Part D: Physical Maps

CYTOLOGICAL MAP OF THE NUCLEAR GENOME

The map positions of cloned genes and polymorphisms as determined by in situ hybridization are shown below each linkage group; the bars represent the distribution of sites recorded by measuring the percent distance from the left end of the chromosome. Positions determined by genetic methods are shown above the line as they would appear on the genetic map. Cloned genes are designated by three lowercase letters. The core histone gene clusters have been named "HIS." Genetically mapped polymorphisms are represented by the laboratory allele designation followed by P and a number or by names in uppercase letters, often including the name of a gene known to map near the polymorphism (e.g., TCUNC52). The scale represents the percent length from the left end of the chromosome.

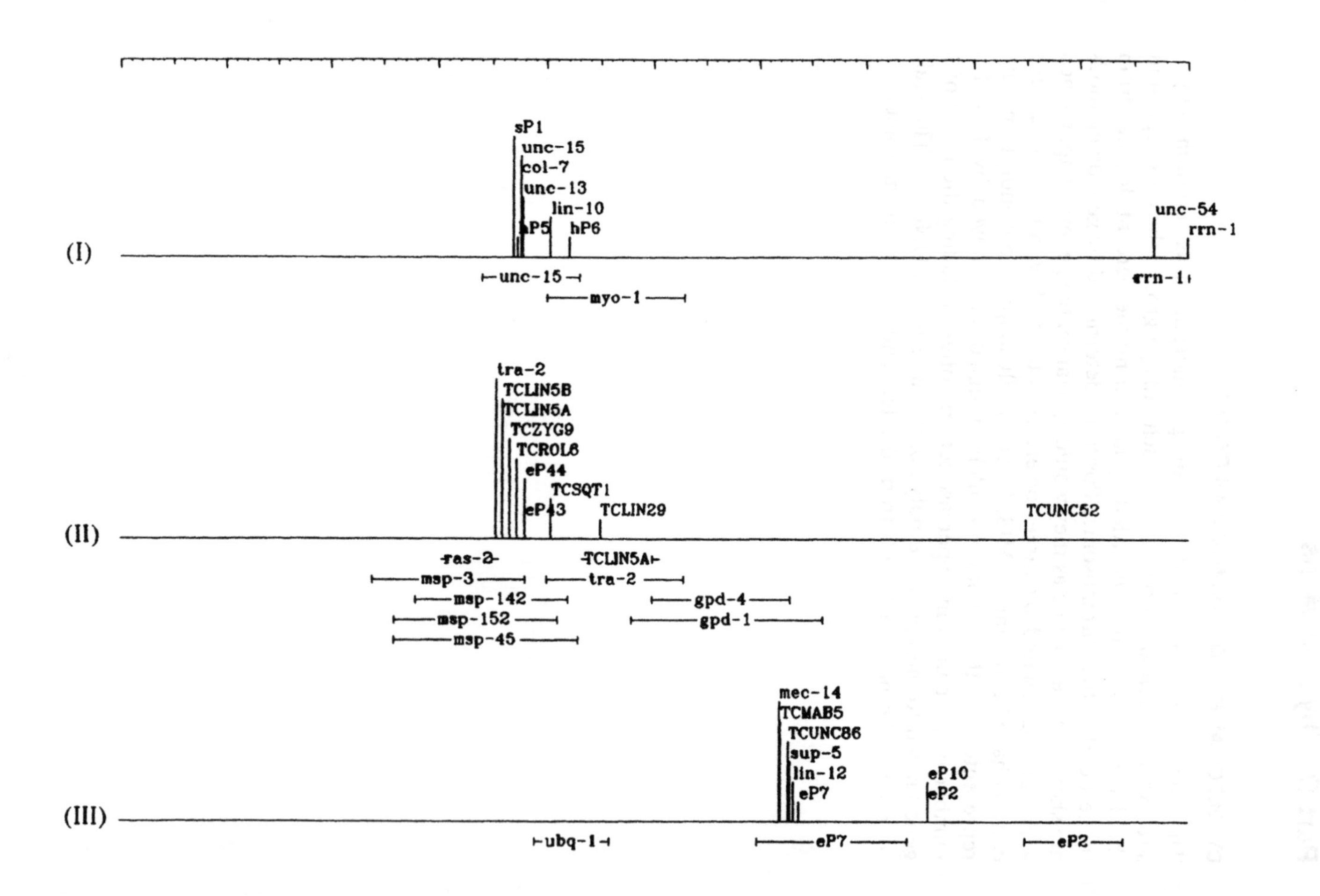
(I)
sP1
unc-15
col-7
unc-13
lin-10
hP5
hP6
unc-54
rrn-1
unc-15
myo-1
rrn-1
(II)
tra-2
TCLIN5B
TCLIN5A
TCZYG9
TCROL8
eP44
TCSQT1
eP43
TCLIN29
TCUNC52
ras-2
TCLIN5A
msp-3
tra-2
msp-142
gpd-4
msp-152
gpd-1
msp-45
(III)
mec-14
TCMAB5
TCUNC86
sup-5
lin-12
eP7
eP10
eP2
ubq-1
eP7
eP2

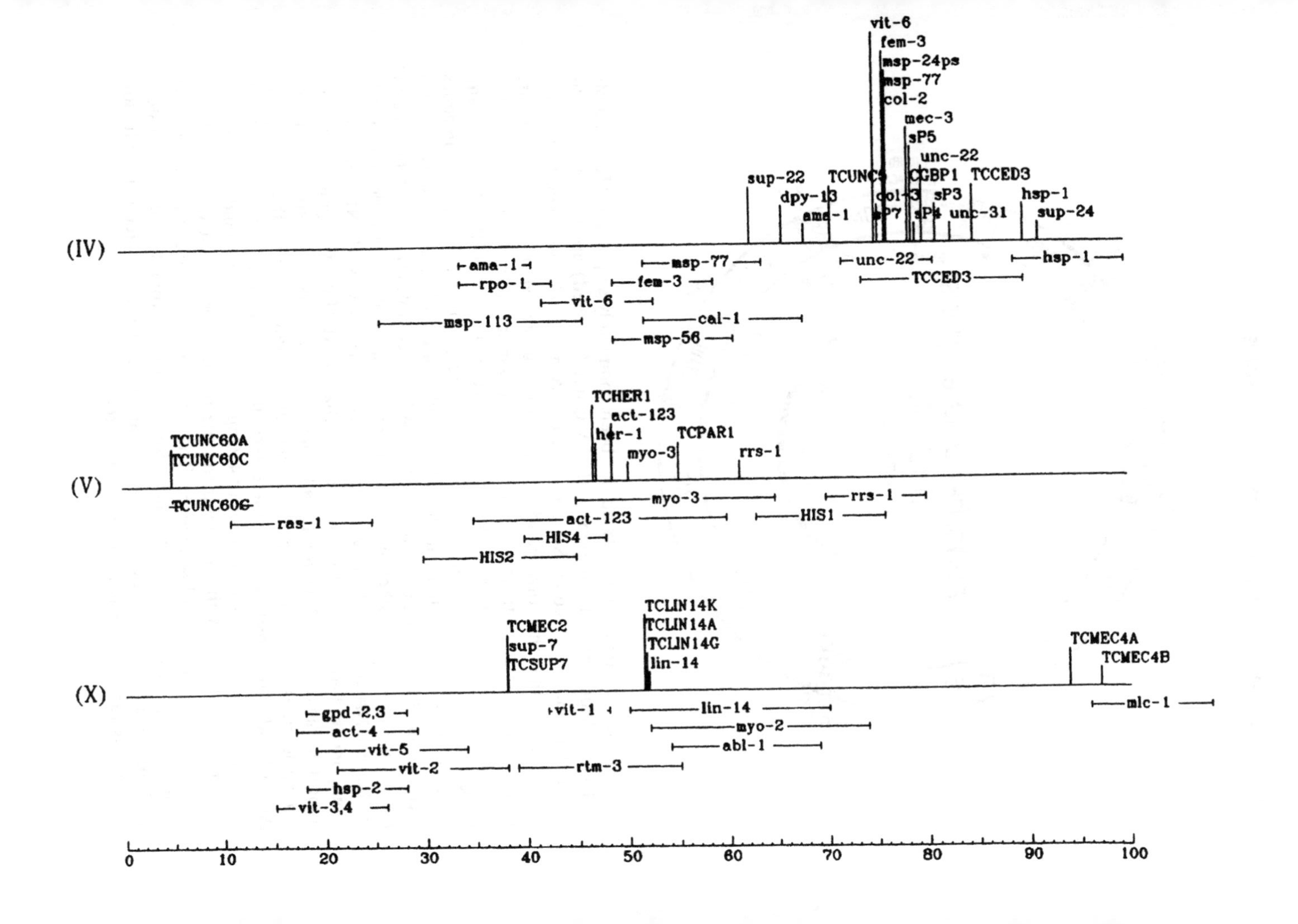
(IV)
vit-6
fem-3
msp-24ps
msp-77
col-2
mec-3
sP5
unc-22
CCBP1
TCCED3
sup-22
TCUNC5
dpy-13
col-3
sP3
hsp-1
ama-1
sP7
sP4
unc-31
sup-24
ama-1
msp-77
unc-22
hsp-1
rpo-1
fem-3
TCCED3
vit-6
msp-113
cal-1
msp-56
(V)
TCHER1
act-123
her-1
TCPAR1
myo-3
rrs-1
TCUNC60A
TCUNC60C
TCUNC60C
myo-3
rrs-1
ras-1
act-123
HIS1
HIS4
HIS2
(X)
TCLIN14K
TCLIN14A
TCLIN14G
lin-14
TCMEC2
sup-7
TCSUP7
TCMEC4A
TCMEC4B
gpd-2,3
vit-1
lin-14
mlc-1
act-4
myo-2
vit-5
abl-1
vit-2
rtm-3
hsp-2
vit-3,4
0
10
20
30
40
50
60
70
80
90
100

PHYSICAL MAP OF THE MITOCHONDRIAL GENOME

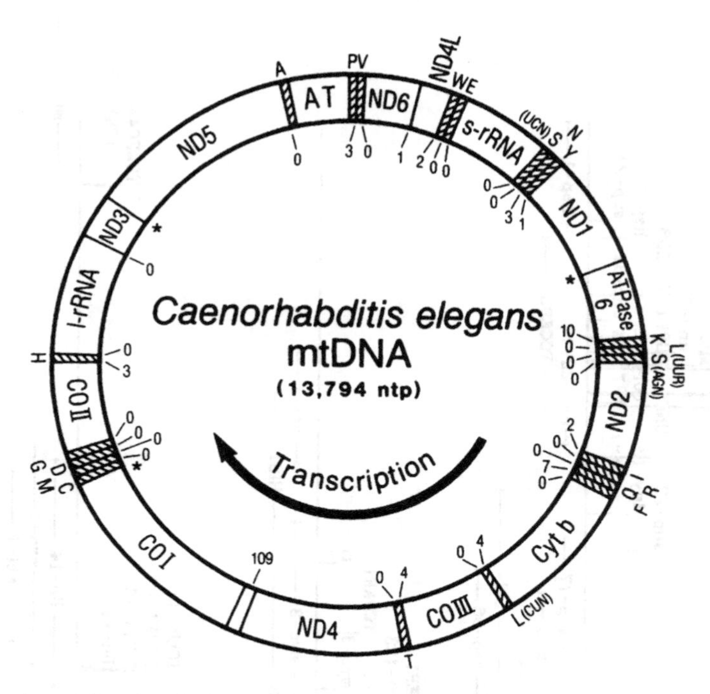

The circular molecule of 13,794 np has been completely sequenced (R. Okimoto and D.R. Wolstenholme, unpubl.). tRNA genes (hatched) are identified by the one-letter amino acid code; Ser and Leu tRNA genes are also identified by the codon family that the corresponding tRNAs recognize. Many of the tRNAs are unusual (e.g., in 20 of them, the TψC arm and variable loop are replaced with a loop of between 6 and 12 nucleotides (see Wolstenholme et al. 1987). s-RNA and l-RNA indicate the small and large rRNA genes, respectively. The 12 protein genes are COI, COII, and COIII (subunits 1, 2, and 3 of cytochrome *c* oxidase), Cyt b (cytochrome *b*), ATPase 6 (subunit 6 of the ATPase complex [note that the subunit 8 gene, identified in all other metazoan mtDNAs sequenced to date, does not occur in the mtDNA of either *C. elegans* or *Ascaris suum*]), and ND1–ND6 and ND4L (components 1–6 and 4L of the respiratory chain NADH dehydrogenase). Arrow indicates transcription direction for all genes. The numbers of apparently noncoding nucleotides between genes are shown on the inner side of the map circle. Asterisks identify possible incomplete termination codons (T or TA). The AT region (93% A and T) is a sequence of 466 np, which may correspond to the region of mammalian mtDNA that contains the site of initiation of first-strand synthesis and transcriptional control sequences. The 109-np region between ND4 and COI includes a sequence with the potential to form a stable hairpin structure with a stem of 12 np and a loop of 9 nucleotides. A similar structure is associated with initiation of second-strand synthesis in mammalian mtDNAs.

Methods

Methods

Compiled by John Sulston and Jonathan Hodgkin

MRC Laboratory of Molecular Biology
Cambridge, England

I. CULTURE TECHNIQUE

A. Stock Handling

Caenorhabditis elegans is maintained on NGM agar plates carrying a lawn of OP50, a leaky uracil-requiring strain of *Escherichia coli* (Brenner 1974). For general purposes, 5-cm plates are suitable; 9-cm plates are used for growing larger quantities, and 2-cm plates are useful for experiments with expensive drugs. For growing numerous clones in parallel for screening, 4 × 6-well microtiter plates are convenient; to prevent cross-contamination, the plates must be of a type in which the wells are separated from one another, and to prevent the worms from burrowing, the agar must be overlaid with 2% top agarose.

Streak OP50 on a plate containing rich medium (e.g., TYE); transfer a single colony to rich broth (e.g., PEN) in a screw-capped bottle or culture tube, and leave at room temperature overnight. The resulting suspension keeps for some months at 4°C.

Keep plates at room temperature for at least 3 days before use, so that excess moisture evaporates and those contaminated with fungi and bacteria can be detected and removed. Seed them in batches sufficient for 1–3-day requirements; as a rough guide, a 5-cm plate receives about 1/40 ml of OP50 suspension, streaked with a 1-ml pipette but not touching the wall. At this and subsequent stages, take care not to damage the agar surface, because the nematodes burrow into the breaks and become difficult to score and pick. Allow the lawn to grow overnight at room temperature before nematodes are added. Transfer the nematodes by means of a platinum wire pick (~32 gauge, sealed into the end of a Pasteur pipette) whose end has been flattened and shaped to your needs. Allow the wire to cool briefly after flaming (e.g., by touching the bacterial lawn) and coat it with sticky bacteria either from the transfer plate or from a plate kept especially for this purpose (the sticky bacteria make it easier to pick up worms). Animals should be active immediately after transfer. The standard wild-type strain N2 can be grown at temperatures between 12°C and 25°C. At 15°C, stock plates can be kept for at least 2 months between transfers; the limit is determined by drying of the plates rather than starvation of the nematodes, which enter the dauer form. Drying of the plates can be slowed by storing them in a covered box. However, cross-contamination between plates has been observed under such conditions because worms can crawl from plate to plate if the ambient humidity is high enough; otherwise, cross-contamination is never a problem.

B. Cleaning

Normal bacterial sterile precautions at the open bench are adequate. With the help of repeated transfers, the nematodes will easily outgrow mycelial fungi. Infestations by mites (which eat agar, worms, and worm eggs and also cause severe contamination problems) can be eliminated by keeping a beaker of moth crystals (paradichlorobenzene) in the incubator: The vapor is toxic to mites but not to worms.

Contamination with foreign bacteria or yeasts, often betrayed by softening of the agar and burrowing of the nematodes, usually requires sterilization of the eggs. A simple procedure is as follows: Take plate with (preferably) many gravid adults and eggs; wash off into about 1 ml of M9 buffer. Add to the suspension one-half volume of alkaline hypochlorite (2 vol 4 M NaOH:3 vol 10–20% NaOCl, preferably freshly mixed), leave 5 minutes, and spin down briefly (30 sec, 1500 rpm). Take up pellet in 0.1 ml of M9 buffer, drop on unseeded side of a half-seeded 9-cm plate. Leave overnight to allow nematodes to hatch and crawl to bacteria; then cut out unseeded half of plate to remove any resistant spores of contaminants.

C. Freezing

The wild-type and mutant strains can be stored indefinitely in the frozen state. For freezing, take one large or two to three small plates with predominantly just starving L1 and L2 larvae (1 day after bacteria are exhausted; longer starvation yields dauer larvae, which do not survive freezing). Wash off into about 1 ml of M9 buffer; add equal volume of freezing solution. Mix, transfer 0.5-ml aliquots to freezing vials, and immediately place either in neck of liquid nitrogen flask or (preferably) in styrofoam boxes at −70°C. The aim is to cool at ~ 1°C/min. Vials can be stored either in liquid nitrogen or at −70°C. For thawing, warm vial between hands until melted, and tip contents onto one side of a seeded 9-cm plate. The next day, pick off young healthy worms onto a fresh plate. Once thawed, nematodes must be regrown before refreezing.

D. Media

NGM Agar

NaCl	3 g
agar	17 g
peptone	2.5 g
cholesterol (5 mg/ml in EtOH)	1 ml
H_2O	975 ml

Autoclave; then, using sterile technique, add the following and mix after each addition.

$CaCl_2$ 1 M	1 ml
$MgSO_4$ 1 M	1 ml
potassium phosphate 1 M pH 6	25 ml

M9 Buffer

KH_2PO_4	3 g
Na_2HPO_4	6 g
NaCl	5 g
$MgSO_4$ 1 M	1 ml
H_2O	1 liter

Freezing Solution

NaCl	5.85 g
KH_2PO_4	6.8 g
glycerol	300 g
1 M NaOH	5.6 ml
H_2O to 1 liter	

Autoclave; then, using sterile technique, add

$MgSO_4$ 0.1 M	3 ml

S Medium

Autoclave the components separately and, using sterile technique, add

S basal	1 liter
1 M potassium citrate pH 6	10 ml
trace metals solution	10 ml
1 M $CaCl_2$	3 ml
1 M $MgSO_4$	3 ml

S Basal

NaCl	0.1 M
potassium phosphate (pH 6)	0.05 M
cholesterol (5 mg/ml in EtOH)	1 ml/liter

Autoclave.

Trace Metals Solution

disodium EDTA	1.86 g	(5 mM)
$FeSO_4 \cdot 7H_2O$	0.69 g	(2.5 mM)
$MnCl_2 \cdot 4H_2O$	0.20 g	(1 mM)
$ZnSO_4 \cdot 7H_2O$	0.29 g	(1 mM)
$CuSO_4 \cdot 5H_2O$	0.025 g	(0.1 mM)
H_2O	1 liter	

Autoclave and keep in the dark.

II. GENETIC METHODS

A. Selfing

The simplest, most common procedure in *C. elegans* genetics is to allow selfing: A single L4 hermaphrodite is picked onto a seeded small NGM plate and allowed to self-fertilize. The hermaphrodite is picked to a fresh plate after a day of egg laying and to successive plates at daily intervals until no more eggs are produced (usually 3 or 4 days at 20°C). This ensures that the entire self-progeny brood is collected and thus can be scored, if necessary, without any possibility of confusion between F1 and F2 progeny. Picking the hermaphrodite at L4 or earlier stages also ensures virginity.

B. Crossing

Crosses between hermaphrodites and males are usually carried out by placing a few young adult virgin hermaphrodites together with an excess of young adult males (e.g., three hermaphrodites and five males) on a small seeded plate. The size and shape of the bacterial lawn are not critical, but the lawn should not touch the edge of the plate because males tend to swim up the plastic wall and die of desiccation. Males are particularly liable to do this if no adult hermaphrodites are present on the plates.

Cross-progeny and self-progeny are usually distinguished by using hermaphrodites homozygous for a recessive visible mutation, so that all self-progeny are marked and all cross-progeny are unmarked. Alternatively, complete outcrossing can usually be guaranteed by picking a single hermaphrodite at L4 stage and placing it on a small plate with five or six adult males.

Single-pair matings (i.e., one hermaphrodite and one male) are sometimes necessary and will result in cross-progeny in at least 90% of crosses under good conditions.

Crosses are facilitated if the hermaphrodite expresses a marker causing it to move more slowly than wild type (e.g., a *dpy* or *unc* mutation). Homozygous males of many strains move poorly and are unable to mate or will only mate successfully with hermaphrodites that move at least as slowly as they do. Therefore, it is frequently necessary to use heterozygous males (*a*/+) to transmit a recessive mutation *a*. For many sex-linked mutations, hemizygous males (*b*/*0*) are unable to mate; in this case, young *tra-1; b*/+ XX males can be used, which can mate successfully, although much less efficiently than XO males. Another solution is to use a male carrying a small duplication of the X chromosome that carries a wild-type copy of the mutant gene, *b*/*0; Dp*(+).

C. Source of Males for Crossing

Males (XO) occur spontaneously in the self-progeny of hermaphrodites (XX) at a frequency of ~0.2%, which is inconveniently low. If the males are capable of efficient mating, a male stock can be established by picking the rare spontaneous males and mating them with a few sibling hermaphrodites. Males and hermaphrodites can then be picked and crossed at each subsequent generation so as to maintain the male stock. If worms of both sexes are not picked to a fresh plate regularly, the frequency of males in the population falls rapidly from the initial 50% to the spontaneous frequency 0.2%, because of the greater growth advantage of a selfing population.

Only a wild-type (N2) male stock need be maintained in this way; heterozygous males of any genotype can then be generated by crossing wild-type males with homozygous mutant hermaphrodites.

If homozygous males are desired from a given hermaphrodite stock, the spontaneous male frequency can be increased at least tenfold by heat shock: Six to ten L4 hermaphrodites are incubated at 30°C for 6 hours on a small seeded NGM plate and then selfed at 20°C. Their progeny will usually include 2–5% males, most of which will be fertile. Alternatively, the self-progeny male frequency can be increased by use of *him* mutations, which cause high levels of X chromosome loss during gametogenesis (Hodgkin et al. 1979). The most useful *him* mutations are *him-5(e1490) V* and *him-8(e1489) IV*, both of which result in male frequencies above 30%.

D. Scoring Crosses

Broods are scored most accurately by physically removing the worms from the plate using a platinum-wire worm pick. If many worms of the same phenotype need to be counted, it is much easier and more convenient to count and remove them by aspiration. The most popular tool is a Pasteur pipette connected to a water aspirator; the end 2 cm of the pipette is bent almost at right angles to the barrel, and the tip is flamed so that it is rounded and slightly narrowed. Such a pipette can be used to suck up worms efficiently without breaking the agar surface of the plate.

E. Two-factor Crosses for Measuring Linkage

The most common method for measuring the recombination distance between two linked recessive mutations *a* and *b*, with phenotypes A and B, is by means of a *cis* two-factor self-cross. Homozygous double-mutant (*ab*/*ab*) hermaphrodites are mated with wild-type males (+ / +), and the resulting hermaphrodite cross-progeny (*ab*/ + + , phenotypically wild type) are selfed. These animals will give four kinds of progeny

	W	A	B	AB
in the proportions:	$\frac{3-2p+p^2}{4}$	$\frac{2p-p^2}{4}$	$\frac{2p-p^2}{4}$	$\frac{(1-p)^2}{4}$

where p is the recombination distance. For most purposes, the most convenient measure of p is the total frequency of recombinant phenotypes

$$\frac{(A + B)}{(W + A + B + AB)} = \frac{2p - p^2}{2}$$

or p for small p.

Sometimes B and AB phenotypes are difficult to distinguish or are subviable, therefore representing less than one quarter of the brood. Under these circumstances, it is better to calculate p from $A/(A + W) = (2p\text{-}p^2)/3$.

In these crosses, p is a recombination frequency P_h that is related to the mean between recombination in hermaphrodite oogenesis (P_{ho}) and recombination in hermaphrodite spermatogenesis (P_{hs}). For sex-linked markers, P_{ho} is sometimes measured directly by crossing *ab*/ + + with wild-type males and scoring male progeny

W	A	B	AB
$\frac{1-p}{2}$	$\frac{p}{2}$	$\frac{p}{2}$	$\frac{1-p}{2}$

In a third type of cross, recombination in male spermatogenesis (P_m) is measured by crossing *ab*/*ab* hermaphrodites with *ab*/ + + males and scoring male progeny (such crosses are convenient when phenotypes A and B can only be recognized in males). This again gives

W	A	B	AB
$\frac{1-p}{2}$	$\frac{p}{2}$	$\frac{p}{2}$	$\frac{1-p}{2}$

Both of these latter crosses therefore measure a recombination frequency that may be different from P_h. It is known that P_{ho}, P_{hs}, and P_m are similar, at least for some intervals (Brenner 1974; Hodgkin 1980), but they may not be identical. Therefore, it is always preferable to measure P_h, because almost all map distances are based on P_h values. On the map, one unit of map distance is equal to 1% recombination ($P_h = 0.01$), without any correction for double events. Consequently, map distances are not strictly additive.

It has been shown that P_h varies with temperature and with the age of the hermaphrodite (Rose and Baillie 1979b); therefore, two-factor crosses should be carried out at 20°C and complete self-progeny broods should always be scored.

F. Ordering Genes by Three-factor Crosses

Two-factor linkage data can sometimes be used to indicate an unambiguous gene order, if the relative linkages for all markers are known and are significantly different. Linkage data are more useful in indicating which additional markers to use in three-factor (or *n*-factor) crosses. Three-factor crosses in which the markers are not fairly tightly linked (all linkages $p < 0.1$) can give misleading results as a consequence of double events and should be avoided.

A typical three-factor cross to order a mutation *c* with respect to two known mutations *a* and *b* would entail the construction and selfing of a hermaphrodite of genotype $a + b/ + c +$. For example, ten recombinants expressing the A phenotype might be picked from among the progeny of this hermaphrodite. If nine of them segregated A, AB, and AC progeny in the next generation of selfing (indicating recombinant genotype $a + b/ac +$, whereas one segregated only A and AB (recombinant genotype $a + b/a + +$), one can conclude that *c* is probably to the left of *b* but close to it. However, it might also be to the right of *b* (the $a + +$ chromosome being generated by a double crossover), so it is important to pick recombinants of both A and B phenotypes whenever possible. If nine B recombinants were also picked and found to be seven $+ + b/a + b$, two $+ cb/a + b$, one would conclude that the order was *acb*, with the approximate relative distances *a* (16/19) *c* (3/19) *b*. It should be noted that it is frequently not possible to distinguish all possible phenotypes in three-factor (and even two-factor) crosses, sometimes because different visible mutants have similar phenotypes but also often because one mutation masks the expression of another in the double mutant. For example, *bli* (blistered) mutations are not expressed in a *dpy* (dumpy) genetic background. Therefore, the genotypes of some multiple heterozygotes must be determined not solely from the phenotypes they segregate but also from the ratios of different phenotypes they produce, or by test crosses.

Three-factor crosses are also often useful in constructing double mutants (e.g., *cb/cb* in the cross above), which may be difficult to make otherwise.

G. Mutagenesis

Worms are usually mutagenized at the L4 stage, when the pool of germ cells in each animal is large but still mostly mitotic. This permits several rounds of replication to intervene between the mutagenic treatment and gamete formation, which may be necessary for the fixation of mutations. Healthy well-fed worms are washed off NGM plates with M9 buffer and washed once or twice to remove excess bacteria before the mutagenic treatment. After treatment, worms are washed again and often allowed to recover overnight on a seeded NGM plate, before picking individuals for selfing or crossing. This allows one to pick the individuals that have best survived the treatment, which may be very toxic.

For radiation mutagenesis, worms are irradiated on the surface of open NGM petri dishes; check that the worms have not burrowed during UV irradiation.

Ethylmethanesulfonate (EMS). This is the most popular and convenient mutagen for the induction of point mutations (GC-to-AT transitions), although small deletions can also be generated (Anderson and Brenner 1984; Dibb et al. 1985). Make a solution 0.1 M in EMS by adding 0.02 ml of liquid EMS to 2 ml of M9 buffer, and then gently agitate until the dense oily liquid has dissolved. Add 2 ml of a suspension of washed worms in M9 buffer to this solution (i.e., final concentration 50 mM). Incubate the suspension in a wide-bore test tube for 4 hours at 20°C, with occasional agitation to increase aeration. Lower doses (5–25 mM) are often used if it is important to avoid double events or to increase the survival of subviable stocks. For a dose-response curve, see Rosenbluth et al. (1983). EMS is a dangerous mutagen/carcinogen, so all contaminated glassware should be rinsed thoroughly with 1 M NaOH after use.

X-rays. The standard high dose, useful for generating deletions, translocations, and duplications, is 7500 R, delivered at a dose rate of approximately 650 R/min. Complex rearrangements have been created by this dose, so it may be preferable to use a lower dose, such as 1500 R, which is still effective in creating deletions.

Gamma Rays. A dose of 750–7500 R is delivered at a rate of 50 R/sec from a ^{60}Co source. As with X-rays, high doses may create complex rearrangements (Rosenbluth et al. 1985).

Ultraviolet Light. 300 seconds irradiation is delivered at a rate of 1 W/m^2 (Greenwald and Horvitz 1980).

Formaldehyde. Formaldehyde is useful for generating deletions. Warm 5 g paraformaldehyde in 50 ml water at 65°C and add NaOH to clear the solution. Adjust to pH 7.2, dilute to 150 ml with water, and then to 500 ml with M9 buffer, making a 1% stock solution. For mutagenesis, add this solution to a suspension of worms in M9 buffer to give 0.07–0.1% and incubate for 4 hours at 20°C (Moerman and Baillie 1981).

Other Mutagenic Treatments. These include diepoxyoctane (Anderson and Brenner 1984), acetaldehyde (Greenwald and Horvitz 1980), ^{32}P (Babu and Brenner 1981), diethylsulfate (Greenwald and Horvitz 1980), ICR 191 (D. Riddle, pers. comm.), and diepoxybutane (C. Trent, pers. comm.). For comparative efficiency of different mutagens, see Greenwald and Horvitz (1980). Also note transposition mutagenesis by Tc1 (Eide and Anderson 1985b).

III. MICROSCOPY OF LIVING ANIMALS

A. Individual Handling

The key to success is a pipette drawn from 1-mm melting point capillary to a diameter of 0.1–0.2 mm. The drawn portion should be as straight sided as possible. The capillary is used with a Micropet-type mouth tube and is kept half full of water or buffer during use. Draw the animal gently in to a point only just behind the tip. There, your prized specimen is protected from desiccation while you hunt for its intended receptacle, and it will never be lost inside the capillary.

B. High-resolution Observation on Plates

For some purposes (e.g., observation of mating), it is desirable to look at the worms under more natural conditions than is possible under a coverslip and yet achieve greater resolution than that of a dissecting microscope. To this end, a plate can be placed directly on the stage of a high-power microscope and viewed through a 6.3 × or 10 × Neofluar objective; higher-power objectives tend to fog because of their shorter working distance. Satisfying photographs can be obtained in this way as well.

C. Mounting for Nomarski Microscopy

Standard Mount. Prepare a 0.4-mm-thick pad of agar by flattening a drop of a hot 5% agar solution on a microscope slide with a second slide; the latter is supported by two spacer slides raised 0.4 mm from the bench by layers of adhesive tape. When set, remove the upper microscope slide and store the pad in a small wet box while preparing the specimens. For worm mounts, precoat the center of a coverslip very thinly with bacteria scraped from a plate. Bacteria are not needed for eggs but are required for worms (even for short periods of observation) in order to confine them and to encourage browsing rather than rapid locomotion. Worms and old eggs can be transferred directly from their plates in a small drop of buffer (e.g., S basal), using a drawn capillary and a mouth tube. For younger eggs, cut several adults in half under water in a watch glass and sort out specimens of roughly the required age, by means of the pipette, before transfer. Remove excess liquid with the pipette and lower the coverslip very gently, pivoting it with fine forceps to avoid the formation of large bubbles. At one time, when PCB immersion oil was available, the agar could be trimmed to the edges of the coverslip and sealed with the same oil that immersed the objective. Current formulations of immersion oil are toxic to nematodes, and it is therefore now advisable to use a coverslip that projects beyond the agar pad and to seal its edges with Voltalef oil or silicone grease.

Quick Mount. Quick mount is used when the specimen is to be recovered. Follow directions for standard mount, but use a 13-mm-diameter circular coverslip and no oil or grease at the edges. Instead, seal with a 25-mm square of Saran Wrap, having an 11-mm-diameter hole made with a cork borer. Use only a tiny drop of oil to immerse the objective, so that the plastic is not wetted by it. The coverslip is then easily removed for retrieval of the specimen.

Anesthetic Mount. This mount is used to immobilize worms for laser microsurgery or drawing. Anesthetize the animal with 0.5–1% 1-phenoxy-2-propanol and mount it on agar containing 0.2% 1-phenoxy-2-propanol as in the Quick mount. No bacteria are required. An alternative anesthetic is 10 mM sodium azide, added freshly to the agar.

Invertible mount. A dreadful technique, but unavoidable at times (Sulston et al. 1980).

D. Laser Microsurgery

Laser microsurgery provides a precise and rapid method for experimental destruction of parts of the animal. A pulsed dye laser is arranged as an epi-illuminator, its beam being deflected downward through the objective by way of a semi-silvered mirror (for full design details, see White and Horvitz 1979). Accurate targeting is achieved by prealignment of the focal point of the laser with a reference graticule in the eyepiece. Except in the case of eggs, the specimen is anesthetized as described above.

With a typical system, a peak laser power of around 100 kW seems to be necessary to produce cell damage when focused down to a spot of 1 μm diameter. Flash-tube-pumped dye lasers with a coumarin dye targeting at 450 nm have been used for many of the single-cell ablation experiments on *C. elegans* that have been described. These lasers typically have a pulse duration of ~1 μsec, and therefore pulse energies of about 100 millijoules are required. Nitrogen-laser-pumped dye lasers have pulse lengths of ~300 psec and so can attain the same peak power levels with considerably less total energy. These lower energy levels allow more precise ablations to be made, but higher energies may be needed in certain situations such as to permeabilize eggshells.

Careful control of the laser pulse power is necessary when performing cell ablations. Too little power has no effect, and too much power causes excessive damage, often bursting open the animal. Generally, the best strategy is to deliver several subthreshold shots within the target area. Usually, cells become refractile after a successful hit, taking on an appearance similar to that seen during programmed cell death. Often, cells that are markedly damaged after ablation will recover and go on to divide; it is therefore advisable to observe the ablated cell for some time to ensure that it dies.

If experimental design permits, a precursor should be killed in preference to a terminally differentiated cell, because it may be difficult to eliminate function in mature cells.

Eggshells may be punctured and permeabilized by laser irradiation. This procedure enables drugs for fixatives to be applied at precisely defined developmental stages. Staining the eggshells with fluorescein isothiocyanate (FITC)-polylysine facilitates this operation (Priess and Hirsh 1986).

IV. MICROSCOPY OF DEAD ANIMALS

A. Subbing Slides

Whole mounts or squashed nematode specimens do not stick well to glass slides; therefore, slides are coated by dipping in a protein solution. For multiple assays, multiwell test slides can be used. The following are alternative subbing protocols.

1. Bovine serum albumin (BSA) (0.03–1 mg/ml H_2O). Dip slides in BSA solution, set in racks to drain, and air-dry.
2. Gelatin. Make 0.1% gelatin by gentle heating (not above 50°C), cool, and add chrome alum to a concentration of 0.01%. Dip slides and air-dry.
3. Poly-L-lysine. Make a fresh solution of 0.1% poly-L-lysine, 0.2% gelatin, 0.01% chrome alum. Dip slides and dry in 100°C oven for 10 minutes. Specimens in M9 buffer or distilled water will stick well, but high-salt solutions should be avoided.

B. Fixation

General Method. Animals are gently washed off plates with M9 buffer or water and allowed to settle away from bacteria. They may be rinsed in buffer or water and pelleted by gentle centrifugation. As a rough guide, one to three small plates of worms will provide sufficient material for five to ten slides. If necessary, embryos may be obtained by alkaline hypochlorite treatment. The specimens may then be fixed in bulk in a centrifuge tube, or a drop containing worms or embryos can be pipetted onto a subbed slide and affixed to the slide by overlaying the drop with a coverslip. At this point, the slide should be inverted and gentle pressure applied if preparing a squash preparation. For whole mounts, it is important that the specimen drop should just fill the area of the coverslip, so that good contact is made with the slide: Too much and the material will float off; too little and the specimen will be squashed. The slide is then frozen by placing on dry ice or plunging into liquid nitrogen. After freezing, the coverslip is flipped off with a razor blade, and the slide is placed in 95% ethanol or directly in a fixative.

Fixatives

1. Carnoy's solution. A good general fixative, particularly of nuclei.

ethanol	3.0 ml
acetic acid	1.5 ml
chloroform	0.5 ml

Fix overnight or for a minimum of 1.5 hours.

2. Ethanol (or methanol)/acetic acid 3:1. Good fixative for nucleic acids. Fix overnight.
3. Methanol and acetone (−20°C). For immunofluorescence, specimens may be fixed first in methanol for 0.5–5 minutes and then in acetone for 4–6 minutes. Optimal times will vary for different antibodies. The fixatives may be maintained at −20°C in a Coplin jar on dry ice. Stir briefly to equilibrate before use.
4. Formaldehyde (~3.0%). Fixative used for immunofluorescence (more trouble than methanol and acetone but may give best results for some antigens). Dissolve paraformaldehyde in buffer with heating in a hood. Cool before use. Fix at 4°C or room temperature for about 1 hour.
5. Heat fixation. Good for some antigens. Place slide with squashed specimen on hot plate at 100°C for 2 minutes.

C. Staining

Visualization of Nuclei and Chromosomes

1. Feulgen. See Sulston and Horvitz (1977).
2. Hoechst 33258 (1 μg/ml in phosphate-buffered saline [PBS]). Rehydrate specimen in PBS and then stain for 5 minutes. Rinse in water for 1 minute. Air-dry and mount in citrate-phosphate buffer (pH 4) or 2% propyl gallate, 80% glycerol (pH 8).
3. DAPI (diamidinophenolindole). Fix in Carnoy's solution; stain with 1 μg/ml DAPI, 1.0 μg/ml phenoxypropanol in M9 buffer (Ellis and Horvitz 1986).

Feulgen may be visualized by bright-field microscopy, whereas Hoechst 33258 and DAPI are fluorescent nuclear stains.

Immunofluorescence. The methods outlined below must be considered only a rough guide, as the nature of the particular antigen or antibody or lectin will necessitate modification (see, e.g., Priess and Hirsh 1986).

1. Rinse slide in PBS or Tween–Tris-buffered saline (TBS) for 10 minutes.
2. Shake off excess buffer and dry slide around specimen.
3. Add antibody. Incubate at 4°C, at room temperature or at 37°C for 1–12 hours in a humidified chamber.
4. Wash slide in two changes of buffer for 5 minutes.
5. Remove excess, add fluorescein- or rhodamine-conjugated second antibody, and incubate 20–30 minutes.
6. Rinse in buffer as in no. 1 above.
7. Drain slides and remove excess buffer.
8. Mount slides in buffered glycerol plus a free radical scavenger to retard fluorescence fading. (a) FITC: 1 mg/ml *p*-phenylenediamine, 10% PBS (pH 8.0), 90% glycerol. (b) Rhodamine: 2% propyl gallate, 80% glycerol buffered to pH 8.
9. Slides are viewed by epifluorescence microscopy using suitable filter sets and may be stored in the dark at 4°C for 1–2 weeks.

PBS

7.31 g	NaCl
2.36 g	Na_2HPO_4
1.31 g	$NaH_2PO_4 \cdot 2H_2O$

H_2O to 1 liter (pH 7.0–7.2)

TBS (4×)

200 ml	1 M Tris (pH 7.4)
600 ml	1 M NaCl
6 ml	4 M NaOH

Tween-TBS

0.5 ml	Tween 20
150 ml	1 M NaCl
50 ml	1 M Tris (pH 7.4)
1.5 ml	4 M NaOH
800 ml	H_2O

Formaldehyde-induced Fluorescence (FIF). For visualizing catecholamine-containing cells (Sulston et al. 1975).

Acetylcholinesterase. See Culotti et al. (1981) and Herman and Kari (1985).

FITC Staining. Certain sensory neurons will take up FITC in living worms and can be visualized in vivo by fluorescence microscopy. For protocols, see Hedgecock et al. (1985).

D. Karyotype

The presence of a free duplication or other unusual chromosomal constitution can be confirmed by visualization of the karyotype.

Meiotic Chromosomes. Free duplications are best scored in the oocyte nucleus in diakinesis. Rinse the individual adult hermaphrodites of the desired genotype in a drop of M9 buffer and then cut just behind the pharynx to allow the gonad to emerge. Transfer several such animals to a subbed slide, overlay with a coverslip, and freeze. Fix the specimen in ethanol and acetic acid and stain with Hoechst 33258 or DAPI after ribonuclease treatment. The free duplication appears as a smaller extra chromosome in the oocyte nuclei.

Metaphase Chromosomes. Wash synchronously growing young adults off three small plates; wash once with water. Suspend in 1 ml of 0.5 M NaOH, 1% sodium hypochlorite for 7 minutes at room temperature without agitation. When the animals start to lyse, make up to 10 ml in a glass centrifuge tube, centrifuge, and wash the pellet with M9 buffer. Place a 40-μl drop of embryos on a gelatin-subbed slide, remove excess buffer, and place an 18 × 18-mm coverslip over the drop. Squash, immediately freeze, and fix in ethanol and acetic acid overnight. Air-dry slides and stain with Hoechst 33258 or DAPI after ribonuclease treatment.

V. ELECTRON MICROSCOPY

A. Osmic Acid Fixation

Rinse worms off plate with 0.1 M HEPES (pH 7.5), centrifuge, remove supernatant; add 1% OsO_4 in 0.1 M HEPES (pH 7.5). Fix for 1 hour. Wash three times with 0.1 M HEPES (pH 7.5). Transfer to a petri dish containing a thin layer of 1% agar (already set). Select suitable animals and cut them into two pieces; place a drop of 1% agar (60°C) on top of selected specimens. Cut blocks from agar containing specimens and dehydrate through

30% alcohol	10 min
50% alcohol	10 min
70% alcohol	10 min
90% alcohol	10 min
absolute alcohol	three changes in 1 hr
absolute alcohol/propylene oxide, 1:1	15 min
propylene oxide	15 min
propylene oxide/araldite, 2:1	30 min
propylene oxide/araldite, 1:2	30 min
araldite (Epon in U.S.)	overnight at room temp.

Transfer to fresh araldite the following morning; embed after 6 hours.

B. Glutaraldehyde Fixation

Pick suitable animals from a plate into a shallow dish containing 3% glutaraldehyde in 0.1 M HEPES (pH 7.5). Cut immediately at a point remote from the region of interest; after 5 minutes, cut again near the region of interest; leave for 1 hour. Wash three times in 0.1 M HEPES, transfer to 1% OsO_4 in 0.1 M HEPES, and proceed as above.

C. Tannic Acid Fixation and Antibody Staining

For tannic acid fixation, see Chalfie and Thomson (1982), and for antibody staining, see Okamoto and Thomson (1985).

VI. BIOCHEMISTRY

A. Bulk Growth on Bacteria

Take a total of 2 liters of 3XD or 2XTY in three 2-liter fluted flasks. To each flask add 3 ml 1 M $MgSO_4$ and 1 ml of a saturated OP50 culture. Shake vigorously overnight at 37°C.

Centrifuge in sterile bottles (5000 rpm, 10 min), drain pellets well, suspend in 1 liter S medium. Transfer to two 2-liter fluted flasks, store at 4°C if nematodes are not quite ready.

To each flask add the nematodes from four 9-cm plates that are just clearing, washing off with a few milliliters of sterile buffer (e.g., M9). Discard any cultures suspected of contamination. Shake at ~250 rpm at 20°C. After about 5 days, the cultures will clear visibly. Pour into 1-liter measuring cylinder; leave to settle overnight at 4°C.

Aspirate most of medium, resuspend sediment in 100 ml of cold 0.1 M NaCl in 4 × 50-ml screw-capped polyethylene tubes. Add equal volume (using calibrations on sides of tubes) of cold 60% w/w sucrose, mix by inversion, and centrifuge *immediately* (2000 rpm, 5 min, cold).

Remove cap of nematodes from each tube (use Pasteur pipette cut at shoulder), resuspend in 50 ml of 0.1 M NaCl, and centrifuge (2000 rpm, 2 min, cold). Resuspend pellet in 0.1 M NaCl; agitate in a shallow layer at room temperature for 30 min, to encourage digestion of bacteria remaining in nematode guts. Check (with dissecting microscope) that debris has been removed satisfactorily; if necessary, repeat sucrose flotation. Wash three times with cold 0.1 M NaCl, centrifuging as before.

Unless nematodes are to be used immediately, aliquot into freezer vials, freeze in liquid nitrogen, and store at −70°C or below. Yield 10–20 g.

Notes

1. Contamination is much more damaging in liquid than on plates; ensure that nematode cultures are clean before beginning protocol.

2. Sucrose flotation works perfectly if done exactly as described; ensure that sucrose is at specified concentrations, and centrifuge within 2 minutes of mixing with nematodes.
3. Nematodes must always be centrifuged cold; otherwise, they wiggle and break up the pellets.
4. Just cleared liquid cultures are devoid of L4s, and the adults are bloated with eggs. For a fully representative population, work up a day earlier; for a mixture of young larvae and dauers, allow to shake for several days after clearing.
5. The method has been scaled up to produce kilogram quantities (C. Johnson et al., pers. comm.). Bacteria are grown in bulk (commercially available bacteria are too contaminated to be used) and fed to the nematodes in fermenters under vigorous aeration.

B. Axenic Growth

Axenic media have been developed for *C. elegans* (for review, see Vanfleteren 1980). However, growth on such media is slower and less convenient than growth on *E. coli*.

C. Isolation of Specific Stages

Eggs. Eggs are obtained by digesting populations containing plenty of adults with alkaline hypochlorite. The method can be adapted to any scale; the following protocol will serve as an example. Suspend the washed nematodes from a 500-ml culture in two volumes of

NaOCl (10–20%)	12 ml
5 N KOH	5 ml
H_2O to 100 ml	

Disperse and agitate the nematodes by forcing them up and down through a Pasteur pipette. Monitor the progress of the reaction by observing aliquots under a dissecting microscope. When the worms begin to break apart (5–10 min), load the suspension into a 50-ml disposable syringe and force it out through a 23-gauge needle onto a 52-μm Nitex screen. Collect eggs by centrifugation of the filtrate. Resuspend the trapped nematodes from the screen in the remainder of the alkaline hypochlorite solution, agitate, and repeat the syringing operation. Wash the combined egg pellets three times by centrifugation from M9 buffer. The eggs are axenic and are ~90% viable.

Larval Stages and Young Adults. *C. elegans* grows more or less continuously, rather than in discrete increments, so complete separation of larval stages by size is not possible. However, it is possible to generate fractions that are highly enriched for particular stages by velocity sedimentation (e.g., Sulston and Brenner 1974) or by sieving through a series of Nitex screens.

For most purposes, the best approach is to establish a large-scale synchronous culture. Purified eggs are allowed to hatch overnight at 20°C in S medium without bacteria; the L1s thus generated neither grow nor die and the population becomes well synchronized. Growth begins upon addition of bacteria, and the culture can be worked up at any desired stage.

Dauer Larvae. Dauer larvae can be isolated from starved liquid cultures by virtue of their high density. For example, start a 5-ml liquid culture with 20–100 worms. After shaking for 12–14 days, layer 0.5 ml of the culture on top of 1 ml of 15% (w/w) Ficoll in 0.1 M NaCl. Dauer larvae settle into the lower solution in 5 minutes, whereas other stages remain at the interface. Wash the dauer larvae with water (Golden and Riddle 1982).

Males. Males can be separated from hermaphrodites by virtue of their tendency to squeeze through meshes that are slightly too small for them. Take a cleared liquid culture of a male-producing strain such as CB1490 (*him-5[e1490]*). Clean the worms by flotation on 1 M sucrose, and remove young larvae by washing on a 20-μm Nitex screen. Layer the worms onto a 35-μm Nitex screen stretched over an embroidery hoop (13–25 cm diam.) and supported in contact with M9 buffer in a petri dish. In the course of an hour, many of the males crawl through the screen and can be collected. Wash them again on 20-μm Nitex to remove any remaining larvae. Males so purified have been used as starting material for the isolation of sperm (Klass and Hirsh 1981; Nelson et al. 1982).

C. Isolation of Nucleic Acids

Bulk Extraction of DNA. Worms are grown in liquid culture as above, washed, quick frozen, and stored in 1-g aliquots. For each DNA preparation, grind one aliquot to a powder in a mortar cooled in liquid nitrogen and then mix very gently in portions into 30 ml of 100 mM EDTA (pH 8), 0.5% SDS, and 50 μg/ml proteinase K. Adding 1% β-mercaptoethanol may assist dissolution of worm carcasses. After incubation at 50°C for 2 hours, cool the solution in ice and extract with phenol by slow rolling at 4°C for 15 minutes. After centrifugation, carefully remove the aqueous layer with a wide-mouth pipette and precipitate with two volumes of ethanol. Wash the pellet with cold 80% ethanol and allow to disperse slowly in TE (10 mM Tris-HCl, 0.5 mM EDTA at pH 8). The DNA can be further purified by CsCl density gradient centrifugation.

Small-scale DNA Extraction. Yields enough DNA for at least 10 Southern lanes. Grow three large (9 cm) plates of worms just to starvation on

NGM plates seeded with OP50. Wash worms thoroughly in M9 buffer, pellet in a 1.5-ml tube, and store frozen at −70°C or below. For extraction, thaw and immediately add 0.5 ml of 100 mM NaCl, 100 mM Tris-HCl (pH 8.5), 50 mM EDTA (pH 7.4), 1% SDS, 1% β-mercaptoethanol, and 100 μg/ml proteinase K. Incubate with occasional agitation for 30 minutes at 65°C. Extract with 0.5 ml phenol, 0.5 ml phenol/chloroform, and 0.5 ml chloroform. To final aqueous layer (0.4 ml), add 1 ml ethanol. Pellet; wash pellet twice with 70% ethanol and once with 95% ethanol. Take up in 50 μl TE (allow at least 1 hr to resuspend).

DNA from worms grown on agar tends to cut poorly with restriction enzymes, but this problem can be circumvented by floating the worms on 1 M sucrose at one of the washing stages and/or running the final DNA solution over a 1-ml Sepharose CL-6B spin column. Adding 5 mM spermidine to the restriction digest may also help. Alternatively, worms can be grown on NGM plates topped with a thin layer of 2% agarose.

RNA Preparation. See MacLeod et al. (1981); but grinding in liquid nitrogen is probably preferable to disruption in a French press.

VII. OTHER METHODS

Behavioral Assays

Chemotaxis	Ward (1973)
Egg laying	Trent et al. (1983)
Mating	Hodgkin (1983)
Mechanosensation	Chalfie and Sulston (1981)
Osmotaxis	Culotti and Russell (1978)
Thermotaxis	Hedgecock and Russell (1975)

Isotopic Labeling

^{32}P	Sulston and Brenner (1974)
	Babu and Brenner (1981)
^{35}S	Epstein et al. (1974)
	Cox et al. (1981b)
$H^{14}CO_3$	Cox et al. (1981c)

Microinjection

See Kimble et al. (1982) and Fire (1986).

In Situ Hybridization

RNA	Hecht et al. (1981a)
	Klass et al. (1982)
	Edwards and Wood (1983)
DNA	Albertson (1984b)

Polarized Light Microscopy

See Sulston and White (1980) and Waterston et al. (1980).

ACKNOWLEDGMENTS

The methods described in this section have been developed by the entire community of *C. elegans* researchers. Particularly helpful advice and explicit contributions were provided by Donna Albertson, Marty Chalfie, Scott Emmons, Ed Hedgecock, Bob Herman, Dick McIntosh, Jim Priess, Don Riddle, Nichol Thomson, Sam Ward, Bob Waterston, and John White.

C. elegans Bibliography

Aamodt, E.J. and J.G. Culotti. 1986. Microtubules and microtubule-associated proteins from the nematode *Caenorhabditis elegans*: Periodic cross-links connect microtubules in vitro. *J. Cell Biol.* **103:** 23–31.

Abdulkader, N. and J.L. Brun. 1978. Induction, detection and isolation of temperature-sensitive lethal and/or sterile mutants in nematodes. I. The free-living nematode *Caenorhabditis elegans*. *Rev. Nematol.* **1:** 27–37.

———. 1979. A temperature-sensitive mutant of *Caenorhabditis elegans* var. Bergerac affecting morphological and embryonic development. *Genetica* **51:** 81–92.

———. 1980. Caractéristiques génétiques et physiologiques de thermosensibilité du developpement embryonnaire chez un mutant à léthalité conditionelle de *Caenorhabditis elegans*. *Rev. Nematol.* **3:** 11–19.

Abdulkader, N., M. Gilbert, J. Starck, C. Bosch, and J. Brun. 1980. Temperature-sensitive mutations in *Caenorhabditis elegans*: A sterile mutation affecting oocyte I core relations. *Rev. Nematol.* **3:** 201–212.

Abi-Rached, M. and J.L. Brun. 1975. Etude ultrastructurale des relations entre ovocytes et rachis au cours de l'ovogenèse du nematode *Caenorhabditis elegans*. *Nematologica* **21:** 151–162.

———. 1978. Ultrastructural changes in the nuclear and perinuclear regions of the oogonia and primary oocytes of *Caenorhabditis elegans*, Bergerac strain. *Rev. Nematol.* **1:** 63–72.

———. 1979. Changes in the synaptonemal complex in the oocyte nucleus in meiotic prophase of *Caenorhabditis elegans*. *C.R. Seances Acad. Sci. Ser. D Sci. Nat.* **288:** 425–428.

Aeby, P., A. Spicher, Y. de Chastonay, F. Müller, and H. Tobler. 1986. Structure and genomic organization of proretrovirus-like elements partially eliminated from the somatic genome of *Ascaris lumbricoides*. *EMBO J.* **5:** 3353–3360.

Agabian, N., H. Goodman, and N. Nogueira, eds. 1987. *Molecular strategies of parasitic invasion. UCLA Symp. Mol. Cell. Biol. New Ser.* **42:** 1–775.

Albert, P.S. and D.L. Riddle. 1983. Developmental alterations in sensory neuroanatomy of the *Caenorhabditis elegans* dauer larva. *J. Comp. Neurol.* **219:** 461–481.

———. 1988. Mutants of *Caenorhabditis elegans* that form dauerlike larvae. *Dev. Biol.* (in press).

Albert, P.S., S.J. Brown, and D.L. Riddle. 1981. Sensory control of dauer larva formation in *Caenorhabditis elegans*. *J. Comp. Neurol.* **198:** 435–451.

Albertson, D.G. 1984a. Formation of the first cleavage spindle in nematode embryos. *Dev. Biol.* **101:** 61–72.

———. 1984b. Localization of the ribosomal genes in *Caenorhabditis elegans* chromosomes by in situ hybridization using biotin-labeled probes. *EMBO J.* **3:** 1227–1234.

———. 1985. Mapping muscle protein genes by in situ hybridization using biotin-labeled probes. *EMBO J.* **4:** 2493–2498.

Albertson, D.G. and J.N. Thomson. 1976. The pharynx of *Caenorhabditis elegans*. *Philos. Trans. R. Soc. Lond. B Biol. Sci.* **275:** 299–325.

———. 1982. The kinetochores of *Caenorhabditis elegans*. *Chromosoma* **86:** 409–428.

Albertson, D.G., O.C. Nwaorgu, and J.E. Sulston. 1979. Chromatin diminution and a chromosomal mechanism of sexual differentiation in *Strongyloides papillosus*. *Chromosoma* **75:** 75–87.

Albertson, D.G., J.E. Sulston, and J.G. White. 1978. Cell cycling and DNA replication in a mutant blocked in cell division in the nematode *Caenorhabditis elegans*. *Dev. Biol.* **63:** 165–178.

Ali, M., A. Wahab, and A.H. El-Kifel. 1973. Nematodes associated with Coleoptera species in Egypt: Part 2. *Parasitol. Hung.* **6:** 169–188.

Ambros, V. and H.R. Horvitz. 1984. Heterochronic mutants of the nematode *Caenorhabditis elegans*. *Science* **226:** 409–416.

———. 1987. The *lin-14* locus of *Caenorhabditis elegans* controls the time of expression of specific postembryonic developmental events. *Genes Dev.* **1:** 398–414.

Anderson, G.L. 1978. Responses of dauer larvae of *Caenorhabditis elegans* (Nematoda: Rhabditidae) to thermal stress and oxygen deprivation. *Can. J. Zool.* **56:** 1786–1791.

———. 1982. Superoxide dismutase activity in dauer larvae of *Caenorhabditis elegans* (Nematoda: Rhabditidae). *Can. J. Zool.* **60:** 288–291.

Anderson, G.L. and D.B. Dusenbery. 1977. Critical oxygen tension of *Caenorhabditis elegans*. *J. Nematol.* **9:** 253–254.

Anderson, P. and S. Brenner. 1984. A selection for myosin heavy-chain mutants in the nematode *Caenorhabditis elegans*. *Proc. Natl. Acad. Sci.* **81:** 4470–4474.

Anderson, R.V. and D.C. Coleman. 1982. Nematode temperature responses: A niche dimension in populations of bacterial-feeding nematodes. *J. Nematol.* **14:** 69–76.

Andrassy, I. 1983. *Caenorhabditis briggsae*: A model of genetics. *Allattani Kozl.* **70:** 113–116.

———. 1985. A dozen new nematode species from Hungary. *Opusc. Zool. (Budap.)* **19:** 3–40.

Andrew, P.A. and W.L. Nicholas. 1976. Effect of bacteria on dispersal of *Caenorhabditis elegans* (Rhabditidae). *Nematologica* **22:** 451–461.

Angstadt, J.D. and A.O.W. Stretton. 1983. Intracellular recordings and Lucifer-yellow fills of interneurons in the nematode *Ascaris*. *Soc. Neurosci. Abstr.* **9:** 302.

———. 1985. Rhythmic membrane potential oscillations in motorneurons of *Ascaris*. *Soc. Neurosci. Abstr.* **11:** 514.

Argon, Y. and S. Ward. 1980. *Caenorhabditis elegans* fertilization-defective mutants with abnormal sperm. *Genetics* **96:** 413–433.

Austin, J. and J. Kimble. 1987. *glp-1* is required in the germ line for regulation of the decision between mitosis and meiosis in *Caenorhabditis elegans*. *Cell* **51:** 589–599.

Axel, R., T. Maniatis, and C.G. Fox, eds. 1979. Eukaryotic gene regulation. *ICN-UCLA Symp. Mol. Cell. Biol.* **14:** 1–661.

Babu, P. 1974. Biochemical genetics of *Caenorhabditis elegans*. *Mol. Gen. Genet.* **135:** 39–44.

Babu, P. and S. Brenner. 1981. Spectrum of ^{32}P-induced mutants of *Caenorhabditis elegans*. *Mutat. Res.* **82:** 269–273.

Baillie, D.L., K.A. Beckenbach, and A.M. Rose. 1985. Cloning within the *unc-43* to *unc-31* interval (LGIV) of the *Caenorhabditis elegans* genome using Tc1 linkage selection. *Can. J. Genet. Cytol.* **27:** 457–466.

Baker, B.S. and J. Belote. 1983. Sex determination and dosage compensation in *Drosophila melanogaster*. *Annu. Rev. Genet.* **17:** 345–393.

Baker, B.S., A.T.C. Carpenter, M.S. Esposito, R.E. Esposito, and L. Sandler. 1976. The genetic control of meiosis. *Annu. Rev. Genet.* **10:** 53.

Bartnik, E., M. Osborn, and K. Weber. 1986. Intermediate filaments in muscle and epithelial cells of nematodes. *J. Cell Biol.* **102:** 2033–2041.

Barton, M.K., T.B. Schedl, and J. Kimble. 1987. Gain-of-function mutants of *fem-3*, a sex-determination gene in *Caenorhabditis elegans*. *Genetics* **115:** 107–119.

Bateson, P.P.G. and R.A. Hinde, eds. 1975. *Growing points in ethology*. Conference, Cambridge, England. Cambridge University Press, New York.

Bazzicalupo, P. 1983. *Caenorhabditis elegans*—A model system for the study of nematodes. In *Molecular biology of parasites* (ed. J. Guardiola et al.). Raven Press, New York.

Beauchamp, R.S., J. Pasternak, and N.A. Strauss. 1979. Characterization of the genome of the free living nematode *Panagrellus silusiae*: Absence of short-period interspersion. *Biochemistry* **18:** 245–251.

Beguet, B. 1972. The persistence of processes regulating the level of reproduction in the herm nematode, *Caenorhabditis elegans*, despite parental aging. *Exp. Gerontol.* **7:** 207–218.

———. 1974. Un exemple de dérivé meiotique chez un nématode libre autofécond *Caenorhabditis elegans*. *C.R. Seances Acad. Sci. Ser. D Sci. Nat.* **279:** 2115–2118.

———. 1975. Génétique de la physiologie ovocytaire chez un nématode libre autofécond *Caenorhabditis elegans*. I. Influence du génotype et sélection. *Genetica* **45:** 405–424.

———. 1976. Cryoconservation de mutants "femelle-stériles" de *Caenorhabditis elegans* à la température de l'azote liquide. *Bull. Soc. Zool. Fr.* **101:** 137.

———. 1978. Etude génétique d'un mutant meiotique dominant chez *Caenorhabditis elegans*, souche Bergerac. *Rev. Nematol.* **1:** 39–45.

Beguet, B. and J.L. Brun. 1972. Influence of parental aging on the reproduction of the F1 generation in a hermaphrodite nematode *Caenorhabditis elegans*. *Exp. Gerontol.* **7:** 195–206.

Beguet, B. and M.-A. Gibert. 1978. Obtaining a self-fertilizing hermaphrodite mutant with a male copulatory bursa in the free-living nematode *Caenorhabditis elegans* var. Bergerac. *C.R. Seances Acad. Sci. Ser. D Sci. Nat.* **286:** 989–992.

Behme, R. and J. Pasternak. 1969. DNA base composition of some free living nematode species. *Can. J. Genet. Cytol.* **11:** 993–1000.

Bejsovec, A., D. Eide, and P. Anderson. 1984. Genetic techniques for analysis of nematode muscle. In *Molecular biology of the cytoskeleton* (ed. G. Borisy et al.), pp. 267–273. Cold Spring Harbor Laboratory, Cold Spring Harbor, New York.

Bender, W. 1985. Homeotic gene products as growth factors. *Cell* **43:** 559–560.

Bennet, M.D., J.S. Heslop-Harrison, J.B. Smith, and J.P. Ward. 1983. DNA density in mitotic and meiotic metaphase chromosomes of plants and animals. *J. Cell Sci.* **63:** 173–179.

Bhat, S.G. and P. Babu. 1980. Mutagen sensitivity of kynureninase mutants of the nematode *Caenorhabditis elegans*. *Mol. Gen. Genet.* **180:** 635–638.

Bird, A.F. 1971. *The structure of nematodes*. Academic Press, New York.

———. 1979. A method of distinguishing between living and dead nematodes by enzymatically induced fluorescence. *J. Nematol.* **11:** 103–105.

———. 1980. The nematode cuticle and its surface. In *Nematodes as biological models*, volume 2: *Aging and other model systems* (ed. B.M. Zuckerman), pp. 213–236. Academic Press, New York.

Bird, A.F., M.V. Jago, and P.A. Cockrum. 1985. Corynetoxins and nematodes. *Parasitology* **91:** 169–176.

Blackburn, E.H. and J.W. Szostak. 1984. The molecular structure of centromeres and telomers. *Annu. Rev. Biochem.* **53:** 163–194.

Blonston, G. 1984. To build a worm. *Science* **5:** 63–67.

Blumenthal, T., M. Squire, S. Kirtland, J. Cane, M. Donegan, J. Spieth, and W. Sharrock. 1984. Cloning of a yolk protein gene family from *Caenorhabditis elegans*. *J. Mol. Biol.* **174:** 1–18.

Boedtker, H., F. Fuller, and V. Tate. 1983. The structure of collagen genes. *Int. Rev. Connect. Tissue Res.* **10:** 1.

Bolanowski, M.A., L.A. Jacobson, and R.L. Russell. 1983. Quantitative measures of aging in the nematode *Caenorhabditis elegans*. II. Lysosomal hydrolases as markers of senescence. *Mech. Ageing Dev.* **21:** 295–319.

Bolanowski, M.A., R.L. Russell, and L.A. Jacobson. 1981. Quantitative measures of aging in the nematode *Caenorhabditis elegans*. I. Population and longitudinal studies of two behavioral parameters. *Mech. Ageing Dev.* **15:** 279–295.

Bollinger, J.A. and J.D. Willett. 1980. A method for synchrony of adult *Caenorhabditis elegans*. *Nematologica* **26:** 491–493.

Bolten, S.L., P. Powell-Abel, D.A. Fischhoff, and R.H. Waterston. 1984. The *sup-7(st5) X* gene of *Caenorhabditis elegans* encodes a transfer RNA-Trp-UAG amber suppressor. *Proc. Natl. Acad. Sci.* **81:** 6784–6788.

Borisy, G., D. Cleveland, and D. Murphy, eds. 1984. *Molecular biology of the cytoskeleton.* Cold Spring Harbor Laboratory, Cold Spring Harbor, New York.

Bornstein, P. and H. Sage. 1980. Structurally distinct collagen types. *Annu. Rev. Biochem.* **49:** 957–1003.

Borst, P. 1986. Discontinuous transcription and antigenic variation in trypanosomes. *Annu. Rev. Biochem.* **55:** 701–732.

Bostock, C.J. and A.T. Sumner. 1978. *The eukaryotic chromosome.* Elsevier/North-Holland, Amsterdam.

Bottjer, K.P., P.P. Weinstein, and M.J. Thompson. 1985. Effects of an azasteroid on growth, development, and reproduction of the free-living nematodes *Caenorhabditis briggsae* and *Panagrellus dividus. Comp. Biochem. Physiol.* **82B:** 99–106.

Boveri, T. 1888. Zellen Studien. *Jena. Z. Naturwiss.* **22:** 685.

———. 1910. Ueber die Teilung centrifugierte Eier von *Ascaris megalocephala. Wilhelm Roux Arch. Entwicklungsmech. Org.* **30:** 101–125.

Brenner, S. 1973. The genetics of behaviour. *Br. Med. Bull.* **29:** 269–271.

———. 1974. The genetics of *Caenorhabditis elegans. Genetics* **77:** 71–94.

———. 1984. Nematode research. *Trends Biochem. Sci.* **9:** 172.

Brown, D.D., ed. 1981. Developmental biology using purified genes. *ICN-UCLA Symp. Mol. Cell. Biol.* **23:** 1–702.

Brown, S.J. and D.L. Riddle. 1985. Gene interactions affecting muscle organization in *Caenorhabditis elegans. Genetics* **110:** 421–440.

Brun, J. 1955. Evolution de la prophase meiotique chez *Caenorhabditis elegans* Maupas 1900, sous l'influence de températures élevées. *Bull. Biol. Fr. Belg.* **89:** 326–346.

———. 1965. Genetic adaptation of *Caenorhabditis elegans* (Nematoda) to high temperatures. *Science* **150:** 1467.

———. 1966a. L'adaptation aux températures élevées chez un nématode *Caenorhabditis elegans* Maupas 1900. II. Stabilité et physiologie de l'adaptation. *Ann. Biol. Anim. Biochim. Biophys.* **6:** 267–300.

———. 1966b. L'adaptation aux températures élevées chez un nématode *Caenorhabditis elegans* Maupas 1900. III. Rôle des facteurs autres que la température. *Ann. Biol. Anim. Biochim. Biophys.* **6:** 439–466.

———. 1966c. L'adaptation aux températures élevées chez un nématode *Caenorhabditis elegans* Maupas 1900. I. L'adaptation et son évolution. *Ann. Biol. Anim. Biochim. Biophys.* **6:** 127–158.

———. 1966d. Influence des conditions de milieu sur la fécondité de *Caenorhabditis elegans* a différentes températures. *Nematologica* **12:** 539–556.

———. 1973. Structure, organisation et variation du matériel génétique du nématode libre hermaphrodite autofécond *Caenorhabditis elegans. Annu. Rep. Univ. CB/Lyon Fr.,* pp. 68–78.

Brun, J.L. and D. Lebre. 1968. Influence of parental aging on the fecundity of 1st generation descendants in a self-fertilizing hermaphrodite nematode: *Caenorhabditis elegans. C.R. Seances Acad. Sci. Ser. D Sci. Nat.* **266:** 2149–2152.

Brun, J., N. Abdulkader, M. Abirached, B. Beguet, M.A. Gilbert, N. Mounier, J. Starck, and C. Bosch. 1978. Contrôle génétique de la gametogenèse et de la différenciation ovocytaire chez *Caenorhabditis elegans* souche Bergerac. *Annu. Rep. Univ. CB/Lyon Fr.,* pp. 25–43.

Brutlag, D.L. 1980. Molecular arrangement and evolution of heterochromatic DNA. *Annu. Rev. Genet.* **14:** 121–144.

Bryant, C., W.L. Nicholas, and R. Jantunen. 1967. Some aspects of the respiratory metabolism of *Caenorhabditis briggsae* (Rhabditidae). *Nematologica* **13:** 197–209.

Buecher, E.J. and E.L. Hansen. 1969. Yeast extract as a supplement to chemically defined medium for auxenic culture of *Caenorhabditis briggsae. Experientia* **25:** 656.

———. 1971. Mass culture of axenic nematodes using continuous aeration. *J. Nematol.* **3:** 199–200.

Buecher, E.J., E.L. Hansen, and T. Gottfried. 1969a. Yeast ribosomes as a source of growth factor for nematodes. *Nematologica* **15:** 619–620.

———. 1970. A nematode growth factor from baker's yeast. *J. Nematol.* **2:** 93–98.

Buecher, E.J., E. Hansen, and E.A. Yarwood. 1966. Ficoll activation of a protein essential for maturation of the free-living nematode *Caenorhabditis briggsae*. *Proc. Soc. Exp. Biol. Med.* **121:** 390–393.

———. 1970. Growth of nematodes in defined medium containing hemin and supplemented with commercially available proteins. *Nematologica* **16:** 403–409.

———. 1971. Cultivation of *Caenorhabditis briggsae* and *Turbatrix aceti* with defined proteins. *J. Nematol.* **3:** 89–90.

Buecher, E.J., G. Perez-Mendez, and E.L. Hansen. 1969b. The role of precipitation during activation treatments of growth factor for *Caenorhabditis briggsae*. *Proc. Soc. Exp. Biol. Med.* **132:** 724–728.

Bull, J.J. 1984. *Evolution of sex determining mechanisms*. Benjamin/Cummings, Menlo Park, California.

Burghardt, R.C. and W.E. Foor. 1978. Membrane fusion during spermatogenesis in *Ascaris*. *J. Ultrastruct. Res.* **62:** 190–202.

Burke, D.J. and S. Ward. 1983. Identification of a large multigene family encoding the major sperm protein of *Caenorhabditis elegans*. *J. Mol. Biol.* **171:** 1–29.

Burr, A.H. 1985. The photomovement of *Caenorhabditis elegans*, a nematode which lacks ocelli. Proof that the response is to light not radiant heating. *Photochem. Photobiol.* **41:** 577–582.

Butler, M.H., S.M. Wall, K.R. Luehrsen, G.E. Fox, and R.M. Hecht. 1981. Molecular relationships between closely related strains and species of nematodes. *J. Mol. Evol.* **18:** 18–23.

Byard, E.H., W.J. Sigurdson, and R.A. Woods. 1986. A hot aldehyde-peroxide fixation method for electron microscopy of the free-living nematode *Caenorhabditis elegans*. *Stain Technol.* **61:** 33–38.

Byerly, L., R.C. Cassada, and R.L. Russell. 1975. Machine for rapidly counting and measuring the size of small nematodes. *Rev. Scientific Instrum.* **46:** 517–522.

———. 1976a. The life cycle of the nematode *Caenorhabditis elegans*. I. Wild type growth and reproduction. *Dev. Biol.* **51:** 23–33.

Byerly, L., S. Scherer, and R.L. Russell. 1976b. The life cycle of the nematode *Caenorhabditis elegans*. II. A simplified method for mutant characterization. *Dev. Biol.* **51:** 34–48.

Campos-Ortega, J.A. 1985. Genetics of early neurogenesis in *Drosophila melanogaster*. *Trends Neurosci.* **8:** 245–250.

Cassada, R.C. 1975. The dauer larva of *Caenorhabditis elegans*: A specific developmental arrest, inducible environmentally and genetically. In *Developmental biology: Pattern formation and gene regulation* (ed. D.M. McMahon and C.F. Fox), pp. 539–547. W.A. Benjamin, Menlo Park, California.

Cassada, R.C. and R.L. Russell. 1975. The dauer larva, a post-embryonic developmental variant of the nematode *Caenorhabditis elegans*. *Dev. Biol.* **46:** 326–342.

Cassada, R., E. Isnenghi, M. Culotti, and G. von Ehrenstein. 1981a. Genetic analysis of temperature-sensitive embryogenesis mutants in *Caenorhabditis elegans*. *Dev. Biol.* **84:** 193–205.

Cassada, R., E. Isnenghi, K. Denich, K. Radnia, E. Schierenberg, and G. von Ehrenstein. 1981b. Genetic dissection of embryogenesis in *Caenorhabditis elegans*. *ICN-UCLA Symp. Mol. Cell. Biol.* **23:** 209–227.

Castillo, J.M., M.J. Kisiel, and B.M. Zuckerman. 1975. Studies on the effects of two procaine preparations on *Caenorhabditis briggsae*. *Nematologica* **21:** 401–407.

Cayrol, J.C. and J. Brun. 1975. Etude, a différentes températures, de l'activité prédatrice de quelques champignons nématophages vis-a-vis du nématode *Caenorhabditis elegans*. *Rev. Zool. Agric. Pathol. Veg.* **74:** 139–146.

Cayrol, J.C. and B. Dreyfus. 1975. Etudes préliminaires sur les relations entre nématodes libres et bactéries dans le sol. *C.R. Seances Soc. Biol. Fil.* **169:** 166–172.

Cech, T.R. and J.E. Hearst. 1975. An electron microscopy study of mouse foldback DNA. *Cell* **5:** 429–446.

Celis, C.E. and J.D. Smith, eds. 1979. Nonsense mutations and tRNA suppressors. In *Proceedings of the EMBO Lab Course and Aarhus University 50 Year Anniversary Symposium July 78*. Academic Press, London.

Certa, U. and G. von Ehrenstein. 1981. Reversed-phase high-performance liquid chromatology of histones. *Anal. Biochem.* **118:** 147–154.

Chalfie, M. 1982. Microtubule structure in *Caenorhabditis elegans* neurons. *Cold Spring Harbor Symp. Quant. Biol.* **46:** 255–261.

———. 1984a. Genetic analysis of nematode nerve-cell differentiation. *Bioscience* **34:** 295–299.

———. 1984b. Neuronal development in *Caenorhabditis elegans*. *Trends Neurosci.* **7:** 197–202.

Chalfie, M. and J. Sulston. 1981. Developmental genetics of the mechanosensory neurons of *Caenorhabditis elegans*. *Dev. Biol.* **82:** 358–370.

Chalfie, M. and J.N. Thomson. 1979. Organization of neuronal microtubules in the nematode *Caenorhabditis elegans*. *J. Cell Biol.* **82:** 278–289.

———. 1982. Structural and functional diversity in the neuronal microtubules of *Caenorhabditis elegans*. *J. Cell Biol.* **93:** 15–23.

Chalfie, M., H.R. Horvitz, and J.E. Sulston. 1981. Mutations that lead to reiterations in the cell lineages of *Caenorhabditis elegans*. *Cell* **24:** 59–69.

Chalfie, M., J.N. Thomson, and J.E. Sulston. 1983. Induction of neuronal branching in *Caenorhabditis elegans*. *Science* **221:** 61–63.

Chalfie, M., J.E. Sulston, J.G. White, E. Southgate, J.N. Thomson, and S. Brenner. 1985. The neural circuit for touch sensitivity in *Caenorhabditis elegans*. *J. Neurosci.* **5:** 956–964.

Chandra, H.S. 1985. Sex determination: A hypothesis based on noncoding DNA. *Proc. Natl. Acad. Sci.* **82:** 1165–1169.

Charnar, Y. and J.L. Brun. 1982. Division and endopolyploidization in intestinal nuclei during postnatal ontogenesis of *Caenorhabditis elegans* (Nematoda). *Rev. Nematol.* **5:** 155–160.

Cheng, A.C., N.C. Lu, G.M. Briggs, and E.L.R. Stokstad. 1979. Effect of particulate materials on population growth of the free living nematode *Caenorhabditis briggsae*. *Proc. Soc. Exp. Biol. Med.* **160:** 203–207.

Chitwood, B.G. 1930. Studies on some physiological functions and morphological characters of *Rhabditis* (Rhabditidae, Nematodes). *J. Morphol.* **49:** 251–275.

Chitwood, B.G. and M.B. Chitwood. 1974. *Introduction to nematology*. University Park Press, Baltimore, Maryland.

Chitwood, D.J., R. Lozano, and W.R. Lusby. 1986. Recent developments in nematode steroid biochemistry. *J. Nematol.* **18:** 9–17.

Chitwood, D.J., W.R. Lusby, R. Lozano, M.J. Thompson, and J.A. Svoboda. 1983. Novel nuclear methylation of sterols by the nematode *Caenorhabditis elegans*. *Steroids* **42:** 311–319.

———. 1984. Sterol metabolism in the nematode *Caenorhabditis elegans*. *Lipids* **19:** 500–506.

Christie, J.R. 1929. Some observations on sex in the Mermithidue. *J. Exp. Zool.* **53:** 59–76.

Ciampi, M.S., D.A. Melton, and R. Cortese. 1982. Site-directed mutagenesis of a tRNA gene: Base alterations in the coding region affect transcription. *Proc. Natl. Acad. Sci.* **79:** 1388–1392.

Ciliberto, G., C. Traboni, and R. Cortese. 1982a. Relationship between the two components of the split promoter of eukaryotic tRNA genes. *Proc. Natl. Acad. Sci.* **79:** 1921–1925.

Ciliberto, G., L. Castagnoli, D.A. Melton, and R. Cortese. 1982b. Promoter of a eukaryotic $tRNA^{pro}$ gene is composed of three noncontiguous regions. *Proc. Natl. Acad. Sci.* **79:** 1195–1199.

Ciliberto, G., G. Raugei, F. Costanzo, L. Dente, and R. Cortese. 1983. Common and interchangeable elements in the promoters of genes transcribed by RNA polymerase III. *Cell* **32:** 725–733.

Cline, T. 1984. Autoregulatory functioning of a *Drosophila* gene product that establishes and maintains the sexually determined state. *Genetics* **107:** 231–277.

Clokey, G.V. and L.A. Jacobson. 1986. The autofluorescent "lipofuscin granules" in the intestinal cells of *Caenorhabditis elegans* are secondary lysosomes. *Mech. Ageing Dev.* **35:** 79–94.

Cohen, C. 1975. The protein switch of muscle contraction. *Sci. Am.* **233(5):** 36–45.

Collins, J., B. Saari, and P. Anderson. 1987. Activation of a transposable element in the germ line but not the soma of *Caenorhabditis elegans*. *Nature* **328:** 726–728.

Colonna, W.J. and B.A. McFadden. 1975. Isocitrate lyase from parasitic and free living nematodes. *Arch. Biochem. Biophys.* **170:** 608–619.

Conklin, E.G. 1905. The organization and cell-lineage of the ascidian egg. *J. Acad. Natl. Sci. Phila.* **13:** 1–119.

Cooper, A.F. and S.D. Van Gundy. 1970. Metabolism of glycogen and neutral lipids by *Aphelenchus avenae* and *C.* sp. in aerobic, microaerobic, and anaerobic environments. *J. Nematol.* **2:** 305–315.

———. 1971a. Senescence, quiescence, and cryptobiosis. In *Plant parasitic nematodes*, volume II: *Cytogenetics, host-parasite interactions, and physiology* (ed. B.M. Zuckerman et al.), pp. 297–315. Academic Press, New York.

———. 1971b. Ethanol production and utilization by *Aphelenchus avenae* and *C.* sp. J. Nematol. **3:** 205–214.

Cooper, A.F., Jr., S.D. Van Gundy, and L.H. Stolzy. 1970. Nematode reproduction in environments of fluctuating aeration. *J. Nematol.* **2:** 182–188.

Cortese, R., R. Harland, and D. Melton. 1980. Transcription of tRNA genes *in vivo*: Single-stranded compared to double-stranded templates. *Proc. Natl. Acad. Sci.* **77:**4147–4151.

Cortese, R., D. Melton, T. Tranquilla, and J.D. Smith. 1978. Cloning of nematode tRNA genes and their expression in the frog oocyte. *Nucleic Acids Res.* **5:** 4593–4611.

Coulson, A., J. Sulston, S. Brenner, and J. Karn. 1986. Towards a physical map of the genome of the nematode *Caenorhabditis elegans*. *Proc. Natl. Acad. Sci.* **83:** 7821–7825.

Cowan, A.E. and J.R. McIntosh. 1985. Mapping the distribution of differentiation potential for intestine, muscle, and hypodermis during early development in *Caenorhabditis elegans*. *Cell* **41:** 923–932.

Cowan, W. and J.A. Ferrendelli, eds. 1977. Approaches to the cell biology of neurons. *Soc. Neurosci. Symp.* **2:** 1–461.

Cox, G.N. and D. Hirsh. 1985. Stage-specific patterns of collagen gene-expression during development of *Caenorhabditis elegans*. *Mol. Cell. Biol.* **5:** 363–372.

Cox, G.N., J.M. Kramer, and D. Hirsh. 1984. Number and organization of collagen genes in *Caenorhabditis elegans*. *Mol. Cell. Biol.* **4:** 2389–2395.

Cox, G.N., M. Kusch, and R.S. Edgar. 1981a. Cuticle of *Caenorhabditis elegans*: Its isolation and partial characterization. *J. Cell Biol.* **90:** 7–17.

Cox, G.N., S. Staprans, and R.S. Edgar. 1981b. The cuticle of *Caenorhabditis elegans*. II. Stage-specific changes in ultrastructure and protein composition during postembryonic development. *Dev. Biol.* **86:** 456–470.

Cox, G.N., S. Carr, J.M. Kramer, and D. Hirsh. 1985. Genetic mapping of *Caenorhabditis elegans* collagen genes using DNA polymorphisms as phenotypic markers. *Genetics* **109:** 513–528.

Cox, G.N., M. Kusch, K. DeNevi, and R.S. Edgar. 1981c. Temporal regulation of cuticle synthesis during development of *Caenorhabditis elegans*. *Dev. Biol.* **84:** 277–285.

Cox, G.N., J.S. Laufer, M. Kusch, and R.S. Edgar. 1980. Genetic and phenotypic characterization of roller mutants of *Caenorhabditis elegans*. *Genetics* **95:** 317–339.

Cristofalo, V.J. 1985. *CRC Handbook of cell biology of aging*. CRC Press, Boca Raton, Florida.

Croll, N.A. 1975a. Components and patterns in the behavior of the nematode *Caenorhabditis elegans*. *J. Zool.* **176:** 159–176.

———. 1975b. Indole alkylamines in the coordination of nematode behavioral activities. *Can. J. Zool.* **53:** 894–903.

———. 1976a. Behavioral coordination of nematodes. In *The organization of nematodes* (ed. N.A. Croll), pp. 343–364. Academic Press, New York.

———. 1976b. When *Caenorhabditis elegans* (Nematoda: Rhabditidae) bumps into a bead. *Can. J. Zool.* **54:** 566–570.

———, ed. 1976c. *The organization of nematodes*. Academic Press, New York.

———. 1977. Sensory mechanisms in nematodes. *Annu. Rev. Phytopathol.* **15:** 75–89.

Croll, N.A. and B.E. Matthews. 1977. *Biology of nematodes*. Halsted Press, New York.

Croll, N.A. and J.M. Smith. 1978. Integrated behaviour in the feeding phase of *Caenorhabditis elegans* (Nematoda). *J. Zool.* **184:** 507–517.

Croll, N.A., J.M. Smith, and B.M. Zuckerman. 1977. The aging process of the nematode *Caenorhabditis elegans* in bacterial and axenic culture. *Exp. Aging Res.* **3:** 175–189.

Crowther, R.A., R. Padron, and R. Craig. 1985. Arrangement of the heads of myosin in relaxed thick filaments from Tarantula muscle. *J. Mol. Biol.* **184:** 429–439.

Cryan, W.S. 1963. A method for axenizing large numbers of nematodes. *J. Parasitol.* **49:** 351–352.

Culotti, J.G. and W.L. Klein. 1983. Occurrence of muscarinic acetylcholine receptors in wild-type and cholinergic mutants of *Caenorhabditis elegans*. *J. Neurosci.* **3:** 359–368.

Culotti, J.G. and R.L. Russell. 1978. Osmotic avoidance defective mutants of the nematode *Caenorhabditis elegans*. *Genetics* **90:** 243–256.

Culotti, J.G., G. von Ehrenstein, M.R. Culotti, and R.L. Russell. 1981. A second class of acetylcholinesterase-deficient mutants of the nematode *Caenorhabditis elegans*. *Genetics* **97:** 281–305.

Curran, J., D.L. Baillie, and J.M. Webster. 1985. Use of genomic DNA restriction fragment length differences to identify nematode species. *Parasitology* **90:** 137–144.

Davidson, E.H. 1976. *Gene activity in early development*. Academic Press, New York.

Davidson, E.H. and R.A. Firtel, eds. 1984. Molecular biology of development. *UCLA Symp. Mol. Cell. Biol. New Ser.* **19:** 1–685.

Davidson, E.H., B.R. Hough, C.S. Amenson, and R.J. Britten. 1973. General interspersion of repetitive with non-repetitive sequence elements in the DNA of *Xenopus*. *J. Mol. Biol.* **77:** 1–23.

Davis, B.O., G.L. Anderson, and D.B. Dusenbery. 1982. Total luminescence spectroscopy of fluorescence changes during aging in *Caenorhabditis elegans*. *J. Biochem.* **21:** 4089–4095.

Davis, B.O., M. Goode, and D.B. Dusenbery. 1986. Laser microbeam studies of the amphid receptors in chemosensory behavior of the nematode *Caenorhabditis elegans*. *J. Chem. Ecol.* **12:** 1339–1347.

Davis, R.E. and A.O.W. Stretton. 1981. Intracellular recordings from identified motorneuron in the nematode *Ascaris*. *Soc. Neurosci. Abstr.* **7:** 745.

———. 1982. Motorneuron membrane constants and signalling properties in the nematode *Ascaris*. *Soc. Neurosci. Abstr.* **8:** 685.

———. 1983. Motorneuron signalling properties and motorneuron-motorneuron synaptic interactions in the nematode *Ascaris*. *Soc. Neurosci. Abstr.* **9:** 301.

deBoer, H.A. and R.A. Kastelein. 1986. Biased codon usage: An exploration of its role in optimization of translation. In *Maximizing gene expression* (ed. W. Reznikoff and L. Gold), pp. 255–285. Butterworth, London.

DeCuyper, C. and J.R. Vanfleteren. 1982a. Nutritional alteration of life span in the nematode *Caenorhabditis elegans. Age* **5:** 42–45.

———. 1982b. Oxygen consumption during development and aging of the nematode *Caenorhabditis elegans. Comp. Biochem. Physiol.* **73A:** 283–289.

Delavault, R. 1952a. Etude cytologique des acides nucléiques chez un nématode libre (*Rhabditis elegans* Maupas 1900). *Arch. Anat. Micro. Morphol. Exp.* **41:** 41–68.

———. 1952b. La teneur en acide desoxyribonucléique des noyaux sexuels chez un Rhabditis hermaphrodite. *C.R. Seances Acad. Sci. Ser. D Sci. Nat.* **234:** 884–885.

———. 1957. Croissance des ovotestis puis des ovaires chez un nématode libre hermaphrodite: *Caenorhabditis elegans* Maupas 1900. *Bull. Soc. Zool. Fr.* pp. 321–325.

———. 1959. Développement, croissance et fonctionnement des glandes génitales chez les nématodes libres. *Arch. Zool. Exp. Gen.* **97:** 109–208.

del Castillo, J., W.C. de Mello, and T. Morales. 1963. The physiological role of acetylcholine in the neuromuscular system of *Ascaris lumbricoides. Arch. Int. Physiol. Biochim.* **71:** 741–757.

———. 1964. Inhibitory action of γ-aminobutyric acid (GABA) on *Ascaris* muscle. *Experientia* **20:** 141–143.

———. 1967. Initiation of action potentials in the somatic musculature of *Ascaris lumbricoides. J. Exp. Biol.* **65:** 773–788.

Denich, K.T.R., E. Schierenberg, E. Isnenghi, and R. Cassada. 1984. Cell-lineage and developmental defects of temperature-sensitive embryonic arrest mutants of the nematode *Caenorhabditis elegans. Wilhelm Roux's Arch. Dev. Biol.* **193:** 164–179.

Deniro, M.J. and S. Epstein. 1978. Influence of diet on the distribution of carbon isotopes in animals. *Geochim. Cosmochim. Acta* **42:** 495–506.

Dente, L., O. Fasano, F. Costanzo, C. Traboni, G. Ciliberta, and R. Cortese. 1982. A prokaryotic tRNA*tyr* gene, inactive in *Xenopus laevis* oocytes, is activated by recombination with an eukaryotic tRNA*pro* gene. *EMBO J.* **1:** 817–820.

Deppe, U., E. Schierenberg, T. Cole, C. Krieg, D. Schmitt, B. Yoder, and G. von Ehrenstein. 1978. Cell lineages of the embryo of the nematode *Caenorhabditis elegans. Proc. Natl. Acad. Sci.* **75:** 376–380.

Deubert, K.H. and B.M. Zuckerman. 1968. The histochemical demonstration of cytochrome oxidase in fresh-frozen sections of nematodes. *Nematologica* **14:** 453–455.

Dibb, N.J., D.M. Brown, J. Karn, D.G. Moerman, S.L. Bolten, and R.H. Waterston. 1985. Sequence analysis of mutations that affect the synthesis, assembly and enzymatic activity of the *unc-54* myosin heavy chain of *Caenorhabditis elegans. J. Mol. Biol.* **183:** 543–551.

Dion, M. and J.L. Brun. 1971. Genetic mapping of the free-living nematode *Caenorhabditis elegans* Maupas 1900, var. Bergerac. I. Study of two dwarf mutants. *Mol. Gen. Genet.* **112:** 133–151.

Doniach, T. 1986. Activity of the sex-determining gene *tra-2* is modulated to allow spermatogenesis in the *C. elegans* hermaphrodite. *Genetics* **114:** 53–76.

Doniach, T. and J. Hodgkin. 1984. A sex-determining gene, *fem-1*, required for both male and hermaphrodite development in *Caenorhabditis elegans. Dev. Biol.* **106:** 223–235.

Doolittle, W.F. and C. Sapienza. 1980. Selfish genes, the phenotype paradigm and genome evolution. *Nature* **284:** 601–603.

Dougherty, E.C. 1950. Sterile pieces of chick embryo as a medium for the indefinite axenic cultivation of *Rhabditis briggsae* Dougherty and Nigon, 1949 (Nematoda : Rhabditidae). *Science* **111:** 258.

———. 1951. The axenic cultivation of *Rhabditis briggsae* Dougherty and Nigon, 1949. II. Some sources and characteristics of "factor Rb." *Exp. Parasitol.* **1:** 34–45.

———. 1953a. The genera of the subfamily Rhabditinae Micolitzky 1922 (Nematoda). *G.S. Thepar. Commem.*, pp. 69–76.

———. 1953b. The axenic cultivation of *Rhabditis briggsae* Dougherty and Nigon, 1949. III. Liver preparation with various supplementation. *J. Parasitol.* **39:** 371–380.

———. 1955. The genera and species of subfamily Rhabditinae: A nomenclatorial analysis—Including an addendum on comparison of family Rhabditidae. *J. Helminthol.* **29:** 105–152.

———. 1959. Introduction to axenic culture of invertebrate metazoa: A goal. *Ann. N.Y. Acad. Sci.* **77:** 27–54.

———. 1960. Cultivation of Aschelminths, especially Rhabditid nematodes. In *Nematology* (ed. J.N. Sasser and W.R. Jenkins), pp. 297–318. University of North Carolina Press, Chapel Hill.

Dougherty, E.C. and H.G. Calhoun. 1948. Possible significance of free-living nematodes in genetic research. *Nature* **161:** 29.

Dougherty, E.C. and E.L. Hansen. 1956. Axenic cultivation of *Caenorhabditis briggsae* (Nematoda : Rhabditidae). V. Maturation on synthetic media. *Proc. Soc. Exp. Biol. Med.* **93:** 223–227.

———. 1957. The folic acid requirement and its antagonism by aminopterin in the nematode *Caenorhabditis briggsae* (Rhabditidae). *Anat. Rec.* **128:** 541–542.

Dougherty, E.C. and D.F. Keith. 1953. The axenic cultivation of *Rhabditis briggsae*. IV. Plasma protein fractions with various supplementation. *J. Parasitol.* **39:** 381–384.

Dougherty, E.C. and V. Nigon. 1953. The effect of "acriflavine" (2,8-diaminoacridine and 2,8-diamino-10-methyl-acridinium-chloride) on growth of *Caenorhabditis elegans*. *Int. Congr. Zool.* **14:** 247–249.

Dougherty, E.C., J.C. Raphael, and C.M. Alton. 1960. The axenic cultivation of *Rhabditis briggsae*. I. Experiments with chick embryo juice and chemically defined media. *Proc. Helminthol. Soc. Wash.* **17:** 1–10.

Dougherty, E.C., E.L. Hansen, W.L. Nicholas, J.A. Mollett, and E.A. Yarwood. 1959. Axenic cultivation of *Caenorhabditis briggsae* with unsupplemented and supplemented chemically defined media. *Ann. N.Y. Acad. Sci.* **77:** 176-217.

Drechsler, C. 1936. A new species of Stylopage preying on nematodes. *Mycologia* **28:** 241–246.

———. 1940a. Three fungi destructive to free-living terricolous nematodes. *J. Wash. Acad. Sci.* **30:** 240–254.

———. 1940b. Three new hyphomycetes preying on free-living terricolous nematodes. *Mycologia* **32:** 448–470.

———. 1941. Some hyphomycetes parasitic on free-living terricolous nematodes. *Phytopathology* **31:** 773–802.

———. 1944. Three hyphomycetes that capture nematodes in adhesive networks. *Mycologia* **36:** 138–171.

———. 1946. A species of *Harposporium* invading its nematode host from the stoma. *Bull. Torrey. Bot. Club* **73:** 557–564.

Dropkin, V.H., W.R. Lower, and J. Acedo. 1971. Growth inhibition of *Caenorhabditis elegans* and *Panagrellus redivivus* by selected mammalian and insect hormones. *J. Nematol.* **3:** 349–355.

Durbin, R.M. 1987. "Studies on the development and organisation of the nervous system of *Caenorhabditis elegans*." Ph.D. thesis, University of Cambridge, England.

Dusenbery, D.B. 1973. Countercurrent separation: A new method for studying behavior of small aquatic organisms. *Proc. Natl. Acad. Sci.* **70:** 1349–1352.

———. 1974. Analysis of chemotaxis in the nematode *Caenorhabditis elegans* by countercurrent separation. *J. Exp. Zool.* **188:** 41–47.

———. 1975. The avoidance of D-tryptophan by the nematode *Caenorhabditis elegans*. *J. Exp. Zool.* **193:** 413–418.

———. 1976a. Chemotactic behavior of mutants of the nematode *Caenorhabditis elegans* that are defective in their attraction to NaCl. *J. Exp. Zool.* **198:** 343–352.
———. 1976b. Attraction of the nematode *Caenorhabditis elegans* to pyridine. *Comp. Biochem. Physiol.* **53C:** 1–2.
———. 1976c. Chemotactic responses of male *Caenorhabditis elegans. J. Nematol.* **8:** 352–355.
———. 1980a. Responses of nematode *Caenorhabditis elegans* to controlled chemical stimulation. *J. Comp. Physiol.* **136:** 327–331.
———. 1980b. Chemotactic behavior of mutants of the nematode *Caenorhabditis elegans* that are defective in osmotic avoidance. *J. Comp. Physiol.* **137:** 93–96.
———. 1980c. Appetitive response of nematode *Caenorhabditis elegans* to oxygen. *J. Comp. Physiol.* **136:** 333–336.
———. 1980d. Behavior of free-living nematodes. In *Nematodes as biological models*, volume 1: *Behavioral and developmental models* (ed. B.M. Zuckerman), pp. 127–158. Academic Press, New York.
———. 1985a. Video camera-computer tracking of the nematode *Caenorhabditis elegans* to record behavioral responses. *J. Chem. Ecol.* **2:** 1239–1247.
———. 1985b. Using a microcomputer and video camera to simultaneously track 25 animals. *Comput. Biol. Med.* **15:** 169–175.
Dusenbery, D.B. and J. Barr. 1980. Thermal limits and chemotaxis in mutants of nematode *Caenorhabditis elegans* defective in thermotaxis. *J. Comp. Physiol.* **137:** 353–356.
Dusenbery, D.B., G.L. Anderson, and E.A. Anderson. 1978. Thermal acclimation more extensive for behavioral parameters than for oxygen consumption in the nematode *Caenorhabditis elegans. J. Exp. Zool.* **206:** 191–198.
Dusenbery, D.B., R.E. Sheridan, and R.L. Russell. 1975. Chemotaxis-defective mutants of the nematode *Caenorhabditis elegans. Genetics* **80:** 297–309.
Eddy, E.M. 1975. Germ plasm and the differentiation of the germ cell line. *Int. Rev. Cytol.* **43:** 229–280.
Edgar, B.A., C.P. Kiehle, and G. Schubiger. 1986. Cell cycle control by the nucleocytoplasmic ratio in early *Drosophila* development. *Cell* **44:** 365–372.
Edgar, L.G. and D. Hirsh. 1985. Use of a psoralen-induced phenocopy to study genes controlling spermatogenesis in *Caenorhabditis elegans. Dev. Biol.* **111:** 108–118.
Edgar, L.G. and J.D. McGhee. 1986. Embryonic expression of a gut-specific esterase in *Caenorhabditis elegans. Dev. Biol.* **114:** 109–118.
———. 1988. DNA synthesis and the control of embryonic gene expression in *Caenorhabditis elegans. Cell* (in press).
Edgar, R.S. 1980. The genetics of development in the nematode *Caenorhabditis elegans*. In *The molecular genetics of development* (ed. T. Leighton and W.F. Loomis), pp. 213–235. Academic Press, New York.
Edgar, R.S. and I. Lielausis. 1968. Some steps in the assembly of bacteriophage T4. *J. Mol. Biol.* **32:** 263–276.
Edgar, R.S. and W.B. Wood. 1977. The nematode *Caenorhabditis elegans*, a new organism for intensive biology study. *Science* **198:** 1285–1286.
Edgar, R.S., G.N. Cox, M. Kusch, and J.G. Politz. 1982. The cuticle of *Caenorhabditis elegans. J. Nematol.* **14:** 248–258.
Edwards, M.K. and W.B. Wood. 1983. Location of specific messenger RNAs in *Caenorhabditis elegans* by cytological hybridization. *Dev. Biol.* **97:** 375–390.
Eide, D. and P. Anderson. 1985a. The gene structures of spontaneous mutations affecting a *Caenorhabditis elegans* myosin heavy chain gene. *Genetics* **109:** 67–79.
———. 1985b. Transposition of Tc1 in the nematode *Caenorhabditis elegans. Proc. Natl. Acad. Sci.* **82:** 1756–1760.
———. 1985c. Novel insertion mutation in *Caenorhabditis elegans. Mol. Cell. Biol.* **5:** 1–6.

Ellis, H.M. and H.R. Horvitz. 1986. Genetic control of programmed cell death in the nematode *Caenorhabditis elegans*. *Cell* **44:** 817–829.

Ellis, R.E., J.E. Sulston, and A.R. Coulson. 1986. The rDNA of *Caenorhabditis elegans*: Sequence and structure. *Nucleic Acids Res.* **14:** 2345–2364.

Emerson, C., D.A. Fischman, B. Nadal-Ginard, and M.A.Q. Siddiqui, eds. 1986. Molecular biology of muscle development. *UCLA Symp. Mol. Cell. Biol. New Ser.* **29**: 1–956.

Emmons, S.W. and L. Yesner. 1984. High-frequency extension of transposable element Tc1 in the nematode *Caenorhabditis elegans* is limited to somatic cells. *Cell* **36:** 599–605.

Emmons, S.W., M.R. Klass, and D. Hirsh. 1979. Analysis of the constancy of DNA sequences during development and evolution of the nematode *Caenorhabditis elegans*. *Proc. Natl. Acad. Sci.* **76:** 1333–1337.

Emmons, S.W., S. Roberts, and K.-S. Ruan. 1986. Evidence in a nematode for regulation of transposon excision by tissue-specific factor. *Mol. Gen. Genet.* **202:** 410–415.

Emmons, S.W., B. Rosenzweig, and D. Hirsh. 1980. Arrangement of repeated sequences in the DNA of the nematode *Caenorhabditis elegans*. *J. Mol. Biol.* **144:** 481–500.

Emmons, S.W., K.S. Ruan, A. Levitt, and L. Yesner. 1985. Regulation of Tc1 transposable elements in *Caenorhabditis elegans*. *Cold Spring Harbor Symp. Quant. Biol.* **50:** 313–320.

Emmons, S.W., L. Yesner, K.-S. Ruan, and D. Katzenberg. 1983. Evidence for a transposon in *Caenorhabditis elegans*. *Cell* **32:** 55–65.

Epstein, H.F. 1986. Differential roles of myosin isoforms in filament assembly. *UCLA Symp. Mol. Cell. Biol. New Ser.* **29:** 653–666.

Epstein, H.F. and J.N. Thomson. 1974. Temperature sensitive mutation affecting myofilament assembly in *Caenorhabditis elegans*. *Nature* **250:** 579–580.

Epstein, H.F., S.A. Berman, and D.M. Miller III. 1982a. Myosin synthesis and assembly in nematode body-well muscle. In *Muscle development: Molecular and cellular control* (ed. M. Pearson and H.F. Epstein), pp. 419–427. Cold Spring Harbor Laboratory, Cold Spring Harbor, New York.

Epstein, H.F., M.M. Isachsen, and E.A. Suddleson. 1976a. Kinetics of movement of normal and mutant nematodes. *J. Comp. Physiol.* **110:** 317–322.

Epstein, H.F., I. Ortiz, and L.A.T. Mackinnon. 1986. The alteration of myosin isoform compartmentation in specific mutants of *Caenorhabditis elegans*. *J. Cell. Biol.* **103:** 985–993.

Epstein, H.F., F.H. Schachat, and J.A. Wolff. 1976b. Molecular genetics of nematode myosin. In *Pathogenesis of the human muscular dystrophies* (ed. L.P. Rowland), pp. 460–467. Excerpta Medica, Amsterdam.

Epstein, H.F., R.H. Waterston, and S. Brenner. 1974. A mutant affecting the heavy chain of myosin in *Caenorhabditis elegans*. *J. Mol. Biol.* **90:** 291-300.

Epstein, H.F., D.M. Miller III, L.A. Gossett, and R.M. Hecht. 1982b. Immunological studies of myosin isoforms in nematode development. In *Muscle development: Molecular and cellular control* (ed. M. Pearson and H.F. Epstein), pp. 7–14. Cold Spring Harbor Laboratory, Cold Spring Harbor, New York.

Epstein, H.F., D.M. Miller III, I. Ortiz, and G.C. Berliner. 1985. Myosin and paramyosin are organized about a newly identified core structure. *J. Cell Biol.* **100:** 904–915.

Epstein, H.F., I. Ortiz, G.C. Berliner, and D.M. Miller III. 1984. Nematode thick-filament structure and assembly. In *Molecular biology of the cytoskeleton* (ed. G. Borisy et al.), pp. 275–286. Cold Spring Harbor Laboratory, Cold Spring Harbor, New York.

Epstein, H.F., H.E. Harris, F.H. Schachat, E.A. Suddleson, and J.A. Wolff. 1976c. Genetic and molecular studies of nematode myosin. *Cold Spring Harbor Conf. Cell Proliferation* **3:** 203–214.

Epstein, J. and D. Gershon. 1972. Studies on aging in nematodes. IV. The effect of antioxidants on cellular damage and life span. *Mech. Ageing Dev.* **1:** 257–264.

Epstein, J., S. Himmelhoch, and D. Gershon. 1972. Studies on aging in nematodes. III. Electronmicroscopical studies on age-associated cellular damage. *Mech. Ageing Dev.* **1:** 245–255.

Epstein, J., J. Castillo, S. Himmelhoch, and B.M. Zuckerman. 1971. Ultrastructural studies on *Caenorhabditis briggsae*. *J. Nematol.* **3:** 69–78.

Evans, A.A.F. and R.M. Perry. 1976. Survival strategies in nematodes. In *The organization of nematodes* (ed. N.A. Croll), pp. 383–424. Academic Press, New York.

Fatt, H.V. 1967. Nutritional requirements for reproduction of a temperature-sensitive nematode, reared in axenic culture. *Proc. Soc. Exp. Biol. Med.* **124:** 897–903.

Fatt, H.V. and E.C. Dougherty. 1963. Genetic control of differential heat tolerance in two strains of the nematode *Caenorhabditis elegans*. *Science* **141:** 266–267.

Fedoroff, N., S. Wessler, and M. Shure. 1983. Isolation of the transposable maize controlling elements *Ac* and *Ds*. *Cell* **35:** 235–242.

Felsenstein, K.M. and S.W. Emmons. 1987. Structure and evolution of a family of interspersed repetitive DNA sequences in *Caenorhabditis elegans*. *J. Mol. Evol.* **25:** 230–240.

———. 1988. Nematode repetitive DNA with *ARS* and segregation function in *Saccharomyces cerevisiae*. *Mol. Cell. Biol.* (in press).

Ferguson, E.L. and H.R. Horvitz. 1985. Identification and characterization of 22 genes that affect the vulval cell lineages of the nematode *Caenorhabditis elegans*. *Genetics* **110:** 17–72.

Ferguson, E., P. Sternberg, and R. Horvitz. 1987. A genetic pathway for the specification of the vulval cell lineages of *Caenorhabditis elegans*. *Nature* **326:** 259–267.

Fiakpui, E.Z. 1967. Some effects of piperazine and methyridine on the free-living nematode *Caenorhabditis briggsae* (Rhabditidae). *Nematologica* **13:**241–255.

Files, J.G. and D. Hirsh. 1981. Ribosomal DNA of *Caenorhabditis elegans*. *J. Mol. Biol.* **149:** 223–240.

Files, J.G., S. Carr, and D. Hirsh. 1983. Actin gene family of *Caenorhabditis elegans*. *J. Mol. Biol.* **164:** 355–375.

Finch, C.E. and E.L. Schneider, eds. 1985. *Handbook of the biology of aging*. Van Nostrand Reinhold, New York.

Findeis, P.M., C.J. Barinaga, J.D. Willett, and S.O. Farwell. 1983. Age-synchronous culture of *Caenorhabditis elegans:* Technique and applications. *Exp. Gerontol.* **18:** 263–275.

Fire, A. 1986. Integrative transformation of *Caenorhabditis elegans*. *EMBO J.* **5:** 2673–2680.

Fixsen, W., P. Sternberg, H. Ellis, and R. Horvitz. 1985. Genes that affect cell fates during the development of *Caenorhabditis elegans*. *Cold Spring Harbor Symp. Quant. Biol.* **50:** 99–104.

Florkin, M. and B.T. Scheer, eds. 1969. *Chemical zoology*, volume III: *Echinodermata, nematoda and acanthocephala*. Academic Press, New York.

Fodor, A. and P. Deak. 1982. Isolation and phenocritical period-analysis of conditional and non-conditional developmental mutants in *Caenorhabditis elegans*. *Acta Biol. Acad. Sci. Hung.* **32:** 229–239.

———. 1983. *Caenorhabditis elegans* as a genetic model. *Allattani Kozl.* **69:** 91–98.

———. 1985. The isolation and genetic analysis of a *Caenorhabditis elegans* translocation (szT1) strain bearing a X-chromosome balancer. *J. Genet.* **64:**143–167.

Fodor, A., P. Deak, and I. Kiss. 1982. Competition between juvenile hormone antagonist precocene II and juvenile hormone analog: Methoprene in the nematode *Caenorhabditis elegans*. *Gen. Comp. Endocrinol.* **46:** 99–109.

Fodor, A., D.L. Riddle, F.K. Nelson, and J.W. Golden. 1983. Comparison of a new wild-type *Caenorhabditis briggsae* with laboratory strains of *Caenorhabditis briggsae* and *Caenorhabditis elegans*. *Nematologica* **29:** 203–217.

Foor, W.E. 1970. Spermatozoan morphology and zygote formation in nematodes. *Biol. Reprod.* (suppl.) **2:** 177–202.

Francis, G.R. and R.H. Waterston. 1985. Muscle organization in *Caenorhabditis elegans*: Localization of proteins implicated in thin filament attachment and I-band organization. *J. Cell Biol.* **101:** 1532–1549.

Friedman, P.A., E.G. Platzer, and J.E. Eby. 1977. Species differentiation in *Caenorhabditis briggsae* and *Caenorhabditis elegans*. *J. Nematol.* **9:** 197–203.

Fuchs, A.C. 1937. Neue parasitische und halbparasitische Nematoden bei Borkenkäfern und einige andere Nematoden, I. Teil. *Zool. Jb.* **70:** 291–380.

Fujita, H., N. Ishii, and K. Suzuki. 1984. Effects of 8-methoxypsoralen plus near-ultraviolet light on the nematode *Caenorhabditis elegans*. *Photochem. Photobiol.* **39:** 831–834.

Fuller, G. and W.D. Nes, eds. 1985. Ecology and metabolism of plant lipids. *ACS Symp. Ser.* **325.**

Gabius, H.G., G. Graupner, and F. Cramer. 1983. Activity patterns of amino acyl-tRNA synthetases, tRNA methylases, arginyltransferase and tubulin-tyrosine ligase. *Eur. J. Biochem.* **131:** 231–234.

Gall, J.G. 1986. Gametogenesis and the early embryo. In *The 44th Symposium of the Society for Developmental Biology*. Alan R. Liss, New York.

Galsky, A.G., H.L. Monoson, and S.A. Williams. 1974. A bioassay for measuring crude nematode extract which induced trap formation in a predaceous fungus. *Nematologica* **20:** 39–42.

Gandhi, S., J. Santelli, D.H. Mitchell, J.W. Stiles, and D.R. Sanadi. 1980. A simple method for maintaining large aging populations of *Caenorhabditis elegans*. *Mech. Ageing Dev.* **12:** 137–150.

Garcea, R.L., F. Schachat, and H.F. Epstein. 1978. Coordinate synthesis of two myosins in wild-type and mutant nematode muscle during larval development. *Cell* **15:** 421–428.

Gehring, W.J., ed. 1978. *Genetic mosaics and cell differentiation*. Springer-Verlag, New York.

———. 1987. Homeo boxes in the study of development. *Science* **236:** 1245–1252.

Gergen, J.P., D. Coulter, and E. Wieschaus. 1986. Sequential pattern and blastoderm cell identities. *Symp. Soc. Dev. Biol.* **44:** 195–220.

Gershon, E.S., S. Matthysse, X.O. Breakefield, and R.D. Ciaranello, eds. 1981. *Genetic research strategies in psychobiology and psychiatry. Psychobiology and psychopathology,* volume I. Boxwood Press, Pacific Grove, California.

Ghiselin, M. 1969. The evolution of hermaphroditism among animals. *Q. Rev. Biol.* **44:** 189–208.

Gibert, M.A., J. Starck, and B. Beguet. 1984. Role of the gonad cytoplasmic core during oogenesis of the nematode *Caenorhabditis elegans*. *Biol. Cell* **50:** 77–85.

Gochnauer, M.B. and E. McCoy. 1954. Responses of a soil nematode *Rhabditis briggsae* to antibiotics. *J. Exp. Zool.* **125:** 377–406.

Goddard, J.M., J.J. Weiland, and M.R. Capecchi. 1986. Isolation and characterization of *Caenorhabditis elegans* DNA sequences homologous to the *v-abl* oncogene. *Proc. Natl. Acad. Sci.* **83:** 2172–2176.

Golden, J.W. and D.L. Riddle. 1982. A pheromone influences larval development in the nematode *Caenorhabditis elegans*. *Science* **218:** 578–580.

———. 1984a. The *Caenorhabditis elegans* dauer larva: Developmental effects of pheromone, food, and temperature. *Dev. Biol.* **102:** 368–378.

———. 1984b. A *Caenorhabditis elegans* dauer-inducing pheromone and an antagonistic component of the food supply. *J. Chem. Ecol.* **10:** 1265–1280.

———. 1984c. A pheromone-induced developmental switch in *Caenorhabditis elegans*: Temperature-sensitive mutants reveal a wild-type temperature-dependent process. *Proc. Natl. Acad. Sci.* **81:** 819–823.

———. 1985. A gene affecting production of the *Caenorhabditis elegans* dauer-inducing pheromone. *Mol. Gen. Genet.* **198:** 534–536.

Goldstein, P. 1982. The synaptonemal complexes of *Caenorhabditis elegans*: Pachytene karyotype analysis of male and hermaphrodite wild-type and him mutants. *Chromosoma* **86:** 577–593.

———. 1984a. Triplo-X hermaphrodite of *Caenorhabditis elegans*: Pachytene karyotype analysis, synaptonemal complexes, and pairing mechanisms. *Can. J. Genet. Cytol.* **26:** 13–17.

———. 1984b. Sterile mutants in *Caenorhabditis elegans*: The synaptonemal complex as an indicator of the stage-specific effect of the mutation. *Cytobios* **39:** 101–108.

———. 1984c. The synaptonemal complexes of *Caenorhabditis elegans*—Pachytene karyotype analysis of the *rad-4* radiation-sensitive mutant. *Mutat. Res.* **129:** 337–343.

———. 1985a. The synaptonemal complexes of *Caenorhabditis elegans*: The dominant *him* mutant mnT6 and pachytene karyotype analysis of the X-autosome translocation. *Chromosoma* **93:**256–260.

———. 1985b. The synaptonemal complexes of *Caenorhabditis elegans*: Pachytene karyotype analysis of the *Dp 1* mutant and disjunction regulator regions. *Chromosoma* **93:** 177–182.

———. 1986a. The synaptonemal complexes of *Caenorhabditis elegans*: Pachytene karyotype analysis of hermaphrodites from the recessive *him-5* and *him-7* mutants. *J. Cell Sci.* **82:** 119–127.

———. 1986b. Nuclear abberations and loss of synaptonemal complexes in response to diethylstilbestrol (DES) in *Caenorhabditis elegans* hermaphrodites. *Mutat. Res.* **174:** 99–107.

Goldstein, P. and D.E. Slaton. 1982. The synaptonemal complexes of *Caenorhabditis elegans*: Comparison of wild-type and mutant strains and pachytene karyotype analysis of wild-type. *Chromosoma* **84:** 585–597.

Goldstein, P. and A.C. Triantaphyllou. 1980. The ultrastructure of sperm development in the plant-parasitic nematode *Meloidogyne hapla*. *J. Ultrastruct. Res.* **71:** 143–153.

Golovin, E. 1901. Nabliudeniia nad nematodami (Okonchanie). *Uch. Zap. Kazan* **68:** 1–50.

Gossett, L.A. and R. Hecht. 1980. A squash technique demonstrating embryonic nuclear cleavage of the nematode *Caenorhabditis elegans*. *J. Histochem. Cytochem.* **28:** 507–510.

Gossett, L., R.M. Hecht, and H.F. Epstein. 1982. Muscle differentiation in normal and cleavage-arrested mutant embryos of *Caenorhabditis elegans*. *Cell* **30:** 193–204.

Gould, S.J. 1977. *Ontogeny and phylogeny*. Belknap, Cambridge, Massachusetts.

Graham, D.E., B.R. Neufeld, E.H. Davidson, and R.J. Britten. 1974. Interspersion of repetitive and non-repetitive DNA sequences in the sea urchin genome. *Cell* **1:** 127–137.

Grasse, P.P., ed. 1965. *Traité de zoologie*, tome IV. Masson et Cie, Paris, France.

Greenwald, I. 1985a. The genetic analysis of cell lineage in *Caenorhabditis elegans*. *Philos. Trans. R. Soc. Lond. B Biol. Sci.* **312:** 129–137.

———. 1985b. *lin-12*, a nematode homeotic gene, is homologous to a set of mammalian proteins that includes epidermal growth factor. *Cell* **43:** 583–590.

Greenwald, I.S. and H.R. Horvitz. 1980. *unc-93(e1500)*: A behavioral mutant of *Caenorhabditis elegans* that defines a gene with a wild-type null phenotype. *Genetics* **96:** 147–164.

———. 1982. Dominant suppressors of a muscle mutant define an essential gene of *Caenorhabditis elegans*. *Genetics* **101:** 211–225.

———. 1986. A visible allele of the muscle gene *sup-10* X of *Caenorhabditis elegans*. *Genetics* **113:** 63–72.

Greenwald, I. and A. Martinez-Arias. 1984. Programmed cell death in invertebrates. *Trends Neurosci.* **7:** 179–181.

Greenwald, I.S., P.W. Sternberg, and H.R. Horvitz. 1983. The *lin-12* locus specifies cell fates in *Caenorhabditis elegans*. *Cell* **34:** 435–444.

Greenwald, I., A. Coulson, J. Sulston, and J. Preiss. 1987. Correlation of the physical and genetic map in the *lin-12* region of *Caenorhabditis elegans*. *Nucleic Acids Res.* **15:** 2295–2307.

Grell, R.F. 1976. Distributive pairing. In *The genetics and biology of* Drosophila (ed. M. Ashburner and E. Novitski), vol. 1a, pp. 436–486. Academic Press, New York.

Guardiola, J., L. Luzzato, and W. Trager, eds. 1983. *Molecular biology of parasites*. Raven Press, New York.

Guerrero, J. and P.H. Silverman. 1971. *Ascaris suum*: Immune reactions in mice. II. Metabolic and somatic antigens from in vitro cultured larvae. *Exp. Parasitol.* **29:** 110–115.

Haase, C.E., J.W. Golden, and D.L. Riddle. 1983. Mutants of *Caenorhabditis elegans* unresponsive to the dauer-inducing pheromone. *Genetics* **104:** s32–33.

Haight, M., J. Frim, J. Pasternak, and H. Frey. 1975. Freeze-thaw survival of the free-living nematode *Caenorhabditis briggsae*. *Cryobiology* **12:** 497–505.

Hall, D.H. 1977. "The posterior nervous system of the nematode *Caenorhabditis elegans*." Ph.D. thesis, California Institute of Technology, Pasadena.

Hall, S.S. 1985. The fate of the egg. *Science* **85:** 40–49.

Hansen, E.L. and A.K. Berntzen. 1969. Development of *Caenorhabditis briggsae* and *Hymenolepsis nana* in interchanged media. *J. Parasitol.* **55:** 1012–1017.

Hansen, E.L. and E.J. Buecher. 1970. Biochemical approach to systematic studies with axenic nematodes. *J. Nematol.* **2:** 1–6.

———. 1971. Effect of insect hormones on nematodes in axenic culture. *Experientia* **27:** 859–860.

Hansen, E.L. and W.S. Cryan. 1966. Continuous axenic culture of free-living nematodes. *Nematologica* **12:** 138–142.

Hansen, E.L. and G. Perez-Mendez. 1970. Large scale preparation of liver growth factor for cultivation of nematodes. *Proc. Soc. Exp. Biol. Med.* **135:** 487–489.

Hansen, E.L., E.J. Buecher, and E.A. Yarwood. 1964. Development and maturation of *Caenorhabditis briggsae* in response to growth factor. *Nematologica* **10:** 623–630.

Hansen, E.L., G. Perez-Mendez, and E.J. Buecher. 1971. Glycogen as a supplement in media for axenic cultivation of nematodes. *Proc. Soc. Exp. Biol. Med.* **137:** 1352–1354.

Hansen, E.L., E.A. Yarwood, W.L. Nicholas, and F.W. Sayre. 1959. Differential nutritional requirements for reproduction of two strains of *Caenorhabditis elegans* in axenic cultures. *Nematologica* **5:** 27–31.

Harrington, W.F. 1979. Contractile proteins in muscle. In *The proteins* (ed. H. Neurath), vol. 3, pp. 246–409. Academic Press, New York.

Harris, H.E. and H.F. Epstein. 1977. Myosin and paramyosin of *Caenorhabditis elegans*: Biochemical and structural properties of wild-type and mutant proteins. *Cell* **10:** 709–719.

Harris, H., M.-Y. Tso, and H.F. Epstein. 1977. Actin and myosin-linked calcium regulation in *Caenorhabditis elegans*. Biochemical and structural properties of native filaments and purified protein. *Biochemistry* **16:** 859–865.

Harris, L.J. and A.M. Rose. 1986. Somatic excision of transposable element TC1 from the Bristol genome of *Caenorhabditis elegans*. *Mol. Cell. Biol.* **6:** 1782–1786.

Harrison, F.W. and R.R. Cowden, eds. 1982. *Developmental biology of freshwater invertebrates*. Alan R. Liss, New York.

Hartman, P.E. and J.R. Roth. 1973. Mechanisms of suppression. *Adv. Genet.* **17:** 1–105.

Hartman, P.S. 1984a. Effects of age and liquid holding on the UV-radiation sensitivities of wild-type and mutant *Caenorhabditis elegans* dauer larvae. *Mutat. Res.* **132:** 95–99.

———. 1984b. UV irradiation of wild type and radiation-sensitive mutants of the nematode *Caenorhabditis elegans*: Fertilities, survival, and parental effects. *Photochem. Photobiol.* **39:** 169–175.

———. 1985. Epistatic interactions of radiation-sensitive (*rad*) mutants of *Caenorhabditis elegans*. *Genetics* **109:** 81–93.

Hartman, P.S. and R.K. Herman. 1982a. Somatic damage to the X chromosome of the nematode *Caenorhabditis elegans* induced by gamma radiation. *Mol. Gen. Genet.* **187:** 116–119.

———. 1982b. Radiation-sensitive mutants of *Caenorhabditis elegans*. *Genetics* **102:** 159–178.

Hartwell, L., J. Culotti, J. Pringle, and B. Reid. 1974. Genetic control of the cell division cycle in yeast. *Science* **183:** 46–51.

Hazelbauer, G.L., ed. 1978. *Receptors and recognition. Series B. Taxis and behavior. Elementary sensory systems in biology*. Halsted Press, New York.

Hecht, R.M., L.A. Gossett, and W.R. Jeffery. 1981a. Ontogeny of maternal and newly transcribed mRNA analyzed by *in situ* hybridization during development of *Caenorhabditis elegans*. *Dev. Biol.* **83:** 374–379.

Hecht, R.M., D.F. Schomer, J.A. Oro, A.H. Bartel, and E.V. Hungerford III. 1981b. Simple adaptations to extend the range of flow cytometry five orders of magnitude for DNA analysis of uni- and multicell systems. *J. Histochem. Cytochem.* **29:** 771–774.

Hecht, R.M., S.M. Wall, D.F. Schomer, J.A. Oro, and A.H. Bartel. 1982. DNA replication may be uncoupled from nuclear and cellular division in its embryonic lethal mutants of *Caenorhabditis elegans*. *Dev. Biol.* **94:** 183–191.

Hedgecock, E.M. 1976a. The mating system of *Caenorhabditis elegans:* Evolutionary equilibrium between self- and cross-fertilization in a facultative hermaphrodite. *Am. Nat.* **110:** 1007–1012.

———. 1976b. "GABA metabolism in *Caenorhabditis elegans.*" Ph.D. thesis, University of California, Santa Cruz.

———. 1985. Cell lineage mutants in the nematode *Caenorhabditis elegans. Trends Neurosci.* **8:** 288–293.

Hedgecock, E.M. and R.L. Russell. 1975. Normal and mutant thermotaxis in the nematode *Caenorhabditis elegans. Proc. Natl. Acad. Sci.* **72:** 4061–4065.

Hedgecock, E.M. and J.N. Thomson. 1982. A gene required for nuclear and mitochondrial attachment in the nematode *Caenorhabditis elegans. Cell* **30:** 321–330.

Hedgecock, E.M. and J.G. White. 1985. Polyploid tissues in the nematode *Caenorhabditis elegans. Dev. Biol.* **107:** 128–133.

Hedgecock, E.M., J.E. Sulston, and J.N. Thomson. 1983. Mutations affecting programmed cell deaths in the nematode *Caenorhabditis elegans. Science* **220:** 1277–1279.

Hedgecock, E., J. Culotti, D. Hall, and B. Stern. 1987. Genetics of cell and axon migration in *Caenorhabditis elegans. Development* **100:** 365–382.

Hedgecock, E.M., J.G. Culotti, J.N. Thomson, and L.A. Perkins. 1985. Axonal guidance mutants of *Caenorhabditis elegans* identified by filling sensory neurons with fluorescein dyes. *Dev. Biol.* **111:** 158–170.

Heine, U. and T. Blumenthal. 1986. Characterization of regions of the *Caenorhabditis elegans* X chromosome containing vitellogenin genes. *J. Mol. Biol.* **188:** 301–312.

Herman, R.K. 1978. Crossover suppressors and balanced recessive lethals in *Caenorhabditis elegans. Genetics* **88:** 49–65.

———. 1984. Analysis of genetic mosaics of the nematode *Caenorhabditis elegans. Genetics* **108:** 165–180.

———. 1987. Mosaic analysis of two genes that affect nervous system structure in *Caenorhabditis elegans. Genetics* **116:** 377–388.

Herman, R.K. and H.R. Horvitz. 1980. Genetic analysis of *Caenorhabditis elegans.* In *Nematodes as biological models*, volume 1: *Behavioral and developmental models* (ed. B.M. Zuckerman), pp. 227–262. Academic Press, New York.

Herman, R.K. and C.K. Kari. 1985. Muscle-specific expression of a gene affecting acetylcholinesterase in the nematode *Caenorhabditis elegans. Cell* **40:** 509–514.

Herman, R.K., D.G. Albertson, and S. Brenner. 1976. Chromosome rearrangements in *Caenorhabditis elegans. Genetics* **83:** 91–105.

Herman, R.K., H.R. Horvitz, and D.L. Riddle. 1980. The nematode *Caenorhabditis elegans. Genet. Maps* **1:** 183–193.

Herman, R.K., C.K. Kari, and P.S. Hartman. 1982. Dominant X-chromosome nondisjunction mutants of *Caenorhabditis elegans. Genetics* **102:** 379–400.

Herman, R.K., J.E. Madl, and C.K. Kari. 1979. Duplications in *Caenorhabditis elegans. Genetics* **92:** 419–435.

Hieb, W.F. and M. Rothstein. 1968. Sterol requirement for reproduction of a free-living nematode. *Science* **160:** 778–780.

Hieb, W.F., E.L.R. Stokstad, and M. Rothstein. 1970. Heme requirement for reproduction of a free-living nematode. *Science* **168:** 143–144.

Higgins, B.J. and D. Hirsh. 1977. Roller mutants of the nematode *Caenorhabditis elegans. Mol. Gen. Genet.* **150:** 63–72.

Himmelhoch, S. and B.M. Zuckerman. 1978. *Caenorhabditis briggsae*: Aging and the structural turnover of the outer cuticle surface and the intestine. *Exp. Parasitol.* **45:** 208–214.

———. 1982. *Xiphinema index* and *Caenorhabditis elegans*: Preparation and molecular labeling of ultrathin frozen sections. *Exp. Parasitol.* **54:** 250–259.

———. 1983. *Caenorhabditis elegans*: Characters of negative charged groups on the cuticle and intestine. *Exp. Parasitol.* **55:** 299–305.

Himmelhoch, S., M.J. Kisiel, and B.M. Zuckerman. 1977. *Caenorhabditis briggsae*: Electron microscope analysis of changes in negative surface charge density of the outer cuticular membrane. *Exp. Parasitol.* **41:** 118–123.

Hirsh, D. 1975. Patterns of gene expression during development of the nematode *Caenorhabditis elegans*. In *Microbiology—1975* (ed. D. Schlessinger), pp. 508–514. American Society for Microbiology, Washington, D.C.

———. 1979. Temperature sensitive maternal effect mutants of early development in *Caenorhabditis elegans*. In *Determinants of spatial organization* (ed. S. Subtelny and I.R. Konigsberg), pp. 149–166. Academic Press, New York.

Hirsh, D. and R. Vanderslice. 1976. Temperature-sensitive developmental mutants of *Caenorhabditis elegans*. *Dev. Biol.* **49:** 220–235.

Hirsh, D., J.G. Files, and S.H. Carr. 1982. Isolation and genetic mapping of the actin genes of *Caenorhabditis elegans*. In *Muscle development: Molecular and cellular control* (ed. M. Pearson and H.F. Epstein), pp. 77–86. Cold Spring Harbor Laboratory, Cold Spring Harbor, New York.

Hirsh, D., D. Oppenheim, and M. Klass. 1976. Development of the reproductive system of *Caenorhabditis elegans*. *Dev. Biol.* **49:** 200–219.

Hirsh, D., S.W. Emmons, J.G. Files, and M.R. Klass. 1979. Stability of the *Caenorhabditis elegans* genome during development and evolution. *ICN-UCLA Symp. Mol. Cell. Biol.* **14:** 205–218.

Hirsh, D., K.J. Kemphues, D.T. Stinchcomb, and R. Jefferson. 1985a. Genes affecting early development in *Caenorhabditis elegans*. *Cold Spring Harbor Symp. Quant. Biol.* **50:** 69–78.

Hirsh, D., G.N. Cox, J.M. Kramer, D. Stinchcomb, and R. Jefferson. 1985b. Structure and expression of the collagen genes of *Caenorhabditis elegans*. *Ann. N.Y. Acad. Sci.* **460:** 163–171.

Hirsh, D., W.B. Wood, R. Hecht, S. Carr, and R. Vanderslice. 1977. Expression of genes essential for early development in the nematode, *Caenorhabditis elegans*. *ICN-UCLA Symp. Mol. Cell. Biol.* **8:** 347–356.

Hirumi, H., D.J. Raski, and N.O. Jones. 1971. Primitive muscles of nematodes: Morphological aspects of platymyarian and shallow coelomyarian muscles in two plant nematodes *Trichodorus christiei* and *Longidorus elongatus*. *J. Ultrastruct. Res.* **34:** 517–543.

Hitcho, P.J. and R.E. Thorson. 1972. Behavior of free-living and plant-parasitic nematodes in a thermal gradient. *J. Parasitol.* **58:** 599.

Hodgkin, J. 1980. More sex-determination mutants of *Caenorhabditis elegans*. *Genetics* **96:** 649–664.

———. 1983a. Male phenotypes and mating efficiency in *Caenorhabditis elegans*. *Genetics* **103:** 43–64.

———. 1983b. Two types of sex determination in a nematode. *Nature* **304:** 267–268.

———. 1983c. X chromosome dosage and gene expression in *Caenorhabditis elegans:* Two unusual dumpy genes. *Mol. Gen. Genet.* **192:** 452–458.

———. 1984. Switch genes and sex determination in the nematode *Caenorhabditis elegans*. *J. Embryol. Exp. Morphol.* (suppl.) **83:** 103–117.

———. 1985a. Males, hermaphrodites and females: Sex determination in *Caenorhabditis elegans*. *Trends Genet.* **1:** 85–88.

———. 1985b. Novel nematode amber suppressors. *Genetics* **111:** 287–310.

———. 1986. Sex determination in the nematode *Caenorhabditis elegans*: Analysis of *tra-3* suppressors and characterization of *fem* genes. *Genetics* **114:** 15–52.

———. 1987a. Primary sex determination in the nematode *Caenorhabditis elegans*. *Development* (suppl.) **101:** (in press).

———. 1987b. A genetic analysis of the sex-determining gene, *tra-1*, in the nematode *Caenorhabditis elegans*. *Genes. Dev.* **1:** 731–745.

Hodgkin, J.A. and S. Brenner. 1977. Mutations causing transformation of sexual phenotype in the nematode *Caenorhabditis elegans*. *Genetics* **86:** 275–287.

Hodgkin, J., T. Doniach, and M. Shen. 1985. The sex determination pathway in the nematode *Caenorhabditis elegans*: Variations on a theme. *Cold Spring Harbor Symp. Quant. Biol.* **50:** 585–594.

Hodgkin, J., H.R. Horvitz, and S. Brenner. 1979. Nondisjunction mutants of the nematode *Caenorhabditis elegans*. *Genetics* **91:** 67–94.

Hofsten, A.V., D. Kahan, R. Katznelson, and T. Bar-el. 1983. Digestion of free-living nematodes fed to fish. *J. Fish Biol.* **23:** 419–428.

Hofstetter, H., A. Kressmann, and M.L. Birnstiel. 1981. A split promoter for a eukaryotic tRNA gene. *Cell* **24:** 573–585.

Hogger, C.H. and R.H. Estey. 1977. Cryofracturing for scanning electron microscope observations of internal structures of nematodes. *J. Nematol.* **9:** 334–337.

Hogger, C.H., E.H. Estey, M.J. Kisiel, and B.M. Zuckerman. 1977. Surface scanning observations of changes in *Caenorhabditis briggsae* during aging. *Nematologica* **23:** 213–216.

Honda, B.M., R.H. Devlin, D.W. Nelson, and M. Khosla. 1986. Transcription of class III genes in cell-free extracts from the nematode *Caenorhabditis elegans*. *Nucleic Acids Res.* **14:** 869–881.

Honda, H. 1925. Experimental and cytological studies on bisexual and hermaphroditic free-living nematodes with special reference to problems of sex. *J. Morphol.* **40:** 191–233.

Horvitz, H.R. 1982. Factors that influence neural development in nematodes. In *Repair and regeneration in the nervous system* (ed. J.G. Nicholls), pp. 41–55. Springer-Verlag, New York.

Horvitz, H.R. and P.W. Sternberg. 1982. Nematode postembryonic cell lineages. *J. Nematol.* **14:** 240–248.

Horvitz, H.R. and J.E. Sulston. 1980. Isolation and genetic characterization of cell-lineage mutants of the nematode *Caenorhabditis elegans*. *Genetics* **96:** 435–454.

Horvitz, H.R., H.M. Ellis, and P.W. Sternberg. 1982a. Programmed cell death in nematode development. *Neurosci. Comment.* **1:** 56–65.

Horvitz, H.R., S. Brenner, J. Hodgkin, and R.K. Herman. 1979. A uniform genetic nomenclature for the nematode *Caenorhabditis elegans*. *Mol. Gen. Genet.* **175:** 129–133.

Horvitz, H.R., M. Chalfie, C. Trent, J. Sulston, and P.D. Evans. 1982b. Serotonin and octopamine in the nematode *Caenorhabditis elegans*. *Science* **216:** 1012–1014.

Horvitz, H.R., P.W. Sternberg, I.S. Greenwald, W. Fixsen, and H.M. Ellis. 1983. Mutations that affect neural cell lineages and cell fates during the development of the nematode *Caenorhabditis elegans*. *Cold Spring Harbor Symp. Quant. Biol.* **48:** 453–463.

Hosono, R. 1978a. Age dependent changes in the behavior of *Caenorhabditis elegans* on attraction to *Escherichia coli*. *Exp. Gerontol.* **13:** 31–36.

———. 1978b. Sterilization and growth inhibition of *Caenorhabditis elegans* by 5-fluorodeoxyuridine. *Exp. Gerontol.* **13:** 369–374.

———. 1978c. Sinusoidal movement of *Caenorhabditis elegans* in liquid phase. *Zool. Mag.* **87:** 191–196.

———. 1980. A study of morphology of *Caenorhabditis elegans*: A mutant of *Caenorhabditis elegans* with dumpy and temperature-sensitive roller phenotype. *J. Exp. Zool.* **213:** 61–67.

Hosono, R. and S. Kuno. 1985. Properties of the *unc-52* gene and its related mutations in the nematode *Caenorhabditis elegans*. *Zool. Sci.* **2:** 81–88.

Hosono, R., S. Kuno, and M. Midsukami. 1985. Temperature-sensitive mutations causing reversible paralysis in *Caenorhabditis elegans*. *J. Exp. Zool.* **235:** 409–421.

Hosono, R., K. Hirahara, S. Kuno, and T. Kurihara. 1982a. Mutants of *Caenorhabditis elegans* with Dumpy and Rounded head phenotype. *J. Exp. Zool.* **224:** 135–144.

Hosono, R., Y. Sato, S.I. Aizawa, and Y. Mitsui. 1980. Age-dependent changes in mobility and separation of the nematode *Caenorhabditis elegans*. *Exp. Gerontol.* **15:** 285–289.

Hosono, R., Y. Mitsui, Y. Sato, S. Aizawa, and J. Miwa. 1982b. Life span of the wild and mutant nematode *Caenorhabditis elegans*. *Exp. Gerontol.* **17:** 163–172.

Howell, M.J., ed. 1986. Parasitology—Quo vadit? In *Proceedings of the Sixth International Congress of Parasitology*. Australian Academy of Science, Canberra.

Huang, L., G. Albers-Schonberg, R.L. Monaghan, K. Jakubas, S.S. Pong, O.D. Hensens, R.W. Burg, D.A. Ostlind, J. Conroy, and E. O. Stapley. 1984. Discovery, production and purification of the Na^+, K^+ activated ATPase inhibitor L-681,110 from the fermentation broth of *Streptomyces SP. MA-5038*. *J. Antibiot.* **37:** 970–975.

Huang, S.-P., T.A. Tattar, R.A. Rohde, and B.M. Zuckerman. 1982. *Caenorhabditis elegans*: Effects of 5-hydroxytryptophan and dopamine on behavior and development. *Exp. Parasitol.* **54:** 72–79.

Huettel, R.N. 1986. Chemical communicators in nematodes. *J. Nematol.* **18:** 3–8.

Hutzell, P.A. and L.R. Krusberg. 1982. Fatty acid compositions of *Caenorhabditis elegans* and *Caenorhabditis briggsae*. *Comp. Biochem. Physiol.* **73B:** 517–520.

Hwang, S.W. 1970. Freezing and storage of nematodes in liquid nitrogen. *Nematologica* **16:** 305–308.

Illmensee, K. and A.P. Mahowald. 1974. Transplantation of posterior polar plasm in *Drosophila*. Induction of germ cells at the anterior pole of the egg. *Proc. Natl. Acad. Sci.* **71:** 1016–1020.

Isnenghi, E., R. Cassada, K. Smith, K. Denich, K. Radnia, and G. von Ehrenstein. 1983. Maternal effects and temperature-sensitive period of mutations affecting embryogenesis in *Caenorhabditis elegans*. *Dev. Biol.* **98:** 465–480.

Ito, K. and J.D. McGhee. 1987. Parental DNA strands segregate randomly during embryonic development of *Caenorhabditis elegans*. *Cell* **49:** 329–336.

Jakstys, B.P. and P.H. Silverman. 1969. Effect of heterologous antibody on *Haemonchus contortus* development in vitro. *J. Parasitol.* **55:** 486–492.

Jameel, S. and B.A. McFadden. 1985. *Caenorhabditis elegans*: Purification of isocitrate lyase and the isolation and cell-free translation of $poly(A)^+$RNA. *Exp. Parasitol.* **59:** 337–346.

Jansson, H.B., A. Jeyaprakash, R.A. Damon, and B.M. Zuckerman. 1984. *Caenorhabditis elegans* and *P. redivivus*—Enzyme-mediated modification of chemotaxis. *Exp. Parasitol.* **58:** 270-277.

Jansson, H., A. Jeyaprakash, N. Marban-Mendoza, and B.M. Zuckerman. 1986. *Caenorhabditis elegans*: Comparisons of chemotactic behavior from monoxenic and zyxenic culture. *Exp. Parasitol.* **61:** 369–372.

Jantunen, R. 1964. Moulting of *Caenorhabditis briggsae* (Rhabditidae). *Nematologica* **10:** 419–424.

Jeffrey, W.R. and R.A. Raff, eds. 1983. *Time, space, and pattern in embryonic development*. A.R. Liss, New York.

Jelinek, W.R. and C.W. Schmid. 1982. Repetitive sequences in eukaryotic DNA and their expression. *Annu. Rev. Biochem.* **51:** 813–844.

Jeyaprakash, A., H.-B. Jansson, N. Marban-Mendoza, and B.M. Zuckerman. 1985. *Caenorhabditis elegans*—Lectin-mediated modification of chemotaxis. *Exp. Parasitol.* **59:** 90–97.

Johnson, C.D. and R.L. Russell. 1975. A rapid, simple radiometric assay for cholinesterase, suitable for multiple determinations. *Anal. Biochem.* **64:** 229–238.

———. 1983. Multiple molecular forms of acetylcholinesterase in the nematode *Caenorhabditis elegans*. *J. Neurochem.* **41:** 30.

Johnson, C.D. and A.O.W. Stretton. 1980. Neural control of locomotion in Ascaris: Anatomy, electrophysiology, and biochemistry. In *Nematodes as biological models*, volume 1: *Behavioral and developmental models* (ed. B.M. Zuckerman), pp. 159–196. Academic Press, New York.

———. 1985. Localization of choline acetyltransferase within identified motoneurons of the nematode *Ascaris*. *J. Neurosci.* **5:** 1984–1992.

———. 1987. GABA-immunoreactivity in inhibitory motoneurons of the nematode *Ascaris*. *J. Neurosci.* **7:** 223–235.

Johnson, C.D., J.G. Duckett, J.G. Culotti, R.K. Herman, P.M. Meneely, and R.L. Russell. 1981. An acetylcholinesterase-deficient mutant of the nematode *Caenorhabditis elegans*. *Genetics* **97:** 261–279.

Johnson, K. and D. Hirsh. 1979. Patterns of proteins synthesized during development of *Caenorhabditis elegans*. *Dev. Biol.* **70:** 241–248.

Johnson, T.E. 1983a. *Caenorhabditis elegans*: A genetic model for understanding the aging process. In *Intervention in the aging process* (ed. W. Regelson and F.M. Sinex), pp. 287–305. Alan R. Liss, New York.

———. 1983b. Aging in *Caenorhabditis elegans*. In *Review of biological research in aging*, (ed. M. Rothstein et al.), vol. 1, pp. 37–49. Alan R. Liss, New York.

———. 1983c. Analysis of the biological basis of aging in the nematode with special emphasis on *Caenorhabditis elegans*. In *Invertebrate models in aging research* (ed. D.H. Mitchell and T.E. Johnson), pp. 37–49. CRC Press, Boca Raton, Florida.

———. 1985. Aging in *Caenorhabditis elegans*: Update 1984. In *Review of biological research in aging* (ed. M. Rothstein), vol. 2, pp. 45–60. Alan R. Liss, New York.

———. 1986. Molecular and genetic analyses of a multivariate system specifying behavior and life span. *Behav. Genet.* **16:** 221–235.

———. 1987. Aging can be genetically dissected into component processes using long-lived lines of *Caenorhabditis elegans*. *Proc. Natl. Acad. Sci.* **84:** 3777–3781.

Johnson, T.E. and G. McCaffrey. 1985. Programmed aging or error catastrophe? An examination by two-dimensional polyacrylamide gel electrophoresis. *Mech. Ageing Dev.* **30:** 285–297.

Johnson, T.E. and V.J. Simpson. 1985. Aging studies in *Caenorhabditis elegans* and other nematodes. In *CRC handbook of cell biology* (ed. V.J. Cristofalo), pp. 481–495. CRC Press, Boca Raton, Florida.

Johnson, T.E. and W.B. Wood. 1982. Genetic analysis of life-span in *Caenorhabditis elegans*. *Proc. Natl. Acad. Sci.* **79:** 6603–6607.

Johnson, T.E., D.H. Mitchell, S. Kline, R. Kemal, and J. Foy. 1984. Arresting development arrests aging in the nematode *Caenorhabditis elegans*. *Mech. Ageing Dev.* **28:** 23–40.

Jones, D., R.H. Russnak, R.J. Kay, and E.P.M. Candido. 1986. Structure, expression, and evolution of a heat-shock gene locus in *Caenorhabditis elegans* that is flanked by repetitive elements. *J. Biol. Chem.* **261:** 12006–12015.

Kahn, R.F. and B.A. McFadden. 1980. A rapid method of synchronizing developmental stages of *Caenorhabditis elegans*. *Nematologica* **26:** 280–282.

Kampfe, L. 1978. The consumption of oxygen by nematodes as a characteristic of zootic activity. *Pedobiologia* **18:** 355–365.

Kampfe, L., V. Kreil, and H.L. Ullrich. 1986. Effects of lathyrogenous compounds and of the synergist piperonly butoxide (PBO) on the free-living nematode species of *Caenorhabditis briggsae* and *Rhabditis oxycerca*. *Zool. Jahrb. Abt. Zool. Physiol. Tiere* **90:** 257–271.

Karn, J., S. Brenner, and L. Barnett. 1983. Protein structural domains in the *Caenorhabditis elegans unc-54* myosin heavy chain gene are not separated by introns. *Proc. Natl. Acad. Sci.* **80:** 4253–4257.

Karn, J., N.J. Dibb, and D.M. Miller. 1985. Cloning nematode myosin genes. In *Cell and muscle motility* (ed. J. Shay), vol. 6, pp. 185–237. Plenum Press, New York.

Karn, J., A.D. McLachlan, and L. Barnett. 1982. *unc-54* Myosin heavy-chain gene of *Caenorhabditis elegans*: Genetics, sequence, structure. In *Muscle development: Molecular and cellular control* (ed. M. Pearson and H.F. Epstein), pp. 129–142. Cold Spring Harbor Laboratory, Cold Spring Harbor, New York.

Karn, J., S. Brenner, L. Barnett, and G. Cesareni. 1980. Novel bacteriophage lambda cloning vector. *Proc. Natl. Acad. Sci.* **77:** 5172–5176.

Kass, I.S., A.O.W. Stretton, and C.C. Wang. 1984. The effects of avermectin and drugs related to acetylcholine and 4-aminobutyric acid on neurotransmission in *Ascaris suum*. *Mol. Biochem. Parasitol.* **13:** 213–225.

Kass, I.S., C.C. Wang, J.P. Walrond, and A.O.W. Stretton. 1980. Avermectin B_1a, a paralyzing anthelmintic that affects interneurons and inhibitory motoneurons in *Ascaris*. *Proc. Natl. Acad. Sci.* **77:** 6211–6215.

Kaufman, T.D., J.R. Bloom, and F.L. Lukezic. 1983. Effect of an extract from saprozoic nematode-infested compost on the mycelial growth of *Agaricus brunnescens*. *J. Nematol.* **15:** 567–571.

Kaufman, T.D., F.L. Lukezic, and J.R. Bloom. 1984. The effect of free-living nematodes and compost moisture on growth and yield in *Agaricus brunnescens*. *Can. J. Microbiol.* **30:** 503–506.

Kay, R.J., R.J. Boissy, R.H. Russnak, and E.P.M. Candido. 1986. Efficient transcription of *Caenorhabditis elegans* heat shock gene pair in mouse fibroblasts is dependent on multiple promoter elements which can function bidirectionally. *Mol. Cell. Biol.* **6:** 3134–3143.

Kemphues, K.J. 1987. Genetic analysis of embryogenesis in *Caenorhabditis elegans*. In *Developmental genetics of higher organisms* (ed. G.M. Malacinski). MacMillan, New York. (In press.)

Kemphues, K.J., J.R. Priess, D.G. Morton, and N. Cheng. 1988. Identification of genes required for cytoplasmic localization in early *Caenorhabditis elegans* embryos. *Cell* (in press).

Kemphues, K.J., N. Wolf, W.B. Wood, and D. Hirsh. 1986. Two loci required for cytoplasmic organization in early embryos of *Caenorhabditis elegans*. *Dev. Biol.* **113:** 449–460.

Kensler, R.W., R.J.C. Levine, and M. Stewart. 1985. Electron microscopic and optical diffraction analysis of the structure of scorpion muscle thick filaments. *J. Cell. Biol.* **101:** 395–401.

Kenyon, C.J. 1983. Pattern, symmetry and surprises in the development of *Caenorhabditis elegans*. *Trends Biochem. Sci.* **8:** 349–351.

———. 1985a. Heterochronic mutations of *Caenorhabditis elegans*—Their developmental and evolutionary significance. *Trends Genet.* **1:** 2.

———. 1985b. Cell lineage and the control of *Caenorhabditis elegans* development. *Philos. Trans. R. Soc. Lond. B Biol. Sci.* **312:** 21–38.

———. 1986. A gene involved in the development of the posterior body region of *Caenorhabditis elegans*. *Cell* **46:** 477–487.

Kermarrec, A. 1973. Recherches sur les ennemis du champignon de couche. I. Zoocenose des composts de champignonnière à *Agaricus bisporus* Lange. *Ann. Zoologie Ecol. Anim.* **5:** 425–464.

Khan, F.R. and B.A. McFadden. 1982a. *Caenorhabditis elegans*: Decay of isocitrate lyase during larval development. *Exp. Parasitol.* **54:** 47–54.

———. 1982b. A rapid method of synchronizing developmental stages of *Caenorhabditis elegans*. *Nematologica* **26:** 280–282.

Kimble, J. 1978. "The post-embryonic cell lineages of the hermaphrodite and male gonads in *Caenorhabditis elegans*." Ph.D. thesis, University of Colorado, Boulder.

———. 1981a. Alterations in cell lineage following laser ablation of cells in the somatic gonad of *Caenorhabditis elegans*. *Dev. Biol.* **87:** 286-300.

———. 1981b. Strategies for control of pattern formation in *Caenorhabditis elegans*. *Philos. Trans. R. Soc. Lond. B Biol. Sci.* **295:** 539–551.

Kimble, J. and D. Hirsh. 1979. The post-embryonic cell lineages of the hermaphrodite and male gonads in *Caenorhabditis elegans*. *Dev. Biol.* **70:** 396–417.

Kimble, J. and W.J. Sharrock. 1983. Tissue-specific synthesis of yolk proteins in *Caenorhabditis elegans*. *Dev. Biol.* **96:** 189–196.

Kimble, J.E. and J.G. White. 1981. On the control of germ cell development in *Caenorhabditis elegans*. *Dev. Biol.* **81:** 208–219.

Kimble, J., L. Edgar, and D. Hirsh. 1984. Specification of male development in *Caenorhabditis elegans*: The *fem* genes. *Dev. Biol.* **105:** 234–239.

Kimble, J., J. Sulston, and J. White. 1979. Regulative development in the post-embryonic lineages of *Caenorhabditis elegans*. In *Cell lineage, stem cells and cell determination* (ed. N. LeDouarin), pp. 59–68. Elsevier, New York.

Kimble, J., J. Hodgkin, T. Smith, and J. Smith. 1982. Suppression of an amber mutation by microinjection of suppressor tRNA in *Caenorhabditis elegans*. *Nature* **299:** 456–458.

Kimble, J., M.K. Barton, T.B. Schedl, T.A. Rosenquist, and J. Austin. 1986. Controls of postembryonic germ line development in *Caenorhabditis elegans*. In *Gametogenesis and the early embryo* (ed. J.G. Gall), pp. 97–110. Alan R. Liss, New York.

Kimpinski, J. 1975. Nematodes associated with vegetables in Prince Edward Island Canada. *Plant Dis. Rep.* **59:** 37–39.

Kisiel, M.J. and B.M. Zuckerman. 1978. Effects of centrophenoxine on the nematode *Caenorhabditis briggsae*. *Age* **1:** 17–20.

Kisiel, M.J., K.H. Deubert, and B.M. Zuckerman. 1976. Biogenic amines in the free-living nematode *Caenorhabditis briggsae*. *Exp. Aging Res.* **2:** 37–44.

Kisiel, M.J., S. Himmelhoch, and B.M. Zuckerman. 1974. *Caenorhabditis briggsae*: Effects of aminopterin. *Exp. Parasitol.* **36:** 430–438.

Kisiel, M.J., B. Nelson, and B.M. Zuckerman. 1969. Influence of a growth factor from bacteria on the morphology of *Caenorhabditis briggsae*. *Nematologica* **15:** 153.

———. 1972. Effects of DNA synthesis inhibitors on *Caenorhabditis briggsae* and *Turbatrix aceti*. *Nematologica* **18:** 373–384.

Kisiel, M.J., J.M. Castillo, L.S. Zuckerman, B.M. Zuckerman, and S. Himmelhoch. 1975. Studies on ageing *Turbatrix aceti*. *Mech. Ageing Dev.* **4:** 81–88.

Klass, M.R. 1977. Aging in the nematode *Caenorhabditis elegans*: Major biological and environmental factors influencing life span. *Mech. Ageing Dev.* **6:** 413–429.

———. 1983. A method for the isolation of longevity mutants in the nematode *Caenorhabditis elegans* and initial results. *Mech. Ageing Dev.* **22:** 279–286.

———. 1984. Biology and cell-specific gene expression in *Caenorhabditis elegans*. In *Invertebrate models in aging research* (ed. D.H. Mitchell and T.E. Johnson), pp. 45–58. CRC Press, Boca Raton, Florida.

———. 1986. Cell-specific gene expression in the nematode. *Int. Rev. Cytol.* **102:** 1–28.

Klass, M. and D. Hirsh. 1976. Nonaging developmental variant of *Caenorhabditis elegans*. *Nature* **260:** 523–525.

———. 1981. Sperm isolation and biochemical analysis of the major sperm protein from *Caenorhabditis elegans*. *Dev. Biol.* **84:** 299–312.

Klass, M.R. and T.E. Johnson. 1985. *Caenorhabditis elegans*. *Interdiscip. Top. Gerontol.* **21:** 164–187.

Klass, M., D. Ammons, and S. Ward. 1988. Conservation of the 5′-flanking sequences of transcribed members of the *Caenorhabditis elegans* major sperm protein gene family. *J. Mol. Biol.* **199:** 14–22.

Klass, M., B. Dow, and M. Herndon. 1982. Cell-specific transcriptional regulation of the major sperm protein from *Caenorhabditis elegans*. *Dev. Biol.* **93:** 152–164.

Klass, M.R., S. Kinsley, and L.C. Lopez. 1984. Isolation and characterization of a sperm-specific gene family in the nematode *Caenorhabditis elegans*. *Mol. Cell. Biol.* **4:** 529–537.

Klass, M., P.N. Nguyen, and A. Dechavigny. 1983. Age correlated changes in the DNA template in the nematode *Caenorhabditis elegans*. *Mech. Ageing Dev.* **22:** 253–263.

Klass, M., N. Wolf, and D. Hirsh. 1976. Development of the male reproductive system and sexual transformation in the nematode *Caenorhabditis elegans*. *Dev. Biol.* **52:** 1–18.

———. 1979. Further characterization of a temperature-sensitive transformation mutant in *Caenorhabditis elegans*. *Dev. Biol.* **69:** 329–335.

Klefenz, H.F. and B.M. Zuckerman. 1978. Review: A comparative analysis of studies of enzyme changes with age, with comments on possible sources of error. *Age* **1:** 60–67.

Kolson, D.L. and R.L. Russell. 1985a. A novel class of acetylcholinesterase, revealed by mutations in the nematode *Caenorhabditis elegans*. *J. Neurogenet.* **2:** 93–110.

———. 1985b. New acetylcholinesterase-deficient mutants of the nematode *Caenorhabditis elegans*. *J. Neurogenet.* **2:** 69–91.

Korner, H. 1954. Die Nematoden-Fauna des vergehenden Holzes und ihre Beziehungen zu den Insekten. *Zoolyst* **82:** 245–353.

Kramer, J.M., G.N. Cox, and D. Hirsh. 1982. Comparisons of the complete sequences of two collagen genes from *Caenorhabditis elegans*. *Cell* **30:** 599–606.

———. 1985. Expression of the *Caenorhabditis elegans* collagen genes *col-1* and *col-2* is developmentally regulated. *J. Biol. Chem.* **260:** 1945–1951.

Krause, M. and D. Hirsh. 1984. Actin gene expression in *Caenorhabditis elegans*. In *Molecular biology of the cytoskeleton* (ed. G. Borisy et al.), pp. 287–292. Cold Spring Harbor Laboratory, Cold Spring Harbor, New York.

———. 1987. A trans-spliced leader sequence on actin mRNA in *Caenorhabditis elegans*. *Cell* **49:** 753–761.

Kreis, H.A. 1929. Freilebende terrestrische Nematoden aus der Umgebung von Peking (China). *Zool. Anz.* **84:** 283–294.

———. 1930. A. Freilebende terrestrische Nematoden aus der Umgebung von Peking (China) II. *Zool. Anz.* **87:** 67–87.

Kreis, H.A. and E.C. Faust. 1933. Two new species of Rhabditis (*R. macrocerca* and *R. clavopapillata*) associated with dogs and monkeys in experimental Stronglyoides studies. *Trans. Am. Microsc. Soc.* **52:** 162–172.

Krieg, C., T. Cole, U. Deppe, E. Schierenberg, D. Schmitt, B. Yoder, and G. von Ehrenstein. 1978. The cellular anatomy of embryos of the nematode *Caenorhabditis elegans*. *Dev. Biol.* **65:** 193–215.

Kumazaki, T., H. Hori, S. Osawa, N. Ishii, and K. Suzuki. 1982. The nucleotide sequence of 5S rRNA's from a rotifer, *Brachionus plicatilis*, and two nematodes, *Rhabditis tokai* and *Caenorhabditis elegans*. *Nucleic Acids Res.* **10:** 7001–7004.

Kunz, P. and J. Klingler. 1976. A method for direct or microscopic observation and photography of nematode tracks during orientation behaviour studies. *Nematologica* **22:** 477–479.

Kusch, M. and R.S. Edgar. 1986. Genetic studies of unusual loci that affect body shape of the nematode *Caenorhabditis elegans* and may code for cuticle structural proteins. *Genetics* **113:** 621–639.

Landel, C.P., M. Krause, R.H. Waterston, and D. Hirsh. 1984. DNA rearrangements of the actin gene-cluster in *Caenorhabditis elegans* accompany reversion of 3 muscle mutants. *J. Mol. Biol.* **180:** 497–513.

Landolt, P. and H. Tobler. 1980. The somatic DNA of *Ascaris lumbricoides* shows short-period interspersion. *Experientia* **36:** 750.

Laskey, R., P. Lawrence, and E. De Robertis. 1977. Genes in development. *Nature* **270:** 477–478.

Laufer, J.S. and G. von Ehrenstein. 1981. Nematode development after removal of egg cytoplasm: Absence of localized unbound determinants. *Science* **211:** 402–404.

Laufer, J.S., P. Bazzicalupo, and W.B. Wood. 1980. Segregation of developmental potential in early embryos of *Caenorhabditis elegans*. *Cell* **19:** 569–577.

Lawrence, P.A. 1981. The cellular basis of segmentation in insects. *Cell* **26:** 3–10.

Lawrence, P.A. and S.M. Green. 1979. Cell lineage in the developing retina of *Drosophila*. *Dev. Biol.* **71:** 142–152.

Le Douarin, N., ed. 1979. *Cell lineage, stem cell and cell determinations*. Elsevier, New York.

Leighton, T. and W.F. Loomis, eds. 1980. *The molecular genetics of development*. Academic Press, New York.

Lew, K.K., S. Chritton, and P. Blumberg. 1982. Biological responsiveness to the phorbol esters and specific binding of ^{3}H phorbol 12,13-dibutyrate in *Caenorhabditis elegans*, a manipulable gene system. *Teratog. Carcinog. Mutagen.* **2:** 19–30.

Lewin, B. 1974. *Gene expression*, vol. 2. *Eukaryotic chromosomes*. John Wiley, New York.

Lewin, R. 1984a. Why is development so illogical? *Science* **224:** 1327–1329.

———. 1984b. The continuing tale of a small worm. *Science* **225:** 153–156.

Lewis, J.A. 1980. Commentary on the uses of small nematode worms. *Neuroscience* **5:** 961–966.

Lewis, J.A. and J.A. Hodgkin. 1977. Specific neuroanatomical changes in chemosensory mutants of the nematode *Caenorhabditis elegans*. *J. Comp. Neurol.* **172:** 489–510.

Lewis, J.A. and I. Paterson. 1984. Preparation of tritium-labelled meta-aminolevamisole of high specific radioactivity by catalytic dehalogenation. *J. Labelled Compd. Radiopharm.* **21:** 945–959.

Lewis, J.A., C.-H. Wu, H. Berg, and J.H. Levine. 1980a. The genetics of levamisole resistance in the nematode *Caenorhabditis elegans*. *Genetics* **95:** 905–928.

Lewis, J.A., C.-H. Wu, J.H. Levine, and H. Berg. 1980b. Levamisole-resistant mutants of the nematode *Caenorhabditis elegans* appear to lack pharmacological acetylcholine receptors. *Neuroscience* **5:** 967–989.

Lewis, J.A., J.S. Elmer, J. Skimming, S. McLafferty, J. Fleming, and T. McGee. 1987. Cholinergic receptor mutants of the nematode *Caenorhabditis elegans*. *J. Neurosci.* **7:** 3059.

Liao, L.W., B. Rosenzweig, and D. Hirsh. 1983. Analysis of a transposable element in *Caenorhabditis elegans*. *Proc. Natl. Acad. Sci.* **80:** 3585–3589.

Lifschytz, E. and R. Falk. 1969. A genetic analysis of the Killer-prune (*K-pn*) locus of *Drosophila melanogaster*. *Genetics* **62:** 353–358.

Link, C.D., J. Graf-Whitsel, and W.B. Wood. 1987. Isolation and characterization of a nematode transposable element from *Panagrellus redivivus*. *Proc. Natl. Acad. Sci.* **84:** 5325–5329.

Lints, F.A., ed. 1985. *Non-mammalian models for research on aging. Interdisciplinary topics in gerontology*. Karger, Basel.

———, ed. 1987. *Non-mammalian models for research on aging: Interdisciplinary topics in gerontology*. Karger, Basel. (In press.)

Liu, A. and M. Rothstein. 1976. Nematode biochemistry XV. Enzyme changes related to glycerol excretion in *Caenorhabditis briggsae*. *Comp. Biochem. Physiol.* **54B:** 233–238.

Long, E.O. and I.B. Dawid. 1980. Repeated genes in eukaryotes. *Annu. Rev. Biochem.* **49:** 727–764.

LoPresti, V., E.R. Macagno, and C. Levintahl. 1973. Structure and development of neuronal connections in isogenic organisms: Cellular interactions in the development of the optic lamina of *Daphnia*. *Proc. Natl. Acad. Sci.* **70:** 433–437.

Lower, W.R. and E.J. Buecher. 1970. Axenic culturing of nematodes: An easily prepared medium containing yeast extract. *Nematologica* **16:** 563–566.

Lower, W.R., E.L. Hansen, and E.A. Yarwood. 1966a. The use of nematodes cultured axenically on defined medium for biological studies. *Am. Nat.* **100:** 367–370.

———. 1966b. Assay of a proteinaceous growth factor required for maturation of the free-living nematode, *Caenorhabditis briggsae*. *J. Exp. Zool.* **161:** 29–36.

———. 1968. Selection for adaptation to increased temperatures in free-living nematodes. *Life Sci.* **7:** 139–146.

Lozano, R., W.R. Lusby, D.J. Chitwood, and J.A. Svoboda. 1985a. Dealkylation of various 24-alkylsterols by the nematode *Caenorhabditis elegans*. *Lipids* **20:** 102–107.

Lozano, R., Y. Ninomiya, H. Thompson, and B.R. Olsen. 1985b. A distinct class of vertebrate collagen genes encodes chicken type IX collagen polypeptides. *Proc. Natl. Acad. Sci.* **82:** 4050–4054.

Lozano, R., W.R. Lusby, D.J. Chitwood, M.J. Thompson, and J.A. Svoboda. 1985c. Inhibition of C-28 and C-29 phytosterol metabolism by N,N-dimethyldodecanamine in the nematode *Caenorhabditis elegans*. *Lipids* **20:** 158–166.

Lozano, R., D.J. Chitwood, W.R. Lusby, M.J. Thompson, J.A. Svoboda, and G.W. Patterson. 1984. Comparative effects of growth-inhibitors on sterol-metabolism in the nematode *Caenorhabditis elegans*. *Comp. Biochem. Physiol.* **79C:** 21–26.

Lu, N.C., A.C. Cheng, and G.M. Briggs. 1983. A study of mineral requirements in *Caenorhabditis elegans*. *Nematologica* **29:** 425–434.

Lu, N.C., W.F. Hieb, and E.L.R. Stokstad. 1974. Accumulation of formimino-L-glutamic acid in the free-living nematode *Caenorhabditis briggsae* as related to folic acid deficiency. *Proc. Soc. Exp. Biol. Med.* **145:** 67–69.

———. 1976. Effect of vitamin B12 and folate on biosynthesis of methionine from homocysteine in the nematode *Caenorhabditis briggsae*. *Proc. Soc. Exp. Biol. Med.* **151:** 701–706.

Lu, N.C., C. Newton, and E.L.R. Stokstad. 1977. The requirement of sterol and various sterol precursors in free-living nematodes. *Nematologica* **23:** 57–61.

Lu, N.C., G. Hubenberg, Jr., G.M. Briggs, and E.L.R. Stokstad. 1978. The growth-promoting activity of several lipid-related compounds in the free-living nematode *Caenorhabditis briggsae*. *Proc. Soc. Exp. Biol. Med.* **158:** 187–191.

Lyons, J.M., A.D. Keith, and I.J. Thomason. 1975. Temperature-induced phase transitions in nematode lipids and their influence on respiration. *J. Nematol.* **7:** 98–104.

Mackenzie, J.M., Jr. and H.F. Epstein. 1980. Paramyosin is necessary for determination of nematode thick filament in vivo. *Cell* **22:** 747–755.

———. 1981. Electron microscopy of nematode thick filaments. *J. Ultrastruct. Res.* **76:** 277–285.

Mackenzie, J.M., Jr., F. Schachat, and H.F. Epstein. 1978a. Immunocytochemical localization of two myosins within the same muscle cells in *Caenorhabditis elegans*. *Cell* **15:** 413–420.

Mackenzie, J.M., Jr., R.L. Garcea, J.M. Zengel, and H.F. Epstein. 1978b. Muscle development in *Caenorhabditis elegans* mutants exhibiting retarded sarcomere construction. *Cell* **15:** 751–762.

MacLeod, A.R., J. Karn, and S. Brenner. 1981. Molecular analysis of the *unc-54* myosin heavy-chain gene of *Caenorhabditis elegans*. *Nature* **291:** 386–390.

MacLeod, A.R., R.H. Waterston, and S. Brenner. 1977a. An internal deletion mutant of a myosin heavy chain in *Caenorhabditis elegans*. *Proc. Natl. Acad. Sci.* **74:** 5336–5340.

MacLeod, A.R., J. Karn, R.H. Waterston, and S. Brenner. 1979. The *unc-54* myosin heavy chain gene of *Caenorhabditis elegans*: A model system for the study of genetic suppression in higher eukaryotes. In *Nonsense mutations and tRNA suppressors* (ed. C.E. Celis and J.D. Smith), pp. 301–312. Academic Press, London.

MacLeod, A.R., R.H. Waterston, R M. Fishpool, and S. Brenner. 1977b. Identification of the structural genes for a myosin heavy-chain in *Caenorhabditis elegans*. *J. Mol. Biol.* **114:** 133–140.

Madl, J.E. and R.K. Herman. 1979. Polyploids and sex determination in *Caenorhabditis elegans*. *Genetics* **93:** 393–402.

Malacinski, G.M. and W.H. Klein, eds. 1984. *Molecular aspects of early development*. Plenum Press, New York.

Manning, A. 1975. The place of genetics in the study of behaviour. In *Growing points in ethology* (ed. P.P.G. Bateson and R.A. Hinde), pp. 327–343. Cambridge University Press, New York.

Marks, C.F. 1971. Respiration response of a *C. sp.* and *Aphelencus avenae* to the nematicide 1,2-dibromoethane (EDB). *J. Nematol.* **3:** 113–118.

Marks, C.F. and O. Sorensen. 1971. Measurement of nematode respiration with the biological oxygen monitor. *J. Nematol.* **3:** 91–92.

Maruyama, K. and R. Hori. 1986. Isolation and characterization of metallothionein from the nematode *Caenorhabditis elegans*. *Eisei Kagaku* **32:** 22–27.

Marx, J.L. 1984a. *Caenorhabditis elegans*: Getting to know you. *Science* **225:** 40–42.

———. 1984b. New clues to developmental timing. *Science* **226:** 425–426.

Maupas, E. 1900. Modes et formes de reproduction des nematodes. *Arch. Zool. Exp. Gen.* **8:** 463–624.

———. 1918. Essais d'hybridisation chez les nematodes. *Bull. Biol. Fr. Belg.* **52:** 466–498.

McClure, M.A. and B.M. Zuckerman. 1982. Localization of cuticular binding sites of concanavalin A on *Caenorhabditis elegans* and *Meloidogyne incognita*. *J. Nematol.* **14:** 39–44.

McGhee, J.D. and D.A. Cottrell. 1986. The major gut esterase locus in the nematode *Caenorhabditis elegans*. *Mol. Gen. Genet.* **202:** 30–34.

McLachlan, A.D. 1983. Analysis of gene duplication repeats in the myosin rod. *J. Mol. Biol.* **169:** 15–30.

———. 1984. Structural implications of the myosin amino acid sequence. *Ann. Rev. Biophys. Bioeng.* **13:** 167–189.

McLachlan, A.D. and J. Karn. 1982. Periodic charge distributions in the myosin rod amino acid sequence match cross-bridge spacings in muscle. *Nature* **299:** 226–231.

———. 1983. Periodic features in the amino acid sequence of nematode myosin rod. *J. Mol. Biol.* **164:** 605–626.

McLaren, D. 1976. Nematode sense organs. In *Advances in parasitology* (ed. B. Dawes), vol. 14, pp. 195–265. Academic Press, New York.

McMahon, D.M. and C.F. Fox, eds. 1975. *Developmental biology: Pattern formation and gene regulation*. W.A. Benjamin, Menlo Park, California.

Meheus, L. and J.R. Vanfleteren. 1985. Nematode chromosomal proteins. IV. The nonhistones of *Caenorhabditis elegans*. *Comp. Biochem. Physiol.* **81B:** 377–383.

———. 1986. Nuclease digestion of DNA and RNA in nuclei from young adult and senescent *Caenorhabditis elegans* (Nematoda). *Mech. Ageing Dev.* **34:** 23–34.

Mellanby, H. 1955. The identification and estimation of acetylcholine in three parasitic nematodes (*Ascaris lumbricoides*, *Litomosoides carnii*, and the microfilariae of *Gyrofilaria repens*). *Parasitology* **45:** 287–294.

Melton, D.A. and R. Cortese. 1979. Transcription of cloned tRNA genes and the nuclear partitioning of a tRNA precursor. *Cell* **18:** 1165–1172.

Meneely, P.M. and R.K. Herman. 1979. Lethals, steriles and deficiencies in a region of the X chromosome of *Caenorhabditis elegans*. *Genetics* **92:** 99–115.

———. 1981. Suppression and function of X-linked lethals and sterile mutations in *Caenorhabditis elegans*. *Genetics* **97:** 65–84.

Meneely, P.M. and W.B. Wood. 1984. An autosomal gene that affects X chromosome expression and sex determination in *Caenorhabditis elegans*. *Genetics* **106:** 29–44.

Messner, B. and U. Kerstan. 1978. The histochemical evidence of peroxidase in invertebrate animals nematoda and insecta. *Acta Histochem.* **62:** 244–253.

Meyer, B.J. and L.P. Casson. 1986. *Caenorhabditis elegans* compensates for the difference in X chromosome dosage between the sexes by regulating transcript levels. *Cell* **47:** 871–881.

Michel, J.F. 1974. Arrested development of nematodes and some related phenomena. *Adv. Parasitol.* **12:** 279–366.

Miller, D.M. and I. Maruyama. 1986. The *sup-3* locus is closely linked to a myosin heavy chain gene in *Caenorhabditis elegans*. *UCLA Symp. Mol. Cell. Biol. New Ser.* **29:** 629–638.

Miller, D.M., F.E. Stockdale, and J. Karn. 1986. Immunological identification of the genes encoding the four myosin heavy chain isoforms of *Caenorhabditis elegans*. *Proc. Natl. Acad. Sci.* **83:** 2305–2309.

Miller, D.M., III, I. Ortiz, G.C. Berliner, and H.F. Epstein. 1983. Differential localization of two myosins within nematode thick filaments. *Cell* **34:** 477–490.

Mitchell, D.H. and T.E. Johnson, eds. 1985. *Invertebrate models in aging research*. CRC Press, Boca Raton, Florida.

Mitchell, D.H., J.W. Stiles, J. Santelli, and D.R. Sanadi. 1979. Synchronous growth and aging of *Caenorhabditis elegans* in the presence of fluorodeoxyuridine. *J. Gerontol.* **34:** 28–36.

Miwa, J., E. Schierenberg, S. Miwa, and G. von Ehrenstein. 1980. Genetics and mode of expression of temperature-sensitive mutations arresting embryonic development in *Caenorhabditis elegans*. *Dev. Biol.* **76:** 160–174.

Miwa, J., Y. Tabuse, M. Furusawa, and H. Yamasaki. 1982. Tumor promoters specifically and reversibly disturb development and behavior of *Caenorhabditis elegans*. *J. Cancer Res. Clin. Oncol.* **104:** 81–88.

Moerman, D.G. and D.L. Baillie. 1979. Genetic organization in *Caenorhabditis elegans*: Fine-structure analysis of the *unc-22* gene. *Genetics* **91:** 95–104.

———. 1981. Formaldehyde mutagenesis in the nematode *Caenorhabditis elegans*. *Mutat. Res.* **80:** 273–279.

Moerman, D.G. and R.H. Waterston. 1984. Spontaneous unstable *unc-22 IV* mutations in *Caenorhabditis elegans* var. Bergerac. *Genetics* **108:** 859–877.

Moerman, D.G., G.M. Benian, and R.H. Waterston. 1986. Molecular cloning of the muscle gene *unc-22* in *Caenorhabditis elegans* by Tc1 transposon tagging. *Proc. Natl. Acad. Sci.* **83:** 2579–2583.

Moerman, D.G., S. Plurad, R.H. Waterston, and D.L. Baillie. 1982. Mutations in the *unc-54* myosin heavy chain gene of *Caenorhabditis elegans* that alter contractility but not muscle structure. *Cell* **29:** 773–781.

Moerman, D.G., G.M. Benian, R.J. Barstead, L. Schreifer, and R.H. Waterston. 1988. Identification and intracellular localization of the *unc-22* gene product of *Caenorhabditis elegans*. *Genes Dev.* **2:** 93–105.

Monoson, H.L. and S.A. Williams. 1973. Endoparasitic nematode-trapping fungi of Mason State Forest. *Mycopathol. Mycol. Appl.* **49:** 177–183.

Monoson, H.L., A.G. Galsky, and R.S. Stephano. 1974. Studies on the ability of various nematodes to induce trap formation in a nematode-trapping fungus *Monacrosporium doedycoides*. *Nematologica* **20:** 96–102.

Morata, G. and P.A. Lawrence. 1977. Homeotic genes, compartments and cell determination in *Drosophila*. *Nature* **265:** 211–216.

Morgan, P.G. and H.F. Cascorbi. 1985. Effect of anesthetics and a convulsant on normal and mutant *Caenorhabditis elegans*. *Anesthesiology* **62:** 738–744.

Moritz, K.B. and G.E. Roth. 1976. Complexity of germline and somatic DNA in *Ascaris*. *Nature* **259:** 55–57.

Mounier, N. 1981. Location of neurosecretory-like material in *Caenorhabditis elegans*. *Nematologica* **27:** 160–166.

Mounier, N. and J. Brun. 1980. A cytogenetical analysis of sterile mutants in *Caenorhabditis elegans*. *Can. J. Genet. Cytol.* **22:** 391–403.

Mount, S.M. 1982. A catalogue of splice junction sequences. *Nucleic Acids Res.* **10:** 459–472.

Muller, H.J. 1932. Further studies on the nature and causes of gene mutations. *Proc. Int. Cong. Genet.* **6:** 213–255.

Mulvey, R.H. 1960. Oogenesis in some free-living nematodes. In *Nematology* (ed. J.N. Sasser and W.R. Jenkins), pp. 321–322. University of North Carolina Press, Chapel Hill.

Munakata, N. and F. Morohoshi. 1984. Effects of alkylating-agents on the nematode *Caenorhabditis elegans*. *J. Rad. Res.* **25:** 31.

———. 1986. DNA glycosylase activities in the nematode *Caenorhabditis elegans*. *Mutat. Res.* **165:** 101–107.

Murfitt, R.R., K. Vogel, and D.R. Sanadi. 1976. Characterization of the mitochondria of the free-living nematode, *Caenorhabditis elegans*. *Comp. Biochem. Physiol.* **53B:** 423–430.

Natzle, J.E., J.M. Monson, and B.J. McCarthy. 1982. Cytogenetic location and expression of collagen-like genes in *Drosophila*. *Nature* **296:** 368–371.

Neigauz, B.M. and V.K. Ravin. 1983. The effect of physiologically active substances on life duration of a nematode *Caenorhabditis elegans*. *Zh. Obsch. Biol.* **44:** 835–841.

Nelson, D.W. and B.M. Honda. 1985. Genes coding for 5S ribosomal RNA of the nematode *Caenorhabditis elegans*. *Gene* **38:** 245–251.

———. 1986a. Genetic mapping of the 5S rRNA gene cluster of the nematode *Caenorhabditis elegans*. *Nucleic Acids Res.* **14:** 869–881.

———. 1986b. Genetic mapping of the 5S rRNA gene cluster of the nematode *Caenorhabditis elegans*. *Can. J. Genet. Cytol.* **28:** 545–553.

Nelson, F.K. and D.L. Riddle. 1984. Functional study of the *Caenorhabditis elegans* secretory–excretory system using laser microsurgery. *J. Exp. Zool.* **231:** 45–46.

Nelson, F.K., P.S. Albert, and D.L. Riddle. 1983. Fine structure of the *Caenorhabditis elegans* secretory–excretory system. *J. Ultrastruct. Res.* **82:** 156–171.

Nelson, G.A. and S. Ward. 1980. Vesicle fusion, pseudopod extension and amoeboid motility are induced in nematode spermatids by the ionophore monensin. *Cell* **19:** 457–464.

Nelson, G.A., K.K. Lew, and S. Ward. 1978. Intersex, a temperature-sensitive mutant of the nematode *Caenorhabditis elegans*. *Dev. Biol.* **66:** 386–409.

Nelson, G.A., T.M. Roberts, and S. Ward. 1982. *Caenorhabditis elegans* spermatozoan locomotion: Amoeboid movement with almost no actin. *J. Cell Biol.* **92:** 121–131.

Neuschulz, N. 1977. Axenic culture of free living nematodes: A literature review. *Z. Versuchstierk.* **19:** 241–247.

Neuschulz, N. and L. Kampfe. 1983. The influence of selected abiotic factors on the population development of axenic cultivated *Caenorhabditis briggsae* (Nematoda). *Zool. Jahrb. Abt. System. Oekol. Geogr. Tiere* **110:** 333–344.

Newmeyer, D. and D.R. Galeazzi. 1977. The instability of *Neurospora* duplication *Dp(IL→ IR)H4250*, and its genetic control. *Genetics* **85:** 461–487.

Newport, J. and M. Kirshner. 1982. A major developmental transition in early *Xenopus* embryos. I. Characterization and timing of cellular changes at the midblastula stage. *Cell* **30:** 675–686.

Nicholas, W.L. 1959. The cultural and nutritional requirements of free-living nematodes of the genus *Rhabditis* and related genera. *Tech. Bull. Min. Agric. Lond.* **7:** 161–168.

———. 1975. *The biology of free-living nematodes*. Clarendon Press, Oxford, England.

———. 1984. *The biology of free-living nematodes*, 2nd edition. Clarendon Press, Oxford, England.

Nicholas, W.L. and J. Jantunen. 1963. A biotin requirement for *Caenorhabditis briggsae* (Rhabditidae). *Nematologica* **9:** 332–336.

———. 1964. *Caenorhabditis briggsae* (Rhabditidae) under anaerobic conditions. *Nematologica* **10:** 409–418.

———. 1966. The effect of different concentrations of oxygen and of carbon dioxide on the growth and reproduction of *Caenorhabditis briggsae*. *Nematologica* **12:** 328–336.

Nicholas, W.L. and M.G. McEntegart. 1957. A technique for obtaining axenic cultures of rhabditid nematodes. *J. Helminthol.* **31:** 135–144.

Nicholas, W.L. and A.C. Stewart. 1978. The calorific value of *Caenorhabditis elegans* (Rhabditidae). *Nematologica* **24:** 45–50.

Nicholas, W.L. and S. Viswanathan. 1975. A study of the nutrition of *Caenorhabditis briggsae* (Rhabditidae) fed on carbon-14 and phosphorus-32 labelled bacteria. *Nematologica* **21:** 385–400.

Nicholas, W.L., E.C. Dougherty, and E.L. Hansen. 1959. Axenic cultivation of *Caenorhabditis briggsae* with chemically undefined supplements; comparative studies with related nematodes. *Ann. N.Y. Acad. Sci.* **77:** 218–236.

Nicholas, W.L., A. Grassia, and S. Viswanathan. 1973. The efficiency with which *Caenorhabditis briggsae* (Rhabditidae) feeds on the bacterium, *Escherichia coli*. *Nematologica* **19:** 411–420.

Nicholas, W.L., E.L. Hansen, and E.C. Dougherty. 1962. The B-vitamins required by *Caenorhabditis briggsae* (Rhabditidae). *Nematologica* **8:** 129–135.

Nicholas, W.L., E.C. Dougherty, E.L. Hansen, O. Holm-Hansen, and V. Moses. 1960. The incorporation of ^{14}C from sodium acetate-2-^{14}C into the amino acids of the soil-inhabiting nematode, *Caenorhabditis briggsae*. *J. Exp. Biol.* **37:** 435–443.

Nicholls, J.G. 1981. Repair and regeneration of the nervous system. In *Dahlem Workshop Reports Series Proceedings, Berlin*, vol. 24. Springer-Verlag, New York.

Nickle, W.R. and G.L. Ayre. 1966. *C. dolichura* in the head glands of the ants *Camponotus herculeanus* (L.) and *Acanthomyops claviger* (Roger) in Ontario. *Proc. Entomol. Soc. Ont.* **96:** 96–98.

Nigon, V. 1947. Le déterminisme du sexe chez un nématode libre hermaphrodite *Rhabditis dolichura* Schneider. *Bull. Soc. Zool. Fr.* **71:** 78–86.

———. 1949a. Les modalités de la reproduction et le déterminisme de sexe chez quelques nématodes libres. *Ann. Sci. Nat. Zool. Biol. Anim.* (ser. 11) **2:** 1–132.

———. 1949b. Effets de la polyploidie chez un nématode libre. *C.R. Seances Acad. Sci. Ser. D Sci. Nat.* **228:** 1161–1162.

———. 1951a. La gamétogenèse d'un nématode tetraploide obtenu par voie expérimentale. *Bull. Soc. Hist. Nat. Toulouse* **86:** 192–200.

———. 1951b. Polploidie expérimentale chez un nématode libre, *Rhabditis elegans* Maupas. *Bull. Biol. Fr. Belg.* **85:** 187–225.

———. 1954. Contributions à la critique expérimentale des théories de la continuité germinale et de la lignée pure. *Acta Biol. Acad. Sci. Hung.* **5:** 96–117.

———. 1965. Développement et reproduction des nématodes. In *Traité de zoologie* (ed. P.P. Grasse), vol. 4. Masson et Cie, Paris, France.

Nigon, V. and R. Arcel. 1951. Effets d'une élévation de température sur la prophase meiotique d'un nématode libre. *C.R. Seances Acad. Sci. Ser. D Sci. Nat.* **232:** 1032–1034.

Nigon, V. and J. Brun. 1955. L'évolution des structures nucléaires dans l'ovogenèse de *Caenorhabditis elegans* Maupas 1900. *Chromosoma* **7:** 129–169.

———. 1967. Génétique et évolution des nématodes libres. Perspectives tirées de l'étude de *Caenorhabditis elegans*. *Experientia* **23:** 161–170.

Nigon, V. and E.C. Dougherty. 1949. Reproductive patterns and attempts at reciprocal crossing of *Rhabditis elegans* Maupas 1900, and *Rhabditis briggsae* Dougherty and Nigon. *J. Exp. Zool.* **112:** 485–503.

———. 1950. A dwarf mutation in a nematode. A morphological mutant of *Rhabditis briggsae*, a free-living soil nematode. *J. Hered.* **41:** 103–109.

Nigon, V., P. Guerrier, and H. Monin. 1960. L'Architecture polaire de l'oeuf et movements des constituants cellulaires au cour des premières étapes du développement chez quelque nématodes. *Bull. Biol. Fr. Belg.* **94:** 132–201.

Niki, Y. 1986. Germline-autonomous sterility of P-M dysgenic hybrids and their application to germline transfers in *Drosophila melanogaster*. *Dev. Biol.* **113:** 255–258.

Ninomiya, Y. and B.R. Olsen. 1984. Synthesis and characterization of cDNA encoding a cartilage-specific short collagen. *Proc. Natl. Acad. Sci.* **81:** 3014–3018.

Nolan, R.A. 1972. *Asteromyces cruciatus* from North America. *Mycologia* **64:** 430–433.

Nomarski, G. 1955. Microinterféromètre differentiel á ondes polarisées. *J. Phys. Radium* **16:** 9–13.

Nonnemacher-Godet, J. and E.C. Dougherty. 1964. Incorporation of tritiated thymidine in the cells of *Caenorhabditis briggsae* (Nematoda) reared in axenic culture. *J. Cell Biol.* **22:** 281–290.

Noonan, D. 1985. Beyond the double helix. *Esquire* **104:** 196–210.

Nuccitelli, R. and D.W. Deamer. 1982. Intracellular pH: Its measurement, regulation and utilization in cellular functions. *Kroc Found. Ser.* **15.**

O'Farrell, P.H. 1975. High-resolution two-dimensional electrophoresis of proteins. *J. Biol. Chem.* **250:** 4007–4021.

O'Hare, K. and G.M. Rubin. 1983. Structure of P transposable elements and their sites of insertion and excision in the *Drosophila melanogaster* genome. *Cell* **34:** 25–35.

Ohba, K. and N. Ishibashi. 1981. Effects of procaine on the development, longevity and fecundity of *Caenorhabditis elegans*. *Nematologica* **27:** 275–284.

———. 1982. A factor inducing dauer juvenile formation in *Caenorhabditis elegans*. *Nematologica* **28:** 318–325.

———. 1984. A nematode, *Caenorhabditis elegans*, as a test organism for nematicide evaluation. *J. Pest. Sci.* **9:** 91–96.

Okai, Y. 1986a. Possible regulating factors for chromatin-dependent RNA polymerase II reaction in *Caenorhabditis elegans*. *Zool. Sci.* **3:** 97–102.

———. 1986b. A low-molecular-weight inhibitory peptide for DNA-dependent RNA polymerase II reaction in a nematoda, *Caenorhabditis elegans*. *Zool. Sci.* **3:** 103–108.

Okamoto, H. and J.M. Thomson. 1985. Monoclonal antibodies which distinguish certain classes of neuronal and support cells in the nervous tissue of the nematode *Caenorhabditis elegans*. *J. Neurosci.* **5:** 643–653.

Oppenheim, R.W. 1981. Neuronal cell death and some related regressive phenomena during neurogenesis: A selective historical review and progress report. In *Studies in developmental neurobiology: Essays in honor of Viktor Hamburger* (ed. W.M. Cowan), pp. 74–133. Oxford University Press, New York.

Orgel, L.E. and F.H.C. Crick. 1980. Selfish DNA: The ultimate parasite. *Nature* **284:** 604–607.

Osche, G. 1952. Systematik und Phylogenie der Gattung *Rhabditis* (Nematoda). *Zool. Jb.* **81:** 190–280.

Otsuka, A.J. 1986. *sup-3* Suppression affects muscle structure and myosin heavy chain accumulation in *Caenorhabditis elegans*. *UCLA Symp. Mol. Cell. Biol. New Ser.* **29:** 619–628.

Ouazana, R. 1978. A morphological temperature-sensitive mutant of the nematode *Caenorhabditis elegans* var. Bergerac. *Experientia* **34:** 170–171.

———. 1981. Cuticle collagen during the post-embryonic development of the nematode *Caenorhabditis elegans*: Comparison between 1st stage larvae and adults. *C.R. Seances Acad. Sci. Ser. III Sci. Vie* **293:** 467–470.

———. 1982. Structure and chemical composition of the cuticular integument of the nematodes. *Bull. Soc. Zool. Fr.* **107:** 419–426.

Ouazana, R. and J.L. Brun. 1975. Intracistronic recombination at a dwarfing locus in the free living self fertilizing nematode *Caenorhabditis elegans*. *C.R. Seances Acad. Sci. Ser. D Sci. Nat.* **280:** 1895–1898.

———. 1978. Effect of dumpiness on the development and functioning of eutelic and noneutelic organs in the nematode *Caenorhabditis elegans*. *Genetica* **49:** 45–52.

Ouazana, R. and M.A. Gilbert. 1979. Cuticular composition of the nematode *Caenorhabditis elegans*, Bergerac wild-type strain. *C.R. Seances Acad. Sci. Ser. D Sci. Nat.* **288:** 911–914.

Ouazana, R. and D. Herbage. 1981. Biochemical characterization of the cuticle collagen of the nematode *Caenorhabditis elegans*. *Biochim. Biophys. Acta* **669:** 236–243.

Ouazana, R., R. Garrone, and J. Godet. 1985. Characterization of morphological and biochemical defects in the cuticle of a dumpy mutant of *Caenorhabditis elegans*. *Comp. Biochem. Physiol.* **80B:** 481–483.

Ouazana, R., D. Herbage, and J. Godet. 1984. Some biochemical aspects of the cuticle collagen of the nematode *Caenorhabditis elegans*. *Comp. Biochem. Physiol.* **77B:** 51–56.

Ouweneel, W. 1976. Developmental genetics of homeosis. *Adv. Genet.* **18:** 179–248.

Ozerol, N.H. and P.H. Silverman. 1972. Exsheathment phenomenon in the infective-stage larvae of *Haemonchus contortus*. *J. Parasitol.* **58:** 34–44.

Pak, W.L. and L.H. Pinto. 1976. Genetic approach to the study of the nervous system. *Ann. Rev. Biophys. Bioeng.* **5:** 397–448.

Park, E.C. and H.R. Horvitz. 1986a. Mutations with dominant effects on the behavior and morphology of the nematode *Caenorhabditis elegans*. *Genetics* **113:** 821–852.

———. 1986b. *Caenorhabditis elegans unc-105* mutations affect muscle and are suppressed by other mutations that affect muscle. *Genetics* **113:** 853–867.

Patel, T.R. and B.A. McFadden. 1976. A simple spectrophotometric method for measurement of nematode populations. *Anal. Biochem.* **70:** 447–453.

———. 1977. Particulate isocitrate lyase and malate synthase in *Caenorhabditis elegans. Arch. Biochem. Biophys.* **183:** 24–30.

———. 1978a. *Caenorhabditis elegans* and *Ascaris suum*: Fragmentation of isocitrate lyase in crude extracts. *Exp. Parasitol.* **44:** 72–81.

———. 1978b. *Caenorhabditis elegans* and *Ascaris suum*: Inhibition of isocitrate lyase by itaconate. *Exp. Parasitol.* **44:** 262–268.

———. 1978c. Axenic and synchronous cultures of *Caenorhabditis elegans. Nematologica* **24:** 51–62.

Pearson, M.L. and H.F. Epstein, eds. 1982. *Muscle development: Molecular and cellular control.* Cold Spring Harbor Laboratory, Cold Spring Harbor, New York.

Pelham, H. 1985. Activation of heat-shock genes in eukaryotes. *Trends Genet.* **1:** 31–35.

Perkins, L.A., E.M. Hedgecock, J.N. Thompson, and J.G. Culotti. 1986. Mutant sensory cilia in the nematode *Caenorhabditis elegans. Dev. Biol.* **117:** 456–487.

Perlman, S., C. Phillips, and J. Bishop. 1976. A study of foldback DNA. *Cell* **8:** 33–42.

Perrimon, N., D. Mohler, L. Engstrom, and A.P. Mahowald. 1986. X-linked female sterile loci in *Drosophila melanogaster. Genetics* **113:** 695–712.

Pertel, R. 1964. Axenic cultivation of two species of rhabditid nematodes on a commercial medium. *Nematologica* **10:** 343.

Pertel, R. and S.H. Wilson. 1974. Histamine content of the nematode, *Caenorhabditis elegans. Comp. Gen. Pharmacol.* **5:** 83–85.

Pertel, R., N. Paran, and C.F.T. Mattern. 1976. *Caenorhabditis elegans*: Localization of cholinesterase associated with anterior nematode structures. *Exp. Parasitol.* **39:** 401–414.

Pinnock, C.B. and E.L.R. Stokstad. 1975. The effect of heme source on growth of *Caenorhabditis briggsae* in peptide and carbohydrate supplemented chemically defined medium. *Nematologica* **21:** 258–260.

Pinnock, C.B., W.F. Hieb, and E.L.R. Stokstad. 1975a. A mass culture bioassay method for *Caenorhabditis briggsae* using population growth rate as a response parameter. *Nematologica* **21:** 1–4.

Pinnock, C., B. Shane, and E.L.R. Stokstad. 1975b. Stimulatory effects of peptides on growth of the free-living nematode *Caenorhabditis briggsae. Proc. Soc. Exp. Biol. Med.* **148:** 710–713.

Poinar, G.O., Jr. 1983. *The natural history of nematodes.* Prentice-Hall, Englewood Cliffs, New Jersey.

Politz, J.C. and R.S. Edgar. 1984. Overlapping stage-specific sets of numerous small collagen polypeptides are translated in vitro from *Caenorhabditis elegans* RNA. *Cell* **37:** 853–860.

Politz, S.M., J.C. Politz, and R.S. Edgar. 1986. Small collagenous proteins present during the molt in *Caenorhabditis elegans. J. Nematol.* **18:** 303–310.

Pong, S.S., C.C. Wang, and L.C. Fritz. 1980. Studies on the mechanism of action of avermection B1a: Stimulation of release of GABA from brain synaptosomes. *J. Neurochem.* **34:** 351–358.

Popham, J.D. and J.M. Webster. 1978. An alternative interpretation of the fine structure of the basal zone of the cuticle of the dauer larva of the nematode *Caenorhabditis elegans. Can. J. Zool.* **56:** 1556–1563.

———. 1979a. Aspects of the fine structure of the dauer larva of the nematode *Caenorhabditis elegans. Can. J. Zool.* **57:** 794–800.

———. 1979b. Cadmium toxicity in the free living nematode *Caenorhabditis elegans. Environ. Res.* **20:** 183–191.

———. 1979c. The use of osmium mixtures in localizing ions in *Caenorhabditis elegans. Nematologica* **25:** 67–75.

———. 1982. Ultrastructural changes in *Caenorhabditis elegans* (Nematoda) caused by toxic levels of mercury and silver. *Ecotoxicol. Environ. Saf.* **6:** 183–189.

Potten, C.S., ed. 1983. *Stem cells, their identification and characterization*. Churchill Livingston, London.
Potts, F.A. 1910. Notes on the free-living nematodes. I. The hermaphrodite species. *Q. J. Microsc. Sci.* **55:** 433–484.
Priess, J. and D. Hirsh. 1986. *Caenorhabditis elegans* morphogenesis: The role of the cytoskeleton in the elongation of the embryo. *Dev. Biol.* **117:** 156–173.
Priess, J.R. and J.N. Thomson. 1987. Cellular interactions in early *Caenorhabditis elegans* embryos. *Cell* **48:** 241–250.
Priess, J.R., H. Schnabel, and R. Schnabel. 1987. The *glp-1* locus and cellular interactions in early *Caenorhabditis elegans* embryos. *Cell* **51:** 601–611.
Pruss, R.M., R. Mirsky, M.C. Raff, R. Thorpe, A.J. Dowding, and B.H. Anderton. 1981. All classes of intermediate filaments share a common antigenic determinant defined by a monoclonal antibody. *Cell* **27:** 419–428.
Quinn, W.G. and J.L. Gould. 1979. Nerves and genes. *Nature* **278:** 19–23.
Raff, R.A. and T.C. Kaufman. 1983. *Embryos, genes, and evolution*. Macmillan, New York.
Rammeloo, J. 1973. *Harposporium anguillulae* Lohde (moniliales), espèce nématophage, trouvé en Belgique. *Biol. Jaarb.* **41:** 180–182.
Rand, J.B. and C.D. Johnson. 1981. A single-vial biphasic liquid extraction assay for choline acetyltransferase using [^{3}H]choline. *Anal. Biochem.* **116:** 361–371.
Rand, J.B. and R.L. Russell. 1984. Choline acetyltransferase-deficient mutants of the nematode *Caenorhabditis elegans*. *Genetics* **106:** 227–248.
———. 1985a. Properties and partial purification of choline acetyltransferase from the nematode *Caenorhabditis elegans*. *J. Neurochem.* **44:** 189–200.
———. 1985b. Molecular basis of drug-resistance mutations in *Caenorhabditis elegans*. *Psychopharmacol. Bull.* **21:** 623–630.
Regelson, W. and F.M. Sinex, eds. 1983. *Intervention in the aging process*, part B. Alan R. Liss, New York.
Riddle, D.L. 1977. A genetic pathway for dauer larva formation in *Caenorhabditis elegans*. *Stadler Genet. Symp.* **9:** 101–120.
———. 1978. The genetics of development and behavior in *Caenorhabditis elegans*. *J. Nematol.* **10:** 1–16.
———. 1980. Developmental genetics of *Caenorhabditis elegans*. In *Nematodes as biological models*, volume 1: *Behavioral and developmental models* (ed. B.M. Zuckerman), pp. 263–284. Academic Press, New York.
———. 1982. Developmental biology of *Caenorhabditis elegans*: Symposium introduction. *J. Nematol.* **14:** 238–239.
———. 1986. Post-embryonic development in *Caenorhabditis elegans*. Parasitology—Quo vadit? In *Proceedings of the Sixth International Congress of Parasitology* (ed. M.J. Howell), pp. 223–231. Australian Academy of Science, Canberra.
Riddle, D.L. and A.F. Bird. 1985. Responses of *Anguina agrostis* to detergent and anesthetic treatment. *J. Nematol.* **17:** 165–168.
Riddle, D.L. and S. Brenner. 1978. Indirect suppression in *Caenorhabditis elegans*. *Genetics* **89:** 299–314.
Riddle, D.L., M.M. Swanson, and P.S. Albert. 1981. Interacting genes in nematode dauer larva formation. *Nature* **290:** 668–671.
Rinker, D.L. and J.R. Bloom. 1982. Phoresy between a mushroom-infesting fly and two free-living nematodes associated with mushroom culture. *J. Nematol.* **14:** 599–602.
Roberts, S.B., M. Sanicola, S.W. Emmons, and G. Childs. 1987. Molecular characterization of the histone gene family of *Caenorhabditis elegans*. *J. Mol. Biol.* **196:** 27–38.
Roberts, T.M. 1983. Crawling *Caenorhabditis elegans* spermatozoa contact the substrate only by their pseudopods and contain 2-nm filaments. *Cell Motil.* **3:** 333–347.
Roberts, T.M. and G. Steitmatter. 1984. Membrane-substrate contact under the sper-

matozoon of *Caenorhabditis elegans*, a crawling cell that lacks filamentous actin. *J. Cell Sci.* **69:** 117–126.

Roberts, T.M. and S. Ward. 1982a. Centripetal flow of pseudopodial surface components could propel the amoeboid movement of *Caenorhabditis elegans* spermatozoa. *J. Cell Biol.* **92:** 132–138.

———. 1982b. Directed membrane flow on the pseudopods of *Caenorhabditis elegans* spermatozoa. *Cold Spring Harbor Symp. Quant. Biol.* **46:** 695–702.

———. 1982c. Membrane flow during nematode spermiogenesis. *J. Cell Biol.* **92:** 113–120.

Roberts, T.M., F.M. Pavalko, and S. Ward. 1986. Membrane and cytoplasmic proteins are transported in the same organelle complex during a nematode spermatogenesis. *J. Cell Biol.* **102:** 1787–1796.

Robertson, A.M.G. and J.N. Thomson. 1982. Morphology of programmed cell death in the ventral nerve cord of *Caenorhabditis elegans* larvae. *J. Embryol. Exp. Morphol.* **67:** 89–100.

Rockstein, M., ed. 1974. *Theoretical aspects of aging*. Academic Press, New York.

Rogalski, T.M. and D.L. Baillie. 1985. Genetic organization of the *unc-22 IV* gene and the adjacent region in *Caenorhabditis elegans. Mol. Gen. Genet.* **201:** 409–414.

Rogalski, T.M., D.G. Moerman, and D.L. Baillie. 1982. Essential genes and deficiencies in the *unc-22 IV* region of *Caenorhabditis elegans. Genetics* **102:** 725–736.

Rose, A.M. and D.L. Baillie. 1979a. A mutation in *Caenorhabditis elegans* that increases recombination frequency more than threefold. *Nature* **281:** 599–600.

———. 1979b. The effect of temperature and parental age on recombination and nondisjunction in *Caenorhabditis elegans. Genetics* **92:** 409–418.

———. 1981. Genetic organization of the region around *unc-51(I)*, a gene affecting paramyosin in *Caenorhabditis elegans. Genetics* **96:** 639–648.

Rose, A.M. and T.P. Snutch. 1984. Isolation of the closed circular form of the transposable element Tcl in *Caenorhabditis elegans. Nature* **311:** 485–486.

Rose, A.M., D.L. Baillie, and J. Curran. 1984. Meiotic pairing behavior of two free duplications of linkage group I in *Caenorhabditis elegans. Mol. Gen. Genet.* **195:** 52–56.

Rose, A.M., L.J. Harris, N.R. Mawji, and W.J. Morris. 1985. Tcl(Hin): A form of the transposable element Tcl in *Caenorhabditis elegans. Can. J. Biochem. Cell Biol.* **63:** 752–756.

Rose, A.M., D.L. Baillie, E.P.M. Candido, K.A. Beckenbach, and D. Nelson. 1982. The linkage mapping of cloned restriction fragment length differences in *Caenorhabditis elegans. Mol. Gen. Genet.* **188:** 286–291.

Rosenbluth, J. 1965. Structural organization of obliquely striated muscle fibers in *Ascaris lumbricoides. J. Cell. Biol.* **25:** 495–515.

Rosenbluth, R.E. and D.L. Baillie. 1981. The genetic analysis of a reciprocal translocation, eT1(II;V), in *Caenorhabditis elegans. Genetics* **99:** 415–428.

Rosenbluth, R.E., C. Cuddeford, and D.L. Baillie. 1983. Mutagenesis in *Caenorhabditis elegans*. I. A rapid eukaryotic mutagen test system using the reciprocal translocation eT1 (III;V). *Mutat. Res.* **110:** 39–48.

———. 1985. Mutagenesis in *Caenorhabditis elegans*. II. A spectrum of mutational events induced with 1500 R of gamma-radiation. *Genetics* **109:** 493–511.

Rosenzweig, B., L.W. Liao, and D. Hirsh. 1983a. Target sequences for the *Caenorhabditis elegans* transposable element Tcl. *Nucleic Acids Res.* **11:** 7137–7140.

———. 1983b. Sequence of the *Caenorhabditis elegans* transposable element Tcl. *Nucleic Acids Res.* **11:** 4201–4209.

Rosenzweig, W.D., D. Premachandran, and D. Pramer. 1985. Role of trap lectins in the specificity of nematode capture by fungi. *Can. J. Microbiol.* **31:** 693–695.

Roth, G.E. 1979. Satellite DNA properties of the germ-line limited DNA and the organization of the somatic genomes in the nematodes *Ascaris suum* and *Parascaris equorum. Chromosoma* **74:** 355–371.

Rothstein, M. 1965. Nematode biochemistry. V. Intermediary metabolism and amino acid interconversions in *Caenorhabditis briggsae*. *Comp. Biochem. Physiol.* **14:** 541–552.

———. 1968. Nematode biochemistry. IX. Lack of sterol biosynthesis in free-living nematodes. *Comp. Biochem. Physiol.* **27:** 309–317.

———. 1969. Nematode biochemistry. X. Excretion of glycerol by free-living nematodes. *Comp. Biochem. Physiol.* **30:** 641–648.

———. 1970. Nematode biochemistry. XI. Biosynthesis of fatty acids by *Caenorhabditis briggsae* and *Panagrellus redivivus*. *Int. J. Biochem.* **1:** 422–428.

———. 1974. Practical methods for the axenic culture of the free-living nematodes *Turbatrix aceti* and *Caenorhabditis briggsae*. *Comp. Biochem. Physiol.* **49B:** 669–678.

———. 1980. Effects of aging on enzymes. In *Nematodes as biological models*, volume 2: *Aging and other model systems* (ed. B.M. Zuckerman), pp. 29–46. Academic Press, New York.

———. 1983. Nematode biochemistry. III. Excretion products. *Comp. Biochem. Physiol.* **9:** 51–59.

———, ed. 1985. *Review of biological research in aging*, volume 2. Alan R. Liss, New York.

Rothstein, M. and E. Cook. 1966. Nematode biochemistry. IV. Conditions for axenic culture of *Turbatrix aceti*, *Panagrellus redivivus*, *Rhabditis anomala* and *Caenorhabditis briggsae*. *Comp. Biochem. Physiol.* **17:** 683–692.

Rothstein, M. and M. Coppens. 1978. Nutritional factors and conditions for the axenic culture of free-living nematodes. *Comp. Biochem. Physiol.* **61B:** 99–104.

Rothstein, M. and H. Mayoh. 1964a. Nematode biochemistry. IV. On isocitrate lyase in *Caenorhabditis briggsae*. *Arch. Biochem. Biophys.* **108:** 134–142.

———. 1964b. Glycine synthesis and isocitrate lyase in the nematode, *Caenorhabditis briggsae*. *Biochem. Biophys. Res. Commun.* **14:** 43–47.

———. 1966. Nematode biochemistry. VIII. Malate synthetase. *Comp. Biochem. Physiol.* **17:** 1181–1188.

Rothstein, M. and W.L. Nicholas. 1969. Culture methods and nutrition of nematodes and Acanthocephala. In *Chemical zoology* (ed. M. Florkin and B.T. Scheer), vol. 3, pp. 289–328. Academic Press, New York.

Rothstein, M. and G.A. Tomlinson. 1961. Biosynthesis of amino acids by the nematode *Caenorhabditis briggsae*. *Biochim. Biophys. Acta* **49:** 625–627.

———. 1962. Nematode biochemistry. II. Biosynthesis of amino acids. *Biochim. Biophys. Acta* **63:** 471–480.

Rothstein, M., W. Adler, V. Cristofalo, C.E. Finch, J. Florini, and G. Martin, eds. 1983. *Review of biological research in aging*, volume 1. Alan R. Liss, New York.

Rowland, L.P., ed. 1976. *Pathogenesis of the human muscular dystrophies*. Excerpta Medica, Amsterdam.

Ruan, K.-S. and S.W. Emmons. 1984. Extrachromosomal copies of transposon Tc1 in the nematode *Caenorhabditis elegans*. *Proc. Natl. Acad. Sci.* **81:** 4018–4022.

———. 1987. Precise and imprecise somatic excision of the transposon Tc1 in the nematode *Caenorhabditis elegans*. *Nucleic Acids Res.* **15:** 6875–6881.

Rush, M.C. and R. Mishra. 1985. Extrachromosomal DNA in eukaryotes. *Plasmid* **14:** 177–191.

Russell, R.L. 1981. Mutants of neurotransmitter metabolism and action in the nematode *Caenorhabditis elegans*. In *Genetic research strategies in psychobiology and psychiatry. Psychobiology and psychopathology* (ed. E.S. Gershon et al.), vol. I, pp. 113–128. Boxwood Press, California.

Russell, R.L. and L.A. Jacobson. 1985. Some aspects of aging can be studied easily in nematodes. In *Handbook of the biology of aging* (ed. C.E. Finch and E.L. Schneider), pp. 128–145. Van Nostrand Reinhold, New York.

Russell, R.L., C.D. Johnson, J.B. Rand, S. Scherer, and M.S. Zwass. 1977. Mutants of acetylcholine metabolism in the nematode *Caenorhabditis elegans*. *ICN-UCLA Symp. Mol. Cell. Biol.* **8:** 359–371.

Russnak, R.H. and E.P.M. Candido. 1985. Locus encoding a family of small heat-shock genes in *Caenorhabditis elegans*: Two genes duplicated to form a 3.8-kilobase inverted repeat. *Mol. Cell. Biol.* **5:** 1268–1278.

Russnak, R.H., D. Jones, and E.P.M. Candido. 1983. Cloning and analysis of cDNA sequences coding for two 16 kd heat shock proteins in *Caenorhabditis elegans*: Homology with the small hsps of *Drosophila. Nucleic Acids Res.* **11:** 3187–3205.

Rutherford, T.A. and N.A. Croll. 1979. Wave forms of *Caenorhabditis elegans* in a chemical attractant and repellent and in thermal gradients. *J. Nematol.* **11:** 232–240.

Salvato, M., J. Sulston, D. Albertson, and S. Brenner. 1986. A novel calmodulin-like gene from the nematode *Caenorhabditis elegans. J. Mol. Biol.* **190:** 281–290.

Samoiloff, M.R. 1980. Action of chemical and physical agents on free-living nematodes. In *Nematodes as biological models*, volume 2: *Aging and other model systems* (ed. B.M. Zuckerman), pp. 81–98. Academic Press, New York.

Sandler, L. and P. Szauter. 1978. The effect of recombination-defective meiotic mutants on fourth-chromosome crossing over in *Drosophila melanogaster. Genetics* **90:** 699–712.

Sanford, T., M. Golomb, and D.L. Riddle. 1983. RNA polymerase II from wild type and alpha-amanitin-resistant strains of *Caenorhabditis elegans. J. Biol. Chem.* **258:** 12804–12809.

Sanford, T., J.P. Prenger, and M. Golomb. 1985. Purification and immunological analysis of RNA polymerase II from *Caenorhabditis elegans. J. Biol. Chem.* **260:** 8064–8069.

Santoro, C., F. Costanzo, and G. Ciliberto. 1984. Inhibition of eukaryotic transfer RNA transcription by potential Z-DNA sequences. *EMBO J.* **3:** 1553–1559.

Sasser, J.N. and W.R. Jenkins, eds. 1960. *Nematology*. University of North Carolina Press, Chapel Hill.

Satoh, N. and S. Ikegami. 1981. A definite number of aphidicolin-sensitive cell-cyclic events are required for acetylcholinesterase development in the presumptive muscle cells of the Ascidian embryos. *J. Embryol. Exp. Morphol.* **61:** 1–13.

Sayre, F.W., E.L. Hansen, and E.A. Yarwood. 1963. Biochemical aspects of the nutrition of *Caenorhabditis briggsae. Exp. Parasitol.* **13:** 98–107.

Sayre, F.W., M.C. Fishler, G.K. Humphreys, and M.E. Jayko. 1968. Growth factor studies: Changes in reactivity of sulfhydryl groups with activation. *Biochim. Biophys. Acta* **160:** 63–68.

Sayre, F.W., E.L. Hansen, T.J. Starr, and E.A. Yarwood. 1961. Isolation and partial characterization of a growth-control factor. *Nature* **190:** 1116–1117.

Sayre, F.W., T.L. Reiko, R.P. Sandman, and G. Perez-Mendez. 1967. Studies of a growth factor: Fractionation studies and amino acid composition of derived fractions. *Arch. Biochem. Biophys.* **118:** 58–72.

Schachat, F., R.L. Garcea, and H.F. Epstein. 1978a. Myosins exist as homodimers of heavy chains: Demonstration with specific antibody purified by nematode mutant myosin affinity chromatography. *Cell* **15:** 405–411.

Schachat, F., H.E. Harris, and H.F. Epstein. 1977a. Actin from the nematode, *Caenorhabditis elegans*, is a single electrofocusing species. *Biochim. Biophys. Acta* **493:** 304–309.

———. 1977b. Two homogeneous myosins in body-wall muscle of *Caenorhabditis elegans*. *Cell* **10:** 721–728.

Schachat, F., D.J. O'Connor, and H.F. Epstein. 1978b. The moderately repetitive DNA sequences of *Caenorhabditis elegans* do not show short-period interspersion. *Biochim. Biophys. Acta* **520:** 688–692.

Schachat, F.H., H.E. Harris, R.L. Garcea, J.W. LaPointe, and H.F. Epstein. 1977c. Studies on two body-wall myosins in wild type and mutant nematodes. *ICN-UCLA Symp. Mol. Cell. Biol.* **8:** 373–380.

Schiemer, F. 1982a. Food dependence and energetics of free living nematodes. I. Respiration, growth and reproduction of *Caenorhabditis briggsae* at different levels of food supplies. *Oecologia* **54:** 108–121.

———. 1982b. Food dependence and energetics of free living nematodes. II. Life history parameters of *Caenorhabditis briggsae* at different levels of food supply. *Oecologia* **54:** 122–128.

———. 1983. Comparative aspects of food dependence and energetics of free living nematodes. *Oikos* **41:** 32–42.

Schierenberg, E. 1982. Development of the nematode *Caenorhabditis elegans*. In *Developmental biology of freshwater invertebrates* (ed. F.W. Harrison and R.R. Cowden), pp. 249–281. Alan R. Liss, New York.

———. 1984. Altered cell-division rates after laser-induced cell fusion in nematode embryos. *Dev. Biol.* **101:** 240–245.

———. 1985. Cell determination during early embryogenesis of the nematode *Caenorhabditis elegans*. *Cold Spring Harbor Symp. Quant. Biol.* **50:** 59–68.

———. 1987. Reversal of cellular polarity and early cell-cell interactions in the embryo of *Caenorhabditis elegans*. *Dev. Biol.* **122:** 452–463.

Schierenberg, E. and R. Cassada. 1983. Cell division patterns and cell diversification in the nematode *Caenorhabditis elegans*. In *Stem cells, their identification and characterization* (ed. C.S. Potten), pp. 73–92. Churchhill Livingstone, England.

———. 1986. Der Nematode *Caenorhabditis elegans*. *Biol. Unserer Zeit* **16:** 1–7.

Schierenberg, E. and W.B. Wood. 1985. Control of cell-cycle timing in early embryos of *Caenorhabditis elegans*. *Dev. Biol.* **107:** 337–354.

Schierenberg, E., C. Carlson, and W. Sidio. 1985. Cellular development of a nematode: 3-D computer reconstruction of living embryos. *Wilhelm Roux's Arch. Dev. Biol.* **194:** 61–68.

Schierenberg, E., J. Miwa, and G. von Ehrenstein. 1980. Cell lineages and developmental defects of temperature-sensitive embryonic arrest mutants in *Caenorhabditis elegans*. *Dev. Biol.* **76:** 141–159.

Schierenberg, E., T. Cole, C. Carlson, and W. Sidio. 1986. Computer-aided three-dimensional reconstruction of nematode embryos from EM serial sections (technical note). *Exp. Cell Res.* **166:** 247–252.

Schlessinger, D., ed. 1975. *Microbiology–1975*. American Society for Microbiology, Washington, DC.

Schmid, T.M. and H.E. Conrad. 1982. A unique low molecular weight collagen secreted by cultured chick embryo chondrocytes. *J. Biol. Chem.* **257:** 12444–12450.

Schmid, T.M. and T.F. Linsenmayer. 1985. Immunohistochemical localization of short chain cartilage collagen (type X) in avian tissues. *J. Cell Biol.* **100:** 598–605.

Schmidt, G.D. and R.E. Kuntz. 1972. Nematode parasites of Oceanica. XVIII. *C. avicola* sp. n. (Rhabditidae) found in a bird from Taiwan. *Proc. Helminthol. Soc. Wash.* **39:** 189–191.

Schneider, A., ed. 1866. *Monographie der Nematoden*. Berlin.

Schotland, D.M., ed. 1982. *Diseases of the motor unit*. John Wiley, New York.

Searcy, D.G. and A.J. MacInnis. 1970. Measurements by DNA renaturation of the genetic basis of parasitic reduction. *Evolution* **24:** 796–806.

Searcy, D.G., M.J. Kisiel, and B.M. Zuckerman. 1976. Age-related increase of cuticle permeability in the nematode *Caenorhabditis briggsae*. *Exp. Aging Res.* **2:** 293–301.

Sebastiano, M., M. D'Alessio, and P. Bazzicalupo. 1986. Beta-glucuronidase mutants in the nematode *Caenorhabditis elegans*. *Genetics* **112:** 459–468.

Seymour, M.K., K.A. Wright, and C.C. Doncaster. 1983. The action of the anterior feeding apparatus of *Caenorhabditis elegans* (Nematoda:Rhabditida). *J. Zool. Soc. Lond.* **201:** 527–539.

Sharrock, W.J. 1983. Yolk proteins of *Caenorhabditis elegans*. *Dev. Biol.* **96:** 182–188.

———. 1984. Cleavage of two yolk proteins from a precursor in *Caenorhabditis elegans*. *J. Mol. Biol.* **174:** 419–431.

Shay, J.W., ed. 1985. *Cell and muscle motility*, volume 6. Plenum Press, New York.

Shepherd, A.M. 1981. Interpretation of sperm development in nematodes. *Nematologica* **27:** 122–125.

Shulkin, D.J. and B.M. Zuckerman. 1982. Spectrofluorometric analysis of the effect of centrophenoxine on lipofuscin accumulation in the nematode *Caenorhabditis elegans*. *Age* **5:** 50–53.

Siddiqui, S.S. and P. Babu. 1980a. Kynurenine hydroxylase mutants of the nematode *Caenorhabditis elegans*. *Mol. Gen. Genet.* **179:** 21–24.

———. 1980b. Genetic mosaics of *Caenorhabditis elegans*: A tissue-specific fluorescent mutant. *Science* **210:** 330–332.

Siddiqui, S. and J. Culotti. 1984. A neural antigen conserved in different invertebrates. *Ann. N.Y. Acad. Sci.* **435:** 341–343.

Siddiqui, S.S. and G. von Ehrenstein. 1980. Biochemical genetics of *Caenorhabditis elegans*. In *Nematodes as biological models*, volume 1: *Behavioral and developmental models* (ed. B.M. Zuckerman), pp. 285–312. Academic Press, New York.

Sigurdson, D.C., G.J. Spanier, and R.K. Herman. 1984. *Caenorhabditis elegans* deficiency mapping. *Genetics* **108:** 331–345.

Sigurdson, D.C., R.K. Herman, C.A. Horton, C.K. Kari, and S.E. Pratt. 1986. An X-autosome fusion chromosome of *Caenorhabditis elegans*. *Mol. Gen. Genet.* **202:** 212–218.

Simpkin, K.G. and G.C. Coles. 1981. The use of *Caenorhabditis elegans* for anthelmintic screening. *J. Chem. Technol. Biotechnol.* **31:** 66–69.

Simpson, V.J., T.E. Johnson, and R.F. Hammen. 1986. *Caenorhabditis elegans* DNA does not contain 5-methylcytosine at any time during development or aging. *Nucleic Acids Res.* **14:** 6711–6717.

Singh, R.N. and J.E. Sulston. 1978. Some observations on molting in *Caenorhabditis elegans*. *Nematologica* **24:** 63–71.

Smith, R.M., W.H. Peterson, and E. McCoy. 1954. Oligomycin, a new antifungal antibiotic. *Antibiot. Chemother.* **4:** 962–970.

Smith, S.J. 1978. The structural gene for myosin, a closer look. *Nature* **272:** 495–496.

Snutch, T.P. and D.L. Baillie. 1983. Alterations in the pattern of gene expression following heat shock in the nematode *Caenorhabditis elegans*. *Can J. Biochem. Cell Biol.* **61:** 480–487.

———. 1984. A high degree of DNA strain polymorphism associated with the major heat shock gene in *Caenorhabditis elegans*. *Mol. Gen. Genet.* **195:** 329–335.

Soos, S. 1941. *Rhabditis carpathicus* spec nov eine neue in Sphagnum-Mooren lebende Nematode. *Fragm. Faun. Hung.* **4:** 115–119.

Spaull, V.W. 1973. Qualitative and quantitative distribution of soil nematodes of Signy Island South Orkney Islands. *Br. Antarct. Surv. Bull.* **33:** 177–184.

Spence, A.M., K.M.B. Malone, M.M.A. Novak, and R.A. Woods. 1982. The effects of mebendazole on the growth and development of *Caenorhabditis elegans*. *Can. J. Zool.* **60:** 2616–2623.

Spieth, J. and T. Blumenthal. 1985. The *Caenorhabditis elegans* vitellogenin gene family includes a gene encoding a distantly related protein. *Mol. Cell. Biol.* **5:** 2495–2501.

Spieth, J., K. Denison, E. Zucker, and T. Blumenthal. 1985a. The nucleotide sequence of a nematode vitellogenin gene. *Nucleic Acids Res.* **13:** 7129–7138.

Spieth, J., K. Denison, S. Kirtland, J. Cane, and T. Blumenthal. 1985b. The *Caenorhabditis elegans* vitellogenin genes: Short sequence repeats in the promoter regions and homology to the vertebrate genes. *Nucleic Acids Res.* **13:** 5283–5295.

Spradling, A.C. and G.M. Rubin. 1981. *Drosophila* genome organization: Conserved and dynamic aspects. *Annu. Rev. Genet.* **15:** 219–264.

Starck, J. 1977. Radioautographic study of RNA synthesis in *Caenorhabditis elegans* (Bergerac variety) oogenesis. *Biol. Cell* **30:** 181–182.

———. 1984. Synthesis of oogenesis specific proteins in *Caenorhabditis elegans*: An approach to the study of vitellogenesis in a nematode. *Int. J. Invertebr. Reprod. Dev.* **7:** 149–160.

Starck, J., M.-A. Gilbert, J. Brun, and C. Bosch. 1983. Ribosomal RNA synthesis and processing during oogenesis of the free living nematode *Caenorhabditis elegans*. *Comp. Biochem. Physiol.* **75B:** 575–580.

Stent, G.S. and D.A. Weisblat. 1982. The development of a simple nervous system. *Sci. Am.* **246(1):** 100–110.

Sternberg, P.W. and H.R. Horvitz. 1981. Gonadal cell lineages of the nematode *Panagrellus redivivus* and implications for evolution by the modification of cell lineage. *Dev. Biol.* **88:** 147–166.

———. 1982. Postembryonic nongonadal cell lineages of the nematode *Panagrellus redivivus*: Description and comparison with those of *Caenorhabditis elegans*. *Dev. Biol.* **93:** 181–205.

———. 1984. The genetic control of cell lineage during nematode development. *Annu. Rev. Genet.* **18:** 489–524.

———. 1986. Pattern-formation during vulval development in *Caenorhabditis elegans*. *Cell* **44:** 761–772.

Stinchcomb, D.T., C. Mello, and D. Hirsh. 1985a. *Caenorhabditis elegans* DNA that directs segregation in yeast cells. *Proc. Natl. Acad. Sci.* **82:** 4167–4171.

Stinchcomb, D.T., J.E. Shaw, S.H. Carr, and D. Hirsh. 1985b. Extrachromosomal DNA transformation of *Caenorhabditis elegans*. *Banbury Rep.* **20:** 251–263.

———. 1985c. Extrachromosomal DNA transformation of *Caenorhabditis elegans*. *Mol. Cell. Biol.* **5:** 3484–3496.

Stinchcomb, D.T., M. Thomas, J. Kelly, E. Selker, and R.W. Davis. 1980. Eukaryotic DNA segments capable of autonomous replication in yeast. *Proc. Natl. Acad. Sci.* **77:** 4559–4563.

Strehler, E.E., V. Mahdavi, M. Periasamy, and B. Nadal-Ginard. 1985. Intron positions are conserved in the 5′ end region of myosin heavy-chain genes. *J. Biol. Chem.* **260:** 468–471.

Stretton, A.O.W., R.E. Davis, J.D. Angstadt, J.E. Donmoyer, and C.D. Johnson. 1985. Neural control of behaviour in *Ascaris*. *Trends Neurosci.* **8:** 294–300.

Stretton, A.O.W., R.M. Fishpool, E. Southgate, J.E. Donmoyer, J.P. Walrond, J.E.R. Moses, and I.S. Kass. 1978. Structure and physiological activity of the motoneurons of the nematode *Ascaris*. *Proc. Natl. Acad. Sci.* **75:** 3493–3497.

Stromberg, B.E. and E.J.L. Soulsby. 1977. Heterologous helminth induced resistance to *Ascaris suum* in Guinea pigs. *Vet. Parasitol.* **3:** 169–175.

Strome, S. 1986a. Fluorescence visualization of the distribution of microfilaments in gonads and early embryos of the nematode *Caenorhabditis elegans*. *J. Cell Biol.* **103:** 2241–2252.

———. 1986b. Establishment of asymmetry in early *Caenorhabditis elegans* embryos: Visualization with antibodies to germ cell components. In *Gametogenesis and the early embryo* (ed. J.G. Gall), pp. 77–95. Alan R. Liss, New York.

Strome, S. and W.B. Wood. 1982. Immunofluorescence visualization of germ-line-specific cytoplasmic granules in embryos, larvae, and adults of *Caenorhabditis elegans*. *Proc. Natl. Acad. Sci.* **79:** 1558–1562.

———. 1983. Generation of asymmetry and segregation of germ-line granules in early *Caenorhabditis elegans* embryos. *Cell* **35:** 15–25.

———. 1984. Segregation of germ-line-specific antigens during embryogenesis in *Caenorhabditis elegans*. In *Molecular aspects of early development* (ed. G.M. Malacinski and W.H. Klein), pp. 141–165. Plenum Press, New York.

Subtelny, K. and F.R. Konigsberg, eds. 1979. Determinants of spatial organization. In *Meeting of the Society for Developmental Biology*, Madison, Wisconsin. June 1978. Academic Press, New York.

Sudhaus, W. 1974. Zur Systematik, Verbreitung, Ökologie und Biologie neuer und wenig bekannter Rhabditiden (Nematoda), 2. Teil. *Zool. Jb. Syst. Bd.* **101:** 417–465.

Sukul, N.C. and N.A. Croll. 1978. Influence of potential difference and current on the electrotaxis of *Caenorhabditis elegans*. *J. Nematol.* **10:** 314–317.

Sulston, J.E. 1976. Post-embryonic development in the ventral cord of *Caenorhabditis elegans*. *Philos. Trans. R. Soc. Lond. B Biol. Sci.* **275:** 287–298.

———. 1983. Neuronal cell lineages in the nematode *Caenorhabditis elegans*. *Cold Spring Harbor Symp. Quant. Biol.* **48:** 443–452.

Sulston, J.E. and S. Brenner. 1974. The DNA of *Caenorhabditis elegans*. *Genetics* **77:** 95–104.

Sulston, J.E. and J. Hodgkin. 1979. A diet of worms. *Nature* **279:** 758–759.

Sulston, J.E. and H.R. Horvitz. 1977. Post-embryonic cell lineages of the nematode *Caenorhabditis elegans*. *Dev. Biol.* **56:** 110–156.

———. 1981. Abnormal cell lineages in mutants of the nematode *Caenorhabditis elegans*. *Dev. Biol.* **82:** 41–55.

Sulston, J.E. and J.G. White. 1980. Regulation and cell autonomy during postembryonic development of *Caenorhabditis elegans*. *Dev. Biol.* **78:** 577–597.

Sulston, J.E., D.G. Albertson, and J.N. Thomson. 1980. The *Caenorhabditis elegans* male: Postembryonic development of nongonadal structures. *Dev. Biol.* **78:** 542–576.

Sulston, J., M. Dew, and S. Brenner. 1975. Dopaminergic neurons in the nematode *Caenorhabditis elegans*. *J. Comp. Neurol.* **163:** 215–226.

Sulston, J.E., E. Schierenberg, J.G. White, and J.N. Thomson. 1983. The embryonic cell lineage of the nematode *Caenorhabditis elegans*. *Dev. Biol.* **100:** 64–119.

Swanson, M.M. and D.L. Riddle. 1981. Critical periods in the development of the *Caenorhabditis elegans* dauer larva. *Dev. Biol.* **84:** 27–40.

Swanson, M.M., M.L. Edgley, and D.L. Riddle. 1984. The nematode *Caenorhabditis elegans*. In *Genetic maps* (ed. S.J. O'Brien), vol. 3, pp. 286–299. Cold Spring Harbor Laboratory, Cold Spring Harbor, New York.

Tabuse, Y. and J. Miwa. 1983. A gene involved in action of tumor promoters is identified and mapped in *Caenorhabditis elegans*. *Carcinogenesis* **4:** 783–786.

Taghert, P.H., C.Q. Doe, and C.S. Goodman. 1984. Cell determination and regulation during development of neuroblasts and neurones in grasshopper embryos. *Nature* **307:** 163–165.

Tang, C. 1982. Studies on the plant nematodes in south Fujian, China. 2. The species of *Rhabditida*. *Acta Zool. Sin.* **28:** 157–164.

Tanii, I., M. Osafune, T. Arata, and A. Inoue. 1985. ATPase characteristics of myosin from nematode *Caenorhabditis elegans* purified by an improved method. Formation of myosin-phosphate-ADP complex. *J. Biochem.* **98:** 1201–1209.

Tattar, T.A., J.P. Stack, and B.M. Zuckerman. 1977. Apparent nondestructive penetration of *Caenorhabditis elegans* by microelectrodes. *Nematologica* **23:** 267–269.

Tilby, M.J. and V. Moses. 1975. Nematode ageing. Automatic maintenance of age synchrony without inhibitors. *Exp. Gerontol.* **10:** 213–223.

Tiner, J.D. 1958. A preliminary in vitro test for anthelminthic activity. *Exp. Parasitol.* **7:** 292–305.

Tobler, H. 1986. The differentiation of germ and somatic cell lines in nematodes. *Results Probl. Cell Differ.* **13:** 1–69.

Tomlinson, G. and M. Rothstein. 1962. Nematode biochemistry. I. Culture methods. *Biochim. Biophys. Acta* **63:** 465–470.

Tomlinson, G., C.A. Albuquerque, and R.A. Woods. 1985. The effects of amidantel (Bay d 8815) and its deacylated derivative (Bay d 9216) on *Caenorhabditis elegans*. *Eur. J. Pharmacol.* **113:** 255–262.

Traboni, C., G. Ciliberto, and R. Cortese. 1982. A novel method for site-directed mutagenesis: Its application to an eukaryotic $tRNA^{pro}$ gene promoter. *EMBO J.* **1:** 415–420.

———. 1984. Mutations in box B of promoter of a eukaryotic $tRNA^{pro}$ gene affect rate of transcription, processing, and stability of the transcript. *Cell* **36:** 179–187.

Tranquilla, T.A., R. Cortese, D. Melton, and J.D. Smith. 1982. Sequences of four tRNA genes from *Caenorhabditis elegans* and the expression of *Caenorhabditis elegans* $tRNA^{Leu}$ (anticodon IAG) in *Xenopus* oocytes. *Nucleic Acids Res.* **10:** 7919–7934.

Trent, C. 1982. "Genetic and behavior studies of the egg-laying system of *Caenorhabditis elegans*." Ph.D. thesis, Massachusetts Institute of Technology, Cambridge.

Trent, C., N. Tsung, and H.R. Horvitz. 1983. Egg-laying defective mutants of the nematode *Caenorhabditis elegans*. *Genetics* **104:** 619–647.

Triantaphyllou, A.C. 1973. Environmental sex differentiation of nematodes in relation to pest management. *Annu. Rev. Phytopathol.* **11:** 441–464.

Turner, R.H. and C.D. Green. 1973. Preparation of biological material for scanning electron microscopy by critical point drying from water miscible solvents. *J. Microsc.* **97:** 357–363.

van Beneden, E. 1883. *Recherches sur la maturation de l'oeuf, la fécondation et la division cellulaire*. Gand and Leipzig, Paris.

Vanderslice, R. and D. Hirsh. 1976. Temperature-sensitive zygote defective mutants of *Caenorhabditis elegans*. *Dev. Biol.* **49:** 236–249.

Vanfleteren, J.R. 1973. Amino acid requirements of the free-living nematode *Caenorhabditis briggsae*. *Nematologica* **19:** 93–99.

———. 1974. Nematode growth factor. *Nature* **248:** 255–257.

———. 1975a. The nature of a nematode growth factor. II. Growth and maturation of *Caenorhabditis briggsae* on haemin proteins. *Nematologica* **21:** 425–437.

———. 1975b. The nature of a nematode growth factor. I. Growth and maturation of *Caenorhabditis briggsae* on ribonucleoprotein particles. *Nematologica* **21:**413–424.

———. 1976a. The nature of a nematode growth factor. III. Growth and maturation of *Caenorhabditis briggsae* on protein-haemin co-precipitates. *Nematologica* **22:** 103–112.

———. 1976b. Large scale cultivation of a free-living nematode (*Caenorhabditis elegans*). *Experientia* **32:** 1087–1088.

———. 1978. Axenic culture of free-living, plant-parasitic, and insect-parasitic nematodes. *Ann. Rev. Phytopathol.* **16:** 131–157.

———. 1980. Nematodes as nutritional models. In *Nematodes as biological models*, volume 2: *Aging and other model systems* (ed. B.M. Zuckerman), pp. 47–79. Academic Press, New York.

———. 1983. Nematode chromosomal proteins. II. Fractionation and identification of the histones of *Caenorhabditis elegans*. *Comp. Biochem. Physiol.* **73B:** 709–718.

Vanfleteren, J.R. and H. Avau. 1977. Selective inhibition of reproduction in aminopterin-treated nematodes. *Experientia* **33:** 902–904.

Vanfleteren, J.R. and D.E. Roets. 1972. The influence of some anthelmintic drugs on the population growth of the free-living nematodes *Caenorhabditis briggsae* and *Turbatrix aceti*. *Nematologica* **18:** 325–338.

Vanfleteren, J.R. and J.J. Van Beeumen. 1983. Nematode chromosomal proteins. III. Some structural properties of the histones of *Caenorhabditis elegans*. *Comp. Biochem. Physiol.* **76B:** 179–184.

Vanfleteren, J.R., K. Neirynck, and D. Huylebroeck. 1979. Nematode chromosomal proteins. I. Isolation of chromatin and preliminary characterization of the chromosomal proteins of *Caenorhabditis elegans*. *Comp. Biochem. Physiol.* **62B:** 349–354.

Vanfleteren, J.R., S.M. Van Bun, L.L. Delcambe, and J.J. Van Beeumen. 1986. Multiple forms of histone H2B from the nematode *Caenorhabditis elegans*. *Biochem. J.* **235:** 769–773.

Villeneuve, A.M. and B.J. Meyer. 1987. *sdc-1*: A link between sex determination and dosage compensation in *Caenorhabditis elegans*. *Cell* **48:** 25–37.

von Ehrenstein, G. 1973. A model system for developmental and behavioral genetics. *Jahrb. Max-Planck-Gesellschaft*, pp. 38–60.

von Ehrenstein, G. and E. Schierenberg. 1980. Cell lineages and development of *Caenorhabditis elegans* and other nematodes. In *Nematodes as biological models*, volume 1: *Behavioral and developmental models* (ed. B.M. Zuckerman), pp. 1–72. Academic Press, New York.

von Ehrenstein, G., E. Schierenberg, and J. Miwa. 1979. Cell lineages of wild type and temperature sensitive embryonic arrest mutants of *Caenorhabditis elegans*. In *Cell lineage, stem cells and cell determination* (ed. N. LeDouarin), pp. 49–59. Elsevier, New York.

Wahab, A. 1962. Untersuchungen über Nematoden in den Drüsen des Kopfes der Ameisen (Formicidae). *Z. Morphol. Ökol. Tiere* **52:** 33–92.

Walker, J., M. Saraste, M.J. Runswick, and N.J. Gay. 1982. Distantly related sequences in alpha- and beta-subunits of ATP synthase, myosin, kinases and other ATP-requiring enzymes and a common nucleotide binding fold. *EMBO J.* **1:** 945–951.

Walrond, J.P. and A.O.W. Stretton. 1985. Excitory and inhibitory activity in the dorsal musculature of the nematode *Ascaris* evoked by single dorsal excitatory motorneurons. *J. Neurosci.* **5:** 16–23.

Walrond, J.P., I.S. Kass, A.O.W. Stretton, and J.E. Donmoyer. 1985. Identification of excitatory and inhibitory motoneurons in the nematode *Ascaris* by electrophysiological techniques. *J. Neurosci.* **5:** 1–8.

Walton, A.C., B.C. Chitwood, and M.B. Chitwood. 1974. Gametogenesis. In *Introduction to nematology* (ed. B.G. Chitwood and M.B. Chitwood), pp. 191–201. University Park Press, Baltimore, Maryland.

Wang, K. and C.L. Williamson. 1980. Identification of an N2-line protein of striated muscle. *Proc. Natl. Acad. Sci.* **77:** 3254–3258.

Wang, K., J. McClure, and A. Tu. 1979. Titin: Major myofibrillar components of striated muscle. *Proc. Natl. Acad. Sci.* **76:** 3698–3702.

Ward, S. 1973. Chemotaxis by the nematode *Caenorhabditis elegans*: Identification of attractants and analysis of the response by use of mutants. *Proc. Natl. Acad. Sci.* **70:** 817–821.

———. 1975. Genetic studies of chemotaxis mutants in nematodes. *Exp. Neurol.* **48:** 58–59.

———. 1976. The use of mutants to analyze the sensory system of *Caenorhabditis elegans*. In *The organization of nematodes* (ed. N.A. Croll), pp. 365–382. Academic Press, New York.

———. 1977a. Invertebrate neurogenetics. *Annu. Rev. Genet.* **11:** 415–450.

———. 1977b. Use of nematode behavioral mutants for analysis of neural function and development. *Soc. Neurosci. Symp.* **2:** 1–26.

———. 1978. Nematode chemotaxis and chemoreceptors. In *Receptors and recognition series. B. Taxis and behavior. Elementary sensory systems in biology* (ed. G.L. Hazelbauer), pp. 141–168. Halsted Press, New York.

———. 1982. Genetic analysis of the sensory nervous system of *Caenorhabditis elegans*. *Proc. Fifth. Int. Congr. Parasitol.* **2:** 28–31.

———. 1986a. The asymmetric localization of gene products during the development of *Caenorhabditis elegans* spermatozoa. *Symp. Soc. Dev. Biol.* **44:** 55–75.

———. 1986b. Asymmetric localization of gene products during the development of *Caenorhabditis elegans* spermatozoa. In *Gametogenesis and the early embryo* (ed. J.G. Gall), pp. 55–75. Alan R. Liss, New York.

Ward, S. and J.S. Carrel. 1979. Fertilization and sperm competition in the nematode *Caenorhabditis elegans*. *Dev. Biol.* **73:** 304–321.

Ward, S. and M. Klass. 1982. The location of the major protein in *Caenorhabditis elegans* sperm and spermatocytes. *Dev. Biol.* **92:** 203–208.

Ward, S. and J. Miwa. 1978. Characterization of temperature-sensitive, fertilization-defective mutants of the nematode *Caenorhabditis elegans*. *Genetics* **88:** 285–303.

Ward, S., Y. Argon, and G.A. Nelson. 1981. Sperm morphogenesis in wild-type and fertilization-defective mutants of *Caenorhabditis elegans*. *J. Cell Biol.* **91:** 26–44.

Ward, S., E. Hogan, and G.A. Nelson. 1983. The initiation of spermiogenesis in the nematode *Caenorhabditis elegans*. *Dev. Biol.* **98:** 70–79.

Ward, S., T.M. Roberts, G.A. Nelson, and Y. Argon. 1982. The development and motility of *Caenorhabditis elegans* spermatozoa. *J. Nematol.* **14:** 259–266.

Ward, S., N. Thomson, J.G. White, and S. Brenner. 1975. Electron microscopical reconstruction of the anterior sensory anatomy of the nematode *Caenorhabditis elegans*. *J. Comp. Neurol.* **160:** 313–337.

Ward, S., T.M. Roberts, S. Strome, F.M. Pavalko, and E. Hogan. 1986. Monoclonal antibodies that recognize a polypeptide antigenic determinant shared by multiple *Caenorhabditis elegans* sperm-specific proteins. *J. Cell Biol.* **102:** 1778–1786.

Ward, S., D.J. Burke, J.E. Sulston, A.R. Coulson, D.G. Albertson, D. Ammons, M. Klass, and E. Hogan. 1988. The genomic organization of transcribed major sperm protein genes and pseudogenes in the nematode *Caenorhabditis elegans*. *J. Mol. Biol.* **199:** 1–13.

Ware, R.W., D. Clark, K. Crossland, and R.L. Russell. 1975. The nerve ring of the nematode *Caenorhabditis elegans*: Sensory input and motor output. *J. Comp. Neurol.* **162:** 71–110.

Wat, C.-K., S.K. Prasad, E.A. Graham, S. Partington, T. Arnason, G.H.N. Towers, and J. Lam. 1981. Photosensitization of invertebrates by natural polyacetylenes. *Biochem. Syst. Ecol.* **9:** 59–62.

Waterston, R.H. 1981. A second informational suppressor, *sup-7 X*, in *Caenorhabditis elegans*. *Genetics* **97:** 307–325.

Waterston, R.H. and S. Brenner. 1978. A suppressor mutation in the nematode acting on specific alleles of many genes. *Nature* **275:** 715–719.

Waterston, R.H. and G.R. Francis. 1985. Genetic analysis of muscle development in *Caenorhabditis elegans*. *Trends Neurosci.* **8:** 270–276.

Waterston, R.H., H.F. Epstein, and S. Brenner. 1974. Paramyosin in *Caenorhabditis elegans*. *J. Mol. Biol.* **90:** 285–290.

Waterston, R.H., R.M. Fishpool, and S. Brenner. 1977. Mutants affecting paramyosin in *Caenorhabditis elegans*. *J. Mol. Biol.* **117:** 679–697.

Waterston, R.H., D. Hirsh, and T.R. Lane. 1984. Dominant mutations affecting muscle structure in *Caenorhabditis elegans* that map near the actin gene cluster. *J. Mol. Biol.* **180:** 473–496.

Waterston, R.H., K.C. Smith, and D.G. Moerman. 1982a. Genetic fine structure analysis of the myosin heavy chain gene *unc-54* of *Caenorhabditis elegans*. *J. Mol. Biol.* **158:** 1–15.

Waterston, R.H., J.N. Thomson, and S. Brenner. 1980. Mutants with altered muscle structure in *Caenorhabditis elegans*. *Dev. Biol.* **77:** 271–302.

Waterston, R.H., S. Bolton, H.L. Sive, and D.G. Moerman. 1982b. Mutationally altered myosins in *Caenorhabditis elegans*. In *Muscle development: Molecular and cellular control* (ed. M. Pearson and H.F. Epstein), pp. 119–127. Cold Spring Harbor Laboratory, Cold Spring Harbor, New York.

Waterston, R.H., D.G. Moerman, D.L. Baillie, and T.R. Lane. 1982c. Mutations affecting myosin heavy chain accumulation and function in the nematode *Caenorhabditis elegans*. In *Disorders of the motor unit* (ed. D.M. Schotland), pp. 747–760. John Wiley, New York.

Waterston, R.H., D.G. Moerman, G.M. Benian, R.J. Barstead, I. Mori, and R. Francis. 1986. Muscle genes and proteins in *Caenorhabditis elegans*. *UCLA Symp. Mol. Cell. Biol. New Ser.* **29:** 605–617.

Watson, J.E., C.B. Pinnock, E.L.R. Stokstad, and W.F. Hieb. 1974. A nephelometer for measurement of nematode populations. *Anal. Biochem.* **60:** 267–271.

White, J.G. 1985. Neuronal connectivity in *Caenorhabditis elegans*. *Trends Neurosci.* **8:** 277–283.

White, J.G. and H.R. Horvitz. 1979. Laser microbeam techniques in biological research. *Electro-Optical Systems Design AUG.*

White, J.G., D.G. Albertson, and M.A.R. Anness. 1978. Connectivity changes in a class of motoneurone during the development of a nematode. *Nature* **271:** 764–766.

White, J.G., H.R. Horvitz, and J.E. Sulston. 1982. Neurone differentiation in cell lineage mutants of *Caenorhabditis elegans*. *Nature* **297:** 584–587.

White, J.G., E. Southgate, J.N. Thomson, and S. Brenner. 1976. The structure of the ventral nerve cord of *Caenorhabditis elegans*. *Philos. Trans. R. Soc. Lond. B Biol. Sci.* **275:** 327–348.

———. 1983. Factors that determine connectivity in the nervous system of *Caenorhabditis elegans*. *Cold Spring Harbor Symp. Quant. Biol.* **48:** 633–640.

———. 1986. The structure of the nervous system of *Caenorhabditis elegans*. *Philos. Trans. R. Soc. Lond. B Biol. Sci.* **314:** 1–340.

White, M.J.D. 1973. *Animal cytology and evolution*, 3rd edition. Cambridge University Press, England.

Whittaker, J.R. 1982. Muscle lineage cytoplasm can change the developmental expression in epidermal lineage cells of Ascidian embryos. *Dev. Biol.* **93:** 463–470.

Wieringa, B., E. Hofer, and C. Weissmann. 1984. A minimal intron length but no specific internal sequence is required for splicing the large rabbit β-globin intron. *Cell* **37:** 915–925.

Wilcox, G., J. Abelson, and C.G. Fox, eds. 1977. Molecular approaches to eukaryotic genetic systems. *ICN-UCLA Symp. Mol. Cell. Biol.* **8:** 1–449.

Wilkins, A.S. 1986. *Genetic analysis of animal development*. John Wiley, New York.

Willett, J.D. 1980. Control mechanisms in nematodes. In *Nematodes as biological models*, volume 1: *Behavioral and developmental models* (ed. B.M. Zuckerman), pp. 197–226. Academic Press, New York.

Willett, J.D., I. Rahim, M. Geist, and B.M. Zuckerman. 1980. Cyclic nucleotide exudation by nematodes and the effects on nematode growth, development and longevity. *Age* **3:** 82–87.

Wills, N., R.F. Gesteland, J. Karn, L. Barnett, S. Bolten, and R.H. Waterston. 1983. The genes *sup-7 X* and *sup-5 III* of *Caenorhabditis elegans* suppress amber nonsense mutations via altered transfer RNA. *Cell* **33:** 575–583.

Wilson, D.A. and C.A. Thomas, Jr. 1974. Palindromes in chromosomes. *J. Mol. Biol.* **84:** 115–144.

Wilson, E.B. 1896. *The cell in development and inheritance*. Macmillan, New York.

———. 1925. *The cell in development and heredity*. Macmillan, New York.

Wilson, P.A.G. 1976. Nematode growth patterns and the moulting cycle: The population growth profile. *J. Zool.* **179:** 135–151.

Wolf, N., D. Hirsh, and J.R. McIntosh. 1978. Spermatogenesis in males of the free-living nematode, *Caenorhabditis elegans*. *J. Ultrastruct. Res.* **63:** 155–169.

Wolf, N., J. Priess, and D. Hirsh. 1983. Segregation of germline granules in early embryos of *Caenorhabditis elegans*: An electron microscopic analysis. *J. Embryol. Exp. Morphol.* **73:** 297–306.

Wolstenholme, D.R., J.L. Macfarlane, R. Okimoto, D.O. Clary, and J.A. Wahleithner. 1987. Bizarre tRNAs inferred from DNA sequences of mitochondrial genomes of nematode worms. *Proc. Natl. Acad. Sci.* **84:** 1324–1328.

Wood, W.B. 1977. Summary of workshop on nematodes. *ICN-UCLA Symp. Mol. Cell. Biol.* **8:** 357–358.

Wood, W.B., J.S. Laufer, and S. Strome. 1982. Developmental determinants in embryos of *Caenorhabditis elegans*. *J. Nematol.* **14:** 267–273.

Wood, W.B., E. Schierenberg, and S. Strome. 1984. Localization and determination in early embryos of *Caenorhabditis elegans*. *UCLA Symp. Mol. Cell. Biol.* **19:** 37–49.

Wood, W.B., S. Strome, and J.S. Laufer. 1983. Localization and determination in embryos of *Caenorhabditis elegans*. In *Time, space, and pattern in embryonic development* (ed. W.R. Jeffery and R.A. Raff), pp. 221–239. Alan R. Liss, New York.

Wood, W.B., P. Meneely, P. Schedin, and L. Donahue. 1985. Aspects of dosage compensation and sex determination in *Caenorhabditis elegans*. *Cold Spring Harbor Symp. Quant. Biol.* **50:** 575–583.

Wood, W.B., R. Hecht, S. Carr, R. Vanderslice, N. Wolf, and D. Hirsh. 1980. Parental effects and phenotypic characterization of mutations that affect early development in *Caenorhabditis elegans*. *Dev. Biol.* **74:** 446–469.

Woods, R.A., K.M.B. Malone, C.A. Albuquerque, and G. Tomlinson. 1986. The effects of amidantel (Bay d 8815) and its deacylated derivative (Bay d 9216) on wild-type and resistant mutants of *Caenorhabditis elegans*. *Can. J. Zool.* **64:** 1310–1316.

Wright, D.J. and F.A. Awan. 1976. Acetylcholinesterase activity in the region of the nematode nerve ring: Improved histochemical specificity using ultrasonic pretreatment. *Nematologica* **22:** 326–331.

Wright, K.A. 1980. Nematode sense organs. In *Nematodes as biological models*, volume 2: *Aging and other model systems* (ed. B.M. Zuckerman), pp. 237–295. Academic Press, New York.

Wright, K.A. and J.N. Thomson. 1981. The buccal capsule of *Caenorhabditis elegans* (Nematoda: Rhabditoidea): An ultrastructural study. *Can. J. Zool.* **59:** 1952–1961.

Wulf, E., A. Deboben, F.A. Bautz, H. Faulstich, and T. Weiland. 1979. Fluorescent phallotoxin, a tool for the visualization of cellular actin. *Proc. Natl. Acad. Sci.* **76:** 4498–4502.

Yamaguchi, Y., K. Murakami, M. Furusawa, and J. Miwa. 1983. Germline-specific antigens identified by monoclonal antibodies in the nematode *Caenorhabditis elegans. Dev. Growth Differ.* **25:** 121–131.

Yarbrough, P.O. and R.M. Hecht. 1984. Two isoenzymes of glyceraldehyde-3-phosphate dehydrogenase in *Caenorhabditis elegans*. Isolation, properties, and immunochemical characterization. *J. Biol. Chem.* **259:** 14711–14720.

Yarwood, E.A. and E.L. Hansen. 1969. Dauer larvae of *Caenorhabditis briggsae* in axenic culture. *J. Nematol.* **1:** 184–189.

Yeargers, E. 1981. Effect of gamma-radiation on dauer larvae of *Caenorhabditis elegans. J. Nematol.* **13:** 235–237.

Yosida, T.H., T. Sadaie, and Y. Sadaie. 1984. Somatic and meiotic chromosomes of the small free-living nematode, *Caenorhabditis elegans. Proc. Jpn. Acad.* **60B:** 54–57.

Zengel, J.M. and H.F. Epstein. 1980a. Muscle development in *Caenorhabditis elegans*: A molecular genetic approach. In *Nematodes as biological models*, volume 1: *Behavioral and developmental models* (ed. B.M. Zuckerman), pp. 73–126. Academic Press, New York.

———. 1980b. Identification of genetic elements associated with muscle structure in the nematode *Caenorhabditis elegans. Cell Motil.* **1:** 73–97.

———. 1980c. Mutants altering coordinate synthesis of specific myosins during nematode muscle development. *Proc. Natl. Acad. Sci.* **77:** 852–856.

Zuckerman, B.M. 1974. The effects of procaine on aging and development of a nematode. In *Theoretical aspects of aging* (ed. M. Rockstein), pp. 177–186. Academic Press, New York.

———. 1976. Nematodes as models for aging studies. In *The organization of nematodes* (ed. N.A. Croll), pp. 211–241. Academic Press, New York.

———. ed. 1980a. *Nematodes as biological models*, volume 1: *Behavioral and developmental models*. Academic Press, New York.

———. ed. 1980b. *Nematodes as biological models*, volume 2: *Aging and other model systems*. Academic Press, New York.

———. 1983. The free-living nematode *Caenorhabditis elegans* as a rapid screen for compounds to retard aging. In *Intervention in the aging process*, part B (ed. W. Regelson and F.M. Sinex). Alan R. Liss, New York.

Zuckerman, B.M. and K.A. Barrett. 1978. Effects of P-chlorophenoxy acetic acid and dimethylaminoethanol on the nematode *Caenorhabditis briggsae. Exp. Aging Res.* **4:** 133–139.

Zuckerman, B.M. and M.A. Geist. 1983. Effects of vitamin E on the nematode *Caenorhabditis elegans. Age* **6:** 1–4.

Zuckerman, B.M. and S. Himmelhoch. 1980. Nematodes as models to study aging. In *Nematodes as biological models*, volume 2: *Aging and other model systems* (ed. B.M. Zuckerman), pp. 3–28. Academic Press, New York.

Zuckerman, B.M. and J.-B. Jansson. 1984. Nematode chemotaxis and possible mechanisms of host/prey recognition. *Annu. Rev. Phytopathol.* **22:** 95–113.

Zuckerman, B.M. and I. Kahane. 1983. *Caenorhabditis elegans*: Stage specific differences in cuticle surface carbohydrates. *J. Nematol.* **15:** 535–538.

Zuckerman, B.M., S. Himmelhoch, and M. Kisiel. 1973. Fine structure changes in the cuticle of adult *Caenorhabditis briggsae* with age. *Nematologica* **19:** 109–112.

Zuckerman, B.M., I. Kahane, and S. Himmelhoch. 1979. *Caenorhabditis briggsae* and *Caenorhabditis elegans*: Partial characterization of cuticle surface carbohydrates. *Exp. Parasitol.* **47:** 419–424.

Zuckerman, B.M., W.F. Mai, and R.A. Rodhe, eds. 1971a. *Plant parasitic nematodes*, volume II: *Cytogenetics, host-parasite interactions, and physiology*. Academic Press, New York.

Zuckerman, B.M., B. Nelson, and M. Kisiel. 1972. Specific gravity increase of *Caenorhabditis briggsae* with age. *J. Nematol.* **4:** 261–262.

Zuckerman, B.M., J.M. Castillo, K.H. Deubert, and H.B. Gunner. 1969. Studies on a growth supplement for *Caenorhabditis briggsae* from freeze-dried bacteria. *Nematologica* **15:** 543–549.

Zuckerman, B.M., S. Himmelhoch, B. Nelson, J. Epstein, and M. Kisiel. 1971b. Aging in *Caenorhabditis briggsae*. *Nematologica* **17:** 478–487.

Subject Index